fourth edition

ENGINEERING TECHNICAL DRAFTING

J.W. GIACHINO   **Western Michigan University
Kalamazoo, Michigan**

HENRY J. BEUKEMA   **Western Michigan University
Kalamazoo, Michigan**

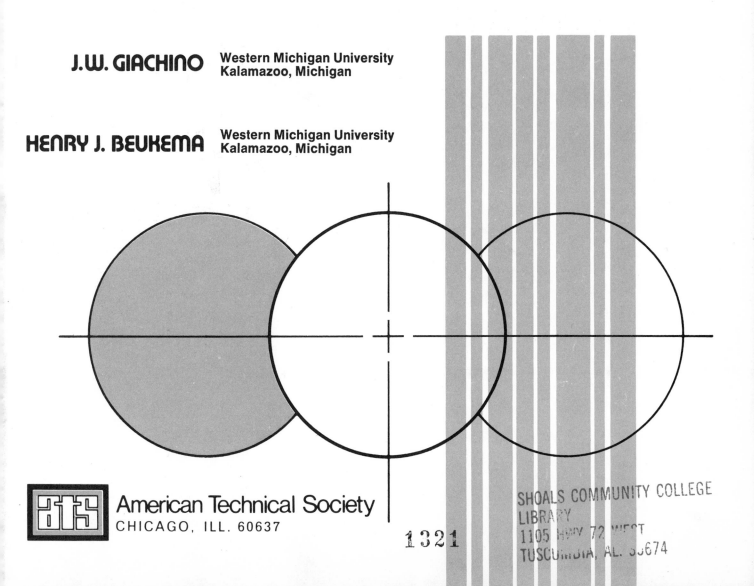

American Technical Society
CHICAGO, ILL. 60637

PREFACE

The fourth edition of *Engineering Technical Drafting* has been completely reorganized to better meet current teaching concepts of modern day drafting. The text is divided into five main sections: Basic Drafting Practices, Representational Drawings, Technical Illustrations, Geometry in Drafting, and Industrial Production Drafting. Each section consists of several units which cover the essential drafting principles and practices recommended by the American National Standards Institute and accepted by industry.

Significant features of the new edition that are particularly noteworthy are:

1. All dimensional values are expressed in decimal-inch or metric units. Metrication is included throughout the text wherever it is applicable.

2. Geometric tolerancing has been expanded and simplified for better understanding.

3. Text material has been streamlined to provide a more realistic approach to the teaching of drafting.

4. Problems are grouped at the end of each section rather than at the end of individual units. With this arrangement drafting practices are integrated for more effective learning.

5. Illustrations have been enlarged and the use of color improved to enhance the reading quality of the text material.

The wide acceptance of previous editions of *Engineering Technical Drafting* has been very gratifying. In preparing the new edition we have endeavored to incorporate the various suggestions made by teachers in the field. We are confident that this fourth edition will continue to serve the needs of those who are interested in and are pursuing a course of instruction in the communication aspects of engineering and technology.

J.W. Giachino
H.J. Beukema

ACKNOWLEDGMENTS

The authors and publisher wish to thank and acknowledge the following companies and organizations for providing numerous illustrations and technical material used in this Fourth Edition:

Addressograph-Multigraph Corporation., Bruning Division
Allis-Chalmers Manufacturing Company
American Chain Associate
American Meter Controls, Incorporated
American Welding Society
American National Standards Institute
Bendix Corporation
Boeing Aircraft
Borg-Warner Corporation
Boston Gear Works
Charles Bruning Company
Chart-Pak, Incorporated
Checkers Motors Corporation
Cincinnati Milacron
Convair Division, General Dynamics
Crane Company
C-Thru Ruler Company
Dake Corporation
Design News
Dodge Manufacturing Company
Eastman Kodak Company
Eaton Corporation, Transmissions Division
Eisler Engineering Company
Eugene Dietzgen Company
Fafnir Bearings Company
Fluid Controls, Incorporated
Foote Bros. Gear & Machine Corporation
Ford Motor Company
General Electric Company
General Electric Plastics Department
General Motors Corporation
Gritzner Graphics
Harrington & King Perforating Company
H.M. Harper Company
Howard W. Sams and Company, Incorporated
International Harvester Company
J.L. Clark Mfg. Company
Keuffel and Esser Company
Klok Institute
Koh-I-Noor Rapidograph, Inc.
Alfred Lant

Library of Congress
Lima Electric Motor Company
Lionel Corporation
Machine Design
Mackinac Bridge Authority
Marbon Division, Borg-Warner Corporation
Michigan Employment Security Commission
Millers Falls Company
National Aeronautics and Space Administration (NASA)
New York Air Brake Company
Perspective Systems, Incorporated
Pierce Corporation
Radio Corporation of America
Rand Corporation, Vickers Incorporated Division of Sperry-Rand Corp.
RapiDesign, Incorporated
Recordak Corporation
Reliance Electrical and Engineering Company
Republic Steel Corporation
Rockwell International—Space Division
Ryan Aeronautical Company
Sandia Corporation
SKF Industries Incorporated
Smithsonian Institute
Socony Mobil Oil Company
J.S. Staedtler, Inc.
Stock Drive Products
T.B. Woods & Sons Company
Teledyne-Post
Tooling & Production Magazine
Tompkins-Johnson Company
U.S. Department of Labor
Universal Drafting Machine Corporation
Varigraph, Inc.—Div. of Sperry-Rand Corporation.
Vickers Incorporated
Weatherhead Company
Weirton Steel Company
Western Michigan University
Wilson Instrument Division, ACCO
Whitlock Manufacturing Company
Woodhams, Blanchard and Flynn

CONTENTS

SECTION I

Basic Drafting Practices

UNIT 1

Introduction to Modern Drafting

> Drafting is a graphic language that is used universally by industry to produce the products of modern civilization. It is a visual system for conveying design ideas or presenting precise instructions for fabrication purposes.
>
> This unit includes material which students of drafting may find informative. Of particular significance are such topics as historical events that contributed to the development of modern drafting, responsibilities and skills required of drafting personnel, and a brief description of recognized professional fields of drafting.

1.1 Historical Developments

Since the dawn of modern civilization, certain events have clearly identified specific eras of development and progress. These events have great historical significance because of their contribution to the world in which we live and work. Perhaps one of the first notable achievements was the invention of the common wheel. Just try to visualize what the wheel has done in easing the burden of man and making possible the vast quantity of modern equipment in use today.

Moving rapidly over many centuries, all of which undoubtedly could claim numerous contributions to civilization, one important event was the development of the spinning jenny, followed by the cotton gin and power loom. See Fig. 1-1. These were instrumental in revolutionizing the making of cloth. Then in 1769-81, James Watt and others ushered in what might be referred to as the Steam Age. The steam power plant made possible the use of power machinery, both for manufacturing purposes and water transportation. See Fig. 1-2.

Fig. 1-1. James Hargreaves' spinning jenny.

A very important period emerged with the patent of the internal combustion engine by Otto in 1876. Not only did this engine pave the way for greater utilization of power machinery but it opened up a whole new mode of travel—the automobile.

3

Fig. 1-2. Fulton's steamboat.

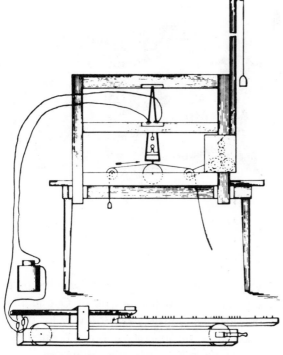

Fig. 1-3. The first telegraphic instrument.

Fig. 1-4. An early working model of the Bell Telephone.

The inventions of the electric telegraph by Morse in 1832, the telephone by Bell in 1876, and the wireless telegraph by Marconi in 1895 provided new avenues of communication which had a great impact in the development of our industrial, economic, and social structure, Figs. 1-3 and 1-4.

Thomas Edison's invention of the incandescent lamp in 1879, Fig. 1-5, added immeasurably to the kind of world we live in today. In 1903 the Wright brothers, with their first controlled air flight, ushered in the Air Age. See Fig. 1-6. The airplane has figuratively reduced the size of the world and made all countries next door neighbors.

During World War II, we saw the beginning of the Atomic Age. Now we are entering what appears to be the most challenging and dramatic period of civilization—exploration of space. Fig. 1-7 shows the space shuttle vehicle, for example.

These are only a few of the many outstanding events which have influenced the course of mankind. But many other inventions and discoveries have also had profound effects on our lives. And it is certain that in the years ahead other epochs will have the same far-reaching impacts on the destinies of people as in the past.

Fig. 1-5. Edison's first successful incandescent lamp.

Fig. 1-6. First controlled air flight by the Wright brothers.

Fig. 1-7. Thousands of engineering drawings were needed in the design and manufacturing of the space mission shuttle vehicle shown here. This vehicle, capable of being launched from a 747 jet aircraft in flight, reenters the earth's atmosphere at transonic speeds.

1.1.1 Origin of drafting. It is difficult to establish an exact date when drafting came into existence. If we accept the premise that drafting is a graphic language, then undoubtedly the first use of drafting was by prehistoric peoples.

When these early cave dwellers wanted to record ideas, they made pictures on stones, skins, and walls of caves. See Fig. 1-8. As civilization developed, more and more implements were conceived and constructed and the graphic form of

Fig. 1-8. Symbols like these were used by ancient peoples to record what they did.

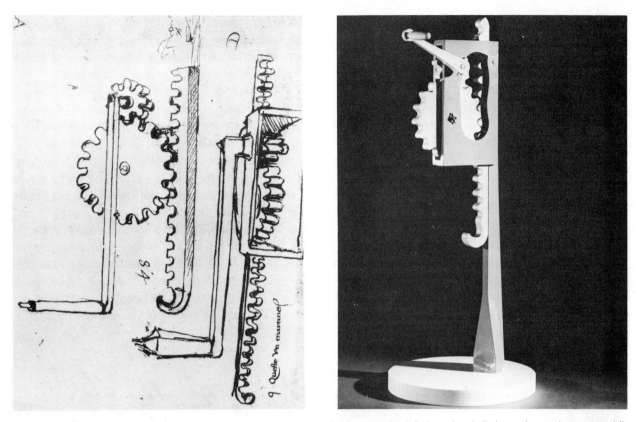

Fig. 1-9. Leonardo da Vinci's sketch of a device for lifting a weight is shown at the left. Note its similarity to the modern automobile jack shown at the right.

expressions gradually improved until they evolved into the refined system we have today.

Leonardo da Vinci is often considered to be the father of modern drafting. His wide range of talent led him to record many engineering ideas by using pictorial and orthographic-like views. See Fig. 1-9.

1.2 Importance of Drafting

At this juncture one might conceivably ask, "Just what do all of these events have to do with drafting?" Actually there is a very direct relationship between drafting and these events. In every instance it is reasonable to assume that some form of graphic communication was involved. Drafting is the graphic language used the world over to represent ideas, design, or construction. Without having had some form of graphic communication, it is questionable if what is still in evidence could have been produced—the fine old buildings, bridges, aqueducts, and other structures, some of which are recognized as "wonders of the world."

Today drafting is the accepted means of communication used by the scientist, engineer, designer, technician, and production worker. Regardless of their specific tasks, all of these people must either make sketches or drawings or be able to read them. Usually an idea starts with a rough sketch. Then the sketch is refined, and eventually it is made into a finished mechanical drawing.

1.3 Modern Concepts of Drafting

Since drafting is a key function in any engineering or development process, a drawing is regarded as a means of instruction, not a work of art. Because time is an important factor in any industrial organization, a drawing must be simple, concise, and accurate, without all of the embellishments of a beautiful picture. The essential factor is understanding—a drawing must clearly convey its intended purpose. Although the execution of a beautiful drawing cannot be justified in terms of time, neither can a sloppy drawing be condoned.

A poorly prepared, incomplete, or careless drawing merely increases time and results in confusion, error, and wasted effort.

Considerable thought is currently being given to practices that eventually may reduce substantially the time required to prepare drawings. Some industries are experimenting with a variety of mechanical devices that can actually make drawings. One such device utilizes an electron beam to produce lines, dimensions, and graphic symbols from information supplied by a computer. Another machine can prepare standard-type drawings on vellum or polyester film with drawing instruments.

At present the greatest limitation of all such equipment is the tremendous cost involved. But in the years ahead there is little doubt that economical means will be found to mechanize the process of making drawings. In spite of what may occur, however, technical personnel will always be needed to read and interpret drawings. The basic fundamentals of drawing will continue to be an essential part of the technician's education.

1.3.1 Metrics in drafting. The United States is in the process of converting its system of weights and measurements to the metric system, which is now in use throughout the world. Some industries such as chemical, photographic, and pharmaceutical have used the metric system for many years. Currently a large number of manufacturing plants, particularly in the automotive and tooling fields, are rapidly converting to metric. However, complete metrication in all walks of life probably will take many years because people have to undergo a new mode of thinking. Our inch, foot, yard, pound, quart, and mile concepts have been with us so long that only through a gradual process of

Fig. 1-10. An increasing number of women are training for careers in drafting and design.

education can we change our thinking from the present system and start thinking only in metric.

During the transition period, dimensioning may often have to be expressed in both the English and metric systems. Drafting professionals, therefore, must learn how to convert measurements from one system to the other.

1.4 Drafting Departments in Industry

Drafting departments vary in size from small units employing only a few people to extremely large departments where several hundred work. Obviously the organization of a small department is not nearly as extensive as that of a department where many people are involved. Regardless of size, however, each department must operate efficiently to produce quality drawings.

A new employee reporting on the job is shown the organizational structure of the drafting department. In a relatively small organization the structure will be comparatively simple, usually a group working together and supervised by one of their number. Drafting departments of considerable size are usually headed by a manager whose principal job is to direct all of the activities involving drafting services. In other cases, depending on the number of people employed, a department may be divided into smaller groups headed by squad bosses or group leaders. Each group is responsible for certain types of work and, in turn, each person is assigned a specific job to do.

The newly employed man or woman (see Fig. 1-10) must learn to understand the existing organizational structure, because the department was developed to produce maximum work efficiency. To maintain this efficiency, each employee must recognize the proper lines of communication and carry out assigned duties in accordance with departmental procedures and specific instructions.

1.4.1 Duties and responsibilities of drafting personnel. The principal function of those employed in drafting departments is to express engineering ideas in the form of mechanical drawings. Drafting personnel are required to produce drawings that are accurate and neat and to be able to complete them in the shortest possible time.

To insure the utmost degree of standardization of method and practice, industries often prescribe certain rules and regulations governing drafting practices. Newly employed personnel have to become familiar with these practices and adhere to them as closely as possible. There is often more than one correct way to make and dimension a drawing. The drafting standard practices adopted by the department are not intended to stifle originality but rather to provide an authoritative guide in the mechanics of drafting. If these recommended drafting practices are followed, fewer mistakes will be made, and more high-quality drawings will be produced.

Those in drafting will have to keep themselves informed of the latest specifications of materials, tools, and processes that are applicable to their assigned specialties. Although experienced checkers will make a final check of drawings, each person is expected to carefully check his or her completed work so that a minimum number of errors will be found by the checker. Above all, all instructions given by a supervisor must be carefully followed. Any deviation from these instructions may seriously misrepresent the design intent and cause unnecessary waste of time and costly delays. On the other hand, if an irregularity is discovered, the discrepancy should be discussed with a supervisor before any changes or corrections are made.

1.4.2 Competencies required of drafting personnel. Drafting personnel are expected to be knowledgeable and skilled in drafting techniques. They must be able to produce quality drawings which include linework, lettering, projection, sectioning, conventional representation, tolerancing, and dimensioning. To be efficient, they must also have a good understanding of manufacturing and assembly operations, materials, castings, forgings, screw threads, fasteners, gears, springs, and other related manufacturing processes and components.

Although the people responsible for preparing drawings may not be directly involved in the original design concept of a product, they must be able to communicate with manufacturing, inspection, and purchasing personnel concerning the design ideas developed by the designer or engineer. In the final analysis the success of a design depends greatly on quality drafting.

1.4.3 Employer-employee invention agreement. Drafting personnel, engineers, and other individuals who in their jobs do creative work are usually required to sign an agreement assigning all inventions to the company. It is accepted policy that if such individuals, while carrying out company work, should discover some new process or device, these processes or devices rightfully belong to the company. See the sample agreement shown in Fig. 1-11.

The agreement that employees sign also stipulates that they will not reveal their discoveries to any other person or competing company and that upon termination of their employment, they will make no attempt to infringe on the patented process or device for a prescribed period of time.

INVENTION ASSIGNMENT AGREEMENT

In part consideration of my Employment by

_____ DIVISION,

BORG-WARNER CORPORATION, an Illinois Corporation, and for the wages now and hereafter paid to me by said Division, I agree in connection with any work assigned to me to use my skill and ability to discover, make and invent new and useful improvements and inventions relating to the art and business of said Division, and to assist other employees in so doing, and to communicate to my immediate superior or to the Patent Department of BORG-WARNER CORPORATION, during the term of my employment, any and all discoveries, improvements and/or inventions relating to said art and business of said Division: and

With respect to any and all such discoveries, improvements, and inventions which I may conceive or make during the term of my employment, and for six (6) months thereafter, either solely or jointly with others, which relate to the art or business of said Division; or relate to the art or business for which BORG-WARNER CORPORATION is wholly or in part equipped and growing out of my contract with said art or business; I agree to assign, and by these presents do hereby assign and transfer, all of my right, title and interest in and to such discoveries, improvements and inventions to BORG-W   name of PORATION, its successes and assigns, and I agree, upon the request of said   by appropri-deliver all documents and do all acts necessary to secure to BO   Patent Law and the cessors and assigns, or its nominee, the right   mon. provements and inventions, including appli countries foreign   ivison or subsidiary of BORG-WARNER then I agree that said for Lett   ary may be included with or substituted for the above identified division and this agree-remain in full force and in effect until superseded by a new agreement or terminated by either party hereto.

This agreement shall be binding upon and the covenants and agreement herein contained shall insure to and be for the benefit of the parties hereto, their heirs, executors, administrators, legal representatives, nominees and assigns.

_____
Signature of Employee

WITNESS: _____

DATED AT _____ 19 _____

Fig. 1-11. Drafting personnel are often required to sign an agreement relinquishing all rights to new inventions which they discover or develop.

Most companies usually do not make any promises that additional compensation will be given to individuals if they invent a new design or make an important discovery. Nevertheless, a company will generally recognize this invaluable service by proper adjustment of salary and advancement in the company.

1.5 Description of Drafting Job Classifications

Although the duties and responsibilities of drafting personnel may vary from one industry to another, certain basic assignments are generally the same. In general, job classifications for drafting personnel include the following:

Chief Draftsperson or Drafting Manager.

1. Administers and coordinates all drafting functions in accordance with company policies.

2. Directs, delegates, and maintains work schedules of all personnel involved in design and drafting activities.

3. Prepares budgets and provides cost estimates on drafting and other documentation.

4. Counsels drafting personnel relative to company policies.

Fig. 1-12. Among other duties a technical clerk produces prints for distribution.

5. Evaluates the efficiency of drafting personnel.

6. Makes certain that all documentation projects are started on time and that they proceed according to established schedules.

7. Prepares progress reports on status of all documentation projects.

8. Makes certain that drafting personnel are kept abreast of new drafting standards.

Senior Designer or Design Group Leader.

1. Prepares or supervises the preparation of all types of drawings, sketches, and diagrams.

2. Maintains design logs.

3. Confers with personnel in engineering, manufacturing, and quality control of design implementation for which drawings are being prepared.

4. Evaluates design proposals as to their feasibility for manufacturing.

5. Performs related tasks as delegated by the drafting manager.

Senior Checker or Chief Checker.

1. Checks the accuracy of all drawings, tolerance accumulations, and adequacy of the design layouts

2. Supervises the correction and revision of drawings.

3. Confers with appropriate personnel relative to quality of drawings.

Senior Draftsperson or Drafting Group Leader.

1. Prepares detail drawings, assembly drawings, and wiring diagrams from design layouts.

2. Supervises the work of other personnel involved in producing required drawings.

3. Makes necessary revisions on drawings.

4. Carries out related assignments.

Junior Draftsperson or Detailer.

1. Prepares complex detail and assembly drawings and wiring diagrams and associated lists, working from design layouts.

2. Revises drawings with minimum instruction.

3. Performs related assignments as required.

Senior Technical Clerk or Supervisor of Engineering Records.

1. Maintains and controls a central file of drawings, prints, microfilm, parts lists, materials lists, and catalogs.

2. Prepares, edits, and correlates engineering documents such as change notices, parts lists, and bills of materials.

3. Distributes drawing standards and other engineering documents to appropriate personnel.

Technical Clerk or Specification Writer.

1. Prepares engineering releases for the purchase or manufacture of parts and assemblies.

2. Prepares engineering change notices.

3. Prepares engineering deviations, which authorizes appropriate personnel to deviate from drawings or specifications.

4. Prepares engineering stop orders, which halts or curtails the manufacture of certain parts or assemblies.

5. Keeps records of drafting assignments.

6. Maintains and operates reproduction equipment. See Fig. 1-12.

7. Distributes drawings and other engineering documents to appropriate personnel.

8. Files all drawings and other engineering documents.

1.6 Standards and Drafting Efficiency

The need for standardization of drafting practices is fully recognized by industry. This is clearly emphasized in the introductory statement of Chrysler Corporation's *Drafting and Design Standards:* "Standardization of methods and practices in the drafting room has always been very necessary due to the dissemination and interchange of drawings and engineering information. Increasing competition is forcing products into a design field of closed limits of positional and decimal tolerancing. Direction toward such standardized drafting room practices cannot help but result in savings of a great deal of time, material and expense, not only in the engineering departments but throughout the entire organization."[1]

To a considerable extent, the standardization of drafting practices has been achieved over the years because of efforts of standardizing groups such as the American National Standards Institute (ANSI) and the Society of Automotive Engineers (SAE); of industries such as the National Electrical Manufacturers Association (NEMA) and the American Gear Manufacturers Association

---

1. *Drafting and Design Standards* (Detroit: Chrysler Corp.).

(AGMA); and of governmental agencies such as the U.S. Department of Defense (DOD) with its Military Standards (MIL).

Drafting standards manual. Most industrial drafting departments have gone even a step further to insure standardization by preparing their own drafting standards manuals. For the most part a company's manual will contain those portions of ANSI, SAE, or MIL standards that are of special significance to the work of the company. In addition, the manual will include many other items, such as company drafting policy, engineering data, specifications of materials, and fabrication processes.

Drafting personnel are expected to become familiar with their company's manual and to use it as the authoritative guide and source for reference whenever a question arises. Those who follow the manual usually make fewer mistakes, thus simplifying the job of the checker and causing less confusion in the shop where the drawings are to be used.

Drafting department library. Almost every modern drafting department maintains a library consisting of reference books, manufacturers' catalogs, trade periodicals, handbooks, and standards. See Fig. 1-13. This fund of information is readily available to all drafting personnel and is intended to help them keep constantly up-to-date.

Personal technical file. Conscientious employees often keep personal files to maintain a

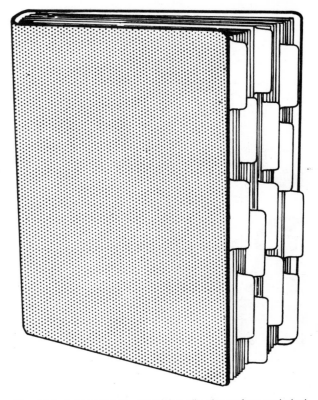

Fig. 1-14. A personal file containing clip sheets from technical magazines and other trade literature keeps a person up-to-date.

current line of communication to new technical information in their field. The file is frequently in the form of a three-ring loose-leaf binder having suitable dividers. See Fig. 1-14. Into this binder clippings from technical magazines, journals, and manufacturers' literature can be inserted. The loose-leaf file makes it easy to add or delete material and has the advantage of expandability and portability. Placing tabs on the divider leaves makes each section of the file readily accessible.

1.7 Professional Classifications of Drafting

The field of drafting encompasses a wide range of specialties. In large industries drafting personnel are usually assigned to specific drafting areas, whereas in smaller industries the work often crosses specialty lines. Even though a person works primarily in one specific area of drafting, a basic knowledge of other drafting fields will often prove

Fig. 1-13. Modern drafting departments maintain libraries of current technical information.

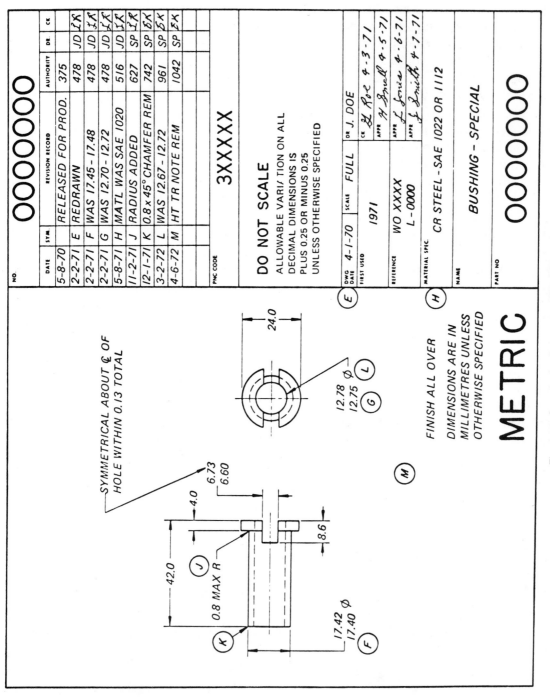

Fig. 1-15. Sample detail drawing. (MIL-STD-100)

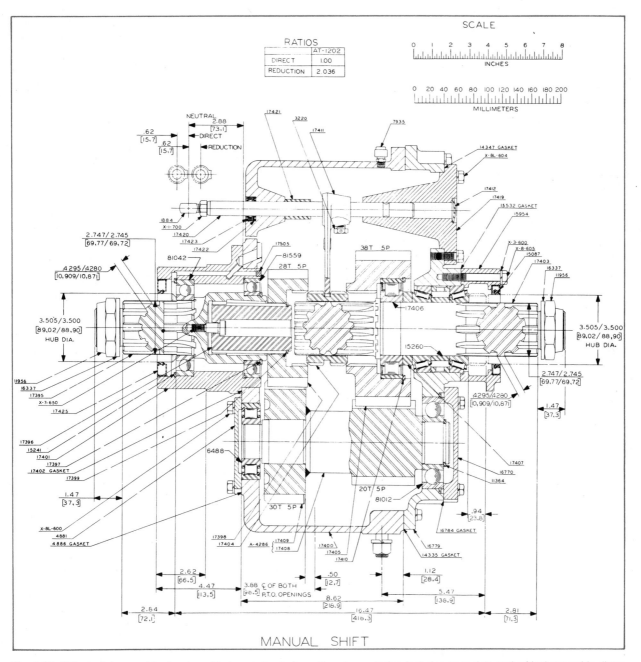

Fig. 1-16. This partial assembly drawing of twin countershaft auxiliary transmission includes gear data and critical assembly dimensions in both decimal-inch and metric sizes.

indispensable. The individual who has a wide range of drafting competencies also has better opportunities for advancement.

The recognized professional specializations in drafting are:

1. Product Manufacturing drafting.
2. Architectural drafting.
3. Structural drafting.
4. Tooling drafting.
5. Electrical drafting.
6. Piping drafting.
7. Fluid power drafting.
8. Map drafting.
9. Patent drafting.

1.7.1 *Product manufacturing drafting.* The area of manufacturing drafting deals with the preparation of drawings that are to be used in the manufacture of a product. The drawings may range from a very simple one-view layout to elaborate detail and assembly documents of complex designs. In all instances the primary consideration is the development of drawings that will show exactly how a part is to be manufactured. For some products only a relatively small number of drawings is required, while for other products a great many drawings may be necessary.

Production drawings can be classified into two broad groups—detail drawings and assembly drawings. *Detail drawings* contain the necessary views and dimensions for producing each individual component part. *Assembly drawings* show how the various parts are to be put together. See Figs. 1-15 and 1-16 for examples of typical detail and assembly drawings.

1.7.2 *Architectural drafting.* Architectural drafting is concerned with the development and preparation of plans for erecting residential and commercial buildings. The actual planning and development of working drawings for a building is done by a certified architect. The mechanical engineer plans the building's plumbing, heating, and ventilating systems, while the electrical engineer plans the necessary circuits and fixtures for power and lighting. For small residential structures the architect or the architect's staff usually does all the planning and layout work.

Actual building planning involves four major stages: (1) analysis of building requirements and costs, (2) preparation of preliminary sketches to show principal room arrangements, (3) preparation of display drawings which contain perspective views of the building, and (4) execution of working plans which cover all essential building construction details, Figs. 1-17, 1-18, and 1-19.

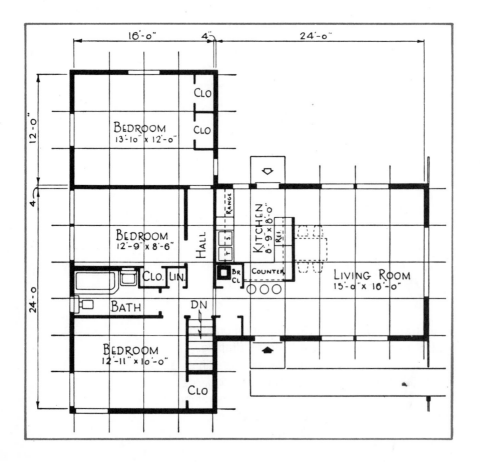

Fig. 1-17. A typical floor plan of a house.

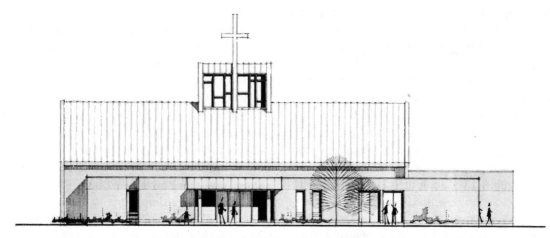

Fig. 1-18. Example of a display drawing.

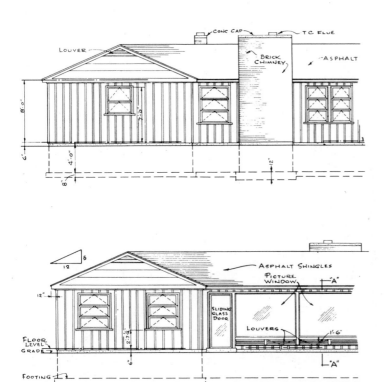

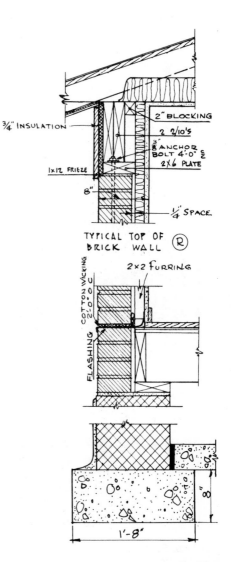

Fig. 1-19. Working plans show all of the construction details of a building.

1.7.3 Structural drafting. Structural drafting deals specifically with the preparation of plans for the construction of commercial buildings, industrial plants, schools, hospitals, bridges, and other large structures. See Fig. 1-20. In the preparation of structural drawings, special techniques are used to present form and shape, and symbols are frequently employed to describe structural elements.

The two most common types of structures are reinforced concrete and steel frame. In reinforced concrete structures the columns, beams, floors, stairways, and so on, are made of reinforced concrete with an exterior facing of brick or other masonry materials backed up by lightweight concrete blocks. Steel frame buildings have steel members as the principal framing elements. The structural members are joined by welding, riveting, or bolting. The exterior facing is also of brick or some other masonry material backed up by concrete blocks.

Steel frame structural planning will usually involve two sets of drawings: design drawings and shop drawings. *Design drawings* include all the essential views required for the erection of a building, such as foundation, room arrangement, elevations, and construction details. An important function of a design drawing is to show the steel framing plan of the building. See Fig. 1-21. *Shop drawings* present details of how various pieces are to be fabricated for assembly. See Fig. 1-22.

Fig. 1-20. Four thousand engineering drawings and eighty-five thousand prints were used in the construction of the Michigan-Mackinac Bridge.

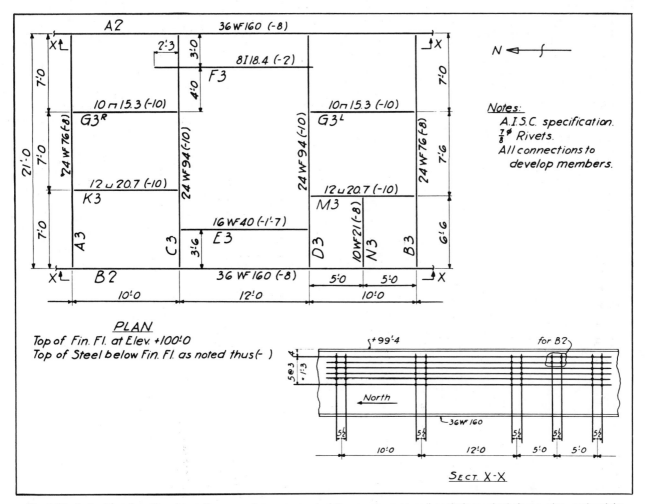

Fig. 1-21. A steel framing plan shows the general arrangement of the members as well as their required shape, size, and weight.

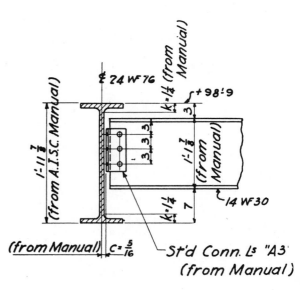

Fig. 1-22. A typical shop drawing showing a square-frame beam unit.

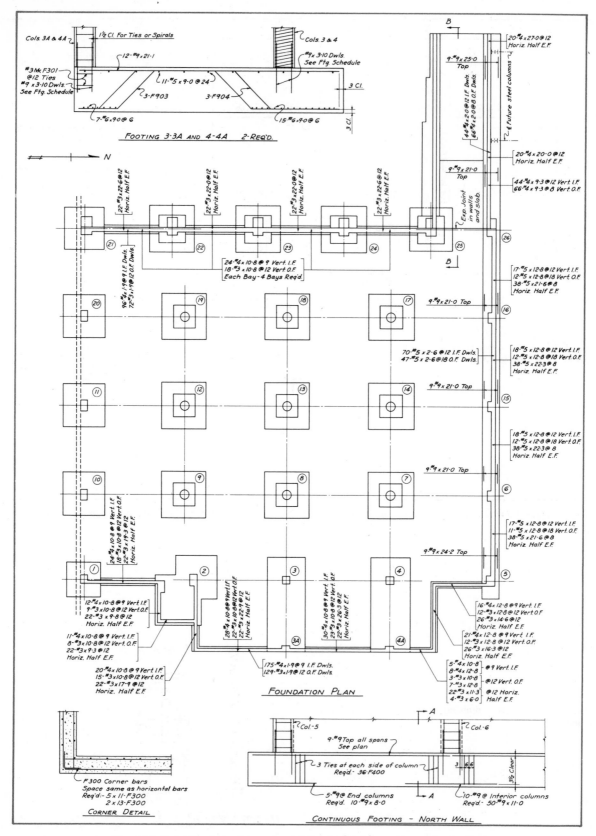

Fig. 1-23. Example of an engineering drawing for a reinforced concrete structure.

Two sets of drawings are also used for reinforced concrete structures, but these are designated as engineering drawings and placing drawings. *Engineering drawings* deal with the general arrangement of the structure, its size, and reinforced members. *Placing drawings* indicate the size, shape, and location of all bars in the structure and, in addition, contain a schedule of beams, joints, columns, and girders. See Fig. 1-23.

1.7.4 Tool drafting. Mass production of finished goods involves the use of countless tools and machines to perform a vast number of operations, many of them complex. To achieve quality products on a low-cost and competitive basis, production tools must be designed that will permit rapid and economical fabrication. The responsibility of producing such equipment rests with the tool designer or tool engineer.

Tool design is a highly specialized field involving the creation of jigs, fixtures, dies, and gages. The individual concerned with this phase of engineering must have a good knowledge of materials, a

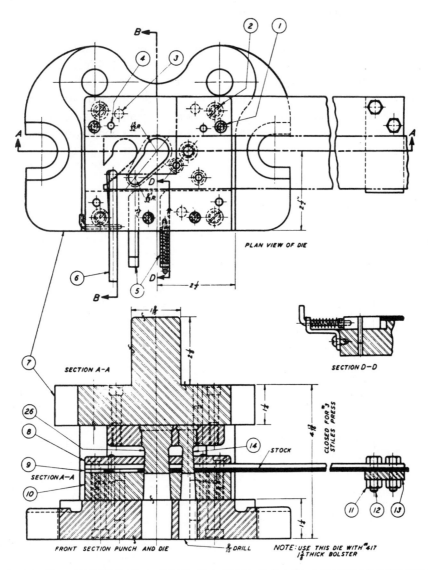

Fig. 1-24. A progressive piercing and blanking die equipped with a stripper to free the metal and a locating stop for correct positioning of the metal.

general understanding of shop practices, mechanical ingenuity, and skill in drafting.

Jigs are devices which are used in performing identical operations easily and rapidly with uniformity of precision. Jigs are specifically adapted for such operations as drilling, reaming, counterboring, and tapping. Their main functions are to hold the workpiece in position and to guide the cutting tool. They are usually movable on the worktable.

Fixtures are also locating and holding devices but, unlike jigs, they are clamped in a fixed position and are not free to move or guide the cutting tool. They are used to perform operations requiring facing, boring, milling, grinding, welding, and so on.

Dies are tools used to produce forgings and sheet-metal stampings. The two main classifications of dies are cutting dies and forming dies. *Cutting dies* perform such operations as blanking, trimming, piercing, shearing, and notching. *Forming dies* are designed to shape metal parts and

are used in operations involving bending, flanging, embossing, beading, and drawing. Very often dies carry out several operations in a single stroke of the press. Thus, the die may be designed to perform blanking and piercing or blanking and forming operations.

Gages are used for checking purposes to insure a uniform degree of accuracy. The many different types of gages are classified as work gages, inspection gages, and master gages. *Work gages* are those employed by an operator for checking a particular operation. *Inspection gages* are used by the inspector for checking the finished product. *Master gages* are reference tools intended primarily to check inspection gages periodically.

Drawings of simple tools are often confined to a single assembly drawing which incorporates all of the essential details. See Fig. 1-24. For more complicated tools both an assembly drawing and parts detail drawings are prepared. In either case standard drafting practices are followed to provide the necessary shape description.

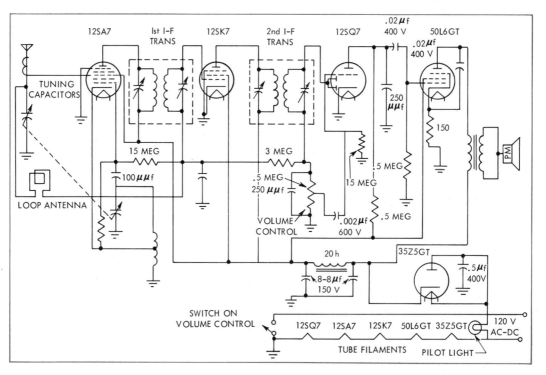

Fig. 1-25. Schematic diagram of a radio receiver.

1.7.5 Electrical drafting. Electrical and electronics drafting involves the preparation of drawings which show the circuitry and fixture arrangement of devices operated by electricity. The drawings are actually a form of schematic diagrams designed for architectural function, power distribution, or communication systems. One characteristic of these diagrams is the extensive use of graphic symbols. See Fig. 1-25.

Electrical and electronics diagrams are often designated as single-line, wiring, highway, baseline, or block. (See Unit 24.) Although each type is intended to serve a particular function, they all have the same basic purpose—to outline the path of electrical flow from its point of origin to the various operational units.

1.7.6 Piping drafting. Piping drawings show the location, identification, and sizes of pipe and fixtures in a piping system. Very often graphic symbols are used to designate pipes, valves, and fittings. Drawings are either one-, two-, or three-view projections. They may also be isometric, oblique, or schematic. See Fig. 1-26.

Single lines or double lines are used to represent piping units. Double-line representation is normally found where the system is made up of large pipe. Where the pipe is relatively small, the single-line representation is more common. (See Unit 22.)

1.7.7 Fluid power drafting. Fluid power drawings are mostly schematic diagrams that show the principal units in a hydraulic system. (See Unit 23.) These diagrams will indicate the piping between components, positions of moving parts, and the flow path of the fluid. Standard graphic symbols are used to depict function and nature of hy-

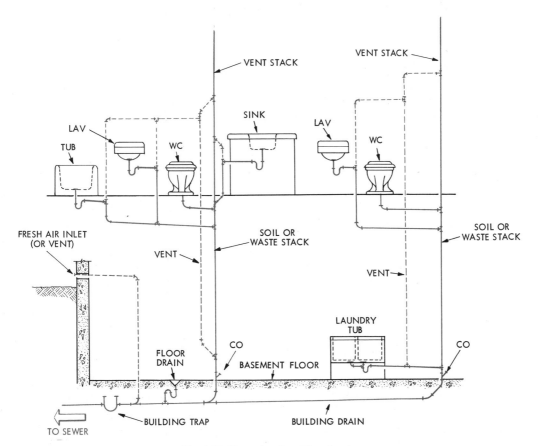

Fig. 1-26. Example of a piping drawing.

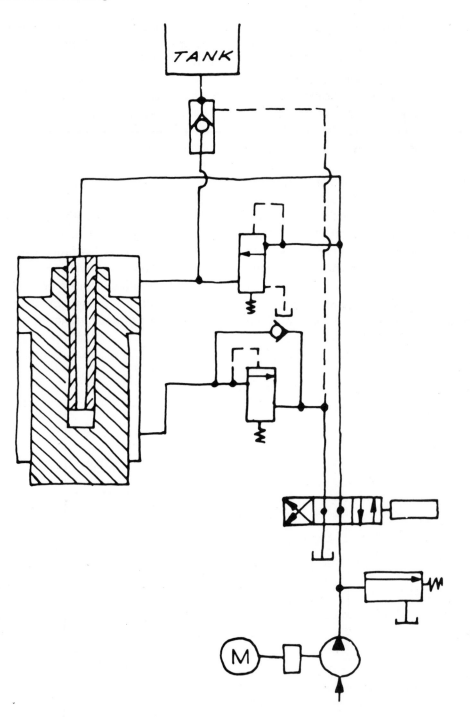

Fig. 1-27. A diagram of an industrial press circuit.

draulic fixtures. A typical fluid power diagram is illustrated in Fig. 1-27.

1.7.8 Map drafting. Map drafting deals with the preparation of such drawings as cadastral, topographic, plat survey, building site, hydrographic, aeronautical, and road maps.

Cadastral maps show layouts of cities, towns, or county districts. These maps serve as records of

Fig. 1-28. This is a portion of a cadastral map.

land ownership and may include locations of gas mains, water lines, sewage systems, steam pipes, and fire hydrants. See Fig. 1-28.

A *plat map* is a land map that contains an accurate description of a tract of land. It indicates the true lengths and bearings of boundary lines and establishes the exact location in reference to section and township divisions. See Fig. 1-29.

Building site (engineering) maps are plot plans that show the orientation of a residential or indus-

trial building and its surrounding area. Included in this layout will be such features as ground elevation contours, property boundaries, utilities, walks, and drives which are identified with the ground. See Fig. 1-30.

A *landscape map* is one which shows how a ground area is to be landscaped. It is used by the architect and landscape gardener to ascertain the most pleasing arrangement of shrubbery, trees, flower beds, and so on.

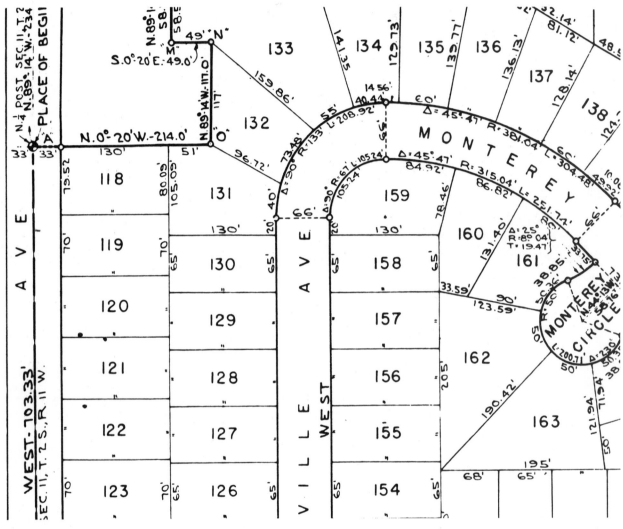

Fig. 1-29. A typical plat map.

Topographic maps are large area maps that show by means of symbols features such as lakes, streams, rivers, hills, valleys, roads, railroads, bridges, cities, towns, villages, electric power lines, and other identifying land shapes. See Fig. 1-31.

Hydrographic maps contain information concerning bodies of water, shorelines, sounding depths, reefs, shoals, and so on. Included in this group of maps are plan charts, which show harbor and anchorage details; coasting charts, which plot outside courses along a coast; approach charts,

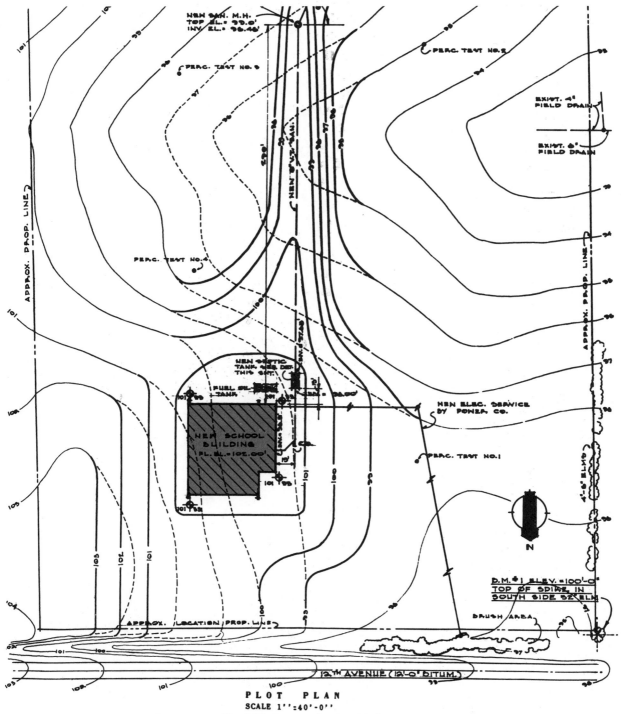

PLOT PLAN
SCALE 1''=40'-0''

Fig. 1-30. An example of a plot plan.

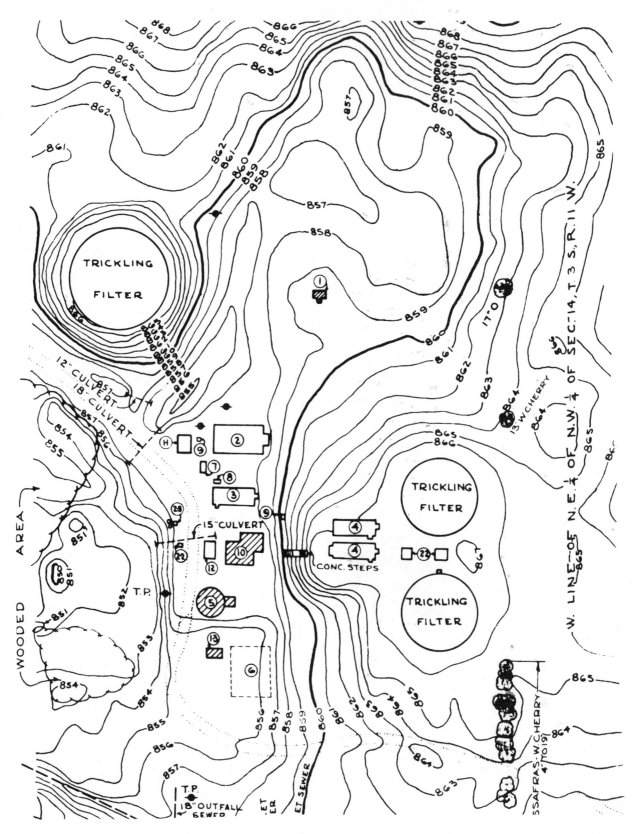

Fig. 1-31. A topographic map depicts the essential features of large ground areas.

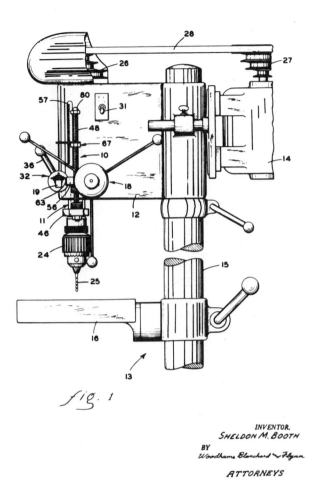

Oct. 20, 19   S. M. BOOTH   2,909,082
AUTOMATIC FEED AND DEPTH CONTROL FOR A DRILL PRESS
Filed Sept. 6, 19   3 Sheets—Sheet 1

Fig. 1-32. An example of a patent drawing.

which present features to enable a ship to make a particular approach to a harbor; and ocean charts for ocean navigation.

Aeronautical maps contain cultural and ground shape features for air navigation purposes. They include details and traffic routes to permit navigation by visual ground references and information to utilize radio and electronic navigation aids.

Road maps are maps used by motorists. They show the network of roads and highways laid out over a given section of land.

1.7.9 Patent drafting. Patent drafting involves the preparation of special drawings that are submitted to the United States Patent Office when an application is made for a patent for an invention. See Fig. 1-32. The drawings must show and identify all features of the invention and be presented in the form prescribed by the patent office. (See Unit 10.) Shape description may include projected views, pictorial views, or a combination of projected and pictorial views.

UNIT 2

Equipment and Materials for Drafting

To have optimum value, industrial drawings must be clear, concise, and subject to but one interpretation. In order to produce drawings that conform to accepted standards and practices, certain types of equipment, drafting materials, and instruments, are used. Since time is an important element in any industrial work, a clear understanding of all the drafting tools and drawing techniques is important to speed up the process of drawing preparation.

2.1 Drafting Table and Board

Most industrial drafting is done on tables similar to those shown in Fig. 2-1. Although the construction details may vary somewhat on different tables, in general they are made either to a standard height or so that they can be adjusted to any desired working height. A turn of a hand knob or lever also permits the top to be regulated to vari-

Fig. 2-1. A modern industrial drafting room.

ous angles; on some tables the top can be moved to full easel position.

The table top is usually covered with green or buff linoleum or vinyl, which not only minimizes

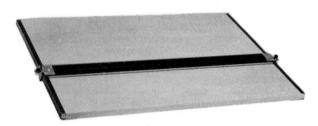

Fig. 2-2. The drawing board is used when drafting tables are not available.

glare but provides a smooth firm working foundation under the drawing sheet, thereby helping to produce sharp clear-cut pencil lines. It also makes erasing easier.

The type of drawing board shown in Fig. 2-2 is used primarily for fieldwork. Sometimes it is found in schools when professional drafting tables are not available. These boards are made of either white pine or basswood and come in a variety of sizes. The 20 inch × 26 inch is the most practical for ordinary drawing and for portability.

2.2 Drafting Machine

A drafting machine consists of a base unit which is controlled by a protractor head. Scales are attached to the protractor head, which makes it pos-

Fig. 2-3. Track type of drafting machine.

sible to draw lines at any angular intervals. On one type of drafting machine, the base unit is fastened to a vertical track which moves over a horizontal track. See Fig. 2-3. On the other type of drafting machine, the base unit is mounted to pivoting arms. See Fig. 2-4.

The drafting machine eliminates the need for separate scales, triangles, protractor, and straightedge. All of the essential drawing elements are combined in such a manner as to permit their in-

stant use. The time-saving value of the machine lies in the fact that many operations, such as drawing horizontal and vertical lines and measuring and laying out angles, can be performed without resorting to individual instruments. The centralized control units allow all of these operations to be accomplished with one hand, leaving the other one free for drawing. Drafting machines are produced both for left-handed and for right-handed persons.

Fig. 2-4. Drafting machine with pivoting arms.

2.3 Straightedges and Triangles

Straight lines are drawn with the aid of a parallel unit, T-square or triangle.

2.3.1 Parallel unit. This type of straightedge is supported at both ends and automatically maintains parallel motion. It may be moved up or down by applying slight pressure at any point along its length. See Fig. 2-5.

The parallel straightedge can be mounted on either the drafting table top or on the drawing board. It is operated by a cord which runs through both ends of the straightedge. The arrangement of the cord and guiding pulleys varies depending on the manufacturer of the equipment.

To prevent smudging of lines, some parallel straightedges have spring-mounted rollers on the underside. These rollers raise the straightedge slightly and keep it off the drawing sheet.

2.3.2 T-square. The T-square, shown in Fig. 2-6, is rarely used in professional drafting rooms. Its greatest use is in fieldwork, where a drafting machine or a parallel straightedge may not be available.

2.3.3 Drawing horizontal lines. Horizontal lines are always drawn from left to right. To use a straightedge to draw horizontal lines, place the pencil point against the working edge of the

Fig. 2-6. The T-square is used mostly for fieldwork.

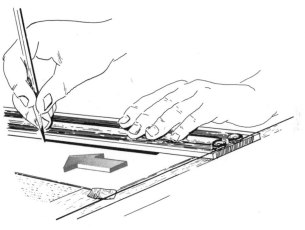

Fig. 2-7. Drawing horizontal lines.

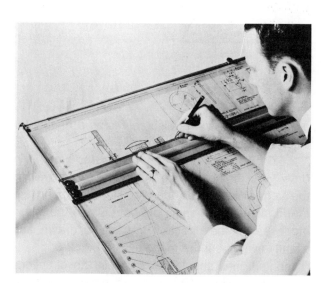

Fig. 2-5. A typical parallel straightedge.

straightedge. Hold the pencil at an angle of approximately 60°, tilting it in the direction of the line being drawn. Also incline the pencil slightly away from the working edge; this will keep the pencil point closer to the straightedge. See Fig. 2-7. It is good practice to rotate the pencil between the thumb and fingers when drawing the line, because such a motion keeps the conical pencil point symmetrical for longer periods and produces more uniform lines.

2.3.4 Triangles. Triangles are used for drawing vertical and slanted lines. The two standard triangles employed for this purpose are known as the 45° triangle and the 30°-60° triangle. They are made of clear plastic and come in a variety of sizes.

However, the adjustable triangle has become the most popular for drafting purposes. See Fig. 2-8. Since it combines the two standard triangles,

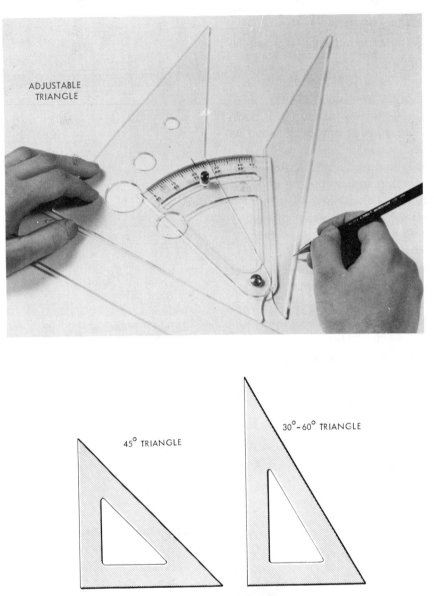

Fig. 2-8. Types of triangles.

it eliminates the need to switch from one triangle to another when lines of different angles must be drawn. Another advantage of the adjustable triangle is that it has a protractor, which permits the drawing of any angle from 0° to 90° direct from the baseline. The protractor has two rows of graduations—the outer row indicates angles from 0° to 45° from the longer base, and the inner row indicates angles from 45° to 90° from the shorter base. The graduations are marked to half degrees, and the adjustable arm is held firmly at any angle by a clamp screw that also serves as a handle for lifting or moving the instrument.

2.3.5 Drawing vertical and slanted lines. To draw vertical lines, move the pencil upward along the vertical leg of the triangle. Hold the pencil so that it slants at an angle of about 60° to the drawing sheet in the direction the line is drawn and slightly away from the triangle. See Fig. 2-9.

Slanted lines should be drawn in the directions illustrated in Fig. 2-10 in order to insure a high degree of accuracy.

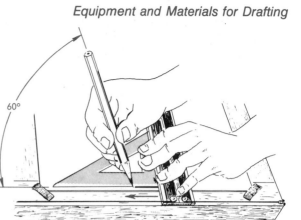

Fig. 2-9. This is how vertical lines are drawn.

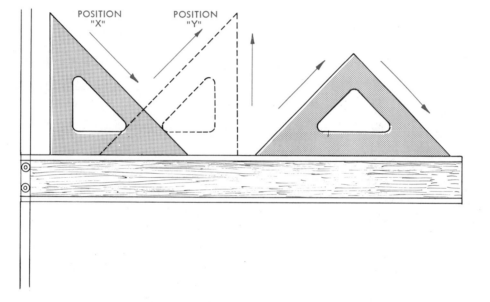

Fig. 2-10. Notice the direction in which the pencil should move in drawing slanted lines with the aid of triangles.

2.4 Scales

The scale has a twofold function: (1) to measure distances accurately and (2) to produce drawings to certain sizes, that is, making drawings to scale. Drawing to scale simply means that objects are represented on paper either in full size, reduced size, or enlarged size. For some parts the dimensional units must be reduced to properly make the drawing on the required size sheet. The size of other objects may have to be increased to clearly show all the necessary details. Scales are constructed to permit production of drawings to any desired size.

Scales are available in either flat or triangular shapes. See Fig. 2-11. An advantage of the triangular scale is that more measuring faces are incorporated on one stick. However, with such a scale, a scale guard, as shown in Fig. 2-12, is often used to keep the required measuring edge in position.

Professionals in drafting use the flat scale almost exclusively, because it is easier to handle and makes the working face more readily available. The only limitation is that several sticks with different graduations must be kept on hand. But since only one scale is used for extended periods,

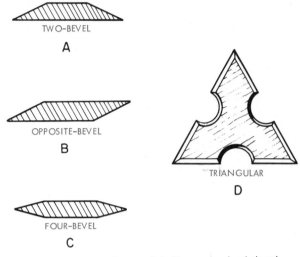

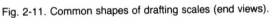

Fig. 2-11. Common shapes of drafting scales (end views).

Fig. 2-12. This scale guard keeps a triangular scale in position for instant use.

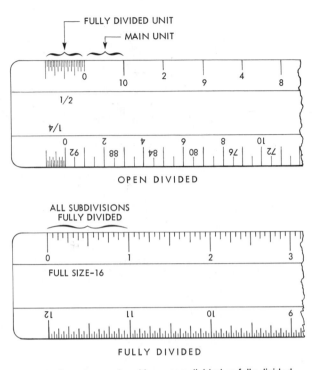

Fig. 2-13. A scale may be either open divided or fully divided.

the need for a number of sticks is not particularly objectionable.

Flat scales are manufactured in several shapes. The *two-bevel* scale has the advantage of a wide base with both scale faces always visible. See Fig. 2-11*A*. The *opposite-bevel* scale is easy to handle and easily lifted from the board by tilting. See Fig. 2-11*B*. The *four-bevel* scale has four faces and is especially convenient as a pocket rule in the 6 inch and 150 mm sizes. See Fig. 2-11*C*.

Scales are either open divided or fully divided See Fig. 2-13. *Open divided* scales are those on which the main units are numbered along the whole length of the edge, with an extra unit fully subdivided in the opposite direction from the zero point. The subdivided unit shows the fractional graduations of the main unit. Open divided scales often have two complete measuring systems on

one face—one double the other and reading in opposite directions.

Fully divided scales have all the subdivisions along the entire length of the stick, so that several values from the same origin can be read without having to reset the stick. They are sometimes double numbered, either to permit both right-to-left and left-to-right reading or to provide two different scales on one face.

2.4.1 Decimal scale. The full-size decimal scale, which is the most common for preparing industrial drawings, has the inch divided into 50 units with each unit being equal to 0.02 inch. See Fig. 2-14. This scale is particularly suited for use as the design unit in the decimal-inch system of measurement.

2.4.2. Metric scale. Metric scales are available in various lengths such as 15, 30, 45, and 60 cm. The most common size is the 30 cm (300 mm). See Fig. 2-15.

Just as there are different inch scales so drawings can be made to different sizes (⅛, ¼, ½, ¾, etc.) so too are metric scales produced for the same purpose. Standard scales for mechanical

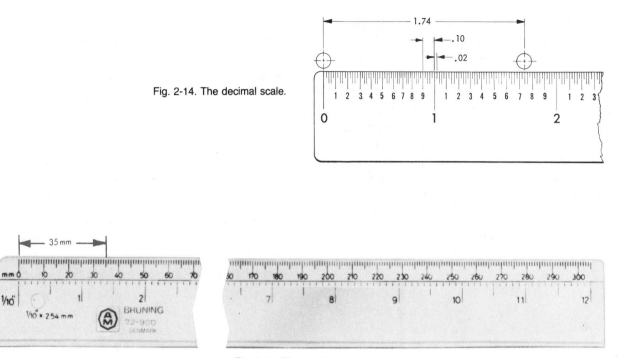

Fig. 2-14. The decimal scale.

Fig. 2-15. The metric scale.

drawings and for tool and die design drawings are usually full-size, ½-size, ¼-size and ⅛-size. Construction and architectural drawings generally are made with metric graduations of 1:1, 1:25, 1:50, and 1:100.

2.4.3 Mechanical Engineer's scale. This scale is used for drawing machine and structural parts. The measuring units are designed to produce drawings that are to be ⅛-size, ¼-size, ½-size, or full-size. The graduations represent inches and fractional parts of inches. Thus, to draw an object to a ¼-size, the ¼-size measuring face would be used. Each main division on this scale represents one inch. The fractional parts of the inch are indicated by small division lines located along the entire edge (fully divided), or only on one unit opposite the zero mark (open divided). See Fig. 2-16.

2.4.4 Architect's scale. The Architect's scale is used for making drawings of buildings as well as of structural elements. This scale provides a wide range of scale reductions. The basic measuring faces are designated as 3, 1½, 1, ¾, ½, ⅜, ¼, ³/₁₆, ³/₃₂, ⅛, and full size. In each case these scales represent the proportions to which drawings can be reduced in terms of feet and inches. Thus,

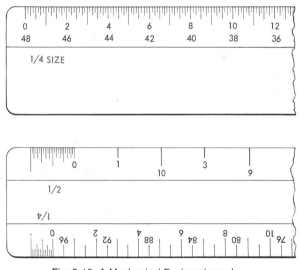

Fig. 2-16. A Mechanical Engineer's scale.

if the measuring edge is marked with a 3 it means *3 inches represents one foot*. The edge labeled 1½ indicates that 1½ inches represents one foot, and so on.

2.4.5 Civil Engineer's scale. The Civil Engineer's scale is used primarily where large reductions are necessary, such as in making maps and charts. The scale is divided into 10, 20, 30, 40, 50, and 60 parts to the inch.

2.4.6 Laying out accurate measurements.
Measurements should not be taken off the scale with dividers and laid off on the drawing. This practice ruins the scale and tends to produce errors. To get accurate results, place the scale on the drawing, and make a small mark with the point of the pencil next to the graduation desired. See Fig. 2-17. When a number of distances are to be laid off end-to-end on the same line, mark off all

the distances without moving the scale. If each distance is laid off individually by moving the scale to a new position each time, there is always a possibility that cumulative errors will be introduced. See Fig. 2-18.

2.5 Protractors

A protractor is a semicircular form divided into units called degrees. Two scales are shown on the protractor, with each scale having units running from 0° to 180°. See Fig. 2-19. The outer scale is designed for laying out angles that extend to the left and the inner scale for angles that must be laid off to the right. This type of protractor is used only when an adjustable triangle is not available.

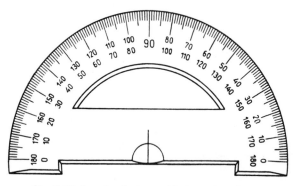

Fig. 2-19. A protractor is used to lay out angles.

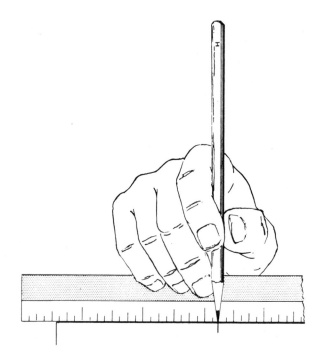

Fig. 2-17. For accurate results lay out dimensions in this manner.

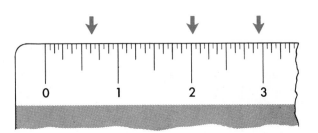

Fig. 2-18. Avoid cumulative error by laying out all distances without moving the scale.

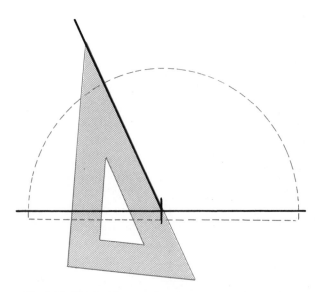

Fig. 2-20. This is the way to lay out angles with a protractor.

To use a protractor, place it on a straightedge with its center point on the mark where the angle is to be located, as shown in Fig. 2-20. Find the desired angle on the protractor, using either the inner or outer scale, and mark it with a point. Remove the protractor and, with the aid of a triangle or straightedge, draw a line connecting the two points.

2.6 Curves

When it is necessary to draw curves other than standard circles or arcs, special devices known as irregular curves are used. These devices, as may be seen from Fig. 2-21, are available in many dif-

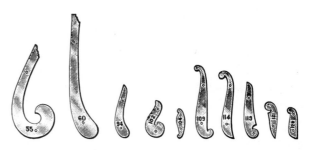

Fig. 2-22. These curves are especially useful to the industrial draftsperson. Numbers identify curves.

ferent shapes and sizes and can produce practically any curvature.

Mechanical Engineer's Curves are a group of irregular curves, usually ten in a set, selected to provide patterns of curves commonly encountered in the work of a mechanical engineer, Fig. 2-22.

Copenhagen Ship Curves are another set of curved patterns which are often used for work requiring more extended curvatures. These curves are not known to be based on any mathematical formula or principle. Undoubtedly they were originally laid out to combine two arcs which had been proven over the years to be of practical curvature for hull design.

Railroad Curves are especially designed to meet highway and road engineering requirements. They are cut to definite radii in which one inch equals 100 feet. Both edges of the curves have the same radius. Each curve is usually stamped with its designated degree of radius and tangency value.

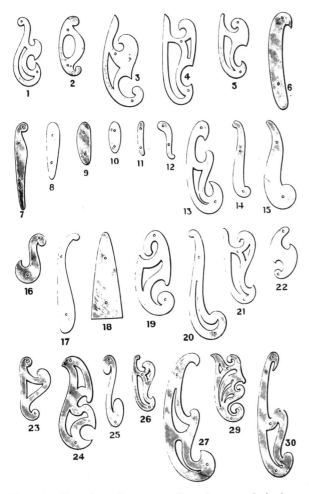

Fig. 2-21. These irregular curves will produce any desired curvature. Numbers identify curves.

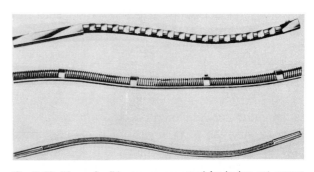

Fig. 2-23. These flexible curves are used for laying out curves not easily located on rigid irregular curves.

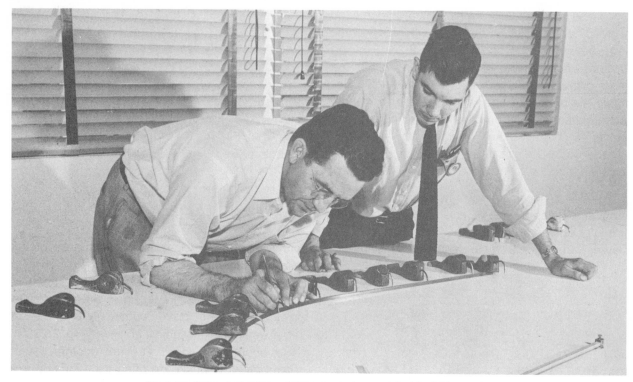

Fig. 2-24. Spline and spline weights are used a great deal for lofting work.

Flexible curves consist of long narrow strips of flexible metal, with either metal or plastic ruling edges, which can be bent to fit any desired curvature. See Fig. 2-23. The spline, as shown in Fig. 2-24, is another type of flexible curve used considerably in lofting work where full-size contours of objects must be laid out. Spline weights or "ducks" hold the spline in any desired position.

2.6.1 Drawing irregular curves. To draw an irregular curve, first lay off a series of points to indicate the shape of the curve, and sketch in a very light line connecting these points. Then select a part of the irregular curve or Mechanical Engineer's curve that fits a portion of the line, as shown in Fig. 2-25. Arrange the curve so that its curvature increases in the direction the curvature of the line increases. Be sure the curve used matches the curved line to be drawn for some distance beyond the point where the curved line and irregular curve appear to coincide. If this is done, each successive portion of the curved line will be tangent to the other without any abrupt breaks in

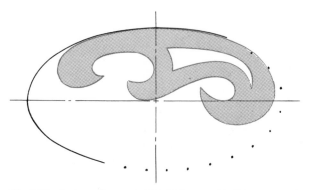

Fig. 2-25. An irregular curve should be used in this manner to secure a smooth line.

the line. Continue to move the irregular curve to new positions until the entire line is completed.

2.7 Compasses

Compasses are used to draw circles and arcs. Several types are available to meet different requirements.

2.7.1 Bow compass. The bow compass is available in several sizes for drawing circles and arcs of different diameters. Most bow compasses are made to hold either a lead or a pen. See Fig. 2-26. If a lead is used, shape the lead to a bevel point as shown in Fig. 2-27. Use a lead that is one grade softer than the pencil employed for straight-line work, because heavy pressure cannot be exerted on the compass to produce lines as dark as those made with a pencil.

To draw circles and arcs, adjust the point of the compass so that it is slightly longer than the lead. After the center of the arc or circle is located, set the compass point at the center mark and adjust

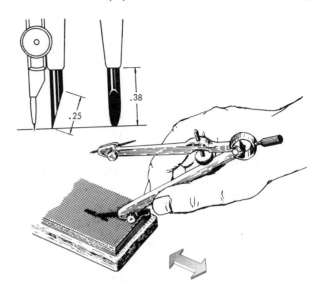

Fig. 2-27. Shape the lead of the compass to a bevel point.

Fig. 2-26. A typical bow compass for drawing circles and arcs.

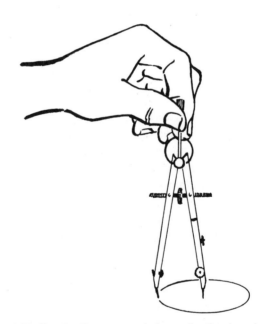

Fig. 2-28. Revolve the compass between the thumb and forefinger to draw the circle.

the compass to the required radius. Hold the stem of the compass between the thumb and forefinger. Draw the circle in a clockwise direction, tilting the compass slightly in the direction in which the circle is being made. Complete the circle in one sweeping motion. See Fig. 2-28.

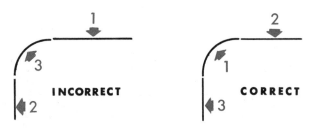

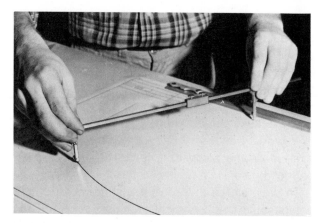

Fig. 2-29. Always draw the arc first and then connect the straight lines to the arc.

Fig. 2-31. Hold the beam compass in this position to draw large circles and arcs.

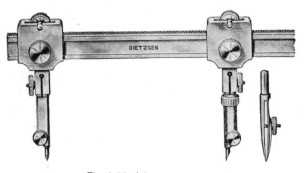

Fig. 2-30. A beam compass.

Fig. 2-32. A drop spring bow compass.

When arcs and straight lines tangent to them are required, it is best to draw the arcs first, since it is easier to connect straight lines to an arc than to adjust the arc to the straight lines, Fig. 2-29.

2.7.2 Beam compass. The beam compass, or trammel, is designed for drawing large circles and arcs that cannot be made with a regular compass or for transferring distances when they are too great to be done with dividers. See Fig. 2-30.

Beam compasses come in various sizes and consist of a bar with two movable point assemblies to hold either lead, divider points, or pen. Some beam compasses are equipped with a micrometer adjustment for accurate settings.

To use a beam compass, set the points to the required radius, employing the micrometer adjustment for accurate settings. Hold the beam compass with two hands as shown in Fig. 2-31 and draw the circle or arc.

2.7.3 Drop spring bow compass. The drop compass is a very useful instrument for drawing very small circles and arcs. Either a pen or a lead assembly can be inserted. See Fig. 2-32. The in-

strument is held by the center pin and the point is raised or lowered with the thumb and third finger. The pen or lead point is turned around the central axis with the third finger.

2.8 Dividers

Dividers, as shown in Fig. 2-33, are used for dividing distances into equal parts, for transferring distances, or for setting off a series of equal spaces.

Fig. 2-33. Dividers.

2.8.1 *Using dividers.* To divide a line into equal parts, adjust the dividers to the approximate space required. Step off the distances, rotating the dividers first to the right and then to the left until the number of units required have been stepped off. See Fig. 2-34. If the line falls short of or beyond the given line, lengthen or shorten the dividers proportionately and repeat the operation. For example, to divide a line into four equal parts, set the dividers by eye to approximately one-quarter of the length of the line. If, after this distance is stepped off, it is found that the distance between the divider points is too small, increase it by one-quarter of the remaining length of the line. Similarly if the last prick of the dividers is beyond the end of the given line, shorten the dividers one-quarter of the extended distance.

To transfer a distance from one part of the drawing to another, simply adjust the dividers to the required length and press the points lightly on the paper in the new position. Avoid too much pressure on the dividers; it will leave unsightly holes in the paper and also may cause the dividers to spread.

To space off equal distances, set the points of the dividers to the required length. Hold the dividers as shown in Fig. 2-34, and set off the spaces by rotating the dividers between the thumb and forefinger. Rotate the dividers first on one side of the line and then on the other to equalize any accumulating error.

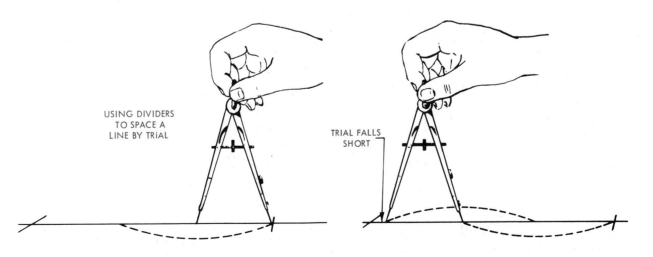

USING DIVIDERS
TO SPACE A
LINE BY TRIAL

TRIAL FALLS
SHORT

Fig. 2-34. Dividers should be used to space off equal distances or to divide a line into equal parts.

Fig. 2-35. Proportional dividers are time-saving devices when making copies of drawings to a specified scale.

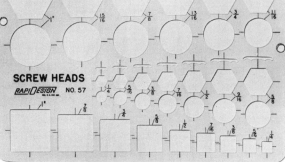

Fig. 2-36. Templates are time-saving devices for drafting.

2.8.2 Proportional divider. When it is necessary to make copies of drawings to an enlarged or reduced scale, a proportional divider is often used. See Fig. 2-35. This instrument permits both reproducing the lines of a drawing so that the lines in the copy are of a known ratio to the original and producing a drawing so that the volume of a solid or area of a plane surface will be in a known proportion to the original.

2.9 Drafting Templates

A variety of templates have been designed to minimize many time-consuming drafting operations. Templates such as these shown in Fig. 2-36 save hours of tedious work in laying out holes, standard symbols, and other figures that are frequently required in plans, drawings, and sketches.

2.10 Drafting Pencils and Lead Holders

Drafting pencils are made with leads of different grades of hardness. The hardness is designated on the pencil by numbers and letters. These symbols range from 7B, which is very soft, through 6B, 5B, 3B, 2B, B, HB, F, H, 2H, 3H, 4H, 5H, 6H, 7H, and 8H, to 9H, which is the hardest.

The mechanical lead holder is often preferred to pencils because its length remains constant and because it can easily be refilled with new leads. See Fig. 2-37.

Pencils having plastic leads for use on polyester film are produced in hardness grades equivalent to HB, F, 2H, 4H, and 6H of regular graphite leads.

Fig. 2-37. The mechanical lead holder is often preferred for drafting.

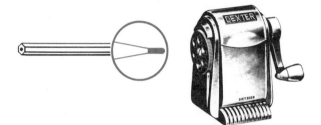

Fig. 2-38. This sharpener is equipped with special cutters which remove only the wood and not the lead on a pencil.

MANUALLY OPERATED

2.10.1 Pencil sharpeners and pointers. Most drafting rooms have mechanical sharpeners equipped with special cutters to simplify the process of sharpening a pencil. See Fig. 2-38. These sharpeners remove the wood only, leaving the lead exposed so that it may be pointed to any desired shape.

The pencil pointer—either manual or electric, as shown in Fig. 2-39—is another piece of standard equipment found in drafting rooms. The pointer is used to shape the lead to a conical point after the wood is removed.

2.10.2 Selecting the correct lead. The hardness of lead to be used depends on such factors as grade of drawing media, kinds of lines required, and the prevailing humidity in the drafting room. Generally a 4H, 5H, or 6H grade lead produces satisfactory lines for layout drawings or for drawings which necessitate heavier lines, a 3H or 4H lead is recommended. A softer lead, such as H or 2H, is

ELECTRICALLY OPERATED

Fig. 2-39. A lead pointer speeds up the process of sharpening the lead of a pencil.

advisable when drawings are to be produced on vellum or cloth. Sketching is best accomplished with an HB, F, or H grade. An F, H, or 2H grade lead is usually considered ideal for lettering.

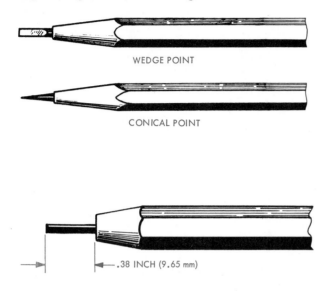

WEDGE POINT

CONICAL POINT

Fig. 2-40. A pencil is sharpened so the point is either conical or wedge-shaped.

.38 INCH (9.65 mm)

Fig. 2-41. The lead can be held closer to the straightedge if a slight shoulder like this is provided.

2.10.3 Sharpening the lead. To produce clear sharp lines, a lead must be sharpened correctly. Two types of points are used for drawing: the conical point and the wedge, or chisel, point. See Fig. 2-40. The conical point is best for all-purpose work. The wedge point is often preferred for drawing straight lines and the conical point for lettering and curved-line work.

In sharpening a wood-case pencil, the end opposite the hardness symbols should be cut so that the grade of the pencil used can always be known. The pencil should first be sharpened in a pencil sharpener and then about 0.38 inch of the lead uncovered with a knife. See Fig. 2-41. Notice the slight shoulder where the lead is exposed. The advantage of shaping the wood section in this manner is that the lead can be held closer to a straightedge or triangle, and the lead may be re-pointed several times without having to use the sharpener. The use of a knife to uncover the lead is not necessary if a sharpener with special cutters is employed.

After the wood is removed, the lead is pointed on a sandpaper pad or a smooth file by a rotating motion until the point assumes the shape of a cone. If a wedge point is desired, the point is rotated slightly, as in shaping a cone, and the operation completed by rubbing both sides of the point flat on the sandpaper or file. See Fig. 2-42. Whenever possible, a lead pointer should be used.

Fig. 2-42. The point is formed by shaping it on a sandpaper pad or file.

2.11 Drafting Media

Most industrial drawings are made directly on vellums or on films, rather than on paper. However, beginning students of drawing usually start their work on drawing paper and then transfer to other media after some skill in drawing is mastered.

2.11.1 Vellums. Vellums are made of 100% pure white rag stock and are particularly noted for their high transparency. They are scientifically processed to provide a non-yellowing finish and do not become brittle with age. The vellums have good pencil- and ink-taking qualities, will withstand repeated erasing without leaving ghost marks, and will allow considerable handling without damage.

2.11.2 Drafting films. Polyester films have brought a new drawing medium into popular use. These materials have numerous advantages— such as phenomenal resistance to tearing, cracking, and the roughest sort of treatment—over other drafting media. Films are also extremely transparent and possess great dimensional stability. They provide maximum contrast between linework and background, and extrasharp reproduction.

The polyester matte surface provides an excellent medium for pencil, ink, or typewriter work. Good line density is obtained with pencils in the H-to-6H range. Pencil smudging is virtually eliminated by using non-graphite plastic films such as Duralar. Erasing is easy, especially with a damp vinyl eraser.

2.12 Appearance of the Drawing

One feature of a high-quality drawing is its clean appearance—one which is free from smudged lines, surface smears, and ghost marks from erasures.

2.12.1 Keeping a drawing clean. To produce clean drawings, the following practices are particularly useful:

1. Always wipe off the table, straightedge, triangle, and other tools before proceeding to draw. Any ordinary soft cloth or cleansing tissue will serve this purpose. Soap and lukewarm water may be used.

2. Be sure the hands are clean at all times. If hands tend to perspire, they should be washed frequently during the drawing period. Sprinkling talcum powder on the hands will often counteract excessive perspiration.

3. Keep hands or sleeves off the penciled area. It is a good idea to roll up the sleeves. If additional work such as lettering or erasing must be done in the penciled area, cover the remaining section with a clean sheet of vellum or other transparent medium.

4. Avoid sliding straightedge, triangles, or other drafting devices over penciled lines. When these

Fig. 2-43. A dusting brush is a necessary item for producing clean drawings.

instruments must be moved over penciled areas, it is better to pick them up. Sliding them will tend to spread tiny dirt or graphite particles, thereby smearing the surface. Sprinkling granulated rubber over the paper will reduce smearing to a minimum.

5. Do not remove erasing particles with the fingers or palm of the hand. Use a dusting brush (Fig. 2-43) or flick the particles off with a clean cloth.

6. Never sharpen a pencil lead over the drawing. After sharpening or pointing, wipe the lead with a clean cloth or stab the point into a styrofoam block to remove small particles of loose graphite.

7. Never store the sandpaper pad in contact with any other drafting equipment unless the pad is completely enclosed.

8. Always cover a drawing at the end of the drawing period.

9. Do not allow drawings to overhang drafting or reference table edges or to be mishandled in other ways that will result in folds or creases.

2.12.2 Erasing a drawing. In the process of making a drawing, corrections and changes that involve some amount of erasing may have to be made. Most pencil marks or lines can be eradicated with an ordinary pencil eraser such as the one shown in Fig. 2-44. Extremely hard or gritty

Fig. 2-44. An eraser of this type simplifies the task of removing lines.

Fig. 2-45. The electric erasing machine speeds up the erasing operation.

erasers should be avoided, since they damage the surface of the drawing medium. Vinyl erasers are recommended for erasing on polyester film.

The electric erasing machine is a common tool in many drafting rooms. See Fig. 2-45. It is built so that the motor can be started and stopped and the erasing operation performed with one hand. A built-in blower blows away the erased particles while the machine is erasing.

Before proceeding with any erasing operation, be sure that the eraser is clean. A simple way to clean an eraser is to rub it on a piece of clean paper. Very often, if a considerable amount of erasing is necessary, it is a good idea to place a triangle under the paper. The triangle provides a hard surface which permits erasing to be accomplished more effectively. To prevent damaging the drawing surface, however, care should be exercised to erase only on the wider sections of the triangle.

To erase, place the fingers of the left hand near the mark that is to be removed and rub the eraser back and forth. Hold the drawing firmly; otherwise it may wrinkle or tear. If a rubber eraser is used, rub slowly, without too much pressure, to avoid overheating the eraser and leaving a stain on the drawing.

If erasing is to be done near lines which are to be left intact, an erasing shield will facilitate the

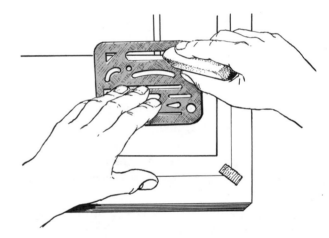

Fig. 2-46. An erasing shield simplifies the task of removing pencil marks.

process. See Fig. 2-46. To use this shield, select an opening that best fits the mark to be removed. Hold the shield firmly over the pencil mark and erase through the opening. If a shield is not available, the same results can be obtained by covering the area not to be touched with a piece of stiff paper.

When the erasing is completed, wipe off the paper with a clean cloth or dusting brush and touch up any lines that might have been damaged

Fig. 2-47. This set contains the instruments for preparing most drawings.

during the erasing process. To help keep a drawing clean, granulated rubber is often sprinkled over the paper. These small particles act as bearing surfaces which keep the straightedge and triangles from actually touching the paper. As a result, dirt and graphite are prevented from smudging the drawing.

Occasionally after a line has been erased, a groove may remain in the paper. The groove, if not too deep, can be removed by rubbing over with a burnisher or even with the thumbnail.

2.13 Drafting Instrument Set

A serviceable set of instruments is essential for producing good drawings with a minimum amount of effort and in the shortest possible time. There are many different kinds of sets. Some contain numerous special accessories, while others include only the basic instruments.

A professional set will usually contain large and small bow compasses with pen and lead attachments, a two-piece beam bar, hairspring dividers, and a technical fountain pen. See Fig. 2-47.

UNIT 3

Line Conventions and Drawing Sheet Formats

In order to insure some uniformity of drafting procedure, certain practices have been standardized. One of these involves the types of lines used on drawings to represent various kinds of features. Another concerns the size and general format of the drawing sheet. Standardized drawings, regardless of point of origin, may be readily exchanged among departments or companies without encountering difficulty of understanding. Having identical sheet sizes, drawings become easier to transmit and file. Also, with uniform formats basic information is expressed in the same way to all users.

3.1 Types of Lines and Line Conventions

Shape descriptions of objects are represented by various types of lines. As shown in Fig. 3-1, two widths of lines are in general use—thick and thin. The actual thickness to be used depends on the size and style of the drawing. All lines should be clean, dense-black, and uniform, with a distinct contrast in thickness between the thick and thin lines. Pencil leads should be hard enough to prevent smudging, but sufficiently soft to produce black lines and be erasable.

The following types of lines are in common use. See Figs. 3-1 and 3-2.

Visible lines. Visible lines are used to show the visible edges or contours of objects. They are the most prominent lines in the field of a drawing.

Hidden lines. These lines are used to show hidden features of an object and consist of short evenly spaced dashes. The length of the dashes may vary to suit the size of the drawing. Hidden lines should begin and end with a dash in contact with a visible line or hidden line except when the dash would appear to be a continuation of the visible line. Dashes should join at corners and start at tangent points of arcs. See Fig. 3-3.

Section lines. Section lines are used to show the cut surfaces of an object in a sectional view. They should be drawn thin to contrast well with visible lines, be equally spaced, and be proportionate to the size or mass of the section.

Center lines. Center lines are used to indicate axes of symmetrical parts and circles, and paths of motion. The long dashes can be made longer on large drawings and shorter on small drawings. Center lines should start and end with the long dashes and should not intersect at spaces between the dashes. The lines should extend a short distance beyond the feature unless a longer extension is necessary for dimensioning purposes. Center lines should not stop at other lines or project through space between views.

Dimension lines, extension lines, and leaders. Dimension lines are used to indicate the extent and direction of dimensions and are terminated with arrowheads. Extension lines are used to indicate the termination of a dimension. There should be a short gap of about 0.12 inch where the extension line joins the object view, so as not to confuse it with the lines of the object. Extension lines should also project a distance of 0.12 inch beyond the dimension line.

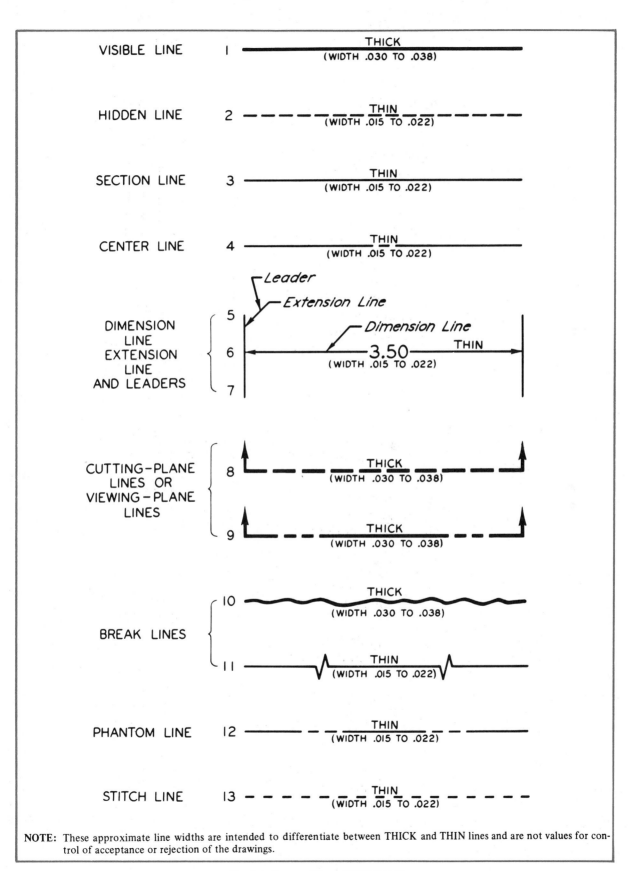

NOTE: These approximate line widths are intended to differentiate between THICK and THIN lines and are not values for control of acceptance or rejection of the drawings.

Fig. 3-1. Types of lines. (ANSI Y14.2)

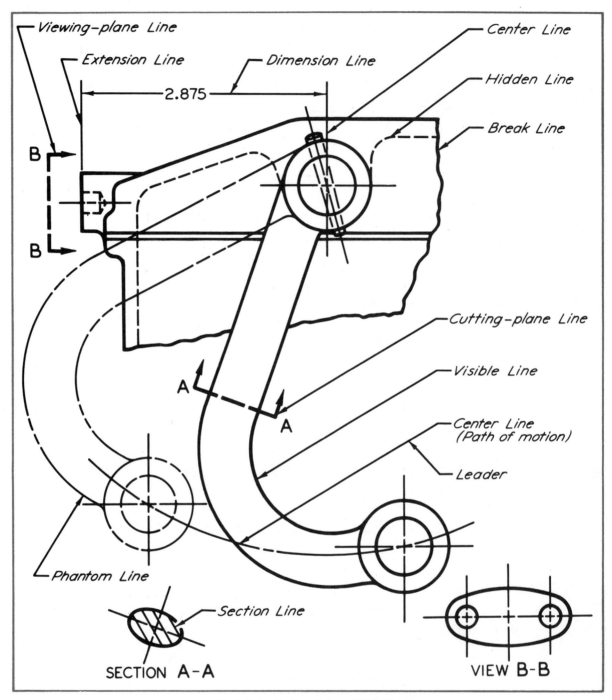

Fig. 3-2. Application of lines. (ANSI Y14.2)

Leaders are used to direct notes, dimensions, or identification symbols to features on the drawing. They should be straight inclined lines and terminated with an arrowhead or a dot. An arrowhead is used when the leader stops on a line and a dot when the leader extends within the outline of the object. The upper end of the leader has a short horizontal line which should run to the center of the height of the first or last letter or digit of the lettering associated with it.

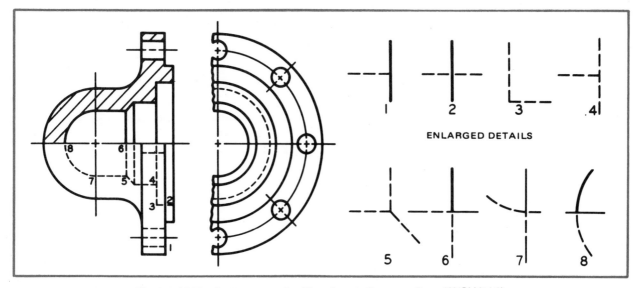

Fig. 3-3. Hidden line junctures should conform to these practices. (ANSI Y14.2)

Cutting-plane lines. These lines are used to indicate the locations of cutting planes in sectioning and the viewing position of removed pieces. Two forms of lines may be used. The first consists of thick alternating long dashes and pairs of short dashes. The long dashes may vary in length, depending on the drawing size.

The second form of line is composed of evenly spaced dashes. The ends of both types of cutting-plane lines are bent at 90° and terminated by arrowheads to indicate the direction of sight for viewing the section.

Break lines. Break lines are used to limit a broken section. For short breaks an uneven freehand line is recommended For long breaks the practice is to use long thin-ruled dashes joined by freehand zigzags.

Phantom lines. Phantom lines are used to show alternate positions of moving parts, lines of motion, adjacent positions of related parts, repeated details, and mold lines of formed metal parts. They consist of long dashes separated by pairs of short dashes. The long dashes may vary in length, depending on the drawing size.

Stitch lines. These lines are used to designate a sewing or stitching process. They are made of short dashes and spaces of equal lengths.

3.2 Line Quality for Microfilming

Because of increased microfilming of industrial drawings, extra precaution becomes necessary to produce high-quality linework. To secure microfilming reliability, lines must be black, consistent, and thick enough not to be lost in reduction. Lines that are gray or light in density will not photograph well. If lines are too thick, they have a tendency to appear smudged.

Best results are achieved when lines are drawn at least 0.06 inch apart. When dimensioning a sheet-metal part or a chamfered surface, the adjacent lines should be broken. See Fig. 3-4*A*. It is also important to prevent leader lines and cross lines from intersecting when dimensioning from a center. See Fig. 3-4*B*.

Finish marks should be broken at intersections as shown in Fig. 3-4*C*.

Extension lines should extend approximately 0.12 inch beyond the dimension line. See Fig. 3-4*D*. In actual practice any line may be broken to improve legibility or to prevent blurring on a reduced print.

To reduce unnecessary blurring, cross-section lining should be eliminated or limited to the extremities of a section. Shading of parts should be avoided entirely.

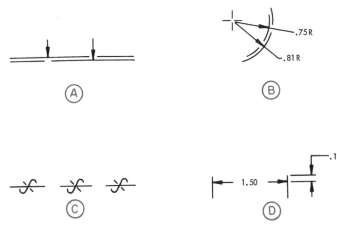

Fig. 3-4. High quality linework is necessary for microfilming.

Standard Drawing Sheet Sizes										
Note: All dimensions are in inches. 1 inch = 25.4 mm										
Flat Sizes					Roll Sizes						
Size Designation	Width (Vertical)	Length (Horizontal)	Margin		Size Designation	Width (Vertical)	Length (Horizontal)		Margin	
			Horizontal	Vertical			Min	Max	Horizontal	Vertical
A (Horiz)	8.5	11.0	0.38	0.25	G	11.0	22.5	90.0	0.38	0.50
A (Vert)	11.0	8.5	0.25	0.38	H	28.0	44.0	143.0	0.50	0.50
B	11.0	17.0	0.38	0.62	J	34.0	55.0	176.0	0.50	0.50
C	17.0	22.0	0.75	0.50	K	40.0	55.0	143.0	0.50	0.50
D	22.0	34.0	0.50	1.00						
E	34.0	44.0	1.00	0.50						
F	28.0	40.0	0.50	0.50						

IS STANDARD

SIZE	MILLIMETERS	INCHES	NEAREST U.S. SIZE
AO	841 x 1189	33.11 x 46.81	E
A1	594 x 841	23.39 x 33.11	D
A2	420 x 594	16.54 x 23.39	C
A3	297 x 420	11.69 x 16.54	B
A4	210 x 297	8.27 x 11.69	A

Fig. 3-5. Standard drawing sheet sizes.

3.3 The Drawing Sheet

The size of sheet used for a drawing is determined by the object to be drawn. Thus, a small size sheet is usually preferable for a small object, while larger size sheets are needed for large parts and assemblies. Actually all industries use a variety of sheet sizes.

3.3.1 Basic sheet sizes. Drawing sheets are either of a flat or roll type. Flat sheets are readily filed in a flat position, whereas roll sheets, because of their size, have to be filed in a roll form. Each size is designated by a letter for identification. Some industries have adopted the International System (IS) standard for flat and roll sheets. The standard sizes of flat and roll drawing sheets are shown in Fig. 3-5.

3.3.2 Sheet margins. The recommended margins allowances will vary for different size drawing sheets. See Figs. 3-5 and 3-6. In addition to the

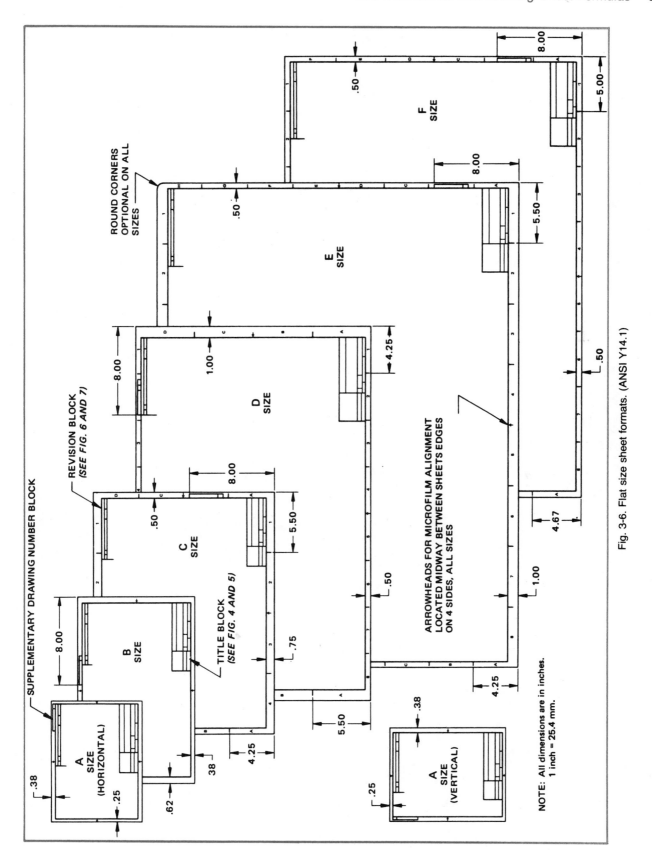

Fig. 3-6. Flat size sheet formats. (ANSI Y14.1)

regular margins, extra margins of 2 to 4 inches on either or both ends are often provided on roll size drawings for protection in handling and storage.

3.3.3 *Fastening sheet to board.* The drawing sheet should be conveniently located on the board so that there is sufficient room to use the straight-edge at the bottom of the sheet. When a drafting machine is employed, place the sheet at an angle as shown in Fig. 3-7. This position reduces the effects of light shadows from the scales and often makes lettering much easier.

The sheet is fastened to the board with drafting tape. The correct procedure is to place a piece of tape across the upper left-hand corner. The

Fig. 3-7. With a drafting machine, the drawing process may be simplified if the sheet is placed in this position on the board.

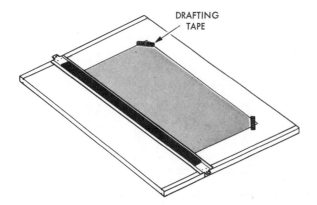

DRAFTING TAPE

Fig. 3-8. This is an example of how the sheet can be fastened to the board.

straightedge is then moved up to the bottom printed border line of the sheet and another strip of tape is fastened on the upper right-hand corner. See Fig. 3-8. This is followed by securing the two bottom corners as well.

3.4 Title Block Format

A title block provides spaces to record essential information concerning the object represented on the drawing. The use of all spaces is not mandatory. Blocks not required are omitted, relabeled for specific requirements, or consolidated with other spaces to provide more room for required information. The title block is located on the drawing in the lower right-hand corner of the sheet. The sizes of the various blocks will vary for different drawing sheets. See Fig. 3-9.

3.4.1 Content of title block. The information in a title block may vary from one manufacturer to another. In general, however, there will be included the following (See Fig. 3-9.):

1. *Block A* Name and address of company, or the design activity.

2. *Block B* Drawing title. (See 3.4.2.)

3. *Block C* Drawing number. (See 17.2.3.)

4. *Block D* Information relative to the preparation of the drawing, such as names of the draftsperson and checker, with relevant dates; approving function; issue date; and contract number.

5. *Block E* Approval of the design activity where a contractor subcontractor exists.

6. *Block F* Approval of an activity other than those described by blocks D and E.

7. *Block G* Scale of drawing.

8. *Block H* Code identification number for identifying the company or design activity.

9. *Block J* Drawing letter size.

10. *Block K* Actual or estimated weight of item.

11. *Block L* Sheet number for multiple drawings.

3.4.2 Drawing title. The title should consist of a basic name and sufficient modifiers to properly establish a clear concept of the item. The basic name should be a noun or noun phrase which describes the part and the use of the part and not the material or method of fabrication. Thus, the words *casting, forging,* and *weldment* would be improper. Instead an appropriate title would be

BRACKET or SUPPORT BRACKET or MIXING VALVE. The noun or noun phrase should be in singular form except where the only form of the noun is plural, as in TONGS or SCISSORS, or when multiple single items appear on the same drawing, such as FUSES or CONNECTORS. An ambiguous noun, or one which designates several classes of items, should not be used alone but as part of a noun phrase. Examples:

ACCEPTABLE	UNACCEPTABLE
Running Board	Board, Running
Slide Rule	Rule, Slide
Soldering Iron	Iron, Soldering
Junction Box	Box, Junction
Cable Drum	Drum, Cable

A noun or noun phrase may require further clarification, in which case a modifier is added. A modifier may be a single word or a qualifying phrase. The first modifier is intended to narrow the area of concept established by the basic name. Subsequent modifiers must continue to narrow the item concept by expressing a different type of characteristic. A word directly qualifying a modifying word should precede the word it qualifies, thereby forming a modifying phrase, for example, BRACKET, UTILITY LIGHT—ENGINEER'S COMPARTMENT. A modifier should be separated from the noun or noun phrase by a comma. The first part of the title should be separated from the second part of the title by a dash.

No abbreviation of any portion of the name (first part of the title) should be made, except those necessarily used trademarked names and the abbreviations ASSY (assembly), SUBASSY (subassembly), and INSTL (installation). Abbreviations may be used in the second part of the title, but, in general, abbreviations should be avoided.

A list of approved item names may be found in the *Cataloging Handbook H6-1, Federal Item Identification Guide for Supply Cataloging.*

3.4.3 Revision block. The block for recording drawing revisions (Fig. 3-10) is placed in the upper right corner of the drawing. When additional space is required, a supplemental block is located to the left of and adjacent to the original revision block. For multiple sheet drawings the revision block may be placed on all sheets or only on the first sheet.

The revision block contains such information as

58

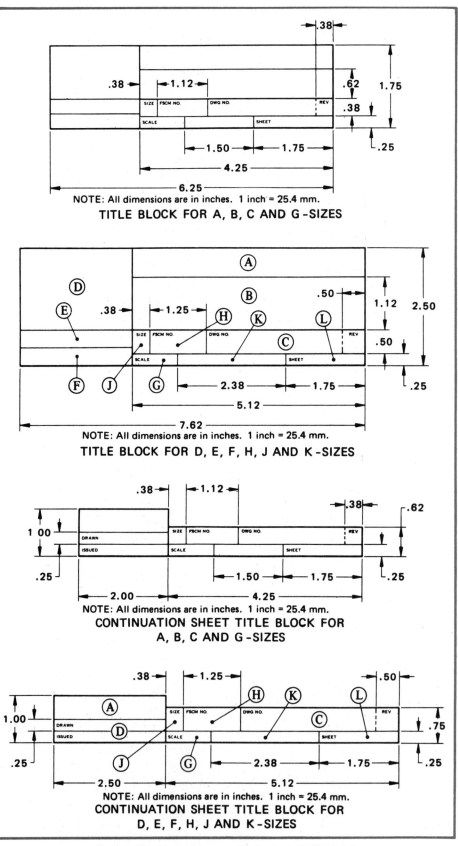

Fig. 3-9. Title block spaces and their sizes. (ANSI Y14.1)

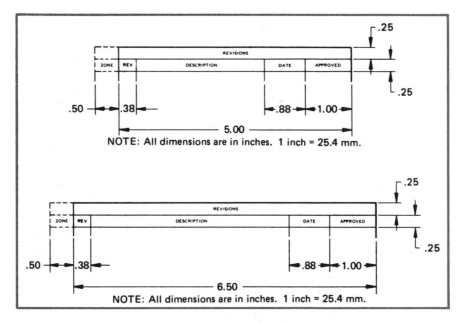

Fig. 3-10. Revision block. (ANSI Y14.1)

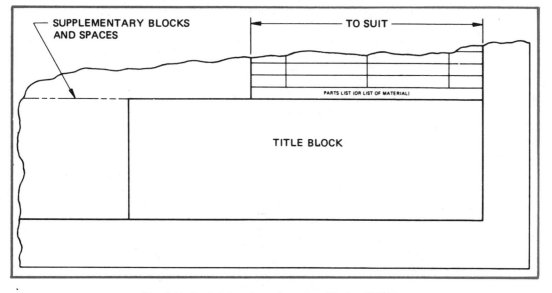

Fig. 3-11. Parts list and supplementary blocks. (ANSI Y14.1)

revision symbol, description or identification of the authorized change, date, and approvals.

3.4.4 Supplementary blocks. Supplementary blocks are used to record such information as dimensioning and tolerancing notes, materials, treatment, and finish. These blocks are normally located to the left of the regular title block. See Fig. 3-11.

3.4.5 Numbering multiple sheets. When a drawing of a part consists of more than one sheet, the second and subsequent sheets are numbered in one of the ways shown in Fig. 3-12.

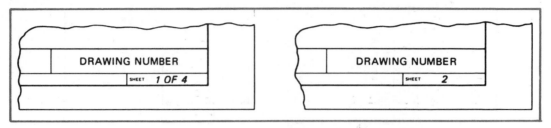

Fig. 3-12. Multiple sheet numbering. (ANSI Y14.1)

3.5 Parts and Materials Lists

An assembly drawing will usually include either a list of materials or a parts list. A list of materials shows the names of the parts with identifying parts numbers, the quantity required, and specification of the materials from which the parts are made. A parts list includes only the names of the parts with their part numbers and the quantity required. No material specifications of the part are shown on this list. See Fig. 3-13.

The lists may be on the drawing where the assembly is shown or on separate sheets. When on the same drawing, the list should be located in the lower right corner above the title block. See Fig. 3-11. Additional lists may be placed at the left of and adjacent to the original block. When lists are placed on separate sheets, these sheets should carry the same number as the related drawings.

3.6 Indicating Scale on a Drawing

As was stated in Unit 2, a drawing is made so that the object is shown in its full or natural size, a reduced size, or an enlarged size. Whenever possible, the practice is to make the drawing full-size.

TITLE	SCAVENGE PUMP ASSY		PL XXXXXXX		
			SHEET X OF X SHEETS		
ITEM NO.	IDENTIFICATION NUMBER	UNITS PER ASSY	DESCRIPTION OR NAME		
1	XXXXXXX	X	GEAR – SCAVENGE PUMP IDLER FRONT		
2	XXXXXXX	X	KEY		
3	XXXXXXX	X	GEAR – SCAVENGE PUMP DRIVE FRONT		
4	XXXXXXX	X	GEAR – SCAVENGE PUMP IDLER		
5	XXXXXXX	X	SCREW – OIL PUMP		
6	XXXXXXX	X	GASKET		
7	XXXXXXX	X	PLUG		
8	XXXXXXX	X	GEAR – SCAVENGE PUMP DRIVE		
9	XXXXXXX	X	SHAFTGEAR – SCAVENGE PUMP DRIVE		
10	XXXXXXX	X	TUBE		
11	XXXXXXX	X	LOCKWIRE		
12	XXXXXXX	X	COVER – SCAVENGE PUMP FRONT		
13	XXXXXXX	X	BODY – SCAVENGE PUMP FRONT		
14	XXXXXXX	X	BODY – SCAVENGE PUMP REAR		
15	XXXXXXX	X	COVER ASSEMBLY – SCAVENGE PUMP		
16	XXXXXXX	X	BOLT – HEX HD DRILLED SHANK		
17					
18					
19					
20					
SYM	REVISIONS				

Fig. 3-13. A typical parts list.

If the size has to be reduced, most industries usually limit the scale to ½, ¼, or ⅛. For parts which are too small to be drawn full-size, the common procedure is to draw them two or four times larger.

The scale of a drawing is always included in the title block. If more than one detail occurs on a drawing and different scales are used, the scale of the principal detail is shown in the title block along

FULL SIZE	OR	1.00 = 1.00	OR	1 = 1	OR	1/1	
HALF SIZE	OR	.50 = 1.00	OR	$\frac{1}{2}=1$	OR	1/2	
QUARTER SIZE	OR	.25 = 1.00	OR	$\frac{1}{4}=1$	OR	1/4	PREFERRED
EIGHTH SIZE	OR	.125 = 1.00	OR	$\frac{1}{8}=1$	OR	1/8	
TWICE SIZE	OR	2.00 = 1.00	OR	2 = 1	OR	2/1	
TEN TIMES SIZE	OR	10.00 = 1.00	OR	10 = 1	OR	10/1	

Fig. 3-14. The scale of a drawing may be shown in any one of these ways.

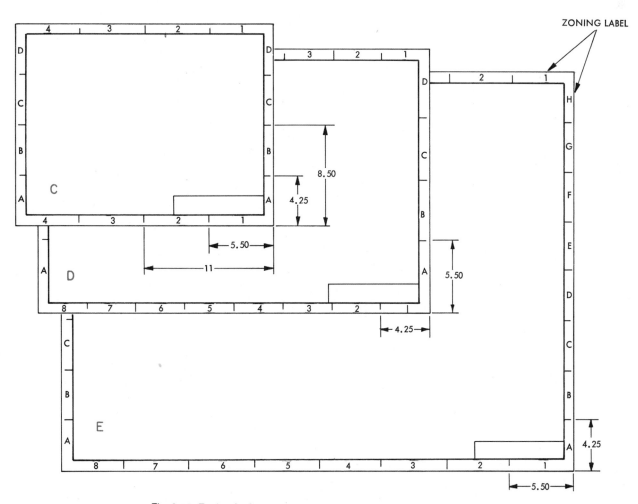

Fig. 3-15. Zoning facilities locating particular information on a large drawing.

with the notation AND NOTED. The other scales are specified under each detail. Fig. 3-14 shows some ways of designating the scale on a drawing.

3.7 Zoning of Drawings

Zoning is used on drawings of C-size and larger to facilitate the location of parts, details, sectional views, notes, and changes in the same manner that numbers and letters on a map help to locate a particular area. Zoning consists of regularly spaced blocks around the edges of the sheet. The upper and lower rows of horizontally spaced blocks are numbered consecutively from right to left. The vertical zone blocks on the right and left edges are designated by letters *A, B, C,* and so on, reading upward. See Fig. 3-15.

3.8 Roll End Marking

Some industries include the following information on the reverse side of each end of roll size drawings for ease in identification:

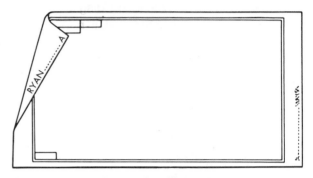

Fig. 3-16. Roll drawings are identified on the ends in this way.

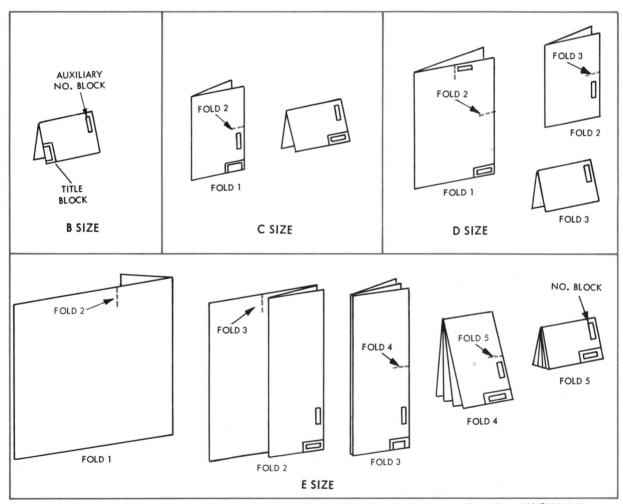

Fig. 3-17. For convenience and ease of filing prints, the practice is to fold them in the manner shown here (ANSI Y14.1)

1. Name of company.
2. Drawing number.
3. Drawing title.
4. Latest revision letter.
5. Security classification when applicable.
6. Sheet number when applicable.

This information is placed on diagonally opposite corners and so located that it may be read from the end of the drawing and may be seen without unrolling the drawing. See Fig. 3-16.

3.9 Folding Prints

To facilitate the handling, mailing, and storing of prints, drafting departments follow a commonly accepted method of folding prints. Regardless of the size of the print, the folding is accomplished so that the top edge has a single fold line. This avoids any possibility of filing one print inside the folds of another when they are filed in a letter-size filing cabinet.

Prints should be folded with the viewing side exposed or unexposed. In either case the sequence of the fold lines remains the same. The advantage of folding with the viewing side unexposed is that it protects the viewing side when prints receive rough handling through constant filing. Fig 3-17 illustrates the directions and number of fold lines for prints of various sizes.

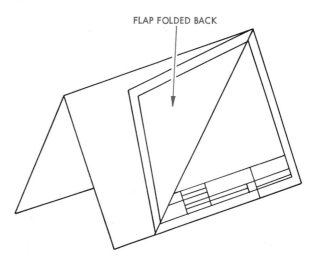

Fig. 3-18. When a print is folded with the viewing side covered, a corner is turned to show the title block.

With the exception of B-size, all prints are folded with the title block in the lower right-hand corner. Since the title block is not visible when the prints are folded with the viewing side unexposed, the practice is to fold back one corner of the print in the form of a triangle in order to expose the title block. See Fig. 3-18. An alternative method is to stamp a duplicate title block on the blank side.

UNIT 4

Lettering a Drawing

A drawing is intended to convey information. If this function is to be executed effectively, two things are necessary: (1) the object must be drawn according to accepted standards and (2) sufficient data must be included to insure adequate understanding of the graphically illustrated object.

The presentation of informational data on a drawing is known as *lettering.* Whether the data is in the form of dimensional sizes, explanatory notes, or the listing of specifications, the lettering must be legible, pleasing in appearance, and simple to execute.

To be able to letter well is a skill that can be acquired only through practice. To acquire this skill, a student of drafting must practice diligently and observe the principles described in this unit.

4.1 Styles of Lettering

There are several different styles of letters, but in actual practice lettering is limited to the type known as the uppercase (capital letters) single-stroke Gothic. Lowercase letters are rarely used on machine drawings. They are used more extensively for maps and other topographical drawings as well as for architectural and structural drawings.

Single-stroke simply means that the width of the lines which form the letters does not vary. It does not imply that letters are begun and completed in one single stroke.

The Gothic style is universally used in industry because it is more legible and can be executed much more rapidly than other styles.

Both the vertical and inclined Gothic letters are acceptable in engineering drawing. The vertical letters are considered by some to possess greater readability. However, they are a little more difficult to produce because a slight variation from the vertical is easily discernible. Although most industries do not prescribe any one form, some prefer that all lettering be done in one style only. Therefore, it is wise for the student of drafting to become proficient in using both the inclined and vertical letters.

4.2 Letter Strokes

To facilitate efficient lettering, letters and numerals should be formed with certain standard strokes. Although there are divergent points of view concerning the number and direction of strokes, an examination of current practices discloses that the strokes shown in Figs. 4-1 and 4-2 are currently favored.

The outstanding features of the strokes illus-

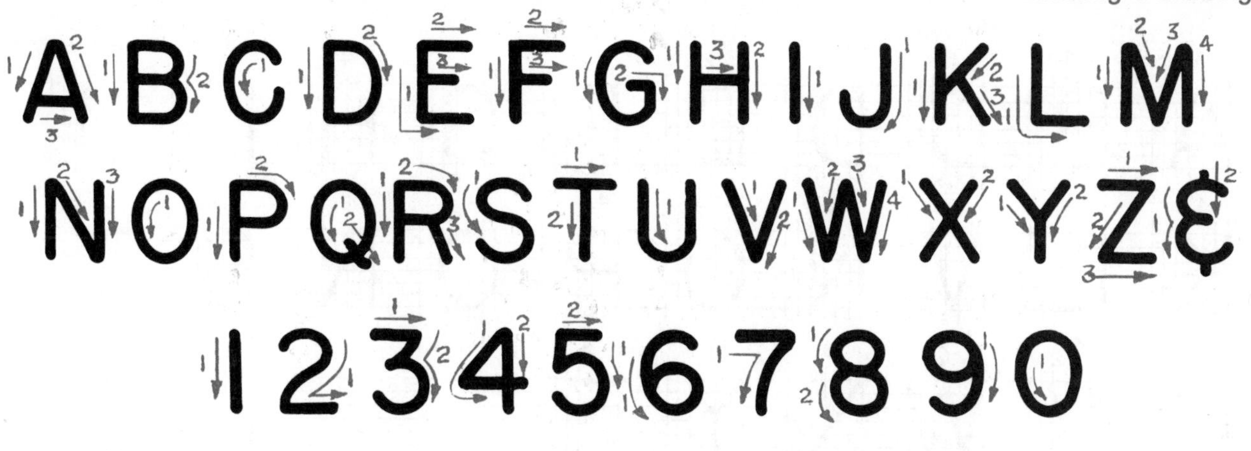

ABCDEFGHIJKLM

NOPQRSTUVWXYZ&

1234567890

Fig. 4-1. The arrows show the order and direction of the strokes made to form letters and numerals properly.

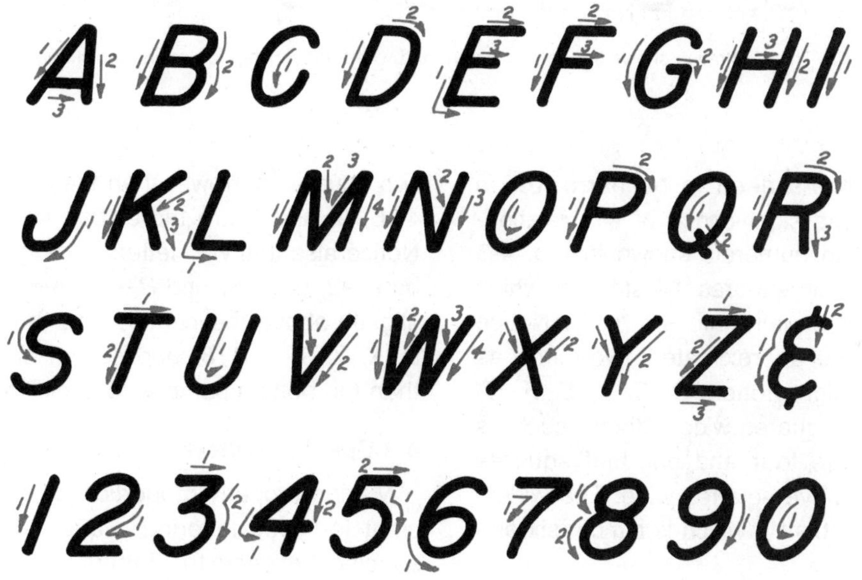

ABCDEFGHI

JKLMNOPQR

STUVWXYZ&

1234567890

Fig. 4-2. Inclined letters are often used instead of vertical letters.

trated are that they permit great speed in the lettering process and provide a natural flow of lines with a minimum change of pencil position. The same strokes are applicable for both inclined and vertical letters.

To keep the strokes of the letters uniform, the pencil should be rotated between the thumb and index finger. The pencil should be held firmly and even pressure applied. As soon as there is evidence that the strokes are broadening, the lead

should be resharpened. If the letters are made rapidly, the strokes will be more uniform in width and height.

4.3 Proportions of Letters and Numerals

To create the effect of balance and of stability, letters and numerals should have a certain relationship between height and width. Some letters and numerals, because of their shapes, must be

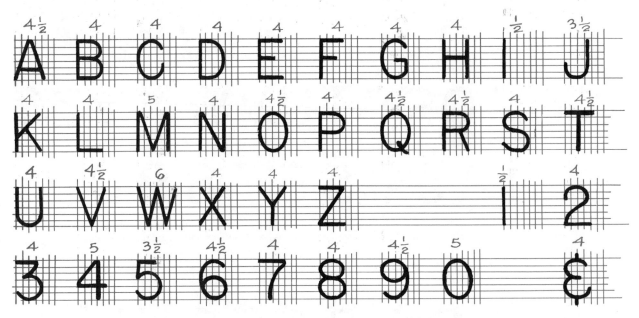

Fig. 4-3. This illustration shows the correct height and width relationship for letters and numbers.

wider, whereas others need to be narrower. In order to illustrate the significance of this relationship, the letters and numerals shown in Fig. 4-3 have been enclosed in squares the sides of which have been divided equally into six parts. Notice that most letters are approximately two-thirds as wide as they are high. Letters *B, C, D, E, F, G,* and so on, are four squares wide. Others, such as *A, O, R,* and *V,* are four and one-half squares wide, whereas *M* is five squares wide. The widest letter of the alphabet is *W,* which is almost six and one-half squares wide.

Fig 4-3 also shows the correct proportions for numerals. The numerals 2, 3, 7, and 8 are about two squares narrower than they are high, whereas 4, 6, 9, and 0 are more than four squares wide. Notice also that with letters having horizontal cross bars—*B, E, F, H,* and *R*—the middle bar is placed slightly above the center, and the top portions of *B, X, Y, 3, 5, 6, 8,* and so on, are made smaller than the bottom portions to avoid top-heaviness.

4.4 Spacing Letters

Making properly proportioned letters is not sufficient for good lettering. Equally important is the spacing between the letters. To give a pleasing appearance, the areas between each letter must appear to be equal. See Fig. 4-4. This uniformity

ALTHOUGH THE LETTERS ARE
SPACED EVENLY, L, T, P AND O
APPEAR TOO FAR APART

MORE PLEASING EFFECT

Fig. 4-4. The area between letters should appear to be equal.

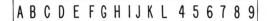

CONDENSED GOTHIC

EXTENDED GOTHIC

Fig. 4-5. Examples of condensed and extended letters.

cannot always be achieved by simply placing the letters the same distance apart. For example, when the letters *L* and *T* are spaced the same as *M* and *E,* an unbalanced effect results. Due to the shape of *L* and *T,* less space is needed between them than for *M* and *E.* The same principle must be observed for some of the other letters in the alphabet. One accepted rule is to place adjacent letters with straight sides farther apart than those with curved sides. Sometimes it may even be necessary to overlap slightly such combinations as *LT* and *AV,* whereas in other cases the width of a letter has to be decreased. The customary practice is to judge the spacing by eye rather than to try to achieve area balance by means of actual measurement.

In the process of lettering composition, the letters and numerals may have to be condensed or extended to meet certain space requirements. Thus, if a space is limited in length, each letter in the composition will have to be correspondingly narrower. Similarly, if a wide space is to be filled with composition, each letter may have to be ex-

tended. See Fig. 4-5. The recommended practice is to space letters and numerals at least 0.06 inch apart for legible reproduction.

4.4.1 Spacing between words, sentences, and lines. The distance between words should be equal to the height of the letters. See Fig. 4-6. As a rule sentences are spaced horizontally a distance equal to approximately twice the distance between words. When several lines of lettering are required, the vertical spacing between them should not be more than the height of the letters. The actual distance will depend on the amount of space available for composition.

The space between two numerals having a decimal point between them should be a minimum of two-thirds the height of numerals.

4.4.2 Height of letters and numerals. The actual height of letters and numerals used on a drawing depends on the function of the composition. Titles and part numbers always have the greatest height, while letters for such items as section designation, part name, notes, dimensions, and list of materials are correspondingly smaller.

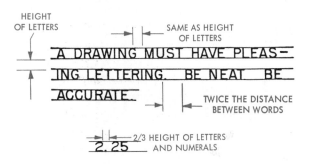

Fig. 4-6. Words, sentences and numerals should be spaced like this.

Use	Letter Heights, In. (Min.)		Drawing Size
	Freehand	Mechanical	
Drawing numbers in title block	5/16 (0.312)	0.350	All
Drawing title	1/4 (0.250)	0.240	
Section and tabulation letters	1/4 (0.250)	0.240	
Zone letters and numerals in border	3/16 (0.187)	0.175	
Dimension, tolerances, limits, notes, subtitles for special views, tables, revisions, and zone letters for the body of the drawing	1/8 (0.125)	0.120	Up to and including 18 x 24 inches
	5/32 (0.156)	0.150	Larger than 18 x 24 inches

Fig. 4-7. Letter heights recommended by ANSI.

Fig. 4-7 shows the letter and numeral heights recommended by ANSI.

4.5 Lowercase Letters

Lowercase letters consist of three parts known as bodies, ascenders, and descenders. See Fig. 4-8. The bodies are made from three-fifths to two-thirds the height of capitals. The ascenders extend to the cap line, while the descenders drop the same distance below to the dropline. Shapes and strokes for lowercase letters are shown in Fig. 4-9.

4.6 Lettering Standards for Microfilming

Hand lettering on A-, B-, and C-size drawings should be at least 0.12 inch high with a spacing of 0.12 inch between lines. For larger size drawings, the height of letters should be 0.16 inch with a

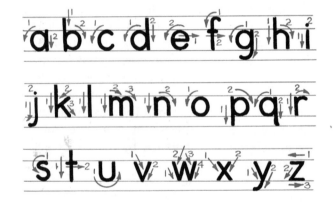

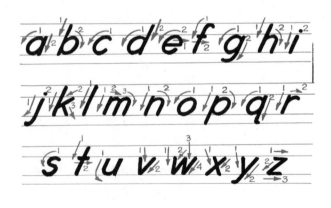

Fig. 4-9. Lowercase letters are formed with these strokes, vertical lettering at top and slant lettering below.

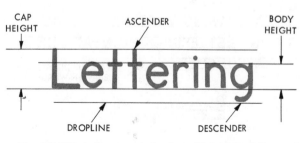

Fig. 4-8. This is the correct structure of lowercase letters.

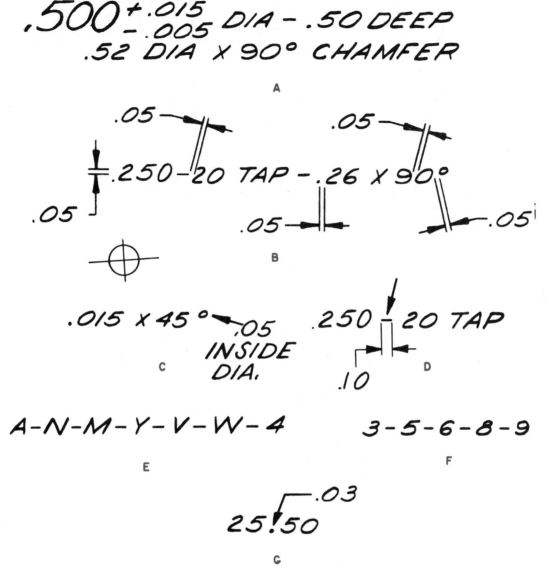

Fig. 4-10. Lettering standards for microfilming (actual size).

spacing of 0.12 inch between lines. See Fig. 4-10A. A minimum distance of 0.05 inch should be maintained between letters, symbols numbers, dashes, and decimals. See Fig. 4-10B. Degree symbols (°) need a minimum inside diameter of 0.05 inches. See Fig. 4-10C.

When dashes are used in notes and dimensions, they should be approximately 0.10 inch long and clearly broken. See Fig. 4-10D. Letters and numerals such as A, N, M, Y, V, W, and 4 must have a clean and full opening at each vertex. See Fig. 4-10E. The numerals 3, 5, 6, 8, and 9 are required to have a full round loop. See Fig. 4-10F. Decimal points should be 0.03 inch in diameter. See Fig. 4-10G.

Revised letter and numeral shapes have been recommended as shown in Fig. 4-11 to minimize the danger of letters filling in or being misread as a result of reduction to microfilm and subsequent blowback to hard copy.

Fig. 4-11. Microfont letters. (ANSI Y14.2)

Guide lines usually will reproduce on microfilm copies regardless of how lightly they are drawn. To avoid ghost lines, it is desirable to use drawing forms with grids or to slip lettering guide sheets under the drawing.

For typewritten information, an electric typewriter equipped with Gothic typeface is recommended. The typeface must be kept clean, and fabric or carbon ribbons should be used.

4.7 Guide Lines

In actual practice, only two guide lines are drawn to represent the height of capital letters. The letters are then visualized to determine their correct spacing in the given area, and the lettering process is carried out.

4.7.1 Drawing guide lines. To draw horizontal guide lines, mark the height of the letters desired and with the aid of a straightedge draw two light lines. If several lines of lettering are required, set the dividers to the correct height and step off the number of lines needed. With a little practice, guide lines eventually can be spaced by eye rather than by measuring them with a scale or dividers. To make erasing unnecessary, all guide lines should be drawn so they are barely visible.

The only other guide lines used occasionally by beginning students are slanted lines for inclined letters. These slanted lines are drawn at random across the horizontal guide lines to help maintain the proper slant of the letters. Inclined letters are drawn at an angle of about 67½° from the horizontal. The proper slope angle may be found by marking off two units on a horizontal line and five units on a vertical line. The termination points of the two lines will produce a slant of approximately 67½°. Then, with a straightedge and triangle held as shown in Fig. 4-12, the necessary lines are drawn.

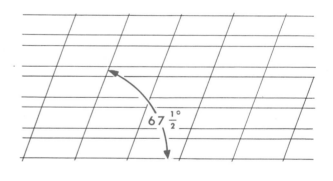

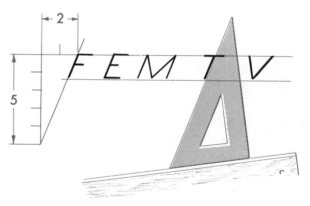

Fig. 4-12. This is how the correct slant for inclined letters can be determined.

4.7.2 Guide line devices. To simplify the process of drawing guide lines, special devices made for this purpose are sometimes used. The two most common are the Braddock-Rowe Lettering Triangle and the Ames Lettering Instrument.

The *Braddock-Rowe Lettering Triangle* is equipped with a series of holes arranged to provide guide lines for lettering and dimensioned figures and for drawing section lines. See Fig. 4-13. The numbers at the base of the triangle designate the spacing for guide lines. If No. 4 holes are used, the guide lines will be 0.12 inch apart. The slot on the left is for drawing slope guide lines for inclined letters. The column of holes on the left is used to draw guide lines for numerals 0.12 inch high and for drawing section lines 0.06 inch apart.

To use the triangle, insert the pencil point through the proper hole and move the instrument back and forth along the straightedge.

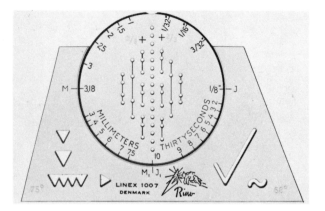

Fig. 4-14. The Ames lettering instrument is another device for drawing guide lines and section lines.

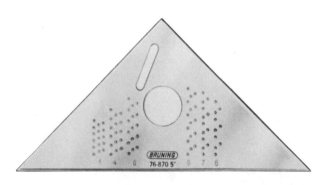

Fig. 4-13. This lettering triangle simplifies the task of drawing guide lines.

The *Ames Lettering Instrument* will produce guide lines for letters varying in height from 0.06 to 1.50 inches. See Fig. 4-14. The correct spacing for guide lines is secured by turning the disk to one of the settings indicated at the bottom of the disk. These numbers designate the height of letters in 0.03 inch. The holes in the center column are equally spaced and are used to draw guide lines for numerals and to draw section lines. The two outer columns of holes are used to draw guide lines for capitals or lowercase letters.

4.8 Lettering Devices

The use of special lettering devices will often help speed up the process of lettering a drawing. The three most frequently used are lettering templates, the LeRoy Lettering Instrument, and the Varigraph.

Fig. 4-15. Lettering templates will speed up the lettering process.

4.8.1 Lettering templates. These templates are plastic strips containing outlines of letters and numerals. Templates are available with vertical or slanted letters of various heights. See Fig. 4-15. The letters and numbers are formed by placing the guide over the portion of the paper on which the lettering is to be done and tracing the outline with a pencil, ballpoint, or fountain pen. The guide is moved back and forth along the edge of a straightedge. No pencil guide lines are necessary.

4.8.2 LeRoy Lettering Instrument. This is basically an ink-lettering device. The instrument consists of a guide or template with grooved letters and numbers. The scriber is equipped with a tracer pin that follows the grooved letters on the template and a pen that forms the letters. See Fig. 4-16. The scriber is either of the fixed type, for reproducing only vertical letters, or adjustable, for both vertical and inclined letters. The guides are available with letters of various heights. The thickness of the letter strokes is governed by the pen size. The instrument is used by moving the template along a straightedge as the letters are formed. See Fig. 4-17.

4.8.3 Varigraph. The Varigraph operates very much like the LeRoy instrument. However, it is designed so that the size of the letters can be changed. With this device it is possible to construct letters so that they are compressed or expanded. As shown in Fig. 4-18, a guide pin in the lower right corner is moved along the grooves in a template. As the guide pin travels, a pen in the upper left corner forms the letters.

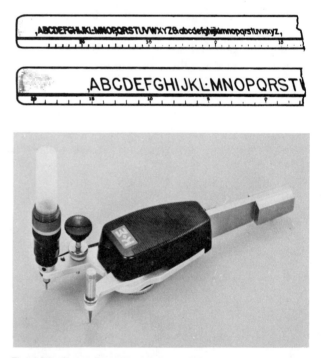

Fig. 4-16. The LeRoy lettering instrument forms perfect letters in ink.

4.8.4 Typing machines. To reduce the time involved in lettering drawings, some industrial drafting rooms are equipped with special lettering machines. These machines are particularly valuable for filling in lists of materials, specifications, schematic diagrams, and notes, as well as for in-

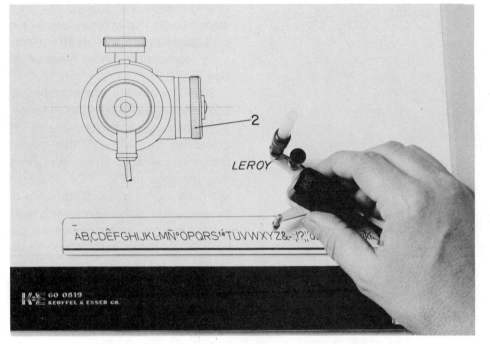

Fig. 4-17. This is how the LeRoy lettering instrument is used.

Fig. 4-18. The Varigraph is a lettering instrument that can vary the size or shape of the letters.

Fig. 4-19. This lettering machine is often used in drafting rooms.

serting dimensional figures. The machines are operated much like ordinary typewriters. Electrically controlled impressions permit uniform, clear, and sharp composition on all types of drawing media. See Fig. 4-19.

Some typewriters are equipped with typefaces comparable to regular drafting lettering styles and are especially useful for lettering drawings which are to be microfilmed. Other typewriters can be placed on the table directly over the drafting sheet. The machine in Fig. 4-20 spaces automatically along an indexing rail. Each strike of a key shifts the typewriter to the next letter position. The indexing rail fits any standard drafting machine or can be held against a parallel bar. The letters are arranged in alphabetical order, making this type of machine easy to use without conventional typing skills.

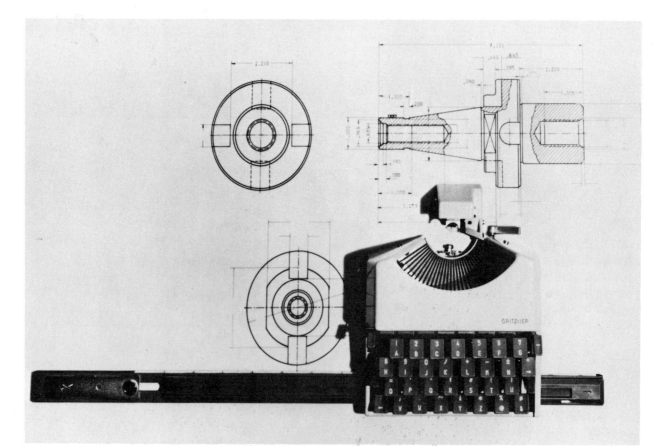

Fig. 4-20. A drafting board typewriter in use with a view of the keyboard.

Fig. 4-21. Transfer lettering is simple to apply.

4.9 Transfer Lettering

Transfer, or pressure, lettering consists of sheets of printed letters and numerals for use in preparing display captions, graphs, and charts, and in a variety of other work requiring special lettering. In application a light pencil line is first ruled to locate the position of the lettering. The backing of the alphabet sheet is removed and the letters placed in position. Next the letters are rubbed with a fingernail, burnisher, or pencil point, and then the transfer sheet is peeled off, leaving the letters on the drafting medium. See Fig. 4-21.

Problems for Section I
Basic Drafting Practices

The problems in this section are intended to help students develop basic understandings and skills dealing with the fundamentals of drafting practices as described in

    Unit 1. Introduction to Modern Drafting
    Unit 2. Equipment and Materials for Drafting
    Unit 3. Line Conventions and Drawing Sheet Formats
    Unit 4. Lettering a Drawing

Problems for Section I include

    Problems 1-12 Lettering
    Problems 13-41 Single View Layout Drawings

Problems 1-12 Lettering

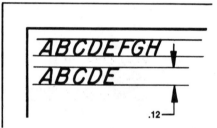

Problem 1
Draw in guide lines and fill an A-size sheet using 0.12″ inclined freehand capital letters.

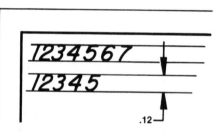

Problem 3
Draw guide lines and complete an A-size sheet with 0.12″ inclined freehand numerals.

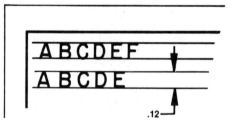

Problem 2
Using freehand lettering, fill an A-size sheet with vertical capital letters 0.12″.

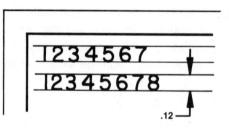

Problem 4
Fill an A-size sheet with 0.12″ vertical freehand numerals.

Problems having decimal-inch values may be converted to equivalent metric sizes by using the millimeter conversion chart or the problems may be redesigned by assigning new millimeter dimensional values more compatible with metric production.

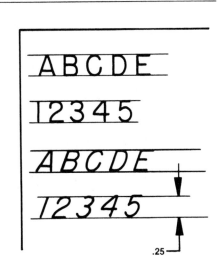

Problem 5

Fill an A-size sheet with 0.25" vertical and inclined letters and numerals, freehand or by lettering template.

Problem 6

Fill an A-size sheet with both vertical and inclined freehand 0.12" lowercase letters.

FUNCTION OF A DRAWING

The function of a drawing is to provide a specification of the required part or assembly. A drawing should contain delineation, dimensions and tolerances, performance requirements, notes and references, as required, in sufficient clarity and detail consistent with the skills and trade practices involved, to insure correct interpretation by any manufacturing organization that may have occasion to use it.

All drawings should provide maximum latitude in the choice of manufacturing methods, processes, etc., consistent with the engineering requirements. References to specific methods of ordering material, detailed instructions for methods of manufacture and assembly, or manufacturing processes are permissible on the drawing, only if they are essential to meet engineering requirements.

Radio Corporation of America

Problem 7

On an A-size sheet, letter the composition shown using 0.12" capital letters, either vertical or inclined, freehand or by using templates.

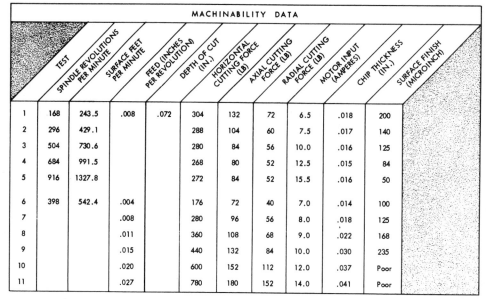

	MACHINABILITY DATA									
TEST	SPINDLE REVOLUTIONS PER MINUTE	SURFACE FEET PER MINUTE	FEED (INCHES PER REVOLUTION)	DEPTH OF CUT (IN.)	HORIZONTAL CUTTING FORCE (LB)	AXIAL CUTTING FORCE (LB)	RADIAL CUTTING FORCE (LB)	MOTOR INPUT (AMPERES)	CHIP THICKNESS (IN.)	SURFACE FINISH (MICROINCH)
1	168	243.5	.008	.072	304	132	72	6.5	.018	200
2	296	429.1			288	104	60	7.5	.017	140
3	504	730.6			280	84	56	10.0	.016	125
4	684	991.5			268	80	52	12.5	.015	84
5	916	1327.8			272	84	52	15.5	.016	50
6	398	542.4	.004		176	72	40	7.0	.014	100
7			.008		280	96	56	8.0	.018	125
8			.011		360	108	68	9.0	.022	168
9			.015		440	132	84	10.0	.030	235
10			.020		600	152	112	12.0	.037	Poor
11			.027		780	180	152	14.0	.041	Poor

WORKPIECE — 1020 HR STEEL — 5.537-IN. DIAMETER (TESTS 1–5)
5.206-IN. DIAMETER (TESTS 6–11)

Problem 8
Lay out a chart similar to the one shown, giving the machinability data indicated. Choose overall sizes, lettering size and style. Use templates for all lettering that you do.

```
                    AUTOMATIC   GRAPH   PLOTTER   PROGRAM

100   FORMAT (15, 15, 15, 15, 15, 15, 15)      2    PUNCH 100, 3, NSFM, NHP, NPHI, NCOEF, NFC
101   FORMAT (F10.2, F10.2, F10.2, F10.2, F10.2)     DO 3 1 = 1, 6
  1   PUNCH 100, 1, 1100, 1600, 280, 178, 0, 0       READ 100, NX, NY, MODE
      PUNCH 100, 2, 34, 10, 10, 10, 1                PUNCH 100, 4, NX, NY, MODE
      DO 2 1 = 1, 5                                  READ 102
      READ 101, SFM, HP, PH1, COEF, FC         3    PUNCH 102
      NSFM = SFM/2.0                         102    FORMAT (20H
      NHP = HP/.02                                  DO 4 1 = 1, 5
      NPHI = PHI/.04                                READ 100, NX, NY, NL, NZ, NI, MODE
      NCOEF = COEF*1000.                       4    PUNCH 100, 5, NX, NY, NL, NZ, NI, MODE
      NFC = FC                                      PUNCH 100.6
                                                    GO TO 1
                                                    END
```

Problem 9
Reproduce the automatic graph plotter program to fit in a 5″ x 7″ space on an A-size sheet. For this problem, use 0.12″ vertical or inclined capital letters and numerals.

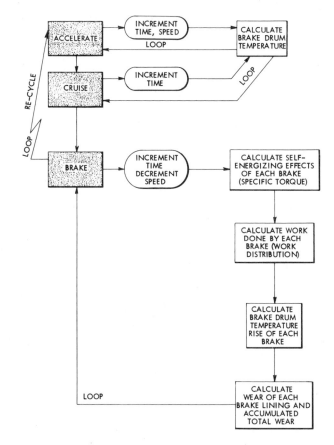

Problem 10
Rearrange the chart shown on an A-size sheet. Use 0.12″ capital letters, either vertical or inclined. Make all letters with a lettering template or a LeRoy lettering instrument.

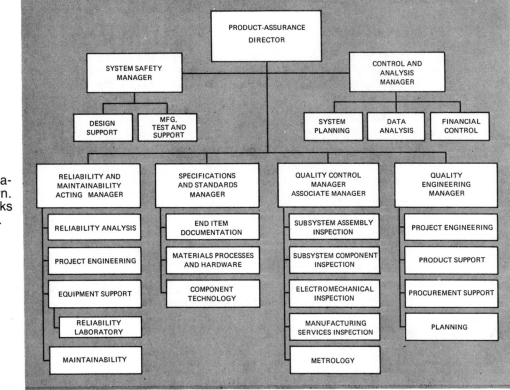

Problem 11
Draw the organizational chart shown. Select size of blocks and size of letters.

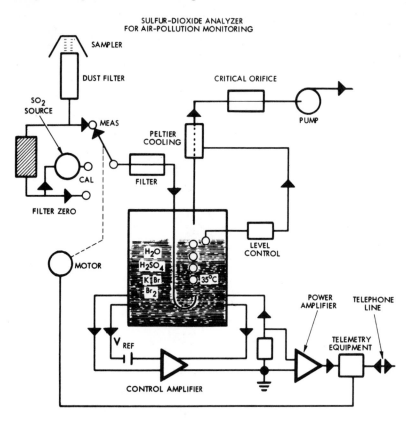

SULFUR-DIOXIDE ANALYZER
FOR AIR-POLLUTION MONITORING

Problem 12
Reproduce the chart of the sulfur dioxide analyzer shown. Select size of chart and lettering.

Problems 13-41 Single View Layout Drawings

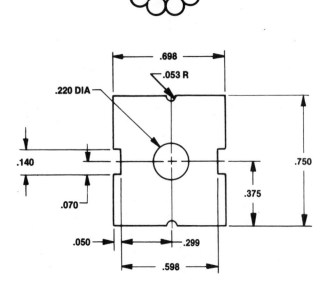

Problem 13
Draw the cross-section 19-strand electrical cable shown. Each strand is No. 3 AWG wire (0.229″ DIA). Make the drawing double size. Determine the minimum inside diameter of a tube to encase this cable.

Problem 14
Make a drawing of the bracket insulation on an A-size sheet. Scale = 4/1. Material: 0.062″ black phenolic plate.

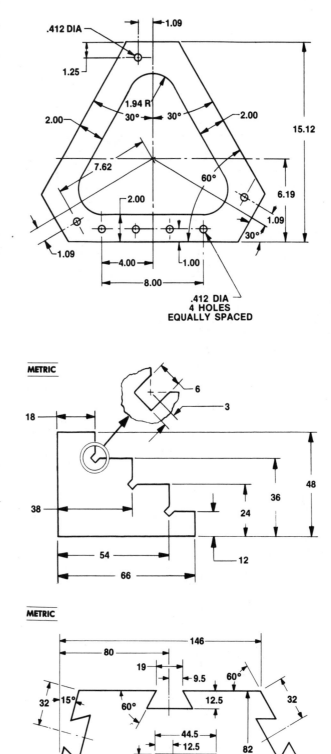

Problem 15

Construct a layout of the lower frame plate as shown. Select any suitable scale.

Problem 16

Make a double-size drawing of the step block shown. Material: tool steel. All dimensions in mm.

Problem 17

Lay out a full-scale profile of the dovetail fixture shown. The three dovetail slots have equal dimensions. All dimensions in mm.

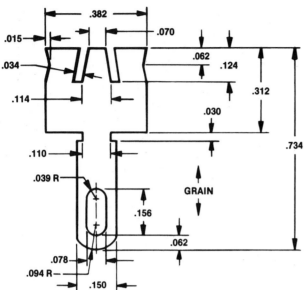

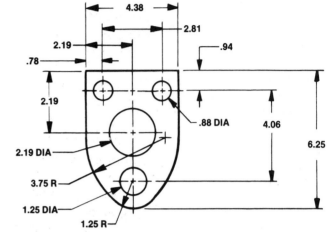

Problem 18
Construct a drawing of the octal contact blank.
Scale = 10/1.

Problem 20
Make a full-size drawing of this nose piece.

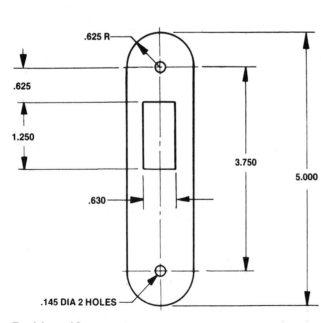

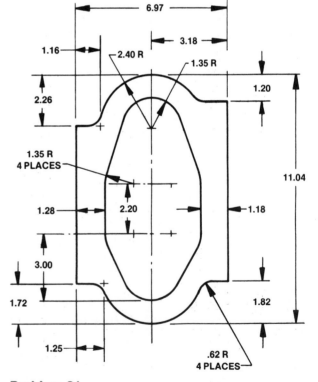

Problem 19
Construct a full-size drawing of the latch front
shown. Material: aluminum, 6063-T5.

Problem 21
Prepare a scaled drawing of the pan seal.

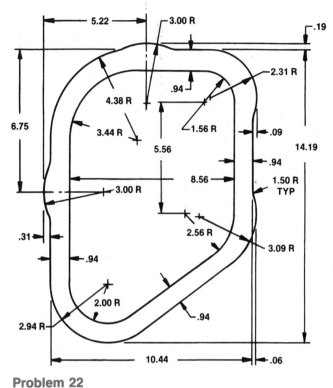

Problem 22
Construct a scaled layout of the oil pan gasket.

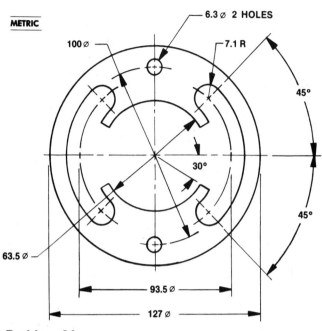

Problem 24
Produce a 1½ size drawing of the gasket shown.
Material: 2.40mm neoprene. All sizes are mm.

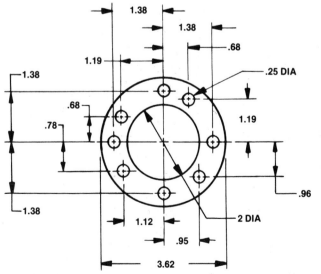

Problem 23
Construct a drawing of the gasket shown. Use any
convenient scale. Material: 0.062″ asbestos fiber.

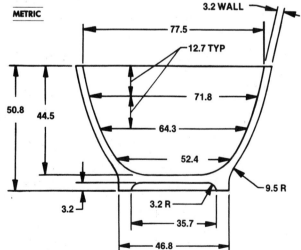

Problem 25
Make a drawing showing the contour of the mixing
bowl. Scale—double size. All sizes are in mm.

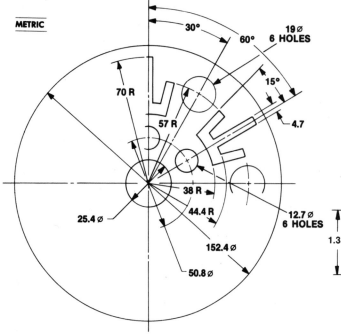

Problem 26
Construct a view of the 0.7mm thick mica shim as shown. Show all six slots and twelve holes plus the center hole. All sizes are in mm.

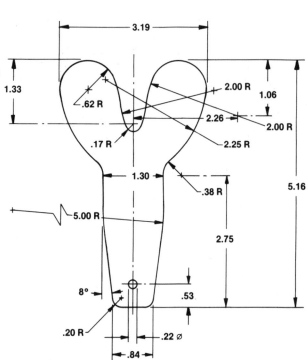

Problem 28
Lay out a full-size drawing of the impression tray blank.

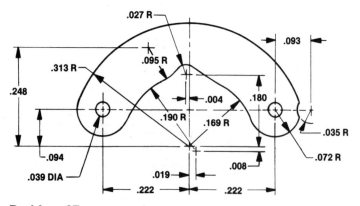

Problem 27
Enlarge the drawing of the lens diaphragm leaf to a suitable scale.

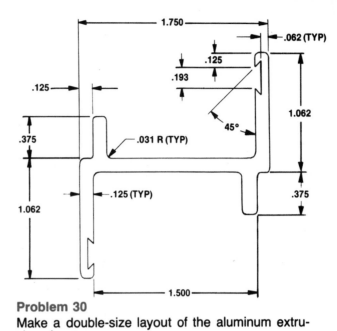

Problem 30

Make a double-size layout of the aluminum extrusion shown.

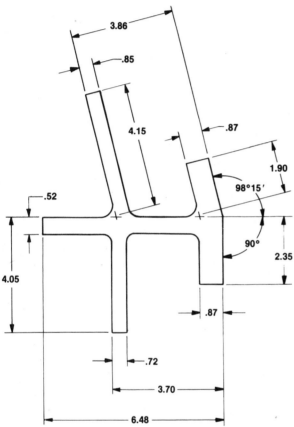

Problem 29

Make an actual size drawing of the cross section of the aluminum extrusion shown.

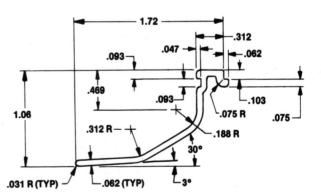

Problem 31

Make a double-size layout of the aluminum extrusion shown.

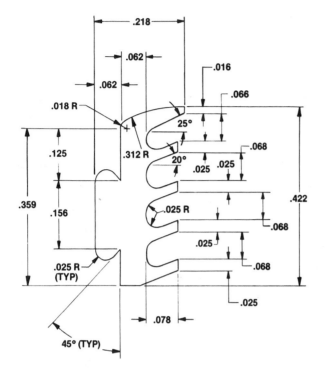

Problem 32

Make an enlarged scale drawing of the polyvinyl weather strip cross-sectional form shown. Include a view showing the actual size of the product.

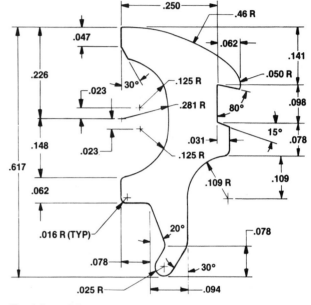

Problem 34

Make an enlarged scale drawing of a polyvinyl weather strip cross-sectional form. Include a view showing the actual size of the product.

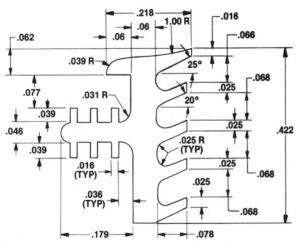

Problem 33

Prepare an enlarged scale drawing of the weather strip cross-sectional form.

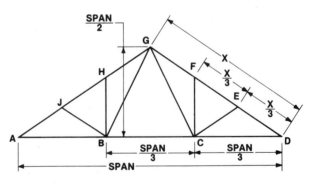

Problem 35

Construct an elevation of the roof truss shown using a span of 32 feet. Measure and record on the drawing the length of the bottom chord AD, the two roof chords DG and GA, and the six webs. Use a scale of ¼" = 1'-0".

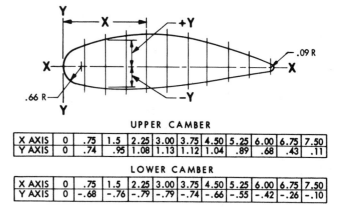

Problem 36
Lay out a pattern for the airfoil section shown.

UPPER CAMBER

X AXIS	0	.75	1.5	2.25	3.00	3.75	4.50	5.25	6.00	6.75	7.50
Y AXIS	0	.74	.95	1.08	1.13	1.12	1.04	.89	.68	.43	.11

LOWER CAMBER

X AXIS	0	.75	1.5	2.25	3.00	3.75	4.50	5.25	6.00	6.75	7.50
Y AXIS	0	-.68	-.76	-.79	-.79	-.74	-.66	-.55	-.42	-.26	-.10

Problem 37
Using any suitable scale, make a layout of the baffle shown.

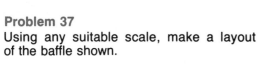

Problem 38
Produce a drawing of the gasket shown. Use any convenient scale. Change all dimensions to two-place decimals.

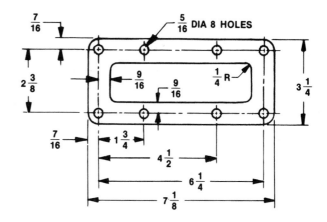

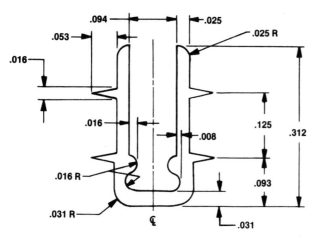

Problem 39
Make an enlarged scale drawing of the polyvinyl weather strip cross-sectional form shown.

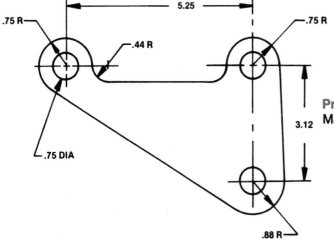

Problem 40
Make a one-view layout drawing of the link shown.

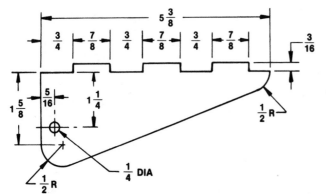

Problem 41
Draw the shelf bracket shown double size. Material: 0.062″ cold rolled steel. Change all dimensions to mm.

SECTION II

Representational Drawings

UNIT 5

Projection Drawing

A representational drawing contains the shape description of an object for manufacturing or construction purposes. The method of producing shape descriptions is based on a system of orthographic projections in which the views are positioned so that they are directly related to one another. The process of preparing projection drawings involves the selection and arrangement of principal views, auxiliary views, and sectional views.

5.1 Orthographic Projection

Orthographic projection is a method of representing the true shape of an object on a single plane. In any orthographic projection process the plane upon which the object is projected is referred to as the *plane of projection.* The position of the eye in viewing the object is called the *point of sight,* which is considered to be an infinite distance from the object. All lines running from the edges or contours of the object to the plane of projection are known as *projectors.* These lines are parallel to one another and perpendicular to the plane of projection. See Fig. 5-1.

Every line of the object must appear as a line or point on the plane of projection. Lines that can be seen are shown as solid lines. Lines that are not visible because they are hidden by some part of the object are represented by dashed lines.

As is seen in Fig. 5-1, the object is assumed to be viewed through a transparent plane from an infinite point. By projecting perpendicular lines from the object to the transparent plane, a true repre-

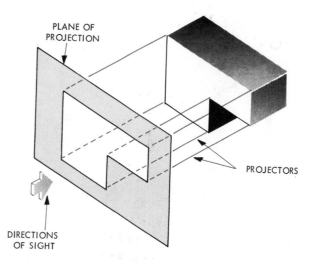

Fig. 5-1. In orthographic projection the shape of an object is projected onto several planes.

sentation of the visible portion of the object appears on the transparent plane.

Since the projection of one face usually does not provide sufficient description of the object, other planes of projection must also be used. To establish its complete shape description, the object is therefore viewed from other positions. The principal projections are called the *front, top,* and *side views.* These projections furnish the necessary views which serve as the foundation for a drawing.

5.2 Planes of Projection

The three planes used in orthographic projection are referred to as the *vertical* or *frontal plane,* the *horizontal plane,* and the *profile plane.* These planes are assumed to revolve about certain axes in order to bring the various views into a single plane. As is shown in Fig. 5-2, the horizontal plane moves about the X-axis and the profile plane moves about the Y-axis. In orthographic projection the horizontal and the profile planes are always revolved into the frontal plane.

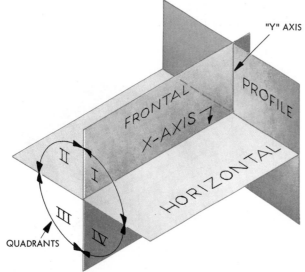

Fig. 5-2. Planes of projection used in orthographic projection.

The intersection of the horizontal and frontal planes generates four quadrants which are known as the first, second, third, and fourth angles. An object may be placed in any one of the four quadrants and its surfaces projected onto the respective planes.

5.2.1 First angle projection.

First angle projection is used principally in European countries in making drawings. In this method of projection the object is located in the first quadrant and is viewed from the positions shown in Fig. 5-3. When the lines of projection are extended to their respective planes, the top view becomes visible on the horizontal plane, the front view on the vertical plane, and the left side view on the profile plane. When the horizontal and profile planes are revolved into the vertical plane, the various views assume the

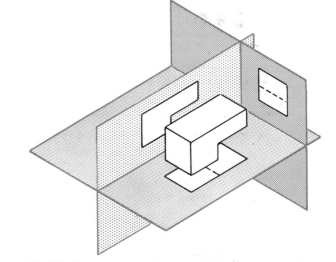

Fig. 5-3. First angle projection is used in European countries.

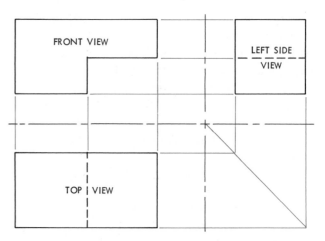

Fig. 5-4. Position of views in first angle projection.

configuration shown in Fig. 5-4. Notice that the front view is directly above the top view and that the left side view is to the right of the front view.

5.2.2 Second and fourth angle projection. If an object is placed either in the second or fourth quadrant and the horizontal and profile planes are revolved into the vertical plane, the top and front views are superimposed. This overlapping of views greatly restricts readability and consequently is ineffective in producing a useful drawing. Hence, the second and fourth angle projections are rarely, if ever, used in the preparation of working drawings.

5.2.3 Third angle projection. The third angle projection is the system of orthographic projection used for drawings in the United States and Canada. As is shown in Fig. 5-5, the object is assumed to be situated in the third quadrant; when the projectors are extended to their respective planes, the top view is shown on the horizontal plane, the front view is on the vertical plane, and the side view is on the profile plane. When the horizontal and profile planes are revolved into the vertical plane, the views fall into the positions shown in Fig. 5-6. Notice that in this arrangement the top view is directly above the front view and the right

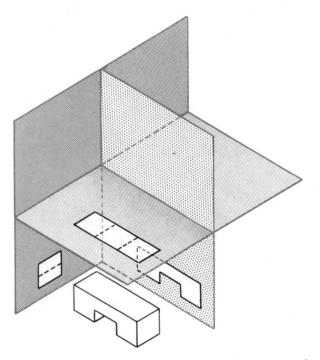

Fig. 5-5. The third angle projection is the system used for drawings in the United States and Canada.

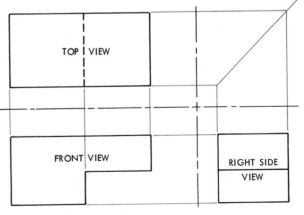

Fig. 5-6. Positions of views in third angle projection.

TOP VIEW

FRONT VIEW

RIGHT SIDE VIEW

side view is to the right of the front view. Compare the third angle projection method with the system used in European countries as described in 5.2.1.

5.2.4 Multiview projection. It was mentioned previously that in order to describe an object completely, it is often necessary to show a number of views. When a series of projections is used, the system is commonly referred to as *multiview projection.* With this system an object can be viewed from six principal positions, namely, top, front, bottom, right side, left side, and rear. See Fig. 5-7. In all cases the plane of projection is between the observer and the object and, except for the rear view, each plane is hinged to the vertical plane and revolved in coincidence with it in a direction away from the object. The rear view may appear

at the extreme right in a reverse position to that shown in Fig. 5-7.

5.3 Visualization of Views

In order to visualize better the positions and the relationship of views, assume that the rectangular box shown in Fig. 5-8 is to be unfolded into a single plane. Assume further that side *A* of the box represents the single plane into which the remaining sides are to be unfolded. For the purpose of swinging the respective surfaces into their correct positions, each panel of the box, except the rear, is considered to be hinged to the front surface. When the panels of the box are swung open as in Fig. 5-9, panel *B* is directly above the front panel

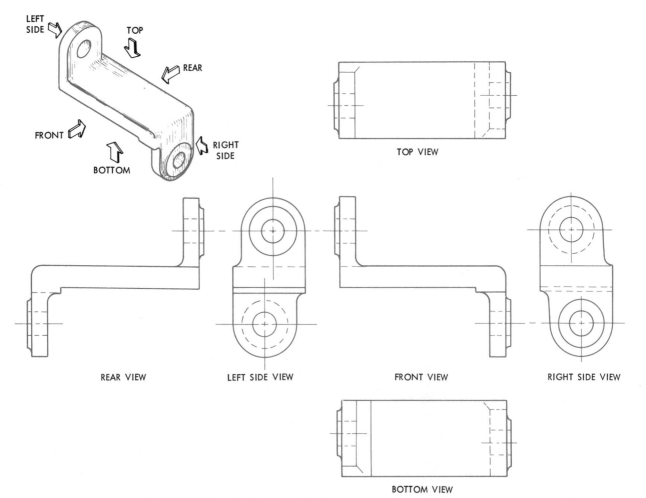

Fig. 5-7. Positions of views in a multiview projection system. (ANSI Y14.5)

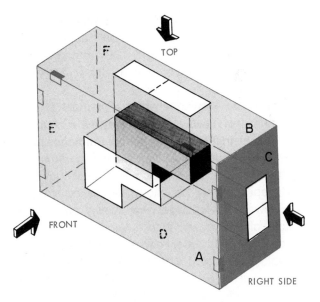

Fig. 5-8. This object is enclosed in a transparent box with projectors extended to the planes.

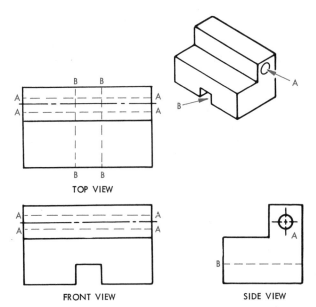

Fig. 5-10. Hidden features of an object are shown by hidden lines.

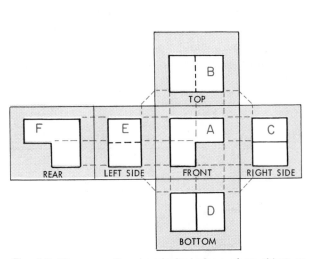

Fig. 5-9. These are the six principal views of an object arranged as they would appear in a drawing.

tion of the observer. The practice is to show these hidden features by means of lines called *hidden lines*. Thus, in the object shown in Fig. 5-10 the hole *A* is visible in the side view, but it cannot be seen when the object is viewed from the front or top. Accordingly, dashed lines *A-A* must be used in the front and top views. Similarly the slot *B* on the bottom of the object can be seen only in the front view. This slot is therefore shown in the top and side views by dashed lines *B-B*.

5.4 Number of Views

Although it has been stated previously that an object has six principal views, it does not necessarily follow that all six views must be shown in order to completely describe the object in a drawing. Quite often some views will be merely duplications of others and therefore will not contribute anything to the drawing. The guiding principle in selecting views is to include only those views which are absolutely necessary to portray clearly the shape of the part. Hence, for some simple objects a single view may be sufficient, whereas for more complicated parts two or more views may be required.

A; panel *C* is to the right of *A;* panel *D* is below *A,* and *E* is to the left of *A.* The rear of the box, panel *F,* unfolds so that it is to the left of *E.* Thus, for any third angle projection the views of an object are always located in these positions.

5.3.1 Showing hidden features. To describe an object completely, a drawing must show all features. In many instances certain edges, intersections, and surfaces cannot be seen from the posi-

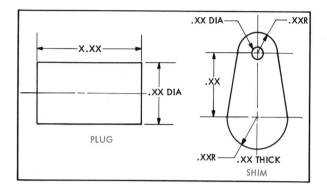

Fig. 5-11. One view drawings are sufficient for cylinders and thin flat pieces having uniform thickness.

5.4.1 One view drawings.
Objects such as rods, shims, gaskets, and thin flat pieces of uniform thickness can often be shown adequately in one view with a note included giving the thickness or diameter. See Fig. 5-11.

5.4.2 Two view drawings.
For many objects, especially those with symmetrical shapes, two views may be sufficient. The most common two view drawings show front and side views or front and top views. Typical examples are shown in Fig.

5-12. In Fig. 5-12*A* only front and side views are needed since a top view would only duplicate the front view. Similarly in Fig. 5-12*B* front and side views are all that is necessary for a complete shape description.

5.4.3 Three view drawings.
For most objects three principal views are necessary—the front, the top, and one side view. The usual practice is to show only the right side view unless the shape of the two ends differs to such an extent that their true shapes cannot be shown clearly by a single side view. If the left side view shows more of the contour of the object, then it should be used instead of the right side view. A typical example is the object illustrated in Fig. 5-13. Notice that by using a left side view instead of the right side view, all hidden lines can be eliminated. This is especially desirable when objects are shown in the positions in which they will appear in a later assembly.

Certain types of objects may need two side views to show all of the details clearly. Observe in Fig. 5-14 that without the right side view it would be virtually impossible to ascertain the true shape of the object.

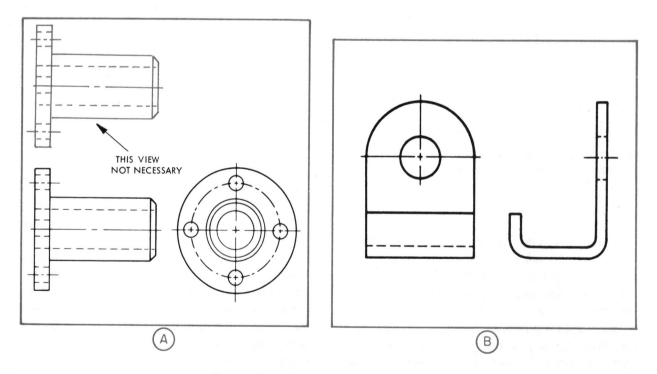

Fig. 5-12. Some objects require only two views.

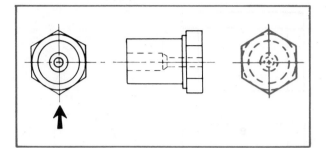

Fig. 5-13. This object is more easily understood if the left side view rather than the right side view is shown.

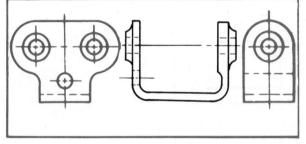

Fig. 5-14. This object requires two side views for a clear representation.

For objects where two side views can be used to advantage, these need not be complete views providing that, when considered together, they will fully describe the shape of the object.

Occasionally there will be objects that will require a bottom view as well as a top view. See Fig. 5-15.

The most common combinations of views are these:

Front, top, right side
Front, top, left side
Front, top, bottom
Front, left side, right side
Front, right side, bottom
Front, left side, bottom
Front, left side, rear
Front, right side, rear

5.5 Orientation of the Object

The object to be drawn can be turned so that its principal surface is parallel to any plane of projec-

RADIO CABINET

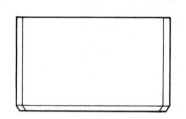

TOP VIEW

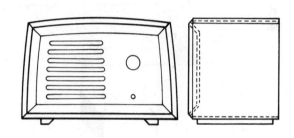

FRONT VIEW   RIGHT SIDE VIEW

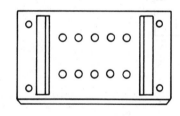

BOTTOM VIEW

Fig. 5-15. This object requires a bottom view as well as a top view.

tion. For example, the object in Fig. 5-16 may be oriented so that it can be viewed in three different ways. It will be seen that the outlines of the resulting views as they would normally appear on the drawing will vary depending on the orientation of the object.

However, the general rule for object orientation is to place the object so that the sides having the

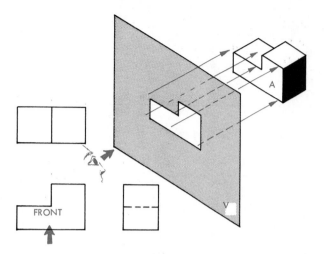

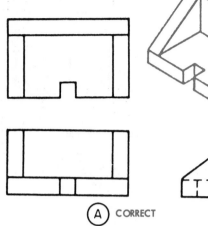

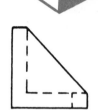

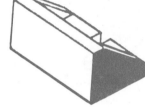

FRONT

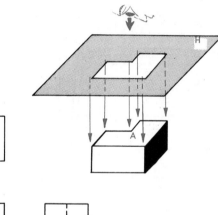

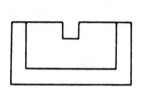

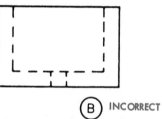

TOP

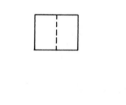

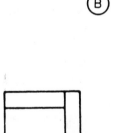

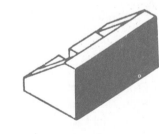

SIDE

Fig. 5-16. The shape of the views will depend on the orientation of the object.

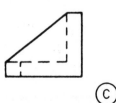

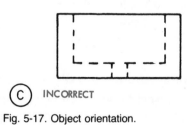

Fig. 5-17. Object orientation.

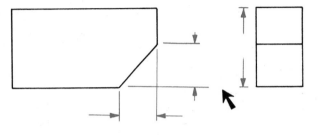

Fig. 5-18. Views should be arranged on the sheet with ample space between them for dimensions.

greatest number of descriptive features are perpendicular to sight lines and parallel to a plane of projection. That is, the object should be situated so that the smallest number of hidden lines will appear in the views. It is apparent in Fig. 5-17 that if the object is oriented as shown in *B,* more hidden lines would be required than if the object were placed in position *A.* It is usually poor practice to have long surfaces serve as side views, as in *C.*

5.5.1 Spacing of views. The views of an object should be arranged so as to present a balanced appearance on the drawing sheet. Ample space must be provided between views to permit the placement of dimensions without crowding and to preclude the possibility of notes overlapping or crowding the views. See Fig. 5-18.

5.5.2 Projecting views. To simplify the positioning of the various views relative to one another, the projection method is often used. This procedure eliminates a great deal of time in measuring distances and locating lines. Thus, in laying out a three view drawing a horizontal line and a 45° line are drawn as in Fig. 5-19. Points can then be readily projected upward, downward, and sideways, thereby reconstructing the shapes of the views in their respective positions.

Many objects have curved surfaces or curved edges tangent to planes. When a circle or curve is parallel to the plane of projection, it is shown as a circle or curve on the parallel plane and is indi-

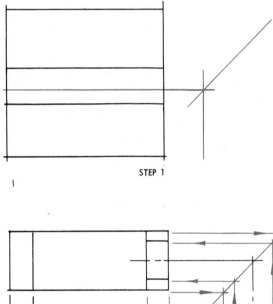

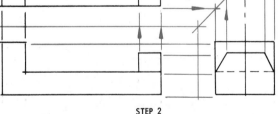

Fig. 5-19. Projecting views simplifies the drawing procedure.

cated on the adjacent planes by straight lines. See Fig. 5-20.

If an object has continuously curved surfaces

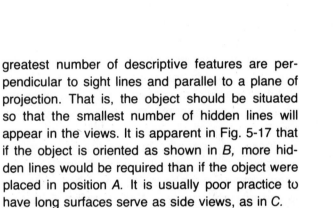

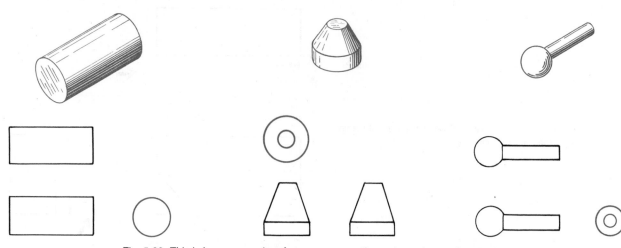

Fig. 5-20. This is how a curved surface appears on the various planes of projection.

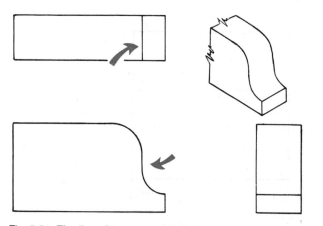

Fig. 5-21. The line of tangency of these two curves appears as a straight line in the top view.

that are perpendicularly tangent to each other, their line of tangency will be perpendicular to the plane of projection and shown as in Fig. 5-21.

If the tangent plane of two curves is at an angle, no line is shown in the plane of projection. See Fig. 5-22.

Where non-circular curves are involved, a number of points on the curve sufficient to show its true shape must be located and then projected to the appropriate view. Notice in Fig. 5-23 that the true shape of the curve in the top view is determined by projecting points from the other two views. Points may be close together or spread out depending on the flatness of the curve. The

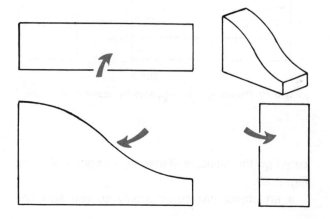

Fig. 5-22. If tangent curves are at an angle, no line is shown in the plane of projection.

coincides with a cutting-plane line the center line is shown. Dimension and extension lines should be placed so that they will not coincide with other lines of the drawing.

5.6 Viewing Special Features

There will be occasions when certain features of an object must be given special treatment for their accurate representation in a drawing. Several common practices governing these special features are as follows:

5.6.1. Partial views. When space is limited, it is permissible to represent symmetrical objects by

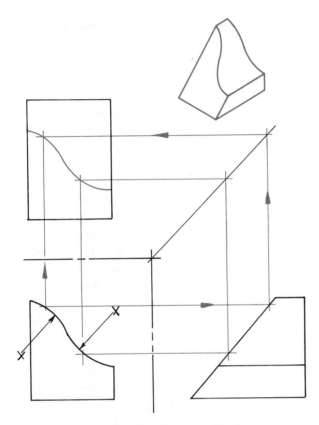

Fig. 5-23. Projection of curved lines.

curved line can be drawn from these points with an irregular curve.

5.5.3 Precedence of lines. In laying out a view there will be instances when hidden lines, visible lines, and center lines will coincide. See Fig. 5-24. Since the essential features of the object are important, visible lines must always take precedence over any other lines. A hidden line always takes precedence over a center line. If a center line

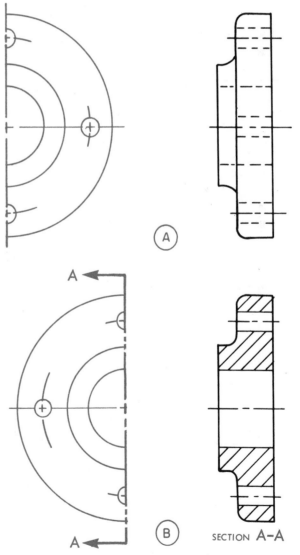

Fig. 5-25. Examples of partial views.

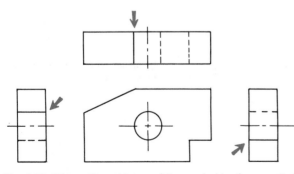

Fig. 5-24. When different types of lines coincide, the accepted precedence of lines must be observed.

partial views. The half view drawn should always be the portion nearest the full view. See Fig. 5-25A. The only exception is when the full view is a sectional view, in which case the partial view is the farthest portion. See Fig. 5-25B. Either the top, front, or side views may be made as partial views.

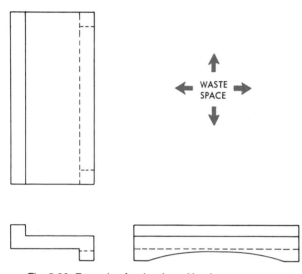

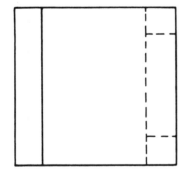

Fig. 5-26. Example of a drawing with a large open area.

5.6.2 Alternate position of side and rear views.

One feature of a well-made drawing is the pleasing balance of the various views. Sometimes proper spacing of views becomes a problem, especially when wide flat objects must be drawn. If the conventional revolution of planes is used for such objects, some parts of the drawing sheet may be crowded while other portions have large open areas. See Fig. 5-26.

To avoid this situation, it is permissible to draw the side view so that it is located to the right of the top view. The right side view is imagined to be hinged to the top view rather than, as is customary, to the front view. The result is a more satisfactory spacing arrangement. See Fig. 5-27. A left side view may also be drawn in an alternate position under similar circumstances.

A rear view may also be drawn in an alternate position rather than in the location shown in Fig. 5-9. In this case the view is considered to be hinged to the top instead of to the left side view. On the drawing sheet the rear view then appears above the top view. See Fig. 5-28.

When views are placed in alternate positions, the views should always be carefully titled to aid in identifying them.

ALTERNATE
RIGHT SIDE VIEW

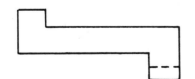

Fig. 5-27. For some objects, better space balance is achieved if the side view projects from the top view.

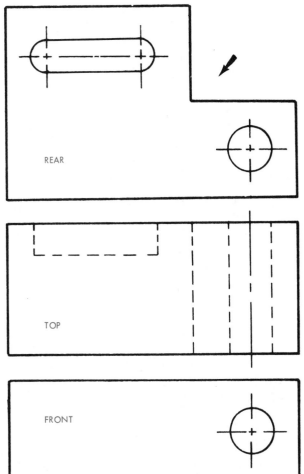

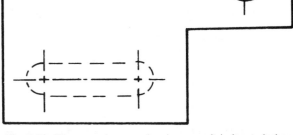

Fig. 5-28. The rear view may be drawn so it is located above the top view.

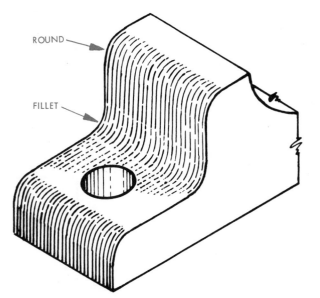

Fig. 5-29. Rounds and fillets provide greater strength in a casting where two surfaces intersect.

rounds. Rounded intersections of surfaces at inside corners are known as *fillets.* See Fig. 5-29.

Fillets and rounds are always shown on a drawing to indicate corners of unfinished castings. If the intersecting surfaces are to be machined, then the corners are drawn square and sharp. See Fig. 5-30. The dimensions of fillets and rounds are des-

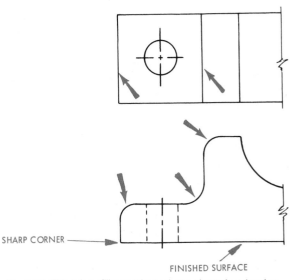

Fig. 5-30. This is how fillets and rounds are shown in a drawing.

5.6.3 Fillets, rounds and runouts. Many parts consist of castings that are made by pouring molten metal into a specially constructed mold. In making castings precautions are always taken to eliminate sharp corners wherever two surfaces intersect. Rounded corners provide greater strength than sharp corners, and they facilitate the casting process. When rounded intersections of surfaces occur at outside edges, they are called

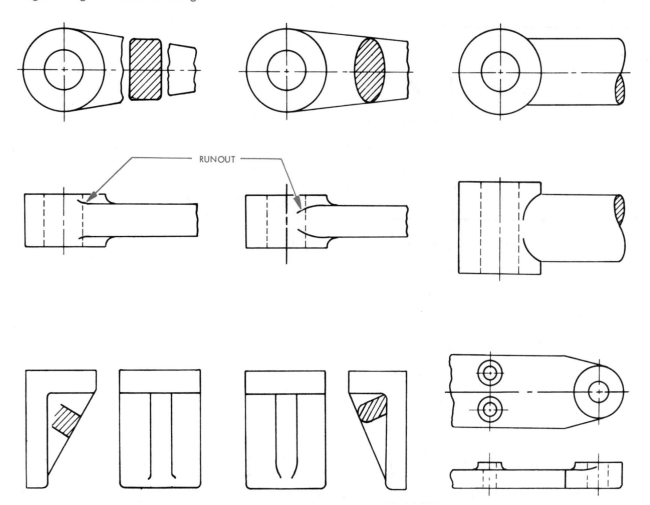

Fig. 5-31. The manner in which runouts are shown is determined by the shape of the intersecting members.

ignated with the note ALL FILLETS AND ROUNDS .XX RADIUS UNLESS OTHERWISE SPECIFIED, which is placed near one of the principal views or included in the area designated for general notes.

When fillets and rounds intersect, the extensions of the curved surfaces are commonly referred to as *runouts.* See Fig. 5-31. Their tangent points should be projected to the related views where they are then shown by an arc or curve having a radius equal to the fillet or round. The required shape and thickness of the filleted area determines whether the arcs are to turn inward or outward. On outside surfaces the theoretical intersections are represented on the related view by solid or dotted lines. See Fig. 5-32.

Arcs to represent fillets and rounds may be drawn freehand or with a compass or irregular curve. The curved segments should be less than a quarter of a circle.

5.6.4 Conventional breaks. Objects which have unusually long features such as rods, shafts, tubes, arms, and bars need not be shown in their entire length. The practice is to break these parts at convenient points with *break lines.* See Fig. 5-33. Whenever break lines are used, the true overall dimension must always be indicated. For short breaks the break line is made freehand and the same thickness as visible lines. Where breaks are particularly long, long thin-ruled lines joined by freehand zigzags are recommended. The appro-

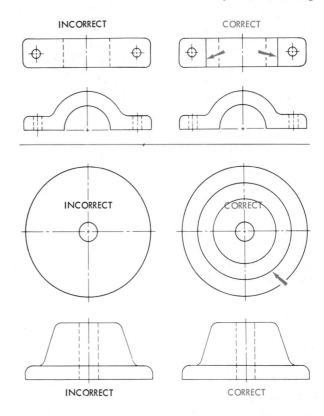

Fig. 5-32. These figures illustrate the conventional method of showing filleted intersections when their true projection may be misleading.

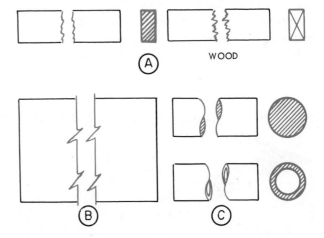

Fig. 5-33. Conventional breaks are used to shorten a view of an elongated object.

priate method of representing breaks on cylindrical surfaces is shown in Fig. 5-33C.

5.6.5 Right hand and left hand parts. On many structures parts frequently operate in pairs, one being on the right and the other on the left. Where this situation occurs, only one part needs to be drawn. However, the part must be marked to indicate which one of the pair it represents. The standard practice is to label the part with the wording RIGHT HAND SHOWN or LEFT HAND OPPOSITE. If the shape of the opposite part is not clearly evident, both right and left versions should be drawn.

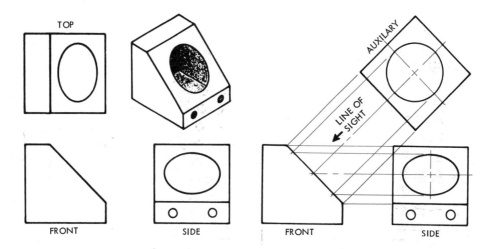

Fig. 5-34. This inclined surface is at an angle to all the principal planes of projection and therefore does not appear in its true size or shape in the front, top, or side views.

5.7 Auxiliary Views

The true shapes of objects that have inclined surfaces cannot always be shown in the regular planes of projection. See Fig. 5-34. Since, to be complete, a drawing must convey the exact shapes of all essential surfaces, it becomes obvious that other means are necessary to describe inclined surfaces. The views employed for such purposes are known as *auxiliary views.*

A primary auxiliary view is one which is pro-

jected to a plane that is perpendicular to one of the three principal planes and inclined to the other two. In Fig. 5-35 the inclined surface is perpendicular to the front plane and inclined to the top and side planes. The true shape of the slanted surface is obtained only by passing a plane parallel to the inclined surface. This auxiliary plane is then considered to be hinged to the plane to which it is perpendicular and revolved into the frontal plane in much the same way that the other views are rotated to their principal planes of projection. Notice

Fig. 5-35. A sloping surface appears as a foreshortened surface in each of the other principal views.

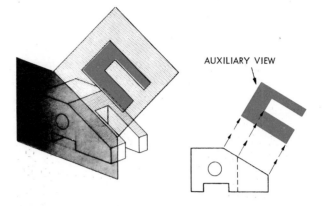

Fig. 5-36. This is the position of the auxiliary view when the inclined surface is revolved into the frontal plane.

in Fig. 5-36 the position the auxiliary view assumes when it is revolved into the frontal plane.

5.7.1 *Types of primary auxiliary views.* The primary auxiliary view may be projected from any of the principal views, that is, from the front, top, or side views. A *front auxiliary view* is one where the inclined surface is perpendicular to the frontal plane and is assumed to be hinged to the frontal plane. See Fig. 5-37A. A *top auxiliary view* is one where the inclined surface is perpendicular to the top plane and is assumed to be hinged to the top plane. See Fig. 5-37B. A *side auxiliary view* is one

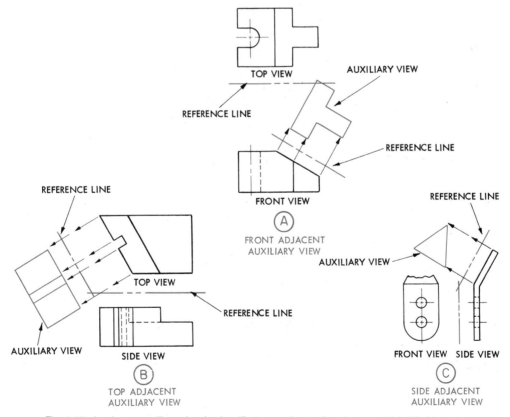

Fig. 5-37. A primary auxiliary view is classified according to the plane to which it is hinged.

where the inclined surface is perpendicular to the side plane and is considered to be hinged to the side plane. See Fig. 5-37C.

5.7.2 Drawing primary auxiliary views.
In making an auxiliary view, the practice is to show the actual contour of only the inclined surface. The projection of the entire view usually tends to confuse rather than to improve the readability of the

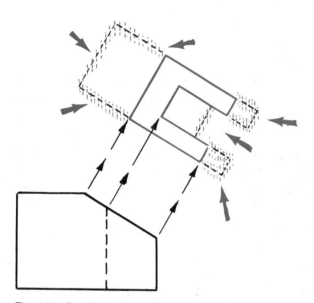

Fig. 5-38. Only the shape of the sloping surface needs to be shown in the auxiliary view.

drawing. Notice in Fig. 5-38 when it is projected alone how much clearer the true shape of the auxiliary surface than if the entire view had been drawn.

Fig. 5-39 illustrates the general procedure for constructing any primary auxiliary view. The basic steps are as follows:

1. Draw two related principal views of the object, such as the front and side views. One of the principal views must always include the edge line of the inclined surface. See step 1 of Fig. 5-39. Here the edge line *AC* of the inclined surface appears in the front view.

2. Draw a reference line parallel to the edge line of the inclined plane. This reference line is assumed to be the hinge line which connects the auxiliary view with the frontal plane. The reference line should be located at some convenient distance from the line of projection so that the auxiliary view falls in a clear space on the drawing sheet. See step 2 of Fig. 5-39.

3. From the principal view containing the edge line of the inclined surface, extend perpendiculars to the reference line.

4. With compass or dividers secure the necessary depth dimensions and transfer them to the auxiliary view. For example, the depth dimensions of the auxiliary view in Fig. 5-39 are obtained from the side view. Thus, the lengths of lines *AB, CD,*

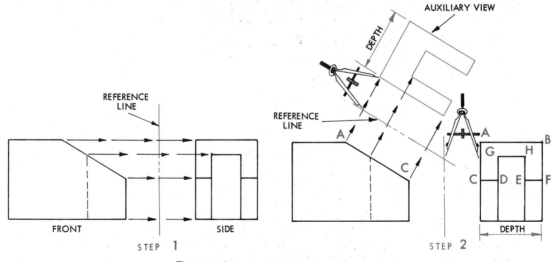

Fig. 5-39. Steps in drawing an auxiliary view.

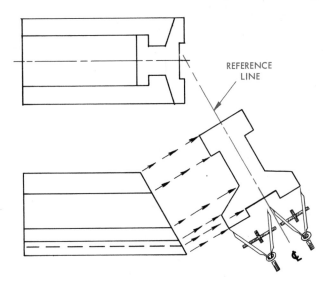

Fig. 5-40. In a symmetrical auxiliary, the view is laid out from the right and left of the reference line.

and *EF* are all transferred from the side view to the auxiliary view.

5. Since both the front and side views of the object shown in Fig. 5-39 contain all the necessary dimensions, no top view is required. A top view in this instance would simply cause an overlapping of projections.

6. Hidden lines are generally omitted from an auxiliary view unless they are needed for clarity.

The same procedure is also followed in projecting an auxiliary view from a top or side view. In all three types of auxiliary views, the shape description of the inclined surface is projected from the view that shows the surface as an oblique line. The distances for the auxiliary are then taken from the other principal views that contain the common sizes for the inclined surface.

5.7.3 Symmetrical auxiliary view. If an auxiliary view is symmetrical, the reference line can serve as a center line, and the auxiliary view can be worked form the right and left of this line. Refer to Fig. 5-40.

In actual practice, in order to save drafting time, only half of the view is usually included. Thus, in the flange shown in Fig. 5-41 only one half is drawn, since the other half is simply a duplication of the portion shown.

5.7.4 Non-symmetrical auxiliary view. A non-symmetrical auxiliary is considered to be unilateral or bilateral. A *unilateral auxiliary* is drawn entirely

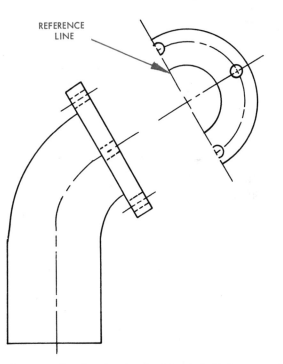

Fig. 5-41. Drawing time is saved if only half of a symmetrical auxiliary is made.

on one side of the reference line. A *bilateral auxiliary* is projected on both sides of the reference line. See Fig. 5-42.

5.7.5 Elimination of principal views. Frequently an auxiliary view permits the elimination of one of the principal views, that is, top, front, or

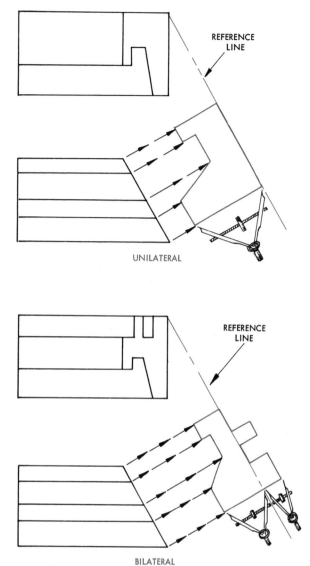

Fig. 5-42. A non-symmetrical auxiliary is considered either unilateral or bilateral.

side views. The general rule is to eliminate a principal view whenever the auxiliary provides sufficient description for a complete understanding of the shape of the part. In Fig. 5-43 the auxiliary furnishes enough information that a top view does not need to be included. Observing this practice facilitates the readability of the drawing and reduces drafting time considerably.

5.7.6. Auxiliary view having curved surfaces. Fig. 5-44 illustrates the steps necessary to draw a curved surface in an auxiliary view. A side view of the curved surface is drawn first and the curve divided into any number of divisions. These points are projected across to the inclined edge of the front view. From the front view the points are projected through the reference line of the auxiliary view. Each distance from the reference line to the curve in the side view is taken and transferred to the auxiliary view as shown.

5.8 Secondary Auxiliary Views

A secondary auxiliary view is one that is projected to define a surface that lies obliquely to all of the principal planes. Whereas a primary auxiliary view is always projected from a principal view, a secondary auxiliary view is always projected from a primary auxiliary view.

The object illustrated in Fig. 5-45 shows the need for a secondary auxiliary view. Notice that the true shape of surface A cannot be shown in

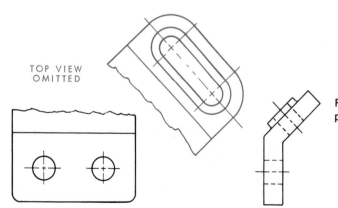

Fig. 5-43. An auxiliary often eliminates the need for one of the principal views.

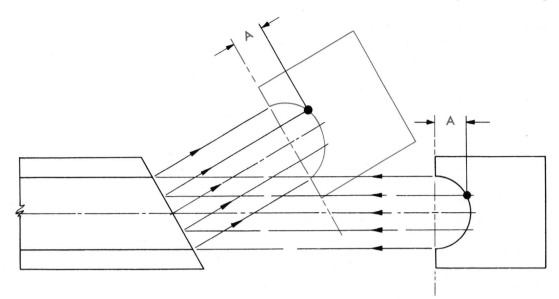

Fig. 5-44. This is how a curved line auxiliary view is drawn.

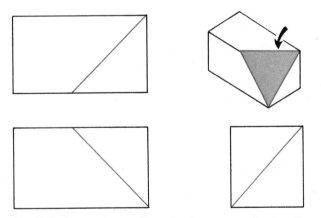

Fig. 5-45. A secondary auxiliary view is necessary to show the true shape of this object.

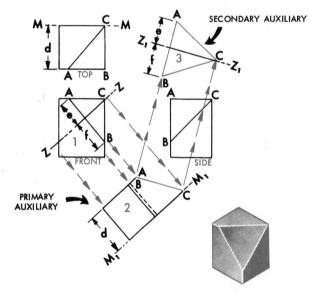

Fig. 5-46. Drawing a secondary auxiliary view which is projected from a front auxiliary view.

the front, side, or primary auxiliary view. The actual shape of the object can be produced only by first drawing a primary auxiliary view and then projecting a secondary auxiliary view from the primary auxiliary view.

A secondary auxiliary view may be projected from a front, top, or side auxiliary view.

5.8.1 Secondary auxiliary view projected from a front auxiliary view. The procedure for drawing a secondary auxiliary view which is to be projected from a front auxiliary view is as follows:

1. Establish reference lines *M-M* in the top view, Fig. 5-46, *Z-Z* in the front view perpendicular to the true length line *AB* of surface *ABC*, and *M₁-M₁* at a convenient distance from the front view parallel to *Z-Z*.

2. Locate points for the primary auxiliary view by transferring distances such as *d* in the top view

to projectors drawn perpendicular to M_1-M_1 from corresponding points in the front view. The surface *ABC* will appear in this view as an edge.

3. Draw reference line Z_1-Z_1 parallel to the edge view of *ABC* at a convenient distance. Complete the secondary auxiliary view by transferring distances *e*, *f*, and so on, from the front view as shown, locating the necessary points on the true view of *ABC*.

5.8.2 Secondary auxiliary view projected from a top auxiliary view. If the arrangement of the view is such that a primary auxiliary view must be projected from a top view, as shown in Fig. 5-47, the secondary auxiliary view is constructed in this manner:

1. Draw a primary auxiliary view by projecting lines parallel to line *AC* in the top view. Locate reference plane *MM* so that it is perpendicular to the projectors from the top view. Distances such as *d* are taken from the front view.

2. Construct the secondary auxiliary view by projecting parallel lines from the edge *CB* in the primary auxiliary view perpendicular to the refer-

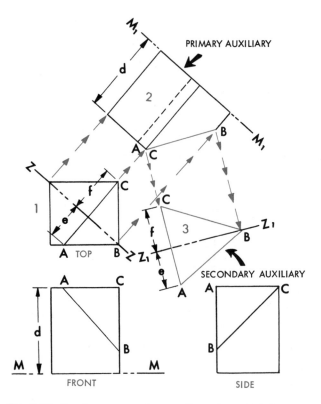

Fig. 5-47. Drawing a secondary auxiliary view which is projected from a top auxiliary view.

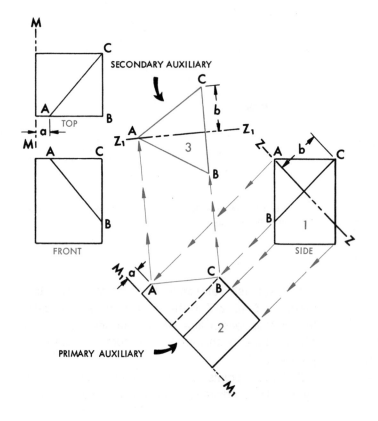

Fig. 5-48. Preparing a secondary auxiliary view which is projected from a side auxiliary view.

ence plane *ZZ*. Draw reference plane *ZZ* in the top view perpendicular to line *AC*. Transfer distances such as *e* and *f* in the top view to the secondary auxiliary view.

5.8.3 Secondary auxiliary views projected from a side adjacent auxiliary view. The procedure for drawing a secondary auxiliary view which is to be projected from a side auxiliary view is essentially the same as described in 5.8.1 and 5.8.2. The primary auxiliary view is completed first and the secondary auxiliary view is then projected from this view. See Fig. 5-48.

5.9 Sectional Views

Many objects have internal shapes which are so complicated in nature that it is virtually impossible to show their true shapes without employing numerous confusing hidden lines. See Fig. 5-49. In situations where interior construction cannot be clearly shown in exterior views, the use of one or more sectional views becomes necessary. A sectional view not only reveals the actual internal shape of an object but it may also retain the significant outline of the external contour.

The cross section of an object is obtained by passing an imaginary cutting plane through the object. The cutting plane is assumed to pass through in some selected position of the object, and the cut part is removed. See Fig. 5-50.

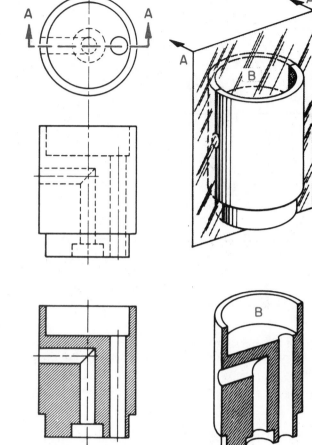

SECTION A–A

Fig. 5-50. A portion can be removed to reveal the internal shape by passing a cutting plane through the object.

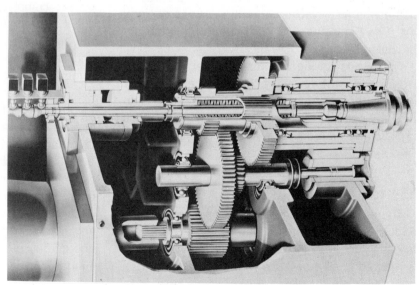

Fig. 5-49. The function of a sectional view is to show the internal construction of an object.

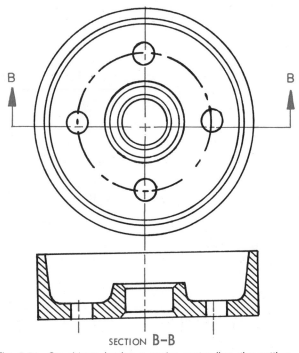

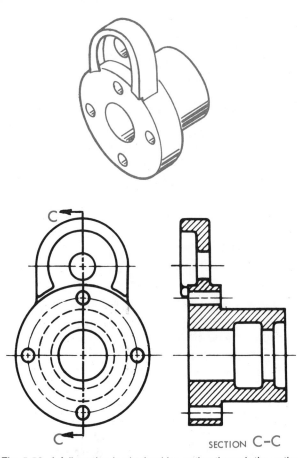

SECTION B-B

Fig. 5-51. On objects having a major center line, the cutting plane may be omitted if the section is taken along that center line.

SECTION C-C

Fig. 5-52. A full section is obtained by cutting through the entire object.

Capital letters *A-A, B-B, C-C,* and so on, are used when it is necessary to identify the section. The letters should be made to read horizontally, should not be underlined, and should be located in front of the arrowheads if the drawing is made to conform to MIL-STD-100, or placed behind the arrows if made to ANSI recommended standards. A notation is also placed under the view as SECTION A-A. See Fig. 5-50. The word SECTION is often abbreviated SECT.

On objects having one major center line and in which the cutting plane is assumed to pass through the axis of symmetry, the practice is to omit the cutting plane line since its position makes it clear that the section is taken along that center line. See Fig. 5-51.

5.10 Types of Sectional Views

There are various types of sectional views, such as full, half, offset, broken-out, revolved, removed, auxiliary, and thin. In each case the view produced is referred to as a sectional view.

5.10.1 Full section. When the cutting plane passes completely through the object, the result is a *full section.* Notice in Figs. 5-51 and 5-52 that the half of the object that is between the observer and the cutting plane is considered to have been removed and the remaining half, in full section, exposed to view. By passing the cutting plane in this position, a full view of the object is seen which can be either a front or side view, depending on the orientation of the object.

5.10.2 Half section. A *half section* results when two cutting planes are passed at right angles to each other along the center lines or symmetrical axes. As is shown in Fig. 5-53, passage of the cutting planes in this manner permits the removal of one-quarter of the object and a half section of the interior is exposed to view.

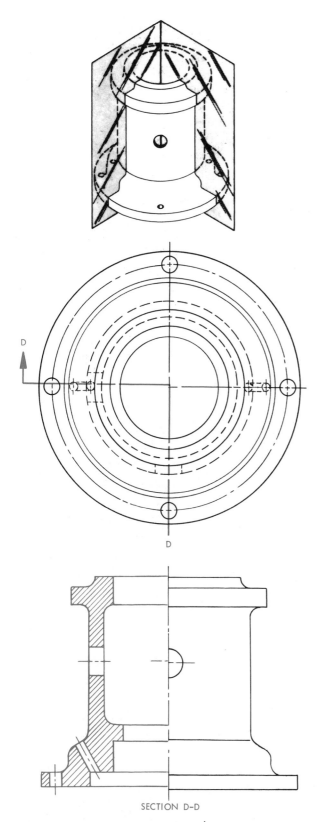

SECTION D-D

Fig. 5-53. When the cutting plane extends halfway through the object, a half section is obtained.

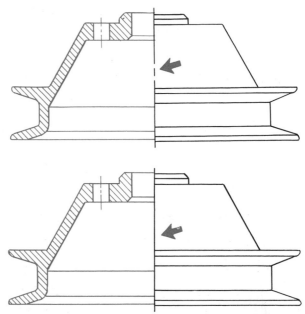

Fig. 5-54. Either a visible line or center line may be used to divide the sectioned parts in a half sectional view, but the center line is preferable.

A half section has the advantage of showing the interior of the object and at the same time maintaining the shape of the exterior. Its usefulness is limited to symmetrical objects. Because it is often difficult to completely dimension the internal shape of a half section, this type of sectional view is not widely used in detail drawings (drawings of single parts). Its greatest value is in assembly drawings where it is necessary to show both internal and external construction on the same view.

Either a visible line or a center line may be used to separate the sectioned half from the unsectioned half. It is generally conceded that a center line is much more realistic for this purpose because removal of a quarter of the object is imaginary only and the actual edge, as implied by a solid line, does not exist. See Fig. 5-54. Hidden lines are usually omitted on the unsectioned part of the view.

5.10.3 Offset section. It is often necessary to change the direction of the cutting plane line from along the main axis in order to include features which are not located in a straight line. The cutting plane is therefore offset to pass through these features, and the resulting section is called an *offset*

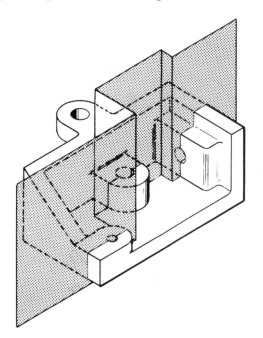

section. Note in Fig. 5-55 that offsetting the cutting plane in several places exposes the shape of the openings and recess which would not be seen if a regular full section were utilized. In making an offset sectional view, the offsets are not included in the sectional view but only in the view showing the cutting plane line.

5.10.4 Broken-out section. For certain types of objects, the removal of only a small portion is necessary to show the interior construction. By removing only a small portion, it is often possible to preserve some detail of the object that otherwise would be eliminated in a full or half section. The sectioned area is limited by a freehand break line. Observe in Fig. 5-56 that in the broken-out sectional view no symbolic cutting plane line is shown.

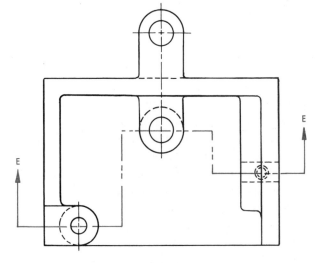

SECTION E-E

Fig. 5-55. In an offset section, the cutting plane line is offset to pass through features not located in a straight line.

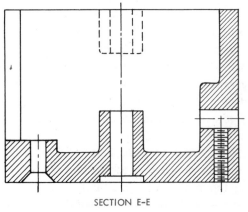

Fig. 5-56. A broken-out section is used to show only a desired feature of the object. No cutting plane line is necessary.

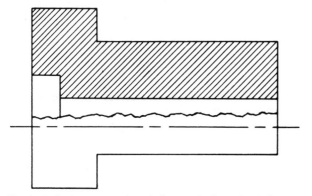

Fig. 5-57. Broken-out section similar to a half sectional view.

Another type of broken-out section that is sometimes used is very similar to the half section. See Fig. 5-57. Notice in this case that an entire segment of the object is removed to reveal the interior but that the removed piece is still outlined with a break line. The advantage of such a section is that

it minimizes any misunderstanding of construction requirements.

5.10.5 Revolved section. Revolved sections are used to show the actual cross sectional shapes of such objects as bars, spokes, propeller blades, arms, ribs, and other elongated parts. The cutting plane is passed perpendicular to the axis of the piece and then revolved in place through 90° into the plane of the sheet. See Fig. 5-58. The visible lines on each side of the adjacent view may be removed and broken lines used so as to leave the revolved section clear. The true shape of the exposed revolved section should always be retained regardless of the direction of the contour lines of the object. Notice that in a revolved section no cutting plane indication is necessary.

5.10.6 Removed section. Greater clarity is often achieved if the section is detached from the projected view and located elsewhere on the sheet. Such a section is known as a *removed section.* See Fig. 5-59. By removing the section, the regular view can be left intact and the removed section drawn to a larger scale to facilitate more complete dimensioning.

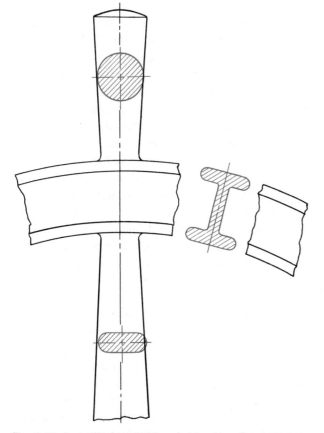

Fig. 5-58. A revolved section is used to show the cross sectional shape of elongated objects.

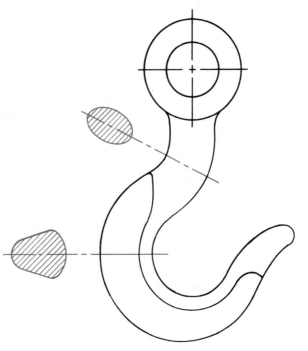

Fig. 5-59. To avoid confusion, place the removed section elsewhere on the sheet.

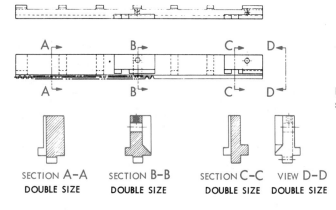

SECTION A–A
DOUBLE SIZE

SECTION B–B
DOUBLE SIZE

SECTION C–C
DOUBLE SIZE

VIEW D–D
DOUBLE SIZE

Fig. 5-60. These are the correct practices in labeling removed sections and a removed view.

The American National Standards Institute recommends the following practices regarding removed sections:[1]

1. A removed section should be labeled in bold lettering, for example, **SECTION B-B**, to identify it with the cutting plane line which is labeled with corresponding letters at the ends. See Fig. 5-60.

2. A removed section should be placed in a convenient location; if possible, on the same sheet with the regular view. On multiple-sheet drawings where it is not practicable to place a removed section on the same sheet with the regular views, identification and zoning references should be indicated on related sheets. Below the section title the sheet number where the cutting plane line will be found should be given as **SECTION B-B ON SHEET 4, ZONE A3**. A similar note should be placed on the drawing where the cutting plane is shown, with a leader pointing to the cutting plane, referring to the sheet where the section will be found. On large drawings having many removed sections, similar cross referencing between cutting plane lines and their corresponding sections should be provided if necessary.

3. If two or more sections appear on the same sheet, they should, if possible, be arranged in alphabetical order from left to right. Section letters should be used in alphabetical order, but to avoid confusing the *I* with the numeral 1, or the *O* with the *Q* or zero, the letters *I, O,* and *Q* should not be used. If more than 23 sections are used, the additional sections should be indicated by double letters in alphabetical order, *AA-AA, BB-BB,* and so on.

4. A removed section may be drawn to a larger scale if necessary, in which case the scale should be shown under the section title.

5. A removed section should not be rotated on the sheet when this would diminish the clarity of the drawing; that is, the edges or center lines of the section should be parallel to the corresponding lines in the normal projected position.

6. Removed sections may be placed on center lines extended from the section cuts.

5.10.7 Auxiliary section. Ocassionally it is necessary to show a sectional view that is not in one of the principal planes. This view is called an *auxiliary section.* An auxiliary section may be full, half, broken-out, removed, or revolved. The section should be shown in its normal auxiliary position and clearly identified with a cutting plane line and appropriate letters. Fig. 5-61 illustrates a partial section at *G-G.*

5.10.8 Section lining. To clearly define the surfaces of a section, thin lines are drawn across the cut area at a 45° angle to the horizontal. See Fig. 5-62. If the section consists of two adjacent parts, such as in an assembly drawing, the section lines should run in opposite directions in order to provide contrast. When three parts form the section, the third part adjacent to the first two should have lines sloping at 30° or 60° to the main outline of

---

1. Extracted from *American Drafting Standards Manual, ANSI Y14.2.*

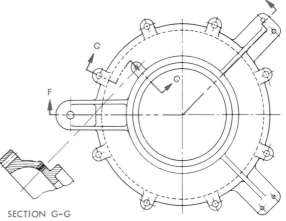

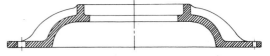

Fig. 5-61. In this drawing section G-G is a typical example of an auxiliary section.

SECTION G-G

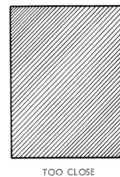

SECTION F-F

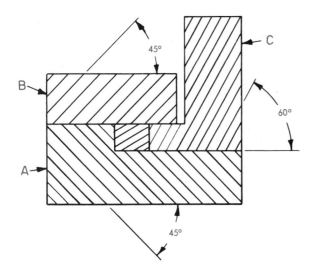

Fig. 5-62. The cut surfaces of adjacent parts in sectional views should be shown with section lines drawn in this manner.

the view. For additional adjacent parts the lines may be drawn at any suitable angle so long as each part stands out clearly.

All lines should be uniformly spaced. The spacing should not be measured but judged by eye. It is also important not to get the lines too heavy or too close together. See Fig. 5-63. Drafting time

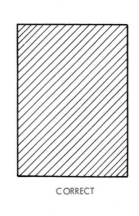

CORRECT

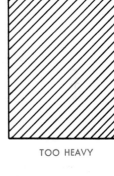

TOO HEAVY

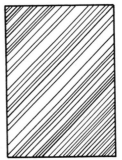

NOT UNIFORM IN SPACING OR LINE WEIGHT

TOO CLOSE

Fig. 5-63. The correct method and some of the common faults that should be avoided in drawing sectional lines.

CORRECT   INCORRECT

Fig. 5-64. If section lines drawn to the conventional angle would be parallel to the contour of the object, they should be drawn at some other angle.

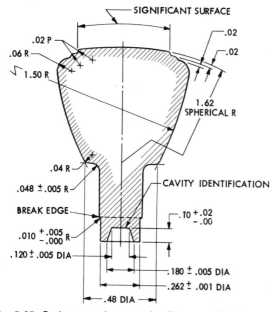

Fig. 5-65. On large sections, section lines need be drawn only near the outside boundary.

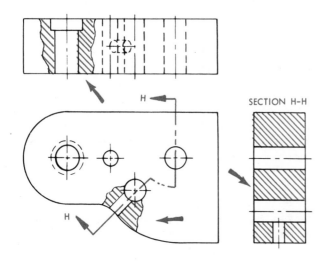

SECTION H-H

Fig. 5-66. If more than one view is sectioned, all section lines must be drawn parallel.

can often be saved by using prelined press-on shading film. See 8.9.5.

If the shape or portion of a sectional area is such that the section lines would be parallel or nearly perpendicular to the dominant visible lines of the section, the section lines should be drawn to some angle other than the standard ones. See Fig. 5-64.

In sectioning very large areas, it is permissible to use section lines only near the outside boundary of the sectioned area, with the interior portion left clear. See Fig. 5-65.

All the sectioned areas of a single part should be lined in the same direction and with the same angle of slope. Thus, if an object has three views, and two or more views are sectioned, the section lines must all be drawn parallel. See Fig. 5-66.

5.10.9 Thin sections. Material such as sheet metal, gaskets, and packings and other thin substances cannot be shown by the ordinary cross sectioning convention. Cross sections of materials of these kinds are drawn solid. See Fig. 5-67. If two or more pieces are adjacent to each other, a white space is left to separate the sections. See Fig. 5-68.

Some industries specify that where two or more adjacent thin sections are used, they should be shown solid, but an additional exploded view of that portion must be included to properly define the arrangement of parts.

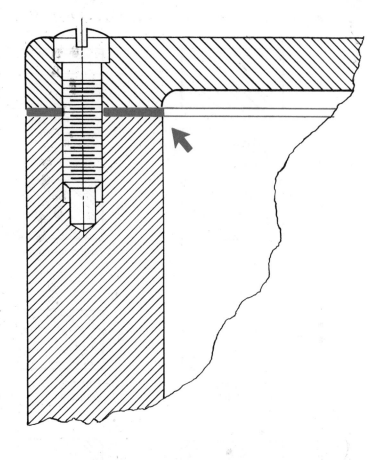

Fig. 5-67. Thin material should be shown solid.

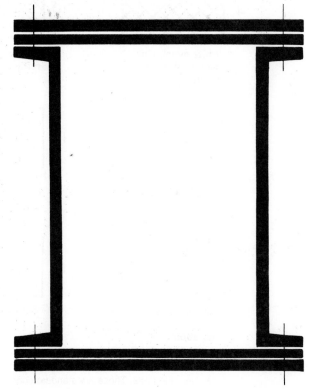

Fig. 5-68. This a method of sectioning adjacent thin parts.

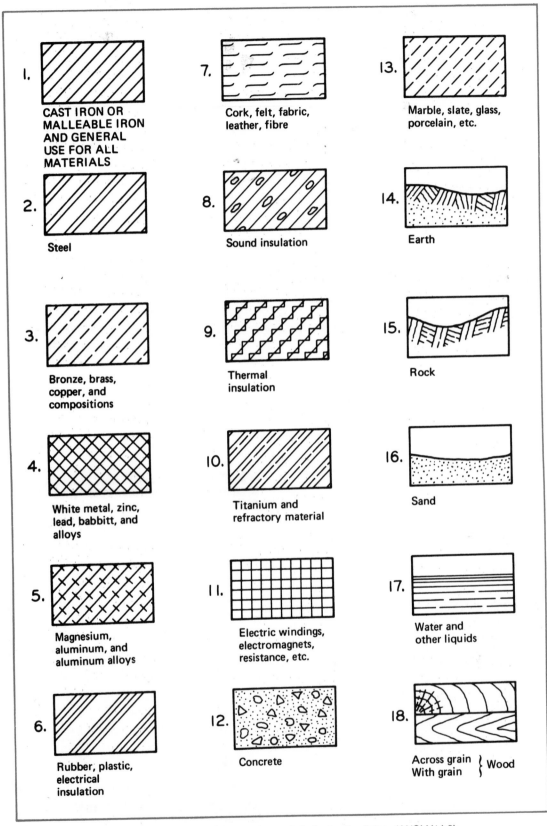

1. CAST IRON OR MALLEABLE IRON AND GENERAL USE FOR ALL MATERIALS

2. Steel

3. Bronze, brass, copper, and compositions

4. White metal, zinc, lead, babbitt, and alloys

5. Magnesium, aluminum, and aluminum alloys

6. Rubber, plastic, electrical insulation

7. Cork, felt, fabric, leather, fibre

8. Sound insulation

9. Thermal insulation

10. Titanium and refractory material

11. Electric windings, electromagnets, resistance, etc.

12. Concrete

13. Marble, slate, glass, porcelain, etc.

14. Earth

15. Rock

16. Sand

17. Water and other liquids

18. Across grain / With grain } Wood

Fig. 5-69. These are standard code symbols for section lining. (ANSI Y14.2)

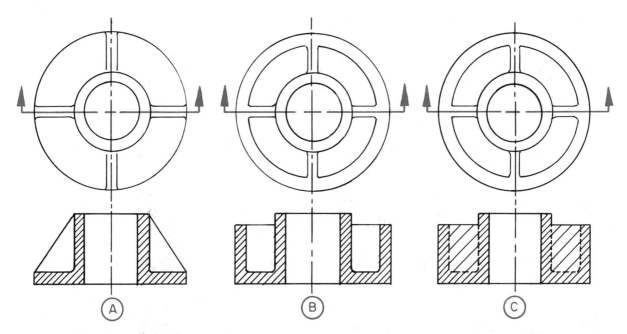

Fig. 5-70. Method of section lining if a cutting plane passes flatwise through a web, rib, or gear tooth.

5.10.10 Section lining symbols. Fig. 5-69 illustrates the various symbols which may be used for sectioning as approved by American National Standards Institute (ANSI). The practice is to employ these symbols on assembly drawings where it is desirable only to distinguish between the different classes of materials without specifying their exact composition. On detail drawings the all-purpose cast-iron symbol is recommended, except for parts made of wood, with the exact specification of the material given in a note near the view or in the title strip or parts list.

5.10.11 Sections through webs or ribs. When the cutting plane passes flatwise through a web, rib, gear tooth, or other similar flat element, to avoid presenting a false impression of thickness or solidity, the element should not be sectioned. See Fig. 5-70. If the cutting plane cuts across elements that are not flatwise, the elements should be section lined in the usual manner. See Fig. 5-71.

Alternate section lining may be used in cases where the actual presence of a flat element is not sufficiently clear without section lining or where clear description of the feature may be improved. For example, in Fig. 5-70B the presence of the ribs is not immediately clear in the sectional view,

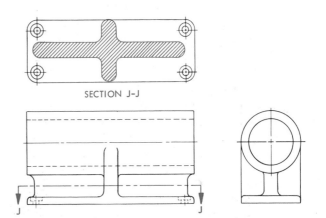

SECTION J-J

Fig. 5-71. If a cutting plane passes across elements that are not flatwise, section lines should be used.

while in Fig. 5-70C the alternate section lining is used to show the ribs. When alternate section lines are drawn as in Fig. 5-70C, the line spacing should be twice as wide as in normal sections.[2]

5.10.12 Sections through shafts, bolts, pins. When the cutting plane contains the center lines of such elements as shafts, bolts, nuts, rods, rivets, keys, pins, spokes, screws, ball or roller bearings,

———————————

2. *Ibid.*

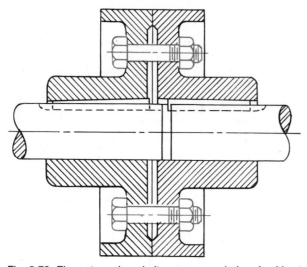

Fig. 5-72. Elements such as bolts, screws, and pins should not be sectioned.

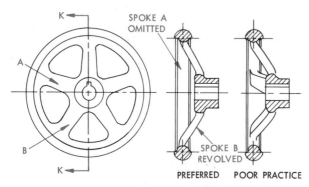

Fig. 5-73. Parts which normally include foreshortened elements should be drawn like this.

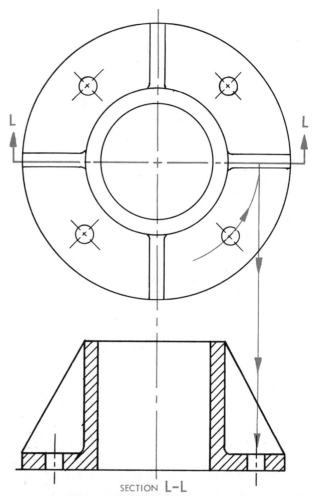

SECTION L–L

Fig. 5-74. Holes in drilled flanges should be rotated so their true distance from the center is shown.

or similar shapes, no sectioning is needed. See Fig. 5-72. However, if the cutting plane cuts across the axes of elongated parts, they should be sectioned in the usual manner.[3]

5.10.13 Foreshortened projections and related features. When the true projection of inclined elements would result either in foreshortening, which might be confusing, or in unnecessary expenditure of drafting time, these elements should be rotated and drawn as in Fig. 5-73.

In drawings of drilled flanges, the holes may be rotated to be shown at their true distance from the center rather than in true projection, if clearness is promoted. See Fig. 5-74.

To include features not located along a straight line, the plane may be bent or changed in direction to pass through these features and the sections drawn as if they were rotated into a plane. See Fig. 5-75. Such sections are called *aligned sections,* whether features are rotated into the cutting plane or the cutting plane is bent to pass through them.[4]

---

3. *Ibid.*

4. *Ibid.*

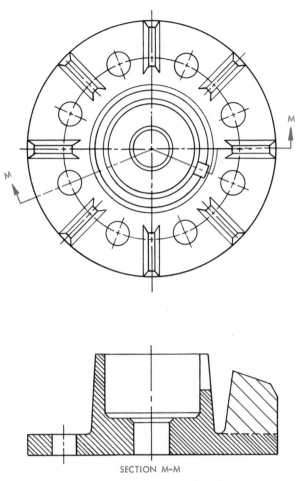

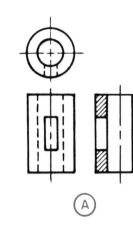

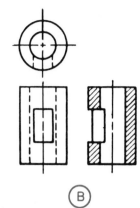

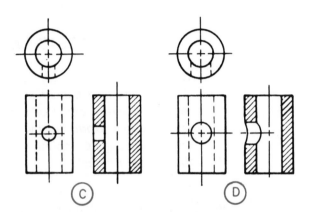

SECTION M–M

Fig. 5-75. Aligned sections should be drawn in this manner.

Fig. 5-76. This is how intersections in sections are shown.

5.10.14 *Intersections in section.* When a section is drawn through an intersection in which the exact figure or curve of intersection is small or of no consequence, the figure or curve of intersection may be simplified as in *A* or *C* of Fig. 5-76. Larger

intersections are usually projected true, as in *B,* or approximated by circular arcs as shown in *D* of Fig. 5-76.[5]

5. *Ibid.*

UNIT 6

Basic Dimensioning

A drawing is expected to convey exact and positive information regarding every detail of the represented part. Without definite specifications expressed by dimensions, it is impossible to indicate clearly any engineering intent or to achieve successful fabrication of the desired product.

An intelligently dimensioned drawing should represent the part as it is to be manufactured. It should permit mass production and interchangeability of parts regardless of the location of the manufacturing plant. A drawing should deal cautiously with the method of production, specifying the results to be obtained rather than the method of fabrication in order to allow the maximum latitude in optional methods of manufacture.[1]

Since the shopworker is expected to follow the instructions which appear on a drawing, those who are responsible for preparing drawings should be familiar with the tools, machines, materials, and manufacturing facilities available. If they have a knowledge of actual manufacturing processes, there is less chance of production error.

Above all, sizes on a drawing should never have to be measured to determine their values. It should never be necessary for the shopworker to have to calculate or assume any dimension in order to fabricate a product.

6.1 Definition of Dimensioning Terms

To fully understand dimensioning practices, a knowledge of basic terms associated with dimensioning is very important. Here are those that are particularly significant:

Dimensions are numerical values that define the exact size and shape of an object.

Basic dimension is a numerical value that describes the theoretically exact shape of an object and serves as a basis from which permissible tolerance variations are established.

Reference dimension is an untoleranced dimension used for informational purposes only and is derived from other given dimensional values.

Datum is a point, line, plane, or cylinder that serves as a base for establishing the true shape of a part.

Feature is any component portion of a part that can be used as a basis for a datum.

---

1. *Chrysler Drafting and Design Standards* (Detroit: Chrysler Corporation).

Nominal size is the given dimensional value that identifies the part.

Allowance is the clearance between mating parts.

Tolerance is the total amount a specific dimension may vary.

Limits are the maximum and minimum values prescribed for a specific dimension.

Unilateral tolerance is a tolerance that is permitted to vary in only one direction from the specified dimension.

Bilateral tolerance is a tolerance that is permitted to vary in both directions from the specified dimension.

Fit defines the range of tightness or looseness of mating parts.

End-product dimensioning is the establishment of all necessary dimensional values to completely show the actual finish shape of a part.

Process dimensioning is the notation that specifies the manufacturing process involved in producing a part.

6.2 Dimensioning Systems

Dimensions on industrial drawings are expressed as decimal-inch or metric values. Fractional dimensions are rarely used. Many industries have adopted a dual dimensioning system in which both decimal-inch and metric units are shown.

6.2.1 Decimal-inch dimensioning system.
Two-place decimals are used where tolerance limits of ±.01 or increments of .01 can be allowed. Decimals to three or more places are used for tolerance limits less than ±.010. No dimension should have fewer than two decimal places.

The second decimal place in a two-place decimal should preferably be an even (rather than an odd) digit, so that dividing by two (as in obtaining a radius from a diameter) will result in an exact two-place decimal. For this reason, .02, .24, .68, and so on are preferred to .01, .23, .69, and so on. However, odd two-place decimals are acceptable when necessary for design reasons (as to provide clearance, strength, smooth curves, and so forth). See decimal chart in Appendix.

The following general practices should be observed in decimal dimensioning:

1. All dimensions should be specified in two-place decimals except where tolerances require decimals to more than two places:

> 1.00 not 1.
> 1.40 not 1.4
> 1.31 not 1.3125

2. When showing dimensions with upper and lower limits, both upper and lower values should contain the same number of decimal places:

CORRECT	INCORRECT
1.212	1.212
1.210	1.21

3. Where tolerances are shown following the nominal dimension, the tolerance and the dimension should have the same number of decimal places:

CORRECT	INCORRECT
6.84 ± .02	6.84 ± .020
1.960 ± .005	1.96 ± .005

4. An endeavor should be made to work with even hundredth increments of an inch. Odd hundredths may be used only where required for accuracy; however, this practice should be avoided as much as possible:

PREFERRED	ACCEPTABLE
1.00	
or	1.01
1.02	
3.12	
or	3.13
3.14	

Care should be exercised in the use of odd hundredths, especially when the dimension will be divided by two, as is necessary to locate a hole or an edge from a center line.

5. Dimensions which are divided into two or more equal lengths (as between equally spaced holes), should be such that the resulting equal dimensions will be in even hundredth increments of the decimal scale, as shown in Fig. 6-1.

6. Where design requires overall dimensions which, when divided into two or more equal spaces, result in dimensions expressed in thousandths of an inch, these dimensions should be rounded off to two-place decimals wherever possible. See Fig. 6-2.

7. Where ordinates are used to dimension curves or irregular lines, all dimensions are ex-

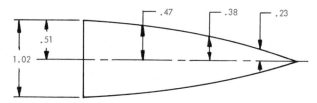

Fig. 6-3. Correct use of decimals for dimensioning curves of irregular lines.

pressed in hundredths as scaled from the layout. No effort is made to use even hundredths. See Fig. 6-3. For greater accuracy on symmetrical parts, dimensions on the overall figure should be in even hundredths to prevent any dimension from center line to side from being a three-place decimal.

8. Nominal sizes of commercially produced materials or of features, such as holes, threads, and keyways, produced and inspected by commercially available tools, gages, and so on should be dimensioned with limits according to the capability of the tool or to the increments of the gage used, for example, .373 — .380 DIA HOLE, .365 — .385 CBORE, .02 — .04 R, or .250 — 20 UNC — 2A.

Wires, cables, sheet stock, and so on that are manufactured to gage or code number should be specified by a decimal indicating the diameter or thickness. The gage or code number is placed in parentheses after the decimal, for example, .0508 (NO. 16 AWG).

6.2.2 Rounding off decimal values. When it is possible to round off decimal values to a lesser number of decimal places, the following procedure should be used:

1. Where the figure following the last number to be retained is greater than five, the number retained is increased by one.

2. Where the figure following the last number to be retained is less than five, the number retained remains unchanged.

3. Where the figure following the last number to be retained is exactly five and the number to be retained is odd, the number retained is increased by one.

4. Where the figure following the last number to be retained is exactly five and the number to be

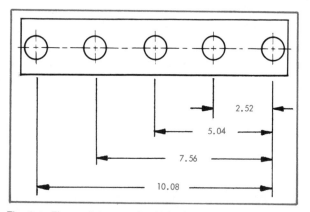

Fig. 6-1. These distances should be in even hundredth increments of the decimal scale.

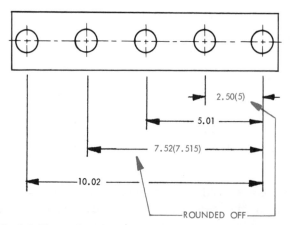

Fig. 6-2. These dimensions should be rounded off to two-place decimals.

retained is even, the number retained remains unchanged.

6.2.3 *Metric dimensioning system.* Metric dimensions on product drawings are expressed in millimeters while those on construction and architectural drawings are shown in meters and centimeters. Other units frequently used on drawings are: kilogram (for mass), newton (for force), kilopascals (for pressure), megapascals (for stress), newtonmeter (for torque), kilowatt (for power), and Celsius (for temperature).

Millimeter dimensioning is based on the use of one-place decimals, that is, one figure to the right of the decimal point. It should be noted that since 0.1 millimeter is about 0.004 inch, two (and occasionally three) figures after the decimal point need only be used where critical tolerances are necessary.

The following are some of the more specific metric dimensioning procedures based on standards of the International Organization for Standardization (ISO):

1. Wherever possible, dimensions should be indicated in even millimeters. A zero should be shown to the left of the decimal point when the millimeter value is less than one:

> 0.2 not .2
> 0.75 not .75

2. When a whole untoleranced number is indicated, neither the decimal point nor a zero is shown:

> 30 not 30.0

3. Where limit dimensioning is used and either maximum or minimum dimensions contain digits to the right of the decimal point, the other value should have zeros added for uniformity:

> 30.0   30
>   not
> 28.4   28.4

4. Where equal plus and minus tolerancing is used, the millimeter dimension and its tolerance need not have the same number of decimal places:

$$42.57 \pm .125$$

5. If a millimeter dimension contains a unilateral tolerance and either the plus or minus tolerance is nil, a single zero is shown without a plus or minus sign:

$$48 \begin{array}{c} 0 \\ -0.02 \end{array} \text{or } 48 \begin{array}{c} +0.02 \\ 0 \end{array}$$

6. Where bilateral tolerancing is used and a millimeter dimension contains an unequal plus and minus tolerance, both the maximum and minimum tolerances should have the same number of decimal places. Zeros are used where necessary:

$$32.00 \begin{array}{c} +0.25 \\ -0.10 \end{array} \text{not } 32.00 \begin{array}{c} +0.25 \\ -0.1 \end{array}$$

7. The decimal point for millimeter dimensions should be the same as the decimal point used for inch dimensions.

8. Commas should not be used to denote thousands in either decimal-inch or metric values:

> 15245 not 15,245

9. To convert a decimal-inch dimension to a millimeter dimension, multiply the inch dimension by 25.4 and round off to one less digit than the inch value according to the rounding off procedures described in 6.2.2:

> 2.456 in. x 25.4 = 62.3824 mm
> Rounded off to 62.38 mm

10. To convert a millimeter dimension to a decimal-inch dimension, divide the millimeter dimension by 25.4 and round off to one more digit than the millimeter value:

> 32.78 mm ÷ 25.4 = 1.2905 in.
> Rounded off to 1.290 in.

6.2.4 *Dual dimensioning system.* Until metrication is fully implemented, most industries that have already adopted the metric system are using dual dimensions on their drawings. In dual dimension-

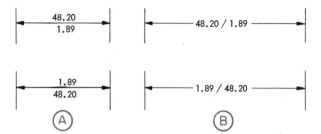

Fig. 6-4. Position method of showing dual dimensions.

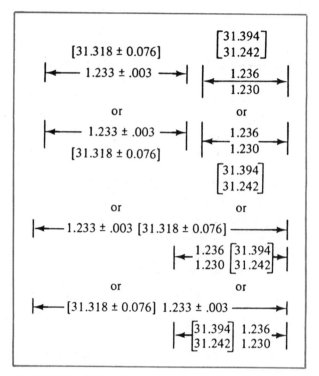

Fig. 6-5. Bracket method of designating dual dimensions. (ANSI Y14.5)

ing both the metric and the decimal-inch values are stated. Two methods are in common use — the position method and the bracket method.

Position method. In the position method the units are shown in one of two ways.

1. Millimeter dimensions are placed above the dimension line and the inch dimensions below, or the inch dimensions are placed above the dimension line and the millimeter below. See Fig. 6-4*A*.

2. The units are placed in a straight line and separated by a slash, with the millimeter dimensions to the left and the inch dimensions to the right, or with the inch dimensions to the left and the millimeter dimensions to the right. (See example in Fig. 6-4*B*.

In the bracket method either the millimeter or inch dimensions are enclosed in a square bracket and placed in the same way as those in the position method. See Fig. 6-5. Computer-prepared drawings may require the use of parentheses instead of brackets because of computer character limitations.

Only one method should be used on a single drawing. Each drawing should illustrate how the inch and millimeter dimensions can be identified. This is done by including one of the following near the title block of the drawing:

| MILLIMETER | or | MILLIMETER/INCH |
| INCH | | |

or

| [MILLIMETER] | or | [MILLIMETER] INCH |
| INCH | | |

A note may also be used as:
DIMENSIONS IN [ ] ARE MILLIMETERS
or
DIMENSIONS IN [ ] ARE INCHES

When a drawing is completely metric, the general note UNLESS OTHERWISE SPECIFIED DIMENSIONS ARE IN MILLIMETERS is included, and the word METRIC is prominently displayed near the title block. See example in Fig. 6-6.

On a metric drawing it is often a good practice to include a conversion chart of all dimensional values. See Fig. 6-6.

6.3 Limits and Tolerances

Limits are the maximum and minimum values prescribed for a specific dimension. The difference between limits is known as a tolerance. Hence, when a tolerance is specified, it represents the total amount a dimension may vary.

There are no specific rules or formulas for establishing tolerances. The amount of tolerance

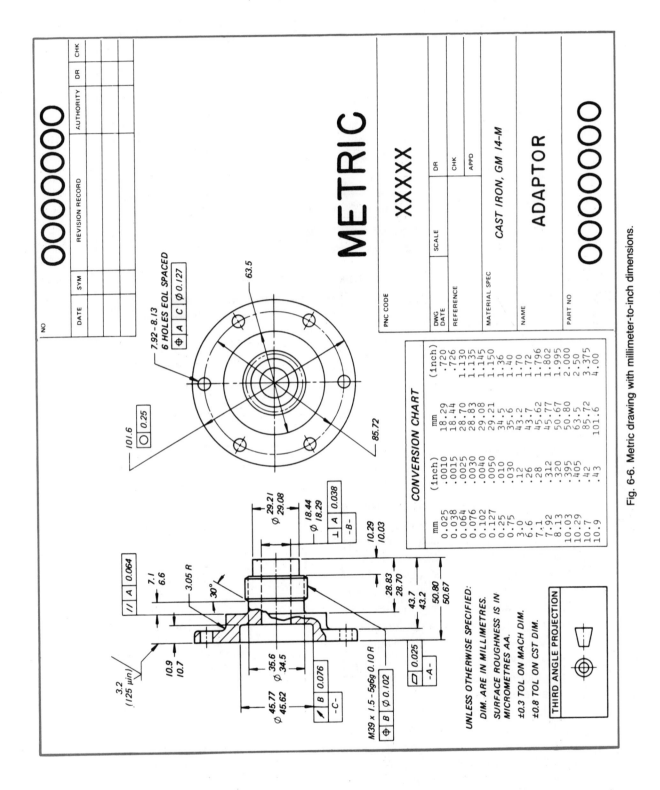

Fig. 6-6. Metric drawing with millimeter-to-inch dimensions.

used for a particular feature is often based on the judgment and experience of the designer or engineer. The primary factors are always how close to

basic dimensions the part must actually be to function adequately and to meet interchangeability criteria, and how accurate the chosen manufactur-

RANGE OF SIZES FROM	TO & INCL	TOLERANCES (TOTAL)								
.000	.599	.00015	.0002	.0003	.0005	.0008	.0012	.002	.003	.005
.600	.999	.00015	.00025	.0004	.0006	.001	.0015	.0025	.004	.007
1.000	1.499	.0002	.0003	.0005	.0008	.0012	.002	.003	.005	.008
1.500	2.799	.00025	.0004	.0006	.001	.0015	.0025	.004	.006	.010
2.800	4.499	.0003	.0005	.0008	.0012	.002	.003	.005	.008	.012
4.500	7.799	.0004	.0006	.001	.0015	.0025	.004	.006	.010	.015
7.800	13.599	.0005	.0008	.0012	.002	.003	.005	.008	.012	.019
13.600	20.999	.0006	.001	.0015	.0025	.004	.006	.010	.015	.025

TOLERANCE RANGE OF MACHINING PROCESSES

LAPPING & HONING

GRINDING, DIAMOND TURNING & BORING

BROACHING

REAMING

TURNING, BORING, SLOTTING, PLANING, & SHAPING

MILLING

DRILLING

Fig. 6-7. A guide for selecting tolerances.

ing methods must be to produce the part. A guide such as the one shown as Fig. 6-7 is frequently used in selecting tolerances.

As a general rule, the greater the permissible tolerance, the less the cost to produce the part. This is primarily due to reduced scrap, lower labor costs, and less expensive tools. Tolerances should never be specified closer than necessary, either by definite specifications or by implication in any general notes. In certain cases it may be advisable to use closer tolerances to facilitate assembly. This course is justified if it can be determined that savings through shorter assembly time will more than offset the increased costs of manufacturing the part to closer tolerances.

6.3.1 Methods of tolerancing. Dimensional tolerances can be expressed in two ways—limit dimensioning and plus and minus tolerancing.

Limit dimensioning. When dimensions are given directly, either the high limit is placed above the low limit, or the high limit is placed to the right of the low limit and is separated from it by a dash. See Fig. 6-8.

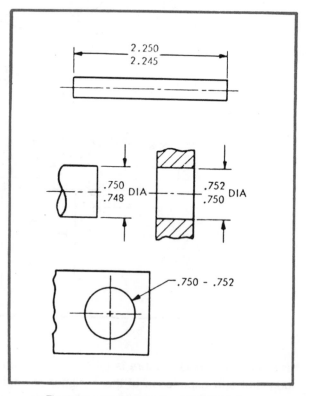

Fig. 6-8. Limit dimensioning. (ANSI Y14.5)

It is not always necessary to state both limits. The abbreviation MIN or MAX may be used after a numeral where other design elements determine the unspecified limit. This usually applies to depth of holes, length of threads, corner radii, chamfers, and so on. See Fig. 6-9.

Plus and minus tolerancing. In this system the tolerance in the form of plus and minus values follows the specified size and can be shown unilaterally or bilaterally. See Fig. 6-10.

6.3.2 Accumulation of tolerances. An accumulation of tolerances in a dimensioning system may have some serious effects on the assembly or function of features or parts. Such an accumulation is most likely to occur in *chain dimensioning,* as shown in Fig. 6-11A. Notice the amount of tolerance accumulation that can result between points X and Y. Thus, in any chain of dimensions with tolerances, the overall variation in position

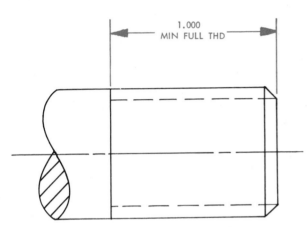

Fig. 6-9. Specifying a single limit.

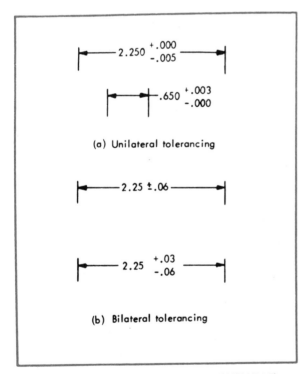

Fig. 6-10. Plus and minus tolerancing. (ANSI Y14.5)

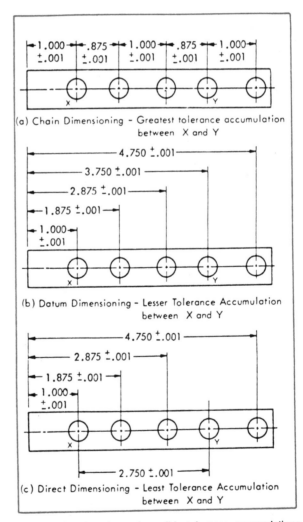

Fig. 6-11. A comparison of possible tolerance accumulations. (ANSI Y14.5)

that may develop is equal to the sum of the tolerances of the intermediate distances.

The *datum dimensioning* method prevents overall accumulations. For example, in Fig. 6-11*B* each hole is individually located from the datum on the left, thereby controlling the positions of the holes between *X* and *Y*.

In *direct dimensioning* the maximum variation between any two features is controlled by the tolerance on each dimension. This produces the least tolerance accumulation. See Fig. 6-11*C*.

6.4 Designating Dimensions

Dimensions on a drawing—regardless of the dimensioning system used—are shown by means of dimension lines, extension lines, leader lines, and notes.

6.4.1 *Use of dimension lines.* Dimension lines with their arrowheads indicate the direction and extent of dimensional values. A break is left in the line for the insertion of numerals that show the units of measurement. When both metric and decimal-inch values are used, the dimension line can be made solid and the units placed above and below the line as in Fig. 6-4*A*, or the dimensions can be inserted in the line with a slash between values as in Fig. 6-4*B*.

Dimension lines should be aligned horizontally and vertically whenever it is possible and grouped uniformly as shown in Fig. 6-12.

Parallel dimension lines should be not less than 0.24 inch apart and should not be closer than 0.40 inch to the outline of the object. See Fig. 6-13. If several parallel dimension lines are necessary, the

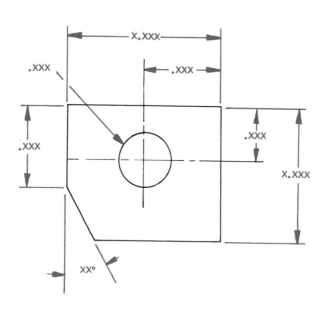

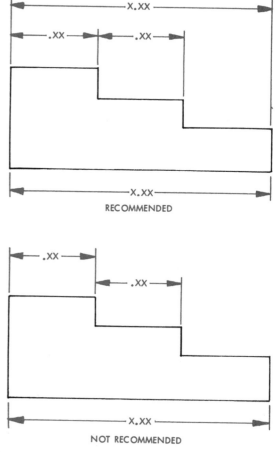

RECOMMENDED

NOT RECOMMENDED

Fig. 6-12. Placement of dimensions.

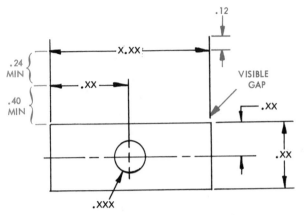

Fig. 6-13. The proper spacing of dimensions.

numerals should be staggered for greater readability. See Fig. 6-14.

Center lines, extension lines, and object lines should never be used as dimension lines. The only exception to this rule is a situation such as the one shown in Fig. 6-15. Every effort should be taken to avoid crossing dimension lines.

Dimensions should not originate at or terminate to hidden lines. If dimensions cannot be placed on views where the actual contour is visible, a sectional view may be required. See Fig. 6-16. Dimensions out of scale should be avoided as much as possible.

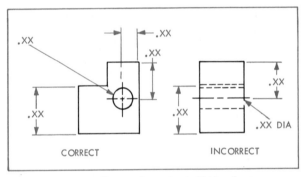

Fig. 6-16. Do not dimension to hidden lines.

Fig. 6-14. Correct placing of staggered dimensions.

Fig. 6-15. Dimension lines may be used as extension lines when situations of this nature are encountered.

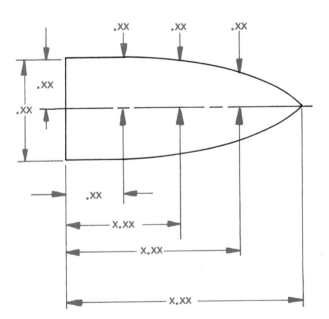

XX.XX

Fig. 6-17. A straight line under a dimension indicates the dimension is not to scale.

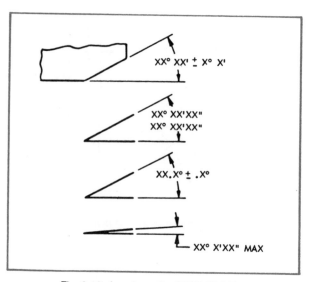

Fig. 6-19. Angular units. (ANSI Y14.5)

However, these may be necessary when a dimension cannot be made to scale because of various types of space limitations, or when changes that do not justify a complete redrawing must be made. Any dimension that is not to scale should be underlined with a straight line. See Fig. 6-17. Often the note NOT TO SCALE or DO NOT SCALE is included in the general notes on the drawing.

6.4.2 Arrowheads. The arrowheads that terminate dimension lines should be made with freehand strokes as shown in Fig. 6-18. The solid type arrowhead is generally preferred to the single-line type. The length of arrowheads should be between 0.12 and 0.19 inch and the width approximately 0.06 inch or one-third the length. See Fig. 6-18.

On any given drawing care should be taken to make all arrowheads the same size and shape.

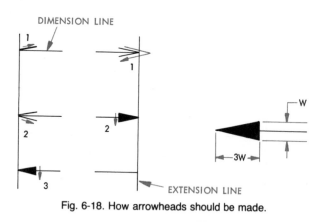

Fig. 6-18. How arrowheads should be made.

6.4.3 Units of measurement. Linear dimensional units are expressed in either decimal inches or millimeters. For decimal-inch values that are less than one, zeros are not placed before the

decimal point. However, zeros are used for metric values. See 1 of 6.2.3.

If a drawing is dimensioned completely in inches or completely in millimeters, the individual units are identified only by a note: UNLESS OTHERWISE SPECIFIED ALL DIMENSIONS ARE IN INCHES (or MILLIMETERS).

Wherever millimeters are used on a primarily inch-dimensioned drawing, the millimeter value should be followed by the abbreviation MM. Wherever inches are used on a millimeter-dimensioned drawing, the inch value should be followed by the abbreviation IN.

Angles should be designated by an arc drawn with the vertex of the angle as a center and the angular dimension inserted in a break in the arc. See Fig. 6-19. As a rule, right angles need not be dimensioned unless this is required for clarity. Angular dimensions should be expressed in ° (degrees), ' (minutes), and " (seconds). When only degrees are used, the numerical value may be followed by the symbol ° alone. If minutes are indicated alone, the value of the minutes should be preceded by 0°. It is also permissible to dimension in degrees and decimal parts of a degree.

6.4.4 Use of extension lines. The ends of a dimension are indicated by extension lines. There must always be a short visible gap between the outline of the part and the beginning of an exten-

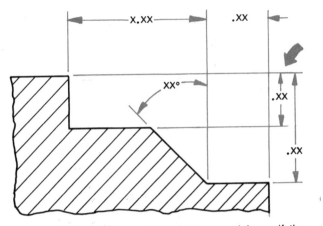

Fig. 6-20. Line crossing can be kept to a minimum if the shortest dimension lines are placed nearest the object outline.

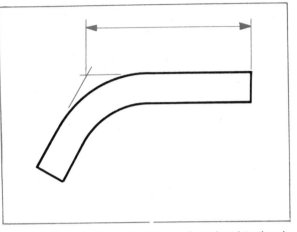

Fig. 6-22. Extension lines should pass through points they locate if the points are not on outlines.

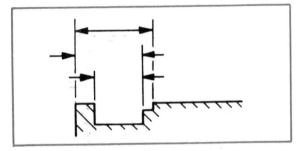

Fig. 6-21. Break extension lines if they cross a dimension line near arrowheads. (ANSI Y14.5)

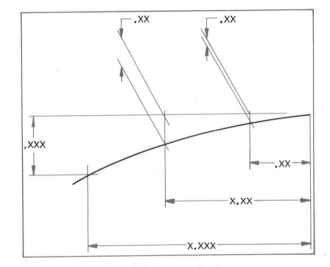

Fig. 6-23. Oblique extension lines.

sion line. The extension line must also extend beyond the arrowhead of the outermost dimension line drawn to it.

Extension lines should, as a rule, be drawn so that they will not cross one another or cross dimension lines. Line crossing can be kept to a minimum if the shortest dimension lines are drawn nearest the outline of the object. Other parallel extension lines then follow in order of their length, with the longest line at the outside. See Fig. 6-20.

When it is impossible to avoid crossing other extension lines, dimension lines, or object lines, the extension lines should not be broken. The exception is where an extension line crosses a dimension line close to arrowheads, in which case a break in the extension line is permitted, Fig. 6-21.

If a point is to be located exclusively by extension lines, as shown in Fig. 6-22, the extension lines should pass through the point.

When it is necessary to use oblique extension lines because of space limitations or object contour, they should be drawn in such a way that dimension lines can run in the directions in which they apply. See Fig. 6-23.

6.4.5 Use of leader lines. A leader is a fine oblique line used to direct attention to a note or to give the size of circles, arcs, and so on. It should be drawn at a convenient angle, usually at least 20° from a vertical or horizontal line or from any line or surface to which it points. If it is directed specifically to a circle or to an arc, it should be

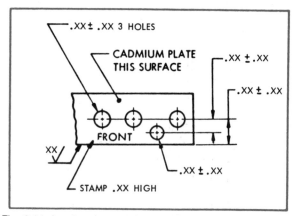

Fig. 6-24. Leaders are used to direct attention to notes or to indicate sizes of arcs, circles, and so on. (ANSI Y14.5)

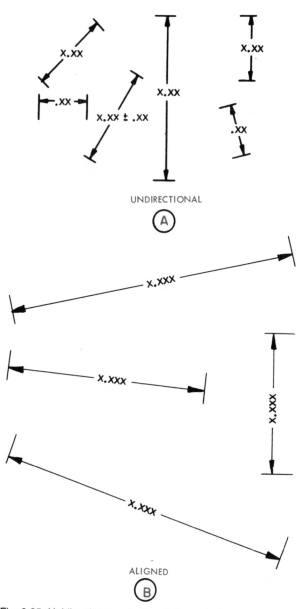

Fig. 6-25. Unidirectional and aligned dimensioning systems. (ANSI Y14.5)

drawn in a radial direction. The leader should terminate with an arrowhead or a dot drawn to the indicated outline or feature. The opposite end should have a short horizontal bar a minimum of 0.25 inch long that should extend to the mid-height of the first or last letter of the note. A minimum space of 0.06 inch should be left between the end of the horizontal bar and the note or dimension.

Leaders should be short and not cross other leaders. Where several leaders are in one group, adjacent leaders are drawn parallel. Care should be taken, however, not to draw leaders so that they are parallel to adjacent dimension or extension lines. See Fig. 6-24.

6.5 Placement of Dimensions

Since dimensions must convey the true geometrical characteristics of an object, their placement on a drawing is very significant. The basic practices described in the following subsections apply to both decimal-inch and millimeter systems.

6.5.1 Reading direction of dimensions. Two systems are used in placing dimensions on a drawing—the unidirectional and the aligned. The unidirectional system is preferred because of the greater ease in reading the figures. It is the principal system used in industry.

Unidirectional system. All dimensions are made to read from the bottom of the sheet. See example in Fig. 6-25*A*.

Aligned system. Each dimension is placed so that it reads either from the bottom or from the right-hand edge of the drawing. With this system, dimension lines that fall in a 45° zone should be avoided. See Fig. 6-25*B*.

6.5.2 Overall dimensions. When overall dimensions are used, they should be placed outside the intermediate dimensions. See Fig. 6-26. When an overall dimension is given, one intermediate dis-

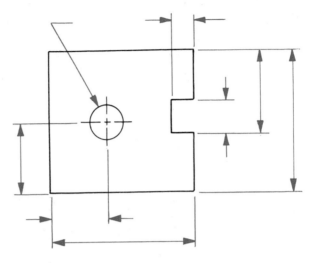

Fig. 6-26. Overall dimension lines should be placed outside the intermediate dimension lines.

tance can be omitted or identified as a reference dimension. Where the intermediate dimensions are more important than the overall dimension, the overall dimension should be shown as a reference dimension. See Fig. 6-30.

6.5.3 Location of dimensions on views. Dimensions should preferably be placed outside the outline of the view. See Fig. 6-27. Only in in-

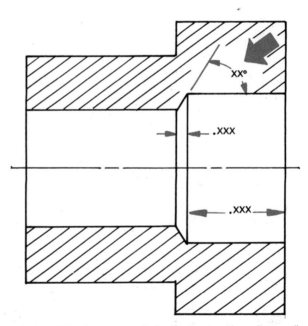

Fig. 6-28. The large arrow indicates the practice when a dimension is placed in a sectional view.

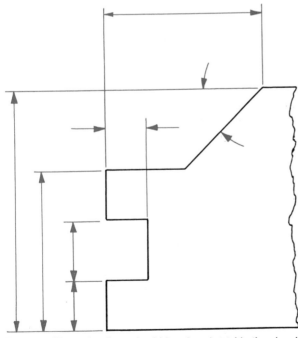

Fig. 6-27. Dimension lines should be placed outside the view if at all possible.

stances when the readability of a drawing is improved, or where extension and leader lines would be excessively long, should dimensions be placed within a view. If a dimension must be placed within a sectional view, a small area around the dimension line should be kept free from the sectional lines. See Fig. 6-28.

6.5.4 Dimensions from datums. Datums are points, lines, planes, or cylinders that are as-

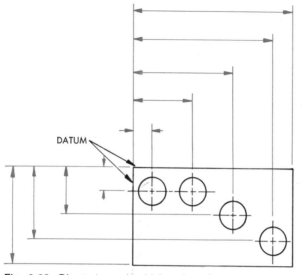

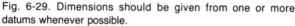

Fig. 6-29. Dimensions should be given from one or more datums whenever possible.

sumed to be exact for purposes of computation or reference and from which the location of features of a part may be established. Where positions are specified by dimensions from a datum, different features of the part are located with respect to this datum and not with respect to one another. See Fig. 6-29. In establishing datums in relation to defining or measuring a part, at least two, and often three, datums must be considered in locating a feature.

Features selected to serve as datums must be clearly identified, easily recognizable, and accessible during manufacture so that measurements can be referred to them readily. Dimensions should not be from rough cast surfaces or other inaccurate reference points.

6.5.5 Reference dimensions. A reference dimension is one without a specified tolerance and is used for informational purposes only. It does not govern manufacturing or inspection operations in any way. Thus, there may be instances when a duplicated dimension on a drawing is not strictly essential in fabricating or inspecting the part but has an important reference value.

The preferred method for indicating reference dimensions is to enclose the dimensions within parentheses. An alternate method is to use the abbreviation REF directly following or under the dimensions. See Fig. 6-30.

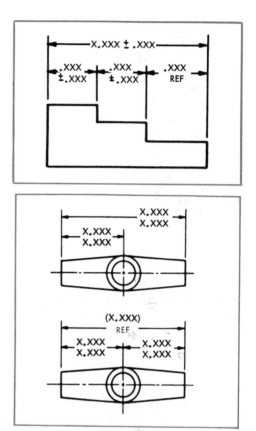

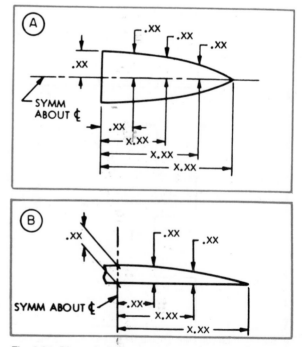

Fig. 6-30. Reference dimensions often have important informational value.

Fig. 6-31. Dimensioning symmetrical outlines. (ANSI Y14.5)

6.5.6 Indicating symmetry. Symmetrical parts are dimensioned as shown in Fig. 6-31A. Where part size or space limitation permits drawing only one-half of the outline, the symmetry is indicated by an appropriate note. See Fig. 6-31B. In situations of this kind, the outline of the part is extended slightly beyond the center line and ends with a break line.

6.6 Dimensioning Special Features

Many objects have features which require more specific dimensional information. Included are such items as cylindrical and spherical surfaces, holes, tapers, keyseats, and knurls.

6.6.1 Diameters. Where the diameters of a number of concentric elements are given, these diameters are dimensioned in the longitudinal view. See Fig. 6-32A.

Where a diameter is dimensioned on a single view that does not illustrate the circularity of the feature, the value of the diameter is followed by the abbreviation DIA or by the diameter symbol, $\varnothing$. See Fig. 6-32B. Where a diameter is dimensioned on a circular view or on a side view that has a reciprocal circular view, neither the abbreviation nor the symbol is necessary. See Fig. 6-32C.

6.6.2 Radii of arcs. A circular arc is dimensioned by giving its radius. Where space permits, a radius dimension line is drawn from the radius center, with the arrowhead ending at the arc and with the numeral between the arrowhead and the center. Each numeral should be followed by the abbreviation R. Where space is limited, as for a small radius, a leader may be used. Where it is inconvenient to place the arrowhead between the center of the radius and the arc, it may be placed outside the arc. See Fig. 6-33.

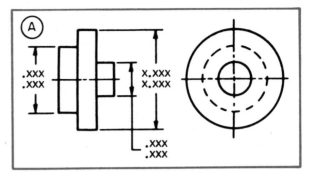

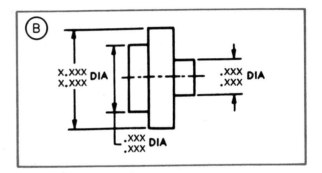

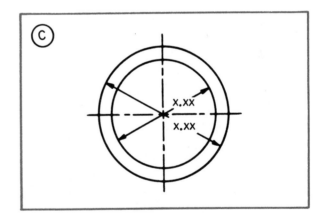

Fig. 6-32. Methods of dimensioning diameters. (ANSI Y14.5)

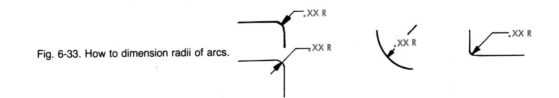

Fig. 6-33. How to dimension radii of arcs.

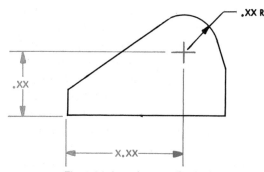

Fig. 6-34. Locating a radius by its center.

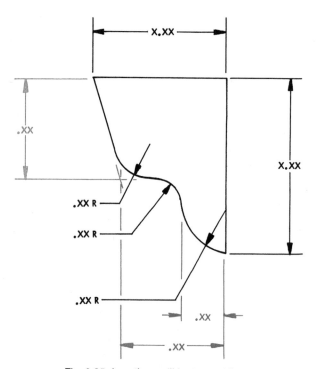

Fig. 6-35. Locating radii by tangent lines.

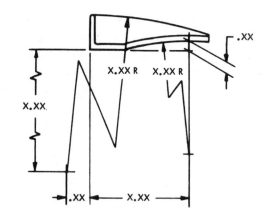

Fig. 6-36. Foreshortened radii save space. (ANSI Y14.5)

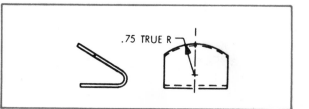

Fig. 6-37. Indicating a true radius.

Where a dimension is given to the center of a radius, a small cross should be drawn at the center. Extension lines are used to locate the radius. See Fig. 6-34. An alternative method for locating radii is by tangent lines. See Fig. 6-35.

Where the center of a radius is outside of the drawing or interferes with another view, the radius dimension line may be foreshortened. See Fig. 6-36. The portion of the line next to the arrowhead should be radial relative to the curved line. Where the radius is foreshortened and the center is lo-

cated by coordinates, the dimension locating the center should be shown.

Where the radius is dimensioned in a view that does not show the true shape of the radius, TRUE R is added after the radius dimension, Fig. 6-37.

Where a part has a number of radii of the same dimension, it is preferable to use a note in lieu of dimensioning each radius separately.

Whenever a portion of a sphere is used in a feature, it is dimensioned by a radius followed by SPHER R. See Fig. 6-38.

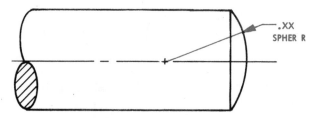

Fig. 6-38. Showing a spherical radius.

The dimensioning of chords, arcs, and angles should be as shown in Fig. 6-39. Where required for clarity, the dimension should be modified with a term such as ARC or CHORD.

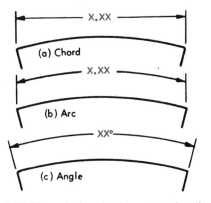

Fig. 6-39. Dimensioning chords, arcs, and angles.

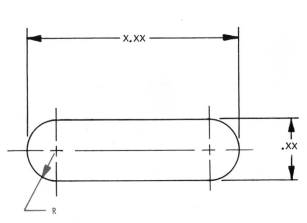

Fig. 6-40. Dimensioning rounded ends.

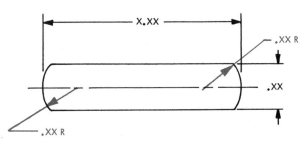

Fig. 6-41. Partially rounded ends. (ANSI Y14.5)

6.6.3 Rounded ends.

Overall dimensions should be used with parts having rounded ends. For fully rounded ends the radius is indicated but not dimensioned. See Fig. 6-40. For parts with partially rounded ends, the radius is dimensioned. See Fig. 6-41. When a hole location is more critical than the location of a radius from the same center, the hole and the radius should be dimensioned and toleranced separately. See Fig. 6-42.

6.6.4 Dimensioning curves.

A curved line composed of two or more circular arcs should be dimensioned by giving the radii and locating their centers or on the basis of their points of tangency. See Fig. 6-43.

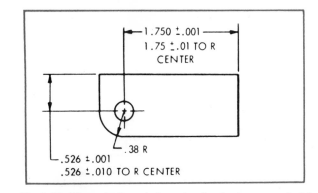

Fig. 6-42. Dimensioning a hole and radius having separate tolerances. (ANSI Y14.5)

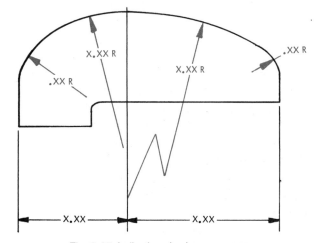

Fig. 6-43. Indicating circular arc curves.

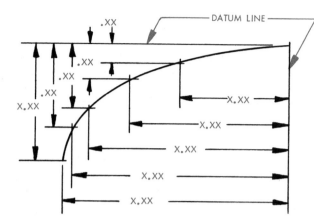

Fig. 6-44. Coordinate dimensioning of an irregular outline.

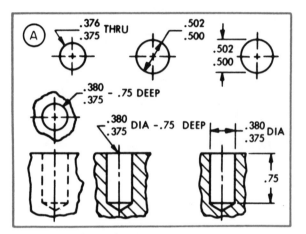

ROUNDED HOLES

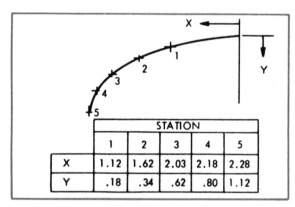

Fig. 6-45. Tabulated outlines. (ANSI Y14.5)

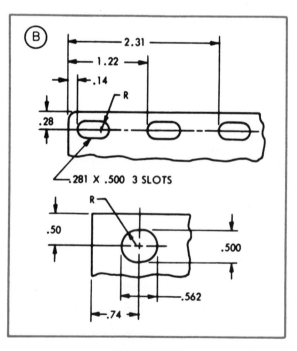

SLOTTED HOLES

Fig. 6-46. Dimensioning holes. (ANSI Y14.5)

Circular or noncircular outlines are dimensioned by the rectangular coordinate, or offset, method. See Fig. 6-44. Where many coordinates are required to describe a contour, the vertical and horizontal coordinate dimensions are tabulated. See Fig. 6-45.

6.6.5 Dimensioning and locating holes. Round holes are dimensioned as shown in Fig. 6-46A. If the hole is to be a through hole, the abbreviation THRU follows the dimension. Blind holes must have the depth specified.

Slotted holes are dimensioned by stating their lengths and widths. Location is indicated by dimensioning to the longitudinal centerplane and to one end. The end radii are shown but not dimensioned. See Fig. 6-46B.

The methods of locating round holes are illlustrated in Figs. 6-47 to 6-51. The same techniques can also be used to locate round pins and other features of symmetrical contour.

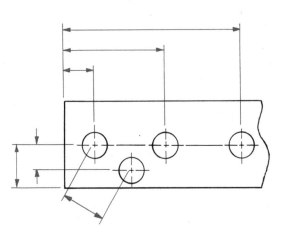

Fig. 6-47. Methods of locating holes by coordinate distances. (ANSI Y14.5)

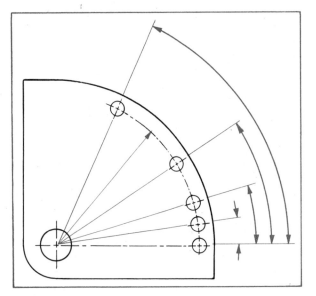

Fig. 6-49. Holes on a circle can be located by polar coordinate dimensions. (ANSI Y14.5)

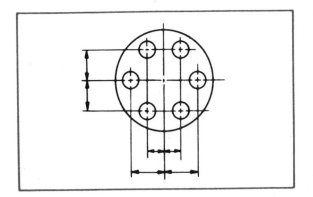

Fig. 6-48. Locating holes by rectangular coordinates. (ANSI Y14.5)

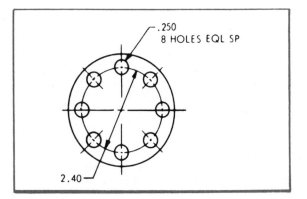

Fig. 6-50. Locating holes on circles by radius or diameter and equal spacing. (ANSI Y14.5)

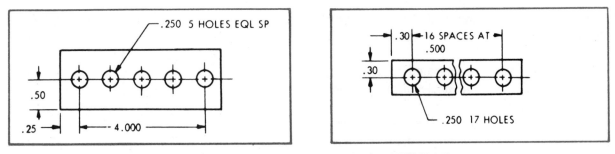

Fig. 6-51. Locating holes equally spaced in a line. (ANSI Y14.5)

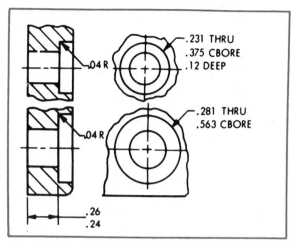

Fig. 6-52. Dimensioning counterbored holes. (ANSI Y14.5)

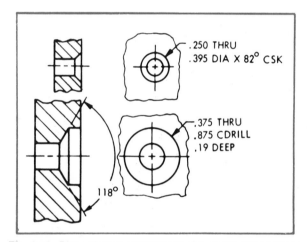

Fig. 6-53. Dimensioning countersunk and counterdrilled holes. (ANSI Y14.5)

Spot-faced holes may be dimensioned by showing the diameter of the faced area and either the depth of the hole or the thickness of the remaining material. See Fig. 6-54.

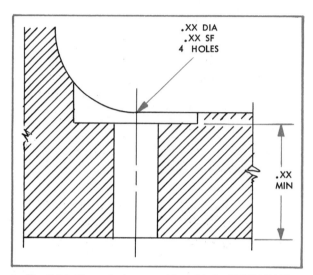

Fig. 6-54. Dimensioning a spot-faced hole. (ANSI Y14.5)

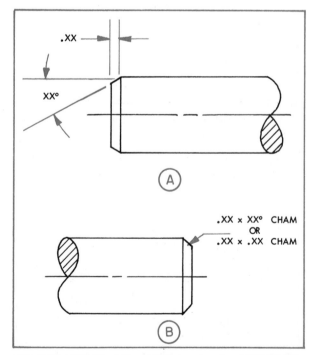

Fig. 6-55. Methods of dimensioning a chamfer. (ANSI Y14.5)

Counterbored holes are specified by a note giving the diameter, depth, and corner radius. See Fig. 6-52. However, if the thickness of the material to remain is more important than the depth of the hole, then that thickness is dimensioned instead of the depth.

Countersunk holes are indicated by specifying the diameter and included angle of the countersink. See Fig. 6-53.

Counterdrilled holes are dimensioned by giving the diameter, depth, and included angle of the counterdrill. See Fig. 6-54.

6.6.6 Dimensioning chamfers, keyseats, and knurls.
Chamfers, keyseats, and knurls are identified on a drawing by dimensioning them in the following ways.

Chamfers. Chamfers should be dimensioned by giving the angle and length, as is shown in Fig. 6-55*A*. When the angle is 45°, the chamfer may be dimensioned as illustrated in Fig. 6-55*B*.

A round hole having a chamfered edge that requires a diameter control is dimensioned as shown in Fig. 6-56.

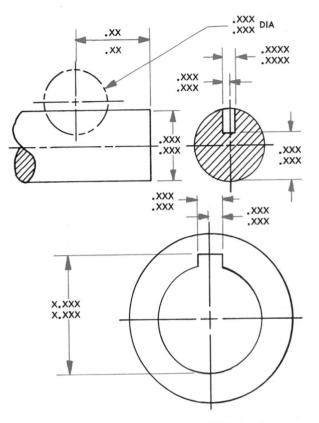

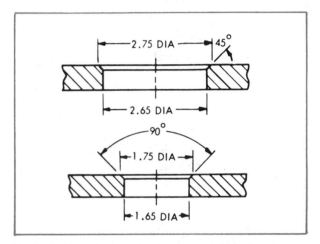

Fig. 6-56. Dimensioning hole chamfers requiring diameter control. (ANSI Y14.5)

Fig. 6-57. Dimensioning keyseats. (ANSI Y14.5)

Keyseats. Keyseats are dimensioned by width, depth, location, and, if necessary, length. The depth is dimensioned from the opposite side of the shaft or hole. See Fig. 6-57.

Knurls. When knurls are used to roughen a surface for the purpose of giving a better grip, the type and pitch of the knurl are indicated, as well as the diameters before and after knurling. When no control is required, the diameter after knurling is omitted. If only a portion of the piece requires knurling, axial dimensioning is required. See example in Fig. 6-58*A*.

When knurls are used to make a press fit between two parts, the original finished surface should be dimensioned with limits. The minimum acceptable diameter of the knurl should be given

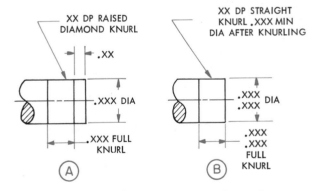

Fig. 6-58. Dimensioning knurls. (ANSI Y14.5)

in a note together with the pitch and type of knurl, such as straight or diamond, depressed or raised. See Fig. 6-58*B*.

6.6.7 Dimensioning tapers.
Conical tapers are defined by a combination of dimensions and toler-

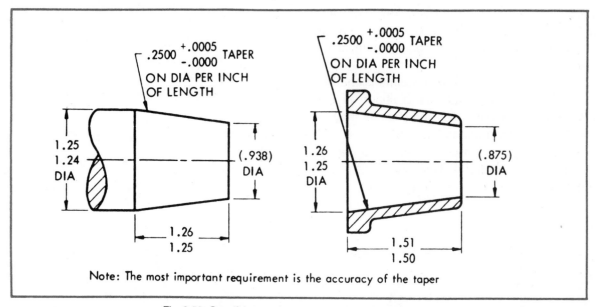

Note: The most important requirement is the accuracy of the taper

Fig. 6-59. Specifying a tolerance on the taper. (ANSI Y14.5)

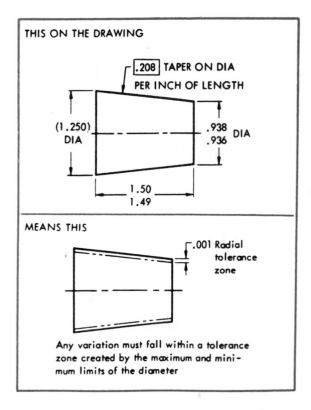

Fig. 6-60. Specifying a basic taper. (ANSI Y14.5)

ances. See Figs. 6-59, 6-60, and 6-61. Included are the following:

1. Diameter at one end.
2. Length of the taper.

3. Diameter at a selected cross-sectional plane within or outside the tapered feature shown with a basic dimension.

4. Dimension with a basic diameter specified.

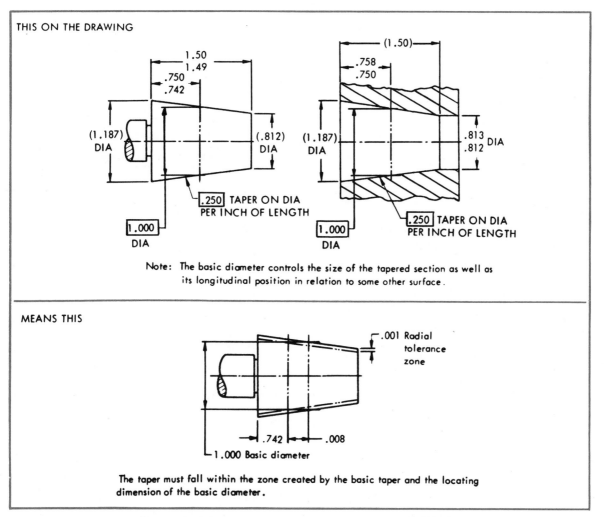

Note: The basic diameter controls the size of the tapered section as well as its longitudinal position in relation to some other surface.

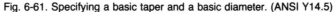

The taper must fall within the zone created by the basic taper and the locating dimension of the basic diameter.

Fig. 6-61. Specifying a basic taper and a basic diameter. (ANSI Y14.5)

Fig. 6-62. Dimensioning a flat taper. (ANSI Y14.5)

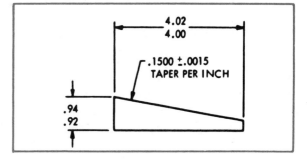

5. Rate of taper.
6. Included angle.

Flat tapers can be dimensioned very much like conical tapers. An example is shown in Fig. 6-62.

6.7 General Notes

Notes are used on drawings to supply information that can be presented more easily in descrip-

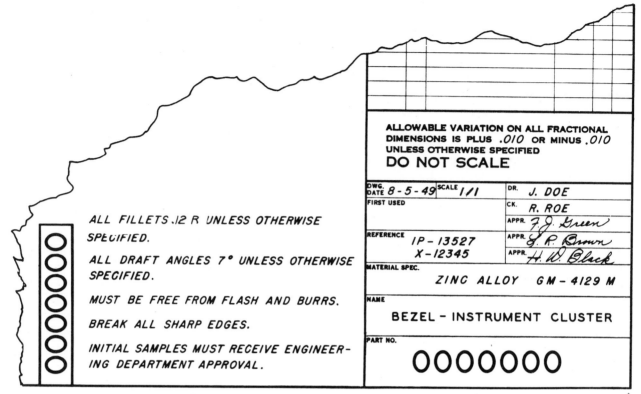

ALL FILLETS .12 R UNLESS OTHERWISE SPECIFIED.

ALL DRAFT ANGLES 7° UNLESS OTHERWISE SPECIFIED.

MUST BE FREE FROM FLASH AND BURRS.

BREAK ALL SHARP EDGES.

INITIAL SAMPLES MUST RECEIVE ENGINEER-ING DEPARTMENT APPROVAL.

ALLOWABLE VARIATION ON ALL FRACTIONAL DIMENSIONS IS PLUS .010 OR MINUS .010 UNLESS OTHERWISE SPECIFIED
DO NOT SCALE

DWG. DATE 8-5-49   SCALE 1/1   DR. J. DOE

FIRST USED   CK. R. ROE

APPR. F. J. Green

REFERENCE   IP-13527   APPR. S. R. Brown
X-12345   APPR. H. W. Black

MATERIAL SPEC.   ZINC ALLOY   GM-4129 M

NAME   BEZEL - INSTRUMENT CLUSTER

PART NO.   **0000000**

Fig. 6-63. Example of a drawing containing general notes.

tive form. They are used for such a variety of purposes that it is not practical to establish a standard note for every condition. The following rules generally apply for all notes.[2]

1. Information conveyed by notes should be clear, accurate, complete, and capable of only one interpretation.

2. Only common shop trade terms and words should be used.

3. Notes should be as simple, brief, and concise as possible. See Fig. 6-63.

6.7.1 Use of abbreviations. Abbreviations are shortened forms of words or expressions and are used to conserve time and space. Abbreviations should be employed sparingly on drawings since the chief purpose of such drawings is to convey manufacturing specifications in unmistakable terms to those who use them. Therefore, abbreviations should not be used where there is any possibility of misunderstanding. The following are a few specific rules which should govern abbreviations:

1. Use abbreviations approved by DOD MIL-STD-100 and ANSI Y1.1. See Appendix.

2. Use uppercase letters for all abbreviations.

3. To abbreviate a combination of words, use the approved abbreviation for each separate word except when common practice has established another form. Thus, the abbreviation for the combination *cubic feet per minute* is CFM, but where each word occurs separately, the abbreviations are CU for *cubic,* FT for *feet,* and MIN for *minute.*

4. Do not use periods with abbreviations except after abbreviations that might otherwise cause confusion. Thus, IN for *inch* should have a period.

6.8 Designating Machined Surfaces

The proper functioning of a machine part depends to a large extent on the quality of its surface. Accordingly, for such parts the surfaces that

---

2. *G.E. Drafting Manual* (Schenectady, N.Y.: General Electric Co.).

are to be machined must be specified, and very often the quality of the surface finish must also be stipulated.

Recent developments and processing and quality requirements have brought greater attention to surface finish. To insure correct interpretation and control of finished surfaces, certain symbols have been standardized and are recognized by all industries. These symbols are placed wherever a surface appears edgewise as a visible line. Hence, the symbols may be repeated in several views. If necessary, they may even be placed on hidden lines.

6.8.1 Noncontrol surface finish. When surface roughness control is not necessary, one of the following three methods may be used to indicate surface quality:

1. A 60° **V** may be drawn with its point touching the line representing the edgewise view of the surface to be machined. Occasionally the letter *R* or *G* is placed in the **V,** the *R* meaning "rough" and the *G* "grind." See Fig. 6-64.

2. A symbol resembling the letter *f* may be used in place of the **V.** See Fig. 6-64. This symbol specifies a smooth machine finish and does not show any control over the quality of the surface. Sometimes a circle with a number enclosed is added to the tail of the *"f"* to better control the type of machine surface. See Fig. 6-64. A note may be used to explain the meaning of the number.

3. A note FINISH ALL OVER, abbreviated FAO, is sometimes used if all surfaces are to be machined.

6.8.2 Controlled surface finish. When the height, width, and direction of surface irregularities must be controlled to exact specifications, the

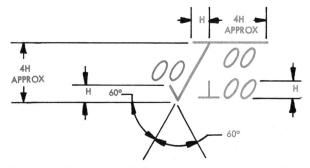

Fig. 6-65. The basic symbol for indicating a control surface finish.

practice is to show the degree of control by a series of roughness symbols adopted by both SAE and ANSI.

The basic symbol used to designate surface irregularities resembles a check mark with a horizontal extension bar. The recommended proportions for construction of the surface symbol are shown in Fig. 6-65. When only roughness height, width, lay, or a combination thereof is indicated, it is permissible to omit the horizontal extension line.

The point of this symbol may be placed on the line representing the surface involved, on the extension line, or on a leader line pointing to the surface. Fig. 6-66 illustrates typical applications of the symbol on a drawing.

The type, direction, and magnitude of the surface irregularities are indicated by additional markings placed about the basic check mark symbol.

Fig. 6-64. These symbols are often used to indicate noncontrol surfaces to be machined.

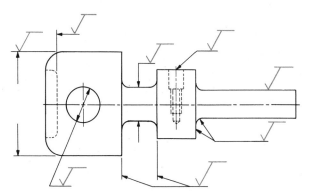

Fig. 6-66. The symbol should be placed in an erect position.

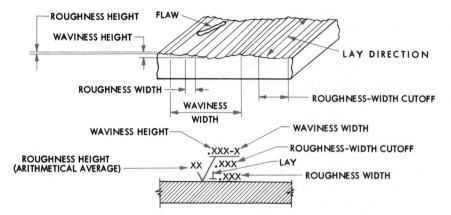

Fig. 6-67. Identification and relation of symbols to surface characteristics.

Roughness value (microinches)	Type of surface	Purpose
1000	Extremely rough	Used for clearance surfaces only where good appearance is not required.
500	Rough	Used where vibration, fatigue, or stress concentration are not critical and close tolerances are not required.
250	Medium	Most popular for general use where stress requirements and appearance are essential.
125	Average smooth	Suitable for mating surfaces of parts held together by bolts and rivets with no motion between them.
63	Better than average finish	For close fits or stressed parts except rotating shafts, axles and parts subject to extreme vibrations.
32	Fine finish	Used where stress concentration is high and for such applications as bearings.
16	Very fine finish	Used where smoothness is of primary importance such as high-speed shaft bearings, heavily-loaded bearings and extreme tension members.
8	Extremely fine finish produced by cylindrical grinding, honing, lapping or buffing	Use for such parts as surfaces of cylinders.
4	Superfine finish produced by honing, lapping, buffing or polishing	Used on areas where packings and rings must slide across the surface where lubrication is not dependable.

Fig. 6-68. Characteristics of roughness height values.

The types of irregularities and how the value of each is indicated are explained below. Fig. 6-67 is a useful reference.

Roughness. Roughness refers to the finely spaced irregularities produced by machining, abrading, extruding, molding, casting, forging, rolling, coating, plating, blasting, or burnishing. The height of irregularities is rated in microinches or micrometers. (A microinch is one millionth of an inch, and a micrometer is one millionth of meter.) The characteristics of roughness-height ratings are shown in Fig. 6-68. These ratings are usually indicated as arithmetical averages in either microinches or micrometers and placed on the left of the long leg of the surface texture symbol. See Fig. 6-67.

In addition to the height, the maximum permissible spacing between repetitive units of the surface pattern is rated. This roughness-width cutoff is given in inches or millimeters and is shown below the extension line of the symbol and to the right of the lay symbol. See Fig. 6-67. The

mm	in.	mm	in.	mm	in.
0.08	(.003)	**0.80**	(.030)	8.0	(.300)
0.25	(.013)	2.50	(.100)	25.0	(1.000)
*Boldface values preferred					

Fig. 6-69. Standard roughness-width cutoff values.

standard roughness-width cutoff values are given in Fig. 6-69.

Waviness. Waviness refers to those surface irregularities spaced too far apart to constitute roughness. See Fig. 6-67. Irregularities of this category result from such factors as machine or work deflections, vibration, heat treatment, or warping strains. Both the height and width of waviness irregularities are rated in inches or in millimeters. The recommended waviness height values are given in Fig. 6-70. Maximum waviness height is placed above the horizontal line of the surface texture symbol. The waviness width is shown to the

TYPICAL SURFACE ROUGHNESS AVERAGE RATING VALUES, MICROMETRES, um AA	
0.025 / Micrometer anvils, mirrors, gages.	0.4 / Compressor blade airfoils, spline shafts, motor-shaft bearings, hydraulic shuttle valves, flanks of gear teeth.
0.05 / Shop-gage faces, comparator anvils.	0.8 / Brake drums, broached holes, bronze bearings, precision parts, ground ball and roller bearings, ablative heat shields, gasket seals for hydraulic fittings.
0.1 / Vernier-caliper faces, wrist pins, hydraulic piston rods, precision tools, honed roller and ball bearings, carbon seal mating surfaces.	1.6 / Gear locating faces, gear shafts and bores, cylinder-head faces, cast iron gear-box faces, piston crowns, Teflon molded parts, waveguide components, turbine blade dovetail pressure faces, rotating labyrinth seals.
0.2 / Crankshaft journals, camshaft journals, connecting-rod journals, valve stems, cam faces, hydraulic cylinder bores, lapped roller and ball bearings.	3.2 / Mating surfaces, no motion.
0.32 / Piston outside diameters, cylinder bores.	6.3 / Clearance surfaces, rough machine parts.

Fig. 6-68 (Continued). Characteristics of roughness height values.

right of the height value and separated from it by a dash. See Fig. 6-67.

Lay. Lay indicates the direction of the predominant pattern of surface irregularities produced by tool marks. These irregularities are specified by a

mm	in.	mm	in.	mm	in.
0.0005	**(.00002)**	0.008	(.0003)	**0.12**	**(.005)**
0.0008	(.00003)	**0.012**	**(.0005)**	0.20	(.008)
0.0012	**(.00005)**	0.020	(.0008)	**0.25**	**(.010)**
0.0020	(.00008)	**0.025**	**(.0010)**	0.38	(.015)
0.0025	**(.00010)**	**0.05**	**(.002)**	**0.50**	**(.020)**
0.005	**(.0002)**	0.08	(.003)	**0.80**	**(.030)**
*Boldface values preferred					

Fig. 6-70. Recommended waviness height values.

lay symbol (Fig. 6-71) placed to the right of the **V** in the surface texture symbol. See Fig. 6-67.

6.8.3 Degree of surface roughness control.

Smoothness and roughness are relative, that is, surfaces are smooth or rough only for the purpose intended. What is smooth for one purpose may be rough for another purpose.

In the mechanical field comparatively few surfaces require any control of roughness beyond that afforded by the processes required to obtain the necessary dimensional characteristics.

Working surfaces such as bearings, pistons, and gears are typical of surfaces for which optimum performance may require control of the surface characteristics. Nonworking surfaces such as the walls of transmission cases, crankcases, or differential housings seldom require any surface con-

LAY SYMBOL	DESIGNATION	EXAMPLE
‖	LAY PARALLEL TO THE LINE REPRESENTING THE SURFACE TO WHICH THE SYMBOL IS APPLIED.	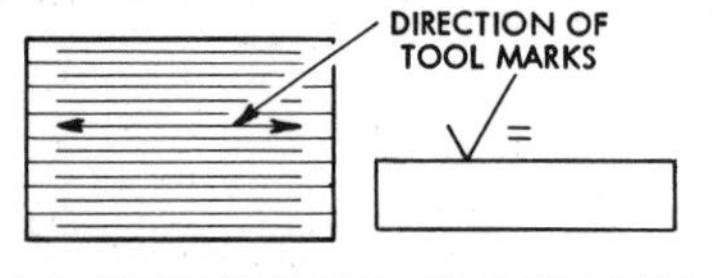 DIRECTION OF TOOL MARKS
⊥	LAY PERPENDICULAR TO THE LINE REPRESENTING THE SURFACE TO WHICH THE SYMBOL IS APPLIED.	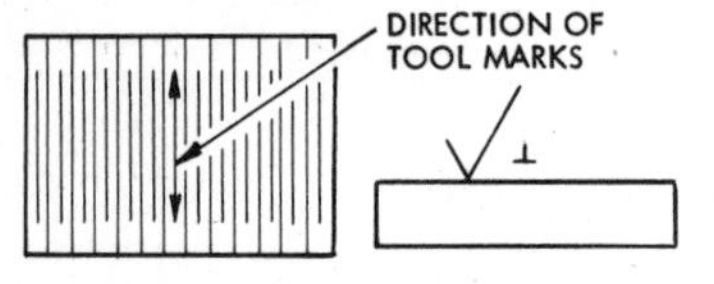 DIRECTION OF TOOL MARKS
X	LAY ANGULAR IN BOTH DIRECTIONS TO LINE REPRESENTING THE SURFACE TO WHICH SYMBOL IS APPLIED.	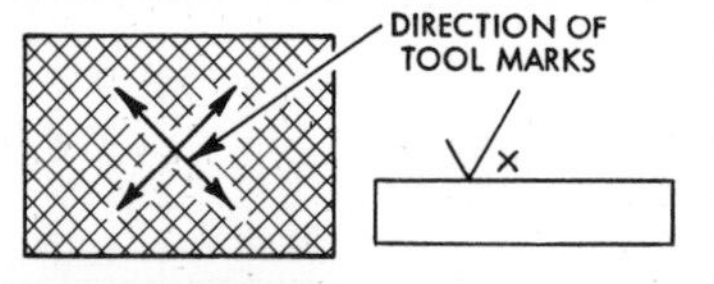 DIRECTION OF TOOL MARKS
M	LAY MULTIDIRECTIONAL.	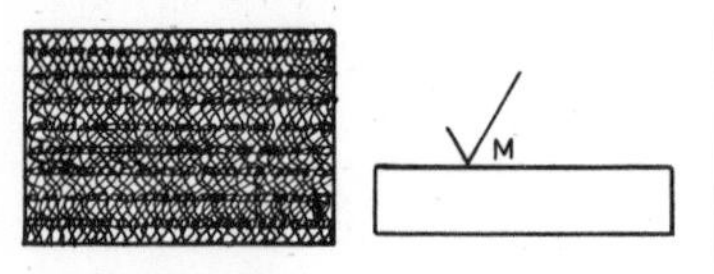
C	LAY APPROXIMATELY CIRCULAR RELATIVE TO THE CENTER OF THE SURFACE TO WHICH THE SYMBOL IS APPLIED.	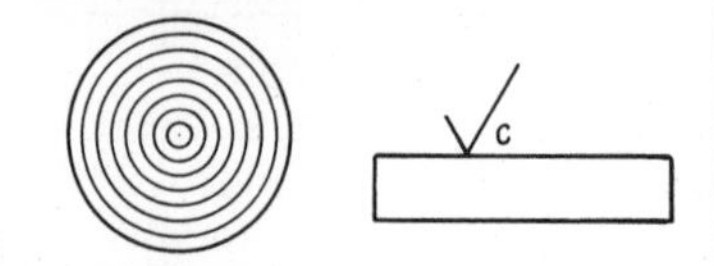
R	LAY APPROXIMATELY RADIAL RELATIVE TO THE CENTER OF THE SURFACE TO WHICH THE SYMBOL IS APPLIED.	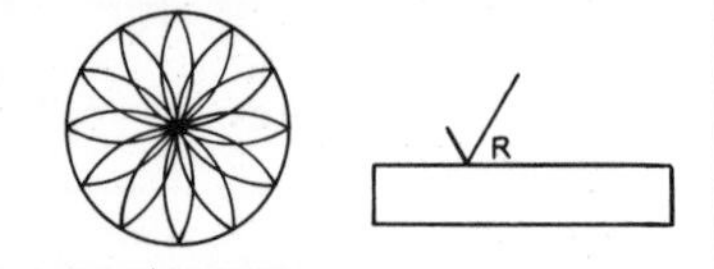

Fig. 6-71. Symbols and their interpretation indicating direction of lay.

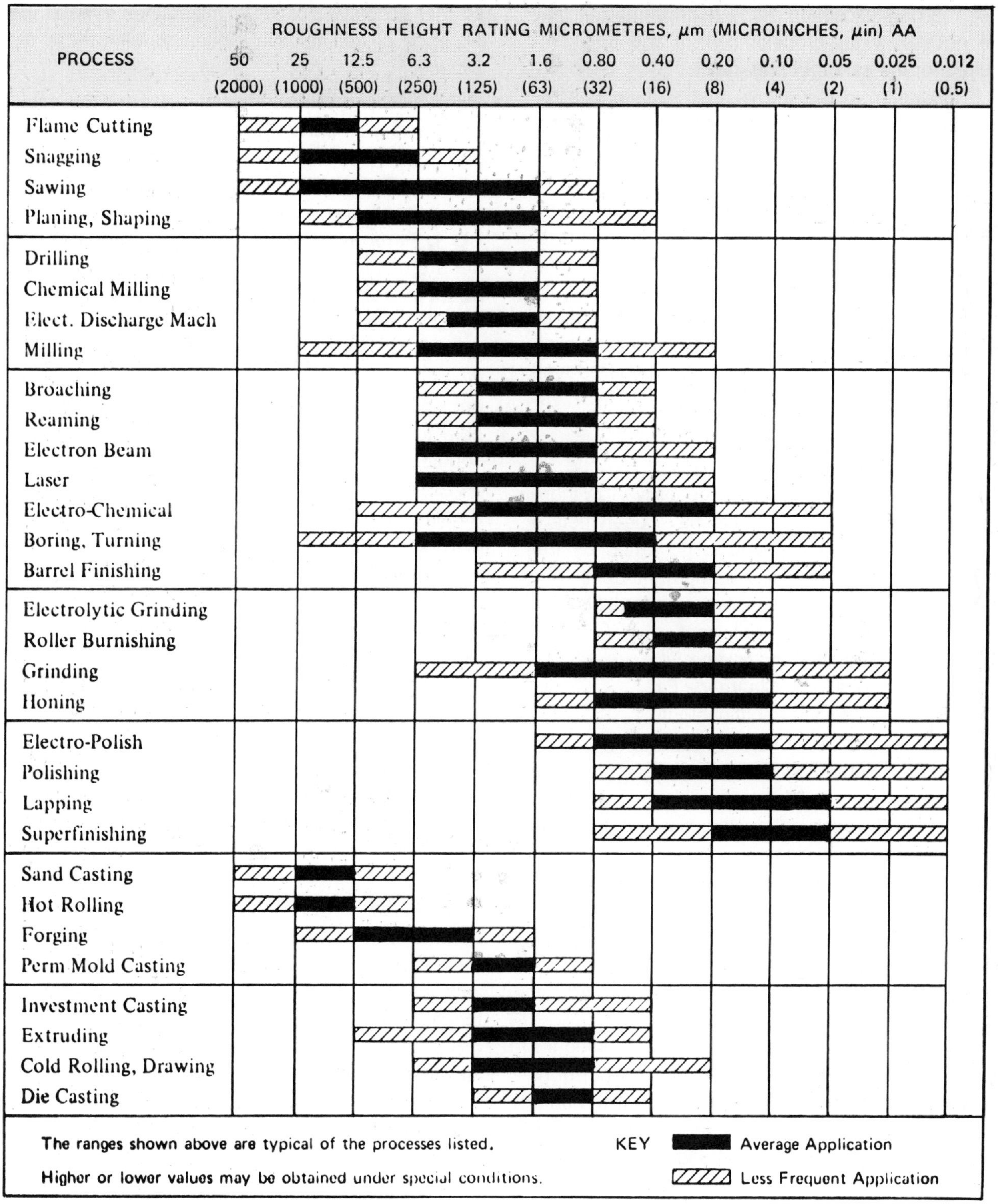
ROUGHNESS HEIGHT RATING MICROMETRES, μm (MICROINCHES, μin) AA
PROCESS
50 (2000) 25 (1000) 12.5 (500) 6.3 (250) 3.2 (125) 1.6 (63) 0.80 (32) 0.40 (16) 0.20 (8) 0.10 (4) 0.05 (2) 0.025 (1) 0.012 (0.5)
Flame Cutting
Snagging
Sawing
Planing, Shaping
Drilling
Chemical Milling
Elect. Discharge Mach
Milling
Broaching
Reaming
Electron Beam
Laser
Electro-Chemical
Boring, Turning
Barrel Finishing
Electrolytic Grinding
Roller Burnishing
Grinding
Honing
Electro-Polish
Polishing
Lapping
Superfinishing
Sand Casting
Hot Rolling
Forging
Perm Mold Casting
Investment Casting
Extruding
Cold Rolling, Drawing
Die Casting
The ranges shown above are typical of the processes listed.
Higher or lower values may be obtained under special conditions.
KEY
Average Application
Less Frequent Application

Fig. 6-72. Surface texture obtained by common production methods.

trol, the only exception being restrictions that may be necessary for process control and finish required for the sake of appearance.

It follows from the above that surface characteristics should not be controlled on a drawing or specification unless such control is essential to appearance or mechanical performance of the product. Imposition of such restrictions when unnecessary may increase production costs and in any event will serve to lessen the emphasis on the control specified for important surfaces.

Fig. 6-72 shows ranges of surface roughness values that may be obtained by various production methods. These values should be considered typical and not absolute limits.

6.9 Standard Cylindrical Fits

To insure precision interchangeability of cylindrical parts, certain types of fits have been established. The type of fit used is governed by the service required from the equipment being designed. The common accepted standard fits and their abbreviations are as follows:

RC   Running and Sliding Fit
LC   Locational Clearance Fit
LT   Transition Locational Fit
LN   Locational Interference Fit
FN   Force and Shrink Fit

These letter symbols are used in conjunction with numbers representing the class of fit; thus, *FN 4* represents a class 4 force fit. Each of the symbols (two letters and a number) represents a complete fit. Generally these symbols are not shown on manufacturing drawings; sizes are specified instead.

6.9.1 *Description of fits*.[3] The following are the basic types of fits and their functions:

Running and sliding fits. Running and sliding fits are intended to provide similar running performances, with suitable lubrication allowance throughout the range of sizes. The clearances for the first two classes, used chiefly as slide fits, increase more slowly with diameter than the other classes,

so that accurate location is maintained even at the expense of free relative motion. Briefly, these fits include:

RC 1 Close sliding fits. Intended for the accurate location of parts which must assemble without perceptible play.

RC 2 Sliding fits. Intended for accurate location but with greater maximum clearance than class RC 1. Parts made to this fit move and turn easily but are not intended to run freely and, in the larger sizes, may seize with small temperature changes.

RC 3 Precision running fits. Intended for precision work at slow speeds and light journal pressures. About the closest fits which can be expected to run freely. Not suitable where appreciable temperature differences are likely to be encountered.

RC 4 Close running fits. Intended chiefly for running fits on accurate machinery with moderate surface speeds and journal pressures, where accurate location and minimum play is desired.

RC 5, RC 6 Medium running fits. Intended for higher running speeds or heavy journal pressure, or both.

RC 7 Free running fits. Intended for use where accuracy is not essential or where large temperature variations are likely to be encountered, or under both these conditions.

RC 8, RC 9 Loose running fits. Intended for use where materials such as cold-rolled shafting and tubing, made to commercial tolerances, are involved.

Locational fits. Locational fits are fits intended to determine only the location of the mating parts; they may provide rigid or accurate location, as with interference fits, or provide some freedom of location, as with clearance fits. Accordingly, they are divided into three groups: clearance fits, transition fits, and interference fits. These fits are

LC Locational clearance fits. Intended for parts which are normally stationary but which can be freely assembled or disassembled. They run from snug fits for parts such as spigots, to the looser fastener fits where freedom of assembly is of prime importance.

LT Locational transition fits. A compromise between clearance and interference fits. For application where accuracy of location is important, but

---

3. Extracted from *Preferred Limits and Fits for Cylindrical Parts. ANSI B4.1* (New York: The American Society of Mechanical Engineers).

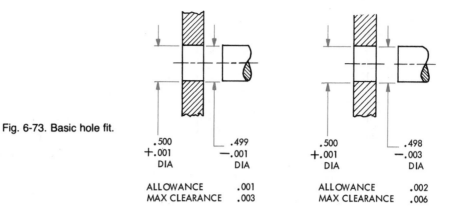

Fig. 6-73. Basic hole fit.

.500
+.001
DIA

.499
—.001
DIA

ALLOWANCE .001
MAX CLEARANCE .003

.500
+.001
DIA

.498
—.003
DIA

ALLOWANCE .002
MAX CLEARANCE .006

(HOLE SIZE IS UNCHANGED)

where a small amount of either clearance or interference is permissible.

LN Locational interference fits. Used where accuracy of location is of prime importance and for parts requiring rigidity and alignment with no special requirements for bore pressure. Such fits are not intended for parts designed to transmit frictional loads from one part to another by virtue of the tightness of fit; these conditions are covered by force fits.

Force fits. Force or shrink fits constitute a special type of interference fit, normally characterized by maintenance of constant bore pressures throughout the range of sizes. The interference therefore varies almost directly with diameter, and the difference between its minimum and maximum value is small, maintaining the resulting pressures within reasonable limits. Briefly, these fits are

FN 1 Light drive fits. Require light assembly pressures and produce more or less permanent assemblies. They are suitable for thin sections or long fits, and in cast iron external members.

FN 2 Medium drive fits. Suitable for ordinary steel parts or for shrink fits on light sections. They are about the tightest fits that can be used with high-grade cast iron external members.

FN 3 Heavy drive fits. Suitable for heavier steel parts or for shrink fits in medium sections.

FN 4, FN 5 Force fits. Suitable for parts which can be highly stressed or for shrink fits where the heavy pressing forces required are impractical.

6.9.2 *Dimensioning for cylindrical fits.*[4] To specify the dimensions and tolerances of an internal and an external cylindrical surface so that these surfaces will fit together as desired, it is necessary to begin calculations based either on a basic hole system, which assumes a minimum hole size, or on a basic shaft system, which assumes a maximum shaft size.

Basic hole system. A basic hole system is a system of fits in which the design size of the hole is the basic size and the allowance is applied to the shaft. Limits for a fit in the basic hole system are determined by (1) specifying the minimum hole size, (2) determining the maximum shaft size by subtracting the desired allowance (minimum clearance) from the minimum hole size for a clearance fit, or adding the desired allowance (maximum interference) for an interference fit, and (3) adjusting the hole and shaft tolerances to obtain the desired maximum clearance or minimum interference. See Fig. 6-73. Tooling economies can often be realized by calculating from the basic hole size, providing the size selected can be produced by a standard tool (reamer, broach, etc.) or gaged with a standard plug gage.

Basic shaft system. A basic shaft system is a system of fits in which the design size of the shaft is the basic size and the allowance is applied to

4. *Ibid.*

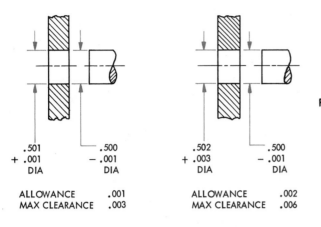

Fig. 6-74. Basic shaft fit.

.501   .500
+ .001   − .001
DIA   DIA

.502   .500
+ .003   − .001
DIA   DIA

ALLOWANCE   .001
MAX CLEARANCE   .003

ALLOWANCE   .002
MAX CLEARANCE   .006

(SHAFT SIZE IS UNCHANGED)

the hole. Limits for a fit in the basic shaft system are determined by (1) specifying the maximum shaft size, (2) determining the minimum hole size by adding the desired allowance (minimum clearance) to the maximum shaft size for a clearance fit, or subtracting for an interference fit, and (3) adjusting hole and shaft tolerances to obtain the desired maximum clearance or minimum interference. See Fig. 6-74. The basic shaft system is recommended only if there is a particular reason for using it; for example, where a standard size of shafting can be used.

6.9.3 Selecting the correct fit limits. Assume that a component must be designed wherein a 2.250 inch diameter shaft having a class RC 8 fit is to slide in a hole with a nominal diameter of 2.250 inches. See Fig. 6-75. Since most limit dimensions are computed on the basic hole system, the limits for the example above can be deter-

mined by converting the nominal size to the basic hole size and adding to or subtracting from the basic size the standard limits specified for hole and shaft sizes. For the illustration shown in Fig. 6-75, the procedure would be as follows:

1. Locate the nominal size range of the hole and shaft in Fig. 6-76. This range is 1.97—3.15.

2. Under the column class *RC 8,* the limit range for the hole size runs from .000 to plus .0045 inch, and for the shaft a minus .006 to a minus .009 inch.

3. The nominal hole and shaft size is 2.250 inches. Therefore, the hole may range from 2.250 to 2.2545 inches and the shaft from 2.244 to 2.241 inches. This is expressed in inches as:

$$\text{Limits on hole:} \quad \frac{2.250 \text{ plus } .0045 = 2.2545}{2.250 \text{ plus } .000 \ = 2.2500}$$

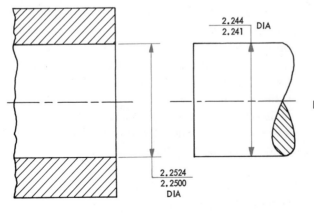

2.244
‾‾‾‾   DIA
2.241

2.2524
‾‾‾‾
2.2500
DIA

Fig. 6-75. Computing hole and shaft size.

RUNNING AND SLIDING FITS

Limits are in thousandths of an inch.

Limits for hole and shaft are applied algebraically to the basic size to obtain the limits of size for the parts.

Data in bold face are in accordance with ABC agreements.

Symbols H5, g5, etc., are Hole and Shaft designations used in ABC System (Appendix I).

Nominal Size Range Inches	Class RC 1			Class RC 2			Class RC 3			Class RC 4		
Over – To	Limits of Clearance	Standard Limits Hole H5	Standard Limits Shaft g4	Limits of Clearance	Standard Limits Hole H6	Standard Limits Shaft g5	Limits of Clearance	Standard Limits Hole H7	Standard Limits Shaft f6	Limits of Clearance	Standard Limits Hole H8	Standard Limits Shaft f7
0 – 0.12	0.1 0.45	+ 0.2 0	− 0.1 − 0.25	0.1 0.55	+ 0.25 0	− 0.1 − 0.3	0.3 0.95	+ 0.4 0	− 0.3 − 0.55	0.3 1.3	+ 0.6 0	− 0.3 − 0.7
0.12 – 0.24	0.15 0.5	+ 0.2 0	− 0.15 − 0.3	0.15 0.65	+ 0.3 0	− 0.15 − 0.35	0.4 1.2	+ 0.5 0	− 0.4 − 0.7	0.4 1.6	+ 0.7 0	− 0.4 − 0.9
0.24 – 0.40	0.2 0.6	+ 0.25 0	− 0.2 − 0.35	0.2 0.85	+ 0.4 0	− 0.2 − 0.45	0.5 1.5	+ 0.6 0	− 0.5 − 0.9	0.5 2.0	+ 0.9 0	− 0.5 − 1.1
0.40 – 0.71	0.25 0.75	+ 0.3 0	− 0.25 − 0.45	0.25 0.95	+ 0.4 0	− 0.25 − 0.55	0.6 1.7	+ 0.7 0	− 0.6 − 1.0	0.6 2.3	+ 1.0 0	− 0.6 − 1.3
0.71 – 1.19	0.3 0.95	+ 0.4 0	− 0.3 − 0.55	0.3 1.2	+ 0.5 0	− 0.3 − 0.7	0.8 2.1	+ 0.8 0	− 0.8 − 1.3	0.8 2.8	+ 1.2 0	− 0.8 − 1.6
1.19 – 1.97	0.4 1.1	+ 0.4 0	− 0.4 − 0.7	0.4 1.4	+ 0.6 0	− 0.4 − 0.8	1.0 2.6	+ 1.0 0	− 1.0 − 1.6	1.0 3.6	+ 1.6 0	− 1.0 − 2.0
1.97 – 3.15	0.4 1.2	+ 0.5 0	− 0.4 − 0.7	0.4 1.6	+ 0.7 0	− 0.4 − 0.9	1.2 3.1	+ 1.2 0	− 1.2 − 1.9	1.2 4.2	+ 1.8 0	− 1.2 − 2.4
3.15 – 4.73	0.5 1.5	+ 0.6 0	− 0.5 − 0.9	0.5 2.0	+ 0.9 0	− 0.5 − 1.1	1.4 3.7	+ 1.4 0	− 1.4 − 2.3	1.4 5.0	+ 2.2 0	− 1.4 − 2.8
4.73 – 7.09	0.6 1.8	+ 0.7 0	− 0.6 − 1.1	0.6 2.3	+ 1.0 0	− 0.6 − 1.3	1.6 4.2	+ 1.6 0	− 1.6 − 2.6	1.6 5.7	+ 2.5 0	− 1.6 − 3.2

Nominal Size Range Inches	Class RC 5			Class RC 6			Class RC 7			Class RC 8			Class RC 9		
Over – To	Limits of Clearance	Standard Limits Hole H8	Standard Limits Shaft e7	Limits of Clearance	Standard Limits Hole H9	Standard Limits Shaft e8	Limits of Clearance	Standard Limits Hole H9	Standard Limits Shaft d8	Limits of Clearance	Standard Limits Hole H10	Standard Limits Shaft c9	Limits of Clearance	Standard Limits Hole H11	Standard Limits Shaft
0 – 0.12	0.6 1.6	+ 0.6 − 0	− 0.6 − 1.0	0.6 2.2	+ 1.0 − 0	− 0.6 − 1.2	1.0 2.6	+ 1.0 0	− 1.0 − 1.6	2.5 5.1	+ 1.6 0	− 2.5 − 3.5	4.0 8.1	+ 2.5 0	− 4.0 − 5.6
0.12 – 0.24	0.8 2.0	+ 0.7 − 0	− 0.8 − 1.3	0.8 2.7	+ 1.2 − 0	− 0.8 − 1.5	1.2 3.1	+ 1.2 0	− 1.2 − 1.9	2.8 5.8	+ 1.8 0	− 2.8 − 4.0	4.5 9.0	+ 3.0 0	− 4.5 − 6.0
0.24 – 0.40	1.0 2.5	+ 0.9 − 0	− 1.0 − 1.6	1.0 3.3	+ 1.4 − 0	− 1.0 − 1.9	1.6 3.9	+ 1.4 0	− 1.6 − 2.5	3.0 6.6	+ 2.2 0	− 3.0 − 4.4	5.0 10.7	+ 3.5 0	− 5.0 − 7.2
0.40 – 0.71	1.2 2.9	+ 1.0 − 0	− 1.2 − 1.9	1.2 3.8	+ 1.6 − 0	− 1.2 − 2.2	2.0 4.6	+ 1.6 0	− 2.0 − 3.0	3.5 7.9	+ 2.8 0	− 3.5 − 5.1	6.0 12.8	+ 4.0 − 0	− 6.0 − 8.8
0.71 – 1.19	1.6 3.6	+ 1.2 − 0	− 1.6 − 2.4	1.6 4.8	+ 2.0 − 0	− 1.6 − 2.8	2.5 5.7	+ 2.0 0	− 2.5 − 3.7	4.5 10.0	+ 3.5 0	− 4.5 − 6.5	7.0 15.5	+ 5.0 0	− 7.0 − 10.5
1.19 – 1.97	2.0 [illegible].6	+ 1.6 − 0	− 2.0 − 3.0	2.0 6.1	+ 2.5 − 0	− 2.0 − 3.6	3.0 7.1	+ 2.5 0	− 3.0 − 4.6	[illegible] 1[illegible]	+ 4.0 0	− 5.0 − 7.5	8.0 18.0	+ 6.0 0	− 8.0 − 12.0
1.97 – 3.15	[illegible].5 [illegible].5	+ 1.8 − 0	− 2.5 − 3.7	2.5 7.3	+ 3.0 − 0	− 2.5 − 4.3	4.0 8.8	+ 3.0 0	− 4.0 − 5.8	6.0 13.5	+ 4.5 0	− 6.0 − 9.0	9.0 20.5	+ 7.0 0	− 9.0 − 13.5
3.15 – 4.73	3.0 6.6	+ 2.2 − 0	− 3.0 − 4.4	3.0 8.7	+ 3.5 − 0	− 3.0 − 5.2	5.0 10.7	+ 3.5 0	− 5.0 − 7.2	7.0 15.5	+ 5.0 0	− 7.0 − 10.5	[illegible]0.0 [illegible].0	+ 9.0 0	− 10.0 − 15.0
4.73 – 7.09	3.5 7.6	+ 2.5 − 0	− 3.5 − 5.1	3.5 10.0	+ 4.0 − 0	− 3.5 − 6.0	6.0 12.5	+ 4.0 0	− 6.0 − 8.5	8.0 18.0	+ 6.0 0	− 8.0 − 12.0	12.0 28.0	+ 10.0 0	− 12.0 − 18.0

Fig. 6-76. Running and sliding fits. (ANSI B4.1)

$$\text{Limits on shaft: } \frac{2.250 \text{ minus } .006 = 2.244}{2.250 \text{ minus } .009 = 2.241}$$

See Appendix for tables covering nominal size range of location fits, transitional fits, interference location fits, and force and shrink fits.

UNIT 7

Geometric Tolerancing

Although the usual methods of tolerancing generally provide sufficient dimensional control of parts, it is often difficult to determine whether such parts will mate satisfactorily with their corresponding components even if the sizes are within the stated tolerances. A more accurate way to achieve greater manufacturing control is by means of geometric tolerancing.

7.1 Meaning of Geometric Tolerancing

Geometric tolerancing deals with three general types of control features—location, form, and runout. Each is intended to control some component portion of a part. Additional treatment of geometric tolerancing will be found in the references listed below.*

7.1.1 Tolerance of location. This is the permissible variation assigned to a dimension that locates one or more features in relation to some other feature. Application of location tolerances is particularly valuable in controlling:

1. center distances between holes, slots, bosses, and tabs.

2. the location of holes, slots, bosses, and tabs relative to datum features such as plane and cylindrical surfaces.

3. coaxiality between features.

4. features with equal center distances about a datum axis or plane.

7.1.2 Tolerance of form. Form tolerances specify the maximum permissible variations of desired surface conditions such as straightness, flatness, roundness, cylindricity, profile of a surface or a line, angularity, parallelism, and perpendicularity. Form tolerancing actually stipulates how far features of a part are allowed to deviate from the perfect geometry implied by the drawing. The use of various geometric forms—cylinders, cones, spheres, and so on—assumes perfect features but since this perfection is impossible to achieve because of variations in the manufacturing process, these variations must be controlled if high quality of interchangeability of parts is to be maintained.

7.1.3 Tolerance of runout. Runout is a composite tolerance used to control the functional relationship of one or more features of a part to a datum axis. Features controlled by runout tolerances include surfaces constructed around a datum axis and those at right angles to a datum axis. See Fig. 7-1.

Dimensioning and Tolerancing. ANSI Y14.5 (New York: The American Society of Mechanical Engineers and Foster, Lowell W. *Geometric Dimensioning and Tolerancing* (Reading, Mass.: Addison-Wesley Publishing Co.).

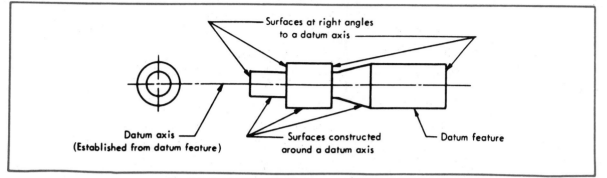

Fig. 7-1. Features applicable to runout tolerancing. (ANSI Y14.5)

7.2 Tolerancing Symbols

Location, form, and runout tolerances are specified by symbols and numerical values enclosed in frames and attached to applicable features of a part. The geometric characteristic symbols are illustrated in Fig. 7-2. The recommended

		CHARACTERISTIC	SYMBOL	NOTES
INDIVIDUAL FEATURES	FORM TOLERANCES	STRAIGHTNESS	—	1
		FLATNESS	▱	1
		ROUNDNESS (CIRCULARITY)	○	
		CYLINDRICITY	⌭	
INDIVIDUAL OR RELATED FEATURES		PROFILE OF A LINE	⌒	2
		PROFILE OF A SURFACE	⌓	2
		ANGULARITY	∠	
		PERPENDICULARITY (SQUARENESS)	⊥	
		PARALLELISM	//	3
RELATED FEATURES	LOCATION TOLERANCES	POSITION	⊕	
		CONCENTRICITY	◎	3,7
		SYMMETRY	≡	5
	RUNOUT TOLERANCES	CIRCULAR	↗	4
		TOTAL	↗	4,6

Note:- 1) The symbol ∿ formerly denoted flatness.

  The symbol ⌒ or — formerly denoted flatness and straightness.

  2) Considered "related" features where datums are specified.

  3) The symbol || and ◎ formerly denoted parallelism and concentricity, respectively.

  4) The symbol ↗ without the qualifier "CIRCULAR" formerly denoted total runout.

  5) Where symmetry applies, it is preferred that the position symbol be used.

  6) "TOTAL" must be specified under the feature control symbol.

  7) Consider the use of position or runout.

Where existing drawings using the above former symbols are continued in use, each former symbol denotes that geometric characteristic which is applicable to the specific type of feature shown.

Fig. 7-2. Geometric characteristic symbols. (ANSI Y14.5)

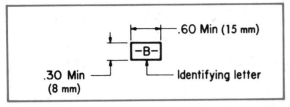

Fig. 7-3. Recommended size of tolerancing frame.

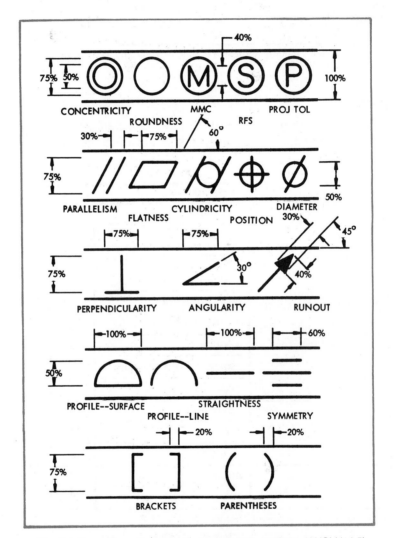

Fig. 7-4. Proportions of geometric characteristic symbols. (ANSI Y14.5)

size of the frame is shown in Fig 7-3. Fig. 7-4 gives the preferred proportions of the various geometric characteristic symbols. These proportions are based on modules of 50% and 75% of the basic frame height.

The number of frames incorporated in any tolerancing system depends on the desired degree of tolerancing control. Generally the data included appears in the following sequence: geometric characteristic symbol, datum, tolerance value, and supplementary symbols. See Fig. 7-5.

7.2.1 Supplementary symbols. Supplementary symbols are used to convey specific information that is needed to clarify some special aspect of

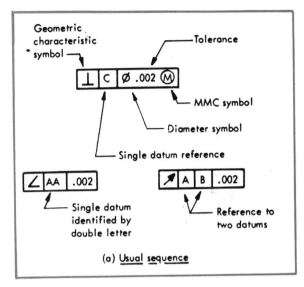

Fig. 7-5. Typical sequence of tolerancing data. (ANSI Y14.5)

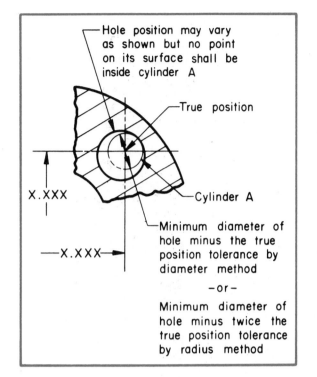

Fig. 7-7. Tolerance zone for surface of hole at MMC. (ANSI Y14.5)

TERM	ABBREVIATION	SYMBOL
Maximum Material Condition	MMC	Ⓜ
Regardless of Feature Size	RFS	Ⓢ
Diameter	DIA	⌀
Projected Tolerance Zone	TOL ZONE PROJ	Ⓟ
Reference	REF	(1.250)
Basic	BSC	3.875

Fig. 7-6. Supplementary tolerancing symbols. (ANSI Y14.5)

the feature. See Fig. 7-6. Types of information conveyed by supplementary symbols are given in the following paragraphs:

Maximum Material Condition (MMC). The maximum material condition, represented by the symbol Ⓜ, refers to the maximum amount of material permitted by the toleranced size dimension for the specified feature. For holes, slots, and other internal features, it means the *minimum* allowable sizes. For shafts, bosses, lugs, tabs, and other external features, it applies to the *maximum* allowable sizes. See Fig. 7-7.

Regardless of Feature Size (RFS). The designation *regardless of feature size,* represented by the symbol Ⓢ, means that a form or positional tolerance must be met irrespective of where the feature lies within its size tolerance. See Fig. 7-8. Thus, where RFS applies to positional tolerance of circular features, the axis of each feature must be located within the specified positional tolerance regardless of the size of the feature. For example, in Fig. 7-8 the six holes shown may vary in size from 0.9994 to 1.0000. To minimize spacing errors, each hole must be located within the specified positional tolerance regardless of the size of that hole.

Diameter. Diameter is designated by the symbol ⌀, which precedes the specified tolerance in a feature control symbol. It is often used on a drawing in place of the abbreviation DIA and is placed after the dimension. See Fig. 7-11.

Projected tolerance zone. Application of this feature control, shown by the symbol Ⓟ, is recommended where variations in perpendicularity of

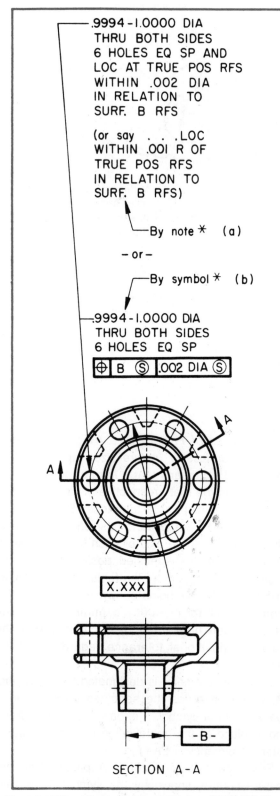

.9994-1.0000 DIA
THRU BOTH SIDES
6 HOLES EQ SP AND
LOC AT TRUE POS RFS
WITHIN .002 DIA
IN RELATION TO
SURF. B RFS

(or say . . . LOC
WITHIN .001 R OF
TRUE POS RFS
IN RELATION TO
SURF. B RFS)

By note ✳ (a)

– or –

By symbol ✳ (b)

.9994-1.0000 DIA
THRU BOTH SIDES
6 HOLES EQ SP

⊕ | B Ⓢ | .002 DIA Ⓢ

X.XXX

-B-

SECTION A-A

Fig. 7-8. RFS applied to feature and datum. (ANSI Y14.5)

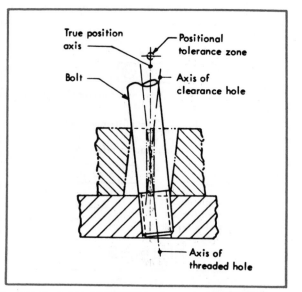

Fig. 7-9. Basis for projected tolerance zone. (ANSI Y14.5)

threaded holes or press fit holes may cause bolts, studs, or pins to interfere with mating parts. See Fig. 7-9.

Reference. When a reference value is required, each value is enclosed with parentheses as shown in Fig. 7-6.

Basic dimension. A basic dimension is a numerical value which is used to describe the theoretically exact size, shape, or location of a feature. It serves as a basis for establishing permissible tolerances. A basic dimension is identified by enclosing it in a separate frame as shown in Fig. 7-10.

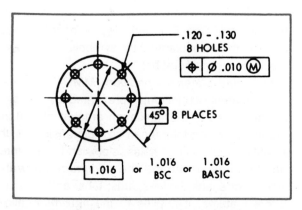

Fig. 7-10. Identifying basic dimensions. (ANSI Y14.5)

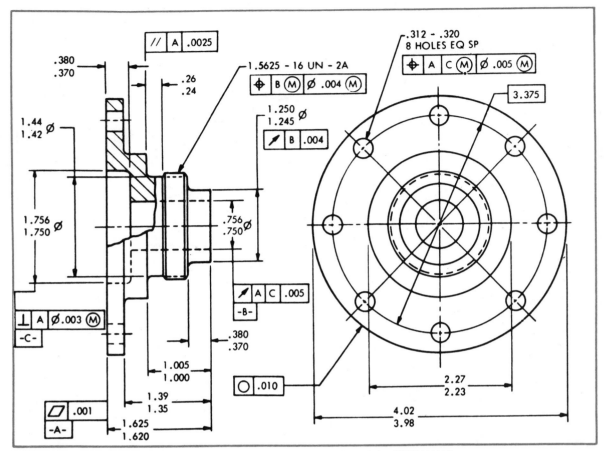

Fig. 7-11. Application of feature control symbols. (ANSI Y14.5)

7.2.2 *Location of feature control symbols.*
Feature control symbols are located on a drawing by any of the following methods: (See Fig. 7-11.)

1. Running a leader from the feature to the symbol.

2. Attaching the symbol frame to the extension line running from the feature.

3. Attaching the symbol frame to the dimension line that pertains to the feature.

4. Adding the symbol to a note or dimension pertaining to the feature.

7.3 Datum Referencing

Datums are points, lines, axes, or planes which serve as origins for dimensions. A sufficient number of features that are most important to the design of the part are selected and positioned to a set of three mutually perpendicular planes. Then all related measurements of the part are made from these planes. See Fig. 7-12.

The actual number of datums used depends on the relationship required. For some objects a single datum reference may be sufficient, whereas for others two or more datums may be necessary.

7.3.1 *Datum identifying symbols.*
A frame with an enclosed letter identifies a datum. See Fig. 7-13. Any letter of the alphabet except *I, O,* and *Q* may be used to identify a reference datum. Each datum feature that requires identification is assigned a different letter. If datum features are so numerous as to exhaust all single letters, double letters AA through AZ are permissible.

In the dimensioning of a part, the three datum planes are referred to as first, second, and third and are so designated on a drawing. See Figs. 7-13 and 7-14. The sequence shown on the draw-

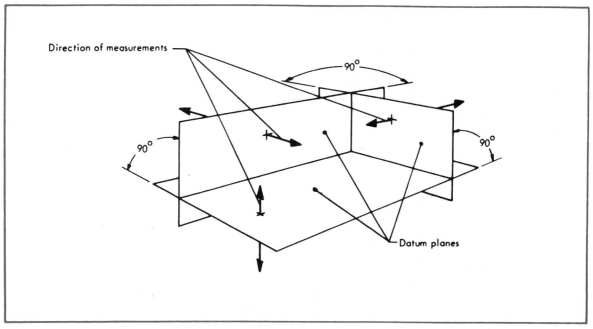

Fig. 7-12. Datum reference frame. (ANSI Y14.5)

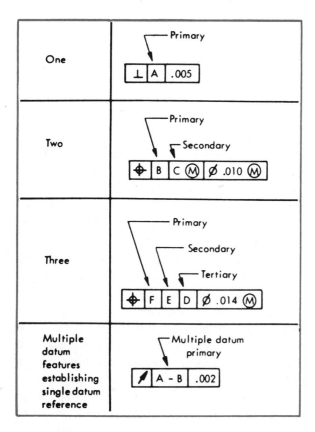

Fig. 7-13. The order of precedence of datum references. (ANSI Y14.5)

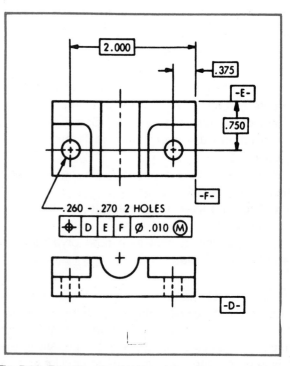

Fig. 7-14. Example of part where datum features are flat surfaces. (ANSI Y14.5)

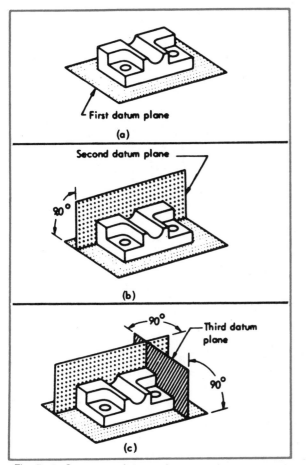

Fig. 7-15. Sequence of datum features relates part to datum reference frame. (ANSI Y14.5)

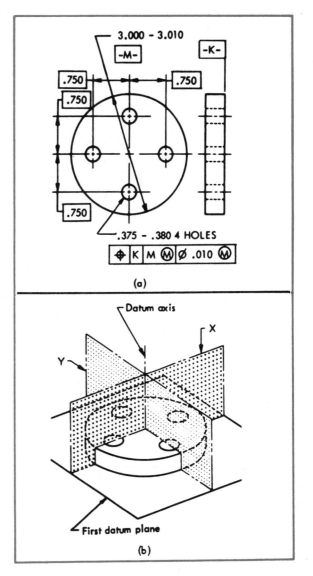

Fig. 7-16. An example of part with cylindrical datum feature. (ANSI Y14.5)

ing depends to a large extent on the design of the part. Notice in Fig. 7-14 that datum features are identified as surfaces D, E, and F. Feature *D* is considered to be primary, *E* secondary, and *F* tertiary. Thus, on a drawing they would appear in that order, meaning that dimensions for the part were established in the given sequence from their respective planes. Each datum reference letter, supplemented by the symbol Ⓜ or Ⓢ if applicable, is entered from left to right in the desired order of preference. See Fig. 7-13.

A further example of positioning features so that they relate to specific datum reference frames is shown in Fig. 7-15. In Fig. 7-15A the primary datum feature is related to the first datum plane. Fig. 7-15B shows two points of the secondary datum feature in contact with the secondary datum

plane. The relationship is completed in Fig. 7-15C by bringing three points into contact with the third datum plane.

The same three-plane concept also applies to cylindrical parts. Notice in Fig. 7-16A that the datum feature *K* is the most applicable reference base and therefore is selected as the primary datum feature. Since the holes are related to the cylindrical datum *M*, they are dimensioned from center lines through the center of *M*. As may be seen in Fig. 7-16B, this part has depth. Hence, to

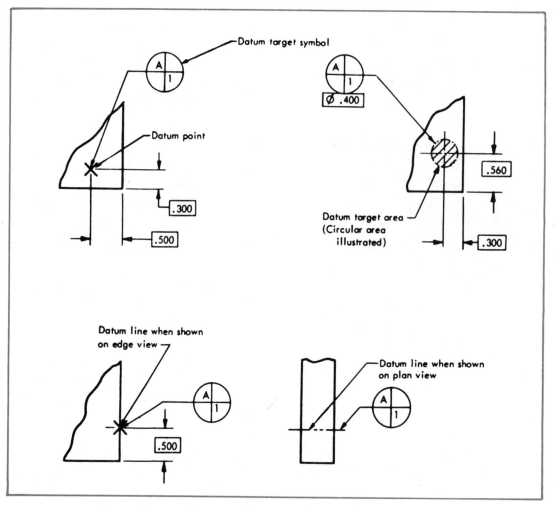

Fig. 7-17. Datum targets. (ANSI Y14.5)

fully describe the part, the second and third datum planes, *X* and *Y,* are necessary. Thus, all measurements for locating clearance holes are made to these three datum planes.

7.3.2 Datum targets. Datum targets are sometimes used to better establish the relationship of points, lines, or areas on an object to datum planes. This technique is particularly helpful in applying the three-plane concept to features that are non-planar or have uneven surfaces resulting from such operations as casting, forging, molding, or welding.

The *datum target symbol* consists of a circle divided into four quadrants. The letter placed in the upper left quadrant identifies the datum feature.

The number in the lower right quadrant identifies the target. Datum targets are located by using basic or toleranced dimensions. See Figs. 7-17 and 7-18.

7.4 Application of Location Tolerances

As previously indicated, tolerance of location controls the relationships of holes, slots, bosses, and tabs to each other and to datum features. These relationships may involve position, concentricity, or symmetry.

7.4.1 Positional tolerancing. A positional tolerance defines a zone within which the center plane of a feature may vary from a true (theoretically

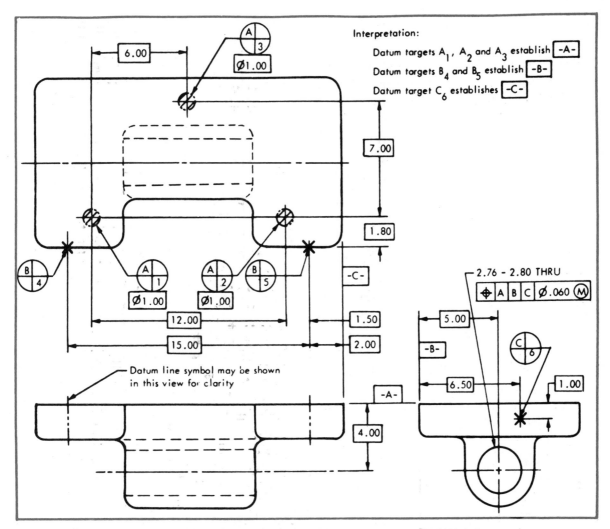

Fig. 7-18. Application of datum point, line and area targets. (ANSI Y14.5)

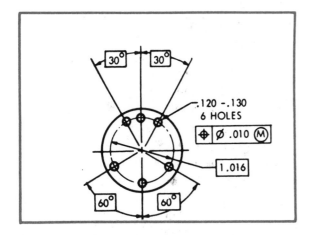

Fig. 7-19. Positional tolerancing. (ANSI Y14.5)

exact) position. Basic dimensions are used to establish the true position, and the permissible variation is expressed as a positional tolerance with its representative geometric characteristic symbol. See Fig. 7-19.

In the traditional coordinate plus and minus system, tolerances are often subject to more than one interpretation. With positional tolerancing, only one interpretation is possible, thereby assuring a greater degree of precision in the assembly of mating parts without interference. A typical example is shown in Fig. 7-20. Notice that in 7-20A the dimensions with coordinate plus and minus tolerances can be interpreted at least in two different

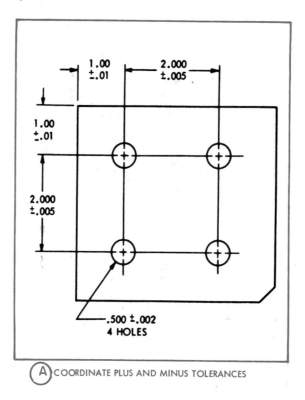

A COORDINATE PLUS AND MINUS TOLERANCES

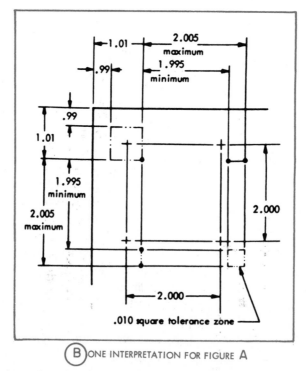

B ONE INTERPRETATION FOR FIGURE A

Fig. 7-20. Coordinate plus and minus tolerancing versus positional tolerancing. (A) Coordinate plus and minus tolerancing. (B) One interpretation for figure (A). (ANSI Y14.5)

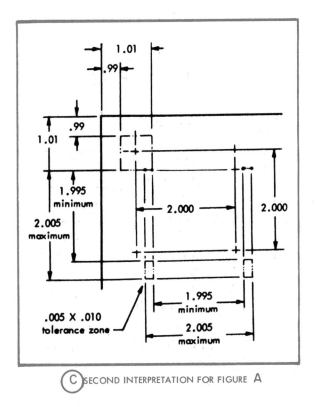

(C) SECOND INTERPRETATION FOR FIGURE A

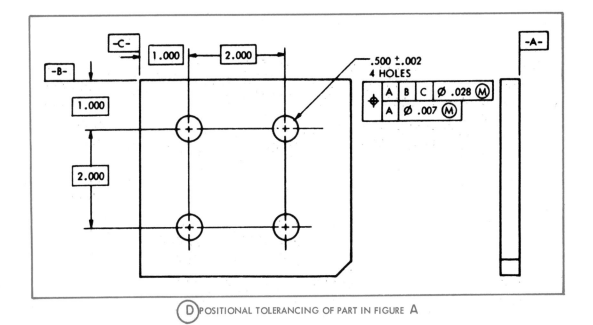

(D) POSITIONAL TOLERANCING OF PART IN FIGURE A

Fig. 7-20 (Continued). Coordinate plus and minus tolerancing versus positional tolerancing. (C) Second interpretation for figure (A). (D) Positional tolerancing of part in figure (A). (ANSI Y14.5)

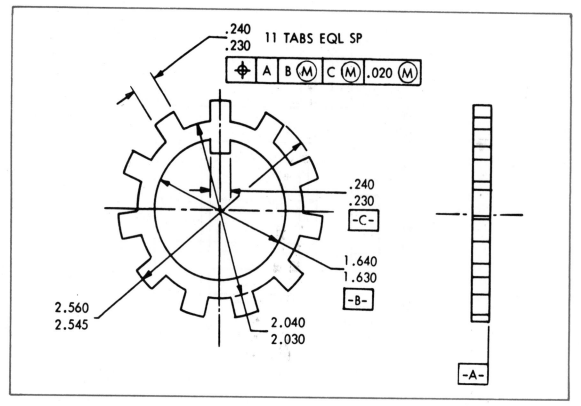

Fig. 7-21. Positional tolerancing of tabs. (ANSI Y14.5)

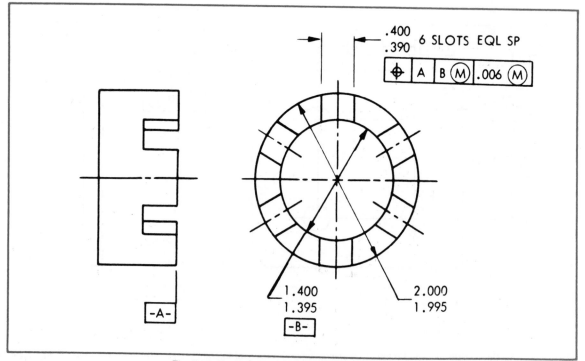

Fig. 7-22. Positional tolerancing of slots. (ANSI Y14.5)

ways. These interpretations are illustrated in Figs. 7-20*B* and 7-20*C.* By specifying this same part with positional tolerancing, as in Fig. 7-20*D,* misinterpretation can be prevented.

In addition to circular holes and bosses, the same principles of positional tolerancing can be applied to noncircular parts, such as slots, tabs, and elongated holes. For these features the positional tolerance is applied to the surfaces related to the center plane of the feature. The designated tolerance value represents a distance between two parallel planes. See Figs. 7-21 and 7-22.

7.4.2 Concentricity tolerancing. Concentricity tolerancing deals with permissible deviation when two or more surfaces of cylinders, spheres, cones, and so on are generated about a common axis. The amount of permissible deviation from such coaxiality is expressed by a concentricity tolerance with its appropriate geometric characteristic symbol. See Fig. 7-23.

A coaxiality tolerance can also be used to control the alignment of two or more holes on a common axis. See Fig. 7-24.

7.4.3 Symmetry tolerancing. Symmetry tolerancing may be called for where one or more features are symmetrically distributed about the center plane of a datum feature. By using this form of control, the desired tolerance may be expressed as MMC or RFS. Symmetry control on a drawing is expressed by a symmetry tolerance with its symmetry symbol. See Fig. 7-25.

7.5 Application of Form Tolerancing

Form tolerancing regulates the permissible deviation of a feature from its true geometric shape where tolerance of size and location do not provide sufficient control of manufacturing. Included are such characteristics as straightness, flatness, roundness, cylindricity, and profile.

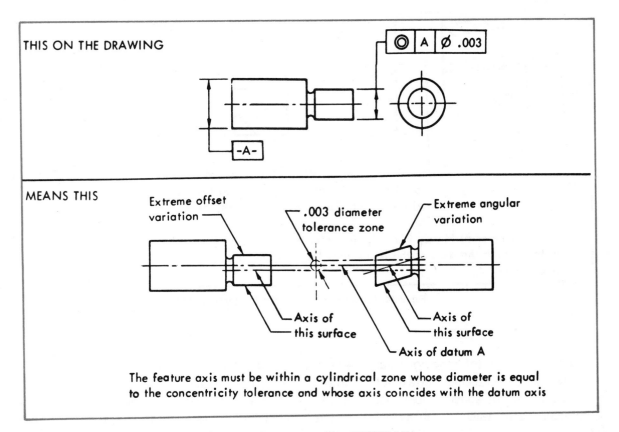

Fig. 7-23. Specifying concentricity. (ANSI Y14.5)

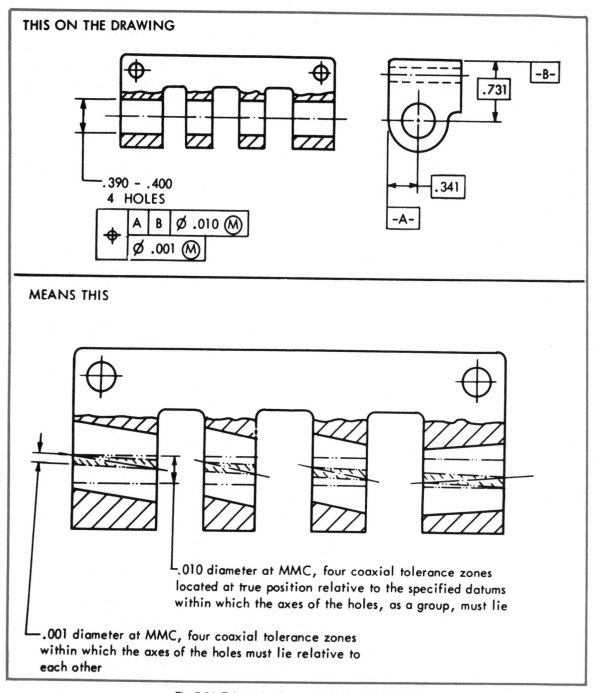

Fig. 7-24. Tolerancing for coaxial holes. (ANSI Y14.5)

7.5.1 Straightness tolerancing. Straightness is a condition where an element or axis of a surface is a straight line. The specified tolerance indicates the tolerance zone within which all points of the element must lie. The straightness tolerance is lo- cated in the view where elements to be controlled are represented by a straight line, Fig. 7-26.

7.5.2 Roundness tolerancing. Roundness is a condition where all surface points of a cylinder or cone, are equidistant from a common axis. A

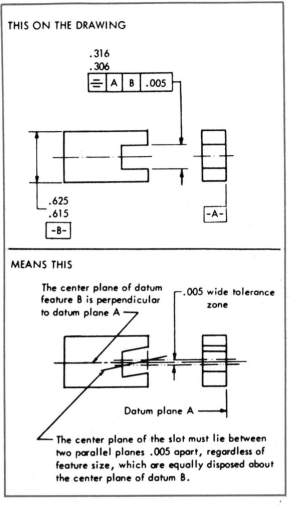

Fig. 7-25. Specifying symmetry. (ANSI Y14.5)

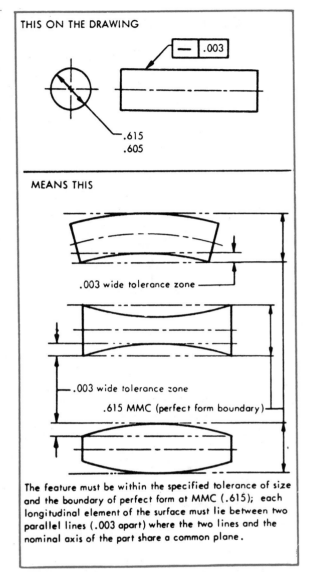

The feature must be within the specified tolerance of size and the boundary of perfect form at MMC (.615); each longitudinal element of the surface must lie between two parallel lines (.003 apart) where the two lines and the nominal axis of the part share a common plane.

Fig. 7-26. Specifying the straightness of surface elements. (ANSI Y14.5)

roundness tolerance specifies a tolerance zone, bounded by two concentric circles, within which each circular element of the surface must lie. See Fig. 7-27.

7.5.3 Flatness tolerancing. Flatness is a characteristic of a surface whose elements are all in one plane. A flatness tolerance specifies a zone, defined by two parallel planes, within which a surface must lie. The tolerance symbol is placed in the view where the surface elements to be controlled are represented by a line. See Fig. 7-28.

7.5.4 Cylindricity tolerancing. Cylindricity involves a feature where all points of a surface are equidistant from a common axis. A cylindricity tol-

erance specifies a zone, bounded by two concentric circles, within which the surface must lie. The identifying symbol is attached to a leader which may be directed to either view. The tolerance value applies to both circular and longitudinal elements of the surface. See Fig. 7-29.

7.5.5 Profile tolerancing. A profile is the outline of any object as shown in a given plane and represents either an entire surface or a line. The true profile is defined by basic dimensions and the tol-

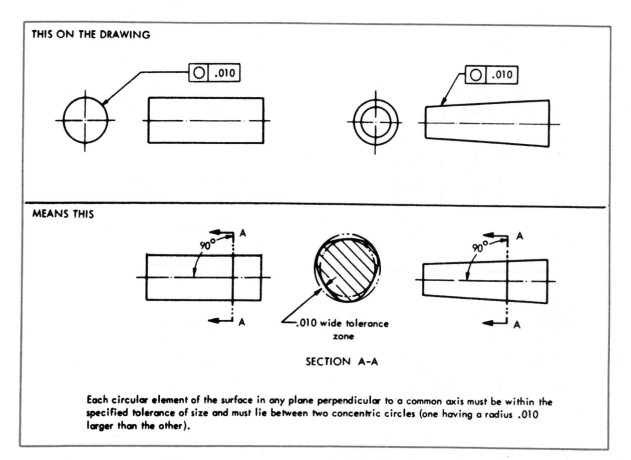

THIS ON THE DRAWING

MEANS THIS

90°

A

A

.010 wide tolerance zone

SECTION A-A

Each circular element of the surface in any plane perpendicular to a common axis must be within the specified tolerance of size and must lie between two concentric circles (one having a radius .010 larger than the other).

Fig. 7-27. Specifying roundness for a cylinder or a cone. (ANSI Y14.5)

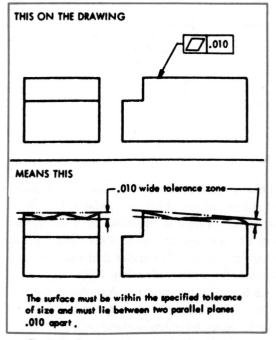

THIS ON THE DRAWING

.010

MEANS THIS

.010 wide tolerance zone

The surface must be within the specified tolerance of size and must lie between two parallel planes .010 apart.

Fig. 7-28. Specifying flatness. (ANSI Y14.5)

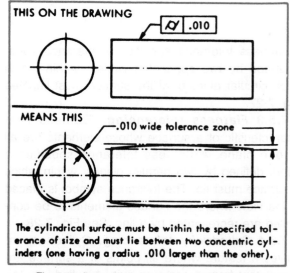

THIS ON THE DRAWING

.010

MEANS THIS

.010 wide tolerance zone

The cylindrical surface must be within the specified tolerance of size and must lie between two concentric cylinders (one having a radius .010 larger than the other).

Fig. 7-29. Specifying cylindricity. (ANSI Y14.5)

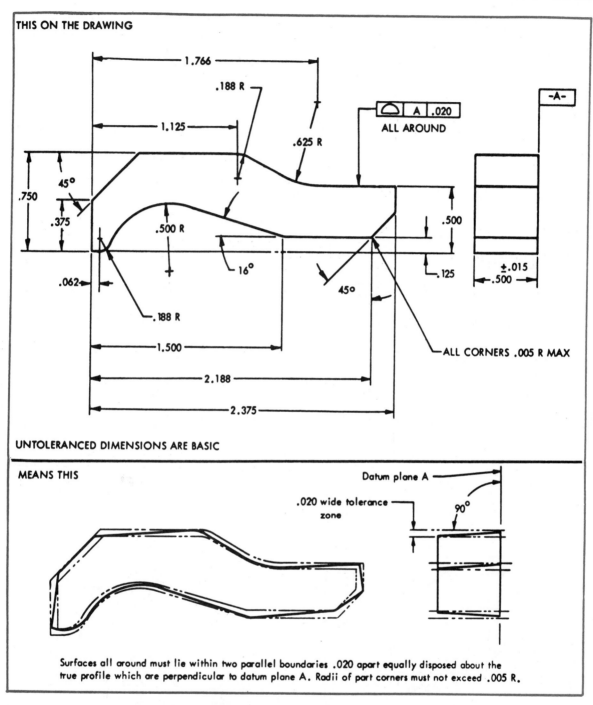

THIS ON THE DRAWING

UNTOLERANCED DIMENSIONS ARE BASIC

MEANS THIS

Surfaces all around must lie within two parallel boundaries .020 apart equally disposed about the true profile which are perpendicular to datum plane A. Radii of part corners must not exceed .005 R.

Fig. 7-30. Specifying profile of a surface. (ANSI Y14.5)

erance specifies a boundary within which the elements of a surface must lie. Refer to Figs. 7-30 and 7-31.

7.5.6 Form tolerancing for related features.

Form tolerancing for related features involves the control of such characteristics as angularity, paral-

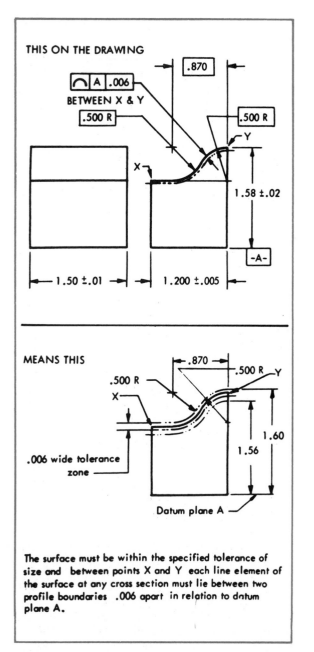

Fig. 7-31. Specifying profile of a line. (ANSI Y14.5)

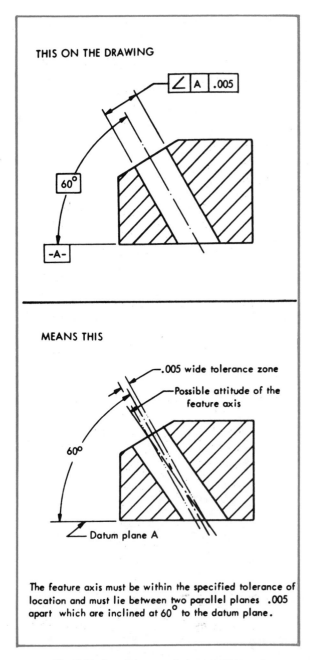

Fig. 7-32. Specifying angularity. (ANSI Y14.5)

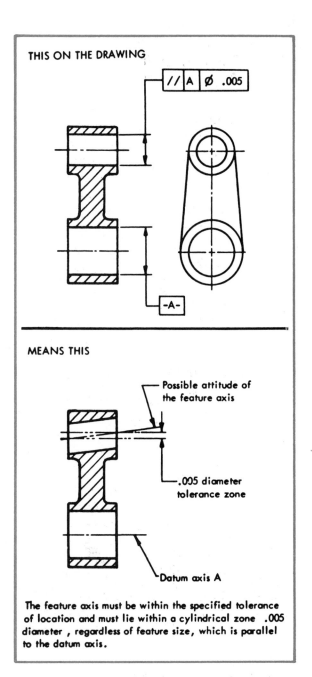

Fig. 7-33. Specifying parallelism. (ANSI Y14.5)

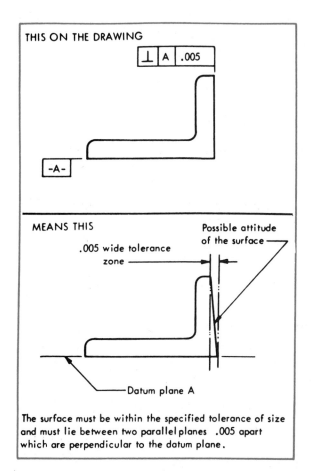

Fig. 7-34. Specifying perpendicularity for a plane surface. (ANSI Y14.5)

lelism, and perpendicularity of an object. Tolerancing of *angularity* governs the condition of a surface or axis at a specified angle from a datum plane or axis. See Fig. 7-32. *Parallelism* refers to the condition of a surface or axis equidistant at all points from a datum. See Fig. 7-33. *Perpendicularity* is the condition of a surface, plane, or axis which is at right angles to a datum. See Fig. 7-34.

7.6 Application of Runout Tolerancing

Two types of runout control are used—circular and total. A *circular runout* provides control of the circular elements of a surface. The specified toler-

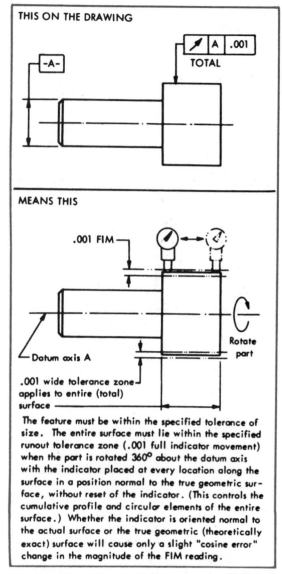

THIS ON THE DRAWING

TOTAL

MEANS THIS

.001 FIM

Datum axis A

.001 wide tolerance zone
applies to entire (total)
surface

The feature must be within the specified tolerance of
size. The entire surface must lie within the specified
runout tolerance zone (.001 full indicator movement)
when the part is rotated 360° about the datum axis
with the indicator placed at every location along the
surface in a position normal to the true geometric surface, without reset of the indicator. (This controls the
cumulative profile and circular elements of the entire
surface.) Whether the indicator is oriented normal to
the actual surface or the true geometric (theoretically
exact) surface will cause only a slight "cosine error"
change in the magnitude of the FIM reading.

Fig. 7-36. Specifying total runout relative to a datum diameter.
(ANSI Y14.5)

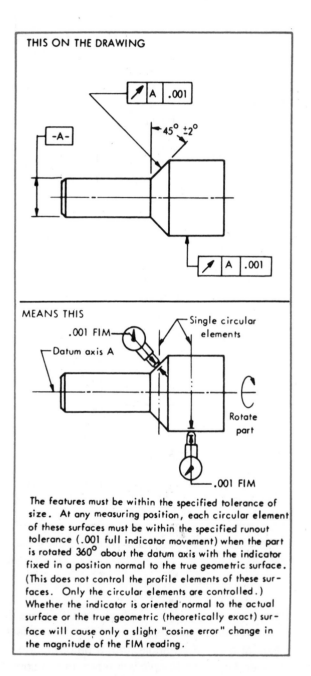

THIS ON THE DRAWING

45° ±2°

MEANS THIS

.001 FIM

Datum axis A

Single circular
elements

Rotate
part

.001 FIM

The features must be within the specified tolerance of
size. At any measuring position, each circular element
of these surfaces must be within the specified runout
tolerance (.001 full indicator movement) when the part
is rotated 360° about the datum axis with the indicator
fixed in a position normal to the true geometric surface.
(This does not control the profile elements of these surfaces. Only the circular elements are controlled.)
Whether the indicator is oriented normal to the actual
surface or the true geometric (theoretically exact) surface will cause only a slight "cosine error" change in
the magnitude of the FIM reading.

Fig. 7-35. Specifying circular runout relative to a datum diameter. (ANSI Y14.5)

ance applies to any circular measuring position as the part is rotated 360°. See Fig. 7-35. *Total runout* governs composite control of all surface elements. This tolerance applies to all circular and profile measuring positions as the part is rotated 360°. Where total runout is used, the word TOTAL is added beneath the feature control symbol. See Fig. 7-36.

Problems for Section II
Representational Drawings

A wide range of problems varying in complexity are included here to give students practice in developing drafting skills involving

> *Unit 5. Projection Drawing*
> *Unit 6. Basic Dimensioning*
> *Unit 7. Geometric Tolerancing*

Problems for Section II include

> **Problems 1-18 Basic Dimensioning Drawings**
> **Problems 19-66 Multiview Drawings**
> **Problems 67-86 Sectional Drawings**
> **Problems 87-104 Auxiliary Drawings**
> **Problems 105-116 Geometric Tolerancing Problems**

Problems having decimal-inch values may be converted to equivalent metric sizes by using the millimeter conversion chart or the problems may be redesigned by assigning new millimeter dimensional values more compatible with metric production.

Problems 1-18 Basic Dimensioning Drawings

Problems 1-6
Draw the views of the problems shown at any convenient scale and place all dimensions in their proper locations. Assume all sizes.

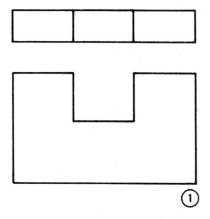

(1)

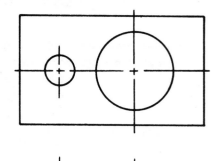

(2)

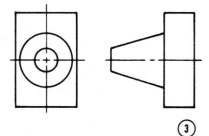

(3)

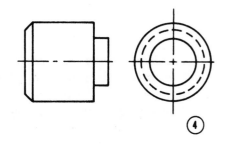

(4)

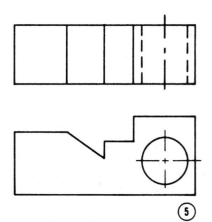

(5)

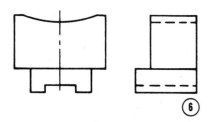

(6)

Problems 7-12
Draw the views of the problems shown at any convenient scale and place all dimensions in their proper locations. Assume all sizes.

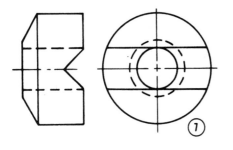

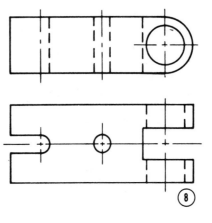

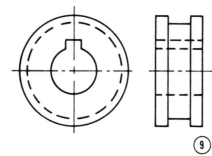

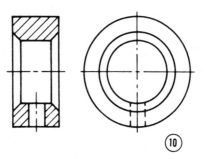

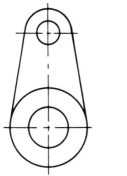

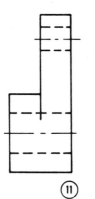

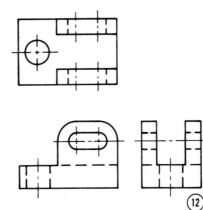

Problems 13-18
Draw the views of the problems shown at any convenient scale and place all dimensions in their proper locations. Assume all sizes.

METRIC

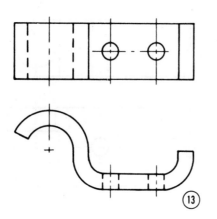

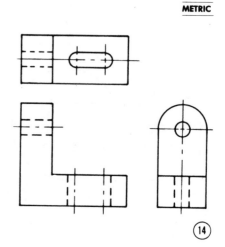

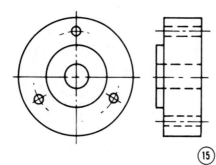

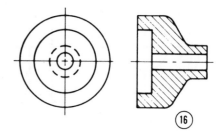

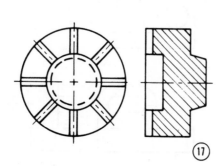

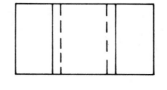

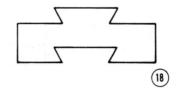

Problems 19-66 Multiview Drawings

Problems 19-53

Prepare multiview dimensioned drawings of the parts shown. Select any convenient scale and assign an appropriate project title to each drawing. Include only the essential views. On cast parts, choose suitable radii for fillets and rounds unless otherwise indicated.

Note to the Instructor:

1. Parts labeled *Metric* are dimensioned in millimeters. All others are dimensioned in decimal-inches. Several problems are dimensioned in fractions, but in those instances sizes are to be converted to millimeters or decimal-inches as indicated. Where applicable, tolerances and surface finish should be shown on the drawings as described in Unit 6.

2. To introduce practice in geometric tolerancing symbology, see Problems 105-116.

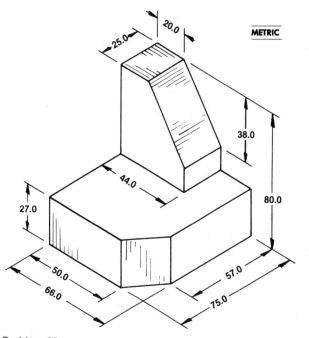

Problem 20.

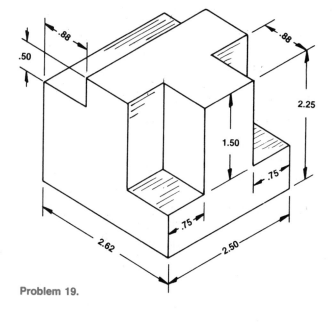

Problem 19.

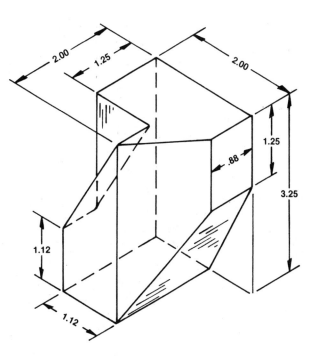

Problem 21.

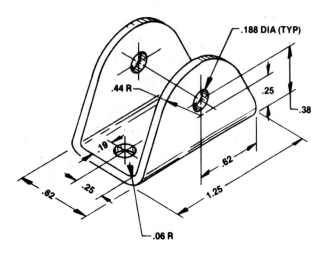

.188 DIA (TYP)

.44 R

.25

.38

.19

.62

1.25

.62

.25

.06 R

Problem 22.

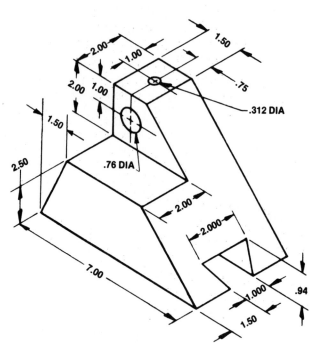

2.00

1.00

1.50

.75

2.00

1.00

.312 DIA

1.50

.76 DIA

2.50

2.00

2.000

7.00

1.000

.94

1.50

Problem 24.

METRIC

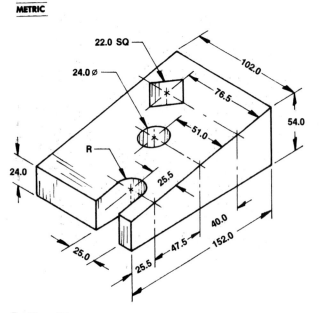

22.0 SQ

24.0 Ø

102.0

76.5

54.0

51.0

R

25.5

24.0

40.0

152.0

25.0

47.5

25.5

Problem 23.

2.50 R

.38

.75

2.12

.88

.50

.75

.50 DIA

1.62

Problem 25.

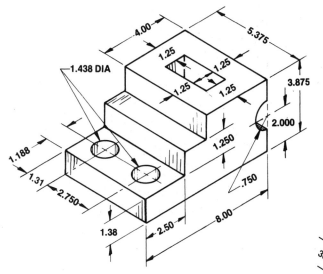

Problem 26.

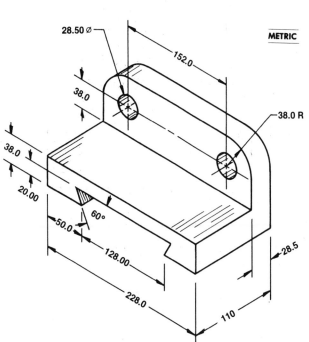

Problem 27.

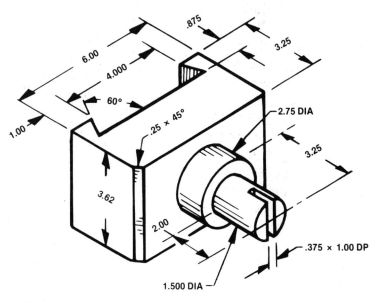

Problem 28.

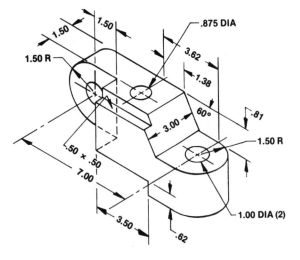

Problem 29.

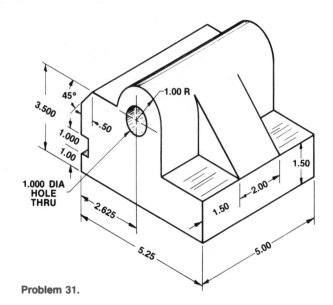

Problem 31.

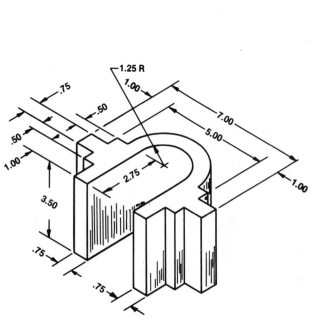

Problem 30.

METRIC

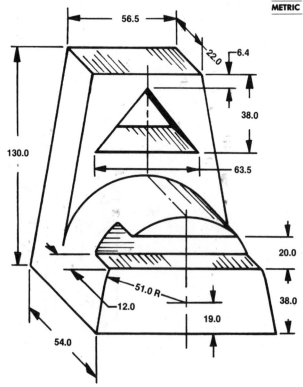

Problem 32.

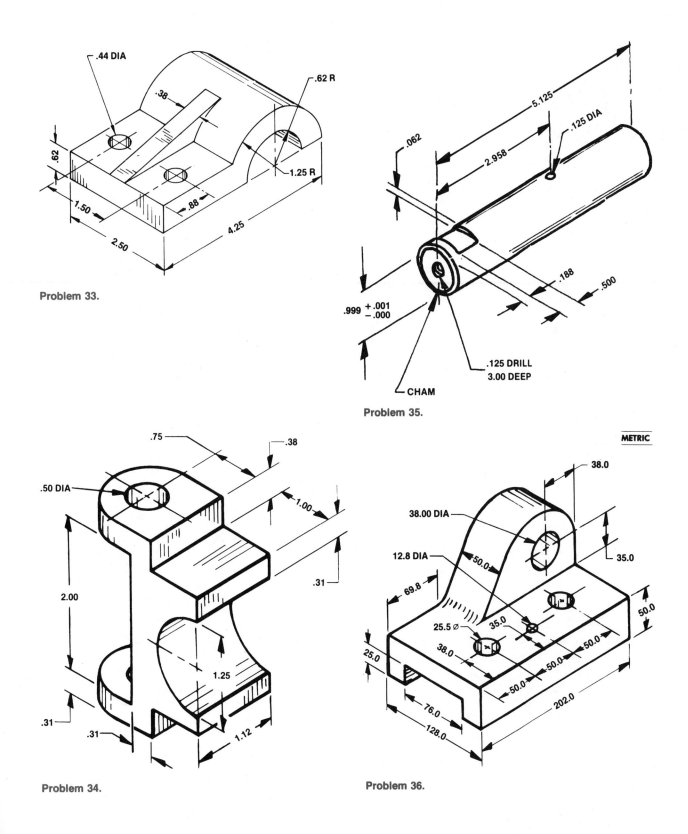

Problem 33.

.44 DIA

.62 R

.38

1.25 R

.62

1.50

.88

2.50

4.25

Problem 35.

.062

5.125

2.958

.125 DIA

.188

.500

.999 +.001 −.000

.125 DRILL
3.00 DEEP

CHAM

Problem 34.

.75

.38

.50 DIA

1.00

2.00

.31

1.25

.31

.31

1.12

Problem 36.

METRIC

38.0

38.00 DIA

35.0

12.8 DIA

50.0

69.8

50.0

25.5 ⌀

35.0

38.0

50.0

50.0

25.0

50.0

76.0

50.0

128.0

202.0

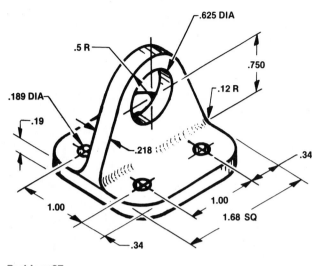

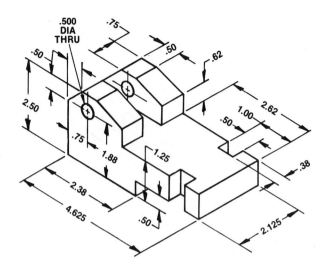

Problem 37.

Problem 40.

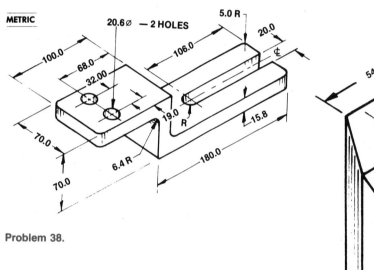

Problem 38.

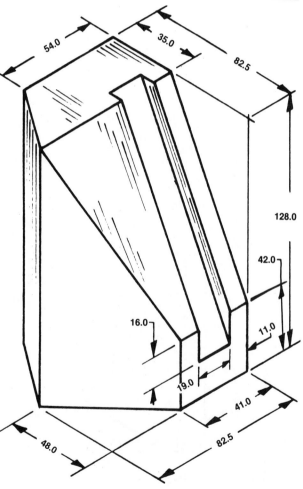

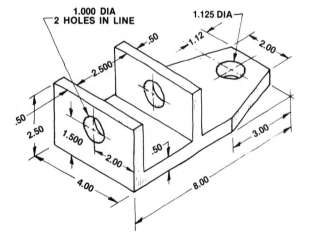

Problem 39.

Problem 41.

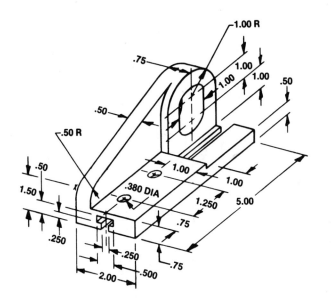

Problem 42.

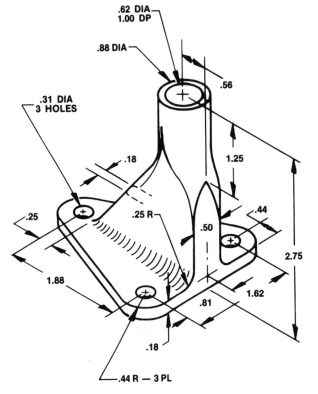

Problem 44.

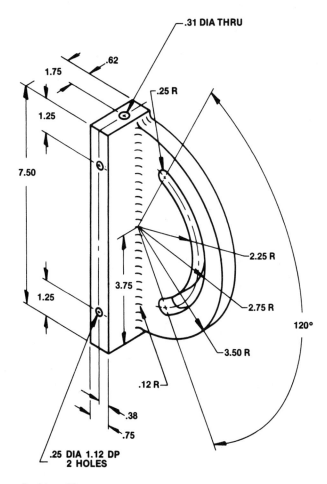

Problem 43.

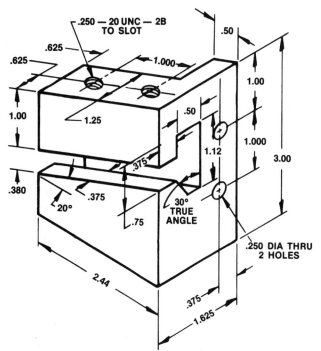

Problem 45.

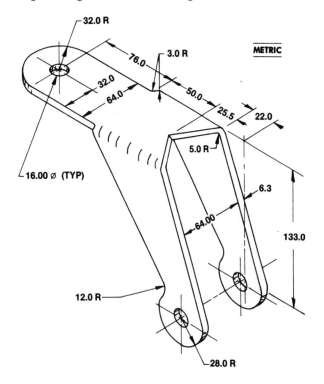

Problem 46.

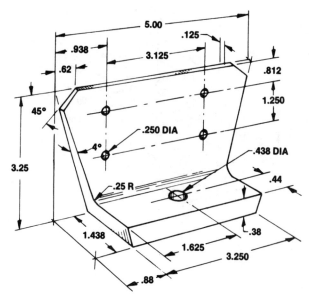

Problem 47.

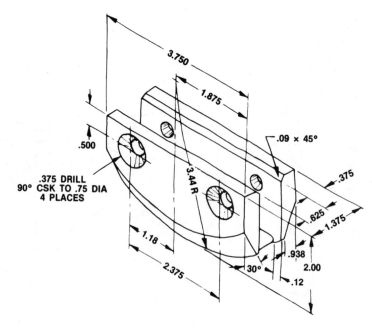

Problem 48.

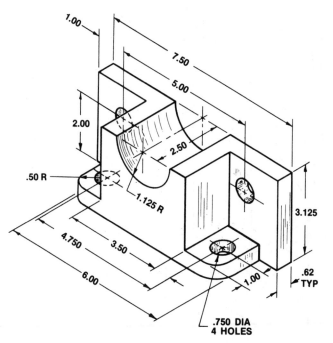

Problem 49.

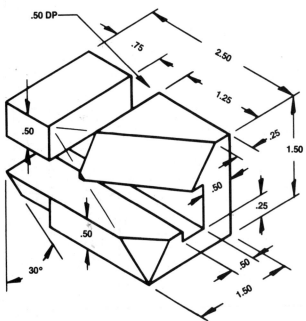

Problem 51.

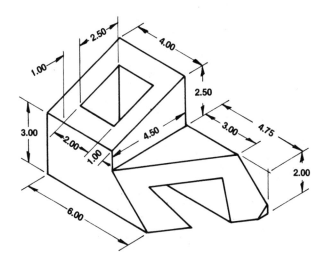

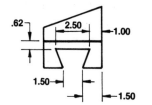

Problem 50.

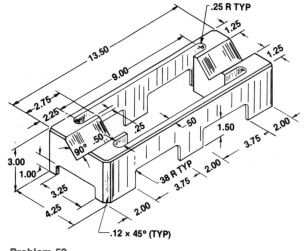

Problem 52.

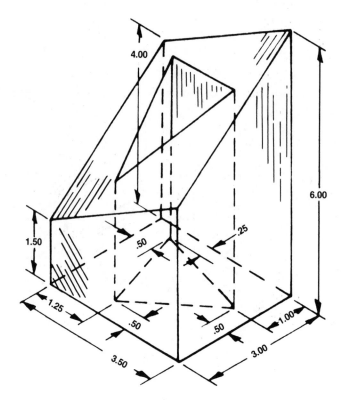

Problem 53.

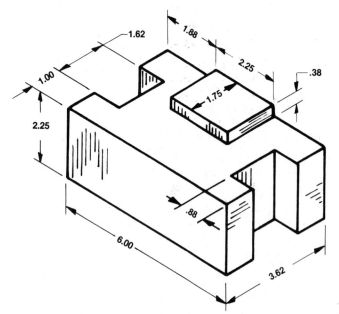

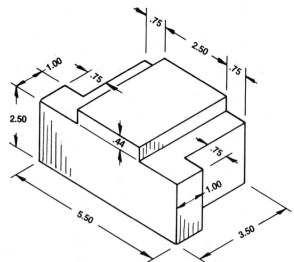

Problem 54
Produce a multiview drawing of the steel block shown. Select a convenient scale and include only essential views.

Problem 55
Make a multiview drawing of the steel block shown. Select appropriate scale and supply missing sizes if required. Indicate machining tolerance of ± 0.002.

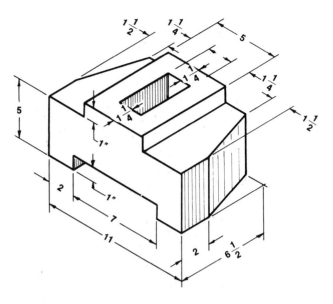

Problem 56
Make a multiview drawing of the steel block shown. Select suitable scale and select appropriate tolerance and roughness finish. Change all sizes to mm.

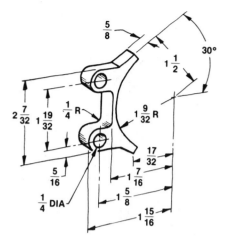

Problem 58
Make a multiview drawing of the guide shown. Change all sizes to metric values. Show appropriate tolerance and surface finish.

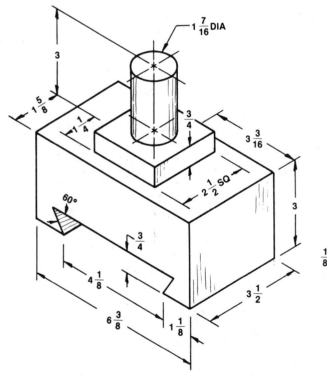

Problem 57
Make a multiview drawing of the steel block shown. Select appropriate scale. Change all sizes to two-place decimals. Show suitable tolerance and surface finish.

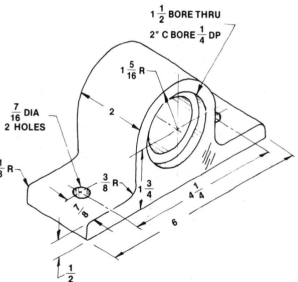

Problem 59
Produce a multiview drawing of the guide shown. Change all sizes to two-place decimals. Choose suitable size radii for fillets and rounds. Specify tolerance and surface finish.

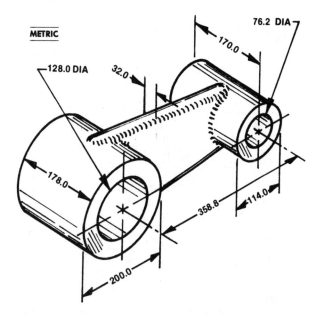

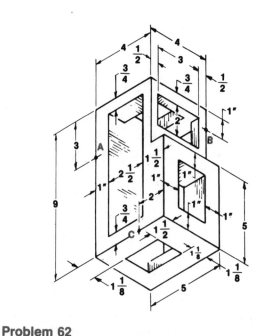

Problem 60

Make a multiview drawing of the steel link shown. Show suitable size radii for fillets and rounds. Designate appropriate tolerance and surface finish. All dimensions are in mm. Use suitable scale.

Problem 62

Make a three-view drawing of the steel block showing portion in front of the plane through points *A, B* and *C* removed. Change all sizes to mm. Assign appropriate machining tolerances and surface finish.

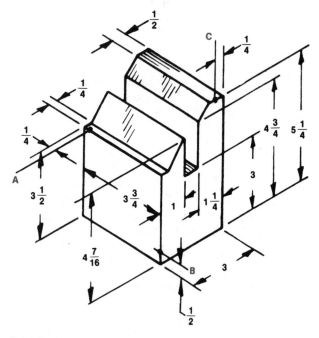

Problem 61

Draw three views of the given steel block showing the portion in front of the plane through points *A, B,* and *C* as having been removed. Select any convenient scale. Change all sizes to two-place decimals. Assign appropriate machining tolerances and surface finish.

TABLE I

Case	Nominal shaft and hole size		Class of fit
1	1.250	DIA	RC 2
2	1.938	DIA	RC 4
3	1.438	DIA	LC 3
4	4.688	DIA	LC 5
5	2.250	DIA	LT 7
6	1.875	DIA	LN 2
7	.625	DIA	LN 3
8	.188	DIA	FN 2
9	1.375	DIA	FN 3
10	2.688	DIA	FN 4

Problem 63

Using the tables in the Appendix, determine the shaft and bore limits for each of the cases in Table I. Make sketches similar to Fig. 6-75 and place the dimensions in proper locations.

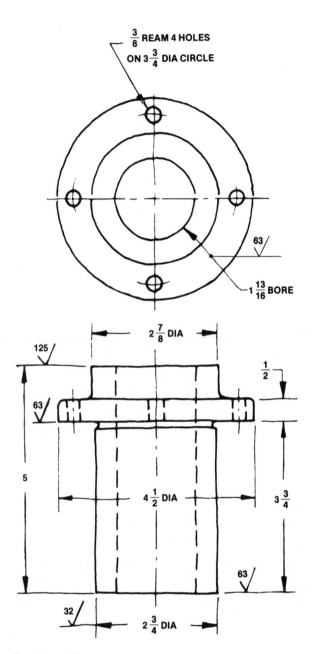

$\frac{3}{8}$ REAM 4 HOLES
ON $3\frac{3}{4}$ DIA CIRCLE

63

$1\frac{13}{16}$ BORE

$2\frac{7}{8}$ DIA

125

63

$\frac{1}{2}$

5

$4\frac{1}{2}$ DIA

$3\frac{3}{4}$

63

32

$2\frac{3}{4}$ DIA

Problem 64
Make a two-view detail drawing of the cast steel flanged hub. Convert all sizes to metric using one-place decimals for cast surfaces and two-place decimals for machined surfaces. Assign tolerances for all machining sizes from Fig. 6-7, Unit 6. Show by symbols that the bore is perpendicular to the machined face of the flange to within .001" and the 2¾" O.D. is concentric with the bore to within .005" TIR. Make the ⅜ reamed holes at true position with respect to the finished surface of the flange and the 2¾" DIA cylinder to within .005 DIA at maximum material condition.

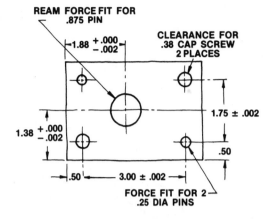

REAM FORCE FIT FOR
.875 PIN

1.88 $\begin{smallmatrix}+.000\\-.002\end{smallmatrix}$

CLEARANCE FOR
.38 CAP SCREW
2 PLACES

1.75 ± .002

1.38 $\begin{smallmatrix}+.000\\-.002\end{smallmatrix}$

.50

.50   3.00 ± .002

FORCE FIT FOR 2
.25 DIA PINS

Problem 65
Make a two-view drawing of the 0.50" x 2.75" x 4.00" steel jig plate shown. Determine hole diameters and tolerances. Use limit dimensions for all sizes.

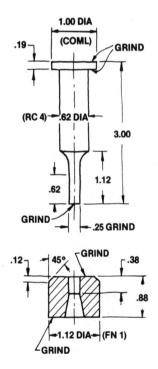

1.00 DIA

(COML)

.19

GRIND

(RC 4) .62 DIA

3.00

.62

1.12

GRIND

.25 GRIND

.12   45°   GRIND   .38

.88

1.12 DIA   (FN 1)

GRIND

Problem 66
Make a complete drawing of the piercing punch and die shown. Determine proper limits for fits specified. (See Appendix.) Assign tolerances to all sizes requiring machining. Use maximum tolerance range in each case. Provide a 0.006" clearance between punch and die.

Problems 67-86 Sectional Drawings

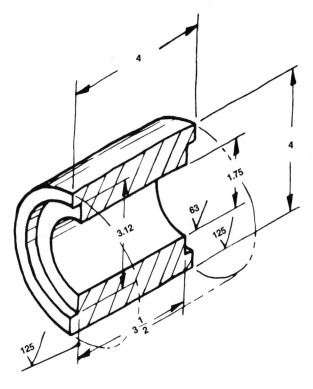

Problem 67
Make a two-view dimensioned drawing of the idler roll shown. Make one view in full section. Material: steel. Change all sizes to one- and two-place metric dimensions.

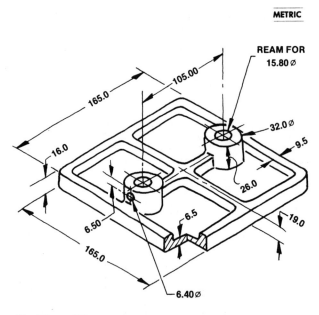

METRIC

Problem 69
Complete the design of the motor bracket shown. Draw the necessary views and sections with dimensions including any omitted from the sketch. Select tolerances for machined surfaces and indicate surface finishes. Make reamed holes perpendicular to bottom surface and cylindrical to within .05 mm. All dimensions in mm. Material: gray cast iron.

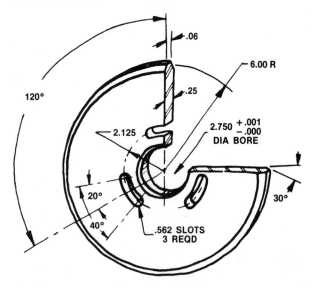

Problem 68
Construct two views of the steel sprocket blank complete with dimensions. Show one view as a half section.

Problem 70
Make a dimensioned drawing of the shear hub shown with one view in full section. Indicate material and provide keyway for a standard square key for a 1.1875 DIA bore. Select tolerances for machined surfaces. Make a bore cylindrical to within .001″ and parallel to shear pin hole to within .002″.

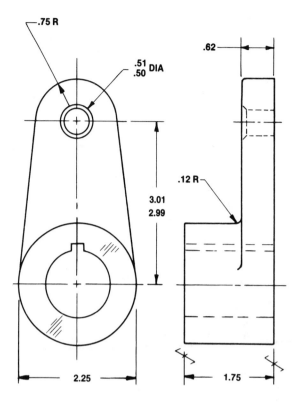

Problem 71
Make a dimensioned drawing of the casting shown. Determine the number of views and necessary section. Material: gray iron. Select machining tolerances and make inside DIA perpendicular to bottom surface to within .002″ and cylindrical to within .003″. Make bottom surface flat to within .002″, and establish the two .312″ DIA holes at true position within .002″ DIA.

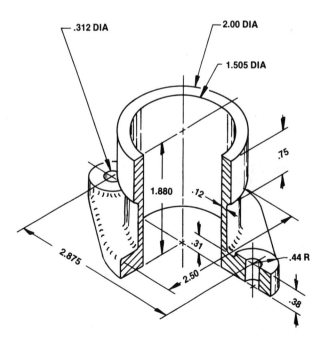

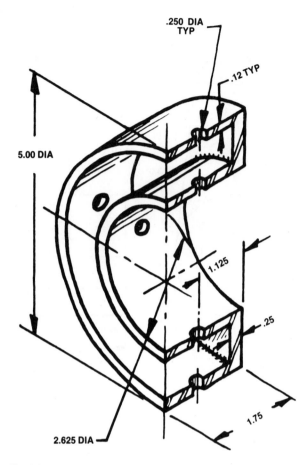

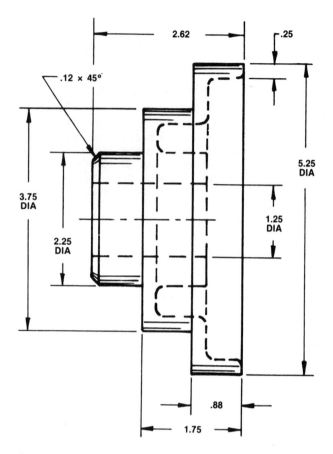

Problem 72
Construct the required views of the ring jacket shown. Make one view a full section. Material: plastic. Include all dimensions.

Problem 74
Draw the idler pulley in two views, one a half-section, with all necessary dimensions and specifications. Assign tolerances as needed. Make all O.D. surfaces concentric with bore to within .003″.

Problem 73
Prepare a two-view dimensioned drawing of the caster wheel with one view in a half section. Indicate all finished surfaces and assign tolerances on dimensions.

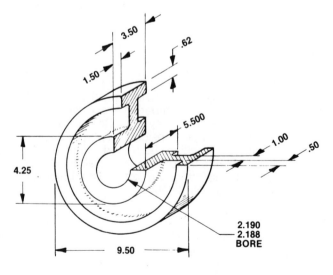

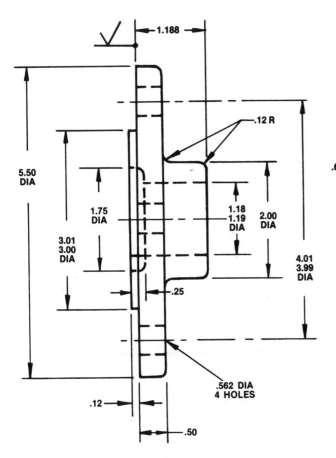

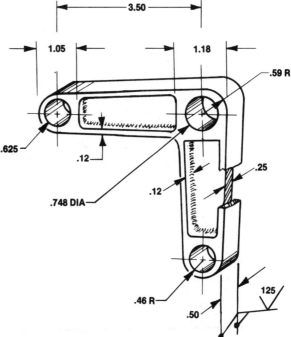

Problem 77
Construct the necessary views of the bellcrank. Include all dimensions and show a revolved section of one arm. Material: aluminum. Make all holes parallel to within .001″.

Problem 75
Produce a two-view drawing completely dimensioned of the flanged coupling shown. Indicate surface roughness for the finished surface shown. Assign an appropriate material specification and feature controls.

Problem 76
Draw two views of the eccentric shown so that the side view is in full section. Show the spoke elliptical in cross section. Provide standard keyway and indicate material. Make bore perpendicular to finished face of hub to within .001″ and flat surfaces of eccentric parallel to within .002″.

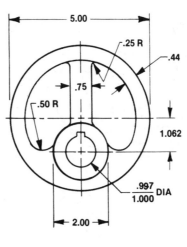

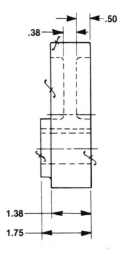

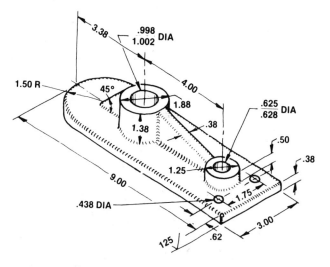

Problem 78

Draw the necessary views of the base shown. Use as many sectional views as are needed to indicate the construction completely. Material: cast iron. Make bottom surface flat to within .002″ and hub bore perpendicular to bottom to within .003″. Make .625″ DIA hole parallel to hub bore to within .001″.

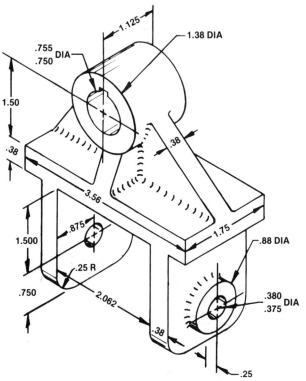

Problem 80

Make a dimensioned drawing with the necessary sections of the rocker. Material: Meehanite, Grade GB. Specify design data to suit, and assign appropriate feature controls so that .750″ bore is perpendicular to .375″ DIA holes and to vertical legs parallel to within .005″.

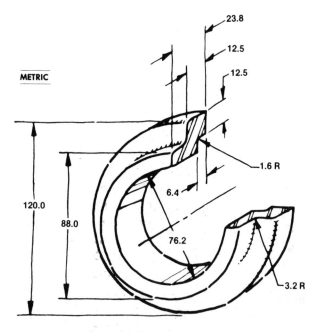

Problem 79

Make a two-view dimensioned drawing of the armature hub shown. Make one view a full section. Material: Meehanite, Grade GA. All dimensions in mm.

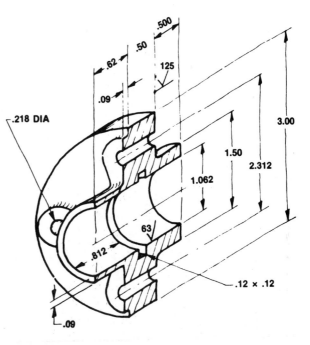

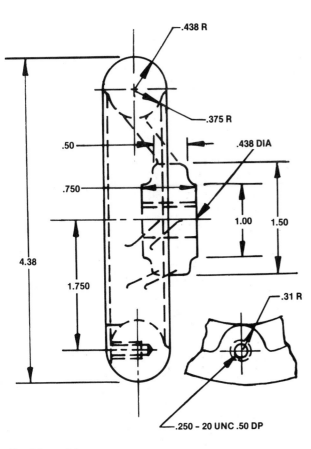

Problem 81
Draw the necessary views including appropriate section of the three-spoke handwheel. Supply design information for features not indicated. Show surface finishes and all dimensions. Material: brass.

Problem 82
Draw the necessary views complete with sizes of the cylindrical shaft housing shown. Assign appropriate feature controls. Make one view a half section. Material: steel casting, SAE 030. Choose suitable fillet radii. Make the .12″ annular groove concentric with bore to within .005″ and .218″ holes at true position to within .002″.

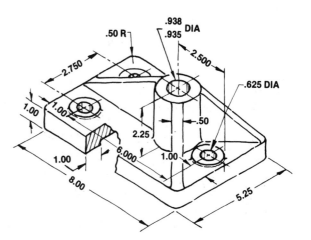

Problem 83
Draw the necessary views including appropriate sections of the base bracket. Supply all missing dimensions and indicate surfaces to be machined. Assign appropriate feature controls. Material: cast iron.

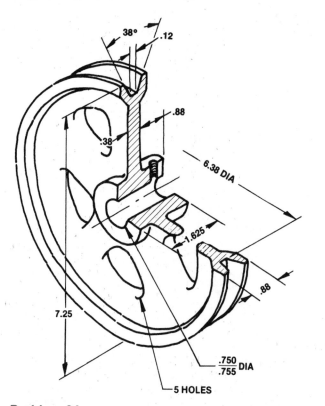

METRIC

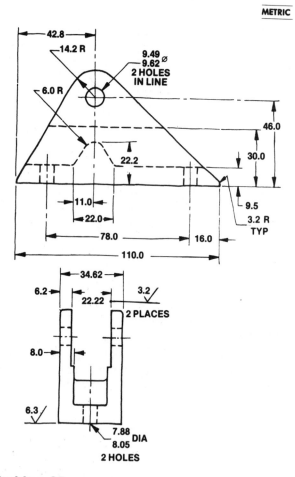

Problem 84

Make the necessary views with dimensions of the V-belt sheave shown. Provide suitable sectional views to show cross-sectional shape details. Include five lightening holes of appropriate size in the design. Supply any omitted design sizes. Material: cast aluminum.

Problem 85

Construct the necessary views of the cast bracket making one view a full section. Show all sizes. Material: steel, Spec. SAE 050-080. Make bottom surface flat to within .12 mm and two holes parallel to bottom within .12 mm. All dimensions in mm.

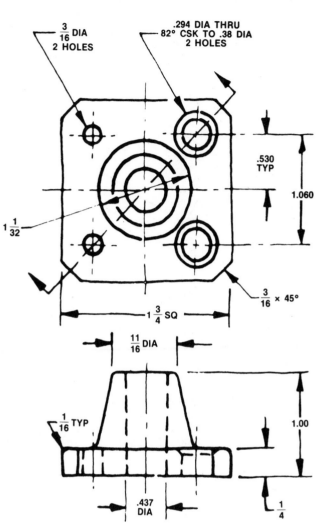

Problem 86

Construct the required views with complete sizes of the filter connector. Assign appropriate feature controls. Material: brass, Spec. QQ-B-626, Comp. 22. Change sizes to appropriate decimals.

Problems 87-104 Auxiliary Drawings

Problem 87

Draw the given front and right side views and add an auxiliary view of problems 1-12 as assigned.

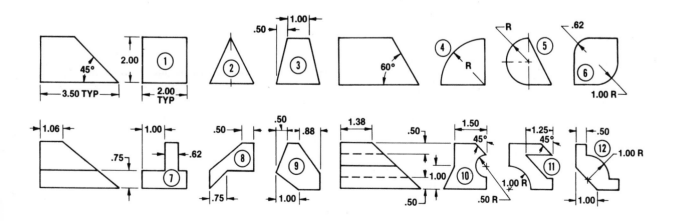

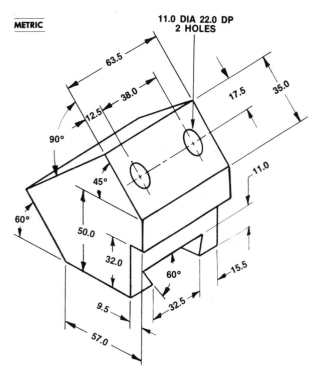

METRIC

Problem 88
Produce a drawing of this sliding gage block complete with the necessary auxiliary views and dimensions. All dimensions are in mm. All surfaces are to be finish machined. Material: tool steel.

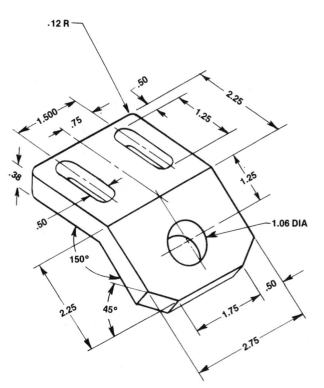

Problem 89
Draw top, front and auxiliary views of the steel adjustable mounting. Show all dimensions.

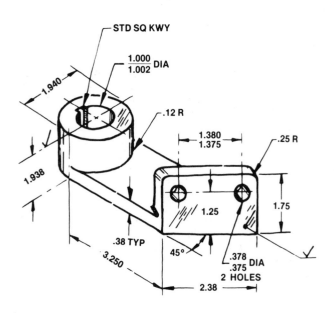

Problem 90
Prepare a dimensioned drawing showing the necessary views of the shaft hanger.

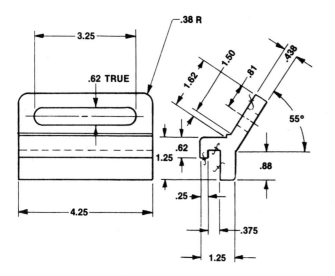

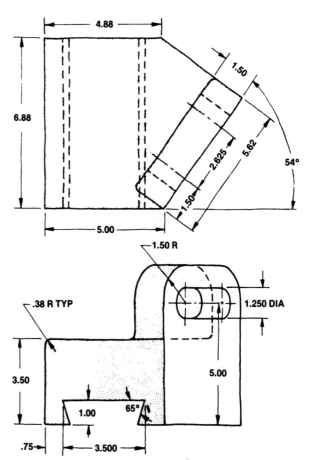

Problem 91
Prepare a completely dimensioned drawing of the sliding support. Show an auxiliary view of the inclined finished surface. Material: brass. Fillets and rounds are to be .12″ R.

Problem 93
Draw the top and front views of the cross slide bracket shown. Add an auxiliary view of the inclined face. Dimension completely. Material: cast steel, Spec. SAE 030.

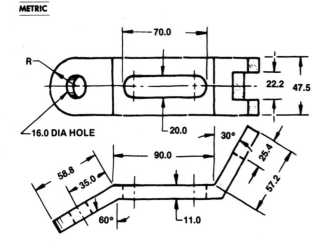

Problem 92
Make a complete, dimensioned drawing of the locating strap. All dimensions in mm.

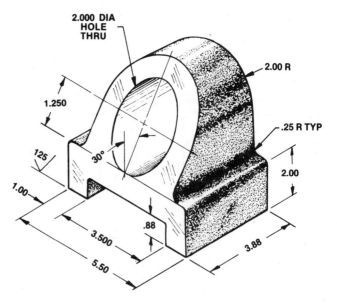

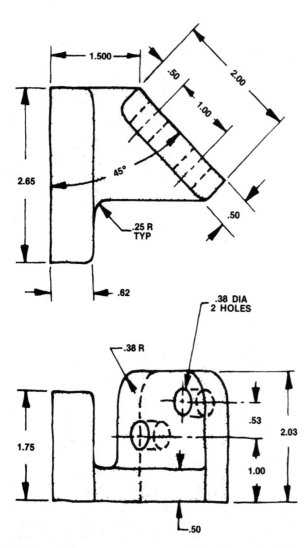

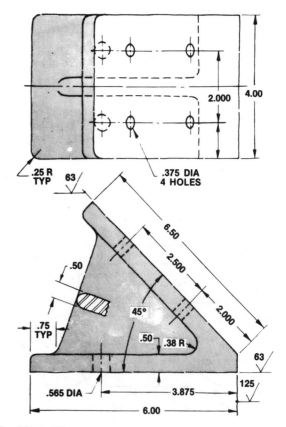

Problem 95
Make a dimensioned drawing of the rod guide showing the necessary principal and auxiliary views. Make the hole parallel to the base surface.

Problem 94
Prepare the necessary principal and auxiliary views of the angle bracket. Show dimensions.

Problem 96
Construct the necessary views of the angle support shown. Dimension properly. Material: Meehanite, Grade GB.

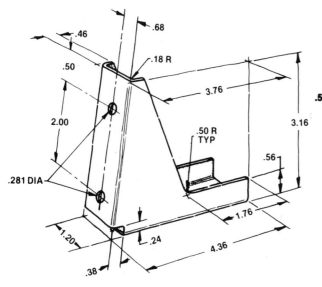

Problem 97
Prepare the necessary views with dimensions of the pulley support. Material: 2020-T4 clad aluminum, .090 thick.

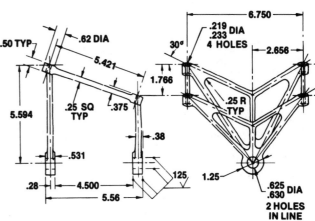

Problem 99
Make a complete dimensioned drawing of the bracket shown. Material: magnesium, Spec. QQ-M-56, Comp. A Z 63, Temp. T6.

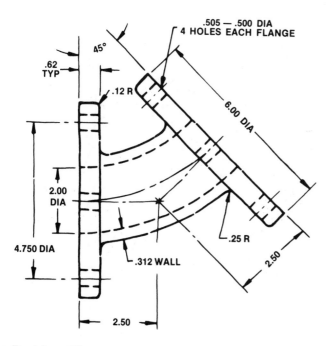

Problem 98
Make a full sectional view and a complete auxiliary view of the tilted flange surface of the cast iron elbow. Show all dimensions and specify a surface roughness for the flange surfaces.

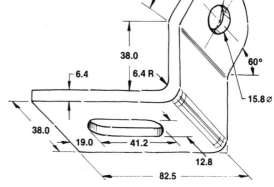

Problem 100
Make a complete dimensioned drawing of the strap showing an auxiliary view of the sloping surface. Material: SAE 1010 steel. All dimensions in mm.

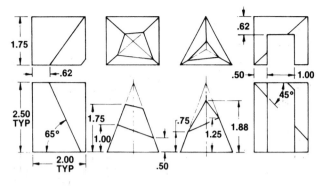

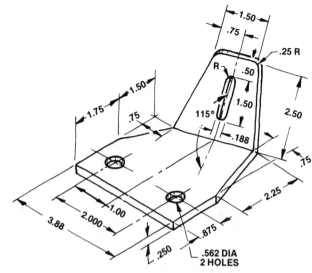

Problem 101

Draw the necessary views including a secondary auxiliary of the problems shown as assigned. Choose an appropriate scale.

Problem 103

Make a dimensioned drawing of the angle mount including an auxiliary view showing the tilted part in its true shape. Material: SAE 1020 HR steel.

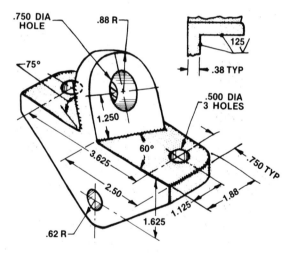

Problem 102

Prepare the necessary views including a secondary auxiliary view of the mounting shown. Material: Zamake, No. 8.

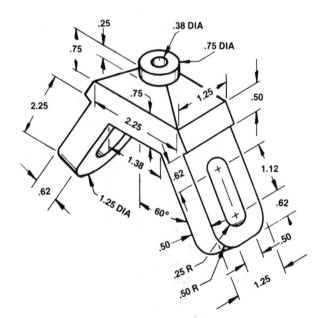

Problem 104

Make a complete dimensioned drawing of the governor mounting shown. Include auxiliary view of the angular legs; supply any design data to suit. Material: bronze.

Problems 105-116 Geometric Tolerancing Problems

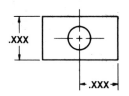

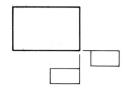

Problem 105
Using geometric tolerance symbols, indicate dimensions to be basic by two methods.

Problem 108
Using geometric tolerance symbols, indicate right surface to be Datum *A,* bottom surface to be Datum *B.*

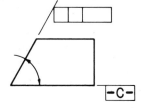

Problem 106
Surface indicated to be at an angle specified within .003″ in relation to surface *C.* Use geometric tolerance symbols.

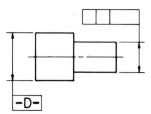

Problem 109
Feature indicated to be concentric to Datum *D* within .002″ dia. Use geometric tolerance symbols.

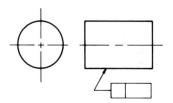

Problem 107
Feature to be cylindrical within a .003″ wide tolerance zone. Use geometrical tolerance symbols.

Problem 110
Surface indicated to be flat within .002″ total. Use geometric tolerance symbols.

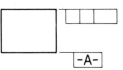

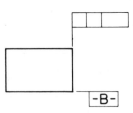

Problem 111
Surface indicated to be parallel to Surface *A* within .002″ total. Use geometric tolerance symbols.

Problem 114
Surface indicated to be perpendicular to Surface *B* within .005″ total. Use geometric tolerance symbols.

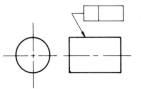

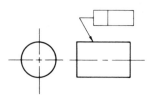

Problem 112
Surface indicated to be round within .002″ wide tolerance zone. Use geometric tolerance symbols.

Problem 115
Feature to be straight within .003″. Use geometric tolerance symbols.

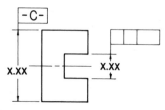

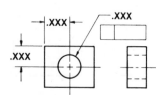
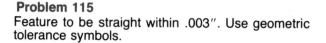

Problem 113
Feature indicated to be symmetrical with Datum *C* within .005″ regardless of feature size. Use geometrical tolerance symbols.

Problem 116
Hole to be at true position within .005″ dia at maximum material condition. Use geometric tolerance symbols.

SECTION III

Technical Illustrations

UNIT 8

Pictorial Drawings

For many years industry relied entirely on conventional multiview drawings to convey ideas and to provide the essential information for manufacturing purposes. Many multiview drawings are complex and require time and special training to interpret. During recent years manufacturers, in order to reduce costly print-reading errors and to clarify engineering multiview drawings, have begun to use pictorial drawings, *often called* graphic illustrations, *more often.*

8.1 Value of Pictorial Drawings

Pictorial drawings have been found to be of particular value to the product designer when explaining design ideas and to the process engineer and tool designer in helping to better visualize the requirements for tooling and assembly procedures. See Fig. 8-1. Drawings of this kind are also being used by quality control and inspection departments

Fig. 8-1. Better visualization of objects is often achieved with pictorial drawings than with orthographic views.

to facilitate checking and inspection. They have become invaluable to shop supervisors and new workers in visualizing with greater ease the sequence of assembly and installation of parts. Graphic illustrations are proving especially helpful in purchasing, advertising, and dealer training as well as for parts catalogs, maintenance and installation manuals, and service bulletins. Extensive use is also made of them in architectural and structural drafting.

Pictorial drawings are classified as axonometric, oblique, and perspective. There are three types of axonometric drawings, known as isometric, dimetric, and trimetric.

8.2 Isometric Drawings

An *isometric drawing* is constructed by using three axes, one of which is vertical, and the other two drawn to the right and left at an angle of 30° to the horizontal. See Fig. 8-2.

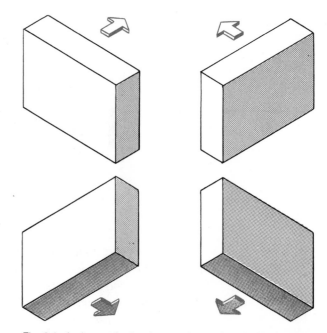

Fig. 8-3. An isometric drawing can be produced with the object placed in any one of these positions.

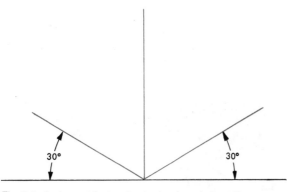

Fig. 8-2. An isometric drawing is developed about three axes.

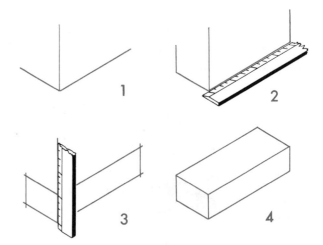

Fig. 8-4. The basic steps in laying out an isometric drawing.

The object to be illustrated can be rotated so that it is tilted either to the right or left with either the top or bottom visible. See Fig. 8-3. The position into which the object is rotated depends entirely upon which is the most advantageous side to show.

The actual width, height, and depth of the object are measured on the three axes and each surface is completed by drawing the necessary lines parallel to the axes. See Fig. 8-4. Hidden lines, as a rule, are omitted on an isometric drawing unless they are absolutely essential for shape description.

8.2.1 Isometric by the box construction method. In making isometric drawings of irregularly shaped objects, the process is often simplified if the box construction method is used. Rectangular or square boxes are drawn having

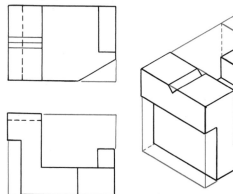

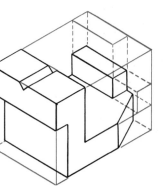

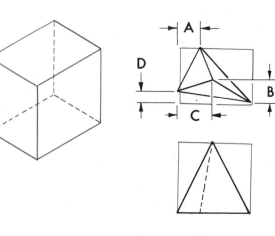

Fig. 8-5. The box construction method often simplifies drawing isometrics of irregularly shaped objects.

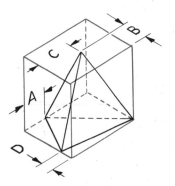

Fig. 8-7. Non-isometric lines are easily located by the box construction method.

sides that coincide with the main faces of the object. The boxes are made with light construction lines, and the irregular features of the object are then drawn within the framework of the boxes. See Fig. 8-5.

8.2.2 Non-isometric lines. When an object has sloping lines that do not run parallel to the isometric axes, the lines are called non-isometric lines. See Fig. 8-6. Since lines of this kind will not appear in their true lengths, they cannot be measured directly, as isometric lines are. To draw non-isometric lines, first locate the ends of the lines.

These locations may be found by the box construction method, that is, the extreme points are located on the regular isometric lines. The slanted lines are then drawn connecting the designated points. See Fig. 8-7.

8.2.3 Angles in isometric. Angles will normally appear in their true sizes only when they lie in surfaces that are parallel to the planes of projection.

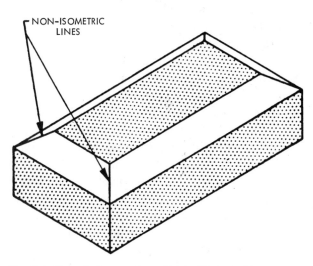

Fig. 8-6. Lines that are not parallel to the isometric axes are called non-isometric.

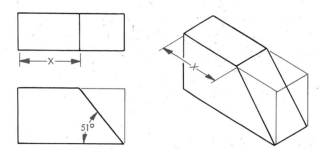

Fig. 8-8. Angles in isometric are drawn by means of coordinate points.

Consequently, they cannot be shown in their true sizes in an isometric drawing. Angles in isometric are laid off by coordinates drawn parallel to the isometric axes. An exact orthographic view drawn to the same scale as the isometric drawing is prepared first. Points locating the angular lines are then transferred from this view to the isometric lines and the points connected to show the lines isometrically. See Fig. 8-8.

8.2.4 Isometric circles. An isometric circle is drawn by the four-center system, which is sufficiently accurate for most work. Fig. 8-9 illustrates

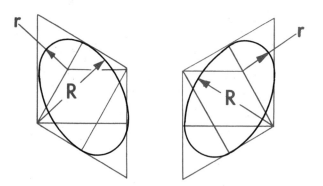

Fig. 8-10. Drawing isometric circles in different positions.

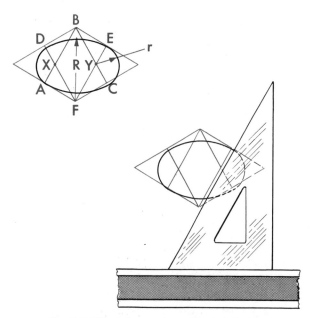

Fig. 8-9. This is how an isometric circle is drawn.

this procedure. First an isometric square of the required size is laid out and the center of each side of the square is located. From the midpoints of these sides, the lines *AB, BC, DF,* and *EF* are constructed. The same slanted lines can be made by means of a 60° triangle. Then with *r* as a radius and *X* and *Y* as centers, arcs *EC* and *AD* are drawn. With *R* as a radius and *B* and *F* as centers, arcs *DE* and *AC* are drawn.

The procedure for drawing circles located in different positions is shown in Fig. 8-10.

Very often a portion of the rear side of a hole must be shown. To show this, lay off the thickness

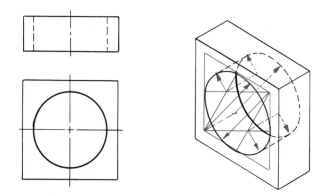

Fig. 8-11. On some isometric drawings a portion of the back side of a hole must be shown.

of the piece or depth of the hole and find the center for the required radius as shown in Fig. 8-11.

8.2.5 Isometric arcs. The same four-center method described for circles is used to draw isometric arcs. However, it is not necessary to draw the entire construction as for a circle. Only the radius needed for drawing the arc is laid out. Fig. 8-12 illustrates how isometric arcs are made.

8.2.6 Irregular curves in isometric. If an object to be drawn in isometric contains an irregular curve, the true shape of the curve is drawn first in an orthographic view as shown in Fig. 8-13. This view must have the same scale as the isometric drawing. The exact shape of the curve is determined by a series of reference lines or coordinates. These coordinates are then drawn in isometric and the points of the curve plotted on

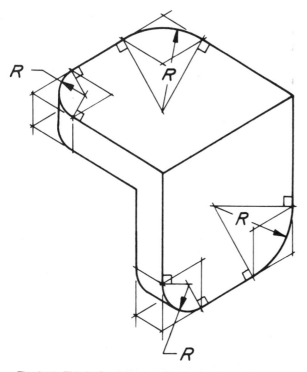

Fig. 8-12. This is the procedure for drawing isometric arcs.

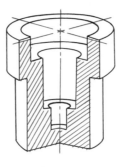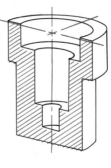

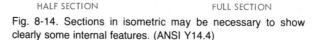

HALF SECTION   FULL SECTION

Fig. 8-14. Sections in isometric may be necessary to show clearly some internal features. (ANSI Y14.4)

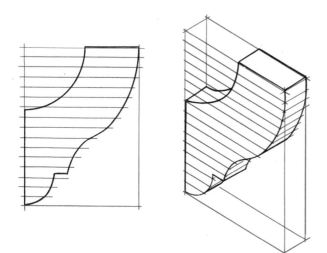

Fig. 8-13. Irregular curves in isometric are produced by a series of reference lines as shown here.

them by transferring lengths measured from the orthographic view.

8.2.7 Isometric sections. Although an isometric drawing is used principally to show the exterior of

an object, occasionally it may be advantageous to illustrate the internal construction of the object in isometric. The cutting plane may be passed in any position, as discussed in Unit 5. Usually it is best to lay out the full shape of the object first and then to remove the required portion. Section lining should be carried out as in multiview drawings. The important consideration is to have the section lines run in the direction that produces the best effect. Fig. 8-14 illustrates two sections in isometric.

8.2.8 Isometric dimensioning. Generally speaking, the rules used for dimensioning multiview drawings apply to isometric drawings. All dimension and extension lines must be parallel to the principal isometric axes. Fig. 8-15 shows how dimension and extension lines should be positioned in the unidirectional and pictorial plane systems of dimensioning pictorial drawings. Note that in the unidirectional system, dimensions read from the bottom of the drawing. In the pictorial plane system, dimensions are perpendicular to the dimension lines and in the same plane as the corresponding extension and dimension lines. Although it is desirable to keep dimensions off the view, this cannot always be done.

8.2.9 Equipment for isometric drawing. To simplify the task of preparing isometric drawings, special isometric graph paper and templates have been developed. Isometric paper consists of grid lines drawn in isometric. The drawing can be made directly on the graph paper or the graph paper can be placed beneath the tracing medium.

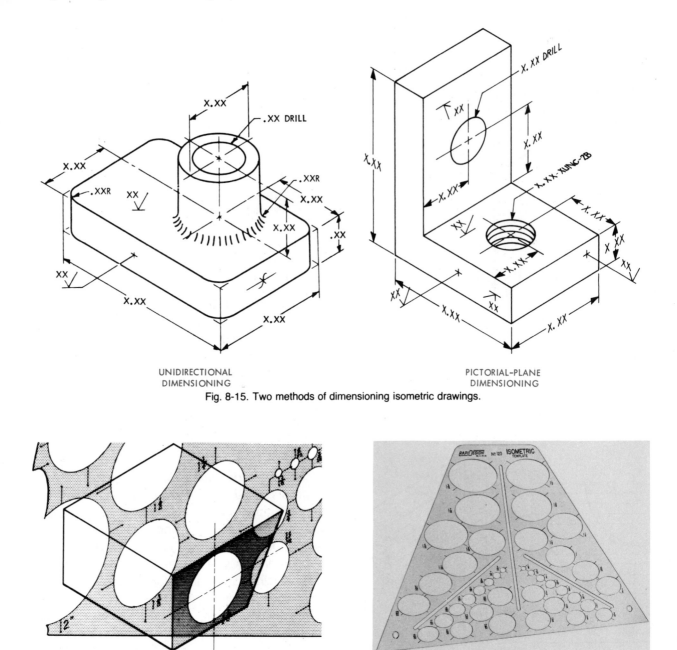

UNIDIRECTIONAL
DIMENSIONING

PICTORIAL-PLANE
DIMENSIONING

Fig. 8-15. Two methods of dimensioning isometric drawings.

Fig. 8-16. Templates of this type speed up the process of preparing isometric drawings.

Fig. 8-16 shows two basic types of isometric templates with which isometric drawings can be produced quickly and accurately.

8.3 Dimetric Projection

An axonometric projection in which two axes make equal angles with a plane of projection while the third axis makes a different angle is called a *dimetric projection.* The edges that are parallel to the first two axes are foreshortened by the same amount, but those edges that are parallel to the third axis are foreshortened to a different value. This necessitates the use of two different scales.

A dimetric drawing is prepared in much the

same way as an isometric. See Fig. 8-17. Its advantage over the isometric is that by using two scale values, the resulting shape has less distortion. The principal limitation of dimetrics is having to use two separate scales for measurement, which is more time-consuming.

The usual practice in preparing a dimetric drawing is to place the object so that one of its main

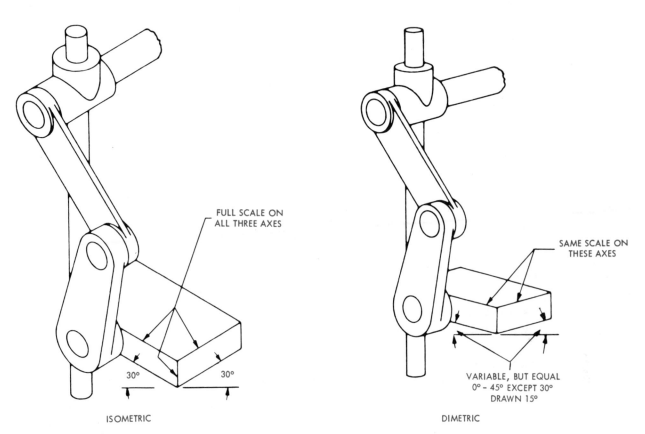

Fig. 8-17. A dimetric projection is similar to an isometric but shows less distortion. (ANSI Y14.4)

Fig. 8-18. The positions and scales illustrated here will be suitable for almost any dimetric drawing.

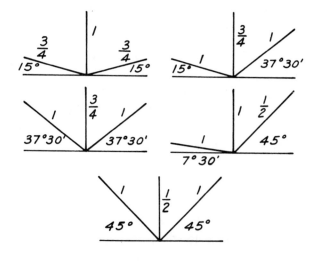

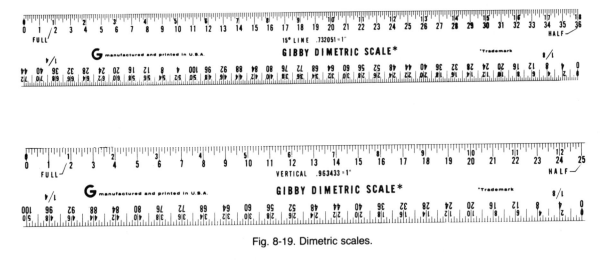

Fig. 8-19. Dimetric scales.

axes is at a 90° angle to the horizontal. Each of the two receding axes can be located at various angles. The scales to be selected may be determined graphically or mathematically, but since true dimetrics are not actually necessary, the usual procedure is to utilize regular scales of assumed ratios. The angles and scales that are shown in Fig. 8-18 generally will be suitable for most dimetric drawings.

Special grid paper is available for making dimetric drawings. Dimetric scales are also available to simplify the task of preparing dimetric drawings. See Fig. 8-19.

8.4 Trimetric Projection

A *trimetric projection* is an axonometric projection in which each of the three axes of an object makes a different angle with the plane of projection. Lines parallel to each of the three axes have different ratios of foreshortening. This means that three different trimetric scales must be used. See Fig. 8-20. The complications resulting from the necessity of three different scales restrict wide use of this form of projection.

8.5 Oblique Projection

An *oblique projection* is one in which the lines of sight are parallel to each other but the projectors are oblique to the plane of projection. The

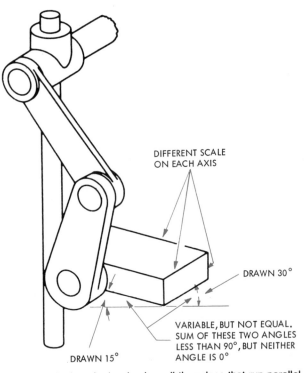

DIFFERENT SCALE ON EACH AXIS

DRAWN 30°

VARIABLE, BUT NOT EQUAL. SUM OF THESE TWO ANGLES LESS THAN 90°, BUT NEITHER ANGLE IS 0°

DRAWN 15°

Fig. 8-20. A trimetric drawing has all the edges that run parallel to the three axes foreshortened to different ratios. (ANSI Y14.4)

principal face is placed parallel to the plane of projection and therefore appears in its true shape.

8.5.1 Types of oblique projections. The three kinds of oblique projections are cavalier, cabinet, and general oblique.

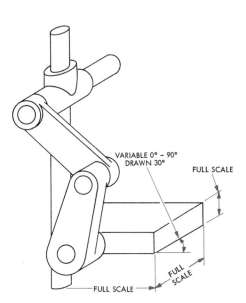

Fig. 8-21. A cavalier drawing. (ANSI Y14.4)

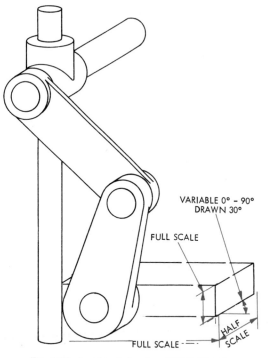

Fig. 8-22. A cabinet drawing. (ANSI Y14.4)

Cavalier projection. When the projectors make any angle from 0° to 90° with the plane of projection and the same scale is used on all axes, the result is a cavalier projection. See Fig. 8-21.

Cabinet projection. An oblique projection in which the receding lines are foreshortened one-half their actual length is called a cabinet projection. The shortening of the receding lines tends to eliminate some of the distortion which is often quite noticeable in cavalier projections. See Fig. 8-22. The receding axes may be drawn at any angle, but they are most commonly made at 30° or 45° to the horizontal.

General oblique projection. A general oblique projection is one in which the projectors make any angle with the plane of projection and the receding axes vary in length from full to one-half scale. For practical purposes the angle of projection is normally kept between 30° and 60°. The angle used will depend on the shape of the object and the effects desired. Thus, a large angle may be selected to obtain a better view of some special surface. The scale on the receding axes will depend on the angle used. The governing principle is to employ a scale that produces that least amount of distortion. See Fig. 8-23.

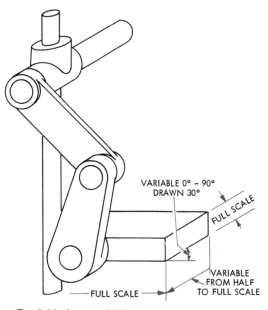

Fig. 8-23. A general oblique drawing. (ANSI Y14.4)

8.5.2 *Position of object.* In making oblique drawings, the object may be placed so that the receding faces assume any number of positions, as

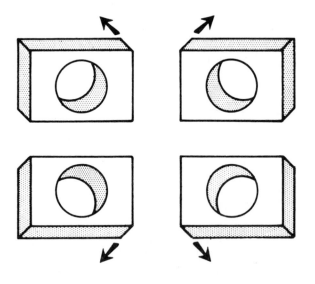

Fig. 8-24. Oblique drawings may be made so the inclined axes assume a variety of positions.

shown in Fig. 8-24. But in every instance the front face, regardless of the position of the inclined axes, remains parallel to the plane of projection.

The principal, or front, face of an oblique drawing should be the view that shows the most essential features of the object. This is especially important for irregular surfaces or those having circles and curves.

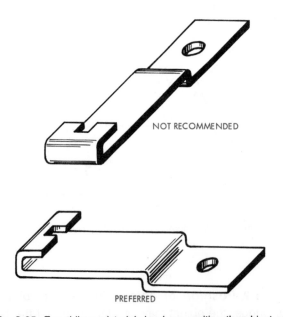

Fig. 8-25. For oblique pictorial drawings position the object so the longest features are parallel to the plane of projection.

Circles and arcs, if drawn on the front face, may be made with a compass since they will appear in their true sizes and shapes. However, if they appear on receding faces, circles and arcs will have to be drawn with an irregular curve or by means of the four-center ellipse method.

The object should also be drawn so the greatest dimensions appear in the front face. Too much distortion results when the long features are placed in receding faces. See Fig. 8-25.

8.5.3 Circles and arcs in oblique projection. When circles and arcs must be made on receding faces, the four-center ellipse system can be used, providing that the receding faces are drawn to the same scale as the principal face. Such a system is applicable in cavalier projections and in some general oblique projections when the receding faces are drawn full scale. If the receding surfaces are foreshortened, then circles and curves may be approximated by using an irregular curve, or by the offset method described in subsection 8.5.4, or with ellipse templates.

The four-center ellipse system for oblique projection will necessitate a flatter parallelogram than in isometric. Accordingly, the perpendicular side bisectors will not intersect in the corners of the parallelogram but will intersect either on the outside or at some point within the parallelogram, depending on the angle of the receding axes. Thus,

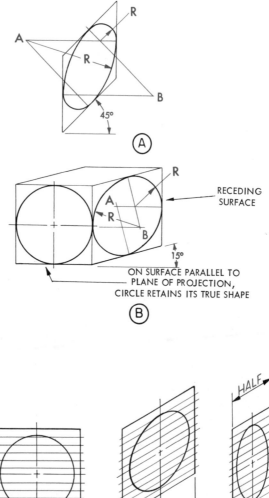

Fig. 8-26. Circles and arcs for oblique projections can be constructed by the four-center ellipse system on receding surfaces.

Fig. 8-27. How to lay off circles, arcs, and irregular curves by the offset measurement system for oblique projection.

if the receding axes are drawn at a 45° angle to the horizontal, the bisectors will intersect on the outside of the parallelogram. See Fig. 8-26*A*. If the angle is less than 45°, as shown in Fig. 8-26*B*, the perpendicular bisectors will intersect within the parallelogram.

To make a circle or curve in an oblique drawing, lay out the parallelogram in the required position and find the perpendicular bisector of each side. The intersection of these perpendicular bisectors will provide the necessary centers for the required radii.

8.5.4 Offset measurements for oblique projection. The offset measuring system is used to draw circles, arcs, and other irregular curves for

oblique projections having foreshortened receding axes. A number of parallel reference lines are first located on a regular orthographic projection as shown in Fig. 8-27, and the curve or circle is located on these lines. The points are then transferred to similar parallel coordinates drawn on the oblique view. The circle or arc is completed with an irregular curve.

8.5.5 Oblique dimensioning. An oblique drawing is dimensioned in much the same way as an isometric drawing. The important factor is to place all dimensions in the planes to which they apply. Dimensions should be kept off the object as much as possible, but they may be placed directly on the object if greater clarity results.

8.6 Perspective Drawings

A *perspective drawing* is one which attempts to present an object as it would appear to the eye or in an actual photo. Perspective drawing is based on the fact that all lines which extend from an observer appear to converge or come together at some distant point. For example, to a person sighting down a long stretch of roadway, light poles and buildings will appear to slope and converge as shown in Fig. 8-28.

In any perspective drawing, the observer assumes that he or she is looking through an imaginary plane of projection called a picture plane or *PP*. See Fig. 8-29. The position of the observer is known as the station point, or *SP*. Lines leading from the observer to the scene are called lines of sight or visual rays. The horizon is an imaginary line in the distance and represents the eye level of the observer. The ground line, or *GL*, is the intersection of the picture plane with the ground plane. The point where all the lines of sight seem to meet on the horizon is referred to as the vanishing point, or *VP*. See Fig. 8-34.

8.6.1 *Types of perspective drawings.* A drawing which has but one vanishing point, that is, one in which two of the principal axes of the object are parallel to the picture plane and the third is at an angle to that plane, is known as a *parallel*, or *one point, perspective.* See Fig. 8-30.

A drawing which has two vanishing points, that is, one where the vertical axis is parallel to the picture plane and the other two axes are inclined to it, is called an *angular,* or *two point perspective.* See Fig. 8-31.

A drawing having three vanishing points, that is, one in which all three principal axes of the object are oblique to the picture plane is called an *oblique,* or *three point, perspective.* See Fig. 8-32.

Fig. 8-28. A perspective drawing represents an object as it actually appears to the eye.

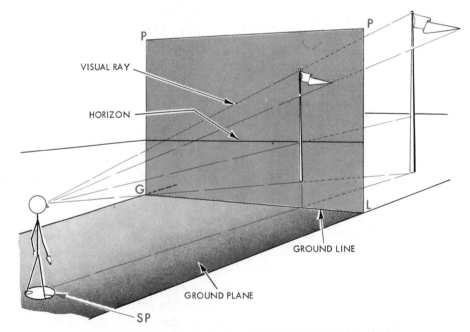

Fig. 8-29. These are the terms used in making a perspective drawing.

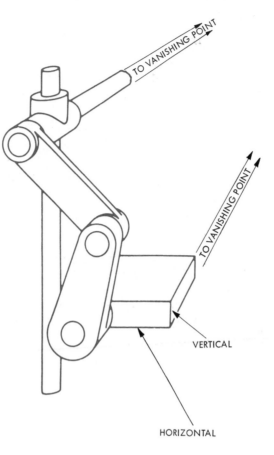

Fig. 8-30. A one point perspective. (ANSI Y14.4)

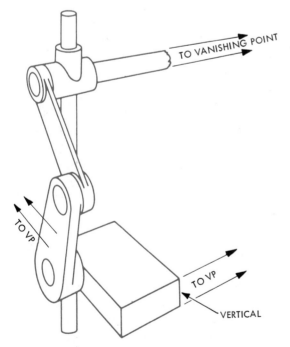

Fig. 8-31. A two point perspective. (ANSI Y14.4)

Fig. 8-32. A three point perspective.

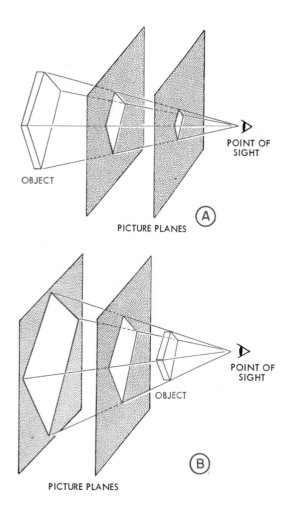

OBJECT

POINT OF SIGHT

PICTURE PLANES Ⓐ

PICTURE PLANES Ⓑ

OBJECT

POINT OF SIGHT

Fig. 8-33. The location of the picture plane affects the size of the perspective drawing.

8.6.2 Location of the picture plane. For most perspective drawings the picture plane is assumed to be between the object and the point of sight. In this position the perspective drawing is smaller than the actual size of the object. As the picture plane is moved further away from the object, the perspective drawing becomes smaller. See Fig. 8-33A.

If the perspective drawing is to be larger than the true size of the object, the object is placed between the observer and the picture plane. See Fig. 8-33B.

8.6.3 Location of the station point. A perspective can be altered to a considerable extent by the location of the point of sight. Thus, by placing the

point of sight above the object, a view of the top will be seen. If the point of sight is below the object, a view of the bottom will be shown. Similarly the point of sight can be stationed to the right or left of the object to reveal either side. See Fig. 8-34. As a rule the point of sight for most small and medium size objects is assumed to be slightly above the horizon. For large objects the horizon is located approximately 5 feet (1.52 m) above the ground.

In preparing a perspective with the least amount of distortion, the station point should be located at a distance so that the cone of visual rays will enclose the entire object at an angle not greater than 30°. See Fig. 8-35. If larger angles are used,

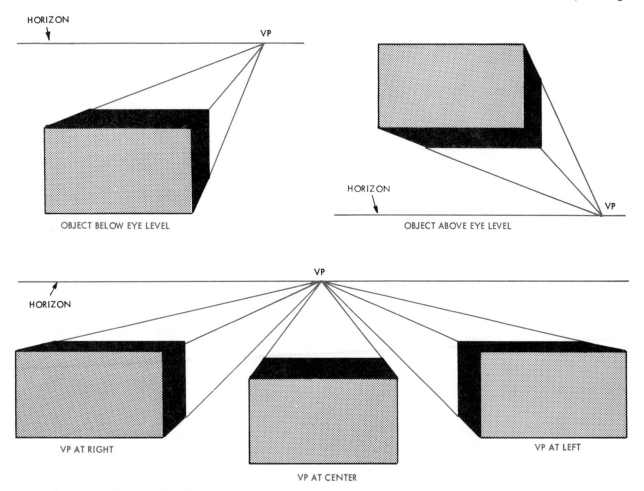

HORIZON

VP

OBJECT BELOW EYE LEVEL

HORIZON

VP

OBJECT ABOVE EYE LEVEL

VP

HORIZON

VP AT RIGHT

VP AT CENTER

VP AT LEFT

Fig. 8-34. The shape of the perspective is determined by the location of the point of sight.

the convergence of horizontal lines will result in shapes that are badly distorted, Fig. 8-36.

8.6.4 *How to draw a parallel or one point perspective.* The steps in making a one point perspective are shown in Fig. 8-37:

1. Draw the ground line *GL* and on it lay out the front view of the object.

2. Locate the picture plane *PP* and on it draw the top view of the object. See subsection 8.6.2.

3. Draw the horizon line at any convenient distance from the ground line.

4. Locate the station point so that the cone of visual rays will enclose the object at an angle not greater than 30°. See subsection 8.6.3.

5. From the station point draw a vertical line to the horizon to provide the vanishing point *VP.*

6. Draw visual ray lines from the station point to all of the top points in the top view.

7. Extend vertical lines downward from the bottom points of the top view as well as from the points where the visual ray lines intersect the picture plane line.

8. Project horizontal lines from the front view to intersect the vertical lines from the top view.

9. Extend lines from the front perspective view to the vanishing point.

The intersections of the horizontal lines from the front view with the vertical lines from the top view and the lines extending from the front perspective view to the vanishing point together will provide the required shape of the one point perspective drawing.

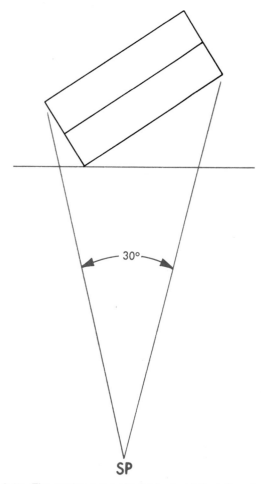

Fig. 8-35. The station point should be located so the visual rays will enclose the object in a cone not greater than 30°.

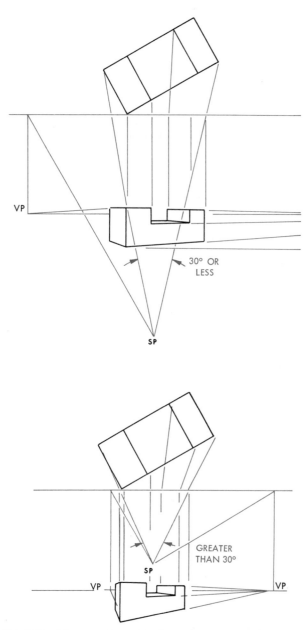

Fig. 8-36. If the cone formed by lines of sight is too great, the drawing will be distorted.

8.6.5 How to draw a two point perspective.
To prepare a two point perspective, proceed as follows: (See Fig. 8-38.)

1. Draw the ground line *GL,* the horizon, and the picture plane *PP.*

2. Draw a front view of the object on the ground line and a top view on the picture plane. Revolve the top view to any convenient angle on the picture plane.

3. Locate the two vanishing points by first drawing two lines from *SP* to *PP* so they are parallel to the edge lines of the top view. Then drop perpendiculars from the picture plane line to the horizon.

4. Draw visual ray lines from *SP* to the essential features of the top view.

5. From the points where the visual ray lines intersect the picture plane line, drop vertical projectors to intersect the lines drawn to the vanishing points.

6. Draw lines from points of intersection until the perspective view is complete.

8.6.6 How to draw a three point perspective.
Fig 8-39 illustrates the steps in preparing a three point perspective. Proceed as follows:

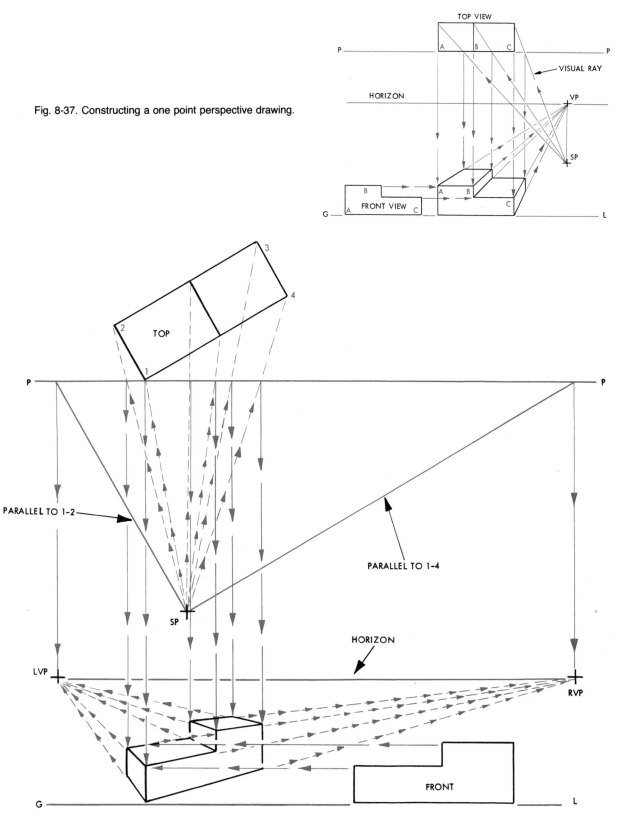

Fig. 8-37. Constructing a one point perspective drawing.

Fig. 8-38. Drawing a two point perspective.

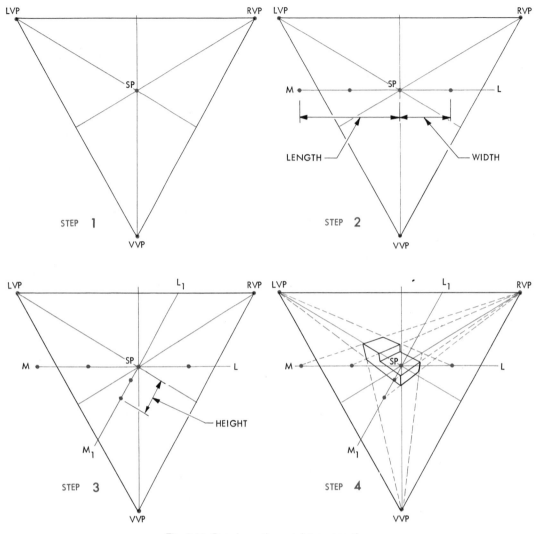

Fig. 8-39. Drawing a three point perspective.

1. Lay out an equilateral triangle and let the corners represent the three vanishing points. Make the triangle as large as can be conveniently drawn on the sheet.

2. From each corner of the triangle construct a perpendicular bisector to the opposite side of the triangle. The point of intersection of the three lines is the required *SP*.

3. Through *SP* draw two measuring lines, one *ML* parallel to *LVP-RVP* and the other M_1L_1 parallel to *RVP-VVP*.

4. On the horizontal measuring line *ML* lay out the actual length of the object from *SP* to the left of *SP* and the true width of the object from *SP* to

the right of *SP*. From these points draw visual rays to the right and left vanishing points respectively.

5. Lay out the height of the object on the measuring line M_1L_1 and draw visual ray lines to the corresponding *VP*.

6. Complete the perspective by drawing the remaining visual ray lines to their respective *VP*s.

8.6.7 Perspective of a circle or arc. If a circle or arc is parallel to the picture plane, it assumes its true shape and can be drawn with a compass. In a one point perspective, when the circle or arc is inclined to the picture plane, its shape may be found as follows:

1. Draw a front view of the circle or arc and di-

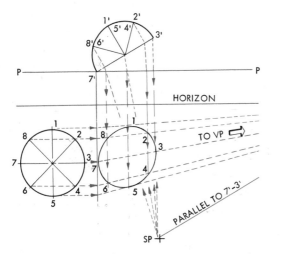

Fig. 8-40. Drawing a circle in a one point perspective.

vide it into any number of equal parts. Refer to Fig. 8-40.

2. Draw horizontal lines through the points and extend them to the vanishing point.

3. Construct a plan view of the circle on the picture plane and lay out the same numbered divisions. From these numbered points draw lines to *SP*. Drop vertical lines from the intersection of the visual ray lines and picture plane. The corresponding intersection of these lines with the vanishing point lines will provide the necessary points for constructing the perspective circle or arc.

Fig. 8-41 illustrates the procedure for drawing a circle in a two point perspective.

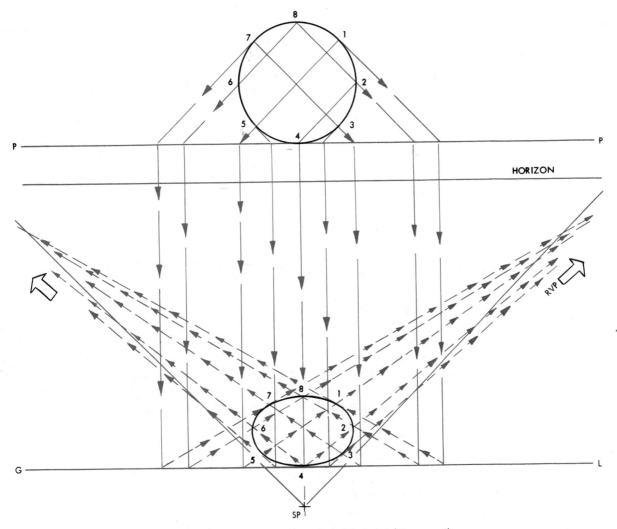

Fig. 8-41. Procedure for drawing a circle in a two point perspective.

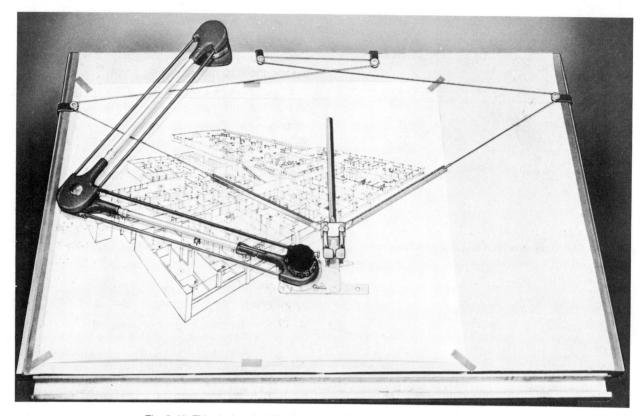

Fig. 8-42. This device simplifies the preparation of perspective drawings.

8.6.8 Preparing perspective drawings with the Perspect-O-Metric.

The Perspect-O-Metic makes it comparatively simple to create complicated perspectives. This device guides the pencil at the proper angle and automatically realigns itself for every position on the drafting table. It provides a special scale which instantly reduces distant portions of the subject to their correct proportions. See Fig. 8-42.

The Perspect-O-Metric consists of an attachment which fits any standard drafting machine. By the addition of a special adapter, it also fits any parallel ruling straightedge. It has three scale arms. The central scale arm is fixed in a position at right angles to the established base line. The left and right scale arms pivot in the plane of the drawing board.

Any movement of the Perspect-O-Metric creates a corresponding angular motion of the floating scale arms. No matter where the Perspect-O-Metric is placed on the drawing board, the scale arms remain oriented to their chosen vanishing points. To prevent the scales from shifting under the pressure of the pencil, left and right brake levers are provided. A pressure of the thumb locks the scale arms in position while the line is being drawn.

Although the Perspect-O-Metric is used to create two point perspective drawings, which covers about 90 percent of ordinary needs, the pulleys can easily be relocated to permit any number of vanishing points.

8.6.9 Klok perspective drawing system.

The Klok perspective drawing system is based on a specially designed drawing board containing a series of graduated scales and vanishing points. See Fig. 8-43. The scales are marked A-1, A-2, B-1, B-2, and so forth. When the letter is followed by 1, it means that the lines running toward the left vanishing point are to be measured on these scales; if

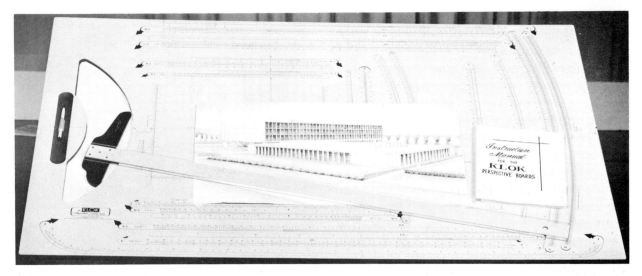

Fig. 8-43. The Klok perspective drawing board.

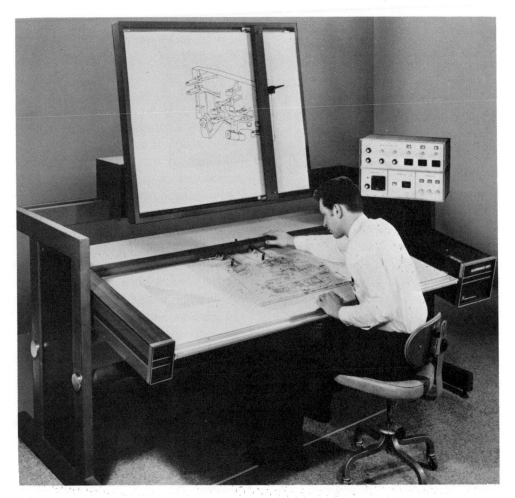

Fig. 8-44. Computerized perspective drawing machine.

the letter is followed by 2, the lines running to the right vanishing point are to be measured on these scales.

When using scales A, B, C, and D, the vanishing point used is on the horizon line 18.5 inches from the right edge. With scales E, F, G, and H the vanishing point on the extreme right edge is used. The other vanishing points are for one-point perspectives.

With this many scales the board may seem complicated and confusing. However, in practice one uses only one set of scales at a time. Thus, scales A-1, A-2, and upright scale A are used to draw perspectives to the scale of 1 inch to 1 foot; scales B-1, B-2, and upright scale B to draw perspectives 1½ inches to 1 foot, and so on. In order to avoid confusion as to which scales to use for a given project, a simple expedient is to place tape at the ends of the particular scales to be used.

8.6.10 Computerized perspective. Perspective Incorporated, has developed a computerized system which simplifies the problem of making perspective drawings. The system uses a machine that produces perspective drawings with accuracy at least four times faster than a skilled illustrator can. An operator simply traces over any two flat orthographic views of the object, and the machine produces the desired three-dimensional line drawing. See Fig. 8-44.

8.7 Exploded Illustrations

Exploded illustrations are pictorial drawings showing the various parts of an assembly in separate, or pulled out, positions but with all of the parts aligned in the correct order for reassembly. See Fig. 8-45. The principal advantage of this type of drawing is that it readily discloses how the individual parts of a mechanism fit together. It is of particular value to those who are unable to read multiview drawings. Exploded illustrations are widely used in design, manufacturing, sales, and service. These drawings may vary from simple sketches to elaborate shaded illustrations.

Drawings of this kind can be prepared either in axonometric, oblique, or perspective form. As a rule the perspective technique is preferred because it gives a more pleasing appearance. When

Fig. 8-45. An exploded drawing shows the parts of an assembly in their corresponding pulled out position.

parts are strung out along a single axis by axonometric or oblique methods, a certain amount of distortion is inevitable.

In making a perspective exploded drawing, a perspective of the assembly is made first. Next a piece of tracing paper is placed over the assembly drawing, and one or more parts are drawn with the perspective axes clearly shown. The overlay sheet is removed and replaced by another sheet, on which some other part is then drawn. When all of the parts have been drawn, the individual sheets are arranged so each part is in its normal spread-out position, and a final drawing is then made.

An experienced illustrator achieves the same results by tracing the main part of the mechanism and then moving the overlay sheet each time to the correct distance and tracing the remaining parts.

It must be remembered that a perspective made with overlay sheets will not produce a final true perspective of each part removed from the original assembly. The overlay method is satisfactory, however, because a true perspective would reduce individual parts to very small proportions.

8.8 Phantom Drawings

Any pictorial drawing that shows the interior parts of a mechanism is called a *phantom drawing*. The contrast between the interior and exterior shape is achieved by shading. The interior sections are rendered in dark tones and outer covering in very light tones. See Fig. 8-46.

8.9 Shading

One of the main purposes of a pictorial drawing is to achieve better visualization of an object than an orthographic view can provide. To make a pictorial drawing appear more natural, shading is often used. By means of shading, depth or distance can be achieved, thereby imparting the desired form or shape to an object. Shading is simply a technique of varying the light intensity on the surfaces by lines or tones. Notice in Fig. 8-47 how shading helps to better visualize the shapes of the objects.

8.9.1 Location of shaded areas. Since shading is a result of light intensity on a surface, the first consideration in producing a shaded effect is to determine the source of light falling on the object. Generally speaking, one can proceed on the basis that the principal source of light is shining over the observer's left shoulder or from the upper corner of the drafting table, as shown in Fig. 8-48. This however, is not a fixed rule, since sources of light from other directions will sometimes produce a better effect. However, if we assume that the

Fig. 8-46. A phantom drawing shows in pictorial form the internal shape of an object.

Fig. 8-48. Direction of the main light source on an object.

Fig. 8-47. Shading aids the imagination in visualizing the shape of the object.

standard direction of light is over the left shoulder, then the top and front surfaces of the object receive the most light. The sides that form smaller angles with the light rays have less light, and the surfaces that are directly opposite the light source are in deep shade. See Fig. 8-49.

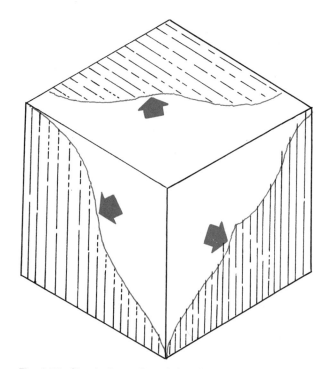

Fig. 8-51. Sketch the outline of the shaded areas with a fine line.

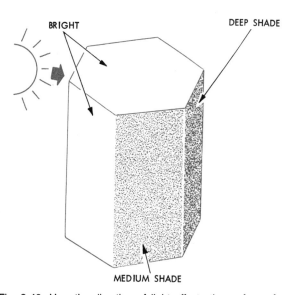

Fig. 8-49. How the direction of light affects the surface of an object.

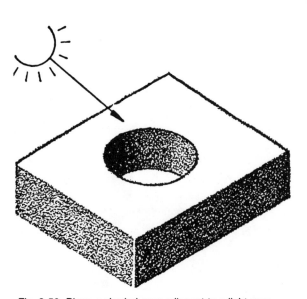

Fig. 8-50. Place a shaded area adjacent to a light area.

Following these guidelines, shading can be achieved by sketching contrasting weights of lines over the surfaces affected by the light. It should be kept in mind, too, that the most pleasing effect is usually obtained by having the shaded area of one plane adjacent to the light area of an adjoining surface. See Fig. 8-50.

Before actually proceeding with any shading, it is good practice to outline with a very fine line the areas that are to be shaded. The line should be light enough so it can be erased after the shading is completed. See Fig. 8-51.

8.9.2 Line shading. The simplest method of producing shading effects is by means of contrasting weights and spacing of lines. Fig. 8-52 shows how such lines are sketched on flat surfaces. The spacing of lines should be judged by eye. Notice that the darker the area, the closer will be the lines. Actually no hard and fast rule can be given as to the amount of space to leave between lines. Practice and judgment will serve as the best guide. Always visualize the intensity of light cast

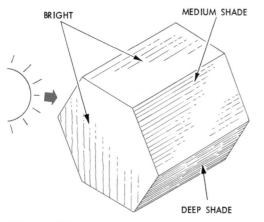

Fig. 8-52. This shows line shading on flat surfaces.

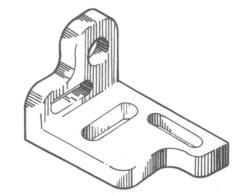

Fig. 8-54. This shows the direction of lines for the best shading effects.

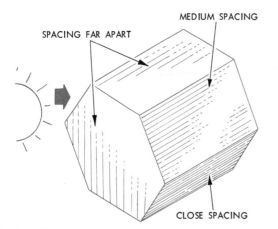

Fig. 8-53 Shading can be produced by varying the spacing of the lines.

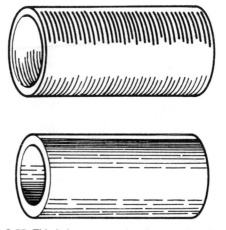

Fig. 8-55. This is how a curved surface can be shaded.

on the object and then space the lines so they will best reflect the effects of this light.

The weight of lines can be achieved by varying the pressure on the pencil. Notice in Fig. 8-52 that the heaviest lines are used where the surface is to have the darkest areas. It is also possible to produce shaded effects by keeping all the lines light and varying the spacing. See Fig. 8-53. As a rule, better results will be achieved by varying both the weight and spacing of the lines.

The direction of the lines must also be taken into consideration. Usually it is best to shade vertical faces with vertical lines and the other faces with lines parallel to one of the edges of the object. See Fig. 8-54.

To shade curved surfaces, the lines may be sketched straight or curved. The practice is to shade approximately one-fifth of the surface nearest the source of light, then leave the next two-fifths white, and shade the two-fifths that are the farthest from the light. See Fig. 8-55. Spheres are shaded by sketching a series of concentric circles, as illustrated in Fig. 8-56.

8.9.3 Stippling. Stippling is another method that can be used to produce shaded areas. This method consists of covering the surface to be

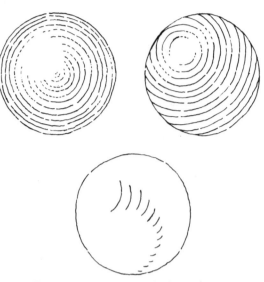

Fig. 8-56. This is how to shade a sphere.

shaded with a series of dots made with the point of a pencil. For dark areas the dots are placed closer together, and for light areas the dots are spaced widely apart. See Fig. 8-57. Although stippling produces a pleasing appearance, the process is much slower than line shading.

8.9.4 Broad stroke and smudge shading. Good shading results can be achieved by rubbing the selected areas with the flattened side of the pencil lead. See Fig. 8-58.

If a smudge effect, as shown in Fig 8-59, is desired, the broad pencil strokes should be rubbed with a paper stump. Once the area is covered, light and dark effects can be achieved by placing

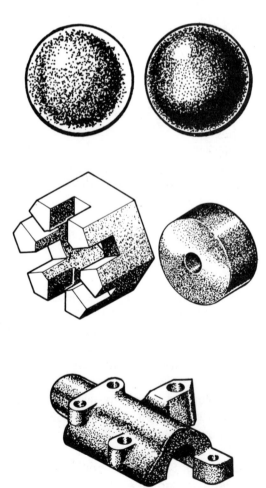

Fig. 8-57. Stippling is used to produce shaded areas.

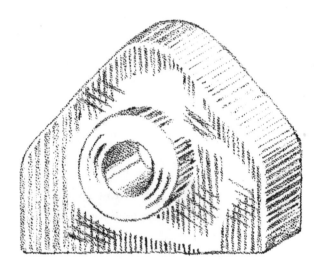

Fig. 8-58. This is an example of broad stroke shading.

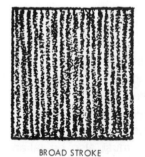 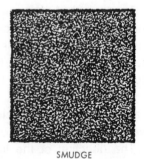

BROAD STROKE   SMUDGE

Fig. 8-59. Broad strokes are first drawn and then rubbed to produce smudge shading.

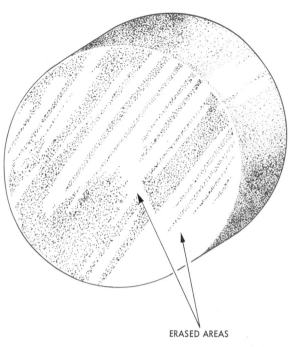

ERASED AREAS

Fig. 8-60. Erasing a shaded area produces light spots.

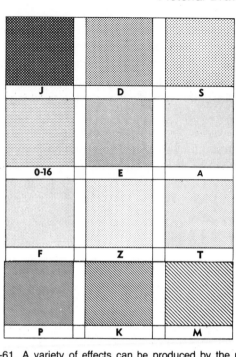

J	D	S
0-16	E	A
F	Z	T
P	K	M

Fig. 8-61. A variety of effects can be produced by the use of shading film. Letters and numbers are for identification.

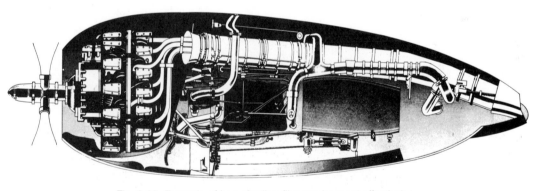

Fig. 8-62. Example of how shading film can be used effectively.

an erasing shield over the shaded part and rubbing with an eraser. See Fig. 8-60.

8.9.5 *Shading film.* For the inexperienced person, the use of shading film will produce excellent shading results. This is a thin transparent film made with a variety of line and dot patterns which will provide a wide scale of shading values. See Figs. 8-61 and 8-62. To use shading film, proceed as follows:

1. Place the entire sheet of shading film over the artwork to determine how much is needed. With a cutting needle, cut out the acetate a quarter of an inch larger all around than the area to be covered. Use enough pressure to cut only the ace-

tate, leaving the backing sheet intact. Before removing the film, be sure that the artwork is clean and free of dust and eraser particles.

2. Remove the backing from the shading film and place the film directly on the drawing, lining up the pattern squarely with the artwork. Rub lightly with a fingertip or a small square of paper to hold the film temporarily in place. The special adhesive on the film acetate does not seize the paper; the pattern can be moved into position quickly and accurately.

3. Use the cutting needle to cut out the pattern exactly where it is wanted. The surplus film is easily removed without damaging the drawing or tracing. The pattern may now be bonded permanently with the drawing by placing a sheet of paper over the acetate and rubbing it thoroughly with a hardwood or bone burnisher.

SECTION III
UNIT 9
Freehand Sketching

The ability to produce a freehand sketch is an important asset to anyone associated with industrial and engineering work. Most ideas are usually expressed first through the medium of a sketch and later translated into finished production drawings. Thus, the engineer or designer will very likely sketch out the preliminary features of a product and then pass them on to the drafting department for detailing. Those who prepare finished drawings frequently resort to a sketch to convey information on a mechanical drawing to the shop supervisor. Even the supervisor may be required to use some kind of technical sketch to show other workers a detail involved in the fabrication process. Any person who can quickly produce a clear and accurate sketch possesses an invaluable means of communication that contributes immeasurably to production efficiency.

9.1 Types of Sketches

Freehand sketches may be either orthographic projections or pictorial, depending on the function for which they are intended. See Fig. 9-1. Orthographic projection sketches are prepared in the same manner as described in Unit 5. All the necessary principal views are drawn and each detail dimensioned so as to present a complete shape and size description of the object.

Fig. 9-1. Sketching is a valuable means of communication in any engineering function.

243

Pictorial sketches are usually isometric, oblique, or perspective. Regardless of the type, they are made in accordance with the rules governing pictorial drawing as discussed in Unit 8.

9.2 Basic Sketching Essentials

Making good sketches requires some degree of practice, but equally important is adherence to a few basic principles.

9.2.1 Paper for sketching. While experienced personnel in design and drafting often use plain unruled paper for sketching, it is usually better to sketch on cross-section paper, since the grids make it easier to draw straight lines and to secure good proportions. Cross-section paper also simplifies laying out exact sizes without employing

a scale. Thus, if the paper has grids of a certain size, each square can be designated to represent a specific linear value. Special ruled paper may be used to make isometric, oblique, and perspective drawings. See Fig. 9-2.

Most engineering departments supply their personnel with sketching pads imprinted with non-reproducible blue line grids. Sketches can then be reproduced as ordinary white prints without the grid lines showing.

9.2.2 Proportions. Keeping correct proportions is a very important feature in freehand sketching. Regardless of how excellent the sketching techniques may be, the final sketch itself will not be a very good one if proper proportions are not maintained.

Proper proportioning is achieved by estimating actual dimensions. Sketches are not made to scale and one has to learn to recognize proportions by comparison. Thus, the process is started by noting the relationships between the width, the length, and the depth. If, for example, the height of the object is twice its width, then this proportion can often be applied to other details.

The problem of proportions is not quite so difficult if grid paper is used. By counting the number of squares, the proper relationships of sizes can be easily maintained. Sometimes the process is simplified if the general area enclosing a view is divided into squares or rectangles having proportions which are the ones to be maintained for the view. Refer to Fig. 9-3.

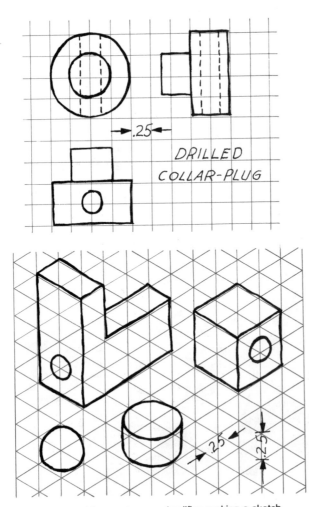

Fig. 9-2. Squared paper simplifies making a sketch.

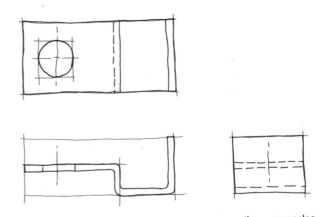

Fig. 9-3. Sketching rectangles or squares over the area enclosing the views will often help to achieve correct proportions.

9.2.3 *Position of the pencil.* Either a medium (F) or soft (HB) pencil is recommended for sketching. Hold the pencil loosely approximately 1.50 to 2.00 inches (38 to 50 mm) away from the point. See Fig. 9-4. As the pencil is used, rotate it slightly. This keeps the point sharp longer and makes clearer lines. The general practice is to slant the pencil at an angle of 50° to 60° from the vertical for drawing straight lines and at about 30°

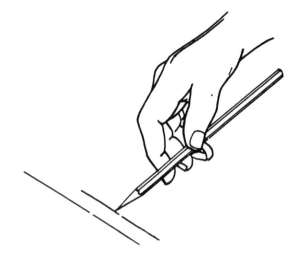
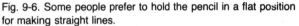

Fig. 9-6. Some people prefer to hold the pencil in a flat position for making straight lines.

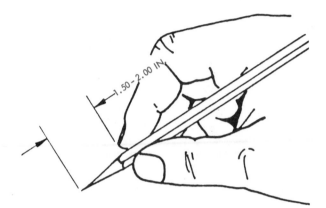

Fig. 9-4. For most sketching purposes hold the pencil in this position.

for circles. See Fig. 9-5. Some people hold the pencil in a flat position for straight lines. In this position the hand is guided on the backs of fingernails. See Fig. 9-6.

9.2.4 *Weight of lines.* The same line conventions used in instrument drawing are also used in technical sketching. However, in freehand sketching, lines are usually made in two weights only—medium weight for outlines and hidden and cutting plane lines; and thin weight for section, center, extension, and dimension lines. The contrast in weight should not be made by degrees of darkness but by difference in thickness. This can be done by varying the amount of pressure on the pencil.

It is a good practice to make all lines as light as possible to start with. The lines can be made heavier after the sketch is completed. In this way lines can easily be corrected without having to erase a great deal.

9.3 Sketching Straight Lines

Sketching straight lines is relatively simple—especially if grid paper is used. Nevertheless, by following a few fundamental rules, good straight lines can be made even without grid guide lines.

9.3.1 *Sketching horizontal lines.* To sketch horizontal lines, first mark off two points to indicate

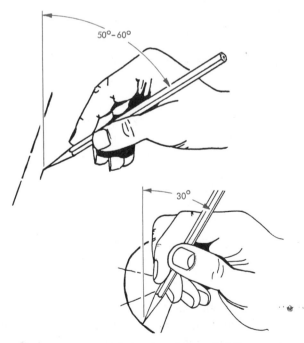

Fig. 9-5. For lines and circles slant the pencil at these angles.

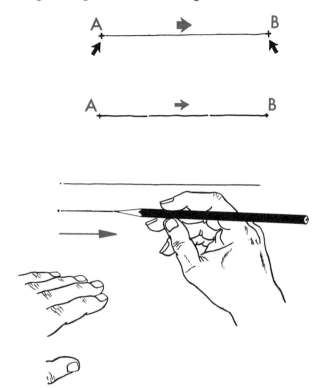

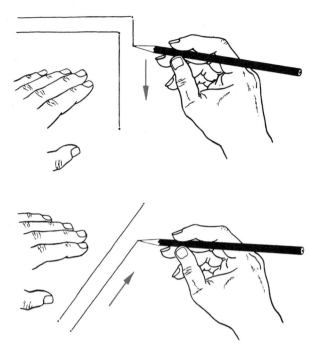

Fig. 9-8. Sketching vertical and slanting lines.

Fig. 9-7. Always pull, rather than push, the pencil when sketching horizontal lines.

the position of the line. See Fig. 9-7. Then sketch the line between the two points, moving the pencil from left to right. For short lines use a finger and wrist movement. With longer lines it is better to use a free arm movement, since using only the fingers and wrist tends to bend the line.

Sketch short lines in a single stroke. Long lines will be more accurate if they are made by a series of short strokes. By using short strokes 1.50 to 2.00 inches long, it is easier to maintain the proper direction than if longer strokes were used. A space of approximately 0.03 inch is left between strokes.

Always pull the pencil in sketching straight lines. If the pencil is pushed, it may catch the surface of the paper.

9.3.2 Sketching vertical and slanted lines.
Sketch vertical lines by starting at the top end of the line and moving the pencil downward. Slanted lines can be sketched better if the pencil is moved

Fig. 9-9. Turning the paper may simplify sketching vertical and slanted lines.

from left to right. See Fig. 9-8. Sometimes it is advisable to turn the paper so the vertical or slanted lines assume a horizontal position as in Fig. 9-9.

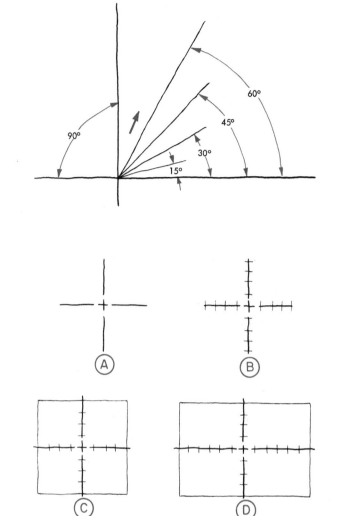

Fig. 9-10. How to sketch angles.

Fig. 9-11. Procedure for sketching a square and a rectangle.

9.3.3 Sketching angles.

To sketch an angle, draw two lines to form a right angle and then divide the angle into equal spaces. Project a line through the point that represents the angle desired. Notice in Fig. 9-10 the direction in which angular lines are made.

9.3.4 Sketching squares and rectangles.

To make a square, sketch a vertical line and a horizontal line as shown in Fig 9-11A. Space off equal distances on these lines as in Fig 9-11B. Sketch light horizontal and vertical lines through the outer points to form the square. Then darken the lines as in C.

A similar procedure can be used to sketch rectangular shaped objects. See Fig. 9-11D.

9.4 Sketching Curved Lines

Compared to straight lines, curved lines are a little more difficult to sketch with any degree of uniformity. But again, if a few basic procedures are used, circles, arcs, or irregular curves can be made with little difficulty.

9.4.1 Sketching circles and arcs.

To sketch a circle, draw a horizontal line and a vertical line through a point marking the center of the desired circle. See Fig. 9-12. On a piece of scrap paper, mark off the desired radius and transfer the distance to the main axes. Sketch diagonal lines at various intervals and locate the outer radial points using the scrap paper. Complete the circle by sketching short arcs through one quadrant at a

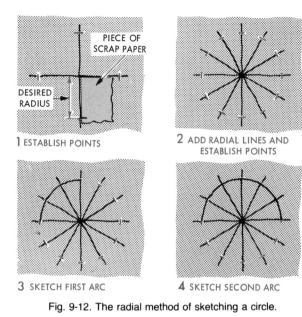

1 ESTABLISH POINTS

2 ADD RADIAL LINES AND ESTABLISH POINTS

3 SKETCH FIRST ARC

4 SKETCH SECOND ARC

Fig. 9-12. The radial method of sketching a circle.

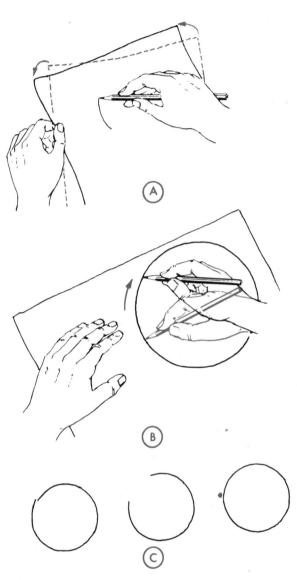

Fig. 9-13. The pivot method of sketching a circle.

time. By rotating the paper, the strokes can all be made in the same direction.

Circles can also be sketched by the pivot method. As shown in Fig. 9-13*A*, the hand is positioned with the second finger resting at the center of the proposed circle. With the pencil lightly touching the paper, rotate the paper with the other hand. Then darken the line by resting the hand on its side and tracing over the light line with the hand pivoting at the wrist, as in Fig. 9-13*B*.

9.4.2 Sketching irregular curves. For any irregular curve, locate a number of points to represent the shape of the required curvature. Complete the curve by sketching a series of arcs through these points. See Fig. 9-14.

9.5 Making a Multiview Sketch

A multiview sketch conforms to all of the practices used in making a mechanical orthographic projection. In proceeding to prepare such a sketch, the first consideration is the number of views which must be shown. As was stated in Unit 5, some objects may be shown in one or two views, whereas for more complicated objects three or more views will be required. In either case it is important that each view be placed in its true plane of projection.

The sketch need not be made to any specific scale, but reasonable proportions should be maintained. If interior features of the object must be shown, a sectional view should be included as described in Unit 5. If necessary, the sketch should be completely dimensioned according to practices discussed in Unit 6.

In general, the following steps should be carried out in making a multiview sketch: (See Fig. 9-15.)

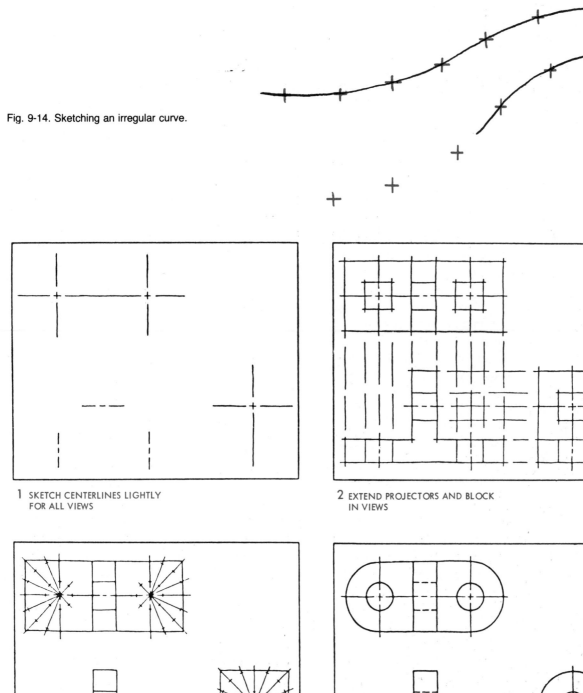

Fig. 9-14. Sketching an irregular curve.

1 SKETCH CENTERLINES LIGHTLY
FOR ALL VIEWS

2 EXTEND PROJECTORS AND BLOCK
IN VIEWS

3 ESTABLISH POINTS AND
DRAW ARCS

4 DARKEN OUTLINES AND DRAW
HIDDEN LINES

Fig. 9-15. Follow these steps in preparing a multiview sketch.

1. Locate the main center lines or base lines of the views.

2. Block in the views, using light construction lines.

3. Locate all radius points and sketch in circles, arcs, and curves.

4. After the views are properly formed, darken all visible and hidden lines. If necessary, sketch in extension lines and dimension lines and include dimensions, notes, and other essential information.

9.6 Making a Pictorial Sketch

Whether a sketch should be isometric, oblique, or perspective will depend on the shape of the object. Since the basic purpose of any pictorial sketch is to convey to the person who has not been trained to read multiview drawings the true shape of the object, the important consideration is to select the kind of sketch that will show the object with the least amount of distortion. For some objects an isometric sketch will prove satisfactory, while for others an oblique or perspective sketch will be more effective.

9.6.1 Isometric sketch. An isometric sketch shows three sides of an object. As was explained in Unit 8, these sides are sketched along one vertical axis and two angular axes. The angular axes extend to the right and left of the vertical axis at angles of 60° (30° to the horizontal). See Fig. 9-16. The object can be rotated so that either the right or left side is visible. See Fig. 9-17. Whether the object is sketched with its main surfaces extending to the right or to the left depends entirely upon which side it is most advantageous to show. As a rule hidden lines are not used on an isometric sketch.

To make an isometric sketch, proceed as follows: (See Fig. 9-18.)

1. Sketch a vertical line. From the top or base of this line, extend two slanted lines at an angle of 30° to the horizontal.

2. Lay out the actual width, length, and height on these three lines.

3. Complete each surface by sketching the necessary lines parallel to the axes.

4. Sketch in the remaining details of the object.

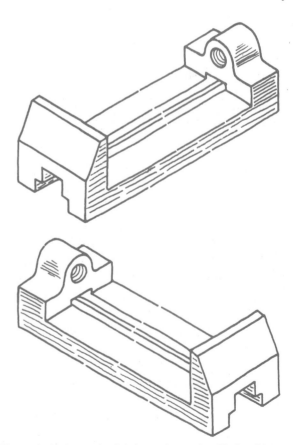

Fig. 9-17. An isometric sketch can be made with the object rotated to the right or to the left.

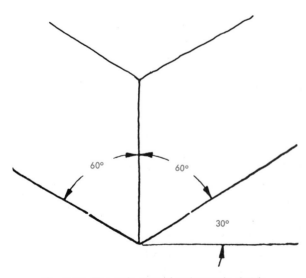

Fig. 9-16. The main axes of an isometric sketch.

Fig. 9-18. Follow these steps in making an isometric sketch.

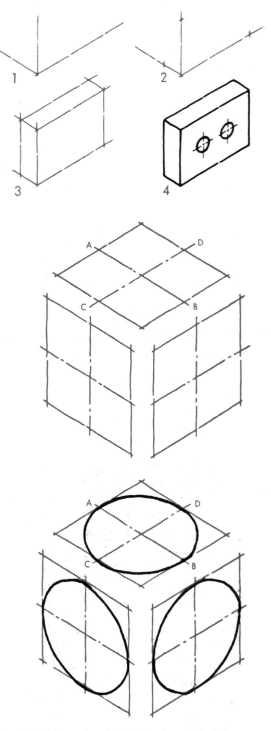

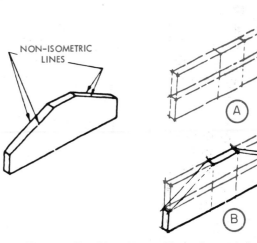

Fig. 9-19. Sketching objects with non-isometric lines.

Fig. 9-20. Procedure in sketching isometric circles.

When an object has slanted lines that do not run parallel to the main axis, these lines are called non-isometric lines. See Fig. 9-19. To sketch non-isometric lines, first lay out one of the views as for a multiview sketch. Then project the points from this view to the isometric lines. The same procedure can be used to sketch angles and irregular curves.

To make an isometric circle or arc, sketch an isometric square of the desired size. See Fig. 9-20. Sketch center lines *AB* and *CD* parallel to the axes of the square. Then sketch short arcs from *A* to *C, C* to *B, B* to *D,* and *D* to *A.*

9.6.2 Oblique sketch. An oblique sketch is similar to an isometric sketch except that the front face is seen in its true shape, much as is the front view of a multiview sketch. The two sides or ends are made to recede at some convenient angle from the horizontal, such as 30° or 45°. See Fig. 9-21. The common practice is to select the face that is most irregular or complicated in shape to serve as

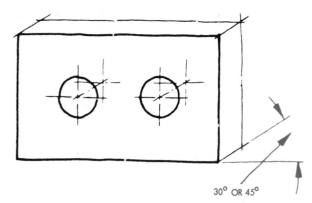

Fig. 9-21. An oblique sketch.

30° OR 45°

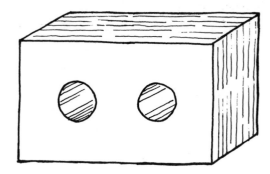

OBLIQUE

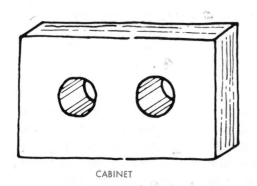

CABINET

Fig. 9-23. A comparison of an oblique and a cabinet sketch of the same object.

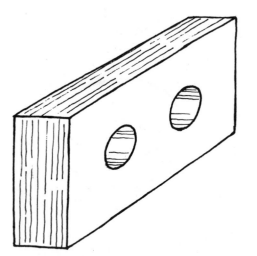

Fig. 9-22. Avoid distortion in oblique sketches by placing the long surface in front.

the front. To avoid undue distortion and to produce a more pleasing appearance, it is advisable to place the longest face in front, as in Fig. 9-22, instead of on a receding axis.

Since the receding faces of an oblique sketch are shown in their true sizes, the result often produces a slightly distorted representation. This distortion can be minimized by foreshortening the receding lines, usually by one-half. A sketch made in such a way is referred to as a cabinet sketch. See Fig. 9-23.

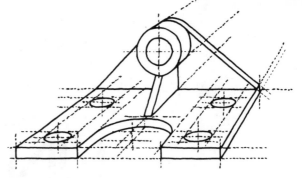

Fig. 9-24. A one point perspective sketch.

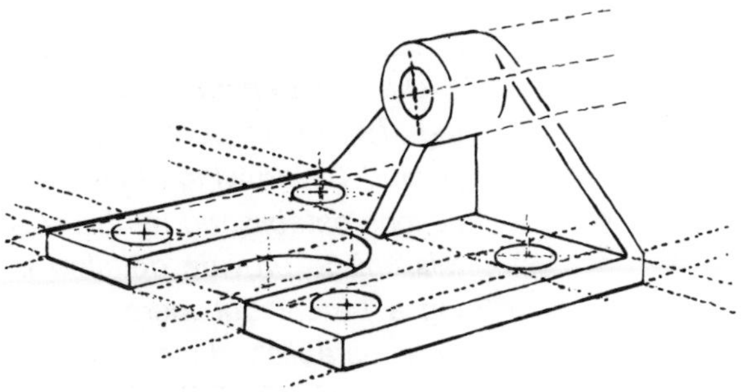

Fig. 9-25. A two point perspective sketch.

9.6.3 Perspective sketch. A perspective sketch is similar in all essentials to a perspective drawing as described in Unit 8. An object can be sketched so that the vanishing point is either to the left or to the right, or directly in the center of vision.

A perspective sketch may be made with one, two, or three vanishing points. The number of vanishing points to be used will depend on the shape of the object as well as the general effects to be achieved. Most freehand perspective sketches are made with one or two vanishing points. To make a perspective sketch, proceed as follows: (See Figs. 9-24 and 9-25.)

1. Assume the location of the horizon line.
2. Locate the position of the vanishing point.
3. For a one point perspective, sketch a front view of the object. If a two point perspective is to be made, sketch a vertical line and on it lay off the full or scaled height of the object.

4. From the front view or vertical line, sketch light construction lines back toward the vanishing points.

5. The points designating the length or depth of the object may be found by the projection method described in Unit 8. However, this procedure is used only for an accurate mechanical perspective. For most purposes location points for depths are simply assumed, that is, the vertical lines representing the ends of the object are placed in positions that produce the most pleasing effect.

6. Darken all outlines. Surfaces that lie in a shaded area may be shaded lightly.

UNIT 10

Patent Drawing

Drafting personnel may sometimes be required to prepare drawings for patent purposes. Any request for a patent must be accompanied by drawings that have been prepared according to specifications established by the U.S. Patent Office. It is not the purpose of this unit to provide detailed information on the procedure for securing a patent but merely to describe the specifications which have been established for making patent drawings.*

10.1 Patents, Copyrights, and Trademarks

A *patent* is an exclusive right granted to an individual to exclude others from making, using, or selling a useful invention in the United States for the term of seventeen years. Any person who invents or discovers any new and useful process, machine, manufacture, or composition of matter, or any new and useful improvements, may obtain a patent. A patent is issued by the United States Patent Office in accordance with rules and regulations (in legal terms, a *statute*) established by the government.

A *copyright* is an exclusive right granted to an individual to prevent a literary, dramatic, musical, or other artistic work from being copied by others. A copyright protects the form of expression rather than the subject matter. Thus, a written description of a machine could be copyrighted, but this would not prevent others from copying the description providing it was rewritten in their particular form of expression. Copyrights are registered in the Copyright Office in the Library of Congress.

A *trademark* refers to a word, letter, device, or symbol used by a manufacturer to indicate the source or owner of goods bearing that mark. Trademark rights are intended to prevent others from using the same name on similar goods, but they do not restrict others from making the same goods. Trademarks used in interstate or foreign commerce may be registered in the Patent Office.

10.2 Patent Representatives

The preparation of an application for patent and the conducting of the proceedings in the Patent Office to obtain the patent are undertakings requiring knowledge of patent law and Patent Office practice as well as knowledge of the scientific or technical matters involved in the invention.

An inventor may prepare his or her own application and file it in the Patent Office and conduct the proceedings in person, but unless the inventor is

---

*Most of the technical information in this unit has been extracted from the U.S. Department of Commerce publications, *Rules of Practice in the United States Patent Office* and *General Information Concerning Patents.*

familiar with these matters or studies them in detail, considerable difficulties may arise. While in many cases a patent may be obtained by persons not skilled in this work, there would be no assurance that the patent obtained would adequately protect the particular invention.

Most inventors employ the services of patent attorneys or patent agents to do the work for them. The statute gives the Patent Office the power to make rules and regulations governing the recognition of patent attorneys and agents to practice before the Patent Office. Persons who are not recognized by the Patent Office for this practice are not permitted by law to represent inventors. The Patent Office maintains a register of attorneys and agents. To be admitted to this register, a person must comply with Patent Office regulations, which now require a showing of the necessary qualifications and the passing of an examination.

The Patent Office registers both attorneys at law and persons who are not attorneys at law. The former are now referred to as patent attorneys and the latter as patent agents. Insofar as the work of preparing an application for patent and conducting the proceedings in the Patent Office is concerned, patent agents are usually just as well qualified as patent attorneys, although patent agents cannot conduct patent litigation in the courts or perform various services which the local jurisdiction considers practicing law. For example, a patent agent could not draw up a contract, such as an assignment or a license, relating to a patent, if the state in which he or she resides considers drawing contracts as practicing law.

10.3 Drawing Standards

A drawing is required by the statute for practically all inventions except processes or compositions of matter.

The drawing must show every feature of the invention specified in the claims and is required by the Office rules to be in a particular form. See Fig. 10-1. The Office specifies the size of the sheet on which the drawing is made, the type of paper, the margins, and other details relating to the making of the drawing. The reason for specifying the standards in detail is that the drawings are printed and published in a uniform style when the patents

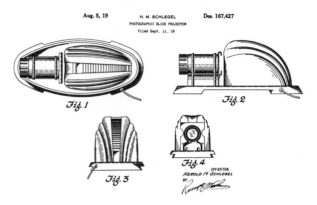

Fig. 10-1. An example of a patent drawing.

are issued, and the drawing must also be such that it can be readily understood by persons using the patent descriptions.

The complete drawing is printed and published when the patent is issued, and a copy is attached to the patent. This work is done by photolithography, the sheets of drawing being reduced about one-third in the process. In addition, a reduction of a selected portion of the drawings of each application is published in the *Official Gazette*. Therefore, it is necessary for these and other reasons that the character of each drawing be brought as nearly as possible to a uniform standard, suited to the requirements of the reproduction process and of the use of the drawing. This high standard will serve the best interests of inventors, of the Office, and of the public.

10.3.1 Paper and ink. Drawings must be made on pure white paper of a thickness corresponding to two-ply or three-ply Bristol board. The surface of the paper must be smooth and of a quality which will permit erasure and correction. To insure perfectly black solid lines, only India ink must be used for pen drawings. The use of white pigment to cover lines is not acceptable.

10.3.2 Size of sheet and margins. The size of a sheet on which a drawing is made must be exactly 8½ by 14 inches (21.6 by 35.6 cm). See Fig. 10-2. One of the shorter sides of the sheet is regarded as its top. The drawing must include a top margin of 2 inches (5.1 cm) and bottom and side margins of ¼ inch (6.4 mm). This leaves a "sight," or framed area, of precisely 8 by 11¾ in-

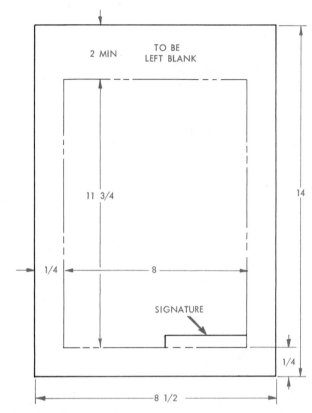

Fig. 10-2. Drawing of inventions must have a sheet format conforming to these specification.

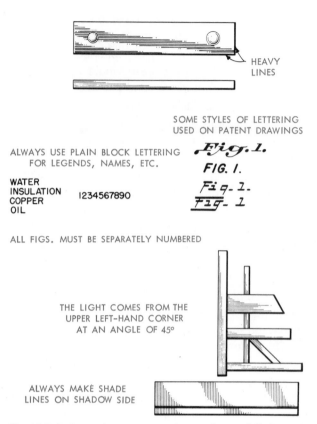

Fig. 10-3. Letters reference numerals must be carefully formed. Several types of lettering and figure marks are shown; any style may be used.

ches (20.3 by 29.8 cm). Margin border lines are not permitted. All work must be included within the sight. The sheets may have two ¼-inch (6.4 mm) diameter holes having their center lines ¹¹/₁₆ inch (17.5 mm) below the top edge and 2¾ inches (7.0 cm) apart, and equally spaced from the side edges.

10.3.3 Character of lines. All drawings must be made with drafting instruments or by a photolithographic process that will give satisfactory reproduction characteristics. Every line and letter, signatures included, must be absolutely black. This direction applies to all lines however fine, to shading, and to lines representing cut surfaces in sectional views. All lines must be clean, sharp, and solid. Fine or crowded lines should be avoided. Solid black should not be used for sectional or surface shading. Freehand work should be avoided wherever it is possible to do so.

10.3.4 Hatching and shading. Hatching should be made by oblique parallel lines not less than about ¹/₂₀ inch apart.

Heavy lines on the shade side of objects should be used except where they tend to thicken the work or obscure reference characters. The light should come from the upper left-hand corner at an angle of 45°. Surface delineations should be shown by proper shading, which should be open. See Figs. 10-3 to 10-12.

10.3.5 Scale. The scale of a drawing ought to be large enough to show the mechanism without crowding when the drawing is reduced in reproduction. Views of portions of the mechanism on a larger scale should be used when necessary to show details clearly. Two or more sheets should be used if a single sheet does not provide sufficient room, but the number of sheets should not be more than is necessary.

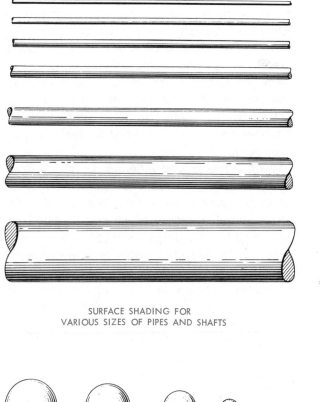

SURFACE SHADING FOR
VARIOUS SIZES OF PIPES AND SHAFTS

SURFACE SHADING FOR SPHERICAL OBJECTS

Fig. 10-4. Surface delineations should be shown by proper shading.

10.3.6 Reference characters.

The different views should be consecutively numbered figures. Reference numerals (letters may be used, but numerals are preferred) must be plain, legible, carefully formed, and not circled. They should, if possible, measure at least one-eighth of an inch in height so that they may bear reduction to one twenty-fourth of an inch; they may be slightly larger when there is sufficient room.

Reference numerals must not be placed in the close and complex areas of the drawing so as to interfere with a complete understanding of the drawing. When necessarily grouped around a certain part, they should be placed at a little distance, at the closest point where there is available space, and connected by lines with the parts to which they refer. They should not be placed on hatched

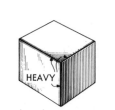

HEAVY

NOTE--THE HEAVY SHADE LINES ARE
PLACED ON THE EDGES CLOSEST
TO THE EYE

SHADING FOR A BLOCK
IN PERSPECTIVE

SURFACE SHADING
ILLUSTRATING A MIRROR

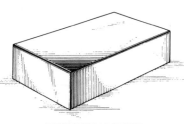

RECTANGULAR BLOCK
IN PERSPECTIVE

ROUND MIRROR

Fig. 10-5. Heavy shaded lines on perspective views are placed on the edges closest to the eye. The rule about light coming from the upper left corner at a 45° angle does not apply to perspective views.

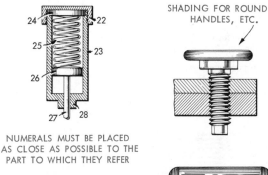

NUMERALS MUST BE PLACED
AS CLOSE AS POSSIBLE TO THE
PART TO WHICH THEY REFER

SHADING FOR ROUND
HANDLES, ETC.

NEEDLE
VALVE

WOOD SCREW

CYLINDRICAL SHADING

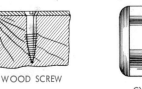

CYLINDRICAL SHADING
HIGH LIGHT

Fig. 10-6. Placing of reference characters, shading, and sectioning methods.

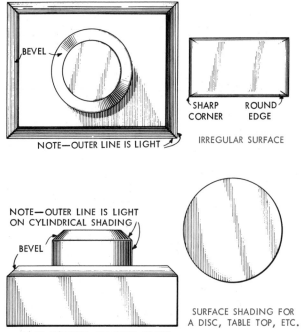

SHARP CORNER   ROUND EDGE

IRREGULAR SURFACE

NOTE—OUTER LINE IS LIGHT

BEVEL

NOTE—OUTER LINE IS LIGHT ON CYLINDRICAL SHADING

BEVEL

SURFACE SHADING ON BEVEL EDGES

SURFACE SHADING FOR A DISC, TABLE TOP, ETC.

Fig. 10-7. Methods of shading various kinds of surfaces.

or shaded surfaces but, when necessary, a blank space may be left in the hatching or shading where the character occurs, so that it will appear perfectly distinct and separate from the work.

Any one part of an invention appearing in more than one view of the drawing must always be designated by the same character, and the same character must never be used to designate different parts.

10.3.7 Symbols and legends. Graphical drawing symbols for conventional elements may be used when appropriate, subject to approval by the Office. The elements for which such symbols are used must be adequately identified in the specification. While descriptive matter on drawings is not permitted, suitable legends may be used, or may be required in cases such as diagrammatic views and flow sheets. The lettering should be as large as, or larger than, the reference characters.

10.3.8 Location of signature and names. The signature of the applicant, or the name of the ap-

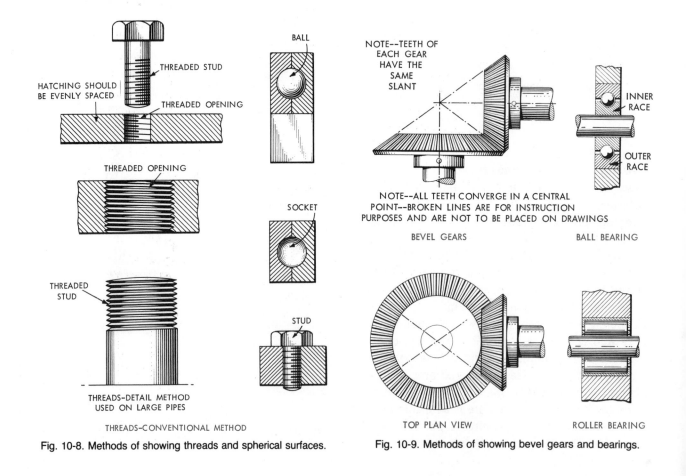

HATCHING SHOULD BE EVENLY SPACED

THREADED STUD

THREADED OPENING

THREADED OPENING

THREADED STUD

THREADS-DETAIL METHOD USED ON LARGE PIPES

THREADS-CONVENTIONAL METHOD

BALL

SOCKET

STUD

Fig. 10-8. Methods of showing threads and spherical surfaces.

NOTE--TEETH OF EACH GEAR HAVE THE SAME SLANT

NOTE--ALL TEETH CONVERGE IN A CENTRAL POINT--BROKEN LINES ARE FOR INSTRUCTION PURPOSES AND ARE NOT TO BE PLACED ON DRAWINGS

BEVEL GEARS

INNER RACE

OUTER RACE

BALL BEARING

TOP PLAN VIEW

ROLLER BEARING

Fig. 10-9. Methods of showing bevel gears and bearings.

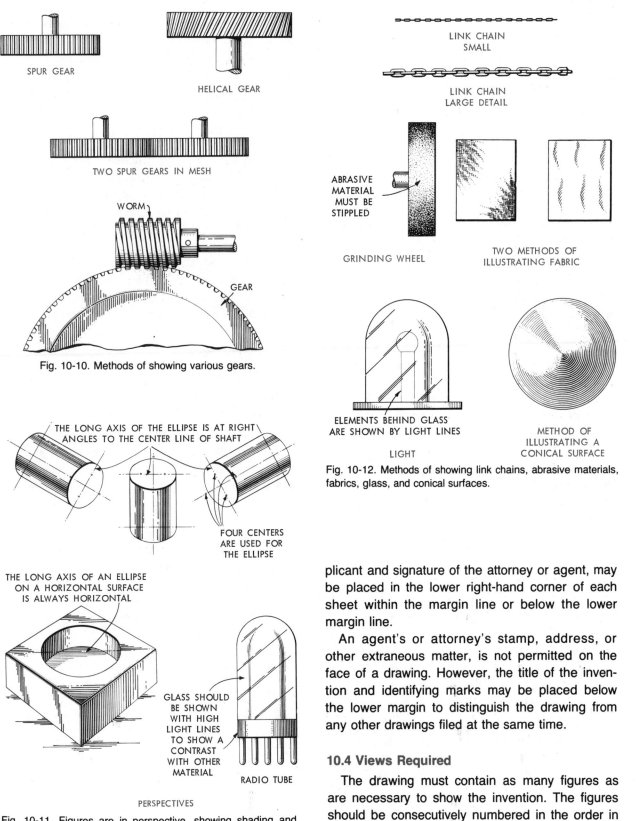

SPUR GEAR

HELICAL GEAR

TWO SPUR GEARS IN MESH

WORM

GEAR

Fig. 10-10. Methods of showing various gears.

LINK CHAIN
SMALL

LINK CHAIN
LARGE DETAIL

ABRASIVE
MATERIAL
MUST BE
STIPPLED

GRINDING WHEEL

TWO METHODS OF
ILLUSTRATING FABRIC

ELEMENTS BEHIND GLASS
ARE SHOWN BY LIGHT LINES

LIGHT

METHOD OF
ILLUSTRATING A
CONICAL SURFACE

Fig. 10-12. Methods of showing link chains, abrasive materials, fabrics, glass, and conical surfaces.

THE LONG AXIS OF THE ELLIPSE IS AT RIGHT ANGLES TO THE CENTER LINE OF SHAFT

FOUR CENTERS
ARE USED FOR
THE ELLIPSE

THE LONG AXIS OF AN ELLIPSE
ON A HORIZONTAL SURFACE
IS ALWAYS HORIZONTAL

GLASS SHOULD
BE SHOWN
WITH HIGH
LIGHT LINES
TO SHOW A
CONTRAST
WITH OTHER
MATERIAL

RADIO TUBE

PERSPECTIVES

Fig. 10-11. Figures are in perspective, showing shading and position of long axis.

plicant and signature of the attorney or agent, may be placed in the lower right-hand corner of each sheet within the margin line or below the lower margin line.

An agent's or attorney's stamp, address, or other extraneous matter, is not permitted on the face of a drawing. However, the title of the invention and identifying marks may be placed below the lower margin to distinguish the drawing from any other drawings filed at the same time.

10.4 Views Required

The drawing must contain as many figures as are necessary to show the invention. The figures should be consecutively numbered in the order in which they appear.

10.4.1 *Types of views.* The figures may be plan, elevation, section, or perspective views. Detail views of portions or elements, on a larger scale if necessary, may also be used. Exploded views are permissible to show the relationship or order of assembly of various parts; the separated parts of the same figure are embraced by a bracket.

When necessary, a view of a large machine or device in its entirety may be broken and extended over several sheets if there is no loss in facility of understanding the view. The different parts should be identified by the same figure number but followed by the letters *a, b, c,* and so on for each part.

The plane on which a sectional view is taken should be indicated on the general view by a broken line, the ends of which should be designated by numerals corresponding to the figure number of the sectional view and have arrows applied to indicate the direction in which the view is taken.

A moved position may be shown by a broken line superimposed on a suitable figure if this can be done without crowding; otherwise, a separate figure must be used for this purpose. Modified forms of construction can only be shown in separate figures. Views should not be connected by projection lines nor should center lines be used.

10.4.2 *Arrangement of views.* All views on the same sheet must read from the same direction and should, if possible, be placed so that they can be read with the sheet held in an upright position. If views longer than the width of the sheet are necessary for the clearest illustration of the invention, the sheet may be turned on its side. The space for a heading must then be reserved at the right and the signatures placed at the left, occupying the same space and position on the sheet as in an upright view, and being horizontal when the sheet is held in an upright position. One figure must not be placed on top of another or within the outline of another.

The drawing should, as far as possible, be planned so that one of the views will be suitable for publication in the *Official Gazette* as the illustration of the invention.

10.5 Transmission of Drawings

Drawings transmitted to the Patent Office can be sent (1) flat, protected by a sheet of heavy binder's board, or (2) rolled in a suitable mailing tube. However, they must never be sent folded. If drawings are received creased or mutilated, new drawings will be required.

UNIT 11

Graphs and Charts

Charts are intended to provide visualization of numerical values or data and to serve as devices for transmitting information. They are useful to scientists, engineers, economists, statisticians, accountants, and others as a rapid and often simplified means to examine, explore, or calculate data, to answer some specific question or questions, or to present material to someone who is not familiar with the subject. Thus, in chart form statistical data becomes much more meaningful. Similarly a technical process can often be made more understandable by a pictorial chart rather than written instructions.

11.1 Line Charts[1]

The line chart is by far the most common of all charts. Its principal purpose is to indicate an amount of change by means of a curve plotted on coordinates. The curve can be in either slope or step form. See Fig. 11-1.

To prepare a line chart, proceed as follows:

1. *Select the paper.* Careful thought must be given to choice of grid proportions, because grid shape can influence the shape of the curve. For example, different grid shapes can make the same data appear to change rapidly or gradually. Grids for time-series charts are usually oblong instead of square, with standard proportions of 2:3 or 3:4.

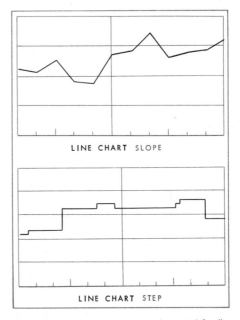

LINE CHART SLOPE

LINE CHART STEP

Fig. 11-1. Types of curves which may be used for line charts. (ANSI Y15.2)

---

1. *ANSI Y15.2.*

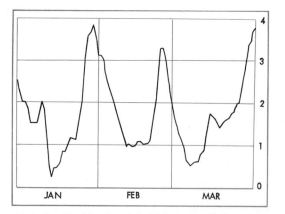

Fig. 11-2. A grid wider than it is high is especially suitable for time series containing many plottings or covering a long period of time. (ANSI Y15.2)

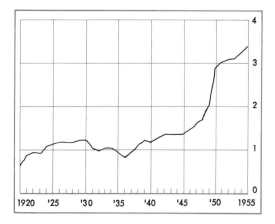

Fig. 11-3. A grid higher than it is wide is sometimes better for series covering a short period or exhibiting rapid change. (ANSI Y15.2)

A grid which is wider than it is high, as in Fig. 11-2, is usually best for:

a. series extending over a long period of time.

b. series extending over a short period but calling for many plottings.

c. series in which the trend or change is moderate or slight.

A grid which is higher than it is wide, as in Fig. 11-3, is preferable when showing:

a. series extending over a very short period of time.

b. series extending over a longer period but calling for only a few plotted points.

c. series with a strong trend or rapid change.

d. overlapping curves that would run together too much on a horizontal grid.

e. series in which important fluctuations would not be noticeable on a horizontal grid.

2. *Draw the coordinate axes.* The horizontal line (abscissa) is known as the X-axis and the vertical line (ordinate) as the Y-axis. Draw the X-axis at the bottom of the sheet and the Y-axis at the left, leaving one-inch margins for lettering.

3. *Select a proper scale.* The independent variables are normally plotted as abscissas and the dependent variables as ordinates. Select a scale that permits the divisions on the coordinate paper to assume values of 1, 2, 3, 4, and so on or a power of ten multiplied by 1, 2, 3, 4, and so on. The scale selected determines the visual change in the dependent variable. A poor choice of scales may give the impression of rapid change even though this is not true. Fig. 11-4 shows the same data plotted using different scales.

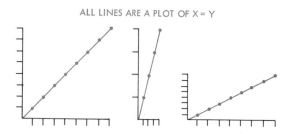

Fig. 11-4. Change in visual impression is possible by altering the scale of a chart.

The intervals of a scale are usually more suitable when they are in even numbers, since the points are easier to plot and read. Intervals of odd numbers are awkward and often are more difficult to visualize.

Normally the scale should start from a base of zero, particularly when the purpose is to compare magnitudes, that is, to show how much or how many there are of something. When the data deals with differences from one period to another or when the shape of one curve is compared with that of another, the scale may start at any convenient value.

4. *Plot the data and draw the curve.* Plot the positions determined by the values of the abscissa and ordinate. Encircle the dots to make the location of each point clear. Then connect the points with a curve.

Data which are not based on scientific theory or mathematical law are known as *discontinuous data.* The curve representing this kind of data is drawn by connecting the plotted points with straight lines as shown in Fig. 11-5A. For *continuous data,* that is, data being related to a theory or a law, a smooth curve should be drawn which distributes the points on either side of the curve and has, as nearly as possible, equal distances alternately on one side and then the other. See Fig. 11-5B.

A single curve should be drawn with a solid line. When more than one curve is required on a chart,

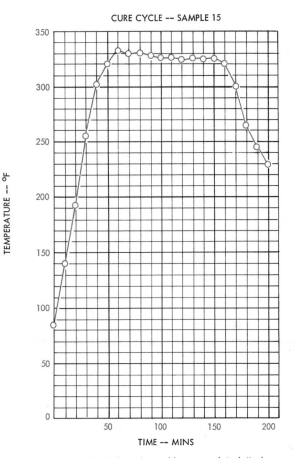

Fig. 11-7. A completed line chart with appropriate lettering.

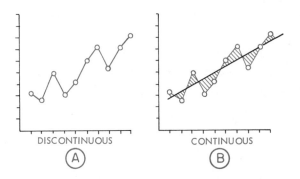

Fig. 11-5. Representing discontinuous and continuous data.

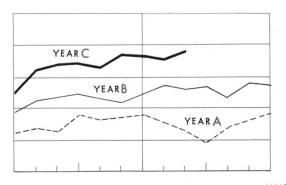

Fig. 11-6. A line chart may contain more than one curve. (ANSI Y15.2)

the type of line may be varied to differentiate the curves. See Fig. 11-6.

5. *Letter the chart.* Each axis should be labeled with an appropriate title. See Fig. 11-7. When more than one curve is used, a key should be incorporated to identify the data represented by each curve. The key should be located away from the curves to avoid confusion. See Fig. 11-8. Each curve should be labeled as illustrated in Fig. 11-9. Although labels are necessary for multiple curves, it is often advantageous to label a single curve, provided it does not repeat what is said in the caption. A curve label should be placed at about one-half to three-quarters of the letter height from the curve and preferably above rather than below the curve.

The chart should carry a suitable title located at the top and usually outside the grid. The practice

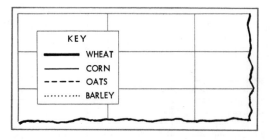

Fig. 11-8. Standard arrangement for a key. (ANSI Y15.2)

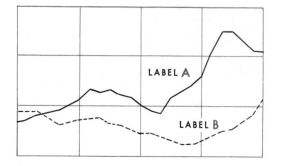

Fig. 11-9. Examples of how curves should be labeled. (ANSI Y15.2)

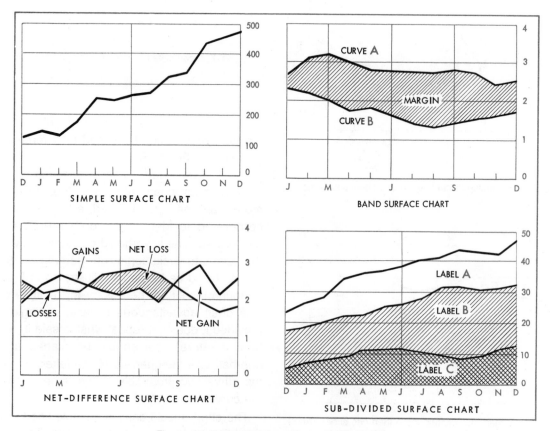

Fig. 11-10. Types of surface charts. (ANSI Y15.2)

is to make the title one size larger than other lettering on the chart. It should be concise but must give the reader a quick and clear understanding of what the chart is about.

11.2 Surface Charts

A surface chart is one in which values are represented by the height of a shaded surface. There are four main types of surface charts as shown in Fig. 11-10.

1. *Simple* surface: emphasizes the total amounts of a single time series.

2. *Band* surface: emphasizes the difference between two series, one of which is always at a higher level than the other.

3. *Net-difference:* emphasizes the difference

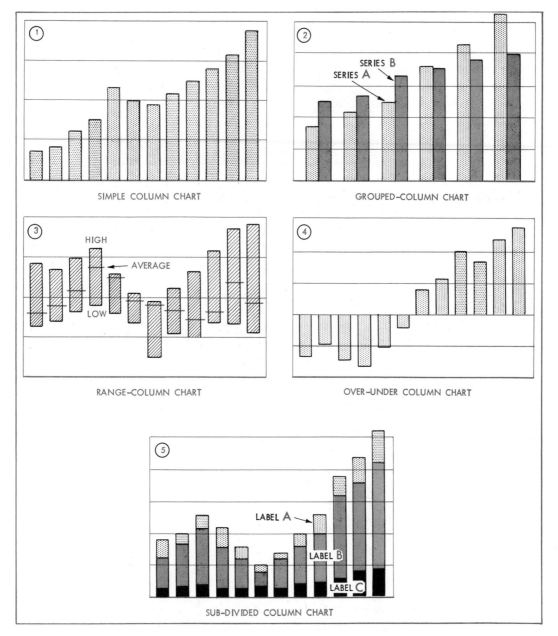

Fig. 11-11. Types of column charts. (ANSI Y15.2)

between two series. It can also be used to measure the difference between a series and a standard against which it is measured.

4. *Subdivided* surface (also called a strata chart): shows the absolute sizes of the component parts of a series of totals.

In general, the layout and design of surface charts should follow the principles and procedures for line charts.

11.3 Column Charts

A column chart is one in which the numerical values are represented by the lengths of bars or columns. The bars can be either vertical or horizontal. The main types of column charts are shown in Figs. 11-11 and 11-12.

1. *Simple column:* emphasizes the total amounts in a single time series.

2. *Grouped-columns:* compares two series.

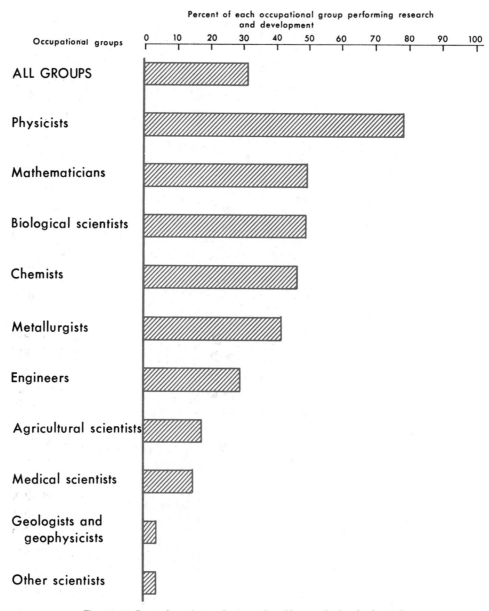

Fig. 11-12. Bars of a column chart can be either vertical or horizontal.

Fig. 11-13. Examples of column spacing. (ANSI Y15.2)

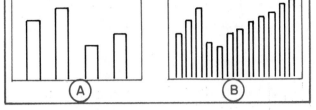

3. *Range-columns:* emphasizes the difference between two time series, one of which is always at a higher level than the other.

4. *Over-under columns:* emphasizes the difference between two series, or the difference between a single series and a standard against which it is measured.

5. *Subdivided columns:* shows the component parts of a series of totals.

Special care must be given in preparing column charts to insure that they have a realistic and balanced appearance. When only a few columns are involved, the general principal is to make the columns narrower than the white space between them. See Fig. 11-13*A*. With many columns the column width should be greater than that of the spaces. See Fig. 11-13*B*.

Correct shading of the columns is also important to achieve the proper effects. Normally shading is darkest at the base and increasingly lighter the further the segments are from the base. Refer to Fig. 11-14.

11.4 Pie Charts

The principal function of a pie chart is to present a graphic comparison of related quantities expressed in percentages. A pie chart is in the form of a circle whose area represents 100 percent. The subdivisions are readily calculated because they are proportionate amounts of the entire circle. A pie chart is illustrated in Fig. 11-15.

To determine the number of degrees of arc that determine each segment, multiply the percentage that the segment will represent by 3.6. Draw the circle and then draw the divisions using a protractor to measure the calculated degrees. Describe the nature of each quantity and its percentage. Shade each segment differently or use colors.

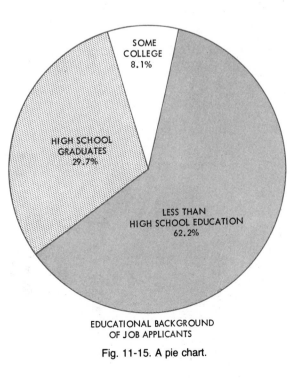

EDUCATIONAL BACKGROUND
OF JOB APPLICANTS

Fig. 11-15. A pie chart.

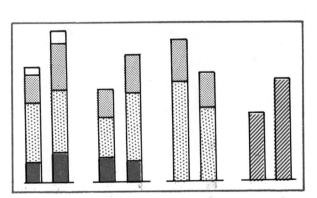

Fig. 11-14. Examples of effective shading. (ANSI Y15.2)

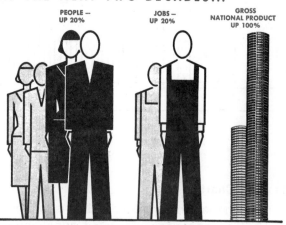

Fig. 11-16. A pictorial chart uses stylized figures to present data.

11.5 Pictorial Charts

The pictorial chart is particularly adapted to show information pertaining to such things as population, costs, resources, and production. The data is represented by symbols depicting humans, money, minerals, animals, and various products. The symbols are made in proportionate sizes or amounts to illustrate the unit to be described. See Figs. 11-16 and 11-38.

11.6 Flow Charts

The flow chart is a form of pictorial chart used to correlate and present data in a simple and understandable manner. See Fig. 11-17. Industrial and chemical engineers use the flow chart to give a concise pictorial description of a process to non-technical persons. A touch of artistry is helpful in selecting meaningful and unique symbols.

The organization chart describes the formal organization of a group of persons, offices, or organization functions. See Fig. 11-18. Drafting personnel are frequently called upon to draw such charts for publication, and they should be aware of the conventions used. The form may differ slightly from one organization to another, but these variations occur in the majority of charts.

The most important office is always centered at the top of the chart. The name of the office and frequently the name of the person performing the duties of the office are lettered inside a rectangular box. Solid lines are drawn from this box to the next echelon of offices. Only those positions reporting directly to the chief officer are included in this line. The names of these offices and officers are boxed and lettered. The process is repeated for the next levels.

Care must be taken to position the boxes in a pattern pleasing to the eye and in the proper relative level in the organization. Dotted lines are sometimes used to denote cooperation and communication between members of the same level.

11.7 Special Charts

Other types of charts in various forms, are frequently used to present detailed technical information. These charts are sometimes referred to as graphs. Several of the more common ones are the

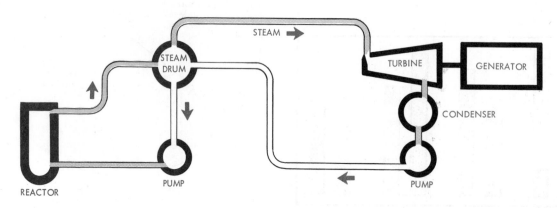

Fig. 11-17. A flow chart conveys a pictorial description of movements, travel, or flow of some particular process or operation.

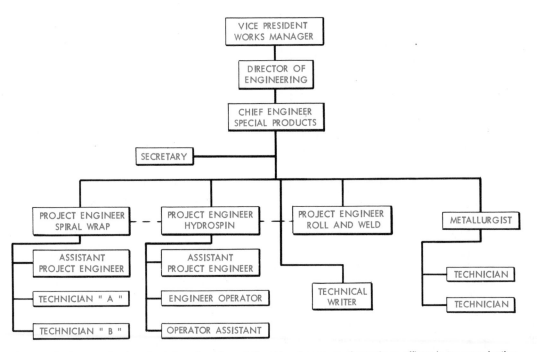

Fig. 11-18. An organization chart describes the relationship of a group of people or offices in an organization.

semi-logarithmic graph, the logarithmic graph, the polar coordinate graph, and the trilinear chart.

11.7.1 Semi-logarithmic graphs. A semi-logarithmic graph is used for presenting data qualitatively and quantitatively to indicate a rate of change rather than an amount of change as on a rectangular coordinate graph. Semi-logarithmic graph paper has equally spaced divisions on one axis and logarithmic scaled divisions on the other and is sometimes known as ratio ruled paper. Measurement of the slope of the curve determines the rate of increase or decrease; a straight line indicates a constant rate of change. This type of graph should be used whenever one variable increases in a geometric progression or other non-linear manner and the other variable increases arithmetically, or if it is necessary to show a percentage of change. See Fig. 11-19.

A semi-logarithmic graph is drawn in the same manner as a rectangular coordinate chart. Care must be exercised to interpolate the locations of plotted points logarithmically on the logarithmic scaled axis.

11.7.2 Logarithmic graphs. A logarithmic graph may be used to present data qualitatively or quan-

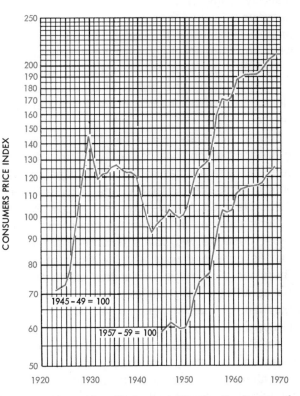

Fig. 11-19. A semi-logarithmic graph showing the Bureau of Labor statistics-price index.

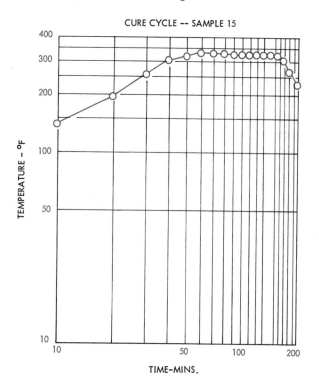

Fig. 11-20. A typical logarithmic graph.

titatively. Some of its advantages include (1) large numerical values are represented with short graphical distances, (2) straight line graphs result when the dependent variables are proportional to a constant power of the independent variables, and (3) the functional errors which may occur in plotting or reading the graph are a constant percentage.

Logarithmic graph paper has both axes scaled in divisions proportional to the logarithms of numbers and is sometimes known as log paper. These printed forms are available in one or more cycles. Part cycle and split cycle forms are also available.

Curves are easily plotted on logarithmic graph paper because two points determine a straight line, as does one point and the slope of a line. The slope of a line is the tangent of the angle it makes with the horizontal axis and is equal to the exponent of the independent variable. See Fig. 11-20.

As a rule it is easier to use logarithmic coordinates and plot the points directly than to use the logarithms of the variables and plot them on rectangular coordinates. All other aspects of graph making explained for rectangular coordinate charts apply to this type. Care must be taken, however, to interpolate logarithmically and not arithmetically, as is done on rectangular coordinate charts.

11.7.3 Polar coordinate graphs. The use of polar coordinates often simplifies the calculation of values. The data is plotted on polar coordinate paper, commonly called by that name. Polar coordinate graphs are frequently used in recording instruments having circular plotting paper and often show intensity of light or heat at various distances from a source. Fig. 11-21 shows a typical polar coordinate graph. It is possible, by using trigonometry, to convert the polar coordinates to rectangular coordinates or vice versa if a more readable graph will result.

11.7.4 Trilinear charts. A trilinear chart is used to present the interrelations between three variable items such as chemical compounds, mixtures, solutions, or alloys. Essentially it is an area chart with the altitude of the triangle equaling 100 percent. It is based upon the premise that the sum of the three perpendiculars from a point equals the altitude. Since the altitude equals 100 percent, the perpendiculars are the proportionate amount of the percentage of the three variables.

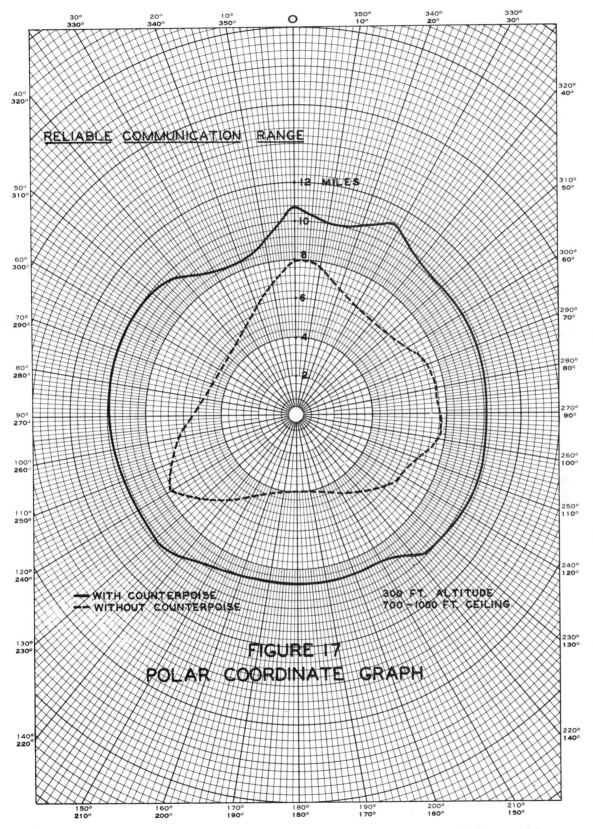

Fig. 11-21. Polar coordinate graphs are used to represent scientific data such as intensity of heat and light.

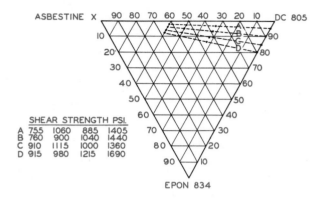

SHEAR STRENGTH PSI.

A	755	1060	885	1405
B	760	900	1040	1440
C	910	1115	1000	1360
D	915	980	1215	1690

Fig. 11-22. Trilinear charts are used to show the interrelations between three variables.

The chart is constructed by dividing two sides of the triangle into equal percentage divisions and then drawing lines through these points parallel to the sides of the triangle. Fig. 11-22 illustrates a trilinear chart.

11.8 Alignment Charts[2]

Alignment charts, or nomographs as they are frequently called, are used extensively as quantitative graphs to easily and quickly determine the numerical values of formulas having three or more variables. Since the majority of nomographs are complicated and difficult to construct, it is unusual to develop one unless its frequency of use is great enough to warrant the effort. Once completed, however, it will justify the time spent, because of the ease and speed with which the values of the variables are determined without resorting to tedious and laborious computations. Each chart must be constructed to fit the particular relationships that are to be shown graphically.

Essentially a nomograph consists of a graphical representation of the relationships that exist between the several variables of an equation. The representation is by means of scaled lengths of the variables along straight or curved axes positioned in such a way that a straight line will intersect the axes at the points satisfying the equation.

It is a graphical equation with three or more variables and only one unknown value.

11.8.1 Nomograph terms. Prior to discussing the design and construction of alignment charts, it is essential that the definitions of various terms and expressions be understood as they apply to this type of graph.

Constant. A symbol which represents a fixed numerical value, incapable of change during a particular discussion.

Variable. A symbol which may assume various values during a discussion.

Function. A dependent variable is said to be a *function* of an independent variable when both are so related in a formula that to each arbitrary value assigned to the first (independent) there corresponds one or more definite values of the other (dependent) variable. This condition is usually written in abbreviated form as *f(x)* which is read "function of *x*" (not as "*f* multiplied by *x*.").

As an example, in the expression $x^2 + 3x + 5$, *x* is the independent variable and the expression is the dependent variable or function of *x*. If the expression were written as the formula $y = f(x) = x^2 + 3 + 5$, then *y* would be a function of *x*.

The equations that will be used in the following charts will have functions of different independent variables, such as $f(x) + f(y) = f(z)$; or *f(u)* multiplied by $f(v) = f(w)$.

11.8.2 Scales. A straight or curved line may have graduations with equal distances between them (uniform scale), or distances corresponding to the values of the function of the variable (functional scale).

---

2. Prepared by Robert Angerman, Industrial Engineer, Progressive Dynamics, Inc.

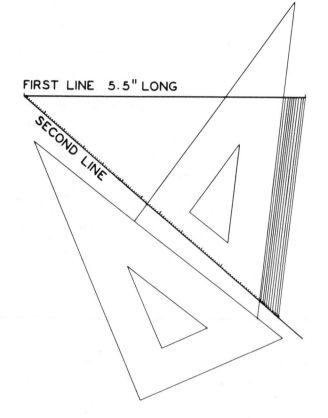

FIRST LINE 5.5" LONG

SECOND LINE

Fig. 11-23. Dividing a line into equal segments to develop a uniform nomograph scale.

Uniform nomograph scale. When a uniform scale is to be constructed, the first step is to solve the function for the upper and lower limits of the independent variable and then determine the length of the scale. For example, when the function $f(x) = x^2 + 3x + 5$ is solved for $f(x)$ with x varying from 0 to 10, the scale must contain 130 units reading from 5 to 135. If this scale must be constructed to 6.5 inches, it is possible to divide the distance (6.5 inches) by 130 units, giving a value of 0.05 inch per unit. Since this is an easily measured value, it is possible to rule off one hundred thirty 0.05-inch increments to make the uniform scale.

If the same function having the same limits had to be drawn to a 5.5-inch long uniform scale, it would not be feasible or possible to measure increments of 0.0423077 inch each. In cases of this sort, it is easier to divide the scale geometrically in the following manner: (See Fig. 11-23.)

1. Draw a straight line 5.5 inches long.

2. Draw another straight line at an angle to the first line passing through the end of the line.

3. On the second line lay out one hundred thirty equally spaced units of any convenient length.

4. Draw a straight line between the opposite end of the first line and the one hundred thirtieth division of the second line.

5. Continue to draw lines from the division marks on the second line to the first line parallel to the one hundred thirtieth division line. The intersection of these parallel lines with the first line will mark the uniform scale desired.

6. To complete the scale, number the intersection points from 5 to 135.

X	0	1	2	3	4	5	6	7	8	9	10
$\frac{X}{2}$	0	0.5	1.0	1.5	2.0	2.5	3.0	3.5	4.0	4.5	5.0
$\left(\frac{X}{2}\right)^2$	0	0.25	1.00	2.25	4.00	6.25	9.00	12.25	16.00	20.25	25.00
$.2\left(\frac{X}{2}\right)^2$	0	0.05	0.20	0.45	0.80	1.35	1.80	2.45	3.20	4.05	5.00

$$\text{PROPORTIONALITY CONSTANT} = \frac{5}{25-0} = 0.2$$

Fig. 11-24. Tabulated calculations of a five inch long functional scale of $f(x) = (x/2)^2$.

Functional nomograph scale. The construction of a functional scale is best accomplished in steps and the data recorded in tabular form. As in the uniform scale, the upper and lower limits of the independent variable and the length of the scale will be known or must be selected. If it is desired to construct a functional scale of the function $f(x) = (x/2)^2$ with values of x ranging from 0 to 10 within a 5-inch space, use the following steps: (See Fig. 11-24.)

1. Record the values of the independent variable in the table.

2. Compute the numerical value of the function for each value of the independent variable.

3. Calculate the proportionality factor by dividing the length of the scale by the difference between the upper and lower values of the function.

4. Multiply the numerical values of the function by the proportionality factor to determine the scaled length of each value.

5. Lay out the computed distances on a straight line and record the value of the independent variable for each distance.

A functional scale based upon logarithmic values can be graduated in the same manner, but a simpler method may be used to reduce the calculation and construction time. Following the geometrical method described in uniform scale construction, draw the first line to the desired length and then graduate this line using a printed log scale. See Fig. 11-25.

11.8.3 Nomograph equations. The general equation on which many nomographs are based is $f_1(x) + f_2(x) \cdot f_3(y) = f_4(z)$, where f_1 and f_2 are different functions of the same independent variable x; f_3 is a function of y and is multiplied by $f_2(x)$; and $f_4(z)$ is a function of z. Since the variables y and z appear in only one term each, they will show on the nomograph as two parallel straight lines. The independent variable x appears in two terms and is normally a curved line. If $f_1(x)$ equals 0, the chart will take the form of an N or a Z when uniform scales are used. If $f_2(x)$ is a constant, the chart will consist of three parallel lines. The forms of the nomographs with their related general equations are shown in Fig. 11-26.

Curved: $f_1(x) + f_2(x) \cdot f_3(y) = f_4(z)$

N or Z: $f_2(x) \cdot f_3(y) = f_4(z)$

Parallel: $f_1(x) + f_3(y) = f_4(z)$

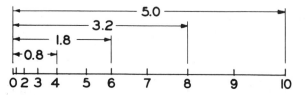

Fig. 11-25. Functional scale of $f(x) = (x/2)^2$ with x ranging from zero to ten.

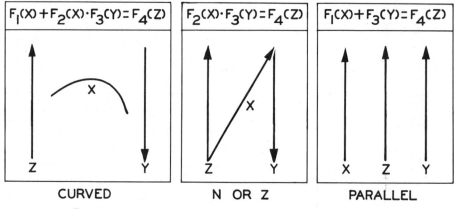

Fig. 11-26. Basic forms of nomographs with their general equations.

11.8.4 Parallel scale nomograph.

The parallel scale nomograph, to depict the values of equation $f_1(x) + f_3(y) = f_4(z)$, is constructed as follows: (See Fig. 11-27.

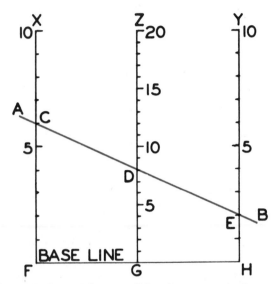

Fig. 11-27. Constructing a parallel scale nomograph where X + Y = Z.

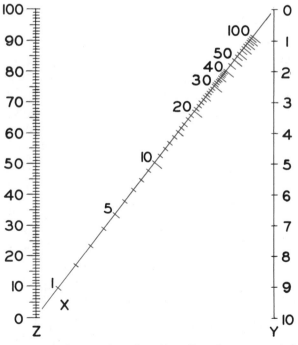

Fig. 11-28. Construction of an N or Z scale nomograph in which X · Y = Z.

1. Draw three parallel lines X, Y, and Z at equal distances apart.

2. Draw a line intersecting all three and call this the base line. The base line does not necessarily have to be drawn perpendicular to the three vertical lines.

3. Lay off a series of equally spaced divisions starting at the base line for the outer lines X and Y. Lay off divisions on line Z using a spacing equal to half that used in the outer lines.

4. Number all divisions as shown.

To illustrate how the nomograph works, draw any straight line, such as AB, intersecting lines X, Z, and Y at C, D, and E, respectively. Lines AB, X, Y and the base line form a trapezoid FCEH. Since line Z is equidistant from CF and EH, the length of DG is equal to half the sum of the lengths of CF and EH. This is the reason for making the divisions on line Z half as long as the divisions on lines X and Y.

It now becomes evident that every straight line intersecting the three lines will solve the equation $f_1(x) + f_3(y) = f_4(z)$, because the value of the function of x plus the value of the function of y can be read as a value of the function of z on line Z.

To use the nomograph for subtraction, convert the equation to the form $f_1(x) = f_4(z) - f_3(y)$. The difference between the values of z and y can be read on line X.

It is obvious that this is too simple a relationship to spend time constructing a nomograph, but the basic concept will be used later.

11.8.5 N or Z scale nomograph.

The N or Z scale nomograph, which solves the equation $f_2(x) \cdot f_3(y) = f_4(z)$, is made in the following manner: (See Fig. 11-28.)

1. Draw the outer vertical lines Y and Z and lay off divisions using uniform spacing. It is necessary to have the origins opposite each other rather than on the same base line as done previously.

2. Draw the diagonal line connecting the origins of lines Y and Z. The diagonal becomes line X.

3. To graduate line X, use any value of Y and align with various values of Z with a straightedge. The intersection of the straightedge with line X is the location of the value of Z divided by Y. In reality, a functional scale is constructed along the diagonal in this manner.

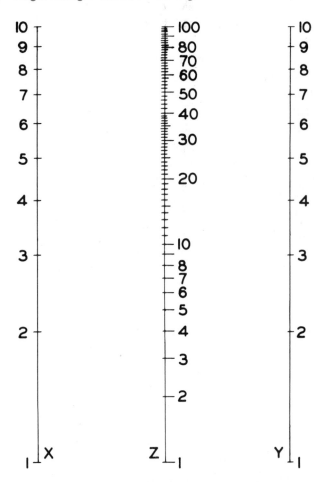

Fig. 11-29. Parallel scale nomograph using logarithmic values.

It is now possible to determine the value of Z by aligning the X and Y values and reading the value at the intersection point with line Z. This method of constructing the nomograph to solve equations involving multiplication has the advantage of using uniform scales on the vertical lines and permitting easy construction of the functional scale.

Another method for constructing a nomograph to solve multiplication equations involves the use of the previously described parallel scale. It is well known that adding the logarithms of numbers will yield the product of the numbers. The equation $f_2(x) \cdot f_3(y) = f_4(z)$ is identical to the equation *log $f_2(x)$ + log $f_3(y)$ = log $f_4(z)$.* It is therefore possible to use the parallel scale form to multiply by using a logarithmic scale instead of uniform spacing. Lines X and Y will have identical scale division spacings starting with one instead of zero at the base line. Line Z has a two-cycle log scale as shown in Fig. 11-29.

11.8.6 Application of a nomograph. While the premises on which nomographs are based are not difficult to understand, applying the basic approach to various equations in designing a nomograph can present practical problems. To avoid the cut-and-try method generally used, it has been found helpful to reduce the equations to simpler forms of expressions. When this is done, it is possible to choose the scale that should be used for each of the vertical lines and the relative distances between the lines. The expressions, choice of scales, and distances between lines are given in Fig. 11-30. The scales, drawn to their proper logarithmic value, are presented in Fig. 11-31.

This method of preparing the nomograph can best be presented by an example. Assume that a nomograph is desired to express the relationship between the volume *(V)* of a right circular cylinder with the cylinder's radius *(r)* and height *(h)*. The formula $V = \pi r^2 h$ is reduced to the expression Z

EXPRESSION	X	Y	Z	RATIO Lx:Ly	EXPRESSION	X	Y	Z	RATIO Lx:Ly	EXPRESSION	X	Y	Z	RATIO Lx:Ly
$Z = XY$	1	1	3	1:1	$Z^2 = XY$	1	1	1	1:1	$Z^3 = XY$	1	3	1	2:1
	1	3	4	2:1		1	4	3	3:1		1	4	2	3:1
	1	4	5	3:1		2	2	2	1:1		3	1	1	1:2
	3	1	4	1:2		3	3	3	1:1		3	3	2	1:1
	3	3	5	1:1		3	5	4	2:1		3	5	3	2:1
	3	5	6	2:1		4	4	4	1:1		4	1	2	1:3
	4	1	5	1:3		4	1	3	1:3		4	4	3	1:1
	4	4	6	1:1		5	3	4	1:2		5	3	3	1:2
	5	3	6	1:2		5	5	5	1:1	$Z^3 = XY^2$	1	1	1	2:1
$Z = XY^2$	1	1	4	2:1	$Z^2 = XY^2$	3	1	3	1:1		2	2	2	2:1
	2	2	5	2:1		3	3	4	2:1		3	1	2	1:1
	3	1	5	1:1		5	1	4	1:2		3	3	3	2:1
	3	3	6	2:1		5	3	5	1:1		5	1	3	1:2
	5	1	6	1:2	$Z^2 = XY^3$	1	1	3	3:1	$Z^3 = XY^3$	1	1	2	3:1
$Z = XY^3$	1	1	5	3:1		3	2	4	2:1		3	2	3	2:1
	3	2	6	2:1		4	1	4	1:1		4	1	3	1:1
	4	1	6	1:1		5	2	5	1:1	$Z^3 = XY^4$	3	1	3	2:1
$Z = XY^4$	3	1	6	2:1	$Z^2 = XY^4$	3	1	4	2:1		5	3	5	2:1
$Z = X^2Y^2$	1	1	5	1:1		5	1	5	1:1	$Z^3 = X^2Y^2$	1	1	2	1:1
	1	3	6	2:1	$Z^2 = X^2Y^2$	1	2	4	2:1		1	3	3	2:1
	3	1	6	1:2		3	2	5	1:1		5	1	3	1:2
$Z = X^2Y^4$	1	1	6	2:1	$Z^2 = X^3Y^3$	1	1	4	1:1	$Z^3 = X^2Y^3$	1	2	3	2:1
$Z = X^3Y^3$	1	1	6	1:1		2	2	5	1:1	$Z^3 = X^2Y^4$	1	1	3	2:1
$Z^4 = XY$	1	4	1	3:1		3	3	6	1:1		3	3	5	2:1
	3	3	1	1:1	$Z^2 = X^3Y^4$	2	1	5	1:1	$Z^3 = X^3Y^4$	2	3	5	2:1
	4	1	1	1:3	$Z^4 = XY^3$	1	1	1	3:1	$Z^4 = XY^2$	3	1	1	1:1
	5	5	3	1:1		2	2	2	3:1		5	3	3	1:1
$Z^4 = X^3Y^3$	2	2	3	1:1		3	3	3	3:1	$Z^4 = XY^4$	5	1	3	1:1

Fig. 11-30. This chart includes expressions, scales, and ratios for designing a variety of nomographs.

$= XY^2$. By referring to Fig. 11-30, it is found there are five combinations of scales to depict this relationship. For ease in reading, it is best to use the scale that has the greatest distance between divisions. If the first set is selected, lines X and Y will be drawn as scale 1 and Z as scale 4. The $L_x:L_y$ ratio column given in Fig. 11-30 establishes the distance of the X and Y lines from the Z line. In the example, locate line Y any desired distance from line Z and then draw line X twice that distance from Z to satisfy the $L_x:L_y$ ratio of 2:1. See Fig. 11-32.

The scales for lines X and Y do not have to be indexed with respect to each other. When these two scales are drawn, a construction line should be drawn at the proper location of line Z. The formula should be solved for V with any value of r and h. Using $r = 2$ inches and $h = 2.5$ inches, V be-

comes 31.416 cubic inches. A straightedge should be used to align 2 on the R scale and 2.5 on the H scale. A mark should be made on the V line at the point of intersecton with the straightedge. A copy of scale 4 should be positioned along the vertical construction line with its 31.416 value on the intersection point that was previously drawn. Locating the height of line Z in this manner makes unnecessary calculating the value of each of the scales at a base line through the three lines to establish their relative heights.

The expressions listed in Fig. 11-30 can be extended to cover an even wider variety of equations by noting the following:

1. Constants do not alter the form of the nomograph; they only shift the location of the Z scale up or down. Z was such a constant in the example just described.

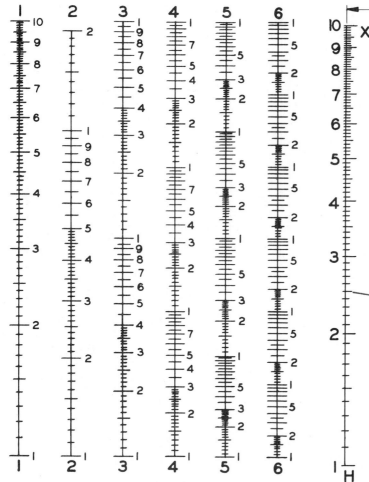

Fig. 11-31. Scales which may be used with Fig. 11-30.

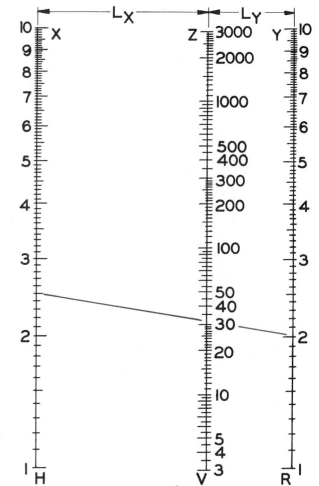

Fig. 11-32. Nomograph for determining volume of a circular cylinder where $V = \pi r^2 h$.

2. If the variables appear as a quotient, as in $Z = X/Y^2$, the expression $Z = XY^2$ should be used with scale Y being drawn upside down.

3. When an equation has more than three variables, $Z = (X) (Y) (W)$, it should be drawn in two stages: first as $A = (X) (Y)$ and then as $Z = (A) (W)$. An example is shown in Fig. 11-33.

4. If the equation has a trigonometric function of the form $z = \sin x/y^2$, the expression $z = x^1/y^2$ should be used, with $x^1 = \sin x$. The scale should be graduated by the method used to develop a functional scale.

11.8.7 Curved scale nomographs. The curved scale nomograph which satisfies the equation $f_1(x) + f_2(x) \cdot f_3(y) = f_4(z)$ is constructed in the following manner: (See Fig. 11-34.)

1. Draw two lines Z and Y parallel to each other.

2. Select the location of the origin on each line and letter them O and O' respectively. Connect these two points with a straight line. It is not necessary for the line OO' to be perpendicular to lines Z and Y.

3. Lay off the scale

$$m_1 \left[f_4(z_{max}) - f_4(z_{min}) \right]$$

on line Z. The total length of the Z scale is equal to OZ.

4. Lay off the scale

$$m_2 \left[f_3(y_{max}) - f_3(y_{min}) \right]$$

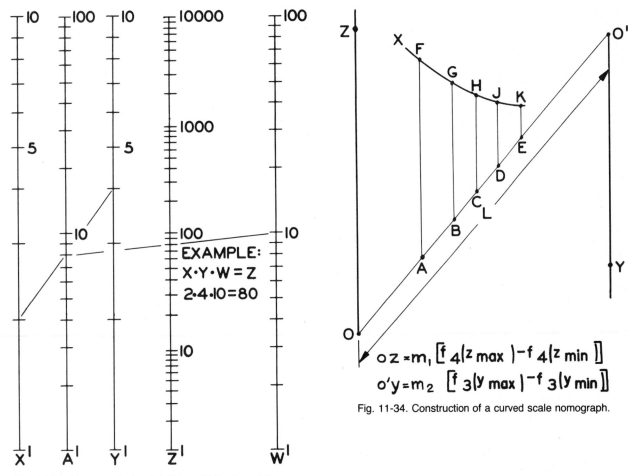

Fig. 11-33. Nomograph showing the multiplication of three variables.

Fig. 11-34. Construction of a curved scale nomograph.

on line *Y*, with the total length equal to *O'Y*. The length of line *OO'* is designated *L*. Fig. 11-34 also includes an imaginary configuration of the curved line *X* and the method of plotting the curve.

5. Use the quotient

$$\frac{m_1 f_2(x) l}{m_1 f_2(x) + m_2}$$

to calculate the distances of points *A*, *B*, *C*, *D*, and *E* from the origin *O* measured along the line *OO'* for various values of *X*.

6. Draw lines through these plotted points parallel to line *Z*. The distances from the line *OO'* to the curved line *X* are determined by the quotient

$$\frac{m_1 m_2 f_1(x)}{m_1 f_2(x) + m_2}.$$

Points *F*, *G*, *H*, *J*, and *K* represent the plotted distance from line *OO'* calculated by the equation. The positive values will be plotted above the line and the negative values below.

7. Draw line *X* using these plotted points as a basis. Any line intersecting the three scales will solve the equation

$$f_1(x) + f_2(x) \cdot f_3(y) = f_4(z).$$

This general background on curved scale nomographs can be best illustrated with an example to show the steps necessary to complete this type of alignment chart. If it is desired to develop a nomograph to determine the chordal height and length of a chord for a circle, it is necessary to put the equation $d^2 = 8rh - 4h^2$ into standard form:

$$f_1(x) + f_2(x) \cdot f_3(y) = f_4(z)$$
$$-4h^2 + 8h \cdot r = d^2.$$

The length of increments for various values of h along the OO' line will be calculated by the quotient

$$\frac{(m_d)\,(8h)\,(l)}{(m_d)\,(8h)\,+\,(m_r)},$$

where $m_d = m_1$, $m_r = m_2$, and $8h = f_2(x)$.

The vertical distances are determined by the quotient

$$\frac{(m_d)\,(m_r)\,(-4h^2)}{(m_d)\,(8h)+(m_r)},$$

where $-4h^2 = f_1(x)$.

It is planned to fit the nomograph into a space four by five inches. The OZ and $O'Y$ lines may both be five inches long with a four-inch perpendicular distance between them. The following ranges of values will provide the desired information: d varies from 0 to 20, r varies from 0 to 10, h varies from 1 to 10.

The value of m_d and m_r should be calculated next using the preceding information:

$$m_d = \frac{5}{(d^2{}_{max} - d^2{}_{min})} = 0.0125$$

$$m_r = \frac{5}{(r_{max} - r_{min})} = 0.5.$$

The value of l should be calculated by the equation $l = \sqrt{(5)^2 + (4)^2}$ because it happens to be the hypotenuse of the triangle OYO'. This calculation yields a value of 6.403 for l. The quotients

$$\frac{(m_d)\,(8h)\,(l)}{(m_d)\,(8h)+(m_r)} \quad \text{and} \quad \frac{(m_d)\,(m_r)\,(-4h^2)}{(m_d)\,(8h)+(m_r)}$$

then become

$$\frac{(0.0125)\,(8)\,(h)\,(6.403)}{(0.0125)\,(8)\,(h)\,+\,(0.5)}$$

and

$$\frac{(0.0125)\,(0.5)\,(-4)\,(h^2)}{(0.0125)\,(8)\,(h)\,+\,(0.5)}$$

which reduce to

$$\frac{0.6403h}{0.1h + 0.5} \quad \text{and} \quad \frac{-0.025\,h^2}{0.1h + 0.5}$$

By applying the desired range of values for h to the above quotients, the distances from the origin on line D can be calculated. In addition, the vertical length of the lines to determine the loci of points on line H can be determined. These values are included in the table of Fig 11-35. Now that all the necessary values for the various lines have been calculated, the chart can be drawn. See Fig. 11-36. Scale D is drawn on the left side using a functional scale because D is squared. Scale R is drawn as a uniform scale. Both of these scales are 5 inches long, and they are 4 inches apart to

EQUATION	1	2	3	4	5	6	7	8	9	10
1	1.067	1.829	2.401	2.846	3.202	3.493	3.735	3.940	4.116	4.269
2	-0.042	-0.143	-0.281	-0.444	-0.625	-0.818	-1.021	-1.231	-1.446	-1.667

Fig. 11-35. Tabulated scale for the construction of the curved scale nomograph in Fig. 11-36.

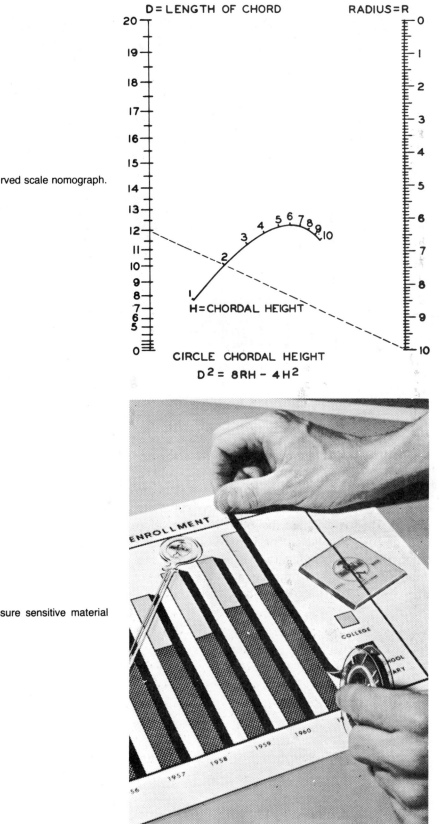

Fig. 11-36. A curved scale nomograph.

D = LENGTH OF CHORD   RADIUS = R

H = CHORDAL HEIGHT

CIRCLE CHORDAL HEIGHT

$$D^2 = 8RH - 4H^2$$

Fig. 11-37. Typical applications of pressure sensitive material for preparing charts.

Fig. 11-38. Some of the many materials available for illustrating charts.

meet the space requirements. The lengths presented in the table are scaled on a light construction line connecting the origins of scales *D* and *R*. The vertical distances are then measured from the construction line and the points for the *H* curve are plotted. These points are the location of the value of *h* given in the table. The curve is then drawn using these points. A straight line through the three scales will solve the equation

$$d^2 = 8rh - 4h^2.$$

11.9 Simplifying Chart Visualization

A variety of commercial materials can be used to simplify the preparation of charts. Printed designs representing lines and bars are available as well as symbols and shading film. See Figs. 11-37 and 11-38. These materials are printed on plastic tapes or sheets with pressure-sensitive adhesive and are relatively easy to use.

Problems for Section III
Technical Illustrations

The problems in this section are designed to provide a knowledge of drafting principles dealing with

> *Unit 8. Pictorial Drawings*
> *Unit 9. Freehand Sketching*
> *Unit 10. Patent Drawing*
> *Unit 11. Drawing Charts*

Problems for Section III include

> **Problems 1-30 Freehand Sketching**
> **Problems 31-42 Isometric Drawings**
> **Problems 43-46 Dimetric Drawings**
> **Problems 47-51 Oblique Drawings**
> **Problems 52-66 Perspective Drawings**
> **Problems 67-73 Patent Drawings**
> **Problems 74-79 Charts and Graphs**

Problems 1-30 Freehand Sketching

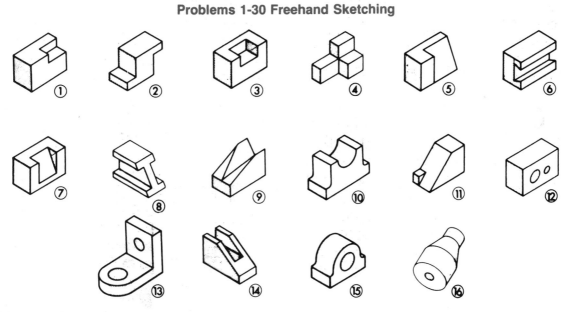

Problems 1-16
Make as many multiview freehand sketches as assigned of the pictorial drawings shown. Draw only the necessary views. Select all required dimensions.

Problems having decimal-inch values may be converted to equivalent metric sizes by using the millimeter conversion chart or the problems may be redesigned by assigning new millimeter dimensional values more compatible with metric production.

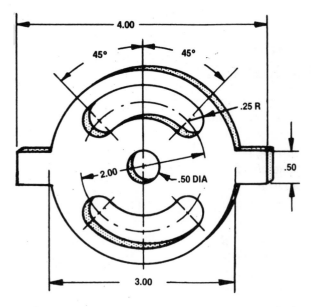

Problem 17
Make a single-view freehand dimensioned sketch of this gasket.

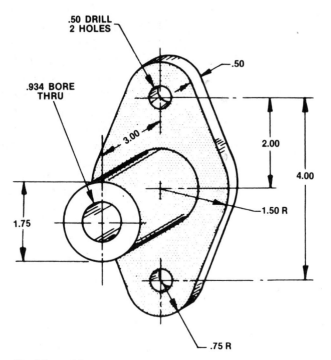

Problem 19
Sketch the support showing required dimensions. Include only the needed views.

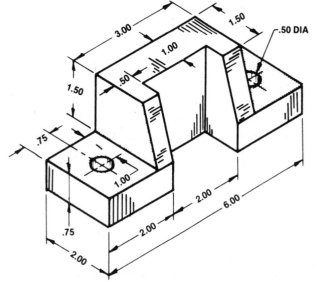

Problem 18
Sketch and dimension the required views of the machined steel block.

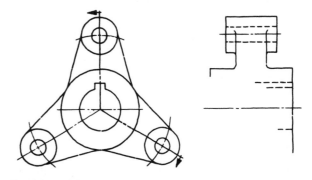

Problem 21
Make a two-view sectional sketch of the object shown. Assume all sizes.

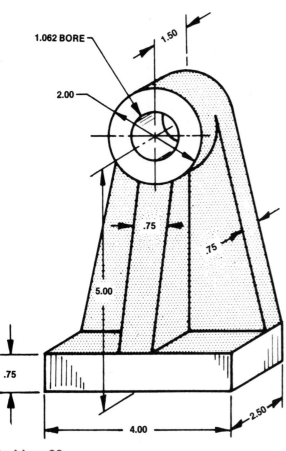

Problem 20
Make a dimensioned sketch showing the necessary views of the bracket.

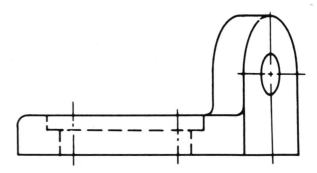

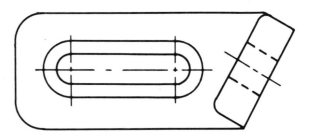

Problem 22
Sketch the necessary views of the object shown. Assume all sizes.

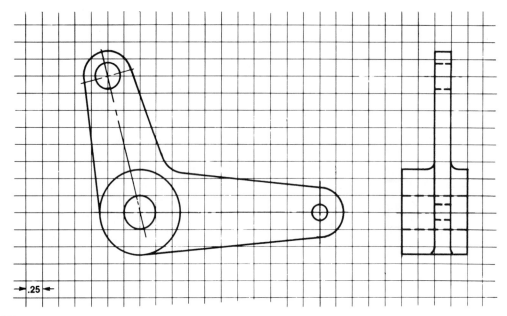

.25

Problem 23
Make a cabinet sketch of this object.

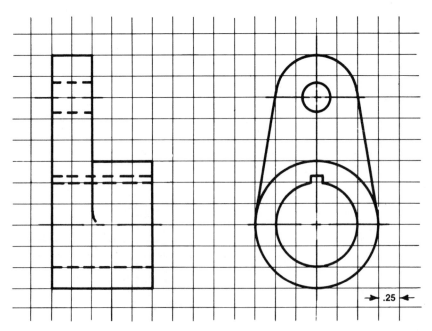

.25

Problem 24
Make an isometric sketch of the object shown.

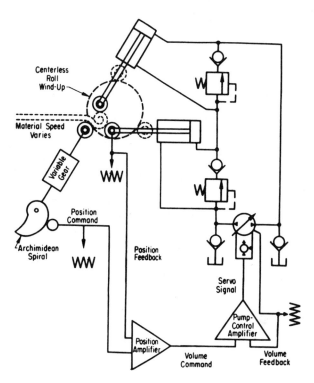

Problem 25
Make a sketch of the illustrated cam-controlled servo-pump circuit.

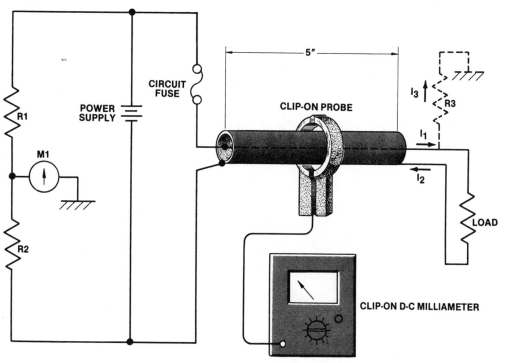

Problem 26
Sketch the diagram for this ground leak detector.

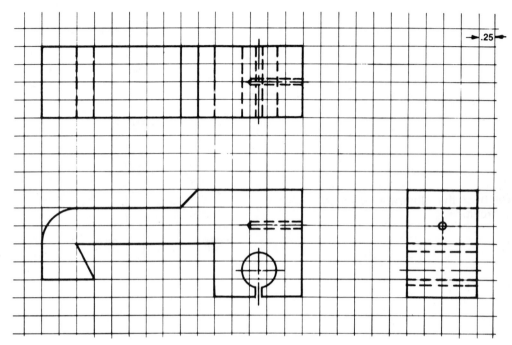

Problem 27
Make a two-point perspective sketch of the object
shown.

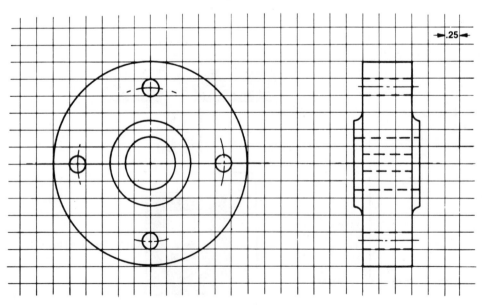

Problem 28
Make an oblique sketch of the object shown.

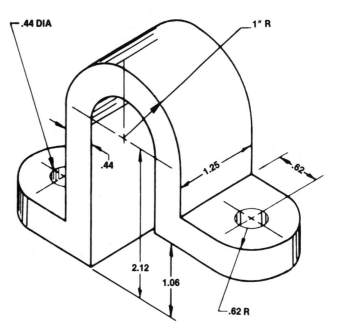

Problem 29
Make a three-view sketch of the support shown.
Dimension completely. Use suitable scale.

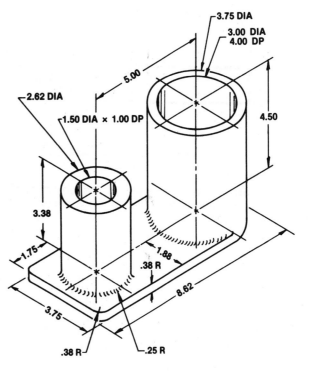

Problem 30
Make a three-view sketch of the holder shown.
Dimension completely. Use convenient scale.

Problems 31-42 Isometric Drawings

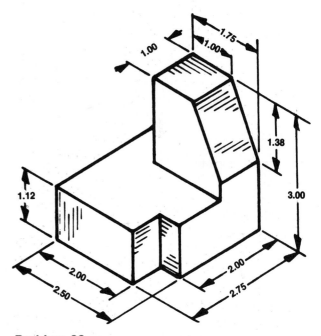

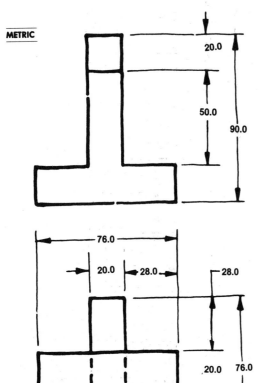

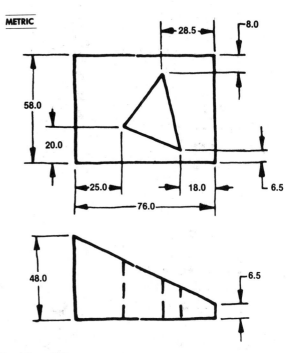

METRIC

Problem 32
Make an isometric drawing of the object shown.
Dimension completely. Use any convenient scale.

Problem 31
Make an isometric drawing of the object shown.
Include all dimensions. Use any convenient scale.
All dimensions in mm.

Problem 33
Make an isometric drawing of the object shown
and include all dimensions. Use any convenient
scale. Dimensions in mm.

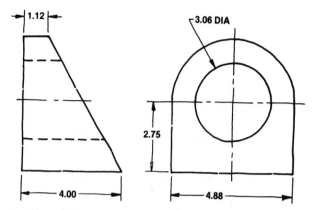

Problem 34
Prepare an isometric drawing of the object shown and include all dimensions. Use any convenient scale.

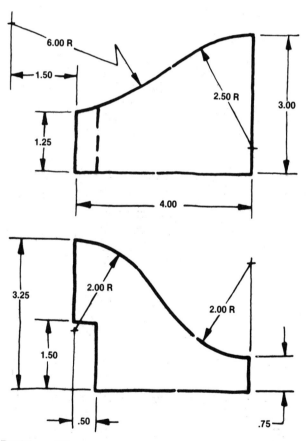

Problem 35
Prepare an isometric drawing of the object shown and include all dimensions. Use any convenient scale. Convert dimensions to mm.

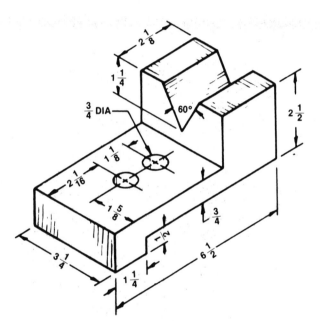

Problem 36
Prepare an isometric drawing of the object shown and dimension completely. Use any convenient scale. Convert sizes to mm.

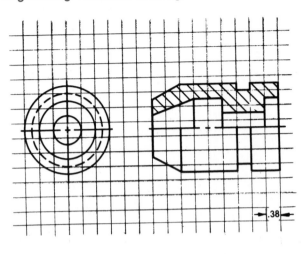

Problem 37
Make an isometric drawing of the object shown and assume all dimensions based on the indicated size of the grid. Use any convenient scale.

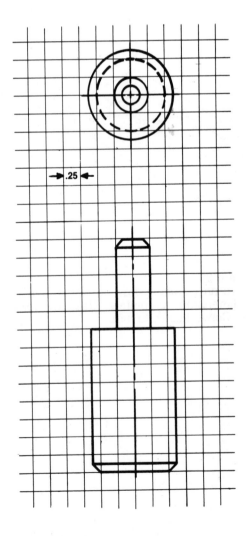

Problem 38
Make an isometric drawing of the object shown and assume all dimensions based on the indicated size of the grid. Use any convenient scale.

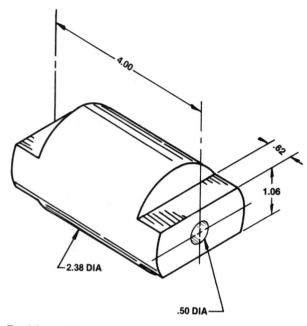

Problem 39
Prepare an isometric drawing of the object shown and dimension. Use any convenient scale and apply suitable shading.

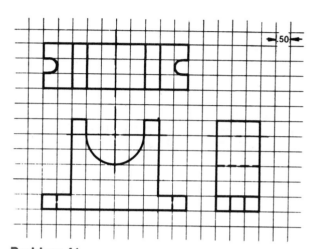

Problem 41
Make an isometric drawing of the object shown and assume all dimensions based on the indicated size of the grid. Use any convenient scale.

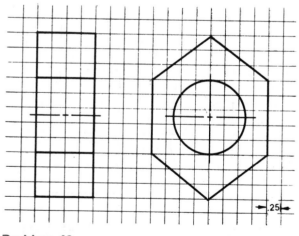

Problem 40
Make an isometric drawing of the object shown and assume all dimensions based on the indicated size of the grid. Use any convenient scale. Apply suitable shading.

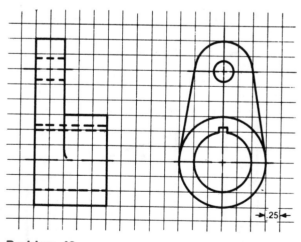

Problem 42
Make an isometric drawing of the object shown and assume all dimensions based on the indicated size of the grid. Use any convenient scale. Apply suitable shading.

Problems 43-46 Dimetric Drawings

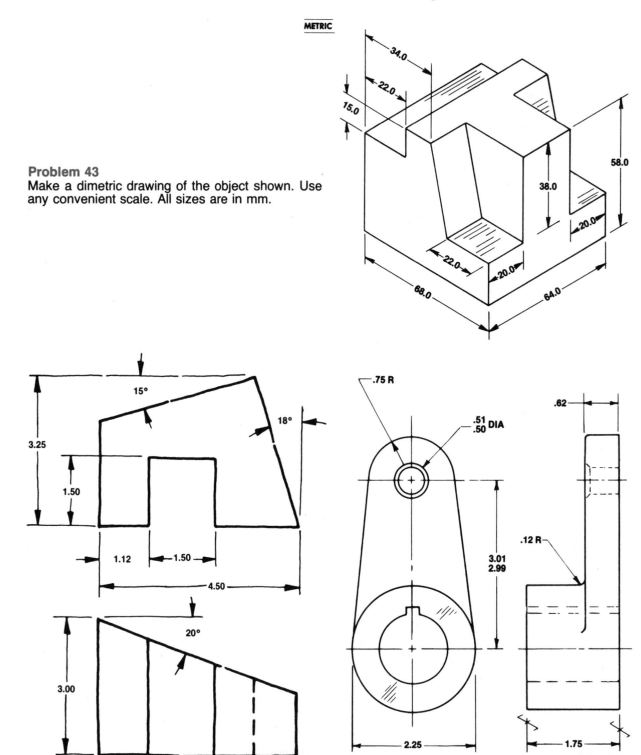

METRIC

Problem 43
Make a dimetric drawing of the object shown. Use any convenient scale. All sizes are in mm.

Problem 44
Make a dimetric drawing of the object shown. Include all dimensions. Use any convenient scale.

Problem 45
Make a dimetric drawing of the shear hub shown. Dimension completely. Use convenient scale. Apply suitable shading.

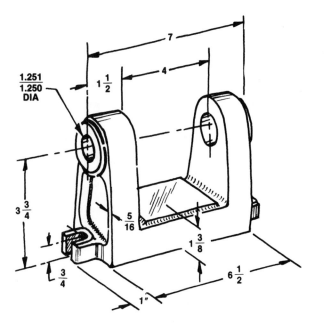

Problem 46
Make a dimetric drawing of the shaft bracket shown. Use convenient scale and dimension completely in metric values. Apply suitable shading.

Problems 47-51 Oblique Drawings

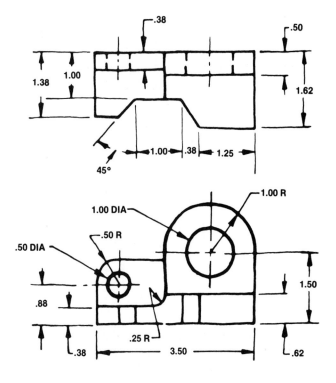

Problem 47
Make a cabinet drawing of the object shown. Include all dimensions. Use any convenient scale.

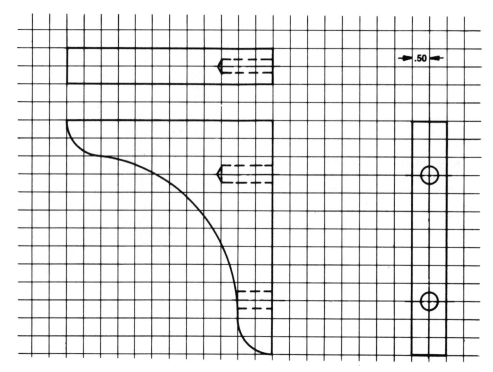

Problem 48
Prepare a cabinet drawing of the object shown
and assume all dimensions based on the indicated
size of the grid. Use any convenient scale.

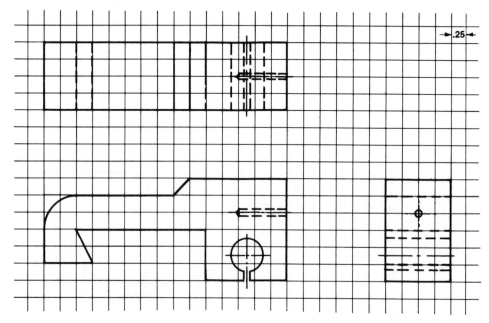

Problem 49
Prepare a cabinet drawing of the object shown.
Assume all dimensions based on the indicated
size of the grid. Use any convenient scale.

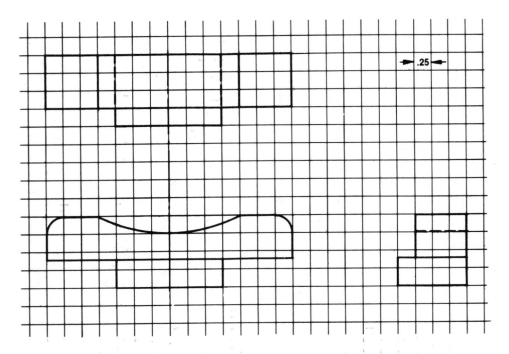

Problem 50
Prepare a cabinet drawing of the object shown.
Assume all dimensions based on the indicated
size of the grid. Use any convenient scale.

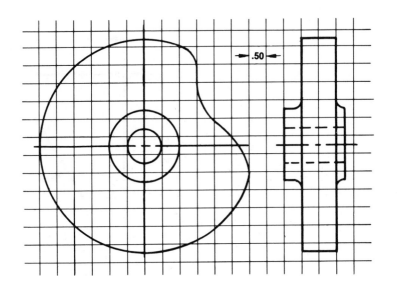

Problem 51
Make a general oblique drawing of the object
shown. Assume all dimensions based on the indicated size of the grid.

Problems 52-66 Perspective Drawings

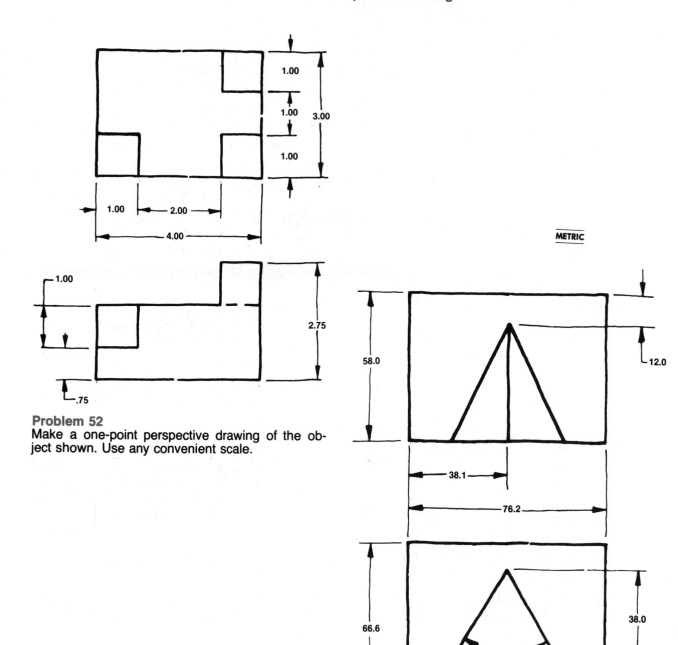

METRIC

Problem 52
Make a one-point perspective drawing of the object shown. Use any convenient scale.

Problem 53
Make a two-point perspective drawing of the object shown. Use any convenient scale. All dimensions in mm.

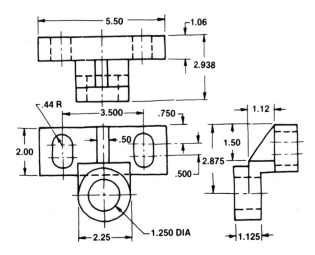

Problem 55
Make a two-point perspective drawing of the object shown. Use any convenient scale. Use proper shading effects.

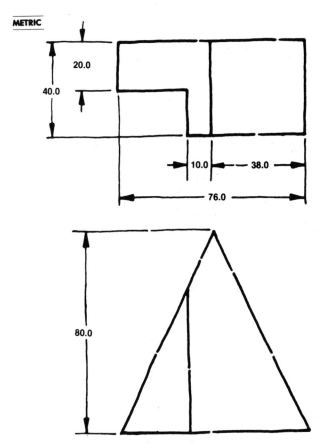

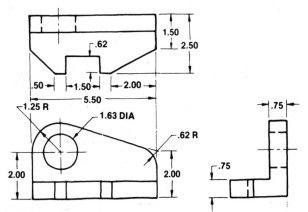

Problem 56
Make a three-point perspective drawing of the object shown and use proper shading effects. Select any convenient scale.

Problem 54
Make a two-point perspective drawing of the object shown. Use any convenient scale. All dimensions in mm.

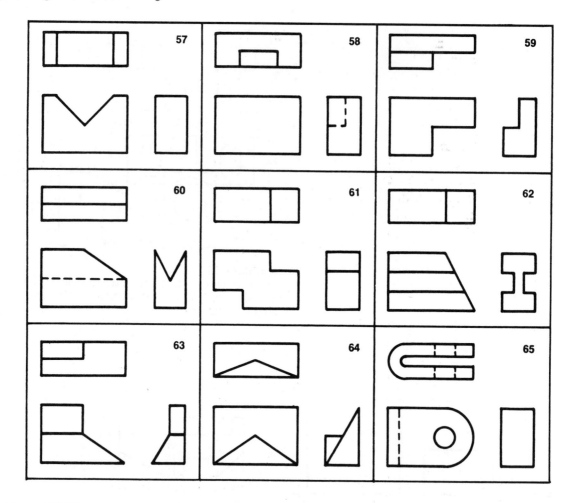

Problems 57-65

First complete views by adding missing lines to objects assigned. Then draw appropriate perspective. Assume all sizes.

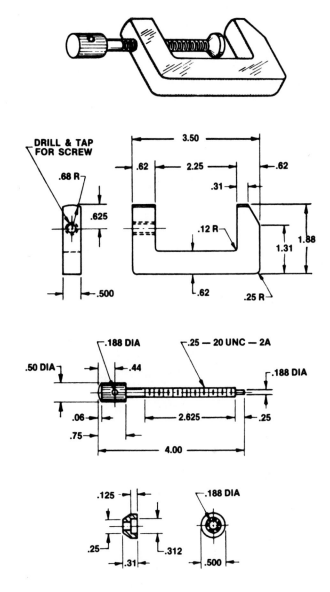

Problem 66

Prepare an exploded assembly drawing of the details shown. Use proper shading effects. Select any convenient scale.

Problems 67-73 Patent Drawings

Problems 67-73

Assume that the products illustrated are patentable. Prepare patent drawings of any selected or assigned detail of the products chosen, incorporating all drawing regulations as prescribed by the Patent Office. Shade the product according to the practices shown in Figs. 10-3 through 10-12. Assume all shape sizes and design features.

FURNITURE LEG

Problem 71.

BROADHEAD
HUNTING ARROW

Problem 67.

POKER CHIP
RACK

Problem 68.

CLOCK
DUST COVER

Problem 72.

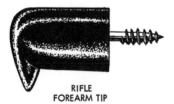

RIFLE
FOREARM TIP

Problem 69.

WHISTLE

Problem 70.

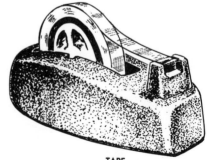

TAPE
DISPENSER

Problem 73.

Problems 74-79 Charts and Graphs

TABLE I.

Elements	Percent
Oxygen	46.43
Silicon	27.77
Aluminum	8.14
Iron	5.12
Calcium	3.62
Sodium	2.85
Potassium	2.60
Magnesium	2.09
Titanium	.63
Phosphorus	.13
Hydrogen	.13
All Others	.49

Problem 74
Using the values shown in Table I showing the occurrence of chemical elements in the earth's crust, draw a percentage bar chart.

TABLE II.

Time (min.)	Temperature	Time (Min.)	Temperature
0	68°F	40	300°F
3	95	45	300
5	115	50	300
10	165	60	300
12	190	65	300
13	199	70	300
14	201	75	301
15	200	76	302
16	200	77	299
17	201	78	300
18	200	79	300
19	210	80	300
20	218	85	300
25	259	86	285
30	292	87	273
31	301	88	260
32	300	90	238
33	300	95	190
35	300	100	152

Problem 75
During a study to determine the optimum curing temperature of a structural adhesive, the temperature of the heated platen press was recorded at various time intervals as shown in Table II.

(a) Using rectangular coordinate graph paper 20 x 20 divisions per inch, construct a graph for a technical report showing the relation between time and the corresponding temperature.

(b) Using logarithmic paper show graphically the relation between time and the corresponding temperature.

(c) Using the graph drawing in (b), determine the rate of increase in temperature per unit time between 68°F and 200°F (0 to 15 minutes) and the decrease in temperature per unit time between 300°F and 152°F (85 to 100 minutes).

Problem 76
Show by a bar chart the assets of the 100 largest manufacturing corporations in the United States.

Petroleum	31.4%
Rubber	3.0%
Steel and Products	13.1%
Textiles	1.6%
Miscellaneous	5.1%
Automobiles	11.5%
Aircraft	1.9%
Chemicals	8.1%
Food, etc.	7.9%
Electrical Machinery	5.9%
Non-electrical Machinery	4.7%
Non-ferrous Metals	5.8%

Problem 77
Show by means of surface chart the unemployment claims of the two governmental units shown.

UNEMPLOYMENT INSURANCE CLAIMS
FOR UNITED STATES AND PAINSVILLE

	U.S. Millions	Painsville Millions
1969	2.5	1.8
1970	3.6	4.4
1971	3.8	3.9
1972	3.4	4.0
1973	2.2	2.5
1974	4.1	4.1
1975	8.2	9.6
1976	7.4	6.4

Problem 78
Draw a pie chart showing the same information given in Problem 74.

Problem 79
Calculate the values of the given equations and plot the curve for quantitive purposes.

(a) $Y = x^3 - 12x + 16$,
x varies from 0 to 5

(b) $Y^2 = 4x$
x varies from 0 to 4

(c) $Y^2 = \dfrac{144 - 9x^2}{16}$
x varies from -4 to $+4$

(d) $Y = \sin x$
x varies from $0°$ to $360°$

(e) $Y = \cos x$
x varies from $0°$ to $360°$

(f) $Y = \log x^2$
x varies from 1 to 10

(g) $Y = \dfrac{1}{x^2}$
x varies from 1 to 10

SECTION IV

Geometry in Drafting

SECTION IV
UNIT 12

Geometric Constructions

In the process of preparing a drawing, there will be many occasions when it will be necessary to utilize one or more of the geometric constructions outlined in this unit. These construction techniques will be helpful in solving problems involving the application of points, lines, planes, and curved and warped surfaces.

12.1 Lines

A line may be defined as a continuous extent of length, straight or curved, without breadth or thickness. Construction problems may involve bisecting a line, dividing a line into equal parts, drawing parallel lines, or drawing lines perpendicular to each other.

12.1.1 Bisecting a line or arc.

Given line or arc *AB.*

Geometric method. (See Fig. 12-1.)

1. Set the compass for any radius greater than one-half of *AB.* Using *A* and *B* as centers, draw two arcs to intersect at *C* and two arcs to intersect at *D.*

2. Draw a line connecting points *C* and *D.* The point at which line *CD* crosses *AB* is the center.

Preferred method. (See Fig. 12-2.)

1. Find the center of line *AB* with dividers by the trial-and-error method.

2. Draw a vertical line through the center.

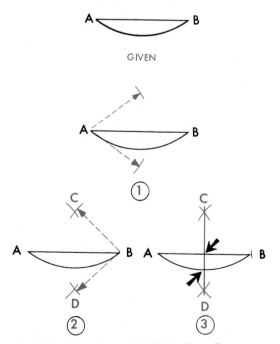

Fig. 12-1. Geometric method of bisecting a line or arc.

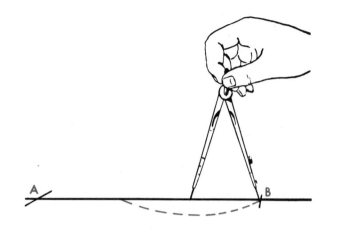

Fig. 12-2. Preferred method of bisecting a line or arc with dividers.

12.1.2 Drawing a straight line parallel to another straight line through a given point or at a given distance.

Given line *AB* and the required point *C*.

Geometric method. (See Fig. 12-3.)

1. With *C* as the center and using any convenient radius, strike the arc *DE* to intersect line *AB* at point *F*.

2. With *F* as the center and using the same radius, draw arc *GH* to intersect line *AB* at *K*.

3. Using *CK* as a radius and *F* as the center, strike an arc intersecting arc *DE* at *L*.

4. Through *L* and *C* draw the required line.

Preferred method. (See Fig. 12-4.)

1. Place a triangle on line *AB* with the base of the triangle resting against a straightedge.

2. Hold the straightedge in position, slide the triangle to point *C* and draw the required line *CD*.

12.1.3 Dividing a line into equal parts.

Given line *AB*.

Geometric method. (See Fig. 12-5.)

1. From point *A* draw a line *AC* at any convenient angle.

2. Starting at *A* on line *AC* lay off the required number of equal spaces (here, six) either with dividers or a scale.

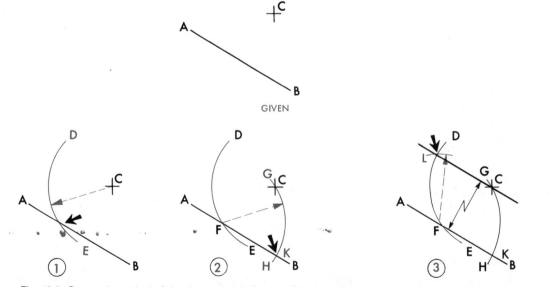

Fig. 12-3. Geometric method of drawing a straight line parallel to another straight line through a given point.

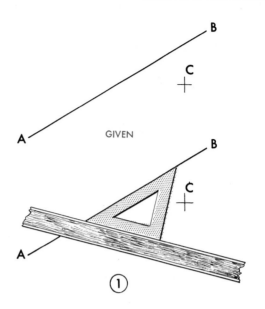

Fig. 12-4. Preferred method of drawing a parallel straight line.

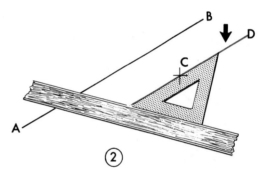

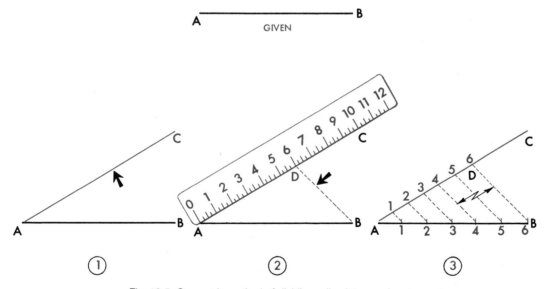

Fig. 12-5. Geometric method of dividing a line into equal parts.

3. From the termination point of the last space *D,* draw a line connecting *D* with *B.*

4. With the edge of a triangle set parallel with line *DB,* draw lines from the points on line AC to line *AB.*

Preferred method A. (See Fig. 12-6.)

1. From point *B* draw a line *BC* perpendicular to *AB.*

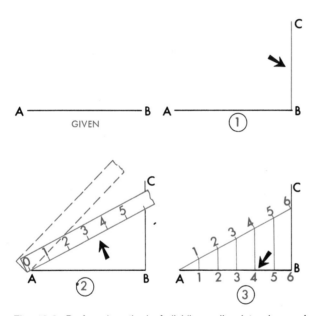

Fig. 12-6. Preferred method of dividing a line into six equal parts.

2. Place a scale so the division point zero coincides with the end of the line *A* and the last required calibration point (here, sixth) falls on the line *BC.*)

3. Lay off the intervening divisions; from these points draw lines parallel to *BC* intersecting *AB.*

Preferred method B. Most professional people use the trial-and-error method of dividing a line. The compass or dividers are set to an approximate spacing, and then adjustments are made in the setting until the proper divisions are secured.

12.2 Angles

An angle is the space formed within two lines or three or more planes diverging from a common point.

Angles may be *straight* (180°), *right* (90°), *acute* (less than 90°), or *obtuse* (more than 90°). If two angles total 90°, they are said to be *complementary,* and if they total 180°, they are *supplementary.* See Fig. 12-7.

12.2.1 Bisecting an angle. (See Fig. 12-8.)

Given angle *BAC.*

1. With *A* as a center and the compass set at any convenient radius, draw an arc cutting line *AB* at *D* and line *AC* at *E.*

2. Set the compass at a radius greater than one-half of *DE.*

3. With *D* and *E* as centers, draw two arcs to intersect at *O.*

4. Draw a line from *O* to *A.* The line *OA* bisects the angle.

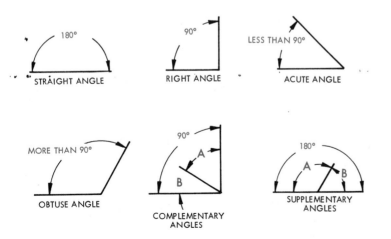

Fig. 12-7. Types of angles.

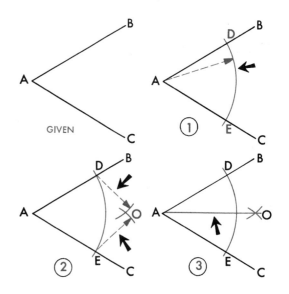

Fig. 12-8. Bisecting an angle.

12.3 Circles and Arcs

Circles and arcs are frequently used in making drawings. Some of the more common construction problems involving circles and arcs are described in the paragraphs to follow. Fig. 12-9 illustrates basic terms associated with circles and arcs.

12.3.1 Locating the center of a circle. (See Fig. 12-10.)

1. Draw any chord *AB* on the given circle.
2. Draw *BC* and *AD* perpendicular to chord *AB*.
3. Draw lines *DB* and *CA*, which are diameters of the circle. The intersection of these diameters at *O* is the center of the circle.

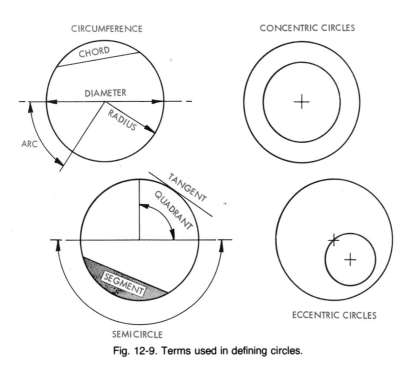

Fig. 12-9. Terms used in defining circles.

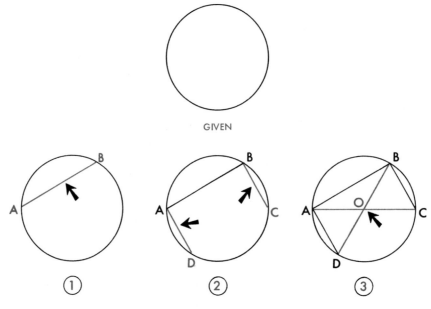

Fig. 12-10. Locating the center of a circle or arc.

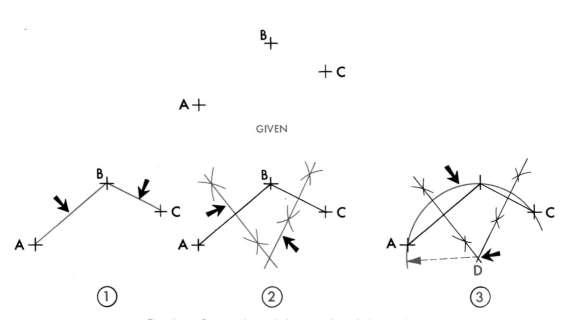

Fig. 12-11. Constructing a circle or arc through three points.

12.3.2 Constructing a circle or arc through three given points. (See Fig. 12-11.)

Given points *A, B,* and *C.*

1. Draw lines *AB* and *BC* connecting the points.

2. Find the bisectors of lines *AB* and *BC.* The intersection of these bisectors at *D* will be the center of a circle or arc through the points.

12.3.3 Drawing an arc of a given radius tangent to two lines at 90°. (See Fig. 12-12.)

Given radius *DE* and lines *AB* and *BC* at right angles.

1. Set the compass to the given radius and with *B* as center, draw an arc cutting line *AB* at *G* and line *CB* at *H.*

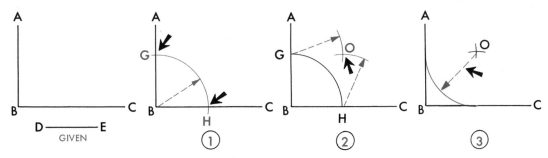

Fig. 12-12. Drawing an arc tangent to two lines at 90°.

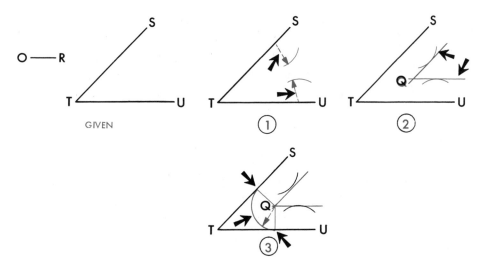

Fig. 12-13. Constructing an arc tangent to two lines not at 90°.

2. Using the same radius and with *G* and *H* as centers, draw two arcs to intersect at *O*.

3. With *O* as a center and the compass set at the same radius, draw the arc tangent to lines *AB* and *BC*.

12.3.4 Drawing an arc of a given radius tangent to two lines not at 90°. (See Fig. 12-13.)

Given a radius *OR* and the intersecting lines *ST* and *TU*.

1. Set the compass to the given radius *OR* and using any points on lines *ST* and *TU*, draw arcs.

2. Draw straight lines parallel to *ST* and *TU* tangent to these arcs. Extend to intersect at *Q*.

3. From *Q* draw perpendiculars to *ST* and *TU*.

4. With *Q* as center and the compass set at the given radius, draw the arc tangent to lines *ST* and *TU*. The perpendiculars from *Q* determine the points of tangency.

12.3.5 Drawing a line tangent to a circle through a point on the circle.

Given point *S* on a circle with center at *O*.

Geometric method. (See Fig. 12-14.)

1. Set the compass to radius *OS* and with *S* as a center, strike arc *OD* to cut circle O at *E*.

2. With *E* as a center and using the same radius, draw an arc to cut arc *OD* at *X*.

3. With *X* and *E* as centers and any convenient radius greater than *OS*, strike arcs intersecting at

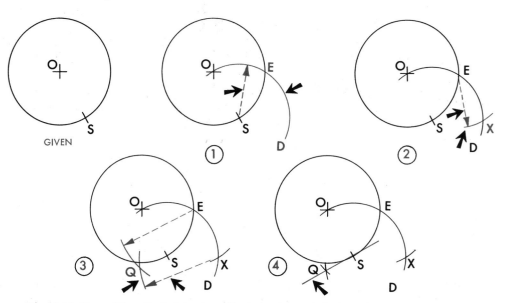

Fig. 12-14. Geometric method of drawing a line tangent to a circle through a point on a circle.

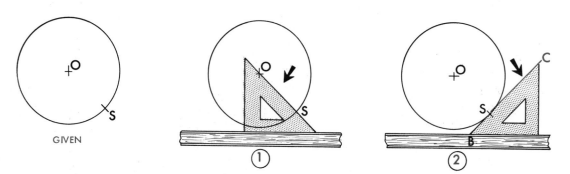

Fig. 12-15. Preferred method of drawing a line tangent to a circle through a point on a circle.

Q. A line drawn through points *Q* and *S* will be tangent to circle *O.*

Preferred method. (See Fig. 12-15.)

1. Place a triangle on a straightedge so the hypotenuse of the triangle passes through the center of the circle at *O* and through point *S.*

2. Hold the straightedge in the same position and turn the triangle so its hypotenuse passes through *S.* The line *BC* drawn along the hypotenuse of the triangle is the required tangent.

12.3.6 Drawing a line tangent to a circle through a point outside the circle. (See Fig. 12-16.)

Given circle *O* and point *X* outside the circle.

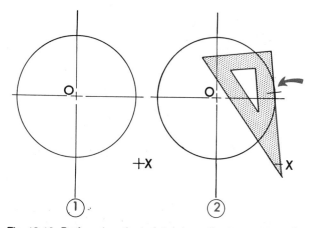

Fig. 12-16. Preferred method of drawing a line tangent to a circle through a point outside the circle.

1. Place a triangle so that one side of the triangle passes through point *X* and is tangent to the circle.

2. Draw the required line from *X* tangent to the circle.

12.3.7 Drawing an arc tangent to both a straight line and an arc. (See Fig. 12-17.)

Given line *EF*, arc *GH* having radius R_1, and radius R_2.

1. Adjust the compass to a distance equal to R_1 plus the specified radius of the connecting arc R_2, and draw arc *JK* using *P* as a center.

2. Draw line *QR* parallel to line *EF* to intersect arc *JK* at *O*. Line *QR* must be a distance equal to R_2 from line *EF*.

3. With *O* as a center and the compass set at the given radius R_2, draw the arc tangent both to arc *GH* and line *EF*.

12.3.8 Drawing tangent arcs externally. (See Fig. 12-18.)

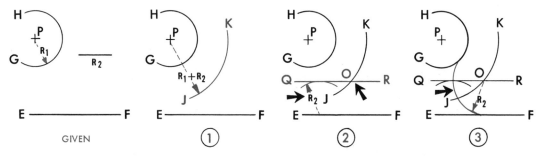

Fig. 12-17. Drawing an arc tangent to a straight line and an arc.

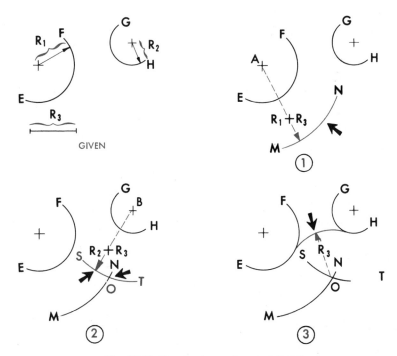

Fig. 12-18. Drawing tangent arcs, external.

Given arc *EF* having radius R_1, arc *GH* with radius R_2, and the radius R_3 of the required tangent arc.

1. Adjust the compass to a distance equal to R_1 plus R_3, and draw arc *MN* using *A* as a center.

2. Set the compass to a distance equal to R_2 plus R_3, and using *B* as a center, draw arc *ST* intersecting arc *MN* at *O*.

3. Using *O* as a center and R_3 as a radius, draw the arc tangent to arcs *EF* and *GH*.

12.3.9 Drawing tangent arcs internally. (See Fig. 12-19.)

Given arc *EF* having radius R_1, arc *GH* with radius R_2, and the radius R_3 of the required tangent arc.

1. Adjust the compass to a distance equal to R_3 minus R_1, and draw arc *MN* using *A* as a center.

2. Set the compass at a distance equal to R_3 minus R_2, and using *B* as a center, draw arc *ST* intersecting arc *MN* at *O*.

3. Using *O* as a center and R_3 as a radius, draw the arc tangent to arcs *EF* and *GH*.

12.3.10 Drawing a straight line tangent to two arcs.

Given arcs *MN* and *RS*.

Geometric method. (See Fig. 12-20.)

1. Find the difference between the radii of arcs *MN* and *RS*. Lay off this difference *OE* on a straight line connecting the centers of arcs *MN* and *RS*.

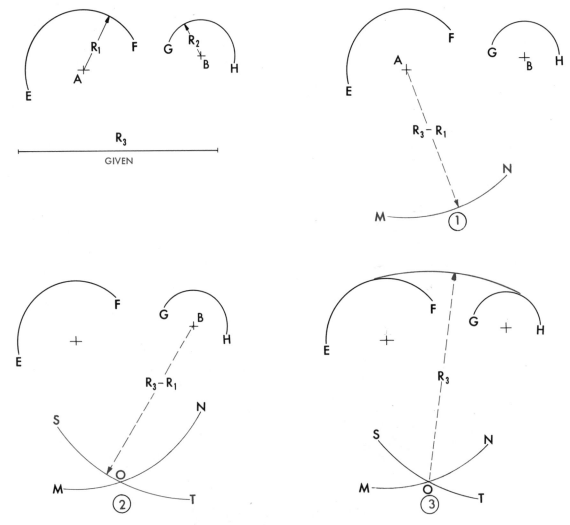

Fig. 12-19. Drawing tangent arcs, internally.

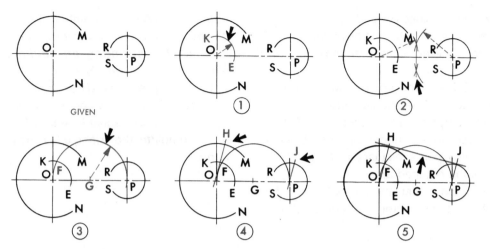

Fig. 12-20. Geometric method of drawing a straight line tangent to two arcs.

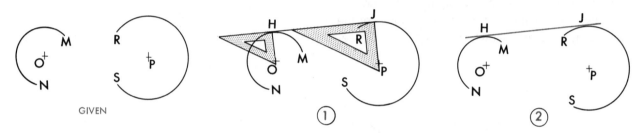

Fig. 12-21. Preferred method of drawing a straight line tangent to two arcs.

2. With *O* as a center and *OE* as a radius, draw the arc *KE.*

3. Find the center *G* of line *OP.*

4. With *OG* as a radius and *G* as a center, draw arc *OP* cutting arc *KE* at *F.*

5. Draw a line through points *O* and *F* cutting arc *MN* at *H. H* becomes the point of tangency of arc *MN.*

6. Draw line *PJ* parallel to line *OH. J* is the point of tangency of arc *RS.*

7. Draw line *HJ,* which is the required tangent to the two arcs.

Preferred method. (See Fig. 12-21.)

1. Place a triangle so that one side is tangent to the two arcs. Draw the required tangent line.

2. Determine the tangent points of each arc by sliding the triangle until the side making 90° with the tangent side intersects the center of the arc.

12.3.11 Drawing a reverse, or ogee, curve. (See Fig. 12-22.)

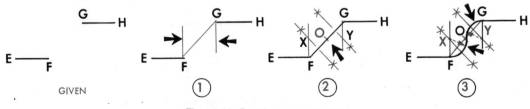

Fig. 12-22. Drawing on ogee curve.

Given two parallel lines *EF* and *GH*.

1. Connect *F* and *G* with a straight line.

2. At *F* and *G* erect lines prependicular to *EF* and *GH*.

3. On line *FG* assume point *O* where it is desired that the ogee curve should cross.

4. Find the perpendicular bisectors of *FO* and *GO*. The intersections *X* and *Y* of these bisectors with the first two perpendiculars at *F* and *G* are the centers of the required arcs.

5. With *X* as the center and the radius *XF*, draw arc *FO*. With *Y* as the center and radius *YG*, draw arc *GO*.

12.4 Polygons

A polygon is a geometric figure enclosed entirely by three or more straight lines. Some of the basic polygons are shown in Fig. 12-23. Procedures for laying out a number of polygons are presented in the following paragraphs.

12.4.1 Constructing an equilateral triangle with one side given. (See Fig. 12-24.)

1. Draw given side *AB* as the base line.

2. With *A* and *B* as centers, draw two intersecting arcs with radii equal to side *AB*.

3. Complete the triangle by drawing lines *AX* and *BX*.

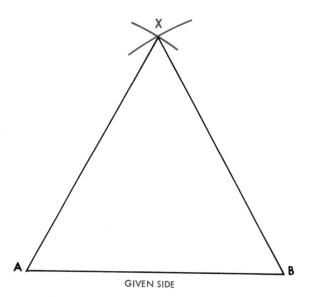

Fig. 12-24. Constructing an equilateral triangle.

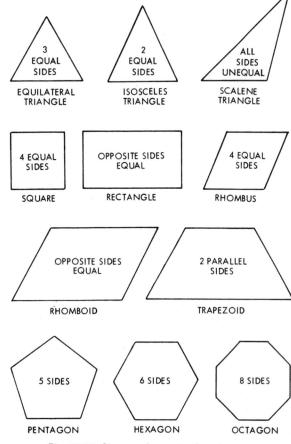

Fig. 12-23. Shapes of some basic polygons.

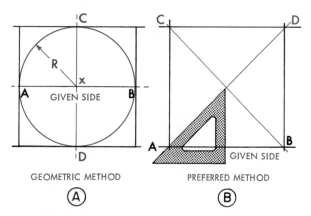

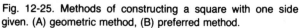

Fig. 12-25. Methods of constructing a square with one side given. (A) geometric method, (B) preferred method.

12.4.2 Constructing a square with one side given.

Geometric method. (See Fig. 12-25A.)

1. Draw line *AB* to represent the width of the square and erect bisector *CD*.

2. With *X* as center and *AX* as radius, draw a circle.

3. Draw lines tangent to the circle at the four bisector points.

Preferred method. (See Fig. 12-25B.)

1. Draw line *AB* as the given side.

2. With a straightedge and a triangle draw lines *AC* and *BD* perpendicular to *AB*.

3. Draw diagonals *AD* and *BC* at 45° to line *AB*.

4. Draw line *CD* with a straightedge.

12.4.3 Constructing a square with a given circle. (See Fig. 12-26.)

1. Draw two diameters at right angles.

2. With *X* as a center, scribe a circle.

3. With a 45° triangle draw lines as shown.

12.4.4 Constructing a regular hexagon with given distance across corners.

Geometric method. (See Fig. 12-27.)

1. Draw a circle with the given distance *AE* as diameter. Draw center lines.

2. With radius *AO* and using *A* and *E* as centers, draw arcs intersecting the circle.

3. Connect with straight lines in sequence the six points formed by the intersecting arcs.

Preferred method. The same hexagon can be

Fig. 12-26. Constructing a square with a given circle.

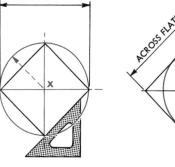
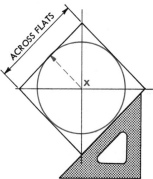

Fig. 12-27. Drawing a hexagon with given distance across corners with a compass.

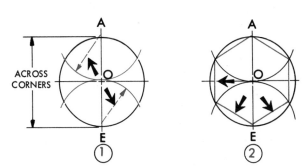

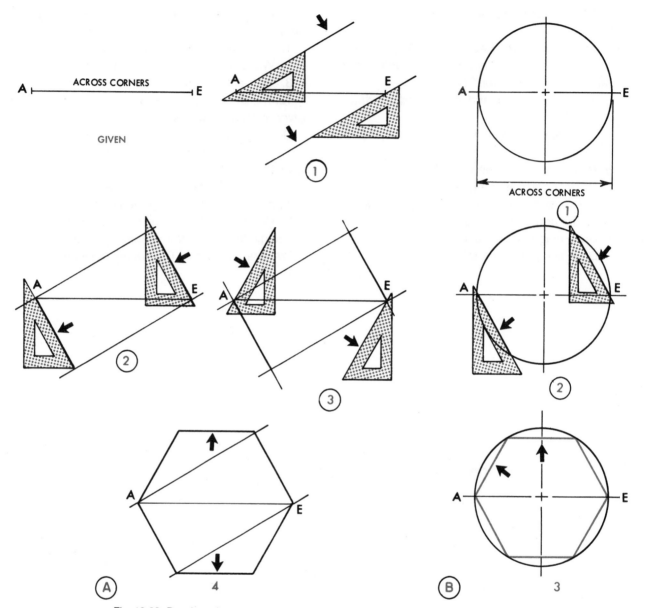

Fig. 12-28. Drawing a hexagon with given distances across corners with 30° - 60° triangle.

drawn with the use of a 30°-60° triangle as shown in Figs. 12-28A and 12-28B.

12.4.5 Constructing a regular hexagon with given distance across flats. (See Fig. 12-29.)

1. With the given distance AE as diameter, draw a circle.

2. Using a 30°-60° triangle, draw tangents to the circle as shown.

12.4.6 Constructing a regular octagon with given distance across flats. (See Fig. 12-30.)

1. Draw a square with sides equal to given distance AE, and its diagonals.

2. Using the corners of the square as centers and a radius equal to one-half of the diagonal, draw arcs intersecting the sides of the square.

3. Connect these points with lines to form the sides of the octagon.

12.4.7 Constructing a regular polygon with any number of sides and one side given. (See Fig. 12-31.)

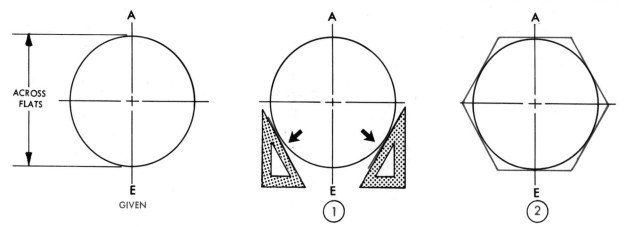

ACROSS
FLATS

GIVEN

①

②

Fig. 12-29. Drawing a hexagon with the given distance across flats.

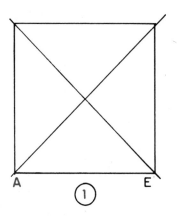

①

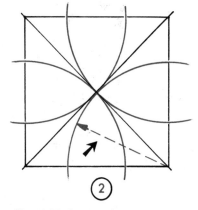

②

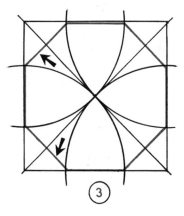

③

Fig. 12-30. Constructing an octagon.

①

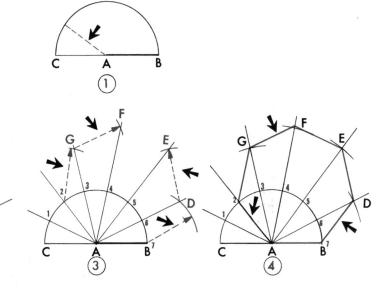

②

③

④

Fig. 12-31. Constructing a regular polygon.

1. With the given side *AB* as a radius and *A* as the center, scribe a semicircle *CB*.

2. Divide the semicircle into as many equal parts as there are to be sides (here, seven) and number the division points as shown.

3. From *A* draw radial lines through points 1–6 inclusive on the semicircle and extend them beyond the semicircle to some convenient distance.

4. With *AB* as a radius and *B* as the center, cut the extended line *A6* at *D*.

5. Using the same radius and with *D* as the center, cut *A5* at *E*.

6. Continue the same procedure to cut radial line *A3* at *G* and *A4* at *F*.

7. Connect the points *2, G, F, E, D,* and *B* with the required lines.

12.5 Ellipses

An ellipse is a plane curve formed by a point moving in such a way that the sum of its distances from two fixed points, called *foci,* is constant and equal to the major axis.

In Fig. 12-32, *AB* is the major axis, and *CD* is the minor axis. Points *F₁* and *F₂* are the foci. Then the sum of the distances from F_1 and F_2 to a point *P* on the curve is equal to the major axis *AB.*

Professionals in drafting often use templates for drawing ellipses. However, if an appropriate template is unavailable, several useful methods of constructing ellipses are described in subsections 12.5.2 through 12.5.6.

12.5.1 Locating the foci of an ellipse. (See Fig. 12-32.)

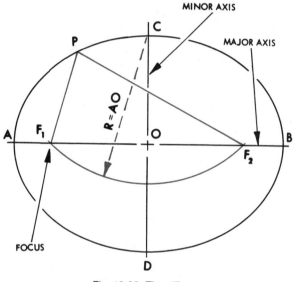

Fig. 12-32. The ellipse.

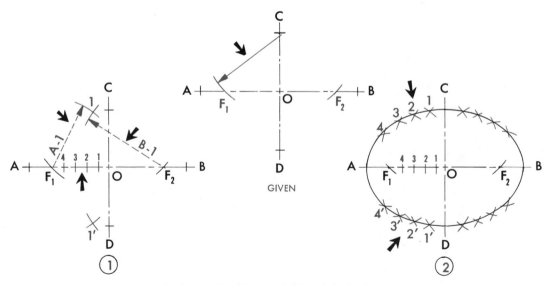

Fig. 12-33. Drawing an ellipse by the foci method.

Given major axis *AB* and minor axis *CD*.

Set the compass to a radius equal to one-half of the major axis *AB*. Using *C* as a center and *AO* as a radius, strike an arc intersecting line *AB*. The points of intersection F_1 and F_2 are the required foci.

12.5.2 Drawing an ellipse. Foci method. (See Fig. 12-33.)

Given major axis *AB* and minor axis *CD*.

1. Locate the foci by cutting axis *AB* with an arc, using *AO* as the radius and *C* as the center. The intersections of the arc with axis *AB* are the foci F_1 and F_2.

2. Between F_1 and *O* lay off a number of equally spaced points (here, five). The larger the ellipse, the more points that should be used to insure a smooth curve.

3. With *A1* as the radius and F_1 and F_2 as centers, scribe arcs at *1* and *1'*.

4. With *B1* as a radius and F_2 and F_1 as centers, draw arcs intersecting the arcs at *1* and *1'*.

5. Proceed in a similar manner with the remaining points to locate intersecting arcs at *2* and *2'*, *3* and *3'*, *4* and *4'*, and so on.

6. Connect the points formed by the intersecting arcs with an irregular curve.

12.5.3 Drawing an ellipse. Trammel method. (See Fig. 12-34.)

Given major axis *AB* and minor axis *CD*.

1. Secure a strip of paper or cardboard having one straight edge.

2. Lay off *EO* equal to one-half the major axis.

3. On the same strip lay off *EF* equal to one-half the minor axis.

4. Place the strip with point *E* on the minor axis and *F* on the major axis. As the strip is moved, the point *O* will provide points for the ellipse.

5. Complete the ellipse by drawing lines through the points with an irregular curve.

12.5.4 Drawing an ellipse. Four-center method. (See Fig. 12-35.)

Given major axis *AE* and minor axis *CD*.

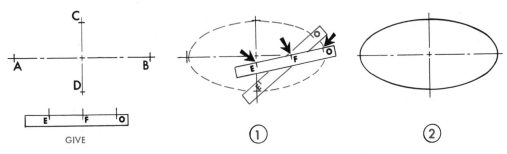

Fig. 12-34. Constructing an ellipse with a trammel.

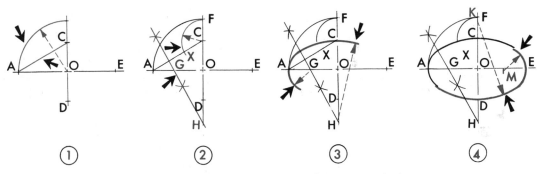

Fig. 12-35. Constructing an ellipse-four-center method.

1. With *O* as a center and radius *OA*, draw an arc intersecting the extension of *OC* at *F*.

2. Draw line *AC*.

3. With *C* as a center and radius *CF*, draw an arc intersecting line *AC* at *X*.

4. Find the perpendicular bisector of *AX* so that it intersects *AE* at point *G* and the extended axis *CD* at *H*.

5. With *G* and *H* as centers, draw two of the required arcs of the ellipse.

6. Lay off *OK* equal to *OH* and *OM* equal to *OG*. With *M* and *K* as centers, draw the other two arcs of the ellipse.

12.5.5 Drawing an ellipse. Concentric circle method. (See Fig. 12-36.)

Given *AE* and *CD* as the principal axes and *O* as center.

1. With *O* as a center and radius *AO*, scribe the outer circle. Draw the inner circle with *CO* as a radius.

2. Divide the outer circle into any number of equal parts.

3. From *O* draw radii intersecting the inner circle to points on the outer circle.

4. From points on the outer circle, draw lines parallel to *CD*.

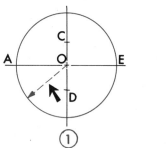
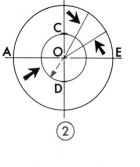
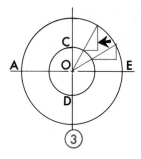
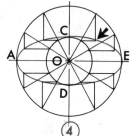

Fig. 12-36. Constructing an ellipse-concentric-circle method.

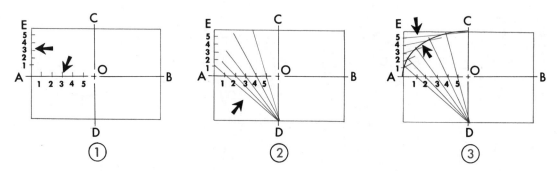

Fig. 12-37. Drawing an ellipse-parallelogram method.

5. From points on the inner circle, draw lines parallel to *AE.* The intersections of the vertical and horizontal lines provide points of the ellipse.

6. Repeat the same procedure for the remaining quadrants and connect the points with an irregular curve.

12.5.6 Drawing an ellipse. Parallelogram method. (See Fig. 12-37.)

Given major axis *AB* and minor axis *CD.*

1. With the two axes, *AB* and *CD,* construct a parallelogram.

2. Divide *AO* into any number of equal parts (here, five) and *AE* into the same number of equal parts. Number the division points as shown.

3. From *D* draw lines through points on *AO.*

4. From *C* draw lines to points on *AE.*

5. The intersections of these lines provide points for the ellipse.

6. Proceed in a similar manner for the remaining quadrants and connect the points with an irregular curve.

12.6 Parabolas

A parabola is a curve that is generated by a point moving along a path equidistant from a fixed point called the *focus* and a straight line called the *directrix.* It can also be defined as a plane curve formed by the intersection of a right circular cone with a plane parallel to a generator of the cone. See Fig. 12-38.

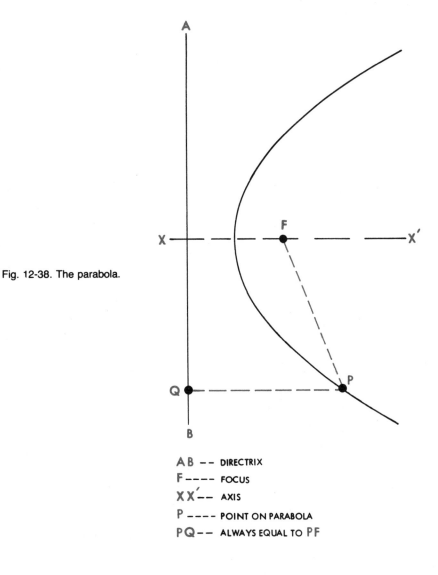

Fig. 12-38. The parabola.

A B -- DIRECTRIX
F ---- FOCUS
X X′-- AXIS
P ---- POINT ON PARABOLA
PQ -- ALWAYS EQUAL TO PF

The principle of the parabola is used in many engineering computations, in constructions such as reflecting surfaces for light and sound and vertical curves on highways and railroads, and for showing bending moment at any point on uniformly loaded beams or girders.

12.6.1 Drawing a parabola. (See Fig. 12-39.)

Given focus *F* and directrix *MN*.

1. Draw line *OR* through *F* and perpendicular to *MN* to intersect *MN* at *Z*.

2. At any point *E* on line *OR*, draw line *XY* parallel to *MN*.

3. With *ZE* as a radius and *F* as the center, draw an arc intersecting line *XY* at *P* and *Q*. These are points on the parabola.

4. Divide distance *ZE* into any number of parts (here, four) and draw lines through the points parallel to *MN*.

5. Repeat step 4 using distances *Z1, Z2, Z3,* and so forth, as radii and *F* as center, locating points on the curve.

6. The vertex *V* is located midway between the directrix *MN* and the focus *F* on *OR*. Draw the curve through the points located.

12.6.2 Drawing a parabola. Parallelogram method. (See Fig. 12-40.)

Given the rise *XY* and the half-span *XZ* of the parabola.

1. Divide *XY* and *XZ* into the same number of equal parts.

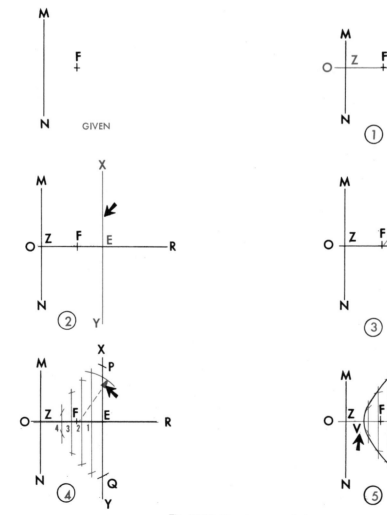

Fig. 12-39. Drawing a parabola.

2. From division points on line *XY,* draw lines converging at *Z.*

3. From the division points on line *XZ,* draw lines parallel to the axis *ZR.* The corresponding intersections of these lines are points on the parabolic curve.

4. Proceed in a similar manner to find the points for the other half of the parabola and connect all the points with an irregular curve.

12.6.3 Drawing a parabola. Tangent method.
(See Fig. 12-41.)

Given the limiting points *R* and *T* of the parabola and the tangents *RS* and *ST.*

1. Divide lines *RS* and *ST* into any number of equal parts and number the division points.

2. Draw lines connecting like-numbered points.

3. The curve drawn tangent to these intersecting lines is the parabolic curve.

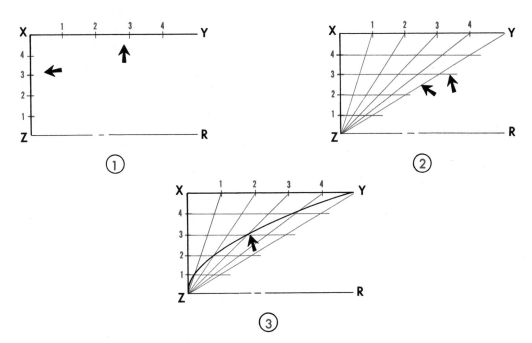

Fig. 12-40. Drawing a parabola-parallelogram method.

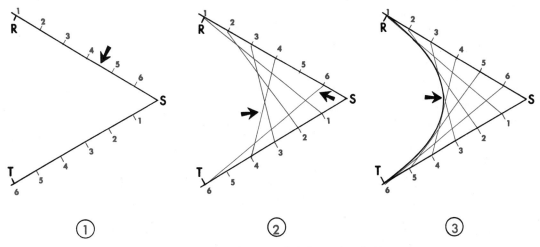

Fig. 12-41. Drawing a parabola-tangent method.

12.7 Special Curves

Several other types of curves are often used in designing and laying out various products and structures. Among these curves are the hyperbola, involute, spiral, cycloid, epicycloid, hypocycloid, and helix. See Fig. 12-42. These curves and methods of constructing them are described in the following subsections.

12.7.1 Hyperbolas. A hyperbola is a curve generated by a point moving in such a way that the difference of the distances from any point of the curve to two fixed points, called the *foci,* is a constant equal to the transverse axis of the hyperbola.

Drawing a hyperbola. (See Fig. 12-43.)

Give foci F_1 and F_2 and transverse axis RS.

1. Extend the transverse axis any convenient distance such as to point Z.

2. Lay off on SZ any number of points.

3. With SZ as a radius and F_2 as the center, strike arcs at M and N.

4. With RZ as a radius and F_1 as the center, strike arcs to intersect the arcs at M and N.

5. Proceed in a similar manner for points *1, 2, 3,* and *4.*

6. Connect the points formed by the intersecting arcs to construct required curve of the hyperbola.

12.7.2 Involutes. An involute is a spiral curve made by a point on a perfectly taut string as it unwinds from around a shape such as a polygon or circle.

Drawing an involute of a polygon. (See Fig. 12-44.)

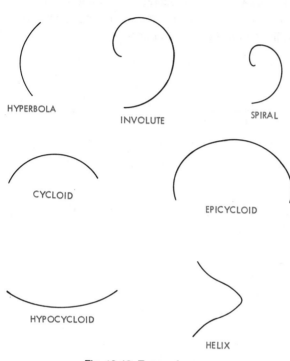

Fig. 12-42. Types of curves.

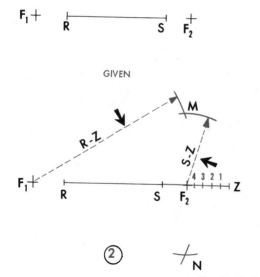

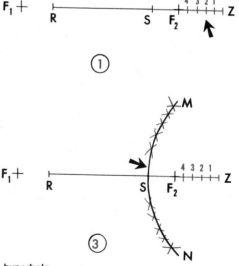

Fig. 12-43. Constructing a hyperbola.

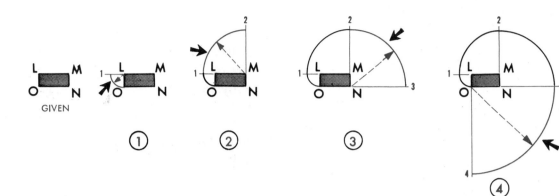

Fig. 12-44. Drawing an involute of a polygon.

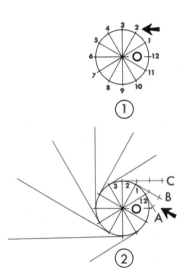

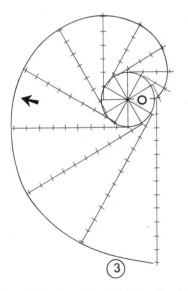

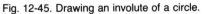

Fig. 12-45. Drawing an involute of a circle.

Given polygon *LMNO.*

1. With *L* as a center and radius *LO,* scribe an arc intersecting the extension of line *LM* at *1.*

2. With *M* as a center and radius *M1,* scribe an arc intersecting the extended line *NM* at *2.*

3. Continue this procedure and determine points *3* and *4* using radii *N2* and *O3.*

4. The connecting arcs determine the involute of the polygon.

Drawing an involute of a circle. (Refer to Fig. 12-45.)

Given circle *O.*

1. Divide the circle into any number of equal parts and number the division points.

2. Draw tangents to the circle at each of these points.

3. On each tangent, step off the length of the corresponding arc divisions such as *1-12* on tangent line *A1,* *2-12* on tangent line *B2,* *3-12* on tangent line *C3,* and so on.

4. Connect the points with an irregular curve.

12.7.3 Spirals. A spiral of Archimedes is a curve generated when a point moves away from a fixed point in such a way that its distance increases uniformly with the angle.

Drawing a spiral. (See Fig. 12-46.)

Given an angle of 30° with vertex *O.*

1. From point *O* lay out a series of equal angles and draw radial lines.

2. Divide one of the radial lines *OG* into the same number of equal parts as there are radial lines.

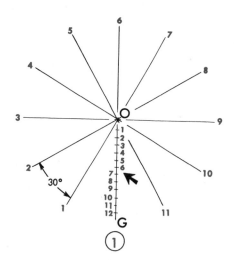

①

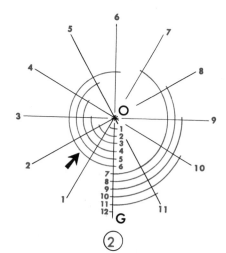

②

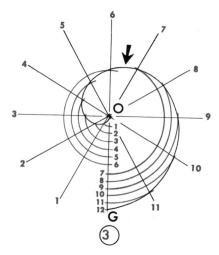

③

Fig. 12-46. Constructing a spiral.

3. Rotate the points on *OG* to the corresponding radial lines.

4. Connect the points of intersection with an irregular curve.

12.7.4 Cycloids. A cycloid is a curve generated by a point in the plane of a circle that rolls along a straight line.

Drawing a cycloid. (See Fig. 12-47.)

Given circle *O* and its path-tangent line *MN* equal in length to the circumference of the circle.

1. Divide the circumference of the circle and line *MN* into the same number of equal parts. Number each point as shown.

2. Draw path *XY* of the center of the circle parallel to *MN*.

3. From points on *MN*, draw perpendiculars intersecting *XY*.

4. From points on the circle, draw lines parallel to *MN*.

5. With the radius of circle *O* and the points of intersection on line *XY* as centers, strike arcs intersecting these horizontal parallel lines.

6. Connect the points of intersection with an irregular curve.

12.7.5 Epicycloids. An epicycloid is a curve traced by a point on the circumference of a generating circle which rolls upon the outside of the circumference of another circle.

Drawing an epicycloid. (See Fig. 12-48.)

Given circumference arc *AE* of a circle *O* and generating circle *P*.

1. Divide small circle *P* into any number of equal parts.

2. From point *Y* on arc *AE*, lay off distances equal to the circumference divisions of circle *P* and number them as shown.

3. With point *O* as the center and radius *OP*, draw the circular center line *CD*.

4. From the center of large circle *O*, draw radial lines through points on arc *AE* to intersect arc *CD*.

5. With *O* as the center, strike arcs through points on the generating circle *P*.

6. Using points on the center line arc *CD* as centers and radius *PY*, draw arcs intersecting the arcs drawn from the points on the small circle in step 5. The intersections of these arcs provide points for the epicycloid.

7. Connect the points with an irregular curve.

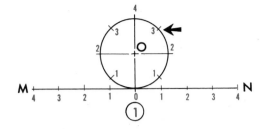

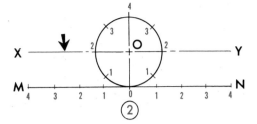

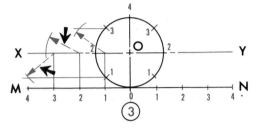

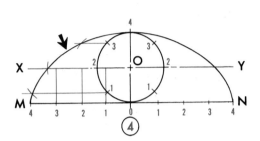

Fig. 12-47. Drawing a cycloid.

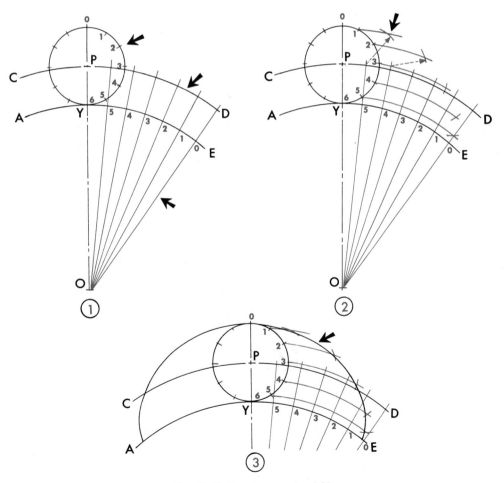

Fig. 12-48. Drawing an epicycloid.

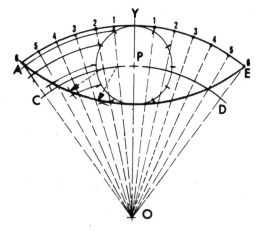

Fig. 12-49. Drawing a hypocycloid.

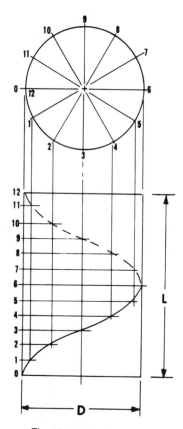

Fig. 12-50. Drawing a helix.

12.7.6 Hypocycloids.

12.7.6 Hypocycloids. A hypocycloid is a curve generated by a point on the circumference of a circle which rolls upon the inside circumference of another circle.

Drawing a hypocycloid. (See Fig. 12-49.) The same procedure that was used to draw an epicycloid is used to construct a hypocycloid.

12.7.7 Helices. A helix is the curve generated by a point moving at a uniform rate around and advancing parallel to, or at varying distances from, an axis. If the generating point is at a fixed distance from the axis, the helix is said to be *cylindrical;* if the distance from the axis varies uniformly, the helix is *conical.* The distance the point moves parallel to the axis in one revolution is called the *pitch,* or *lead.*

The helix is easily recognized in the curves of screw threads, coil springs, and twist drills or auger bits.

Drawing a cylindrical helix. (See Fig. 12-50.)

Given the diameter *D* and lead *L.*

1. Draw the views of the cylinder.

2. Divide the circumference of the circular top view into any number of equal parts (here, twelve). Divide the lead *L* into the same number of equal parts.

3. From each division point on the circumference, project lines parallel to the axis so as to intersect the corresponding division lines of the front view.

SECTION IV
UNIT 13

Practical Descriptive Geometry

In engineering drawing one is continually confronted with problems involving the relationships that exist among geometric elements such as points, lines, and planes. Although many of these problems may be solved mathematically, actual solution by graphic means frequently provides an added check on computations. For practical purposes much of the original layout and development of surfaces is done by using geometric procedures described in this unit. These constructions may also be made by computer graphic techniques.

13.1 Essentials of Descriptive Geometry

While it is not the purpose here to present a complete coverage of descriptive geometry, sufficient material is nevertheless included to solve the more common geometric problems encountered in engineering drawing. Most of these problems can be solved by one or more of the following basic operations involving the use of auxiliary views:

1. Finding the true length of a line.
2. Determining a point view of a line.
3. Constructing an edge view of a plane.
4. Finding the true size of a plane figure.

These and related operations are detailed in the sections that follow.

13.1.1 Notation. In order to understand the graphic solutions of the basic problems included in this section, it is essential that a system of notation be first established. See Fig. 13-1. Points in space are called out as uppercase letters, while projections or views of such points are noted by lowercase letters with subscripts to indicate the view. Thus, a_T will designate a top view of point A, a_F the front view, and a_R the right side view.

Auxiliary views of points will be identified with the subscript letter assigned to the respective views. For example, a_A and a_B designate point A in auxiliary views A and B, respectively.

A line connecting two points is indicated by, for

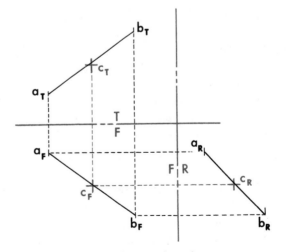

Fig. 13-1. System of notations.

example, *AB,* and the resulting views are shown as a_Tb_T, a_Fb_F, and a_Rb_R.

Reference lines, or datum lines as they are often called, are placed conveniently between views. They are made of lines broken occasionally by two short dashes as shown in Fig. 13-1. Principal views are labeled *T, F,* and *R* to indicate the top, front, and right side views, respectively. These view notations are placed in position on opposite sides of the reference line. Reference lines should be placed "tight," that is, close to the views, to prevent drawings from becoming unnecessarily large.

13.1.2 The views of a point and a line. A discussion of the proper method for projecting the various views of an object is described in Unit 5. Since that unit deals only with surfaces of primary objects, it is now necessary to use these same principles as they apply to a point and a line.

In Fig. 13-1 line *AB* is shown in its three principal views—top, front, and right side. Observe the notation of points *A* and *B* as well as the reference lines. From this drawing it is quite obvious that the top and front views of a point, as well as of a line, are aligned vertically and that the front and side views are aligned horizontally. Notice that point *C* on line *AB* appears on all three views of the line and that the requirements of alignment are true of this point as well as of the line. It follows then that any point on a line must appear on that line in all views of the point and line.

13.1.3 Auxiliary views. The three principal views of a geometric element do not always give the complete information necessary for an engineering drawing. As was explained in Unit 5, other views, called auxiliary views, are often needed to provide the true size and shape description of a part, especially for inclined surfaces or oblique lines that do not appear in their true shapes in the top, front, or side views. Moreover, it is frequently necessary to determine how an object appears from a position other than directly above, front, or side.

In Fig. 13-2 the oblique line *AB* is shown in the top and front views together with two auxiliary views. Auxiliary view *A* is taken by establishing reference line *F-A* parallel to the front view of line *AB.* Distances *X* and *Y,* measured from reference

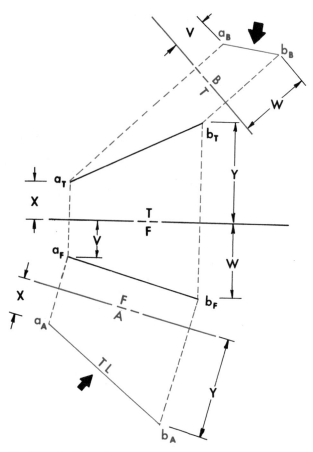

Fig. 13-2. Auxiliary views are used to determine true lengths as well as how a given line appears from some specified position.

line *T-F* to a_T and b_T, are laid out from reference line *F-A* on perpendicular projectors drawn from a_F and b_F. This locates the auxiliary view a_Ab_A. Another auxiliary view *B* is shown by first drawing reference line *T-B.* By transferring distances *W* and *V* taken from reference line *T-F* to a_F and b_F to projectors from a_T and b_T, view a_Bb_B is established. It should be noted that view a_Ab_A shows line *AB* in its true length, whereas view a_Bb_B does not.

13.1.4 A plane surface. A plane surface may be formed by (1) three points, (2) a point and a line, (3) two intersecting lines, or (4) two parallel lines. However, a plane is usually indicated on a drawing by two or more intersecting lines. Although an actual plane surface is limited in size, it is frequently necessary to consider the plane to extend without limits, in order to obtain the so-

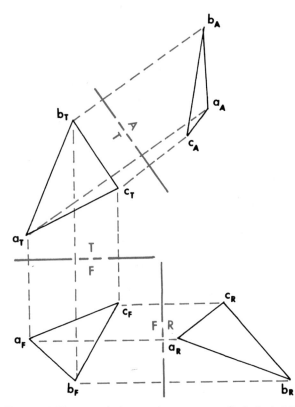

Fig. 13-3. Planes and plane surfaces are usually indicated by two or more intersecting lines.

lutions required. A typical example of a plane surface with its principal views, including an auxiliary view, is shown in Fig. 13-3.

13.2 Finding the True Length of a Line

Any line that is parallel to a reference line or to a plane of projection must necessarily show its true length in the adjacent view. In Fig. 13-4 line

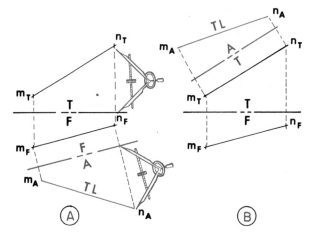

Fig. 13-5. The true length of a line may be found in a front-adjacent or top-adjacent view.

XY is shown in three positions, each parallel to a reference line. This line appears in its true length in the front view at A, in the top view at B, and in the side view at C. Lines in these positions are called *principal lines.* However, most lines to be measured are not principal lines but oblique lines slanted to the three principal planes of projection.

To obtain the true length of an oblique line, a reference line that is parallel to one of the principal views must first be established. In Fig. 13-5A reference line F-A is placed parallel to the front view of line MN. By taking the distances from reference line T-F to m_T and n_T and laying these distances out from reference line F-A on the projectors drawn from $m_F n_F$, the true length of MN is obtained as $m_A n_A$. A similar construction is shown in Fig. 13-5B in which the true length of the line is shown by $m_A n_A$ in the top adjacent auxiliary view.

Fig. 13-4. A principal line is one that shows its true length in a principal view.

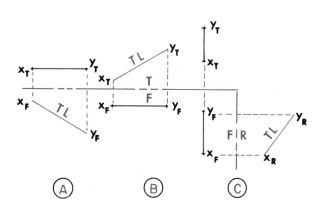

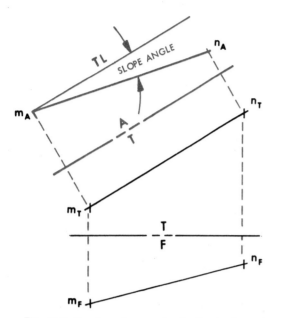

Fig. 13-6. The true slope angle of a line is shown in an elevation view in which the line appears in its true length.

13.3 Finding the True Slope Angle of a Line

The deviation of a line from the horizontal is called *slope angle.* While any elevation view of an oblique line will show that the line slopes from the horizontal, it is only in an elevation view in which the line appears in its true length that the true sloping relationship of the line to the horizontal becomes evident. Thus, in Fig. 13-6 view *A* shows line *MN* in its true length. In addition, because this view is an elevation view, the true angular relationship of $m_A n_A$ to the reference line *A-T* (the horizontal datum) is indicated, and the true slope angle becomes apparent. Whether the line slopes upward or downward is signified by the way it is identified. By reading the line as *MN* it slopes upward, while as *NM* it slopes downward.

13.4 Finding the Point View of a Line

Once the true length of a line is found, the next step in the solution of many problems is to show the same line as a point view. To find this view, a reference line must be drawn perpendicular to the true length view of the line. By transferring distance *X* as described previously, the line can be shown as a point at $m_B n_B$. See Fig. 13-7.

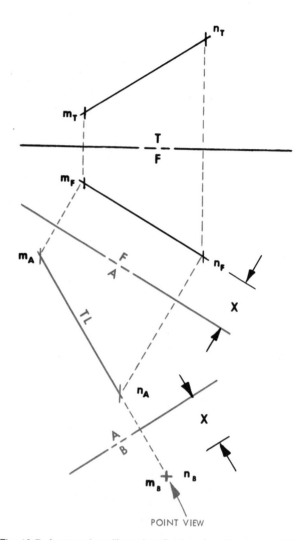

Fig. 13-7. A second auxiliary view *B* taken from the true length view *A* shows line *MN* as a point view.

13.4.1 The shortest distance from a point to a line or between parallel lines. Another very common measurement that must be determined in engineering drawing is the shortest distance from a given point to a given line. In Fig. 13-8 the shortest distance from point *O* to line *RS* is required. To ascertain this length, first obtain a true length view of line *RS* and show point *O* in this view. Then show line *RS* as a point view. By projecting *O* into this view, the actual distance between *O* and *RS* can now be measured.

To find the shortest distance between two parallel lines, the same procedure as above is followed.

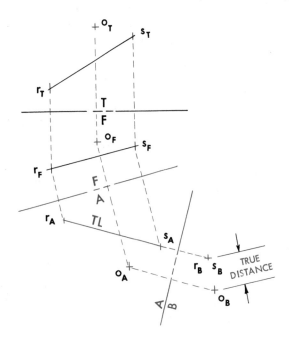

Fig. 13-8. The shortest distance from a point to a line is found in the view in which the line appears as a point.

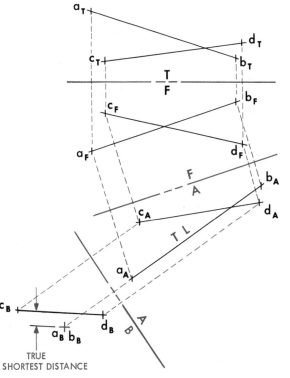

Fig. 13-9. The shortest distance between two skew lines is measured on a perpendicular from the line shown as a point view to the second line.

First, auxiliary views are constructed to determine the true lengths, and then the lines as point views are drawn. This latter view will provide the true distance.

13.4.2 The shortest distance between two skew lines. Lines that are neither parallel to each other nor intersecting are called *skew lines*. The shortest distance between these lines is another measurement often needed. To determine this distance, construct an auxiliary view of the skew lines showing one of them in its true length, as in Fig. 13-9. Draw a second auxiliary view to show the true length line in step 1 as a point. Show the other skew line by normal projection. The perpendicular distance from the point view line to the other skew line can now be measured. This is the shortest distance.

13.5 Plane Measurements

The true size measurements of plane surfaces and of angles formed by intersecting lines, and the true shape description of planes are all frequently required for layouts as well as for finished production drawings. The problem of true plane measurement involves two basic steps. First, the plane, formed by intersecting or parallel lines or by a point and a line, must be drawn as an edge view. From this edge view the true view showing the actual shape can be projected and the required dimensions then determined.

13.5.1 Edge view of a plane. A plane will appear as an edge, that is, as a straight line, in a view showing a true length line in the plane as a point view. In order to establish this view, a line first must be drawn in the plane such that the line will appear in its true length in an adjacent view. This can be done by drawing a line in any one of the positions similar to those shown in Fig. 13-4. Thus, in Fig. 13-10 the line *DC* is drawn in the top view of plane *ABC* parallel to reference line *T-F*. The line *DC* then will show its true length in the front view of the plane *ABC*.

A point view must now be constructed of line *DC* by drawing reference line *F-A* perpendicular to the true length of *DC* in the front view and projecting the distances as shown in Fig. 13-7. By also

projecting points *A* and *B*, the plane *ABC* now appears as an edge.

13.5.2 True size and shape of a plane. The edge view of the plane *ABC*, as previously determined, may now be used as the basis for a second auxiliary view by drawing the reference line *A-B* parallel to the edge view of *ABC*. By projecting the proper distances of points *A*, *B*, and *C*, as shown in Fig. 13-10, the true size and shape of the plane may be drawn. It is obvious that since this view is a true view, both the lengths of the lines and the angles formed by the intersecting lines will show their true measurements.

13.6 Finding the Angle Formed Between Two Planes (Dihedral Angle)

The size of the angle formed by two non-parallel planes, often called a *dihedral angle*, is especially

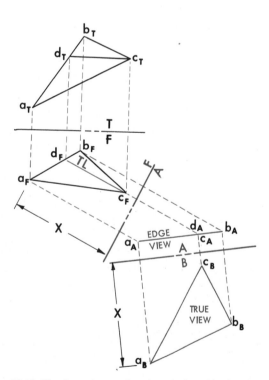

Fig. 13-10. The true shape of a plane is found in the view adjacent to the edge view in which the direction of sight is perpendicular to the plane.

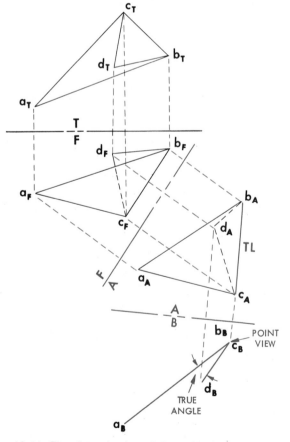

Fig. 13-11. The determination of the angle formed by two planes is a construction often used for sheet-metal fabrication.

important in the fabrication of products that require bending and forming materials in sheet or plate form. To determine the angle between planes *ABC* and *BCD* in Fig. 13-11, first obtain a view of the line of intersection *BC* showing its true length, and then draw a view showing *BC* as a point view. By projecting the other points and lines of the two planes into these auxiliary views, the view containing *BC* as a point will provide the actual angle between the two planes.

13.7 Finding the Point Where a Line Pierces a Plane

Frequently it is necessary to locate the position of an opening in a plane surface. The problem is essentially that of finding the point in a plane at which a line pierces it. The solution to such a problem may be found by the edge view plane method or by the cutting plane method.

13.7.1 The edge view plane method. By observing a plane as an edge, the point at which a line pierces the plane can be found at the intersection of the line and the edge view of the plane. Thus, the edge view of the plane *ABC* first must be shown as in Fig. 13-12, and the line *MN* drawn in this view. The point of intersection of the line and the plane is found at o_A. This piercing point is then projected back into its front and top views on line *MN* at o_T and o_F.

13.7.2 The cutting plane method. If an edge view plane is passed containing line *MN*, it will intersect plane *ABC* at *EF*. Line *MN* will pierce plane *ABC* at *O*, the intersection of *MN*, with the line of intersection *EF*. See Fig. 13-13.

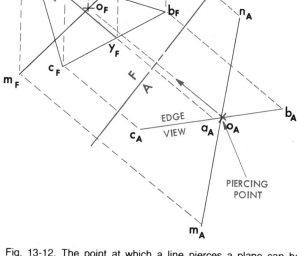

Fig. 13-12. The point at which a line pierces a plane can be found in the view showing the plane as an edge view.

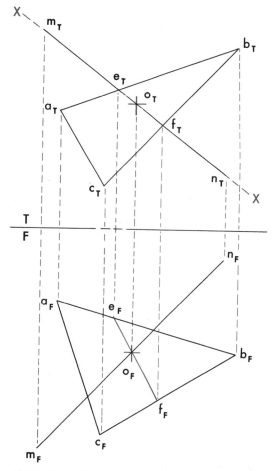

Fig. 13-13. The cutting plane method for locating the point where a line pierces a plane may be more convenient than the edge view plane method.

This method of locating the piercing point of a line and a plane often saves time since the existing views may be used without the addition of auxiliary views. Accuracy is also increased because transfer of distances into auxiliary views is not needed.

13.8 The Angle a Line Makes with a Plane

The true angle between a line and a plane will be shown in the view in which the plane appears as an edge and the line in its true length. Thus, if the angle formed by line *EF* and plane *RST* in Fig. 13-14 is required, the procedure is to find an edge view of plane *RST* and then its true view. Carry the projections of line *EF* into these views. Then from the true view of *RST*, project a view showing line *EF* in its true length. This view will show plane *RST* again as an edge view. The angle formed by the true length of *EF* and the edge view of *RST* is the required angle.

It might be noted that had *RST* been shown as an edge view in one of the given views, only two supplementary views would have been required for the solution.

13.9 Visibility of Lines and Surfaces

As was pointed out in Unit 5, it is extremely important in the delineation of engineering drawings that visible and hidden lines (and surfaces) be correctly distinguished from each other. While the out-

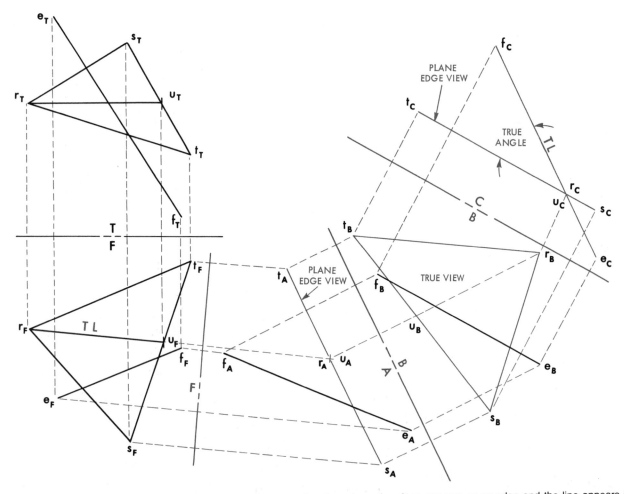

Fig. 13-14. The angle between a line and a plane is shown in the view where the plane appears as an edge and the line appears in its true length.

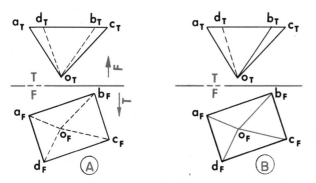

Fig. 13-15. The determination of visibility of lines on an object may be done by inspection.

line of a view will always be visible, lines within this outline may be visible or hidden, depending upon their relative positions with respect to the observer's line of sight.

Whether or not a line is visible in a view often may be ascertained by inspection. In Fig. 13-15A it is obvious that the outline in each view is visible. However, whether or not the lines shown by dashes on the object are visible is determined by the position of point O.

In the top view the point O is shown nearest the reference line T-F. Since the direction of sight for the front view is in the direction of arrow F, it becomes evident that all lines emanating from O are visible for this view. See Fig. 13-15B. Similarly by viewing the object in the direction of arrow T, it can be seen that lines OA and OB, being closest to the observer, must be visible in the top view. In the same way, since point D is farthest from the observer, line OD becomes invisible or hidden in the top view.

Thus, it might be stated that the corner or edge of an object nearest the observer will be visible, and the corner or edge farthest from the observer will usually be hidden if it lies inside the outline of the view.

In many views, crossing edges are often located at approximately the same distance from the viewer. The visibility of lines formed by these edges must be determined independently for each view by testing the visibility of the point where the lines apparently cross. See Fig. 13-16.

For example, to determine the visibility of lines AC and BD, in A of Fig. 13-16A, the apparent

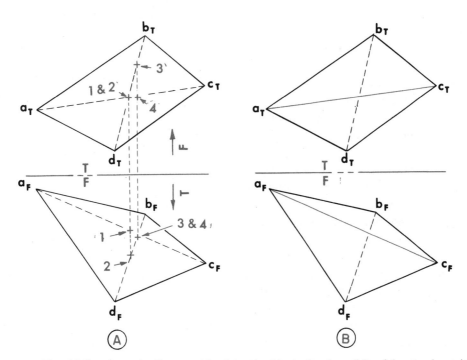

Fig. 13-16. The visibility of crossing lines must be determined by testing the validity of the crossing point.

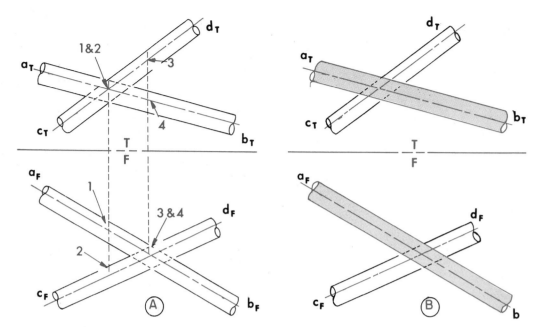

Fig. 13-17. The visibility of non-intersecting linear objects must be determined by the use of the apparent point of crossing.

crossing point of a_Tc_T and b_Td_T is labeled 1 and 2. These points are then located on the lines in the front view; that is, 1 on a_Fc_F and 2 on b_Fd_F. Since point 1 on a_Fc_F is higher and, therefore, nearer the observer, line *AC* in the top view must be visible. See Fig. 13-16*B*. Likewise, in the front view of Fig. 13-16*A*, the apparent intersection of *AC* and *BD* is labeled points 3 and 4. By projecting 3 to line b_Td_T and 4 to line a_Tc_T, point 4 on a_Tc_T is nearer the observer and is visible in the front view a_Fc_F.

The above described procedure must be used in finding the visibility of non-intersecting linear objects such as rods, pipes, and wires. In Fig. 13-17 two non-intersecting pipes are shown. Since they do not intersect, it is obvious that one of the pipes must be above the other at the apparent point of crossing in the top view. Likewise, in the front view one of the pipes must be in front of the other at a similar point of crossing.

If, at the apparent point of crossing in the top view, points 1 and 2 are given, then by projecting these two points to the separate pipes in the front view, it is found that the pipe *AB* is above *CD* at that location. This is shown in Fig. 13-17*B*. By the same method, the apparent point of crossing in the front view is labeled 3 and 4 in Fig. 13-17*A*

and these points projected to the pipes in the top view. Point 4 on a_Tb_T lies in front of point 3 on c_Td_T, and therefore the pipe *AB* is visible in the front view.

This same procedure may be used for any two adjacent views since by rotating the drawing, the views can be oriented to the position of front and top views. It must be remembered that in such cases, visibility must be determined independently for each view.

13.10 Revolution

While the principal views of an object ordinarily represent the object satisfactorily in a fixed position, it is often necessary to revolve an object or its elements for purposes of measurement and true shape description.

If an object is drawn in an oblique position, that is, revolved about an axis which is perpendicular to a principal plane of projection, it is referred to as *simple revolution*. There are three types:

1. Revolution about a horizontal axis perpendicular to a frontal plane.

2. Revolution about a vertical axis perpendicular to a horizontal plane.

3. Revolution about a horizontal axis perpendicular to a profile plane.

13.10.1 Revolution about a horizontal axis perpendicular to a frontal plane. This type of revolution is illustrated in Fig. 13-18. Both the front and top views of the object are shown with the axis of revolution *XY*. The axis may be located at the center of gravity, at an edge, or at any predetermined position. After the object is revolved 30° counterclockwise about its axis, the shape of the front view remains the same since it is still parallel to the frontal plane. The top view, however, is elongated but maintains the same thickness. The top view in this position is merely projected from the front view by regular projection methods.

13.10.2 Revolution about a vertical axis perpendicular to a horizontal plane. In this type of revolution, the front view changes shape while the top view only changes position. See Fig. 13-19.

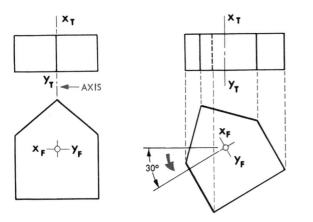

Fig. 13-18. Revolving an object about a horizontal axis perpendicular to a frontal plane changes the front view position while the top view is changed in both size and shape.

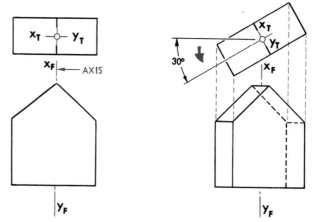

Fig. 13-19. As an object is revolved about a vertical axis perpendicular to a horizontal plane, the size and shape of the front view are changed while only the position of the top view is altered.

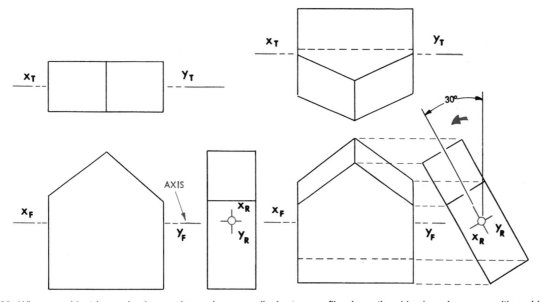

Fig. 13-20. When an object is revolved around an axis perpendicular to a profile plane, the side view changes position while the top and front views are altered in both size and shape.

The top view is drawn first in the revolved position. The front view is then projected from it and the vertical heights measured or projected from the principal front view.

13.10.3 Revolution about a horizontal axis perpendicular to the profile plane. This revolution is shown in Fig. 13-20 and is accomplished by first revolving the side view into its new position and projecting the top and front views from it. Notice that both the front and top views change in size as well as shape, whereas in the side view only the position changes.

13.10.4 True length of a line by revolution. The revolution method, rather than the auxiliary method, is often used to find the true length of a line. An advantage of the revolution method is that it involves only the existing principal views; no additional views are necessary.

Any line will appear in its true length if it is positioned parallel to a plane of projection. Hence, by

revolving a line until it is in such a position, the true length can readily be seen in the adjacent view. In Fig. 13-21A, point A is kept stationary while B is revolved in the top view until the line is parallel to the reference line T-F. Point B thus moves to

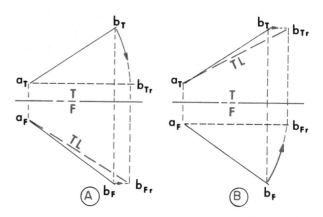

Fig. 13-21. The true length of a line may be found by revolving the line in either a top view or a front view.

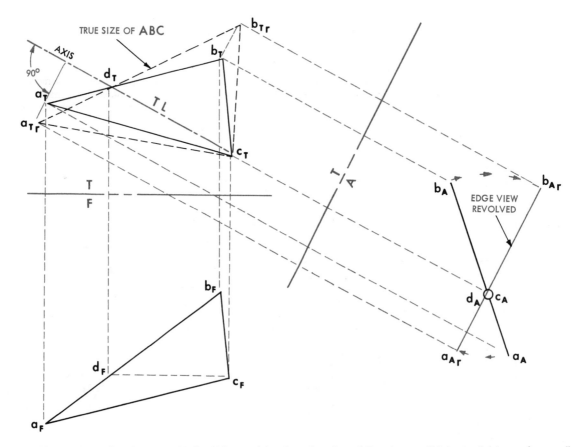

Fig. 13-22. The true size of a plane may be found by revolving the edge view of the plane until it is parallel to a reference line and projecting the revolved points into the adjacent view.

the right in the front view while remaining at the same level. The true length is measured on $a_F b_{Fr}$. A similar solution is shown in Fig. 13-21B, where the line is revolved in the front view, using a_F as the axis, until it is parallel to reference line T-F. The true length of AB is now shown in the revolved position $a_T b_{Tr}$.

13.10.5 True shape of a plane by revolution. Determining the true size and shape of a plane is also commonly done by revolution. To produce revolved views of a plane, an auxiliary view showing the plane as an edge must first be drawn. Refer to Fig. 13-10. Thus, in Fig. 13-22 the axis, line DC in its true length, is shown as a point view at $d_A c_A$. By revolving both ends of $a_A b_A$ so that it is parallel to reference line T-A, the true shape can be drawn by simply projecting the revolved points to the top view.

UNIT 14

Developments and Intersections

Developments involve the construction of full-size layouts of surfaces of various shapes. Intersections deal with the location of lines of intersection between various geometrically shaped objects such as cylinders, cones, and prisms. Application of developments and intersections may be found in fabricating sheet-metal and plate structures, in pattern making, in paper and plastic work, and in numerous other types of industrial processes.

14.1 Classification of Surfaces

A surface is a geometric form generated by a straight or curved line. The two main groups of surfaces are known as ruled surfaces and double-curved surfaces.

A *ruled surface* is one which has straight line elements. Such a surface may be classified as a plane, a single-curved surface, or a warped surface. A plane is a surface generated by a straight line moving in such a manner that one point on the line touches another straight line as the generating line moves parallel to its original position. See Fig. 14-1. A single-curved surface is one which is generated by moving a straight line in contact with a curved line so that any two consecutive positions of the generating line either intersect or are parallel. See Fig. 14-2. A warped surface is one in which no two consecutive elements are in the same plane. See Fig. 14-3.

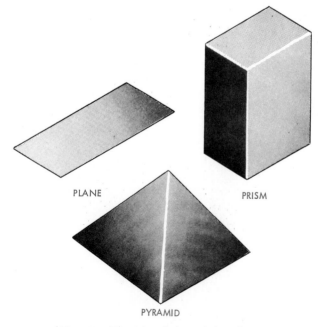

PLANE

PRISM

PYRAMID

Fig. 14-1. Examples of plane-surface figures.

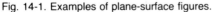

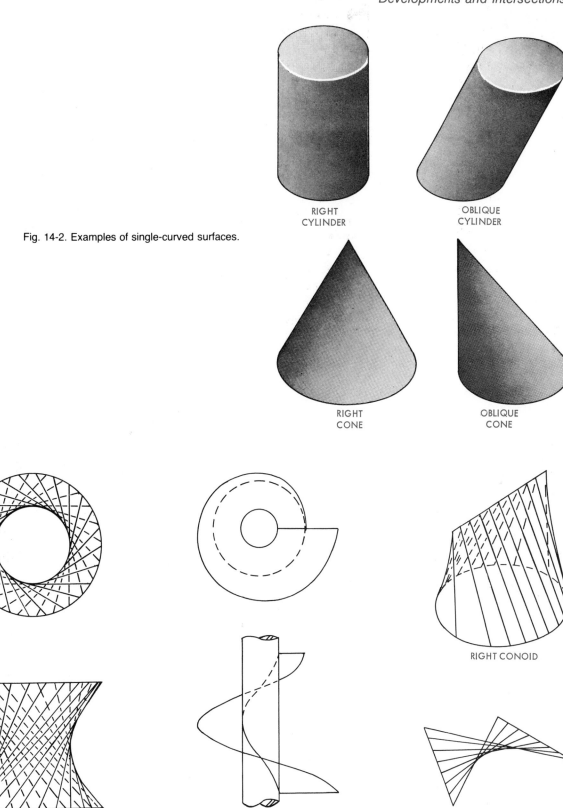

Fig. 14-2. Examples of single-curved surfaces.

RIGHT
CYLINDER

OBLIQUE
CYLINDER

RIGHT
CONE

OBLIQUE
CONE

RIGHT CONOID

HYPERBOLOID OF
REVOLUTION

RIGHT
HELICOID

HYPERBOLIC
PARABOLOID

Fig. 14-3. Examples of warped surfaces.

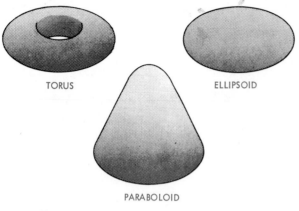

Fig. 14-4. Examples of double-curved surfaces.

A *double-curved surface* is a surface which is formed only by a moving curved line. Examples of such surfaces are the sphere, torus, ellipsoid, and so on. See Fig. 14-4.

14.2 Developments

A surface that can be rolled out or unfolded without distortion is said to be developable. Any object composed of single-curved surfaces is developable. See Fig. 14-5. Warped and double-curved surfaces are not developable, because consecutive elements cannot be brought into a flat plane without distortion. These surfaces can be developed only by approximation, for which triangulation is used.

A developed surface is referred to as a *pattern* or *stretchout*. The pattern is usually prepared with drawing instruments on paper and the stretchout then transferred to the required material.

When a pattern is to be used repeatedly, it is generally made of metal. This type of metal pattern is often referred to as a *template* or *master pattern*.

A developed view for use as a pattern or template must be drawn with accuracy consistent with the function of the part. The development should be plainly marked "Development of _____," "Pattern for _____," or "Template for _____." See Fig. 14-6. If elements of the surface are shown, they should be drawn as phantom lines. Bends should be clearly indicated—the angle of bend as well as whether the angle is to be turned up or down from the flat. Metal for seams should be shown with a phantom line and the seam positions clearly marked.

The four methods used to develop surfaces are known as parallel line development, radial line development, triangulation, and rollout.

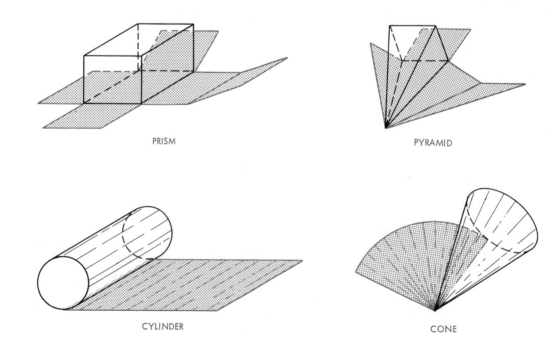

Fig. 14-5. Development refers to the unfolding of the surfaces of an object.

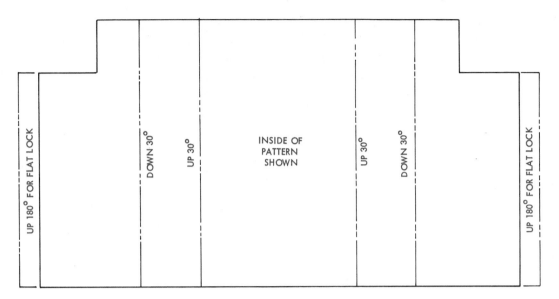

Fig. 14-6. How bends on a development should be marked.

14.3 Parallel Line Developments

Parallel line developments deal with patterns for prisms and cylinders. The constructions which follow are the ones most commonly used.

14.3.1 Truncated rectangular duct. (Shown in Fig. 14-7.)

Draw elevation and plan views of the required duct as shown. Draw the stretchout line *KL* at the same level as *XY* of the elevation view. On this line lay off the perimeter distances *EF, FG, GH,* and *HE* taken from the plan view and construct perpendicular elements representing the edges to be folded at points 1, 2, 3, and 4.

Project a dashed line from *C* on the elevation view to intersect elements 1, 4, and 1. Project another dashed line from *D* to intersect elements 2 and 3. Connect these points with straight lines, completing the pattern.

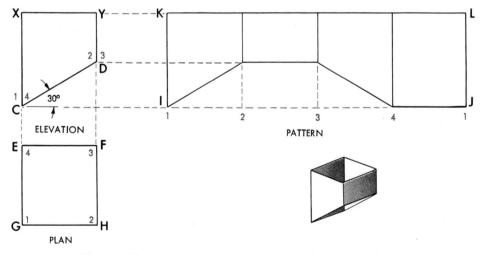

Fig. 14-7. How to develop a pattern for a truncated rectangular duct.

14.3.2 Intersection of two circular ducts at an angle. (See Fig. 14-8.)

Draw an elevation of ducts *A* and *B* with a 45° miter line. Draw a half plan view of duct *A* and divide it into any number of equal parts. Number the division lines as shown. Project the points from the plan view to the miter line on the elevation and across duct *B* to line *CD*.

Draw the stretchout line equal to the circumference of duct *A*, and on it, working to the left of the center line, lay out the points to correspond with the numbered points on the half plan *A*. Project

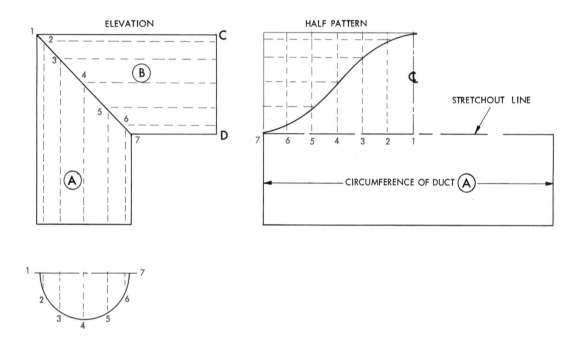

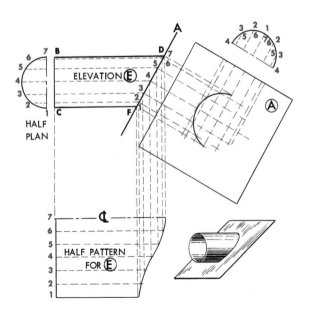

Fig. 14-8. How to lay out a pattern for two intersecting ducts at right angles to each other.

Fig. 14-9. How to develop patterns for a circular duct intersecting a plane surface.

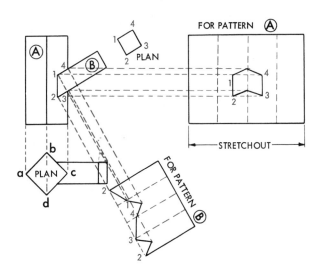

Fig. 14-10. How to develop the patterns of two intersecting square ducts at an angle.

the points from the miter line in the elevation view to the corresponding vertical elements on the stretchout. The intersections of these lines form the points for the contour of the pattern.

Since duct *B* is similar to duct *A,* the patterns for both ducts will be identical.

14.3.3 Circular duct intersecting a flat surface. (See Fig. 14-9.)

Draw an elevation view of duct *E* with line *BC* representing the diameter of the duct and the miter line *DF* having the desired angle for the plane of intersection. Using *BC* as a diameter, draw a semicircle and divide it into any number of equal parts. Project these points to miter line *DF.*

Draw the stretchout line for the duct equal to one-half the circumference and on it lay out the spaces from the plan view *E.* Project the points from the miter line *DF* to the corresponding elements on the stretchout. The intersection of these lines will provide the contour of the half pattern for the duct.

To find the opening in the flat plane *A,* lay out the desired length and width of the plane. Draw a semicircle above the flat plane to represent the plan view of the intersecting duct and divide it into the same number of equal spaces as the plan view of duct *E.* The intersections of lines drawn from these points with the projections from corresponding points on the mitered line *DF* will provide the contour of the opening.

14.3.4 Intersection of two square ducts at an angle. (See Fig. 14-10.)

Draw an elevation and plan view of the two intersecting square ducts as shown. Lay out the pattern for the vertical duct *A* and locate the opening 1-2-3-4 by projecting the intersecting points of the two ducts in the elevation.

Draw the stretchout for the slanted duct *B* and project the intersecting points of the two ducts from the elevation view.

14.3.5 90° T-pipe with like diameters. (See Fig. 14-11.)

Draw an elevation view of two ducts, one at right angles to the other. At one end of each duct, draw a semicircular plan view and divide each semicircle into any number of equal parts, numbering the division points as shown. Project lines through the numbered division points to both cylinders and mark their points of intersection to form the miter lines.

Draw the stretchout line *D* for the pattern of duct *C* and on it lay off the divisions spaced on the semicircle of cylinder *C.* Draw lines through these points. From the points where the corresponding elements of each duct intersect, project lines to intersect the elements on the stretchout. Connect the intersections on the stretchout located with an irregular curve to complete the duct C pattern.

Draw the stretchout line *E* for duct *B* by laying out the true length of the circumference of that cylin-

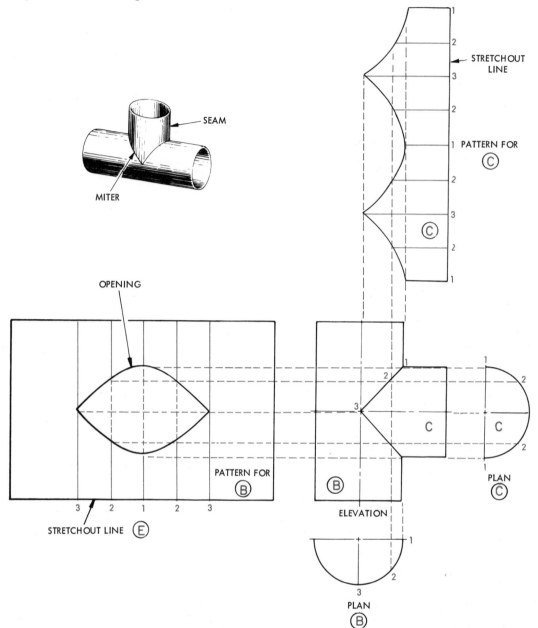

Fig. 14-11. How to develop patterns for a 90° T-pipe with like diameters.

der and locate the element spacings from plan *B*. Then draw lines through these division points. Project lines from the semicircle of plan view *B* until they intersect the corresponding elements of the stretchout. Connect the intersection points with an irregular curve.

14.3.6 90° T-pipe with unlike diameters. (See Fig. 14-12.)

Draw an elevation view of ducts *X* and *Y* with the desired diameters. On line *AB* scribe a semicircle and divide it into any number of equal parts. Since the cylinders have unlike diameters, draw also a side view of the intersecting cylinders. Turn this view of the cylinders around to bring the point numbered 1 on the semicircle to show in the center position.

Draw line *A'B'* for the stretchout and lay out the divisions, numbering them as shown. The true out-

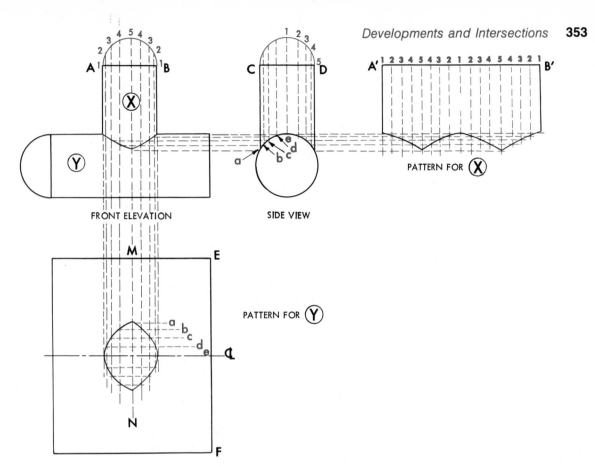

Fig. 14-12. How to develop patterns for a 90° T-pipe of unlike diameters.

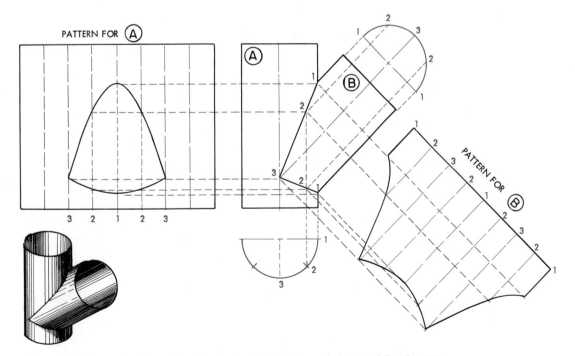

Fig. 14-13. How to develop patterns for an angle intersection of cylinders of like diameters.

line of the intersection is then found by projecting lines from the side view of cylinder *X* to the corresponding elements on the stretchout.

To find the opening of cylinder *Y*, lay out *EF* to represent the true circumference of cylinder *Y*. Project the elements of cylinder *X* to this plane. Starting at the center line on the stretchout plane, lay off distances *ed, dc, cb,* and *ba*, taken along the arc of the side view, on the corresponding projected elements from duct *X*. The intersection of these points will produce the correct opening for the intersecting duct.

14.3.7 Ducts of like diameters intersecting at an angle. (See Fig. 14-13.)

The method for laying out an angle intersection of two cylinders is similar to that for laying out an intersection of cylinders of like diameters. Any desired angle may be used. Details for making the stretchouts are shown in Fig. 14-13.

The only point that needs to be observed is to see that the center line of cylinder *A* intersects the center line of cylinder *B*.

14.3.8 Four piece elbow. (See Fig. 14-14.)

Draw arcs *BC* and *DE*. Divide arc *BC* into three

equal parts and bisect each of these divisions. From these bisectors, which provide the miter lines of the elbow, draw lines to center *A*. Draw the semicircle *EC* and divide it into any number of equal parts. From these points project lines to intersect miter lines *KL* and *MN*.

Draw the stretchout line for section *F* and on it lay off the correct number of divisions to represent the true circumference of the elbow. Draw vertical lines through these points. Project lines from elbow *F* to the corresponding elements on the stretchout and connect the intersections with a curved line.

Follow a similar procedure to lay out the pattern for section *G*. Actually pattern *G* may be laid out by duplicating the miter line contour of pattern *F*.

Develop only *F* and *G* since the remaining two patterns will be identical to those just found.

14.3.9 Twin elbows. (See Fig. 14-15.)

Construct an elevation and plan view of the ducts. Divide the plan view into any number of equal parts and project the points as shown. The stretchouts are developed in the same manner as described in subsection 14.3.8.

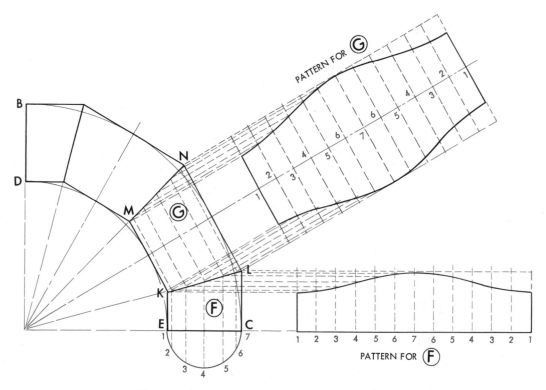

Fig. 14-14. How to develop patterns for a four piece elbow.

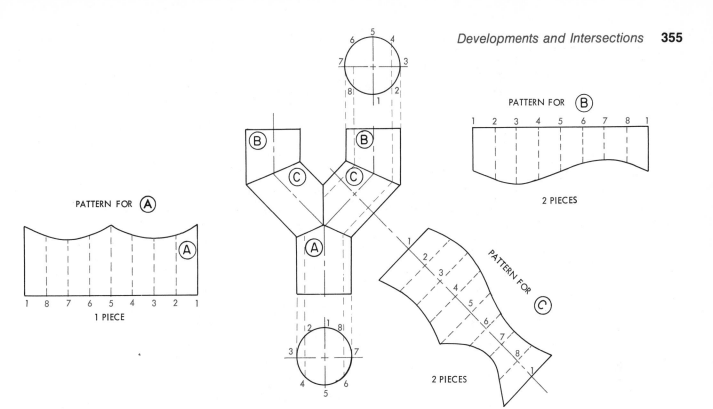

Fig. 14-15. How to develop patterns for twin elbows.

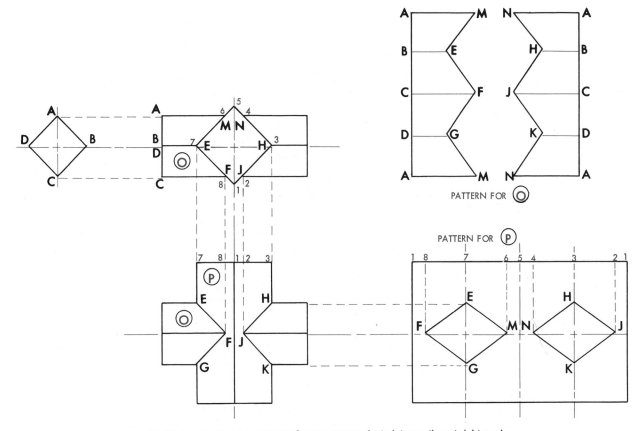

Fig. 14-16. How to develop patterns for two square ducts intersecting at right angles.

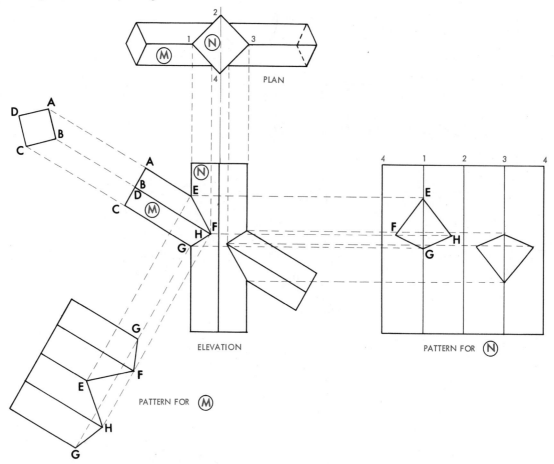

Fig. 14-17. How to develop patterns for two square ducts intersecting at an angle.

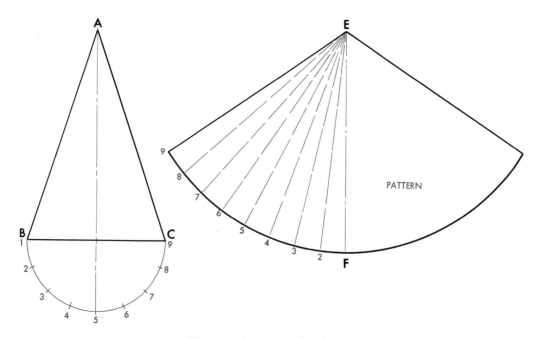

Fig. 14-18. How to develop a cone.

14.3.10 Square ducts intersecting at right angles. (See Fig. 14-16.)

Construct an elevation and plan view of ducts *O* and *P* and label the points as shown. Draw the stretchout line for duct *P* and on it lay out the lateral sides taken from the path view. Find the required openings by projecting the points from duct *O* in the elevation view.

Draw the stretchout lines for duct *O* and lay out the sides with true lengths *AB, BC,* and so on, taken from the plan view of duct *O*. Find the true lengths of the other elements of duct *O* in the plan view (*AM, BE, CF,* and so on) and lay them out on the stretchout lines.

14.3.11 Square ducts intersecting at an angle. (See Fig. 14-17.)

The intersection and development of the two oblique ducts in Fig. 14-17 is very similar to those for the ducts in Fig. 14-16. An elevation and plan view of the ducts are drawn and the sides correctly labeled. The contours for ducts *M* and *N* are obtained by laying out the lateral sides and projecting the corresponding points from the elevation.

14.4 Radial Line Developments

Radial line development is used to develop surfaces of regular tapering forms such as cones and pyramids. Several typical radial developments are described in the paragraphs that follow.

14.4.1 Cone. (See Fig. 14-18.)

To lay out a pattern for a cone, draw a full-size front view of the cone as triangle *ABC*. Using the base *BC* of the triangle as a diameter, draw a semicircle. Divide this semicircle into any number of equal parts, numbering them as shown.

With side *AB* or *AC* of the traingle *ABC* as a radius, draw an arc of unlimited length with *E* as a center to represent one boundary of the stretchout. Then draw line *EF* to intersect this arc at *F*. Beginning at *F*, lay off the distances that were spaced on the semicircle. Since this semicircle represents only half of the circumference of the base of the cone, it will be necessary to lay out an equal length for the opposite half of the same arc. From the end points on the arc, draw straight lines to the center *E*.

14.4.2 Frustum of a right cone. (Shown in Fig. 14-19.)

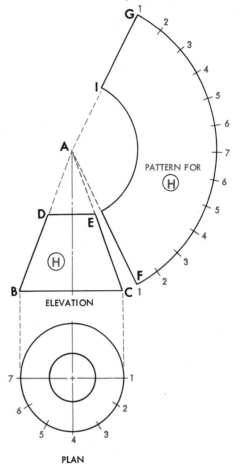

Fig. 14-19. How to develop a frustum of a right cone cut parallel to its base.

The procedure for laying out a frustum of a cone is very similar to that for a regular cone, except that two arcs having radii of *AB* and *AD* are drawn from the apex *A*.

14.4.3 Truncated right cone cut at an angle. (See Fig. 14-20.)

Draw the front view of the cone *ABC* with its truncating line. Bisect line *BC* and using *BF* as a radius, draw the plan view of the base. Divide the circle into any number of equal parts and number them as shown. Project lines from these points to the base line *BC* and extend them to the apex of the cone at *A*. Number the lines at the truncating line as shown.

With *AB* as a radius, draw the stretchout arc *DE*. On this stretchout line lay out the same di-

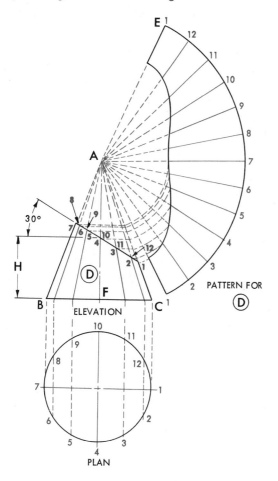

Fig. 14-20. How to develop a pattern for a truncated right cone cut at an angle.

visions as those that were spaced on the base circle and number them in the same order. Connect each point with the apex *A* by means of light construction lines.

To obtain the true lengths of the elements for the pattern, project lines horizontally from the points on the truncating line to *AC*. Then for each point on AC, set the compass equal to the distance from A to that point and scribe arcs to intersect the corresponding element on the pattern. Connect these intersecting points with a curved line.

14.4.4 Pyramid having a square base. (See Fig. 14-21.)

The important consideration in the development of a pyramid is to find the true slant length of the elevation. This is found by constructing the base

of the elevation so that it is parallel to the center line at the widest part of the plan view.

Fig. 14-21 illustrates two methods of developing patterns for a pyramid. In method *A* the plan view is arranged with its widest part parallel to the base of the elevation. Hence, the slant line *LM* appears in its true length.

If the plan view is drawn as in method *B,* the slant line *PR* is not the true length, because the widest part of the plan is not parallel to the base of the elevation. Therefore, to ascertain the actual length of the slant line, first extend the center line *NU* of the plan. With *N* as a center and *N*1 as a radius, scribe an arc to intersect *NU* at *X.* From *X* project a line vertically to the extension of base line *QR* in the elevation. Finally, draw line *PT,* which is also the true length of slant line *PR.*

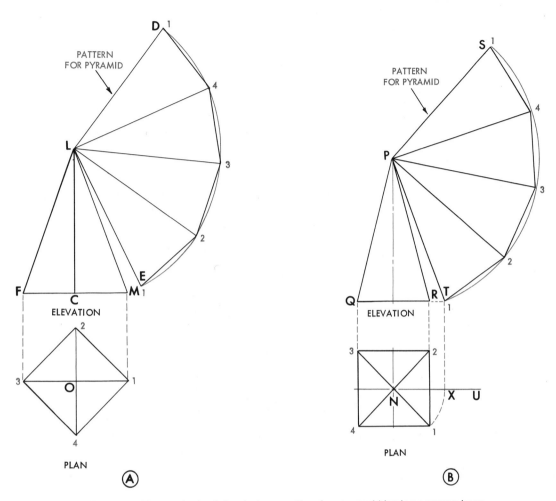

Fig. 14-21. Two methods of developing a pattern for a pyramid having a square base.

14.4.5 Truncated pyramid. (See Fig. 14-22.)

Draw an elevation and plan view of the pyramid with truncating line 1-2.

Find the true length of the slant line as described in subsection 14.4.4 and draw the stretchout arc *XY*.

On the stretchout arc lay off the true sides of the base as *AB, BC, CD,* and *DA.* Project lines through these points to center *E.*

Now find the true lengths of the lateral edges from the elevation view and transfer them to the stretchout as *A*1, *B*2, *C*3, and *D*4.

If a bottom base and top are required, draw them in as shown.

14.4.6 Truncated oblique pyramid. (See Fig. 14-23.) Draw an elevation and plan view of the truncated oblique pyramid and label the edges as shown.

Determine the true lengths of elements *E*1 and *E*2 as described in subsection 14.4.4. With the true lengths of *E*1 and *E*2 as radii, scribe two stretchout arcs from E₁.

Beginning at 1 on the outer arc, lay off the base lengths 1-2, 2-3, and so on taken from the plan view so that the lengths intersect the corresponding stretchout arcs. From these points draw radial lines to E₁. Lay off the true lengths of the lateral edges and connect the points with straight lines.

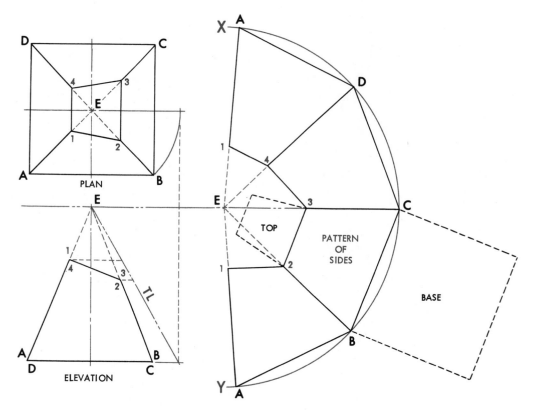

Fig. 14-22. How to develop a truncated right pyramid.

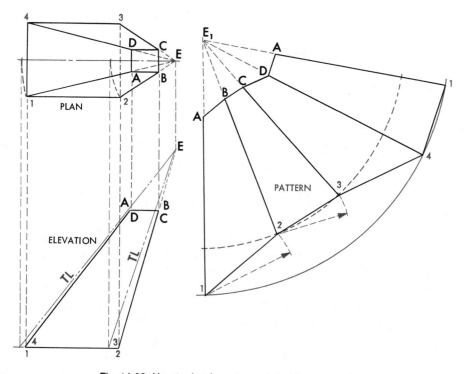

Fig. 14-23. How to develop a truncated oblique pyramid.

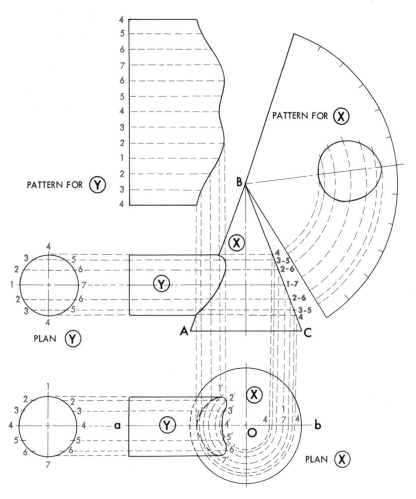

Fig. 14-24. How to develop a cylinder intersecting a cone at right angles.

14.4.7 *Cylinder intersecting a cone at right angles.* (See Fig. 14-24.)

Draw an elevation and plan view of cone *X* and horizontal cylinder *Y* as illustrated. Divide the plan view of cylinder *Y* into equal spaces and number the points as shown. Project these points horizontally to intersect the cone element *BC.*

From the points of intersection on line *BC,* drop vertical lines to intersect the center line *ab* in the plan view. With *O* as the center and radii equal to *O*4, *O*3-5, *O*2-6, and *O*1-7, scribe arcs as shown.

Draw a plan end view of cylinder *Y* on the plan view of cone *X* and number the points as indicated. From these points draw horizontal lines to intersect the corresponding numbered arcs at 1′, 2′, 3′, and so on. Draw vertical lines to intersect

similarly numbered horizontal elements in the elevation view. The intersections found will provide the miter line for the cylinder and cone.

To draw the pattern for the cone, develop the stretchout in the usual manner for a regular cone. To find the opening for the intersecting cylinder, draw arcs on the stretchout of the cone using *B* as a center, and *B*4, *B*3-5, and so on as radii. Lay out the distances taken along the arcs from the center line *ab* to the points of intersection 1′, 2′, and so on in the plan view, on the corresponding arcs in the stretchout of the cone. A line drawn through these points will produce the true opening.

To develop the pattern for cylinder *Y,* lay out a stretchout line equal to the circumference of the cylinder. Project the points from the miter line in the elevation to intersect the corresponding elements on the stretchout.

14.4.8 Cone intersecting a cylinder obliquely.
(See Fig. 14-25.)

Draw an elevation and plan view of the cone and cylinder as shown. Extend the sides of the cone B in the elevation to the base line 1-7. On this base line scribe a semicircle and divide it into any number of equal parts. Project these points to the base line and draw radial lines to center R.

On center line DE draw the plan view for cone B and divide it into equal parts. Project the points on plan view B to plan view A so that they will intersect vertical lines drawn from the base line 1-7 in the elevation. Where these lines intersect, draw radial lines to R'. From the points where the radial lines cut the circle for plan A, draw vertical lines to the corresponding cone elements in the elevation. The intersection of these lines will provide the miter line for cone B.

Using R as a center, draw an arc to represent the stretchout line for cone B as well as additional arcs from the points on miter line 1-7. The intersec-

tion of these arcs with the radial lines in the stretchout will provide the contour of cone B.

Draw the pattern for cylinder A and extend horizontal lines from the miter line 1-7 in the elevation. Find the opening in pattern A by transferring to the stretchout the circular distances 1-2, 1-3, 1-4, and so on from the center line DE on plan view A. The intersection of these points with the corresponding horizontal lines will produce the correct curvature for the opening.

14.5 Triangulation

Triangulation is a procedure used to develop warped and double-curved surfaces. Several triangulation techniques are described in the paragraphs that follow.

14.5.1 Square-to-square transition. (See Fig. 14-26.)

Draw an elevation and plan view of the piece and label the plan as shown. Draw line FG equal in length to the true height of the part in elevation and line GH of indefinite length.

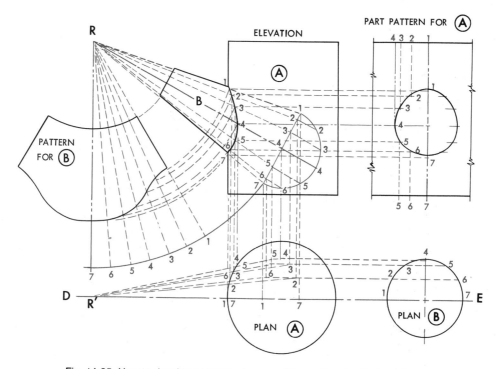

Fig. 14-25. How to develop a pattern of a cone intersecting a cylinder obliquely.

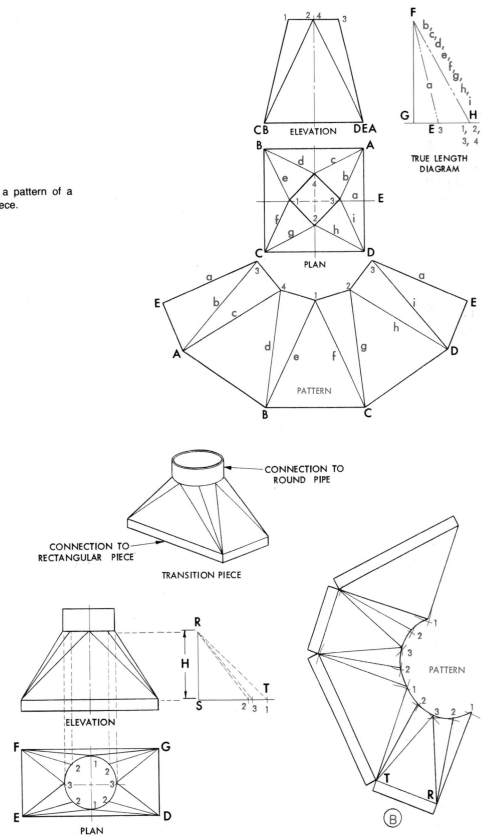

ELEVATION

TRUE LENGTH
DIAGRAM

Fig. 14-26. How to develop a pattern of a
square-to-square transition piece.

PLAN

PATTERN

CONNECTION TO
ROUND PIPE

CONNECTION TO
RECTANGULAR PIECE

TRANSITION PIECE

Fig. 14-27. How to develop
a pattern of a rectangular-
to-round transition piece.

ELEVATION

PLAN

PATTERN

Ⓑ

With *G* as the center, lay out on line *GH* distances equal to *D2, D3, A3, A4,* and so on from the plan view. Draw a line from points 1, 2, 3, 4 to *F*. This line represents the true lengths of elements *b, c, d, e, f, g, h,* and *i* in the plan view. Next lay out on line *GH* the length of *E3* taken from the plan view. Draw a line from this point to *F*. This line *a* is the true length of *E3*.

To develop the pattern, draw line *BC* equal in length to line *CD* in the plan view. With *F1* as the

true length taken from the true length diagram and *B* and *C* as centers, scribe intersecting arcs at 1. Then draw elements from *B* and *C* to the intersecting arcs at 1. With 1 as the center and the distances 1-2 and 1-4 from the plan view, scribe arcs at 2 and 4. With *B* and *C* as centers, *F2* as the true length, and *F4* as a radius, intersect the previously drawn arcs at 2 and 4. Connect these points with elements *C2* and *B4*. Similarly construct triangles *D2C, A4B, D2-3* and *A4-3*. Now

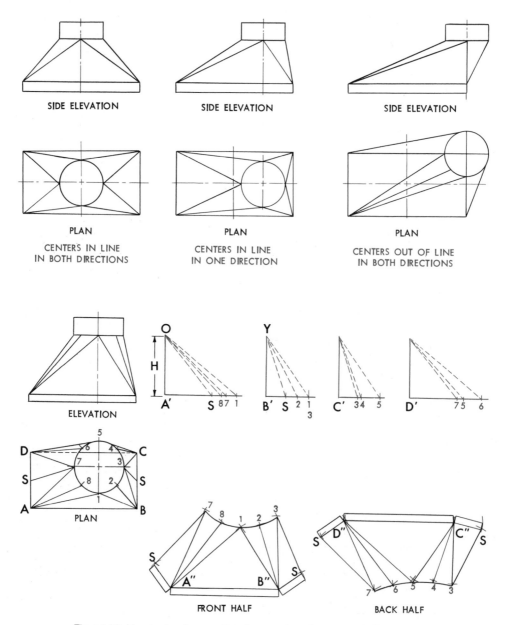

Fig. 14-28. How to develop a pattern for a rectangular-to-round offset transition.

using A and D as centers and the true distance of AE and ED in the plan view as a radius, scribe arcs at E and E. With 3 and 3 as centers and the true distance of E3 taken as *a* in the true length diagram as a radius, draw arcs to intersect at E and E. This completes the pattern.

14.5.2 Rectangular-to-round transition. (See Fig. 14-27.)

Draw an elevation and plan view of the piece as shown. Divide the circle in the plan view into any number of equal parts. Connect the numbered points of each quarter circle to the adjacent corners of the rectangular bases D, E, F, and G.

Lay out the true length diagram by drawing the vertical line RS equal to the true height of the transition in elevation and drawing line ST of any convenient length. Transfer the lengths of elements D1, D2, and D3 from the plan view to the true length diagram line ST. From these points draw lines to R.

To develop the pattern, draw line R1, as shown in B, equal in length to R1 in the true length diagram. Set the compass equal to distance 1-2 in the plan view and, with 1 as center, strike an arc. Intersect this arc with an arc using R as a center and a radius equal to the true length of R2. Connect R and 2 with a line. Strike an arc using 2 as center and a radius 2-3 of the plan view. Intersect this arc using R as center and a radius of the true length R3. Connect points 1, 2, and 3 with a

curve. Set the compass equal to EF of the plan view and, with R as center, strike an arc. Intersect this arc with an arc using 3 as a center and a radius equal to the true length R3. Draw the triangle R3T.

Complete the pattern by adding the remaining parts in a similar order.

14.5.3 Rectangular-to-round offset transition. (See Fig. 14-28.)

Draw an elevation and plan view of the piece. Divide the round end into any number of equal parts and draw element lines to the corners A, B, C, and D. Label each element as shown.

Since this piece contains numerous lines, it is advisable to construct four true length diagrams to facilitate locating the various lines.

Develop the pattern by transferring the true length of each element to the pattern as described in subsections 14.5.1 and 14.5.2.

14.5.4 Oblique cone. (See Fig. 14-29.)

Draw an elevation and plan view of the cone and divide the plan view into any number of equal parts. From the division points draw radial lines to O.

With O as a center, transfer the division points in the plan to line OP. Project these points down to the horizontal line O'P' and draw lines to meet at O. These lines now represent the true lengths of the elements of the oblique cone.

To lay out the pattern, find the true length of

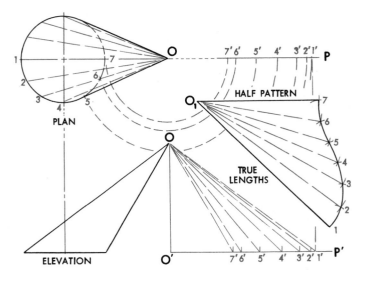

Fig. 14-29. How to develop a pattern for oblique cone.

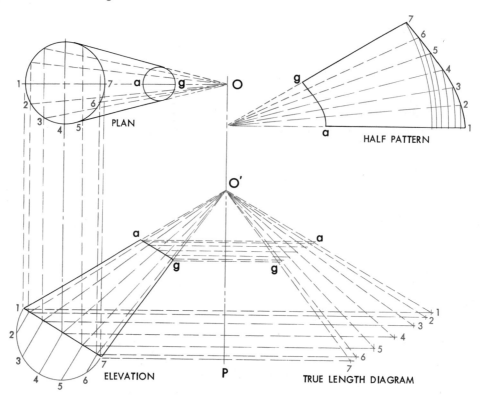

Fig. 14-30. How to develop a pattern for a transition piece between two circular openings of different diameters.

element *O*1′ from the true length diagram and, with *O*₁ in any convenient location, scribe an arc. With 1 as a center and a radius 1-2 obtained from the plan view, draw an arc at 2. With *O*₁ as the center and the true length 2′ as the radius, scribe an arc intersecting the previously drawn arc at point 2. Continue in this manner until all the true elements are laid out on the stretchout. Connect the base points with a smooth curve.

14.5.5 Round-to-round transition. (Shown in Fig. 14-30.)

Draw an elevation and plan view of the transition. Scribe a semicircle on the base of the oblique cone in elevation and divide it into any number of equal parts. Project the division points to the base line 1-7 and to the plan view. From these points draw radial lines to *O* and *O*′.

From the points on base line 1-7 in the elevation view, draw horizontal lines of indefinite length to the true length diagram. Find the true length of each element by transferring the distances *O*1, *O*2, *O*3, and so on in the plan view to the corresponding horizontal lines, measuring from vertical

line *O*′*P*. From each of these points, draw radial lines to *O*′. Draw horizontal lines from the miter line *ag* in the elevation view to intersect the corresponding element in the true length diagram.

To develop the pattern, lay out each element on the stretchout, starting with *a*1 in the true length diagram and the chordal distances 1-2, 2-3, and so on, from the semicircle in elevation.

14.5.6 Y branch. (See Fig. 14-31.)

Draw the elevation and plan views as shown. Divide the half and quarter plans into any number of equal parts and project the divisions to the base lines of the respective cylinders. Number all points, using even numbers for section *B* and odd numbers for section *A*.

Lay out the pattern for section *A* by means of regular parallel line development.

To develop the pattern for part *B*, first find the true lengths of the slant lines by constructing one true length diagram for the dotted lines and another for the solid lines. To determine the true lengths of the solid lines, draw a horizontal line of any length and mark off distance 3-4 from eleva-

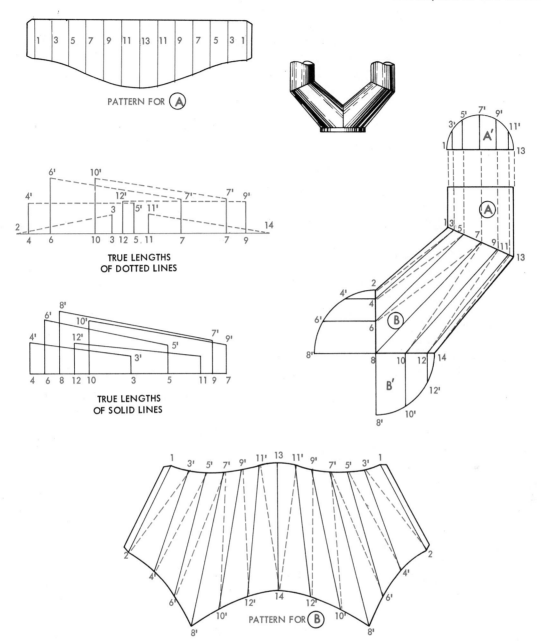

Fig. 14-31. How to develop patterns for a Y branch.

tion *B*. At 3 erect a perpendicular line equal to the distance 3-3′ in profile *A′*. At 4 erect a perpendicular line equal to the distance 4-4′ in profile *B′*. Draw a line connecting points 3′ and 4′. This line is the true length for line 3-4 in the elevation view. Proceed in a similar manner to find the true lengths of the remaining solid lines. Find the true lengths of the dotted lines in a like manner.

To develop the pattern for *B*, draw line 1-2 equal to line 1-2 of the elevation. With point 2 as center and a radius equal to the true length dotted line 2-3′, strike an arc at 3′. Set the compass to space 1-3′ of profile *A′* and, with 1 on pattern *B* as center, intersect the arc at point 3′. With the true length solid line 3′-4′ as a radius and point 3′ as center, strike an arc at 4′. Set the compass to space 2-4′ of *B′* and, with point 2 on the pattern as center, cut an arc at point 4′. Continue in a similar manner until all the true length lines are drawn in their respective positions.

14.5.7 Rectangular-to-rectangular transition. (See Fig. 14-32.)

Draw the necessary views of the piece as shown. Divide the curved edges of the flat side view into any number of equal parts and connect the points with solid and dotted lines. From these points project lines to the plan view. Lay out throat and heel patterns by parallel line development.

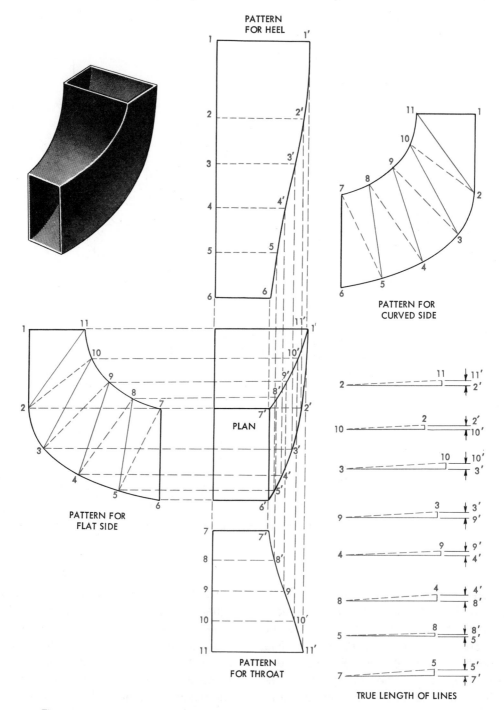

Fig. 14-32. How to develop a rectangular-to-rectangular transition piece with one flat side.

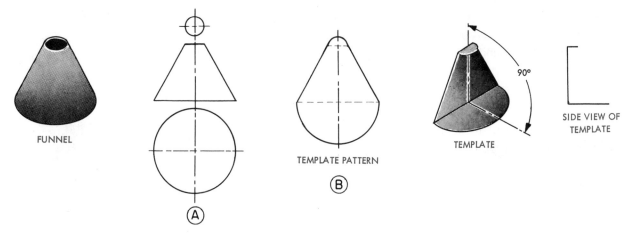

Fig. 14-33. How to prepare a template for a cone by the short method development.

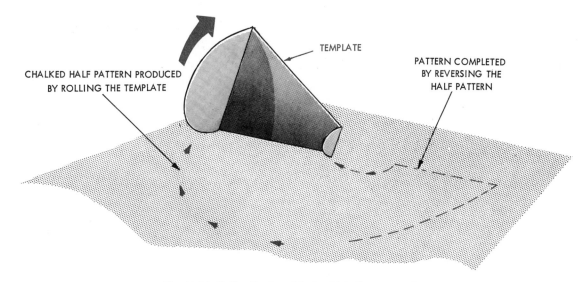

Fig. 14-34. Rolling the template to obtain the cone pattern.

To develop the pattern for the curved side, lay out true length diagrams for the slant lines in the flat side. The base of each triangle is found by measuring between the appropriate vertical lines in the plan view. Lay out the pattern using true length lines as described in subsection 14.5.6.

14.6 Rollout Development

Rollout is a short method of development which involves making a sheet-metal template and then rolling the template on paper.[1]

---

1. Ralph W. Poe, *Short Method of Pattern Development* (Middletown, Ohio: Armco Steel Corp.).

To illustrate this process, assume that a funnel or truncated cone is to be developed.

First, front, bottom, and top views of the cone are drawn, as shown in Fig. 14-33A. Next the cone is laid out with one-half of its top and bottom attached as in B. Then the half circles representing one-half of the top and bottom openings of the cone are bent at right angles to the cross section. The pattern is produced by chalking the edges of the template and rolling it over a sheet of soft black building paper. An alternate method is to brush the edges of the template with oil and roll it over layout paper. In each case the result is distinct lines that provide the stretchout of the pattern as shown in Fig. 14-34.

Figures 14-35, 14-36, 14-37, and 14-38 show the short method of developing representative patterns commonly found in sheet-metal work.

14.7 Bend Allowance

In laying out sheet-metal templates, allowances must be made wherever the metal is to be bent to

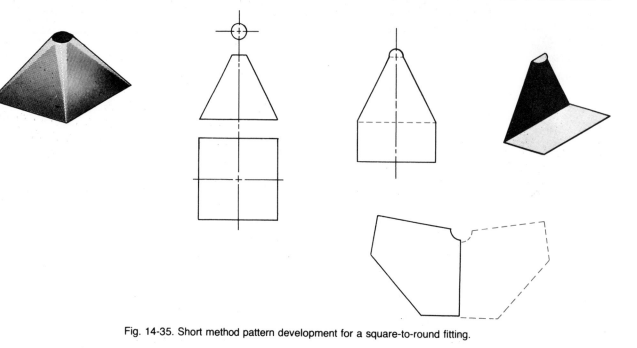

Fig. 14-35. Short method pattern development for a square-to-round fitting.

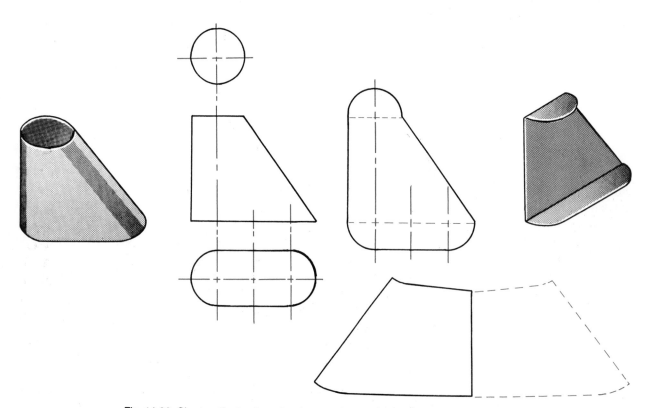

Fig. 14-36. Short method pattern development for an off-center oval-to-round fitting.

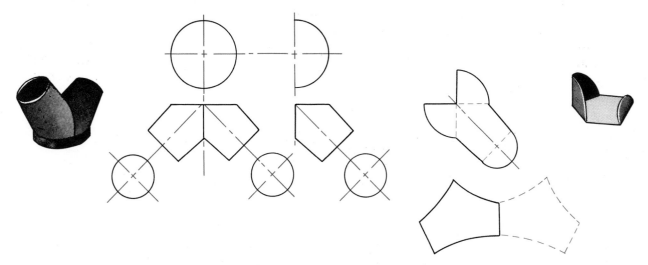

Fig. 14-37. Short method pattern development for a Y-fitting.

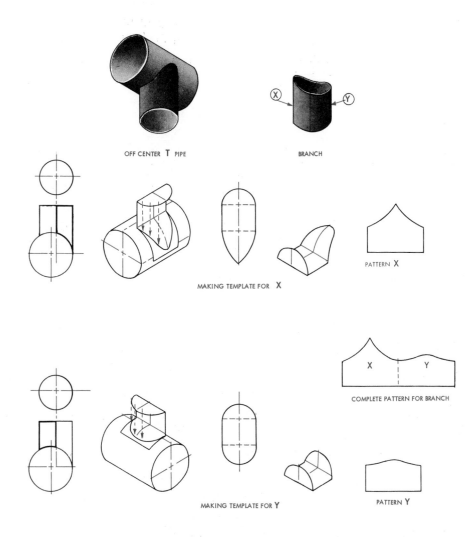

OFF CENTER **T** PIPE

BRANCH

MAKING TEMPLATE FOR **X**

PATTERN **X**

COMPLETE PATTERN FOR BRANCH

MAKING TEMPLATE FOR **Y**

PATTERN **Y**

Fig. 14-38. Short method pattern development for an off-center T-pipe.

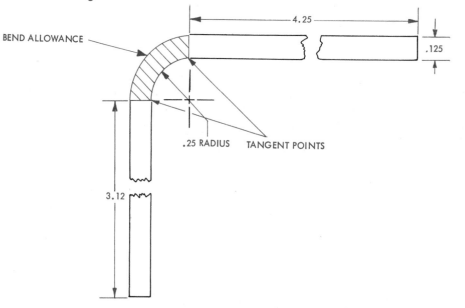

Fig. 14-39. Bend allowance is added to the tangent point dimensions.

Top No. in Each Space = Bend Allowance for 90° Bends   Bottom No. in Each Space = Bend Allowance Constant for 1°

Bend Radii / Thickness	.031	.063	.094	.125	.156	.188	.219	.250	.281	.313	.344	.375	.438	.500	.531	.625	.656	.750	.781	1.000
.013	.058/.00064	.108/.00120	.157/.00174	.205/.00228	.254/.00282	.304/.00338	.353/.00392	.402/.00446	.450/.00500	.501/.00556	.549/.00610	.598/.00664	.697/.00774	.794/.00883	.843/.00937	.991/.01101	1.039/.01155	1.187/.01319	1.236/.01373	1.580/.01755
.016	.060/.00067	.110/.00122	.159/.00176	.208/.00231	.256/.00285	.307/.00342	.355/.00395	.404/.00449	.453/.00503	.503/.00559	.552/.00613	.600/.00667	.699/.00777	.796/.00885	.845/.00939	.993/.01103	1.041/.01157	1.189/.01321	1.238/.01375	1.582/.01758
.019 and .020	.062/.00069	.113/.00125	.161/.00179	.210/.00233	.259/.00287	.309/.00343	.358/.00397	.406/.00452	.455/.00506	.505/.00561	.554/.00616	.603/.00670	.702/.00780	.799/.00888	.848/.00942	.995/.01106	1.044/.01160	1.192/.01324	1.240/.01378	1.584/.01760
.022	.064/.00071	.114/.00127	.163/.00181	.212/.00235	.260/.00289	.311/.00345	.359/.00399	.408/.00453	.457/.00508	.507/.00563	.556/.00617	.604/.00672	.703/.00782	.801/.00890	.849/.00944	.997/.01108	1.046/.01162	1.193/.01326	1.242/.01380	1.586/.01762
.025	.066/.00074	.116/.00129	.165/.00184	.214/.00238	.263/.00292	.313/.00348	.362/.00402	.410/.00456	.459/.00510	.509/.00566	.558/.00620	.607/.00674	.705/.00784	.803/.00892	.851/.00946	.999/.01110	1.048/.01164	1.195/.01328	1.244/.01382	1.588/.01764
.028	.068/.00076	.119/.00132	.167/.00186	.216/.00240	.265/.00294	.315/.00350	.364/.00404	.412/.00458	.461/.00512	.511/.00568	.560/.00622	.609/.00676	.708/.00786	.805/.00894	.854/.00948	1.001/.01112	1.050/.01167	1.198/.01331	1.246/.01385	1.590/.01767
.031 and .032	.071/.00079	.121/.00134	.170/.00189	.218/.00243	.267/.00297	.317/.00353	.366/.00407	.415/.00461	.463/.00515	.514/.00571	.562/.00625	.611/.00679	.710/.00789	.807/.00897	.856/.00951	1.004/.01115	1.052/.01169	1.200/.01333	1.249/.01387	1.593/.01770
.038	.075/.00084	.126/.00140	.174/.00194	.223/.00248	.272/.00302	.322/.00358	.371/.00412	.419/.00466	.468/.00520	.518/.00576	.567/.00630	.616/.00684	.715/.00794	.812/.00902	.861/.00956	1.008/.01120	1.057/.01174	1.205/.01338	1.253/.01392	1.597/.01775
.040	.077/.00085	.127/.00141	.176/.00195	.224/.00249	.273/.00303	.323/.00359	.372/.00413	.421/.00468	.469/.00522	.520/.00577	.568/.00632	.617/.00686	.716/.00796	.813/.00904	.862/.00958	1.010/.01122	1.058/.01176	1.206/.01340	1.255/.01394	1.599/.01776
.050 and .051		.134/.00149	.183/.00203	.232/.00258	.280/.00312	.331/.00368	.379/.00422	.428/.00476	.477/.00530	.527/.00566	.576/.00640	.624/.00694	.723/.00804	.821/.00912	.869/.00966	1.017/.01130	1.066/.01184	1.213/.01348	1.262/.01402	1.606/.01784
.063 and .064		.144/.00160	.192/.00214	.241/.00268	.290/.00322	.340/.00378	.389/.00432	.437/.00486	.486/.00540	.536/.00596	.585/.00650	.634/.00704	.732/.00814	.830/.00922	.878/.00976	1.026/.01140	1.075/.01194	1.222/.01358	1.271/.01412	1.615/.01794
.072			.198/.00220	.247/.00274	.296/.00328	.346/.00384	.394/.00438	.443/.00492	.492/.00546	.542/.00602	.591/.00656	.639/.00710	.738/.00820	.836/.00929	.885/.00983	1.032/.01147	1.081/.01201	1.228/.01365	1.277/.01419	1.621/.01801
.078			.202/.00225	.251/.00279	.300/.00333	.350/.00389	.399/.00443	.447/.00497	.496/.00551	.546/.00607	.595/.00661	.644/.00715	.743/.00825	.840/.00933	.889/.00987	1.036/.01152	1.085/.01206	1.233/.01370	1.281/.01424	1.625/.01806
.081			.204/.00227	.253/.00281	.302/.00335	.352/.00391	.401/.00445	.449/.00499	.498/.00554	.548/.00609	.598/.00664	.646/.00718	.745/.00828	.842/.00936	.891/.00990	1.038/.01154	1.087/.01208	1.235/.01372	1.283/.01426	1.627/.01808
.091			.212/.00235	.260/.00289	.309/.00343	.359/.00399	.408/.00453	.456/.00507	.505/.00561	.555/.00617	.604/.00671	.653/.00725	.752/.00835	.849/.00944	.898/.00998	1.045/.01162	1.094/.01216	1.242/.01380	1.290/.01434	1.634/.01816
.094			.214/.00237	.262/.00291	.311/.00346	.361/.00401	.410/.00456	.459/.00510	.507/.00564	.558/.00620	.606/.00674	.655/.00728	.754/.00838	.851/.00946	.900/.00999	1.048/.01164	1.096/.01218	1.244/.01382	1.293/.01436	1.636/.01818
.102				.268/.00298	.317/.00352	.367/.00408	.416/.00462	.464/.00516	.513/.00570	.561/.00626	.612/.00680	.661/.00734	.760/.00844	.857/.00952	.906/.01006	1.053/.01170	1.102/.01224	1.249/.01388	1.298/.01442	1.642/.01825
.109				.273/.00303	.321/.00357	.372/.00413	.420/.00467	.469/.00521	.518/.00575	.568/.00631	.617/.00685	.665/.00739	.764/.00849	.862/.00958	.910/.01012	1.058/.01176	1.107/.01230	1.254/.01394	1.303/.01448	1.647/.01830
.125				.284/.00316	.333/.00370	.383/.00426	.432/.00480	.480/.00534	.529/.00588	.579/.00644	.628/.00698	.677/.00752	.776/.00862	.873/.00970	.922/.01024	1.069/.01188	1.118/.01242	1.266/.01406	1.314/.01460	1.658/.01842
.156					.355/.00394	.405/.00450	.453/.00504	.502/.00558	.551/.00612	.601/.00668	.650/.00722	.698/.00776	.797/.00886	.895/.00994	.943/.01048	1.091/.01212	1.140/.01266	1.287/.01430	1.336/.01484	1.680/.01867
.188						.427/.00475	.476/.00529	.525/.00583	.573/.00637	.624/.00693	.672/.00747	.721/.00801	.820/.00911	.917/.01019	.966/.01073	1.114/.01237	1.162/.01291	1.310/.01455	1.359/.01510	1.702/.01892
.203								.535/.00595	.584/.00649	.634/.00704	.683/.00759	.731/.00813	.830/.00923	.928/.01031	.976/.01085	1.124/.01249	1.173/.01303	1.320/.01467	1.369/.01521	1.713/.01903
.218								.546/.00606	.594/.00660	.645/.00716	.693/.00770	.742/.00824	.841/.00934	.938/.01042	.987/.01097	1.135/.01261	1.183/.01315	1.331/.01479	1.380/.01533	1.724/.01915
.234								.557/.00619	.606/.00673	.656/.00729	.705/.00783	.753/.00837	.852/.00947	.950/.01055	.998/.01109	1.146/.01273	1.194/.01327	1.342/.01491	1.391/.01545	1.735/.01928
.250								.568/.00631	.617/.00685	.667/.00741	.716/.00795	.764/.00849	.863/.00959	.961/.01068	1.009/.01122	1.157/.01286	1.206/.01340	1.353/.01504	1.402/.01558	1.746/.01940

Fig. 14-40. Bend allowance chart (decimal-inch).

avoid cracking. Two methods are used to calculate bend allowance. One method involves the calculation of the material required to bend around a given radius by means of the empirical formula:

B.A. = *N*(0.01745 × *R* + 0.0078 ×*T*)

where *B.A.* = bend allowance
N = number of degrees in bend
R = radius of bend
and *T* = thickness

The resulting allowance is then added to the tangent point dimensions. See Fig. 14-39.

To eliminate the time-consuming task of making the necessary calculations for each required bend, special bend allowance charts are available that provide the correct allowance for materials of different thicknesses and bends of different radii. See Figs. 14-40 and 14-41.

Example. Find the total length of a fitting to be made of metal 0.12 inch thick, having tangent dimensions of 4.25 and 3.12 inches and which is to be bent 90° over a 0.25-inch radius. See Fig. 14-39. Using the bend allowance chart in Fig. 14-40, the calculation would be made as follows:

B.A. for 1° = .00534
B.A. for 90° = .00534 × 90 = .480
Tan. Dim. = 4.25 + 3.12 = 7.37
Total Length = 7.37 + .480 = 7.850

The second method of determining bend allowance is by means of setback. *Setback* is simply the difference between the sum of the outside dimensions of the angle and the actual distance needed to form the angle. See Fig. 14-42. The amount of setback is found by using a setback chart as shown in Fig. 14-43. Once the amount of setback is determined for a given bend

Min Metal Thickness	0.40	0.50	0.60	0.80	1.0	1.2	1.4	1.6	1.8	2.0	2.5	3.0	3.5	4.0
Rad Deg	Allowance													
0.5 1	0.011	0.012	0.012	0.013	0.015	0.016	0.017	0.018	0.019	0.020	0.023	0.026	0.029	0.032
0.5 90	0.99	1.05	1.10	1.20	1.31	1.41	1.52	1.62	1.73	1.83	2.09	2.36	2.62	2.88
1.0 1	0.020	0.020	0.021	0.022	0.023	0.024	0.026	0.027	0.028	0.029	0.032	0.035	0.038	0.041
1.0 90	1.78	1.83	1.88	1.99	2.09	2.20	2.30	2.41	2.51	2.62	2.88	3.14	3.40	3.66
1.5 1	0.028	0.029	0.030	0.031	0.032	0.033	0.034	0.035	0.037	0.038	0.041	0.044	0.047	0.049
1.5 90	2.56	2.62	2.67	2.78	2.88	2.98	3.09	3.19	3.30	3.40	3.66	3.93	4.19	4.45
2.0 1	0.037	0.038	0.038	0.040	0.041	0.042	0.043	0.044	0.045	0.047	0.049	0.052	0.055	0.058
2.0 90	3.35	3.40	3.46	3.56	3.66	3.77	3.87	3.98	4.08	4.19	4.45	4.71	4.97	5.23
2.5 1	0.046	0.047	0.047	0.048	0.049	0.051	0.052	0.053	0.054	0.055	0.058	0.061	0.064	0.067
2.5 90	4.14	4.19	4.24	4.34	4.45	4.55	4.66	4.76	4.87	4.97	5.23	5.50	5.76	6.02
3.0 1	0.055	0.055	0.056	0.057	0.058	0.059	0.060	0.062	0.063	0.064	0.067	0.070	0.073	0.076
3.0 90	4.92	4.97	5.03	5.13	5.23	5.34	5.44	5.55	5.65	5.76	6.02	6.28	6.54	6.80
3.5 1	0.063	0.064	0.065	0.066	0.067	0.068	0.069	0.070	0.072	0.073	0.076	0.079	0.081	0.084
3.5 90	5.71	5.76	5.81	5.92	6.02	6.12	6.23	6.33	6.44	6.54	6.80	7.07	7.33	7.59
4.0 1	0.072	0.073	0.073	0.074	0.076	0.077	0.078	0.079	0.080	0.081	0.084	0.087	0.090	0.093
4.0 90	6.49	6.54	6.60	6.70	6.80	6.91	7.02	7.12	7.22	7.33	7.59	7.85	8.11	8.38
4.5 1	0.081	0.081	0.082	0.083	0.084	0.086	0.087	0.088	0.089	0.090	0.093	0.096	0.099	0.102
4.5 90	7.28	7.33	7.38	7.49	7.59	7.70	7.80	7.90	8.01	8.11	8.38	8.64	8.90	9.16
5.0 1	0.090	0.090	0.091	0.092	0.093	0.094	0.095	0.097	0.098	0.099	0.102	0.105	0.108	0.111
5.0 90	8.06	8.12	8.17	8.27	8.38	8.48	8.59	8.69	8.79	8.90	9.16	9.42	9.69	9.95
5.5 1	0.098	0.099	0.099	0.101	0.102	0.103	0.104	0.105	0.106	0.108	0.111	0.113	0.116	0.119
5.5 90	8.85	8.90	8.95	9.06	9.16	9.27	9.37	9.47	9.58	9.69	9.95	10.21	10.47	10.73
6.0 1	0.107	0.108	0.108	0.109	0.111	0.112	0.113	0.114	0.115	0.116	0.119	0.122	0.125	0.128
6.0 90	9.63	9.68	9.74	9.84	9.95	10.05	10.16	10.26	10.37	10.47	10.73	10.99	11.26	11.52
6.5 1	0.116	0.116	0.117	0.118	0.119	0.120	0.122	0.123	0.124	0.125	0.128	0.131	0.134	0.137
6.5 90	10.42	10.47	10.52	10.63	10.73	10.84	10.94	11.05	11.15	11.26	11.52	11.78	12.04	12.30
7.0 1	0.124	0.125	0.126	0.127	0.128	0.129	0.130	0.131	0.133	0.134	0.137	0.140	0.143	0.145
7.0 90	11.20	11.26	11.31	11.41	11.52	11.62	11.73	11.83	11.94	12.04	12.30	12.56	12.83	13.09
7.5 1	0.133	0.134	0.134	0.136	0.137	0.138	0.139	0.140	0.141	0.143	0.145	0.148	0.151	0.154
7.5 90	11.99	12.04	12.09	12.20	12.30	12.41	12.51	12.62	12.72	12.83	13.09	13.35	13.61	13.87
8.0 1	0.142	0.143	0.143	0.144	0.145	0.147	0.148	0.149	0.150	0.151	0.154	0.157	0.160	0.163
8.0 90	12.77	12.83	12.88	12.98	13.09	13.19	13.30	13.40	13.51	13.61	13.87	14.13	14.40	14.66

Fig. 14-41. Bend allowance chart (mm).

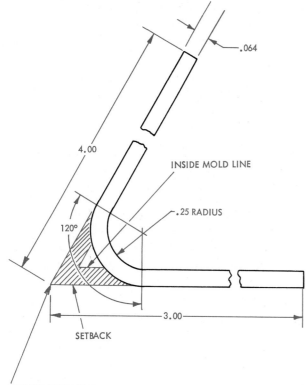

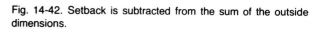

Fig. 14-42. Setback is subtracted from the sum of the outside dimensions.

CLOSED ANGLE

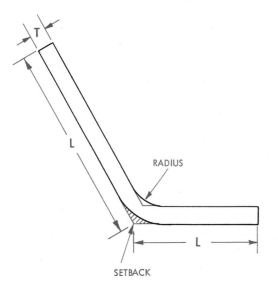

OPEN ANGLE

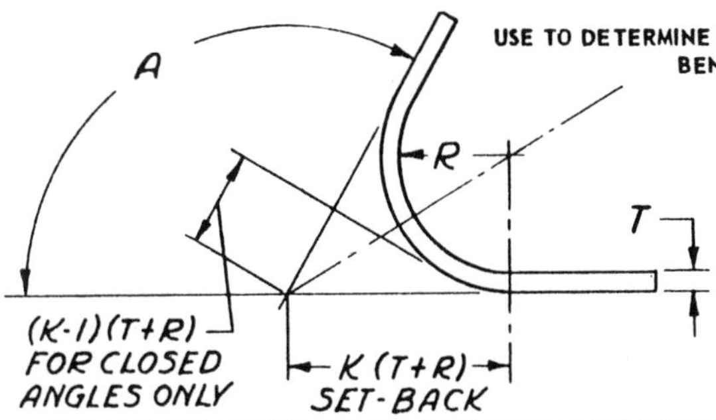

USE TO DETERMINE THE DISTANCE BETWEEN THE MOLD LINE AND THE
BEND LINE OF SHEET METAL FLANGES

A

R

T

$(K-1)(T+R)$
FOR CLOSED
ANGLES ONLY

$K(T+R)$
SET-BACK

EXAMPLE ·

DEGREES IN BEND (A)= 120°
GAGE (T) = .032
RADIUS = .125
K (FROM TABLE) = 1 .732
SET-BACK=1.732 (.032+.125)
1.732 X .157
= .272

A	K	A	K	A	K	A	K
1°	.00873	51°	.47697	101°	1.2131	151°	3.8667
2°	.01745	52°	.48773	102°	1.2349	152°	4.0108
3°	.02618	53°	.49858	103°	1.2572	153°	4.1653
4°	.03492	54°	.50952	104°	1.2799	154°	4.3315
5°	.04366	55°	.52057	105°	1.3032	155°	4.5107
6°	.05241	56°	.53171	106°	1.3270	156°	4.7046
7°	.06116	57°	.54295	107°	1.3514	157°	4.9151
8°	.06993	58°	.55431	108°	1.3764	158°	5.1445
9°	.07870	59°	.56577	109°	1.4019	159°	5.3995
10°	.08749	60°	.57735	110°	1.4281	160°	5.6713
11°	.09629	61°	.58904	111°	1.4550	161°	5.9758
12°	.10510	62°	.60086	112°	1.4826	162°	6.3137
13°	.11393	63°	.61280	113°	1.5108	163°	6.6911
14°	.12278	64°	.62487	114°	1.5399	164°	7.1154
15°	.13165	65°	.63707	115°	1.5697	165°	7.5957
16°	.14054	66°	.64941	116°	1.6003	166°	8.1443
17°	.14945	67°	.66188	117°	1.6318	167°	8.7769
18°	.15838	68°	.67451	118°	1.6643	168°	9.5144
19°	.16734	69°	.68728	119°	1.6977	169°	10.385
20°	.17633	70°	.70021	120°	1.7320	170°	11.430
21°	.18534	71°	.71329	121°	1.7675	171°	12.706
22°	.19438	72°	.72654	122°	1.8040	172°	14.301
23°	.20345	73°	.73998	123°	1.8418	173°	16.350
24°	.21256	74°	.75355	124°	1.8807	174°	19.081
25°	.22169	75°	.76733	125°	1.9210	175°	22.904
26°	.23087	76°	.78128	126°	1.9626	176°	28.636
27°	.24008	77°	.79543	127°	2.0057	177°	38.188
28°	.24933	78°	.80978	128°	2.0503	178°	57.290
29°	.25862	79°	.82434	129°	2.0965	179°	114.590
30°	.26795	80°	.83910	130°	2.1445	180°	infinite
31°	.27732	81°	.85408	131°	2.1943		
32°	.28674	82°	.86929	132°	2.2460		
33°	.29621	83°	.88472	133°	2.2998		
34°	.30573	84°	.90040	134°	2.3558		
35°	.31530	85°	.91633	135°	2.4142		
36°	.32492	86°	.93251	136°	2.4751		
37°	.33459	97°	.94896	137°	2.5386		
38°	.34433	88°	.96569	138°	2.6051		
39°	.35412	89°	.98270	139°	2.6746		
40°	.36397	90°	1.00000	140°	2.7475		
41°	.37388	91°	1.0176	141°	2.8239		
42°	.38386	92°	1.0355	142°	2.9042		
43°	.39391	93°	1.0538	143°	2.9887		
44°	.40403	94°	1.0724	144°	3.0777		
45°	.41421	95°	1.0913	145°	3.1716		
46°	.42447	96°	1.1106	146°	3.2708		
47°	.43481	97°	1.1303	147°	3.3759		
48°	.44523	98°	1.1504	148°	3.4874		
49°	.45573	99°	1.1708	149°	3.6059		
50°	.46631	100°	1.1917	150°	3.7320		

Fig. 14-43. Setback chart.

radius and metal thickness, it is deducted from the outside dimensions of the angle.

Example. Find the setback required to calculate the exact length of a .064-inch metal fitting having a 0.25 inch bend radius of 120° and outside dimensions of 4.00 and 3.00 inches. See Fig. 14-43.

$$
\begin{aligned}
\text{Degrees in bend} &= 120° \\
\text{Thickness} &= .064 \\
\text{Radius} &= .25 \\
K \text{ from Table} &= 1.732 \\
\text{Setback} &= 1.732 \, (.064 + .25) \\
&= 1.732 \times .314 \\
&= .544
\end{aligned}
$$

Problems for Section IV
Geometry in Drafting

The problems in this section deal with drafting practices described in

Unit 12. Geometric Construction

Unit 13. Practical Descriptive Geometry

Unit 14. Developments and Intersections

Problems for Section IV include

Problems 1-17 Geometric Construction

Problems 18-43 Descriptive Geometry

Problems 44-70 Developments and Intersections

Problems 1-17 Geometric Construction

Problem 1

1. Bisect a line 2.25″ long drawn at an angle of 37°.
2. Draw a straight line parallel to another straight line at a distance of .62″.
3. Draw any curved line. Construct a parallel curved line at a distance of .81″ from the first line.
4. Divide a straight line 2.69″ long into five equal parts.
5. Lay out any acute angle and find its bisector.
6. Locate any three points at random. Determine the circular arc that will pass through these points.

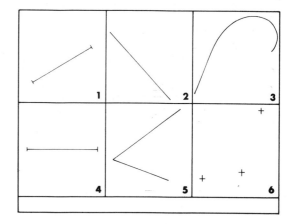

Problem 2

1. Draw a 90° angle with sides of any convenient length. Construct a circular arc tangent to the two lines having a radius of 1.5″.
2. Construct two lines forming any convenient obtuse angle. Draw a tangent arc having a radius of 1.88.″.
3. Draw a 2″ diameter circle and through any point outside the circle, such as *P*, construct a straight line tangent to the circle.
4. Draw a 0.62″ radius arc tangent to a straight line and an arc having a radius of 1″.
5. Construct an arc having a radius of 0.75″ tangent to the two given arcs.
6. Construct a straight line tangent to two arcs. Determine points of tangency.

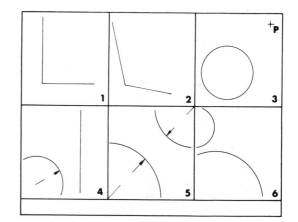

Problems having decimal-inch values may be converted to equivalent metric sizes by using the millimeter conversion chart or the problems may be redesigned by assigning new millimeter dimensional values more compatible with metric production.

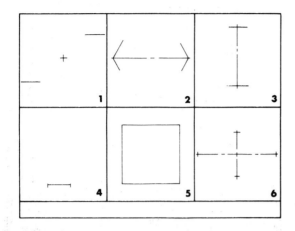

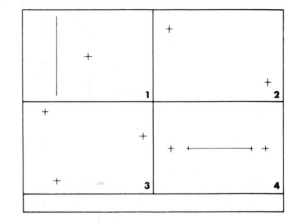

Problem 3

1. Draw an ogee curve tangent to two parallel straight lines spaced 1.06″ apart through a point midway between the lines.
2. Construct a regular hexagon having a distance across corners of 2.88″.
3. Draw a regular hexagon having a distance across flats of 2.09″.
4. Construct a regular hexagon having a side of .81″.
5. Lay out a regular octagon in 2.12″ square.
6. Using a major axis of 3.00″ and a minor axis of 1.62″, produce an ellipse by the foci method.

Problem 5

1. Construct a parabolic curve using a directrix and focus.
2. Construct a parabola through two points using the parallelogram method.
3. Using three points, construct a parabola using the tangent method.
4. Using a transverse axis and two foci, construct a hyperbolic curve.

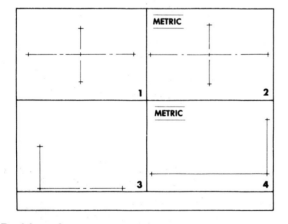

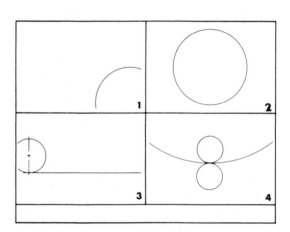

Problem 4

1. Lay out by the trammel method an ellipse having axes of 4.00″ and 2.00″.
2. Using the four-center method, construct an ellipse with a major axis of 114 mm and a minor axis of 55 mm.
3. Construct a quarter of an ellipse using the concentric-circle method having a major axis of 6.25″ and a minor axis of 3.25″.
4. Use the parallelogram method to produce a quarter of an ellipse having a major axis of 228 mm and a minor axis of 100 mm.

Problem 6

1. Construct the involute of an arc of a circle.
2. Construct the spiral of Archimedes in one turn of the circle.
3. Draw a half cycloid formed by a 1.25″ generating circle rolling along a straight line.
4. Draw the epicycloid and hypocycloid formed by a 1.00″ diameter generating circle which rolls on a 4.00″ radius arc.

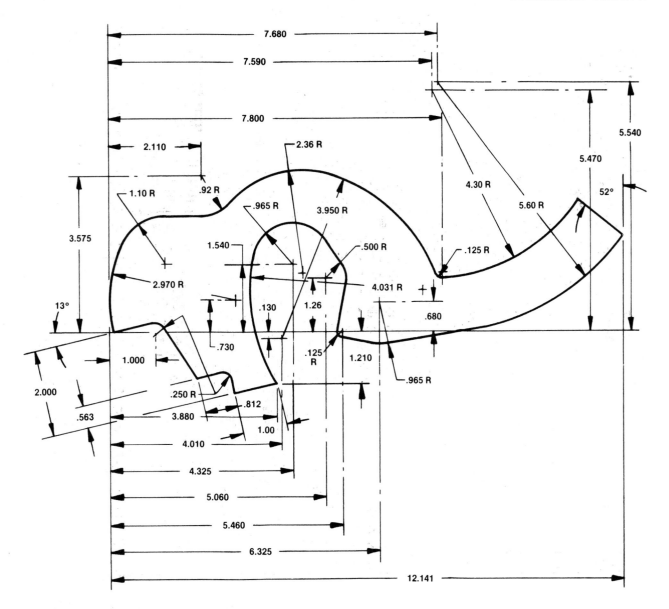

Problem 7
Construct a full-size layout of the seal shown.

Problem 8
The large diameter cylinder of an offset printing press is to be used to transfer ink from a 1″ diameter cylinder to a second 1″ diameter cylinder. Assuming that the surface of the large cylinder is to be tangent to the theoretical center line between the two smaller cylinders, determine the distance between the shaft centers of the small cylinders.

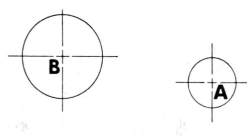

Problem 9

Construct a 2.50″ radius arc so that it is internally tangent to a .75″ diameter circle *A* and externally tangent to a 1.25″ diameter circle *B* when the distance between the center points of circles *A* and *B* is 3.00″. Measure the distance between center point *A* and the center of the 2.50″ arc. Also determine the distance between the center of the 2.50″ arc and the center of circle *B*.

Problem 10

A tubular underground system is to contain three cylindrical tubes each having an outside diameter of 4′-0″ and positioned so that there is a 13″ clearance between each tube. Assuming that they are to be encased in a larger cylinder with its inside wall a distance of 13″ from the outside surface of the 4′-0″ tubes, determine the minimum inside diameter of the encasing cylinder.

Problem 11

Referring to Problem 10, assume that at a later date the three 4′-0″ diameter tubes are to be replaced within the encasing cylinder by seven tubes having equal outside diameters. If these tubes are positioned so there is no clearance between them and the inside diameter of the encasing cylinder, what is the maximum diameter of the seven tubes.

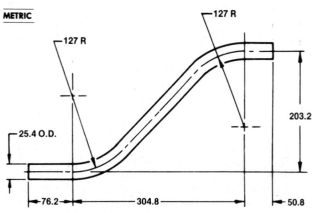

Problem 12

Prepare a half-size drawing to show the given tube bend. Sizes are in mm.

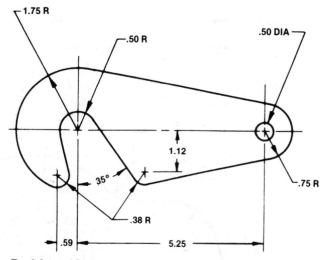

Problem 13

Construct a full-size layout of the latch shown.

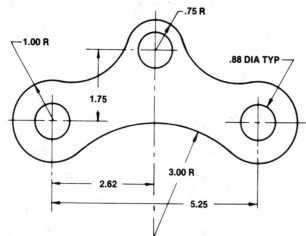

Problem 14

Make a full-size drawing of the link shown.

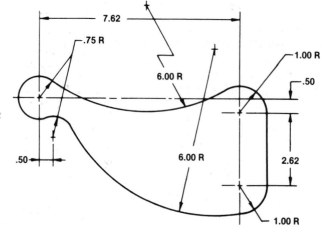

Problem 15

Prepare a drawing of the vane as shown.

Problem 16
Construct a cylindrical helix having a diameter of
4″ and a lead of 6″. (See Fig. 12-50.)

Problem 17
Modify the ink cylinder design in Problem 8 so that
all three cylinders are tangent to a theoretical
straight line. Determine the new distance between
the shaft centers of the smaller cylinders.

Problems 18-43 Descriptive Geometry

Problem 18-23
Lay out these problems on an A-size sheet using
the scale shown.

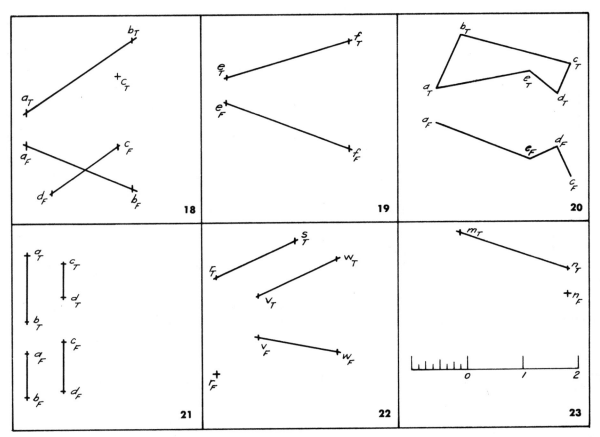

Problem 18
Draw the top view of the line *CD* which intersects
line *AB*.

Problem 19
Determine the true length of the line *EF*.

Problem 20
Complete the front view of surface *ABCDE*.

Problem 21
Determine by a drawing whether or not lines *AB*
and *CD* are parallel.

Problem 22
Line *RS* is parallel to line *VW*. Find the front view
of *RS*.

Problem 23
Line *MN* is 2.44″ long. Find its front view.

Problems 24-28

Lay out these problems on an A-size sheet using the scale shown.

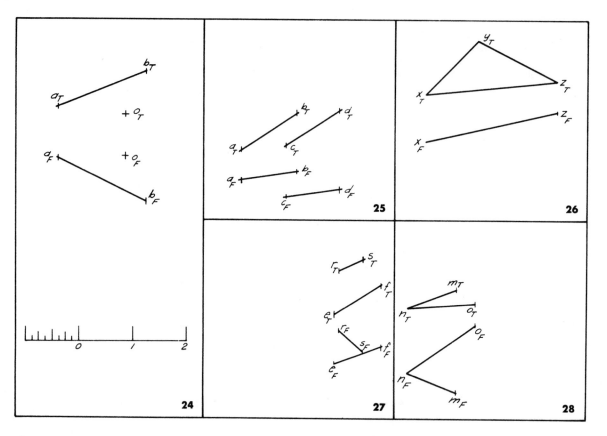

Problem 24

Determine the shortest distance from point *O* to line *AB*.

Problem 25

Determine the shortest distance between lines *AB* and *CD*.

Problem 26

Side *YZ* of triangle *XYZ* is 2.38″ long. Complete the front view of *XYZ*.

Problem 27

Find the shortest line connecting lines *EF* and *RS*.

Problem 28

Find and measure the true angle formed by intersecting lines *NO* and *MN*.

Problems 29-33
Lay out these problems on an A-size sheet using the scale shown.

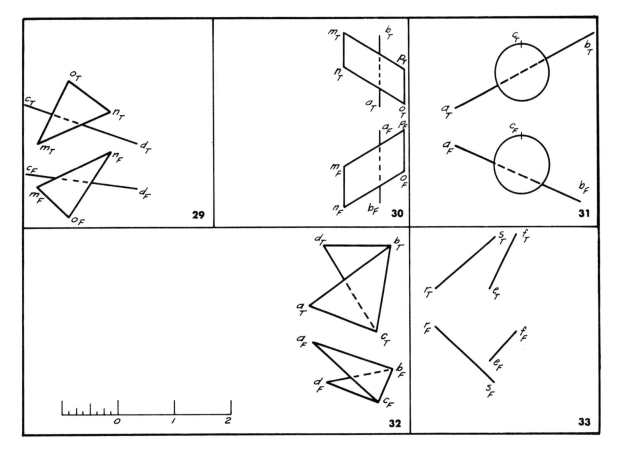

29

30

31

32

33

Problem 29
Find the point at which line *CD* pierces plane *MNO*.

Problem 30
Find the point at which line *AB* pierces *MNOP*.

Problem 31
Determine the piercing point of line *AB* and the elliptical surface.

Problem 32
Find and measure the true angle formed by the intersecting planes *ABC* and *BCD*.

Problem 33
What is the shortest distance between lines *EF* and *RS?*

Problems 34-38

Lay out these problems on an A-size sheet using the scale shown.

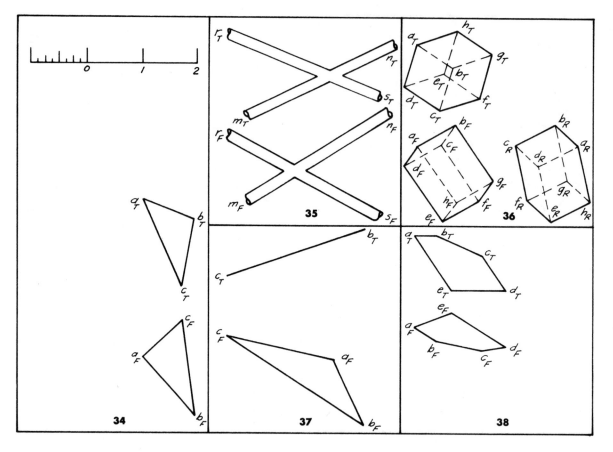

Problem 34
Determine the true size and shape of plane *ABC*.

Problem 35
Complete the views of the two rods, showing visibility.

Problem 36
Determine visibility of the edges of the block.

Problem 37
Line *AB* is 2.12″ long. Complete the top view of triangle *ABC*. Solve by revolution.

Problem 38
Find the true size and shape of plane *ABCD* by revolution.

Problems 39-42
Lay out these problems on an A-size sheet using
the scale shown.

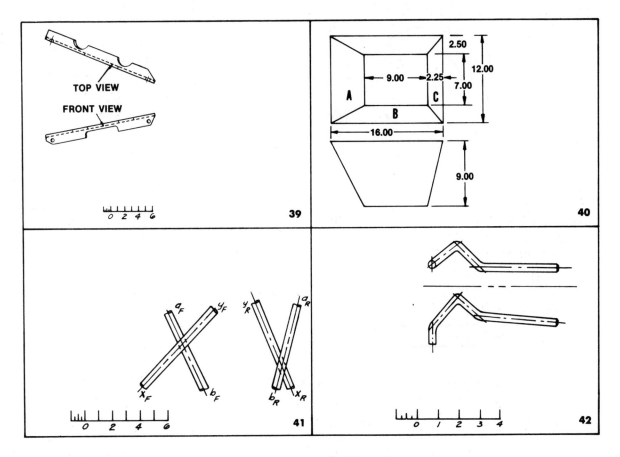

Problem 39
Find the true length of the retainer.

Problem 40
Determine the dihedral angles formed between
plates *A* and *B,* and *B* and *C,* of the chute illustrated.

Problem 41
What is the minimum clearance between the two
support rods?

Problem 42
Find the amount of bend angle for each bend in
the connector rod.

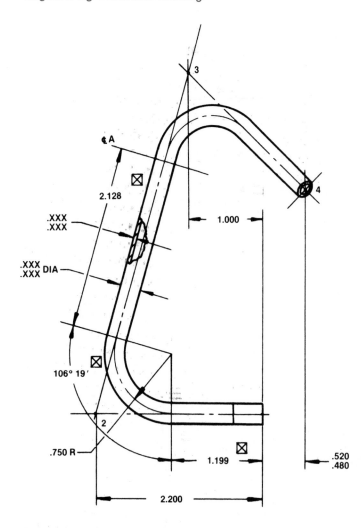

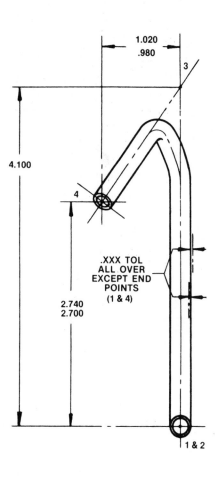

Problem 43

Find the bend angle at point 3 of the tubular member.

Problems 44-70 Developments and Intersections

Problems 44-53
Make stretchout patterns of the lateral surfaces of the problems shown. In Problem 47 include the bottom.

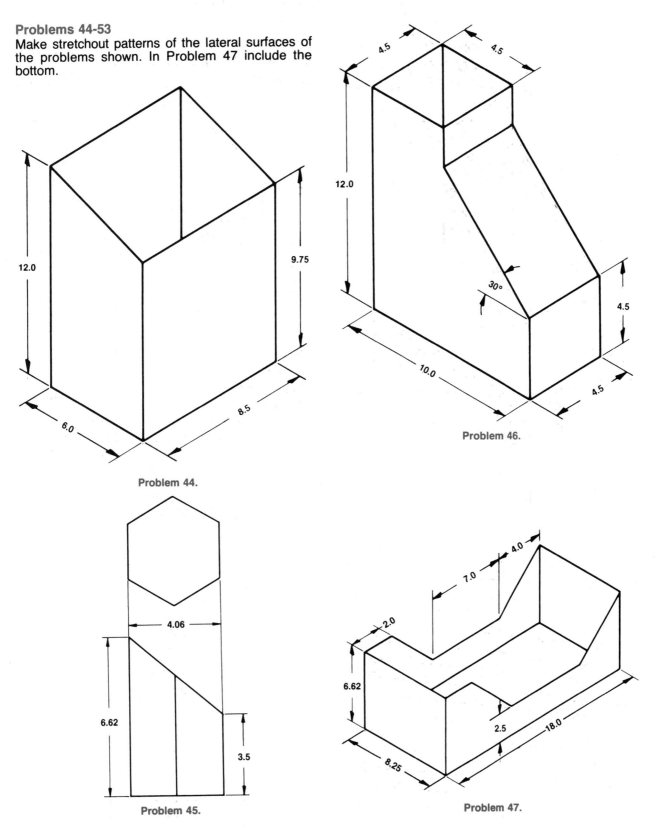

Problem 44.

Problem 45.

Problem 46.

Problem 47.

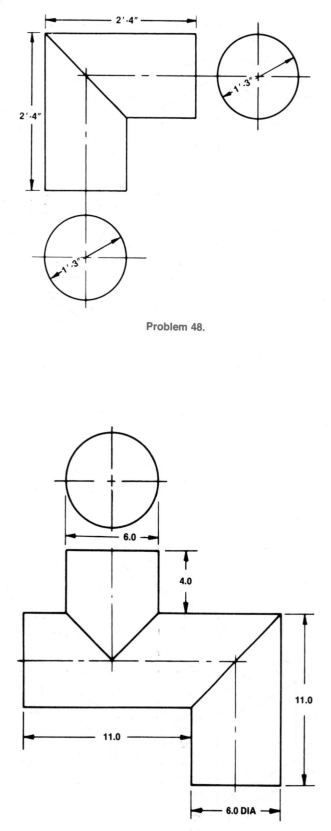

Problem 48.

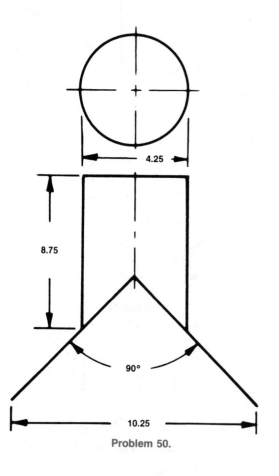

Problem 50.

Problem 49.

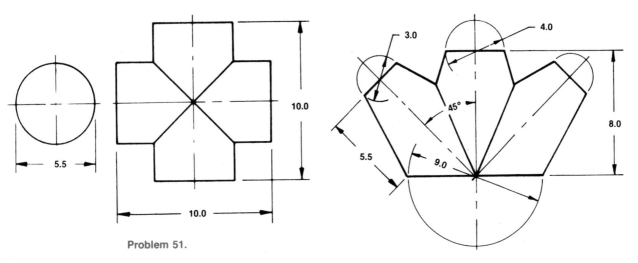

Problem 51.

Problem 53.

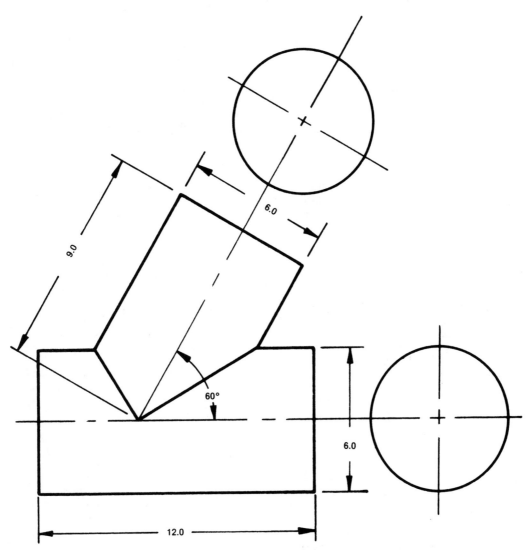

Problem 52.

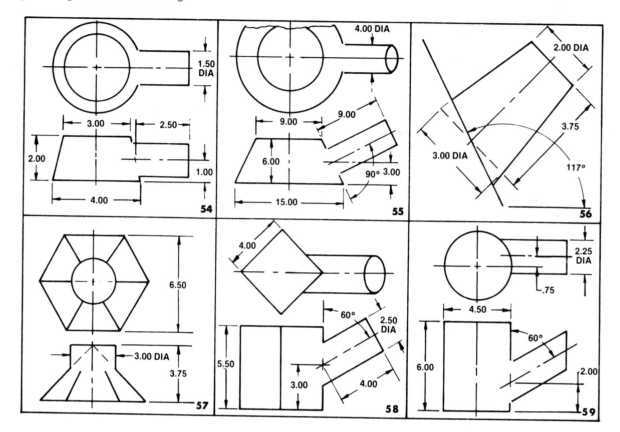

Problems 54-59
Select one of these problems. Determine the intersections and lay out the patterns for the lateral surfaces. Make full size or to scale as assigned.

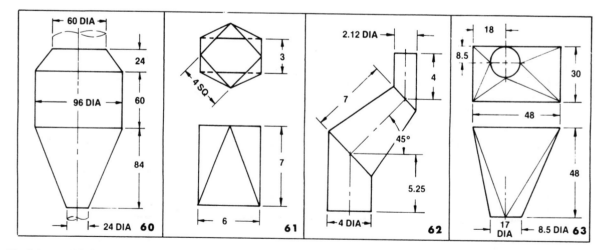

Problems 60-63
Prepare full-size or scale patterns of the lateral surfaces of the items shown as assigned.

Problems 64-65
Using the rollout method for pattern development, prepare stretchouts of the transitions shown. Cut out the patterns and assemble.

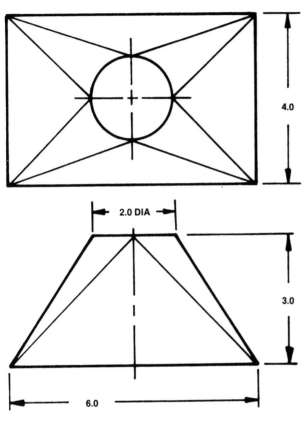

2.5 R

6.0

4.0

6.0

Problem 65.

2.0 DIA

3.0

6.0

Problem 64.

Problem 66
Determine the true shapes of the lateral surfaces of the chute. Assemble a model of the problem from cardboard.

TOP OPENING

10

12

45°

10

6

BOTTOM OPENING

7

Problems 67-68
Determine the length of the cross sections using the bend allowance chart, Fig. 14-40.

Problems 69-70
Determine the lengths of the cross sections shown using the setback chart, Fig. 14-42.

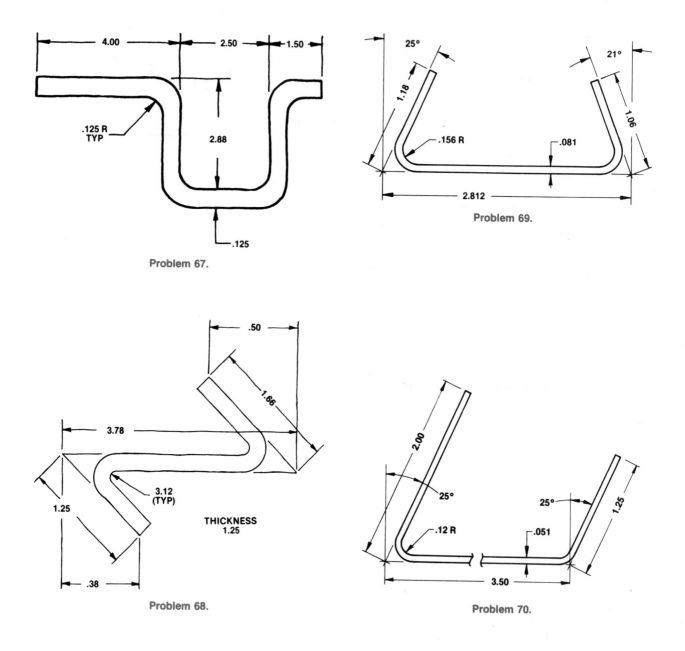

Problem 67.

Problem 69.

Problem 68.

Problem 70.

SECTION V

Industrial Production Drafting

SECTION V
UNIT 15

Production Design and Development

Product design, often referred to as industrial design, has as its primary function the creation of consumer goods that have wide general acceptance by the public and can be readily manufactured on a mass production basis at a reasonable cost. Its simplest manifestation is the production of a quality product that fully meets the needs of the buyer and complies with essential environmental and safety requirements.

15.1 Industrial Designing

The responsibility of creating acceptable products usually rests with the industrial designer. However, the industrial designer does not work alone but in collaboration with other company engineering, manufacturing, and marketing personnel. A design in itself is meaningless unless it can be transformed into a product that the consumer wants, can be efficiently manufactured, and can produce sufficient sales at a reasonable profit. Although the industrial designer must possess artistic talent, a basic understanding of engineering, manufacturing, and merchandising processes is also needed. Without some knowledge of these areas, it is impossible to analyze and to evaluate data provided by other people in the company.

The mechanical engineer may find that although a design looks good, it is not acceptable because it fails to meet necessary strength requirements. Similarly the industrial engineer may find that the design creates costly manufacturing problems. The marketing department may ascertain that the

Fig. 15-1. The designer often checks the final engineering drawings to insure that the design intent is reflected in the manufacturing directions.

design is too radical and, as a result, will not have general appeal.

If, on the other hand, the industrial designer has some basic concepts of manufacturing and marketing and works harmoniously with professional co-workers, the possibility of becoming involved in an impractical design that will prove costly and time-consuming is minimized. See Fig. 15-1.

Industrial designing is a highly specialized profession. Consequently, it is impossible to cover here all of the knowledge, skills, and practices needed for industrial designing. This material primarily is intended to help drafting personnel develop a better understanding of a few basic principles of product development. Areas discussed are

1. merchandizing
2. research and analysis
3. materials and manufacturing processes
4. design elements
5. fabricating design practices
6. strength of materials

15.1.1 Merchandizing.

15.1.1 Merchandizing. Any manufacturer, to stay in business, must regularly improve or add new features to his products. In any system of free industrial enterprise, competition is a dominant factor, and survival means competitive changes. The buying public knows no loyalty; it does not hesitate to switch products or manufacturers if someone else has something better to offer.

Often the sale of a product is not dependent on its basic function but rather on some special appeal it may have to the consumer. See Fig. 15-2. For example, there is no longer a need to convince the public what an automobile, clothes dryer, or electric razor will do. The buyer usually makes a choice based on what the manufacturer can demonstrate are outstanding features that competing products do not have. Hence, each year new models appear on the market that emphasize changes or additions having some special appeal to the buyer.

Frequent design changes not only help a manufacturer maintain a competitive position but also . stimulate greater sales, which contributes to the economy of the social structure in which people live and work. Product changes also help to satisfy the general public's seemingly insatiable desire to have something new. Rarely today will you hear the comment "In the old days they made things to last." As a matter of fact, manufacturers have the capability to make most products last a virtual lifetime. They do not for two reasons: (1) the cost of the product would become prohibitive, and (2) the average person becomes tired of the product and wants something new.

Inevitably the role of merchandising requires constant shifting of sales strategy, changes of pricing, and tactical sales psychology to give a manufacturer a favorable edge over the competition. Although the various strategies of merchandising have little to do with producing a new design, the personnel responsible for product improvement still must understand what is involved in merchandising, because what is produced will have a profound effect on what has to be sold. If the product is to sell, it must be priced right and still possess uniqueness of design that will be attractive to buyers.

15.1.2 Research and analysis.

15.1.2 Research and analysis. Before anything is done to develop a new product or to modify an old product, some attempt is made to find out what the consumer wants. Assuming that market analysis, through various market surveys, has verified the potential sales of a product, the industrial designer needs sufficient data about the buying public to insure that the new design will meet the needs and expectations of the people. For example, the designer should know whether the contemplated product will be used by a man or a woman, what dimensions it must be limited to, what environmental conditions it will be subjected to, what its service requirements will be, and, above all, what the features and shortcomings of competitive products are. Without this kind of information, it is impossible for the designer to understand what the consumer really wants. Failure to accurately perceive the needs and wants of the public is likely to result in disastrous consequences.

As a rule the industrial designer does not personally conduct market surveys to secure essential consumer data. Usually this information is compiled by other departments in the company. The designer's responsibility is to analyze the facts and then take them into consideration as the proposed design is developed. Normally after the designer has made a careful study of the data, an

Fig. 15-2. The sale of a product often depends on some special appeal it has to the consumer.

attempt will be made to formulate some objectives. From them the established specifications necessary to meet the requirements can be developed.

For example, let us assume that a decision has been made to redesign an umbrella which the company has been producing for a number of years. Several other umbrella manufacturers have recently introduced new models that are selling well. The designer may state the problem in the following way.

Problem: To design a more functional umbrella, to include

1. shorter length for greater ease in carrying
2. stronger canopy bracing without increased weight
3. new type of fabric for canopy
4. better means of closing and locking canopy
5. more attractive handle.

With the task clearly outlined, the designer is

able to identify basic specifications and make preliminary sketches. Some of the specifications may have to be modified as the concept of the new design begins to take shape. Nevertheless, by clearly identifying the basic elements of the project early, many pitfalls can be avoided later.

15.1.3 Materials and manufacturing processes.

A very important factor in any design process is knowing what techniques are to be used in manufacturing the product and what kind of materials will be the most appropriate for the job to be done. First, of course, the designer must decide whether the product can be produced with the existing equipment in the plant. Thus, it would be foolish to design a product that calls for plastic molding if there is no plastic molding equipment. Naturally there are circumstances when additional equipment may have to be purchased to insure economy and feasibility of manufacturing the product. However, this is a decision which must be made by management. Insofar as the designer is concerned, the available manufacturing facilities must be a primary consideration.

Since any industrial designer ultimately must consider manufacturing processes, it follows that a practical knowledge of these processes is essential. For example, if the product involves sand casting, it would be extremely difficult to design a proper casting without knowing something about how castings are made in the foundry. A design itself may be excellent, but if the product is not producible, then the value of the design is lost. Often designs have to be reworked simply because the designer failed to understand production techniques.

The industrial designer does not have to possess specific operational skills. For instance, it is not necessary to be a welder to know whether a product can be effectively designed for welding fabrication. On the other hand, by having a basic knowledge of welding, the designer is less likely to specify a joint that does not meet strength requirements.

With some practical knowledge of manufacturing processes, a designer can often supplement his or her knowledge by consulting a specialist in the plant. Through these efforts a design can be translated from paper to an economically produced product.

For similar reasons an industrial designer should have an extensive understanding of materials. Otherwise a material may be specified that cannot be handled with existing plant equipment or is too costly to meet competitive pricing. Sometimes by specifying a certain type of plastic or aluminum alloy, a commanding superiority over a competitive product can be achieved without making extensive alterations in the design. Unless a designer understands materials and manufacturing techniques, possession of artistic talents may prove virtually worthless. Only through a knowledge of manufacturing techniques and materials can one develop cost awareness. Without a deep sense of cost values, products cannot be manufactured economically.

15.2 Elements of Design

To some extent, good design is relative. We all do not see things the same way, nor do we have the same likes and dislikes or the same set of

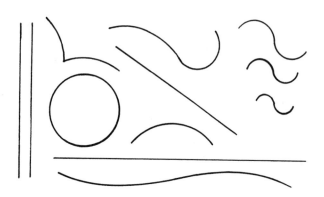

Fig. 15-3. Types of lines used for design purposes.

standards. Moreover, it is difficult to apply any objective measurement or rules to ascertain whether a design is good or bad. Nevertheless, there are certain basic principles that contribute to the production of attractive products. How well the results reflect these principles depends to a large extent on the ingenuity of the designer in blending them into a functional product. The contemplated product should not only carry out its intended function but do so for a reasonable period of time. On the other hand, function in itself must not be the dominant factor. If all of our homes were designed entirely for function, we might have row upon row of identical concrete and glass boxes with no charm, warmth, variety, or originality, with a resulting grim monotony. Good design implies beauty as well as function.

There are many design principles which experienced designers use to obtain desired results. For the purpose here we are concerned primarily with unity and variety, balance and symmetry, propor-

tion, and rhythm. These principles are achieved by blending together lines, planes, forms, and surface qualities.

15.2.1 Lines. A line is the path of a point moving in space. It has but one dimension—length—and its most important characteristic is direction. A carefully drawn line is the first step in transforming a mental image into an actual physical shape. Lines cause the eye to skip from one point of interest to another within a design, thereby adding interest and retaining the attention of the observer.

Lines, when joined together, form the boundaries of planes, produce outlines of shapes, contours, openings, and intersections of planes. They define and give shape to an object and establish size relationships and proportions. Basically there are four types of lines: straight, circular, curved, and S-shaped. See Fig. 15-3.

Lines can be used to express a variety of feelings or motions such as dignity, repose, action, stability, or flight. See Fig. 15-4. They also can in-

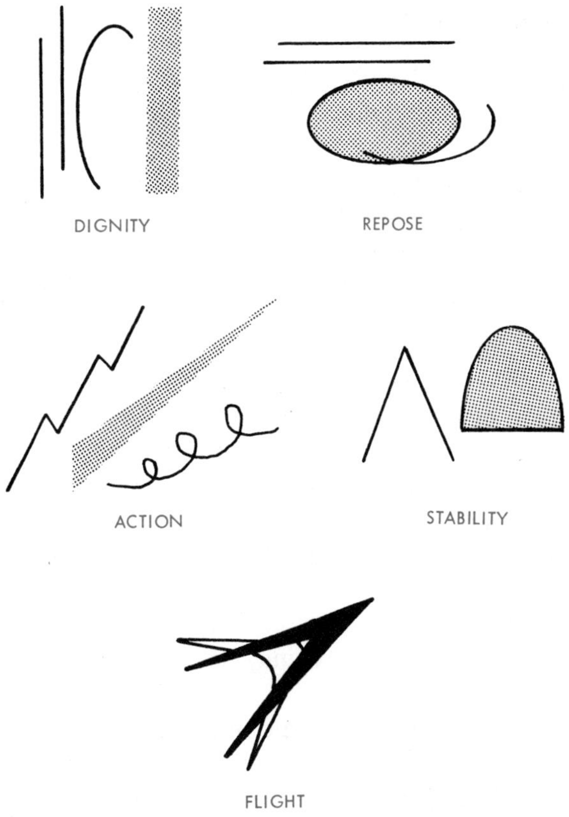

DIGNITY   REPOSE

ACTION   STABILITY

FLIGHT

Fig. 15-4. Lines and shapes may produce feelings of action and emotion.

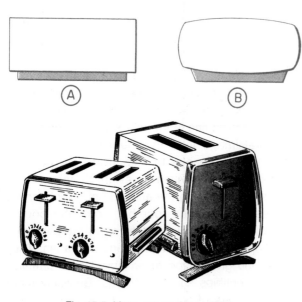

Fig. 15-5. Lines can produce subtlety.

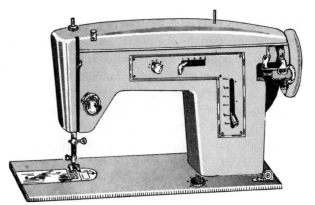

Fig. 15-6. Good design is characterized by its simplicity and smooth flowing lines.

duce a high degree of subtlety or generate a fine and delicate perception of some particular phenomenon. A good example of this is shown in Fig. 15-5. See how box *A* with its straight lines has little artistic impact compared with box *B*. Simply by swelling the surfaces slightly, the convex curvature reduces the monotonous flatness of the box and creates greater attention-holding power for the viewer.

The graceful movement of lines can be found in the design of many products. Compare any article made several decades ago with its counterpart today. In most cases modern design has eliminated massiveness, sharp edges, true geometric contours, and has concentrated instead on simplicity and smooth flowing lines. See Fig. 15-6.

One of the greatest impacts on modern design has been streamlining. The original purpose of streamlining was to reduce friction on objects moving through water or air. See Fig. 15-7. But the general contour achieved by streamlining was found to be so pleasing to the eye that it has been applied to the design of countless products even though speed or friction may no longer be a concern.

15.2.2 Planes and solids. A plane is a flat level surface having only two dimensions—length and width. Common plane shapes are the square, rec-

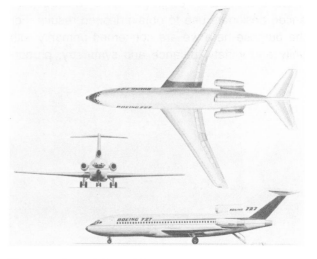

Fig. 15-7. Streamlining, originally used in aircraft design, has had great impact on the design of many consumer products.

tangle, circle, and triangle. If thickness is added to planes, three-dimensional solids such as the cube, ellipsoid, sphere, cylinder, cone, and pyramid result. See Fig. 15-8.

These planes and solids serve as the bases from which the shape of a product is developed. As design elements, they give direction, depth, and interest. However, to the designer, the concept of planes is not limited to flat surfaces. Ingenious ways are used to enrich a design and reduce the dull effects of pure geometric shapes.

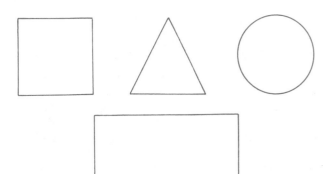

PLANES

The monotony in geometric figures can be minimized by changing slightly some feature of their basic shapes. With some solids this can be done more readily than with others. Take the sphere, for example. Here you have a solid where every point on its surface is equidistant from its center. If any part of it is flattened or pushed out, the resulting shape produces a distorted effect. See Fig. 15-9. Combining several spheres, however, often produces useful results. By grouping them in various patterns, different design effects can be highlighted. Pattern variety is achieved by varying the sizes of the spheres or changing their relative positions. See Fig. 15-10.

The cube can be manipulated over a wide range of pleasing ways without altering its basic character. This is done by combining several cubes to

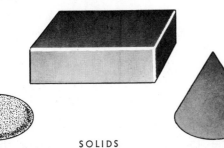

SOLIDS

Fig. 15-8. Planes and solids are the main foundations of a product design.

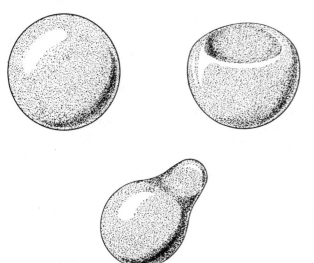

Fig. 15-9. Changing the basic contour of a sphere may result in a less pleasing shape.

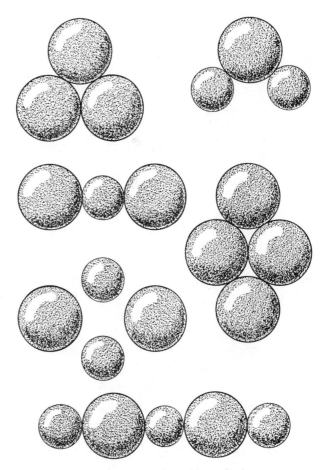

Fig. 15-10. Varying the sizes and positions of spheres can produce pleasing designs.

produce an infinite number of symmetrical and non-symmetrical shapes, many of which are readily translated into commercial products. See Fig. 15-11.

Fig. 15-13. Various shapes are used in modern design of consumer products.

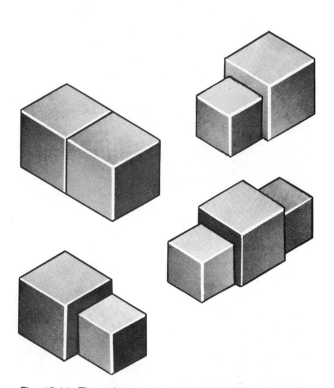

Fig. 15-11. The cube can be manipulated in many ways to achieve pleasing results.

Fig. 15-12. The ellipsoid and the cylinder provide the bases for many product designs.

Unlike the sphere, the ellipsoid provides possibilities for three-dimensional contour changes, since its major and minor axes can be changed without compromising its basic configuration. The same is true with the cylinder. Although the cylinder has a fixed radius, its length is variable. Hence, its shape can be altered without affecting its cylindrical appearance. See Fig. 15-12.

The pyramid is probably the most difficult shape to work with in terms of effective design for a functional commercial product. Because of this fact, it is not widely used in industrial design. The cone, on the other hand, is more adaptable, particularly for such products as doorknobs, controls, and switches. See Fig. 15-13.

15.2.3 Surface quality. The way an object appears to the observer is often influenced by its surface quality. There are three factors that affect surface quality—light value, texture, and color.

Light value is the ability of a surface to reflect light. See Fig. 15-14. Theoretically a white surface reflects all the light striking it and is placed at the

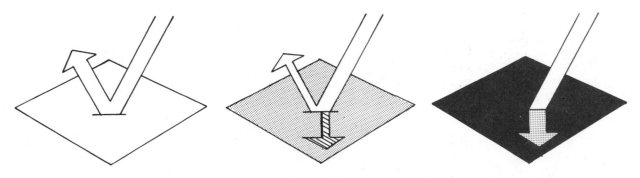

Fig. 15-14. Reflection and absorption of light by white, gray, and black surfaces.

Fig. 15-15. Values ranging from white to black determine the amount of reflected light.

top of the value scale. Black is considered to have no reflecting ability and is at the bottom of the value scale. Thus, all colors and tones fall somewhere between these two extremes. See Fig. 15-15. Furthermore, surface value also varies with the intensity and distance of the light.

Surface texture refers to the cavities, peaks, plateaus, and valleys that make up a surface. Since these are physical elements of a surface, they are most easily identified by our sense of touch and usually described as rough, smooth, coarse, and velvety.

Whenever possible, good design takes advantage of the natural texture that each material possesses. Some textures are the result of manufacturing processes. Other textures are specifically imparted to surfaces as a design element. See Figs. 15-16 and 15-17.

Surface color involves adding colors to surfaces in order to achieve specific results. We know that colors can affect our emotions and feelings. Some colors are relaxing and soothing while others are irritating. Colors can give us a feeling of warmth or make us cold and uncomfortable.

Correctly used, color can produce many optical illusions. Combinations of certain colors can

Fig. 15-16. Surface texture is important in good design.

change the apparent size relationships of adjacent areas. They can cause a surface to appear to be advancing or receding and give the illusion of a

Fig. 15-17. A variety of surface textures are easily produced to create good design features.

three-dimensional object. Designers use a variety of color schemes to make products look more interesting and exciting.

15.3 Design Principles

If lines, planes, solids, and surface quality are to be used successfully in a meaningful and pleasing design, they must be bound together by such principles as unity, variety, balance, proportion, and rhythm.

15.3.1 Unity and variety. A design that has everything woven together according to a definite plan is said to have unity. Such a design has its functional and aesthetic relationships combined into a complete and harmonious unit. Notice in Fig. 15-18 how the form of the vacuum cleaner projects an attractive aesthetic quality while remaining completely functional. Consequently, the first premise of unity is to make sure that the basic

Fig. 15-18. This vacuum cleaner is not only functional but also has appealing contour.

Fig. 15-19. Even an electric drill can be designed to have greater eye appeal.

function of the product is maintained. If an electric hand drill is to be designed, it must be able to perform effectively as a drill. But at the same time its contour can be made to look less mechanical and more appealing to the eye. See Fig. 15-19.

Variety refers to the use of contrasting elements to gain and hold the attention of the viewer. It is synonymous with interest and the opposite of monotony. A simply way to achieve variety is to avoid repetition of form and, at the same time, to use elements that make the object graceful, striking, or otherwise interesting. See Fig. 15-20.

15.3.2 Balance and symmetry. Balance and symmetry are design characteristics that cause corresponding parts to be equal or appear to be equal in weight or size. See Fig. 15-21. If a line is drawn through the center of a design and each half appears to have equal weight or area, the design is said to have balance. Most objects in nature—leaves, flowers, trees, animals, and people—are balanced. Articles that are out of balance appear unstable and look as if they may tip over, or are distorted in shape. There are two kinds of balance—formal and informal. *Formal balance* exists when both halves of a design are exactly alike and equal. Designs organized radially

Fig. 15-20. Variety is achieved by using contrasting elements.

Fig. 15-21. Products that have balance and symmetry.

around a center point are symmetrical, and symmetrical designs always have formal balance. See Fig. 15-22. *Informal balance* is achieved when the two parts appear to be equal but are not. See Fig. 15-23. Tone, texture, color, and different materials can be used to achieve this illusion. For example, a large area of light color on one side may balance a smaller area of dark color on the other, or a small solid area may be balanced by a larger open area. See Fig. 15-24.

15.3.3 Proportion. Proportion refers to the size relationship of one part to another or of one part to the whole. Properly used, proportion is an effective means of creating unity among the various components of a design. Since proportion relates to quantities and ratios of quantities, it may be expressed by mathematical formulas. Although famous designers have developed many useful mathematical relationships, proportion is still subjective and only effective when all factors of a design are balanced. The most famous mathematical design formula was devised by the ancient Greeks. It was called the golden rectangle, or golden oblong. The sides of this rectangle have a ratio of 1:1.618, or about 5:8. See Fig. 15-25. This pleasing proportion can be used for many products, such as serving trays, tabletops, and picture frames. Another proportion that is effective in dividing space is 5:7:9. See Fig. 15-26.

Proportion is extremely important in establishing balance and stability in a design. For example, a large massive lamp base with a small shade will look awkward and out of proportion, or a large

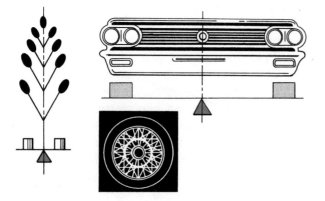

Fig. 15-22. Designs with formal balance.

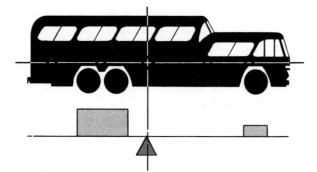

Fig. 15-23. An example of informal balance.

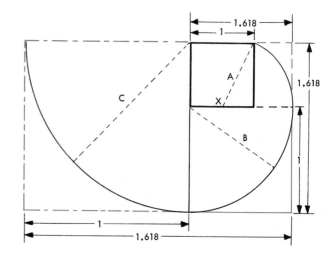

Fig. 15-24. Light and dark areas can help achieve balance.

Fig. 15-25. The golden rectangle.

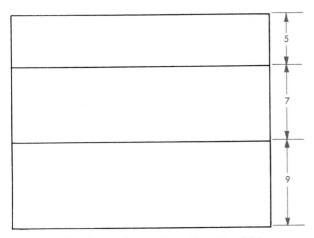

Fig. 15-26. Good proportion can be achieved by dividing space on the basis of a 5:7:9 ratio.

Fig. 15-27. Using overlapping sheets is a timesaver in product design sketching.

bowl with a small spindly base will be top-heavy and unstable.

15.3.4 Rhythm. Rhythm in design is similar to rhythm in music. In music, rhythm is the repetition or regular occurrence of time intervals. In design the repetition may be of line patterns, size, shape, texture, or color. In many designs the patterns are easily recognized, while in others careful analysis may be necessary to discover and understand them.

15.4 Product Design Processes

It is impossible to present a detailed step-by-step procedure for producing product designs, because no two people will work alike. However, there are several basic processes which are fundamental to most product design ventures. Briefly these are visualization, model study, and rendering. However, it does not necessarily follow that each product design will encompass all of these stages of development.

15.4.1 Visualization. Visualization is simply a means of getting a three-dimensional idea of what the product will look like. This is done by sketching the design pictorially. The sketch may be a perspective, isometric, or oblique, depending on the complexity of the design and the amount of detail that needs to be shown. (See Unit 9.)

The sketching is generally done on tracing or layout paper made up into pads. The use of semi-transparent paper is advisable because it permits placing sheets over each other and reworking a design without having to redraw the entire design each time some variation is to be made. This is a labor-saving technique because, once the basic proportions of the design are developed, many variations can be devised in a very short time. See Fig. 15-27.

15.4.2 Model study. Model study follows once the design is completed. This involves making a model of some kind. It is only by means of a model that accurate clearances, dimensions, fixture locations, and other construction details can be decided. In general, model study is centered around four basic types: clay, scale, mockup, and prototype.

A clay model is used a great deal to study form relationship and other design factors discussed previously. Modeling clay is indispensable for this purpose because it can be worked over and over again. The model can offer solutions to design problems that cannot easily be worked out on paper. See Fig. 15-28.

A scale model of a new design product is sometimes made when a very accurate presentation of the article is required. The scale model is an actual replica of what the finished product will look like except that it is usually made to some smaller size, although for some products the scale model is often actual size. Most scale models are made

Fig. 15-28. Modeling with clay is a practical means of working out solutions to design problems.

of plaster and are finished so they will have a surface appearance exactly like the actual product.

A mock-up is a full-size replica of the product made of wood or plaster. Its main purpose is to help solve detailed construction problems, such as flow of electrical circuits, seating arrangements, clearances between moving parts, locations of switches, and operation of moving panels. From the mock-up, final specifications are obtained for

preparing production drawings and compiling tooling data for production. See Fig. 15-29.

A prototype is an actual full-size unit of the product to be manufactured. The unit is made of the same material to be used for the final product and is built exactly like it in every detail, both inside and outside.

15.4.3 Rendering. A rendering is a finished drawing of a product scheduled for manufacture. It is used primarily to sell the product to the public. A rendering is generally drawn in perspective with a station point located so the full effects of the product are visible.

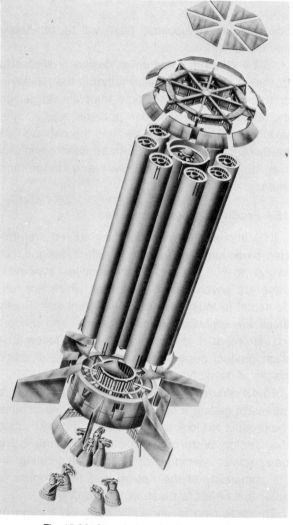

Fig. 15-30. A rendering of a Saturn rocket.

Fig. 15-29. Full size mock-up of an automobile.

Since the drawing must represent the actual object as it appears to the eye, care is taken to draw it in its true proportions. Usually a rendering is laid out from regular mechanical drawings.

One of the features of any rendering is the coloring used to make the product impressive and interesting to the viewer. Although coloring effects can be achieved in various ways, best results are obtained with an airbrush. There are no hard and fast rules to govern the scale of the drawing. Some small objects are drawn full-size, whereas large objects may be one-quarter, one-half, or three-quarter size. Fig. 15-30 shows a small-scale rendering.

15.5 Fabrication Design Practices

Many standard practices have been established in the field of product design. Most of these practices are based on results obtained through experimentation, experience, safety studies, and cost analyses. Some of the most important fabricating design practices deal with weldments, castings, forgings, and stampings.

First of all, it should be noted that, to reach the goal of producing a part at the lowest overall cost regardless of the fabricating process used, it is necessary to evaluate the design to make certain that optimum use is made of the metallurgical and physical properties of the materials. These are some basic guidelines:

1. The design should satisfy the strength and rigidity requirements of the product. Designing beyond these requirements means that more material and labor will be used, thereby adding to the costs.

2. The safety factor designated for the design should be adequate to meet all contingencies. Using too large a safety factor also means added material and labor.

3. The appearance of the product should be pleasing, but where areas are hidden from view, appearance should not be a critical factor. Therefore, it is often more expedient to use less expensive grades of material where appearance is unimportant. For example, hot-rolled steel will frequently prove as effective as cold-rolled steel.

4. Since more weight means higher costs, members should be analyzed to see if stiffeners could be employed for rigidity as a means of reducing weight.

5. Wherever possible, non-premium steel should be specified. A high grade steel, however, is best where the added strength reduces the cost.

6. Standard sizes and shapes should be specified wherever it is practical.

7. When a hard surface is required, the use of hard surfacing, heat treating, or plating should be noted.

15.5.1 Design of weldments. In any product involving weldments, joint design is a primary consideration. Joint design is greatly influenced by the cost of preparing the joint, the accessibility of the weld, the adaptability of the joint for the product being designed, and the type of loading the weld is required to withstand.

The basic joint configurations which are applicable for shielded metal-arc, gas metal-arc, and submerged-arc welding are broadly classified as grooved and fillet. Each group incorporates several variations to provide for different service requirements.

The *square butt joint* is intended primarily for materials that are up to 0.38 inch thick and require full and complete fusion for optimum strength. The joint is reasonably strong in static tension but is not recommended when tension due to bending is concentrated at the root of the weld. The square butt joint should never be used when the joint will be subjected to fatigue or impact loads, especially at low temperatures. The preparation of the joint is relatively simple since it requires only matching edges of the plates. Consequently the cost is low. See Fig. 15-31A.

The *single vee butt joint* is used on plate 0.38 inch or greater in thickness. Preparation is more costly because a special beveling operation is required and more filler material is necessary. The joint is strong in static loading but, like the square butt joint, is not particularly suitable when tension due to bending is concentrated at the root of the weld. See Fig. 15-31B.

The *double vee butt joint* is best for all load conditions. It is often specified for stock that is heavier than stock used for a single vee. For maximum strength the penetration must be complete on both sides. The cost of preparing the joint is

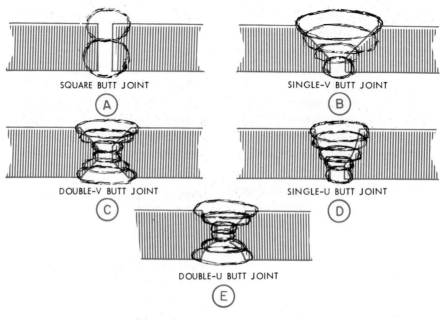

SQUARE BUTT JOINT
Ⓐ

SINGLE-V BUTT JOINT
Ⓑ

DOUBLE-V BUTT JOINT
Ⓒ

SINGLE-U BUTT JOINT
Ⓓ

DOUBLE-U BUTT JOINT
Ⓔ

Fig. 15-31. Types of butt joint designs.

higher than for the single vee, but less filler material is required. To keep the joint symmetrical and warpage to a minimum, the weld bead must be alternated, welded first on one side and then the other. See Fig. 15-31*C*.

The *single U butt joint* readily meets all ordinary load conditions and is used for work requiring high quality. It has greatest applications for joining plates 0.50 to 0.75 inch thick. See Fig. 15.31*D*.

The *double U butt joint* is suitable for 0.75 inch or heavier plate where welding can readily be accomplished on both sides. Although the preparation cost is higher than for the single U butt joint, less weld metal is needed. The joint meets all regular loading conditions. See Fig. 15-31*E*.

The *square tee joint* requires a fillet weld, which can be made on one or both sides. It can be used for light to reasonably thick materials where loads subject the weld to longitudinal shear. Since its stress distribution is not uniform, care must be taken in specifying this joint where severe impact or heavy transverse loads may be encountered. For maximum strength, considerable weld metal is required. See Fig. 15-32*A*.

The *single bevel tee joint* will withstand more severe loadings than the square tee joint, due to better distribution of stresses. It is generally confined to plates 0.50 inch or less in thickness where welding can be done from one side only. See Fig. 15-32*B*. .

The *double bevel tee joint* is intended for use where heavy loads are applied in both longitudinal and transverse shear and where welding can be done on both sides. See Fig. 15-32*C*.

The *single J tee joint* is used on plates 1.00 inch or more in thickness where welding will be limited to one side. It is especially suitable where severe loads are encountered. See Fig. 15-32*D*.

The *double J tee joint* is suitable in heavy plates 1.50 inches or more in thickness where unusually severe loads must be absorbed. Joint location should permit welding on both sides. See Fig. 15-32*E*.

The *double fillet lap joint* can withstand more severe loads than the single fillet lap joint. It is one of the most widely used joints in welding. See Fig. 15-33.

The *flush corner joint* is designed primarily for

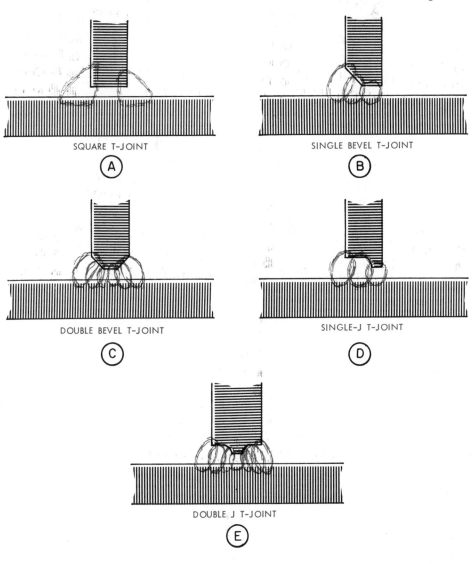

Fig. 15-32. Types of tee joint designs.

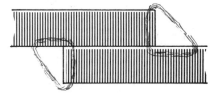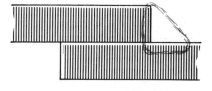

DOUBLE FILLET LAP-WELD JOINT

SINGLE FILLET LAP-WELD JOINT

Fig. 15-33. Types of fillet lap joint designs.

welding sheet 12 gage and lighter. It is restricted to lighter materials because deep penetration is sometimes difficult, and it supports only moderate loads. See Fig. 15-34A.

The *half open corner joint* is usually more adaptable than the closed corner joint for materials heavier than 12 gage, and penetration is better. See Fig. 15-34B.

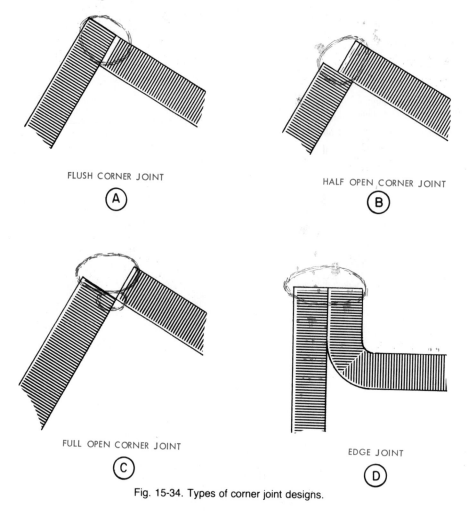

FLUSH CORNER JOINT
(A)

HALF OPEN CORNER JOINT
(B)

FULL OPEN CORNER JOINT
(C)

EDGE JOINT
(D)

Fig. 15-34. Types of corner joint designs.

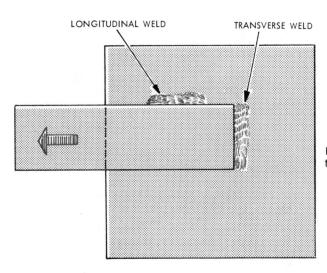

LONGITUDINAL WELD

TRANSVERSE WELD

Fig. 15-35. Transverse welds are stronger than welds parallel to lines of stress.

The *full open corner joint* is used where welding can be done from both sides. Plates of all thicknesses can be welded. This joint is strong and capable of carrying heavy loads. It is also recommended for fatigue and impact applications, because of good stress distribution, Fig. 15-34C.

The *edge joint* is suitable for plate 0.25 inch or less in thickness and can sustain only light loads. See Fig. 15-34D.

Weld location. Welding is generally considered a low-cost method of joining parts, but, unless careful thought is given to joint design and location of welds, much of the cost saving can be nullified. Weld location is an integral part of joint design and, as such, its value must not be minimized. There are many factors that affect proper weld location. Only a few can be illustrated here.

1. The location of the weld in its relationship to the members being joined has a decided effect on the strength of the joint. Normally welds having linear dimensions in a transverse direction to the stress lines are much stronger than welds with linear dimensions parallel to stress lines, Fig. 15-35.

2. A member that is subjected to a moment (turning load) at a joint has greater resistance to turning motions if the welds are spaced apart instead of being close together. See Fig. 15-36.

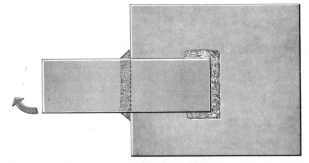

Fig. 15-36. Example of proper placement of welds to resist turning effect of one member at the joint.

3. When joining special shape sections such as channels and angles, consideration must be given to proper distribution of loads on the welds. Notice in Fig. 15-37 that the length of the weld at the heel of the angle is greater than at the toe. This eliminates the tendency for the angle to turn and generate highly eccentric loads at the joint.

4. Wherever possible, welded members should be located to avoid bending or shearing actions. Greater strength is achieved if joints are kept symmetrical. Load distribution is more uniform on symmetrical than on non-symmetrical joints.

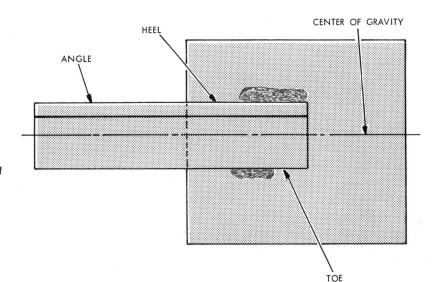

Fig. 15-37. Example of correct lengths of weld for equal load distribution.

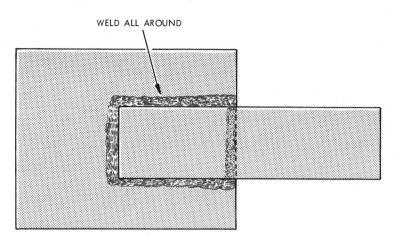

Fig. 15-38. Load distribution in linear welds is better if the joint is welded around all sides.

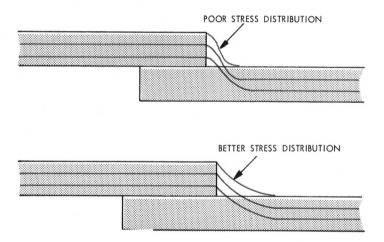

Fig. 15-39. For even stress distribution, avoid any abrupt surface changes in a joint.

5. To attain uniform distribution of loads in linear welds, it is better to weld all around the joint than to weld only on two sides. See Fig. 15-38.

6. Abrupt surface changes in a weld joint should be avoided because greater local stresses will be concentrated at the point of surface change. There is a more uniform transfer of stress through a weld if the surface flow is gradual. See Fig. 15-39.

7. Weld joints should be located so they are readily accessible and welding can be carried out with comparative ease. Notice that the weld in Fig. 15-40*A* is almost impossible to make, whereas the weld at *B* is easily accessible.

8. When an angle is used as a stiffener where it is joined to a plate that serves as a locating surface, it provides greater economy if positioned as in Fig. 15-41*A* than as in 15-41*B*. The weld location at *B* is more expensive, since both beveling and machining the weld are required.

15.5.2 Casting design practices. In the design of castings, particular attention should be given to the considerations below.[1]

---

1. *General Motors Drafting Standards* (Detroit: General Motors Corp.).

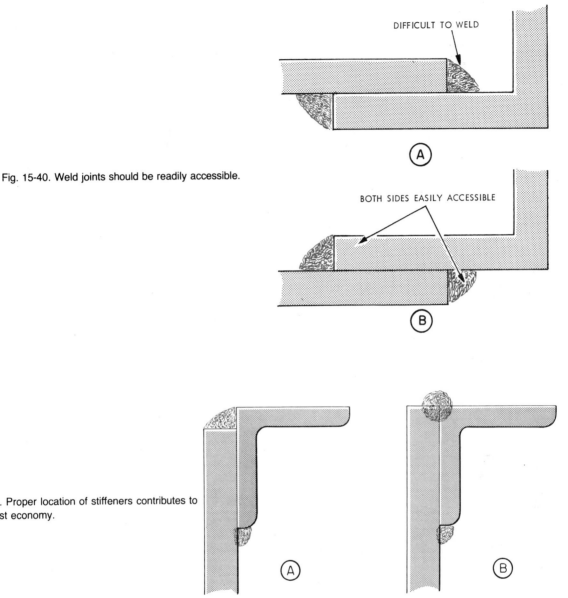

Fig. 15-40. Weld joints should be readily accessible.

DIFFICULT TO WELD

Ⓐ

BOTH SIDES EASILY ACCESSIBLE

Ⓑ

Fig. 15-41. Proper location of stiffeners contributes to greater cost economy.

Ⓐ   Ⓑ

Section thickness. Sections should be of minimum thickness consistent with good foundry practice and should provide adequate strength and stiffness. The least expensive design is one that is light enough but still presents the fewest hazards and difficulties in manufacturing. Metal flows in various directions when entering a mold, and as the mold fills, this metal must join together. If the walls are too thin or if the metal must travel too far, it will not be hot enough to join together properly when it meets. The result is a "cold shut," a seam giving a weak spot in the casting.

Wall thickness in castings should be uniform wherever the design permits.

Walls of gray iron castings and aluminum sand castings should be not less than 0.16 inch thick. Walls of malleable iron and steel castings should be not less than 0.19 inch thick. Walls of brass, bronze, or magnesium castings should be not less than 0.16 inch thick.

Sections of unequal size that do not blend gradually in thickness cause severe internal stresses and frequently produce actual cracking of the metal. This strain is due to more rapid cooling of

Fig. 15-42. These casting sections show some typical webs, clearances for machine cuts, and blending of sections of varying thicknesses.

the thin sections which results in non-uniform contraction of the poured metal. Example sections in Fig. 15-42 are designed to avoid high cooling strains in the metal. Notice the proper blending of thickness between walls and bosses, which is very important in reducing casting stresses to a minimum.

Ribs and bosses. Ribs are used primarily as stiffeners and reinforcing members. In certain castings the tendency of large flat areas to distort when cooling from casting temperature may be eliminated by properly designed ribs. The ribs solidify more quickly than the section which they adjoin and act as a bond and as conductors of heat to promote cooling of the section involved. The relation of the rib or boss section to the main section should permit, as far as possible, a uniformly blended metal section.

Fillets and rounded corners. Adequate fillets at all intersections materially increase the strength and soundness of castings. Sizes and fillets depend upon the metal used and the shape and thickness of the wall section. Too large a fillet in a rib or spoke causes localized heavy sections resulting in weakness at this point.

In general, the fillet radius should be equal to the wall section thickness as shown in Fig. 15-43. On casting drawings, fillets and rounded corners are indicated by a note such as ALL FILLETS .XX R AND ALL ROUNDED CORNERS .XX R UNLESS OTHERWISE SPECIFIED.

Casting tolerances. There are a great many factors which contribute to the dimensional variations of castings. However, the standard drawing tolerances specified in Table I can be satisfactorily attained in the production of fairly simple castings.

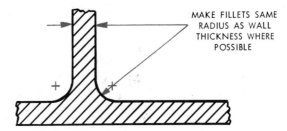

MAKE FILLETS SAME RADIUS AS WALL THICKNESS WHERE POSSIBLE

Fig. 15-43. Fillet radius on a casting should be equal to the wall section thickness.

TABLE I BASIC TOLERANCES FOR CASTINGS

TYPE OF CASTING	STANDARD DRAWING TOLERANCE ±	NORMAL MINIMUM TOLERANCE ±
SAND CASTING	.03 USE NORMAL MIN. TOLERANCES FOR DIMENSIONS GREATER THAN 8".	.03 (UP TO 8") .05 (8" TO 16") .06 (16" TO 24")
PERMANENT MOLD CASTING (SEMI-PERMANENT MOLD)	.03 USE NORMAL MIN. TOLERANCES FOR DIMENSIONS GREATER THAN 12".	.02 (UP TO 5") .03 (5" TO 12") .05 (12" TO 24")
PLASTER MOLD CASTINGS	.02 USE NORMAL MIN. TOLERANCES FOR DIMENSIONS GREATER THAN 8".	.010 (UP TO 4") .02 (4" TO 8") .03 (8" TO 12")
CENTRIFUGAL PRECISION CASTINGS	.02	.005 (UP TO 1/2") .010 (1/2" TO 5") .015 (OVER 5")

The somewhat closer tolerances specified in the normal minimum tolerance column may be applied to critical dimensions.

15.5.3 Die casting design hints. The careful consideration of the factors discussed below will facilitate die casting design.[2]

Section thickness. All sections should be of minimum thickness consistent with reasonable ease of casting and with adequate strength and stiffness. Thin sections are stronger in proportion to thickness than are thicker ones. Thick sections are apt to be more porous. A smoother surface is likely to be attained on thin sections. Sections should be made as uniform in thickness as related features permit. Where variations in thickness of sections are necessary, the transition from thinner to thicker sections should be gradual.

The possibility of adding ribs to thin walls for strength and casting control while hot should be considered. "Shadow marks" and "sinks," caused by variation in cooling on faces opposite to ribs, can be avoided by making the width of the rib not over 80 percent of the adjacent wall thickness.

Use of cores. Cores, or metal lighteners, should be provided for wherever consistent with other requirements. They save metal and machine work, result in sounder castings of reasonably uniform section thickness, and usually lower costs. The greater the number of castings required, the more important it is that cores be provided.

Cores can be of a variety of shapes, providing they are designed to permit withdrawal after the casting is formed. In general, cores are less costly and easier to operate if the core lies parallel to the axis of the die itself. A core may, however, be applied at almost any angle, but it will require special means of activation and will increase the cost of the die and parts.

Cores should be strong enough to withstand shrinkage stresses resulting from the casting solidifying around them. Allow ample draft to facilitate their withdrawal.

Cores of extreme length—especially slender cores—are to be avoided unless their usage is completely justified by some special condition. In general, the length of cores supported at one end only should not exceed three to four times the core diameter, particularly in sizes of less than 0.50 inch in diameter.

Flash and its removal. Castings should be designed in such a way that the cost of flash removal is minimized. Flash always occurs at die partings and where joints in the die form crevices in the wall of the die cavity, as at slides, movable cores, and ejector pins. Location of the parting line on significant surfaces is to be avoided if possible.

Parting flash on a single plane, preferably at right angles to the motion of the die, is easily removed by shearing. If the parting is not in a single

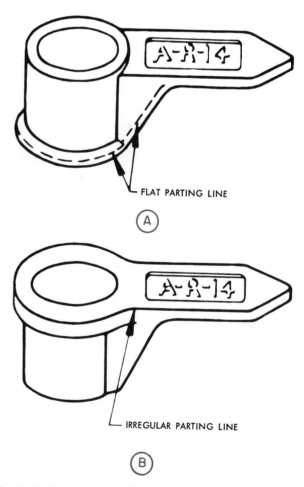

FLAT PARTING LINE

Ⓐ

IRREGULAR PARTING LINE

Ⓑ

Fig. 15-44. Die casting showing an irregular parting line and a flat parting line.

---

2. *Ibid.*

plane, greater cost of flash removal and trim dies is incurred. The casting shown in Fig. 15-44A requires an irregular parting line, while the design shown in Fig. 15-44B permits a flat parting with lower cost for the die and tools for flash removal.

Minimizing machine work. The casting should be designed so that the cost of machine work is minimized without undue sacrifices in precision, appearance, or function.

Fillets and rounds. To minimize stress risers, sharp corners should be avoided. Generous radii should be provided at both inside corners and outside corners.

General shape of castings. The shape of the castings should be kept as simple as conditions permit. Unnecessary irregularities are to be avoided. To attain smooth surfaces for finishing, large flat surfaces should also be avoided. Curvature adds to strength and stiffness. If the surface has a glossy finish, highlights created by the curvature will tend to mask surface defects such as waviness.

The thin die casting shown in Fig. 15-45, if cast flat with cored holes in the bosses, is inexpensive. If made of a ductile alloy, the casting can be bent for limited application.

Location of ejector pins. Permissible ejector pin areas should be indicated on the part drawing to avoid their location on significant surfaces or in other objectionable areas.

If the part is of such a nature that ejection is not permissible on any surface of the part, other means for ejection can be provided outside the part, as shown in Fig. 15-46, and trimmed off later.

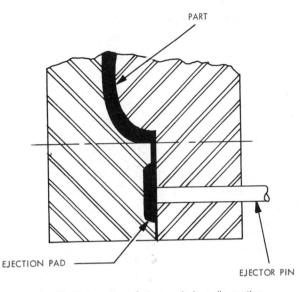

Fig. 15-46. Location of ejector pin in a die casting.

Machining location points. Die castings are generally subject to porosity and center line shrinkage. Consideration should be given to machining location points to avoid finished sections which are too thin. Generally the best location is on surfaces formed by the ejector portion of the die, since this portion usually forms the interior surfaces and carries the movable cores.

Castings having large quantities of cored holes to be drilled or having widely separated surfaces to be machined occasionally require locators. These equalize the machining stock in order to avoid drill breakage or deep machine cuts.

Locating points for machining should be selected to avoid ejector pin and core flash. Areas around gates are subject to rapid die wear and should also be avoided.

Drafts. Ample draft on cavity walls and cores should be allowed. Castings tend to shrink away from die walls. Should cores or projections come at points between side walls, shrinkage may bind the casting to or between these elements and make it difficult to eject. A good commercial high production casting should have at least 2° draft angles.

Integral fastening provisions. Integral cast fastening parts should be employed wherever their

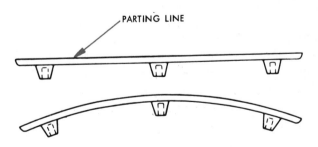

Fig. 15-45. A thin die casting with cored holes is less expensive if cast flat.

use will reduce assembly cost yet meet other requirements. Fig. 15-47 illustrates a typical provision.

Threads. Cast threads should be specified wherever their use reduces cost to less than that for cut threads. Most die casting alloys are easily and rapidly tapped or threaded.

Use of undercuts. The use of undercut castings that require movable cores in the die should be avoided.

The internal lugs shown in Fig. 15-48*A* extend to the lower wall. This is more economical than the undercuts shown in Fig. 15-48*B*, which require expensive collapsible cores, or "knockouts."

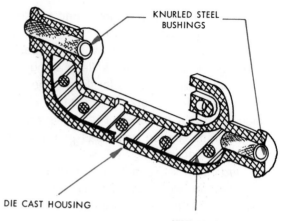

Fig. 15-49. Inserts in a die casting are used for reinforcement purposes.

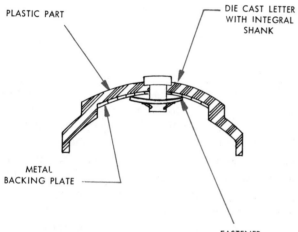

Fig. 15-47. Provisions should be made for fastening parts to reduce assembly costs.

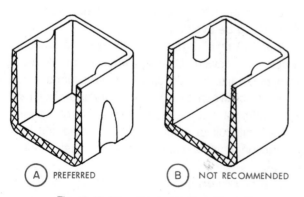

Fig. 15-48. Use of undercuts in a die casting.

Inserts and their utility. Inserts should be employed wherever their use results in economy or in meeting requirements that cannot be realized at equal cost by other means.

Fig. 15-49 shows how bushings and a reinforcing plate may be cast into a pipe cutting frame to provide hardness and strength.

15.5.4 Forging design practices. A forged part should be designed as simply and practically as possible and must serve its purpose in the assembly to the best advantage. To accomplish this function, the factors below should be considered.[3]

Draft angle. Draft is defined as the angle of taper given the side walls of the die impression in order to permit easy removal of the forging from the die. Draft must always be provided on parts produced by ordinary drop die forging. The normal amount of draft for exterior contours is 7°, and for interior contours 10°.

Die draft equivalent is the amount of offset that results when it is necessary to apply draft to a forging. See Table II. Fig. 15-50 illustrates the manner in which draft may be applied to some of the more common sections of parts.

Parting line. The surfaces of dies that meet in forging are the striking surfaces. The line of meeting is the parting line. The parting line of the forg-

3. *Ibid.*

TABLE II DIE DRAFT EQUIVALENT

DIE DRAFT EQUIVALENT			
Depth of Draft	Draft Equivalent for Angle of		
	5°	7°	10°
.04	.0034	.0049	.0070
.08	.0069	.0098	.0141
.12	.0104	.0147	.0211
.16	.0139	.0196	.0282
.20	.0174	.0245	.0352
.24	.0209	.0294	.0423
.28	.0244	.0343	.0493
.32	.0279	.0392	.0564
.36	.0314	.0442	.0634
.40	.0349	.0491	.0705
.44	.0384	.0540	.0775
.50	.0437	.0626	.0881
.60	.0524	.0736	.1057
.70	.0612	.0859	.1234
.80	.0699	.0982	.1410
.90	.0787	.1105	.1586
1.00	.0874	.1227	.1763

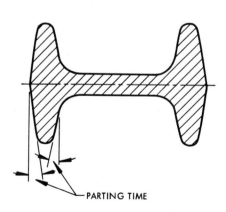
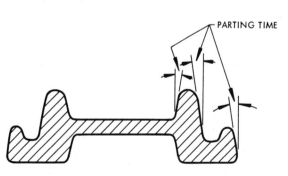

Fig. 15-50. Draft for a forging may be applied as shown here.

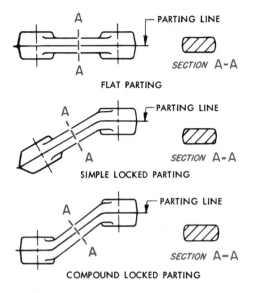

Fig. 15-51. Typical locations and types of parting lines for simple forgings are represented here.

ing must be established in order to determine the amount of draft and its location.

The locations and types of parting as applied to simple forgings are shown in Fig. 15-51. The parting may be either flat or locked where the dies have their faces in two or more planes. Flat partings are generally the most economical, but some forgings must be parted by locked partings.

Fillets and corner radii. An important aspect of forging design is the use of correct radii where two surfaces meet. Corner and fillet radii on forgings should be as large as possible to assist the flow of metal for sound forgings, and to promote economical manufacture.

Stress concentrations due to abrupt changes in section thickness or direction are minimized by fillet and corner radii of the correct size. Any radius larger than recommended will increase die life. Any radius smaller than recommended will decrease die life. Fig. 15-52 has recommendations.

Grain direction. Grain direction is one of the most important properties of forged metal. Grain, or fiber, direction is defined as the extension of alignment of metal grains in the direction of working. In forging, tests have shown that metal may be as much as 5 percent stronger in tension along the grain fibers than across them and have the

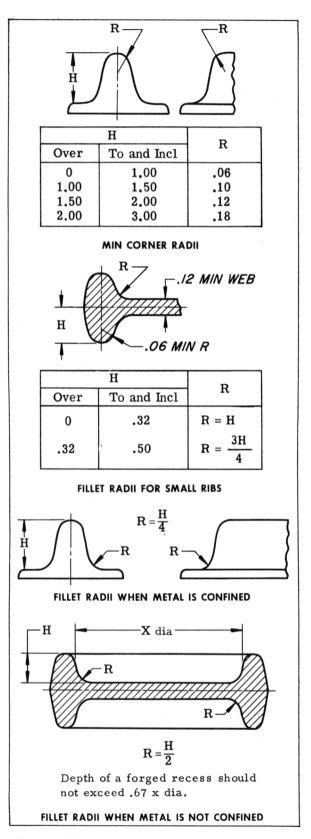

MIN CORNER RADII

H		R
Over	To and Incl	
0	1.00	.06
1.00	1.50	.10
1.50	2.00	.12
2.00	3.00	.18

FILLET RADII FOR SMALL RIBS

H		R
Over	To and Incl	
0	.32	$R = H$
.32	.50	$R = \dfrac{3H}{4}$

FILLET RADII WHEN METAL IS CONFINED

Depth of a forged recess should not exceed .67 x dia.

FILLET RADII WHEN METAL IS NOT CONFINED

Fig. 15-52. These corner radii are recommended for forgings.

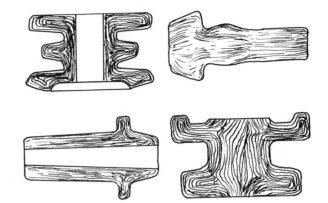

Fig. 15-53. Forgings should be designed to take advantage of the grain direction of the forging metal.

ability to resist shock and impact to a much greater degree than across grain direction.

Forgings should be designed to take advantage of the grain direction of the forging metal. Fig. 15-53 shows a macro-etched cross section of a forging. Grain structure and fiber direction have been directionally worked to produce a strong and tough forging to meet the bending and torsional stresses to which the part will be subjected.

Tolerance. Due allowance must be made in the design of a forged part for variation in outline dimensions and weight of the finished part. These variations are caused by die wear and mismatch or lateral misalignment of the two halves of the die. Die wear causes the greatest variation in contours and planes that are parallel to the parting line. In many parts, variation in contours measured perpendicular to the parting line will cause large variations in weight, which is an important factor in the design of such parts as engine connecting rods. These variations can be corrected only by resinking the die. Since mismatch is a matter of die alignment, it can be controlled by set-up and general machine condition. It is very noticeable in the case of a gear forging with a web and rim design. Mismatch will cause one side of the rim to be eccentric to the other, making the finished part very difficult to balance dynamically. Therefore, tolerances for allowable variations in forging contours and mismatch should be clearly stated on the drawings of forged parts. These tolerances should be as generous as possible consistent with good design of the part. Close forging tolerances

can be held only through increased die maintenance, which adds to the cost of producing the part in large quantities.

Allowance for machining. When a forging is to be machined, allowance must be made for additional metal to be removed. Machine finish will vary with the size of the forging. All surfaces which are to be machined should be indicated by conventional finish marks to signal manufacturers to provide material for finish where indicated. Forging details of normal size should provide allowance for finish of 0.06 inch, while large and intricate shapes should allow 0.12 inch for finish, depending upon tooling requirements. If surfaces carry draft, the draft is additional, and dimensions should be given as shown in Fig. 15-54.

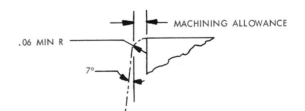

Fig. 15-54. Allowance for machining and draft should be shown on a forging drawing.

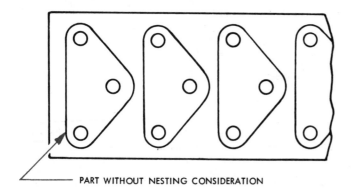

PART WITHOUT NESTING CONSIDERATION

Fig. 15-55. Blanks must be nested on a sheet to keep scrap to a minimum.

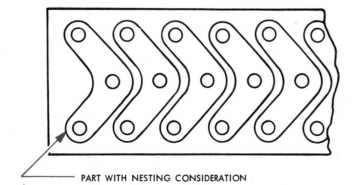

PART WITH NESTING CONSIDERATION

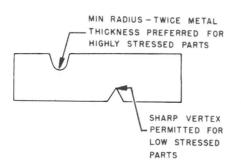

MIN RADIUS — TWICE METAL THICKNESS PREFERRED FOR HIGHLY STRESSED PARTS

SHARP VERTEX PERMITTED FOR LOW STRESSED PARTS

Fig. 15-56. Notches in stampings are provided either to facilitate forming or for clearance, attachment, or locating.

15.5.5 *Stamping design considerations.* In the design of stampings, special attention should be given to the details discussed below.[4]

Dimensions. The general practice is to dimension stampings either on the punch side or die side and not by placing part of the dimensions on one side and part on the other. Wherever possible, dimensions should be given to intersection or tangent points instead of to a locus of a radius.

Nesting of blanks. Consideration must be given to the arrangement of the blanks on the sheet to minimize scrap. See Fig. 15-55.

Notches. V-shaped notches should never be used on highly stressed parts because the sharp

4. *Ibid.*

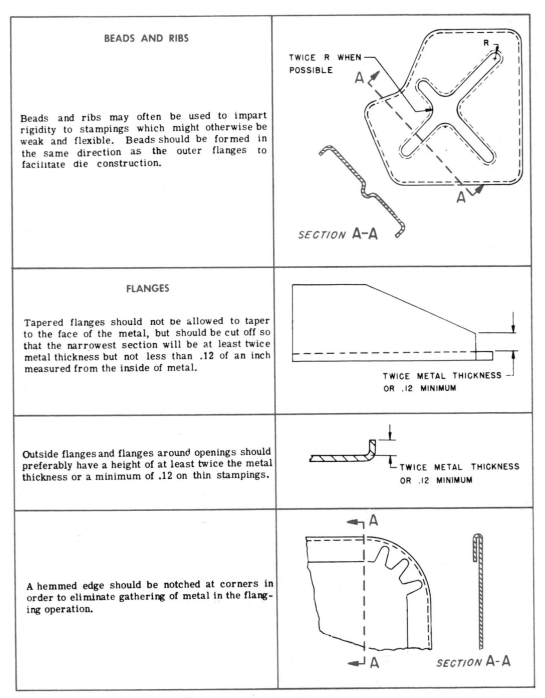

BEADS AND RIBS

Beads and ribs may often be used to impart rigidity to stampings which might otherwise be weak and flexible. Beads should be formed in the same direction as the outer flanges to facilitate die construction.

TWICE R WHEN POSSIBLE

R

SECTION A-A

FLANGES

Tapered flanges should not be allowed to taper to the face of the metal, but should be cut off so that the narrowest section will be at least twice metal thickness but not less than .12 of an inch measured from the inside of metal.

TWICE METAL THICKNESS OR .12 MINIMUM

Outside flanges and flanges around openings should preferably have a height of at least twice the metal thickness or a minimum of .12 on thin stampings.

TWICE METAL THICKNESS OR .12 MINIMUM

A hemmed edge should be notched at corners in order to eliminate gathering of metal in the flanging operation.

SECTION A-A

Fig. 15-57. Design hints for stampings.

vertex of the notch might result in the starting point of a tear. Notches for highly stressed parts should be specified with a minimum of twice the metal thickness. Sharp notches may be used on low-stressed parts. See Fig. 15-56.

Design hints. The design hints given in Figs. 15-57 to 15-61 illustrate some of the important points that should be considered in the design of stampings. Attention to these practices will facilitate standardization of approved detail design.

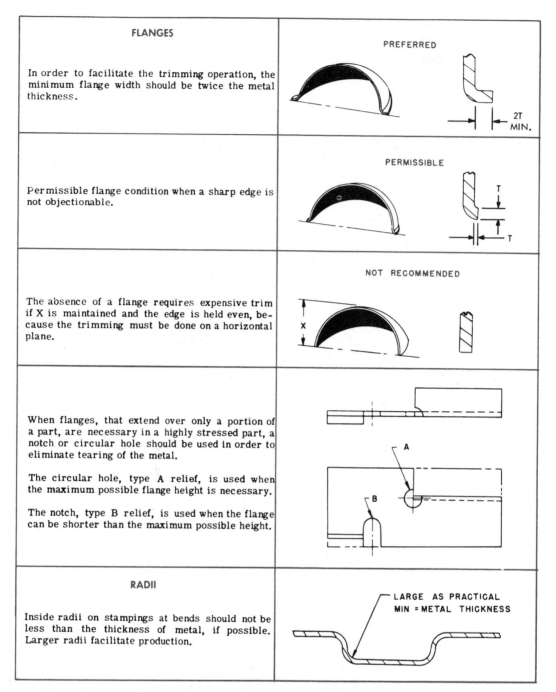

FLANGES	
In order to facilitate the **trimming** operation, the minimum flange width should be **twice** the metal thickness.	PREFERRED
Permissible flange condition when a sharp edge is not objectionable.	PERMISSIBLE
The absence of a flange requires expensive trim if X is maintained and the edge is held even, because the trimming must be done on a horizontal plane.	NOT RECOMMENDED

When flanges, that extend over only a portion of a part, are necessary in a highly stressed part, a notch or circular hole should be used in order to eliminate tearing of the metal.

The circular hole, type **A** relief, is used when the **maximum** possible flange height is necessary.

The notch, type **B** relief, is used when the flange can be shorter than the **maximum** possible height.

RADII

Inside radii on stampings at bends should not be less than the thickness of metal, if possible. Larger radii facilitate production.

LARGE AS PRACTICAL
MIN = METAL THICKNESS

Fig. 15-58. Design hints for stampings.

15.6 Strength of Materials*

Strength of materials is the branch of mechanics that is concerned with the action of forces and

*This section prepared with the assistance of G. Stewart Johnson, Associate Professor of Mechanical Engineering, Western Michigan University.

their effect on structural members. In the design and manufacture of any product, size and strength of the materials must be considered. Material size and strength are significant because they are directly related to the serviceability of the product. Component and structural members are subjected to various kinds of loads. If the materials are to

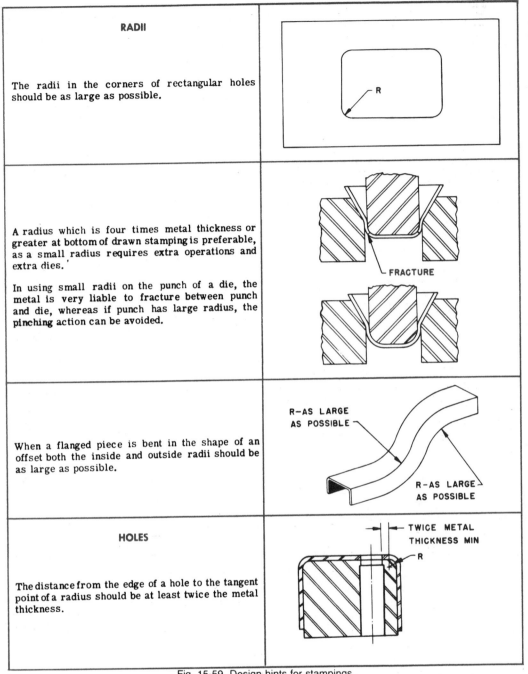

RADII The radii in the corners of rectangular holes should be as large as possible.	
A radius which is four times metal thickness or greater at bottom of drawn stamping is preferable, as a small radius requires extra operations and extra dies. In using small radii on the punch of a die, the metal is very liable to fracture between punch and die, whereas if punch has large radius, the pinching action can be avoided.	FRACTURE
When a flanged piece is bent in the shape of an offset both the inside and outside radii should be as large as possible.	R—AS LARGE AS POSSIBLE R—AS LARGE AS POSSIBLE
HOLES The distance from the edge of a hole to the tangent point of a radius should be at least twice the metal thickness.	TWICE METAL THICKNESS MIN R

Fig. 15-59. Design hints for stampings.

withstand these loads, they must be of a specific size and strength.

Although the subject of mechanics is basically an engineering discipline, drafting people frequently encounter design problems involving strength of materials. The information in this section is confined to basic principles that are usually of significance to drafting personnel.

15.6.1 Mechanical properties. Mechanical properties are measures of how materials behave under applied loads. These properties are described in terms of the kinds of forces materials have to resist and the manner in which these forces are resisted.

Common types of loads are tensile, compressive, torsional, direct shear, impact, or a combina-

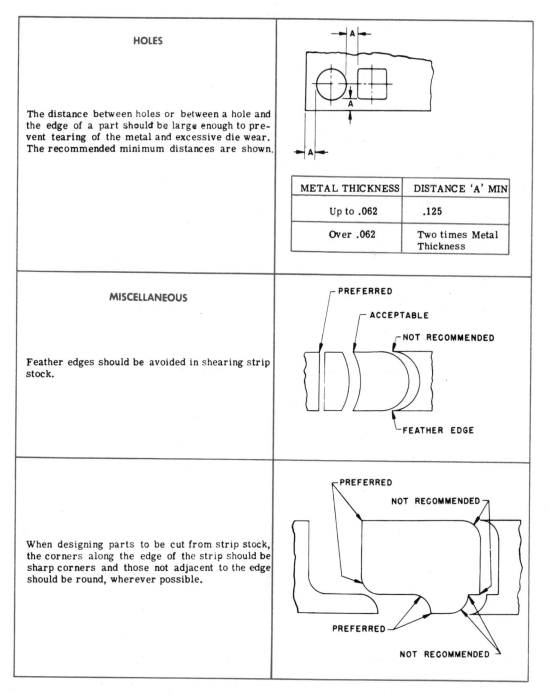

HOLES

The distance between holes or between a hole and the edge of a part should be large enough to prevent tearing of the metal and excessive die wear. The recommended minimum distances are shown.

METAL THICKNESS	DISTANCE 'A' MIN
Up to .062	.125
Over .062	Two times Metal Thickness

MISCELLANEOUS

Feather edges should be avoided in shearing strip stock.

When designing parts to be cut from strip stock, the corners along the edge of the strip should be sharp corners and those not adjacent to the edge should be round, wherever possible.

Fig. 15-60. Design hints for stampings.

tion of these. The following are some of the common terms associated with mechanical properties of materials.

Stress is the internal resistance a material offers to being deformed.

Strain is the deformation resulting from stress.

Modulus of elasticity is the ratio of stress to strain within the elastic limit. The less a material deforms under a given stress, the higher the modulus of elasticity. By checking the modulus of

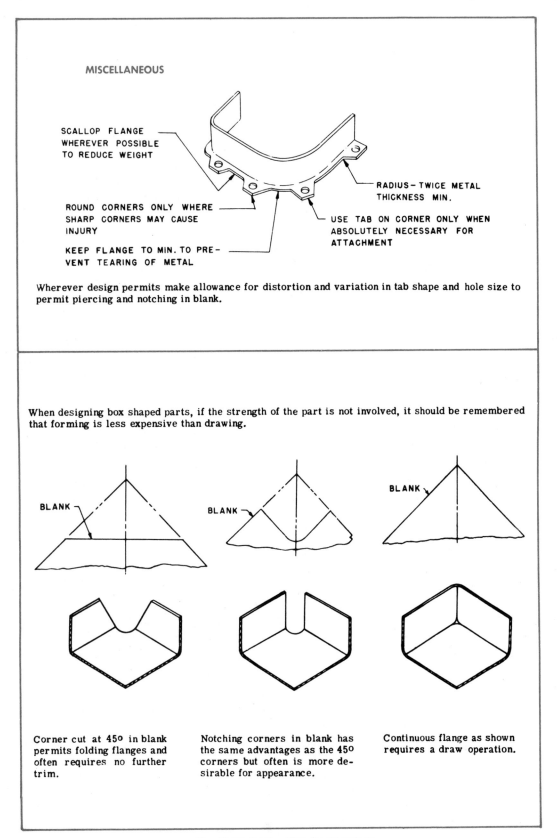

MISCELLANEOUS

SCALLOP FLANGE
WHEREVER POSSIBLE
TO REDUCE WEIGHT

RADIUS – TWICE METAL
THICKNESS MIN.

ROUND CORNERS ONLY WHERE
SHARP CORNERS MAY CAUSE
INJURY

USE TAB ON CORNER ONLY WHEN
ABSOLUTELY NECESSARY FOR
ATTACHMENT

KEEP FLANGE TO MIN. TO PRE-
VENT TEARING OF METAL

Wherever design permits make allowance for distortion and variation in tab shape and hole size to permit piercing and notching in blank.

When designing box shaped parts, if the strength of the part is not involved, it should be remembered that forming is less expensive than drawing.

BLANK

BLANK

BLANK

Corner cut at 45° in blank permits folding flanges and often requires no further trim.

Notching corners in blank has the same advantages as the 45° corners but often is more desirable for appearance.

Continuous flange as shown requires a draw operation.

Fig. 15-61. Design hints for stampings.

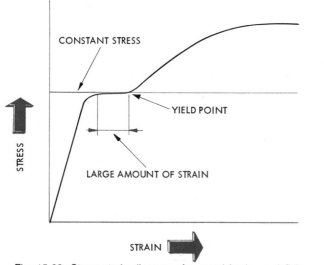

Fig. 15-62. Stress-strain diagram of a metal having a definite yield point.

elasticity, the comparative stiffness of different materials can readily be ascertained.

Proportional limit is the final point at which stress remains proportional to strain. Within the elastic limit most materials exhibit a straight line relationship between stress and strain.

Tensile strength is the ability of a material to resist being pulled apart.

Compressive strength is the ability of a material to resist being crushed.

Yield point is the point on the stress-strain diagram where there is a marked increase in strain with no corresponding increase in stress. Refer to Fig. 15-62.

Elasticity is the ability of a material when deformed under a load to return to its original or undeformed condition after the load is released.

Elastic limit is the last point at which a material may be stretched and still return to its undeformed condition upon release of the stress.

Fatigue strength is the ability of a material to resist various kinds of rapidly alternating stresses.

Impact strength is the ability of a material to resist suddenly applied loads. The higher the impact strength of a material, the greater the energy required to break it.

Hardness is the property of a material to resist permanent indentation.

Ductility is the ability of a material to deform appreciably without rupture.

Toughness is the ability of a material to absorb large amounts of energy without breaking. It is found in materials which exhibit a high elastic limit and good ductility.

Cryogenic properties represent behavior characteristics under stress environments of very low temperatures.

15.6.2 Determining simple stresses. The internal stress produced on a straight bar of a constant cross section by a central load may be assumed to be uniformly distributed over the cross-sectional area. Since unit stress is the internal stress over unit of area, it is constant due to the central load.

If the intensity of a tensile and compressive stress is equally distributed over an area, then the following relationship exists:

Equation 1

$$P = \sigma A$$

where
P = total external load (lbs)
σ = average stress (psi) (Greek *sigma*)
A = stressed area (sq in.)

A corresponding equation involving shear stress is:

Equation 2

$$P = \tau A$$

where
P = total external load (lbs)
τ = average unit shear stress (psi) (Greek *tau*)
A = area being sheared (sq in.)

A study of equations 1 and 2 indicates that given a load and area, the stress may be determined; given a stress and area, the load may be obtained; and given a stress and load, the required area may be found.

Example. A shaft with a diameter of 2 inches is subjected to a tensile load of 10,000 pounds. Determine the average stress.

Solution. From equation 1,

$$\sigma = \frac{P}{A} = \frac{10,000}{\pi \times 1^2} = 3,183 \text{ psi.}$$

15.6.3 Allowable stresses. A unit stress which safely may be used in design is designated as an *allowable stress*. This stress represents the maximum load which should be applied to a material, according to the judgment of some competent authority. Table III includes the allowable stresses for some of the more common materials. They represent average values where the usage of the member being analyzed is not unduly severe.

In the design of any component, it is important that the entire structure be analyzed to insure optimum use of materials and methods of fabrication. Likewise, it is well to recall the basic engineering principle that a machine or structure is no stronger than its weakest member. Since design is based on the premise that materials will not fail, the allowable stress provides a margin of safety.

15.6.4 Factor of safety. A stress greater than the elastic limit will produce a permanent deformation of the material. Repeated stresses of this magnitude usually cause failure.

In order to avoid exceeding certain values that might cause failure, a safety margin is used. Safety margin is referred to as *factor of safety* (FS) and represents the ratio of ultimate stress to allowable stress. Items involved in determining the proper factor of safety are

1. Uniformity of material. The greater the potentials of inclusions, blowholes, corrosion, and so on in a material, the greater should be the FS.

2. Danger to human life. The greater the possibility of personal injury, the greater should be the factor of safety.

3. Type of load. The greater the unpredictability of the load, the greater should be the FS.

4. Permanency of design. The longer the life of a product or component, the greater should be the factor of safety.

Table IV lists factors of safety for common materials when their treatment is not too severe.

15.6.5 Strains. The changes in size that occur due to stresses within a body are called *deformation* or *strains*. The total change of length is called deformation and is denoted by δ (Greek *delta*). The unit change of length is called strain and is

TABLE III ALLOWABLE STRESS

MATERIAL	TENSION PSI	COMPRESSION PSI	SHEAR PSI
STRUCTURAL STEEL	20, 000	20, 000	12, 000
CAST IRON	3, 000	15, 000	3, 000
ALUMINUM ALLOY	15, 000	15, 000	10, 000
BRASS	12, 000	12, 000	8, 000

TABLE IV FACTOR OF SAFETY

MATERIAL	STEADY STRESS	REPEATED STRESS
STRUCTURAL STEEL	4	10
HARD STEEL	6	12
CAST IRON	6	18
TIMBER	10	15

indicated by ϵ (Greek *epsilon*). Therefore, the equation relating deformation to strain is:

Equation 3

$$\epsilon = \frac{\delta}{L}$$

where

ϵ = strain (in./in.)
δ = deformation (in.)
L = total length (in.)

Example. A bar 10 inches long has a total change in length of 0.075 inch when a load is applied. Determine the resulting strain.

Solution. Using equation 3,

$$\epsilon = \frac{\delta}{L} = \frac{.075}{10} = .0075 \text{ in./in.}$$

The relationship between the stress applied to a member and the resulting strain is plotted to give a stress-strain diagram as shown in Fig. 15-62.

The slope of the stress-strain diagram up to its proportional limit is the *modulus of elasticity*. Table V shows the moduli of elasticity of some of the more commonly used materials. The formula showing this relationship is:

Equation 4

$$E = \frac{\sigma}{\epsilon}$$

where

E = modulus of elasticity (psi)
σ = stress (psi)
ϵ = strain (in./in.)

From equations 1, 3, and 4, the following relationship can be established:

Equation 5

$$\delta = \frac{PL}{AE}$$

where

δ = deformation (in.)
P = total external load (lbs)
L = length (in.)
A = stressed area (sq in.)
E = modulus of elasticity (psi)

Example. A cylindrical steel bar having a length of 10 inches is subjected to a tensile force of 8,000 pounds. Determine the required diameter if the stress is not to exceed 18,000 psi or an elongation of 0.005 inch.

Solution. Since this problem involves two conditions, both must be solved and the final diameter must satisfy both conditions. (Note in Table V that E = 30,000,000 psi for steel.)

From equation 1,

$$A = \frac{P}{\sigma} = \frac{8,000}{18,000} = .445 \text{ sq in.}$$

TABLE V MODULUS OF ELASTICITY

MATERIAL	MODULUS OF ELASTICITY (PSI)
STEEL	30, 000, 000
CAST IRON	15, 000, 000
BRASS, BRONZE.	14, 000, 000
ALUMINUM	10, 000, 000
MAGNESIUM	6, 500, 000

From equation 5,

$$A = \frac{PL}{\delta E} = \frac{8,000 \times 10}{.005 \times 30,000,000}$$
$$= .533 \text{ sq in.}$$

Using the larger of the two areas, the result is

$$A = \frac{\pi D^2}{4}$$

$$D = \sqrt{\frac{4A}{\pi}} = \sqrt{\frac{4 \times .533}{\pi}} = .823 \text{ in.}$$

Therefore, by using a shaft with a diameter of 0.823 inch, the elongation will be limited to 0.005 inch, and the stress will be less than the limit of 18,000 psi.

15.6.6 Thermal expansion. If a member is subjected to changes in temperature, it will expand when the temperature increases and contract when it decreases. The change of length per unit of length for each degree of temperature change, indicated by ΔT (Greek capital *delta*), has been measured through experiments. This change is called *thermal expansion coefficient,* and is represented in the formula as α (Greek *alpha*). Table VI lists the coefficients of some of the more commonly used materials.

The equation showing the total change of length in a member is:

Equation 6

$$\delta = \alpha \, (\Delta T) \, (L)$$

where
   δ = total deformation (in.)
   α = thermal expansion coefficient (in./in./°F)
   ΔT = temperature change (°F)
   L = length of member (in.)

Example. New steel rails 30 feet in length were positioned during a temperature of 50°F. In the summer the temperature rose to 100°F. Determine the total elongation of the rails. (Note from Table VI that α = .0000067 for steel.)

Solution. Using equation 6,
   $\delta = a \, (\Delta T) \, (L)$
      $= (.0000067) \, (100 - 50) \, (30 \times 12)$
      $= .121 \text{ in.}$

15.6.7 Beams. A beam is a structural member which is subjected to loads acting transversely to its longitudinal axis. Various kinds of beams are in use. Loads on beams may be concentrated loads, distributed loads, or a combination of both. A concentrated load is supported on an area so small that it is assumed to be at a point. A distributed load, as the term indicates, is distributed over a larger area.

Beams may be classified according to the type of support. Fig. 15-63A shows a simple beam with a concentrated load. Fig. 15-63B shows an overhanging beam with a uniformly distributed load.

TABLE VI THERMAL EXPANSION COEFFICIENTS

MATERIAL	COEFFICIENT - IN./IN. (PER ° F)
STEEL	.0000067
CAST IRON	.0000056
BRASS, BRONZE	.0000102
ALUMINUM	.0000128
MAGNESIUM	.0000145

Fig. 15-63C shows a double overhanging beam with a combination of two concentrated loads and a uniformly distributed load. Fig. 15-63D shows a cantilever beam with a distributed load.

The calculation of the reactions R_1 and R_2 shown in Fig. 15-63 is a problem in statics involving the equilibrium of forces. Equilibrium is assured in these cases since the member is not in motion. The equations involved in equilibrium are

Equation 7

$$\Sigma P_x = 0,$$
$$\Sigma P_y = 0,$$
$$\Sigma M_o = 0$$

where

Σ = sum of

P_x = forces in X direction (lbs)

P_y = forces in Y direction (lbs)

M_o = moments (lb-ft about point 0)

A moment is defined as the product of a force and the perpendicular distance from a point to the line of action of that force. Therefore, M_o means the moment with respect to point 0.

Example 1. Determine the reactions at the supports of a simple beam 15 feet long which carries a uniform load (including its own weight) of 100 pounds per linear foot and a concentrated load of

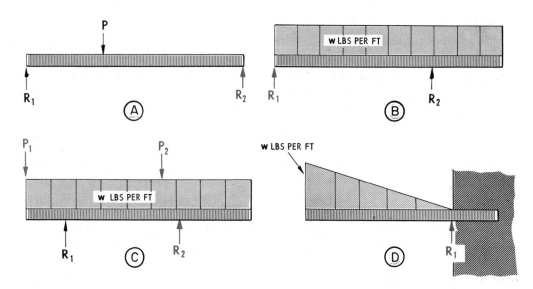

Fig. 15-63. Types of beams.

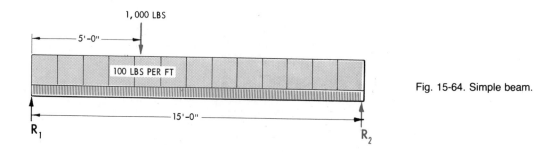

Fig. 15-64. Simple beam.

1,000 pounds at a point 5 feet from the left support. (See Fig. 15-64.)

Solution.

1. The beam being 15 feet long, its total weight is $100 \times 15 = 1500$ lb. Since the center of gravity of a uniform load is at the middle of the load, the distance to it is 7.5 feet from either support.

2. $\Sigma M_1 = 0$
$$15R_2 - 1500 \times 7.5 - 1,000 \times 5 = 0$$
$$R_2 = 1083 \text{ lbs.}$$

3. $\Sigma M_{R2} = 0$
$$15R_1 - 1,000 \times 10 - 1500 \times 7.5 = 0$$
$$R_1 = 1417 \text{ lbs.}$$

4. Check $\Sigma y = 0$ (To see if errors were made in steps 2 and 3.)
$$1,000 + 1500 - 1083 - 1417 = 0$$
$$0 = 0$$

Example 2. Determine the reaction and moment at the wall of a cantilever beam 10 feet long which carries a uniform load (including its own weight) of 60 pounds per linear foot and a concentrated load of 800 pounds at the free end of the beam. (See Fig. 15-65.)

Solution.

1. With the beam 10 feet long, its total weight is $60 \times 10 = 600$ pounds. Since the center of gravity of a uniform load is at the middle of the load, the distance to it is 5 feet from the wall.

2. $\Sigma M_R = 0$
$$M - (600 \times 5) - (800 \times 10) = 0$$
$$M = 11,000 \text{ lb-ft}$$

3. $\Sigma Y = 0$
$$R - 800 - 600 = 0$$
$$R = 1400 \text{ lbs.}$$

15.6.8 Vertical shear and bending moment.

When a beam is loaded, the loads and the reactions cause it to bend. The bending is resisted by forces which are set up within the beam. For you to visualize this phenomenon, the beam is cut as shown in Fig. 15-66 and the force and moment equivalent to the portion removed are applied to the remaining section.

Since the entire beam is in equilibrium, the cut portion of the beam is also in equilibrium. Using equation 7,

$$\Sigma y = 0$$
$$R_1 + V_c = P$$
$$V_c = P - R_1$$

$$\Sigma M_A = 0$$
$$R_1 (X) - P (d) - M_c = 0$$
$$M_c = R_1 (X) - P (d)$$

The force V_c at an arbitrary section is caused by the adjoining portion of the beam and acts as a shearing force on the section. The moment M_c is likewise caused by the adjoining portion and acts to bend the beam and so is called the bending moment.

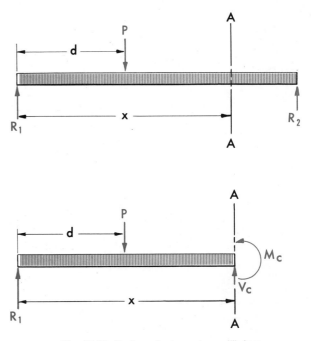

Fig. 15-66. Portion of a beam in equilibrium.

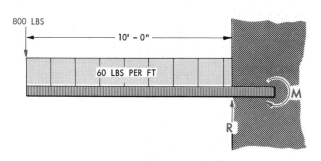

Fig. 15-65. Cantilever beam.

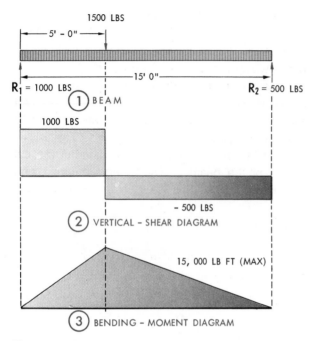

Fig. 15-67. Diagrams for a simple beam with concentrated load.

Vertical-shear and bending-moment diagrams are graphical representations of the variation of the vertical-shear and bending-moment along the full length of the beam. In order to understand the construction of these diagrams, check the examples shown in Figs. 15-67, 15-68, and 15-69. Observe these diagrams and note the following:

1. Where there is no loading, the shear diagram is horizontal and the moment diagram is a straight diagonal line.

2. Where there is a concentrated load, the shear diagram is vertical.

3. Where there is a uniform load, the shear line is a straight diagonal line, and the moment diagram is a parabolic curve.

4. Maximum and minimum bending moments occur where the shearing force is zero.

5. The bending moment is zero at the ends of the beam.

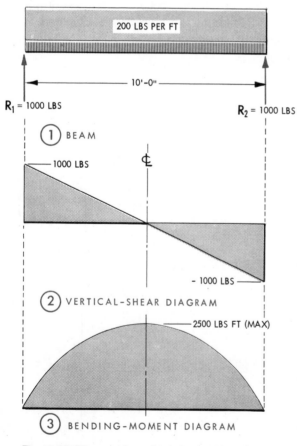

Fig. 15-68. Diagrams for a simple beam with uniform load.

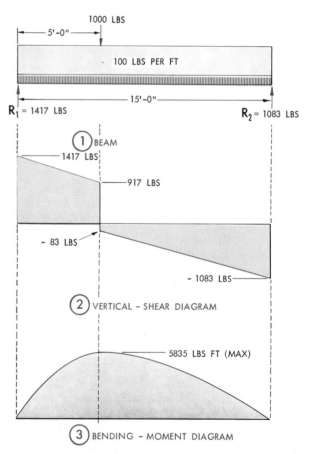

Fig. 15-69. Diagrams for a simple beam with combination loads.

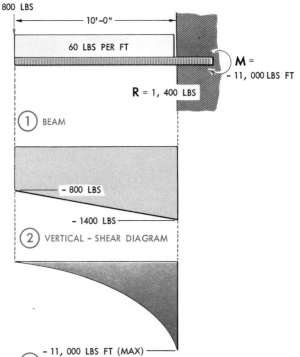

800 LBS

10'-0"

60 LBS PER FT

M = - 11, 000 LBS FT

R = 1, 400 LBS

① BEAM

- 800 LBS

- 1400 LBS

② VERTICAL - SHEAR DIAGRAM

- 11, 000 LBS FT (MAX)

③ BENDING - MOMENT DIAGRAM

Fig. 15-70. Diagram for a cantilever beam with combination loads.

In vertical-shear and bending-moment diagrams for cantilever beams, it should be noted that there is only one reaction. The reactive force is the summation of the loads. The bending moment is zero at the free end and reaches its maximum at the support. See Fig. 15-70.

15.6.9 Stresses in beams. A beam subjected to loads tends to bend. The resisting bending moment in the beam must be analyzed to determine the effect on the material. In Fig. 15-66 the moment was shown to be a twisting action. A mo-

ment can also be shown as two parallel forces opposite in direction and not along the same line as in Fig. 15-71A. Tests have indicated that a force is not concentrated as shown in Fig. 15-71A, but is distributed as in Fig. 15-71B. Therefore, the stress is zero at the neutral axis and reaches its maximum at the outer fibers. In this case the upper fiber is in compression and the lower in tension. The formula used to determine this stress is:

Equation 8

$$\frac{M}{\sigma} = \frac{I}{c}$$

where
M = bending moment (lb-in.)
σ = stress (psi)
I = moment of inertia of the section (in.^{4})
c = distance from neutral axis to outer fiber (in.)

Example. Determine the maximum tensile stress in a 10-foot-long simple beam with a uniform load of 200 pounds per linear foot. The cross section of the beam is 2 inches thick and 4 inches high.

Solution. In Fig. 15-68 the maximum bending moment was determined to be 2,500 lb-ft.

$$I = \frac{bh^3}{12} = \frac{2(4)^3}{12} = 10.67 \text{ in.}^4$$

The distance from the neutral axis to the lower fiber (to give the maximum tension) is 2 inches.

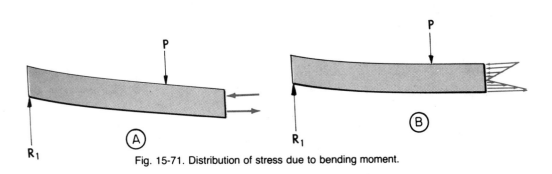

P

R₁

Ⓐ

P

R₁

Ⓑ

Fig. 15-71. Distribution of stress due to bending moment.

Using equation 8,

$$\sigma = \frac{cM}{I}$$

$$= \frac{2(2500 \times 12)}{10.67}$$

$$= 5623 \text{ psi}$$

The maximum shearing stress on the cross section of a rectangular beam is given by

Equation 9

$$\tau = \frac{3}{2} \times \frac{V}{A}$$

where

τ = shear stress (psi)
V = vertical shearing force (lbs)
A = area of cross section (in.²)

The corresponding formula for a beam with a circular cross section is:

Equation 10

$$\tau = \frac{4}{3} \times \frac{V}{A}$$

Example. Refer to Fig. 15-69 and determine the maximum shear stress (a) when the beam has a cross section 2 inches square and (b) when the cross section is 2 inches in diameter.

Solution.

(a) Using equation 9,

$$\tau = \frac{3}{2} \times \frac{V}{A}$$

$$= \frac{3}{2} \times \frac{1417}{2 \times 2}$$

$$= 531 \text{ psi}$$

(b) Using equation 10,

$$\tau = \frac{4}{3} \times \frac{V}{A}$$

$$= \frac{4}{3} \times \frac{1417}{\pi 1^2}$$

$$= 601 \text{ psi}$$

15.6.10 Torsion. External forces which cause a member to twist are called torsional loads. The product of an external force and the distance to the axis of the member is known as twisting moment, or torque. A torque is normally applied in a plane perpendicular to the longitudinal axis of the member.

When the shaft shown in Fig. 15-72 is held stationary at the left end and a torque applied to the right end, the right end is rotated until line *OB* as-

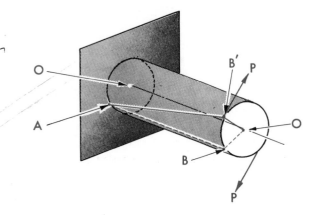

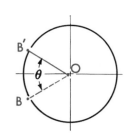

Fig. 15-72. A shaft in torsion.

sumes the position OB'. The relative movement of the sections of the shaft is resisted by the material, thereby setting up internal shear stresses known as torsional stress.

The formula relating the torque and stress for a solid circular shaft is:

Equation 11

$$T = .196\ \tau d^3$$

where

T = torque (lb-in.)
τ = shear stress at outer fiber (psi)
d = shaft diameter (in.)

Similarly for a hollow shaft the formula is

Equation 12

$$T = .196\ \tau \times \frac{(d_o^4 - d_i^4)}{d_i}$$

where

T = torque (lb-in.)
τ = shear stress at outer fiber (psi)
d_o = outside diameter (in.)
d_i = inside diameter (in.)

It was noted in Fig. 15-72 that when the shaft was subjected to a torque, the line OB rotated through an angle θ (Greek *theta*). The formula relating torque and angle of twist for a solid circular shaft is:

Equation 13

$$\theta = 584 \left(\frac{TL}{Gd^4} \right)$$

where

θ = angle of twist (degrees)
T = torque (lb-in.)
L = length (in.)
G = modulus of elasticity in shear (psi)
d = shaft diameter (in.)

Similarly for a hollow shaft the formula is

Equation 14

$$\theta = 584 \left(\frac{TL}{G(d_o^4 - d_i^4)} \right)$$

where

θ = angle of twist (degrees)
T = torque (lb-in.)
L = length (in.)
G = modulus of elasticity in shear (psi)
d_o = outside diameter (in.)
d_i = inside diameter (in.)

Example. Calculate the minimum diameter of a steel shaft 4 feet long subjected to a torque of 100,000 lb-in. if the maximum shearing stress is not to exceed 10,000 psi and the angle of twist is not to exceed 1°. (For steel, G = 12,000,000 psi)

Solution.

1. Using equation 11,

$$T = .196\ \tau\ d^3$$
$$100,000 = .196 \times 10,000 \times d^3$$
$$d = 3.71 \text{ in.}$$

2. Using equation 13,

$$\theta = 584 \left(\frac{TL}{Gd^4} \right)$$

$$L = 584 \left(\frac{100,000 \times 4 \times 12}{12,000,000 \times d^4} \right)$$

$$d = 3.91 \text{ in.}$$

Therefore, to satisfy both conditions, the shaft should be at least 3.91 inches in diameter.

15.6.11 *Horsepower relationships.* Shafts are generally used to transmit power from a motor to a machine. The following formula shows the relationship between horsepower, torque, and revolutions per minute (RPM).

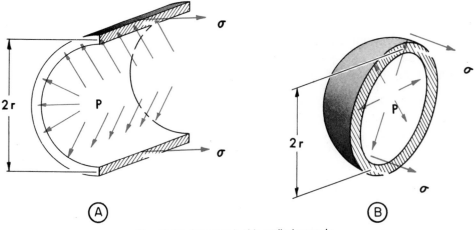

Fig. 15-73. Stresses in thin-walled vessels.

Equation 15

$$HP = \frac{TN}{63,000}$$

where

HP = horsepower
T = torque (in.-lb)
N = revolutions per minute

Example. If a 5-horsepower motor is turning at 3,000 RPM, determine the torque.

Solution. Using equation 15,

$$T = \frac{63,000\,(HP)}{N}$$

$$= \frac{63,000 \times 5}{3,000} = 105 \text{ in.-lb}$$

15.6.12 Thin-walled pressure vessels. A pressure vessel is described as thin-walled when the ratio of the wall thickness to the radius of the vessel is less than 1:10.

In order to obtain the stress in the wall of a cylinder subjected to an internal pressure, the equation used is:

Equation 16

$$\sigma = \frac{Pr}{t}$$

where

σ = stress (psi)
P = pressure (psi)
r = radius (in.)
t = wall thickness (in.)

(See Fig. 15-73A.)

The corresponding equation for obtaining the stress in the wall of a hemisphere is:

$$\sigma = \frac{Pr}{2t}$$

(See Fig. 15-73B.)

Example. A steel pipe is made by wrapping a .063-inch plate around a mandrel and welding it parallel to the axis of the mandrel. If the pipe has a diameter of 2 inches and is subjected to a pressure of 100 psi, determine the stress at the weld.

Solution. Using equation 16,

$$\sigma = \frac{Pr}{t} = \frac{100 \times 1}{.063} = 1587 \text{ psi}$$

UNIT 16

Manufacturing Processes and Materials

Manufactured parts may be produced by casting, forging, stamping, welding, machining, or some combination of these. The processes used will depend on construction material, size and shape of the part, degree of accuracy required, quality of the finished product, and cost of manufacturing. All of these factors must be considered not only during the design stages of the part but also while various working drawings are made. The engineer and drafting personnel, therefore, must have a reasonably good knowledge of basic manufacturing operations. Furthermore, since most design and drafting assignments also involve materials of various kinds, an understanding of basic materials used in manufacturing consumer products is equally important. See Fig. 16-1.

16.1 Casting

A casting is made by pouring molten metal into a mold. The principal methods are sand mold casting, plaster mold casting, permanent mold casting, investment casting, centrifugal casting, and die casting.

Sand mold casting. In this method a wood or metal pattern is used to make a mold. The mold is prepared by placing the pattern in a wood or metal

Fig. 16-1. Both drafting personnel and engineer should have a basic knowledge of manufacturing operations and materials.

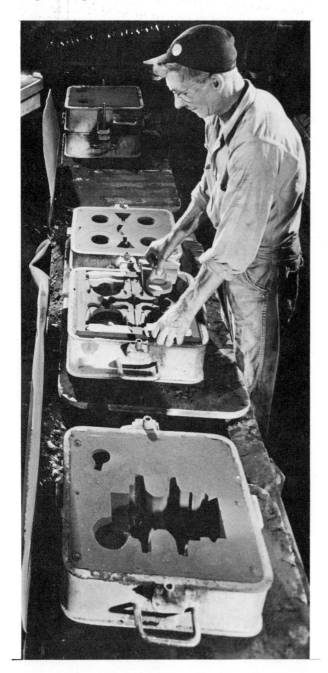

Fig. 16-2. A sand mold is the form into which molten metal is poured to produce a casting.

frame called a flask and packing sand around the pattern. The pattern is then removed and metal is poured into the mold cavity. See Fig. 16-2.

Plaster mold casting. The plaster mold process is similar to sand mold casting, except that the mold material is plaster or a combination of plaster and sand. This process is confined to the casting of non-ferrous metals. Plaster mold castings have a smoother finish and greater dimensional accuracy than sand mold castings.

Permanent mold casting. Unlike sand and plaster mold casting, where a new mold has to be prepared for each casting operation, the permanent mold process uses a metal mold which can

Fig. 16-3. In permanent mold casting a metal mold is used to receive the metal.

be utilized repeatedly. Because of their greater precision, metal molds produce more accurate castings than sand molds. This type of casting is employed when high production warrants the additional cost of equipment. See Fig. 16-3.

Investment casting. Investment casting, sometimes referred to as the lost wax process, is used to produce small and intricate parts requiring a high degree of surface smoothness and dimensional accuracy. This process is particularly adaptable to the production of parts for aircraft, ordnance, and radar.

The pattern is prepared by forcing molten wax or plastic into a metal die. The resulting pattern is used to make a sand mold, after which the mold is fired at a high temperature to remove the wax or plastic. Molten metal is then fed into the cavity either by centrifugal force or by gravity pouring.

Centrifugal casting. Centrifugal casting uses a permanent mold which is rotated rapidly while a measured amount of molten metal is poured into the mold cavity. The process is applicable to cylindrical castings made either of ferrous or nonferrous metals. Centrifugal force holds the metal in the mold, and the volume of metal poured controls the wall thickness of the casting. The advantage of centrifugal casting is that it produces smoother outside surfaces, thereby reducing a great deal of machining. See Fig. 16-4.

Die casting.[1] Die casting is a process of forcing metal under pressure into metal dies. It is especially applicable for casting soft alloys of zinc, aluminum, magnesium, and copper. Castings formed by this method are extremely accurate and require little or no machining. The process is adaptable to almost unlimited shapes, without expensive supplementary operations.

---

1. *General Motors Drafting Standards* (Detroit: General Motors Corp.).

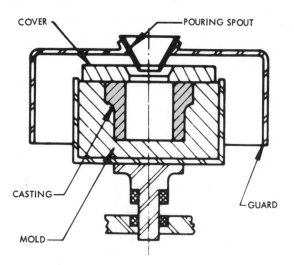

Fig. 16-4. Centrifugal casting utilizes a permanent mold that is rotated while a measured amount of molten metal is poured into the cavity.

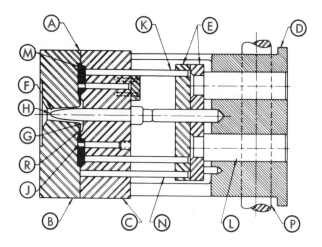

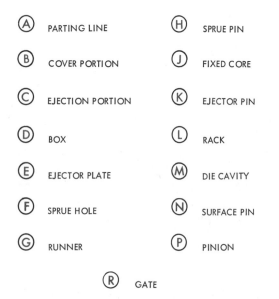

Ⓐ	PARTING LINE	Ⓗ	SPRUE PIN
Ⓑ	COVER PORTION	Ⓙ	FIXED CORE
Ⓒ	EJECTION PORTION	Ⓚ	EJECTOR PIN
Ⓓ	BOX	Ⓛ	RACK
Ⓔ	EJECTOR PLATE	Ⓜ	DIE CAVITY
Ⓕ	SPRUE HOLE	Ⓝ	SURFACE PIN
Ⓖ	RUNNER	Ⓟ	PINION
Ⓡ	GATE		

Fig. 16-5. A typical die for a die casting process.

Dies are constructed in two sections that come together at the parting. This is preferably a plane surface, but it often has to be irregular. Since all casting is done under pressure, the die sections are securely locked while casting and subsequently opened to remove the part. The front, or cover portion, of the die, *B* in Fig. 16-5, is generally fixed to the front plate of the machine on the side towards the metal pot or cold chamber. The rear or ejector portion *C* is arranged to be drawn away from the cover portion when the die is opened. It usually contains the major part of the die cavity. The ejector *E* is placed in the hollow of the box *D,* spaced away from the movable plate of the machine. The ejector plate *E* moves with the ejector portion *C* until the die is partially open and then stops. The casting is held stationary by ejector pins *K*, while the ejector portion continues to withdraw. The surface pins *N* return the ejector plate to casting position when the die is closed.

Many dies have a single cavity for making one casting per "shot." But when parts compatible in size and shape are required in sufficient quantities, several cavities are used. If the cavities are all duplicates, the die is commonly referred to as a multiple-cavity die; if the die has cavities of different shapes, it is usually referred to as a combination die. It is frequently found that several parts of the same assembly can be economically cast simultaneously. It is in such circumstances that the combination die is employed.

Powder metallurgy. Although powder metallurgy is not an actual casting process, parts made by

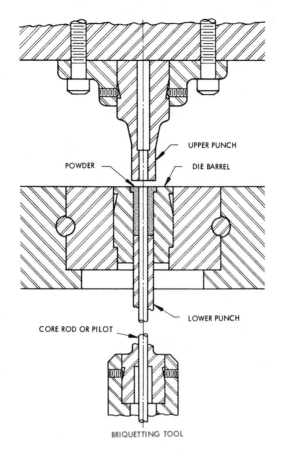

UPPER PUNCH

POWDER

DIE BARREL

LOWER PUNCH

CORE ROD OR PILOT

BRIQUETTING TOOL

Fig. 16-6. *Top,* powder metallurgy makes parts by compacting metal powder in a form, heating the mixture, and then finishing the part. The briquetting tool shown compacts the powder into the die. *Bottom,* typical products made by the powder metallurgy process.

this method require the use of specially made dies. Metal powders are compressed into a form under extremely high pressures varying from 15,000 to 100,000 pounds per square inch. The powder metals most commonly used are copper and tin to produce bronze for bearings, and brass and iron for structural parts.

The first operation involves the mixing of the powders to obtain a homogeneous blend. The powder is then compressed into the form by means of briquetting tools with pressure supplied either by mechanical or hydraulic presses. See Fig. 16-6. The briquetted compacts are next passed through a furnace where heating bonds the particles firmly together. Upon cooling, the piece is ejected from the die and subjected to various treatments.

16.1.1 Patterns. Before a casting can be produced, a pattern, which is a duplicate of the part to be cast, is made. If only a few castings are required, the pattern is constructed of wood; if a large number of castings are needed, the pattern is generally made of metal. Below are explanations of some of the principal terms associated with patterns.

Draft. Draft is the amount of taper incorporated in a pattern to permit its removal from the sand without tearing the walls of the mold. The amount of draft provided on a pattern is governed by the shape and size of the casting and the method of production. Thus, machine molding requires less draft than hand molding, and interior surfaces often need more draft than exterior surfaces.

Parting line. The parting line represents the point where the pattern is divided for molding or the surface where sections of the mold separate. The location of the parting line on a pattern depends on the shape of the casting. In general, it is placed at the largest part of the pattern and where there are no projections or undercut faces. See Fig. 16-7.

Cores. When part of a casting is to be hollow, a form must be inserted in the mold to shape the interior. The piece that occupies the hollow volume is called a core. A core is made by packing sand in a core box having an impression of the internal shape to be produced. See Fig. 16-8. The sand is treated with a bonding agent to achieve cohesion. After the packing is completed, the unit is baked or cured. This hardens the core so that it can be handled and placed in the mold. The core is supported in the mold by projections known as core prints.

Shrinkage. When metal poured into the mold solidifies, a certain amount of contraction takes

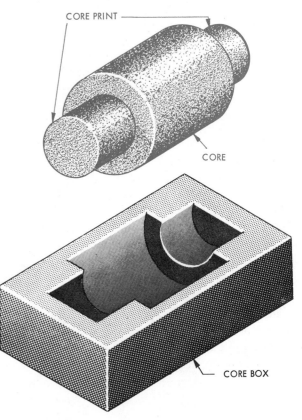

Fig. 16-8. A core is made in a core box and is used to form the internal shape of a casting.

Fig. 16-7. The parting line divides the pattern for molding.

place. To obtain a casting of the required size, the pattern is made slightly oversize to compensate for this shrinkage. The allowance is based on the shrinkage characteristics of the particular metal used.

Machining allowances. A pattern must provide for certain finishing operations. The amount of machine finish allowance depends upon the size and kind of casting, type of surface, method of machining, and the accuracy required of the finished product. This allowance will vary anywhere from 0.12 to 0.38 inch or more.

16.1.2 Pouring a casting.[2] Molten metal is poured into a mold that has been formed in a container called a *flask.* See Fig. 16-9. A flask consists of two or more sections. The upper section is known as the *cope* and the lowest section as the *drag.* Intermediate sections called *cheeks* are sometimes placed between the cope and drag for complicated castings. Molten metal is poured into the mold cavity through vertical passages in the sand called *sprues.* From the sprue the metal flows into the cavity through horizontal passages called *gates.* A *feeder,* or *runner,* is an opening formed in the molding sand to supply additional metal to the casting during the cooling and shrinkage period to eliminate voids and hollows in the cast part. A small opening called a *vent* is provided through the sand to permit the escape of gases generated during the pouring process.

The oldest and most generally used—and the most economical—type of furnace for melting iron is the *cupola,* which is a continuously melting furnace. Fundamentally it consists of a vertical steel cylinder lined with a refractory material and provided with openings for the air under pressure to enter the cylinder. Alternate charges of coke, iron, and a suitable flux, generally limestone, are placed in the refractory-lined cylinder in properly predetermined quantities. Air forced through the charges causes combustion of the coke and melts the metal, which then drips down through the incandescent fuel to a hearth at the bottom of the cylindrical shaft. It is then withdrawn, either continuously or as desired, and poured into the molds.

2. *Ibid.*

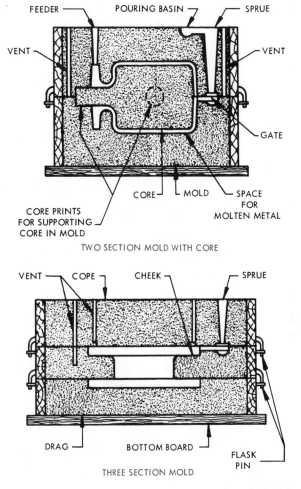

TWO SECTION MOLD WITH CORE

THREE SECTION MOLD

Fig. 16-9. A mold is made by placing a pattern in a flask and packing sand around it. After the pattern is removed, molten metal is poured into the cavity through passages called sprues and gates.

The *electric furnace* is being used to an increasing extent in melting iron, because the composition can be controlled quite accurately and there is flexibility in the temperature control. However, the melting cost is normally higher than that of cupola melting.

The molten metal, when drawn from the furnace, is transferred in ladles to the molds, where it is poured into the sprues. See Fig. 16-10. Where the mold material is a moist refractory sand and the metal is poured into the moist mold, the process is known as *casting in green-sand molds.* Where the mold material is dried before the casting is poured, the process is known as *casting in*

Fig. 16-10. Pouring a casting.

Fig. 16-11. Mechanical forging machine 31 feet high.

dry-sand molds. The rate of pouring is determined by the size of the casting and its metal section. After the mold is poured and the metal has cooled sufficiently, the castings are removed from the molds and taken to the cleaning room.

In the cleaning of castings, adhering sand is removed by brushing, by tumbling in a revolving barrel, or by abrasive or water blasting. In a few cases castings are pickled in acid solutions to remove the adhering sandy scale or oxide. When the casting is cleaned, the gates, risers, and fins not already broken off in handling are removed, and the rough places on the casting are smoothed by chipping or grinding.

16.2 Forging

Forging is a process of producing machine or structural parts that must withstand shock or sudden impacts and cannot be fabricated by ordinary casting operations. Forgings are made by any one of the following methods. (See Fig. 16-11.)

Drop forging. Drop forging is the process of forming the desired shape by placing a heated bar or billet on the lower half of a forging die and pounding the top half of the die into the metal by means of a power-driven machine called a drop hammer. See Fig. 16-12.

Press forging. In this process the heated billet is squeezed between dies. The pressure is applied by a forging press which completes the operation in a single stroke.

Rolling. Rolling involves the passing of a heated bar between revolving rolls that contain an impression of the required shape. It is a process designed chiefly to reduce short thick sections to long slender pieces.

Upsetting. Upsetting is the process of increasing the area of the forging metal by pressure applied between dies on a power-driven machine called an upsetter, or forging machine. This process is particularly applicable to the manufacture of bolts, forming cavities in the upset part of a forging, or in piercing holes.

Extruding. Extruding consists of forcing metal under pressure through a die having the same cross section as the aperture in the die. The resulting shapes are then cut to their proper lengths, straightened, and heat treated if necessary.

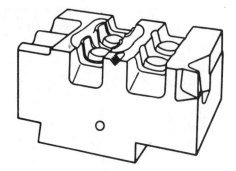

UPPER FORGING DIE (INVERTED)

FORGED PART

Fig. 16-12. This is an example of a hammer forging die.

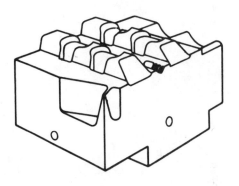

LOWER FORGING DIE

16.3 Stampings

Stampings are parts which have been formed, punched, or sheared from flat sheet-metal stock. The following are the specific operations used in making stampings. See Fig. 16-13.

Blanking is the process of cutting out a piece in the desired shape from flat stock.

Punching is the operation of forming a hole or opening in a sheet-metal part.

Forming involves bending, flanging, folding, offsetting, or twisting metal to the needed shape.

Drawing is the process of stretching metal over a form to produce the required shape.

Trimming is the process of cutting off superfluous metal around the edges of drawn pieces or cutting strips of metal to produce blanks.

Coining is the process of forcing metal to flow from an area which decreases its thickness into an adjacent area which increases its thickness.

16.4 Welding

There are many different welding methods. The type used depends on the kind of material to be joined, operating cost, shape and size of components to be welded, and strength and appearance of the seam. A brief description of these various

Fig. 16-13. Hydraulic press is used here to stamp out automotive body panels.

welding processes is included here for general familiarization purposes.

16.4.1 Oxyacetylene welding. The oxyacetylene process is a form of welding in which fusion of metal is achieved by a gas flame burning a mixture of oxygen and acetylene. Combustion of these two gases produces a temperature of approximately 6300°F, which can melt and effect fusion of weldable metals. Most commercial metals can be welded by the oxyacetylene process.

The welding operation is performed by directing a lighted torch over the seam of a joint. As the metal melts, filler rod is added to the molten pud-

dle to strengthen the weld and form a bead of the required shape and size.

16.4.2 Metallic arc welding. In the metallic arc process an electric arc formed between the work and the electrode liberates the necessary heat to effect fusion of the joint. The arc is produced by a DC generator, rectifier, or transformer. The intense heat that is developed by the arc instantly brings to a melting point a small portion of the work to be welded. The tip of the flux-coated electrode is simultaneously melted and the tiny globules of molten metal are deposited into the molten pool of the parent metal. See Fig. 16-14.

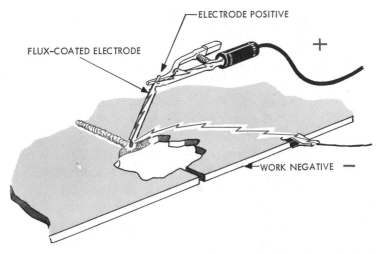

ELECTRODE POSITIVE

FLUX-COATED ELECTRODE

WORK NEGATIVE —

Fig. 16-14. The metallic arc is widely used for manual welding of structural parts.

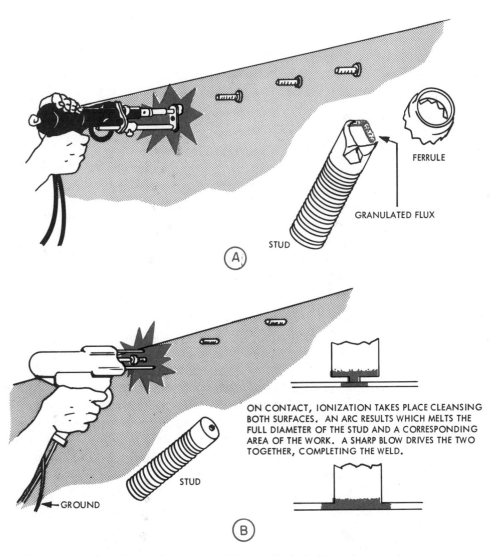

FERRULE

GRANULATED FLUX

STUD

(A)

ON CONTACT, IONIZATION TAKES PLACE CLEANSING BOTH SURFACES. AN ARC RESULTS WHICH MELTS THE FULL DIAMETER OF THE STUD AND A CORRESPONDING AREA OF THE WORK. A SHARP BLOW DRIVES THE TWO TOGETHER, COMPLETING THE WELD.

STUD

← GROUND

(B)

Fig. 16-15. Stud welding is a form of arc welding used in fastening studs to metal components.

16.4.3 Stud welding.[3] Stud welding is a form of electric arc welding used in fastening studs to metal components. Two methods of stud welding have been developed, each with different principles of operation.

In the first method a stud is loaded into the chuck of a gun and a ferrule is positioned over the stud. When the trigger is depressed, the current energizes a solenoid, which lifts the stud away from the plate, causing an arc which melts the end of the stud and the area on the plate. A timing device shuts off the current at the proper time. The solenoid releases the stud, and a spring action plunges the stud into the molten pool and the weld is made. See Fig. 16-15*A*.

Another method is characterized by a small cylindrical tip on the joining face of the stud. See Fig. 16-15*B*. The diameter and length of this tip vary with the diameter of the stud and the material being welded. This method operates on alternating current; a source of about 85 pounds of air pressure is also required. The gun is air-operated with a collet to hold the stud attached to the end of a piston rod. Constant air pressure holds the stud from the metal until the weld is ready to be made; then air pressure drives the stud against the work. When the small tip touches the workpiece, a high-amperage low-voltage discharge results, creating an arc that melts the entire area of the stud and the corresponding area of work. The stud is driven at a velocity of about 31 inches per second, and the explosive action as it meets the workpiece cleanses the area to be welded.

16.4.4 Gas-shielded arc welding. There are two types of gas-shielded arc welding processes: tungsten-inert gas (Tig) and metallic-inert gas (Mig). Each has certain advantages, but both produce welds that are deep penetrating and relatively free from atmospheric contamination.

Tungsten-inert gas (Tig). This process uses a virtually non-consumable tungsten electrode to provide the arc for welding. During the welding cycle a shield of inert gas, such as argon, helium,

or a mixture of both, pushes the air away from the welding area and prevents oxidation of the electrode, weld puddle, and surrounding heat-affected zone. On joints where filler rod is needed, a rod is fed into the puddle in the same way that it is added in welding with the oxyacetylene flame. See Fig. 16-16*A*.

Metallic-inert gas (Mig). In this welding process a continuously consumable wire electrode is used. The molten puddle is completely covered with a shield of inert gas. The wire electrode is fed through the torch at controlled speeds. The shielding gas also is fed through the torch. The welding can be fully automatic or semi-automatic. If fully automatic, the welding unit is arranged so it can travel over the workpiece and is operated entirely by controls. With the semi-automatic the flow of wire and shielding gas is pre-set, but the torch is manually operated. See Fig. 16-16*B*.

Most industrially used metals can readily be welded with either the Tig or Mig process. These include aluminum, magnesium, low-alloy steel, carbon steel, stainless steel, copper, nickel, Monel, Inconel, and titanium.

16.4.5 Submerged arc welding. In this method an AC or DC electric arc buried in a protective layer of granular mineral material provides the heat of welding. Filler metal serves as the electrode; the work is grounded. Automatically controlled mechanisms must be used in this type of welding. See Fig. 16-17.

Essentially the operation consists of a unit that moves at a controlled speed over the weld area. This unit contains a feeding hopper which deposits the granulated flux ahead of the filler rod. The filler rod, which is also the electrode, is automatically fed into the flux so that a constant distance between the melting end of the electrode and the pool of molten metal is maintained. That portion of the granular flux immediately around the arc fuses and covers the molten metal. A means of reclaiming the unused flux—usually a suction tube—follows this.

The arc is not visible, since it is buried in the flux. Thus, there is neither flash nor splatter.

Due to the nature of this operation, welding must be done on a horizontal, or nearly horizontal, plane. In some setups the welding head moves

---

3. *Welding, Brazing, Soldering and Hot Cutting Republic Stainless Steels* (Cleveland: Republic Steel Corp.).

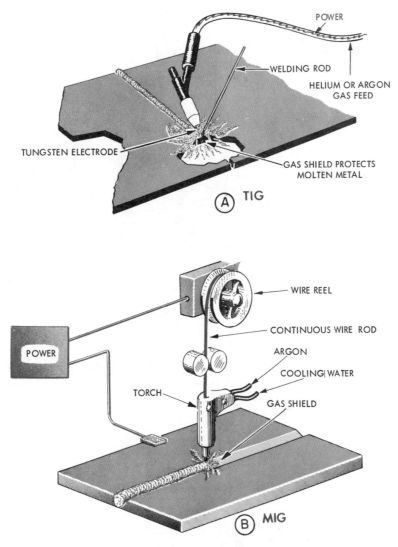

Fig. 16-16. Gas-shielded arc welding processes.

and the work remains stationary. In others the head is stationary and the work moves, as in joining sections of large-diameter pipe that can be rotated under the welding head.

16.4.6 Spot welding. This method, with its various forms, is probably the most common and generally used of the resistance welding methods. Two or more layers of material can be joined simultaneously. These layers are placed between two electrodes, pressure is applied, and a quick "shot" of electricity is sent from one electrode through the material to the other. The pressure is

continued momentarily, and the weld is completed. The method is adaptable to material as thick as one inch. See Fig. 16-18.

Spot welding machines are also designed with multiple sets of electrodes. The current is passed through each set of electrodes separately which, in rotation, provide a number of spot welds with one setting operation.

16.4.7 Pulsation welding. Pulsation welding is merely a form of spot welding in which the current is regulated to go on and off a given number of times during the making of one weld. It is claimed

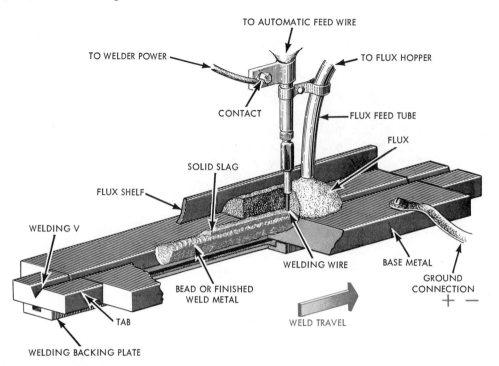

Fig. 16-17. Submerged arc welding is an automatic welding process in which heat is provided by an arc buried in a protective layer of granular mineral material.

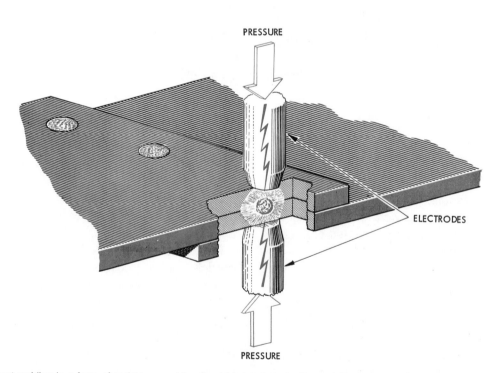

Fig. 16-18. Spot welding is a form of resistance welding in which two layers of material are placed between two electrodes, a shot of electricity is sent through, and a small spot of fusion occurs as pressure is applied momentarily.

to have advantages over straight spot welding in that (1) welding of thicker materials is possible, (2) electrode life is increased, since the interrupted current tends to keep the electrodes cooler, thus minimizing electrode distortion, and (3) there is less tendency for the weld to spit or spark.

16.4.8 Projection welding. This method is similar to the spot method, except that projections are formed on one of the sheets being welded—usually on the thicker sheet when sizes vary. The two electrodes contact the material in line with the projection, the projection itself acting as a sharp-pointed electrode which localizes the heat. Other characteristics of the welding procedure are the same as for spot welding.

Projection welding can be set up in such a manner as to complete more than one "spot" at a time. See Fig. 16-19.

16.4.9 Seam welding. Two types of set-ups are common to seam welding: the *lap method* and the *line method.* In both cases rolling electrodes are used.

Lap seam welding. In this method, which is most frequently used on flat pieces, the materials are placed one on top of the other with roller-type electrodes on opposite sides. As the work is fed between these rollers, a controlled timing of the current flow produces a series of spot welds. The faster the cycle of current flow (or the slower the work moves), the closer the spot welds will be. Watertight seams are possible by controlling the process so that each weld slightly overlaps the former.

Line seam welding. In line seam welding, two roller-type electrodes contact the work, one on each side of the seam, as shown in Fig. 16-20. The current cycle and speed of work feed can be adjusted as in lap seam welding to produce a watertight weld. In both cases the constantly rolling electrodes and moving work leave less contact time between the electrodes and hot weld.

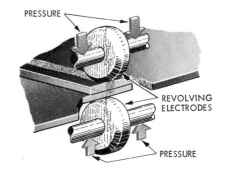

Fig. 16-20. Seam welding is a production-type process in which the work to be welded is fed between two revolving electrodes.

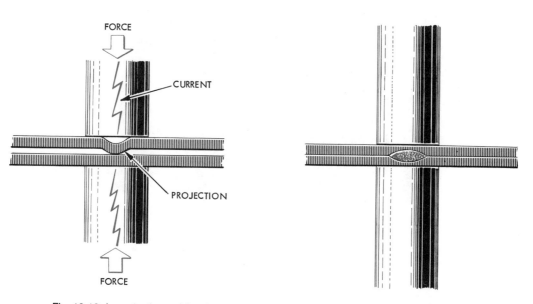

Fig. 16-19. In projection welding fusion occurs over the preformed projections on the sheet.

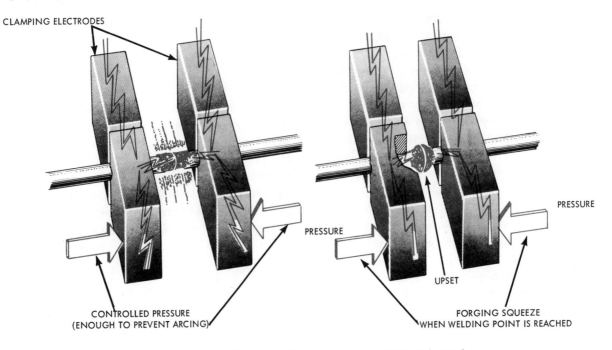

CLAMPING ELECTRODES

PRESSURE

PRESSURE

CONTROLLED PRESSURE
(ENOUGH TO PREVENT ARCING)

UPSET

FORGING SQUEEZE
WHEN WELDING POINT IS REACHED

Fig. 16-21. Butt welding is used to join bars, rods, or wire end to end.

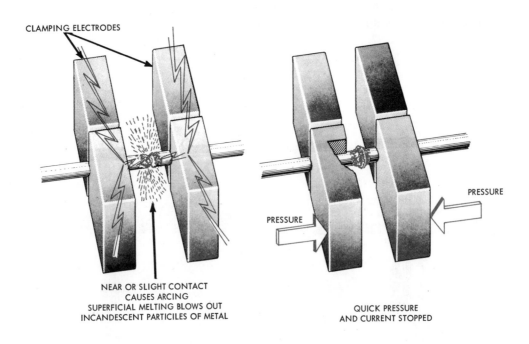

CLAMPING ELECTRODES

PRESSURE

PRESSURE

NEAR OR SLIGHT CONTACT
CAUSES ARCING
SUPERFICIAL MELTING BLOWS OUT
INCANDESCENT PARTICILES OF METAL

QUICK PRESSURE
AND CURRENT STOPPED

Fig. 16-22. In flash welding, an arc melts the edges to be joined, and they are forced together while in a molten state.

Thus, less cooling effect is realized from the electrode, and a large amount of coolant is required.

16.4.10 Butt welding. This form of welding is frequently used for joining bars, rods, or wires end to end. The two ends are butted together by use of clamps which also serve as electrodes. See Fig. 16-21. Pressure is applied, and a high amperage current passes through one electrode to the other. A melting at the joining faces results, and the continuous pressure forces any oxidized metal out of the joint and provides the necessary pressure for fusion of the two ends when the proper temperature is reached.

There is, of course, no arcing, and thus no flash or weld splatter. Even so, this method is largely being supplanted by the faster and less power-consuming flash welding.

16.4.11 Flash welding. Flash welding differs from butt welding in that the pieces being joined do not touch initially. As the current is turned on and the two edges are brought into proximity, intense arcing occurs. Incandescent particles of metal are blown out of the joint by the extremely rapid superficial melting that takes place; thus, the name flash welding. At the proper moment the edges are forced together; the molten metal, slag, and impurities are forced out of the joint; and a very solid weld is produced between the two plastic edges. See Fig. 16-22.

16.4.12 Inertia welding. Inertia welding relies on stored kinetic energy to generate the heating required for fusion. In this process one member to be welded is placed in a stationary chuck or fixture, while the other member is securely clamped in a rotating spindle attached to a flywheel. See Fig. 16-23.

The flywheel is rotated by an external energy source. When a predetermined RPM is reached, the drive source is disconnected, and the members are brought into contact under a precomputed constant thrust load. The kinetic energy contained in the rotating mass converts into frictional heat. Welding occurs as the rotation ceases.

This welding process is particularly effective for joining many dissimilar or exotic metals, as well as for similar metals. Welds are exceptionally strong and free of defects. Weld strength is equal to that of the original metals.

16.4.13 Electron beam welding. Electron beam welding is a fusion process in which coalescence is achieved by focusing a high power density beam of electrons on the area to be joined. Upon striking the metal, the kinetic energy of the high-velocity electrons changes to thermal energy and causes the metal to melt and fuse. Welding is usually done in a vacuum chamber. Refer to Fig. 16-24.

16.4.14 Laser welding. Laser welding consists of directing a highly concentrated light beam to a spot about the diameter of a human hair. This light beam has a higher energy concentration than the electron beam. The laser is generated by exciting atoms in a synthetic ruby rod by means of an external light source. To excite the atoms to a high energy state, the parallel ends of the rod are mirrored, which bounces the atoms back and forth.

Fig. 16-23. In inertia welding, stored energy is converted into heat to produce high-quality welds.

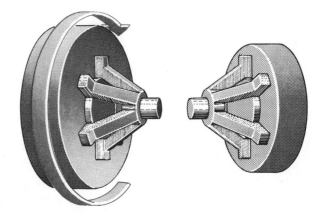

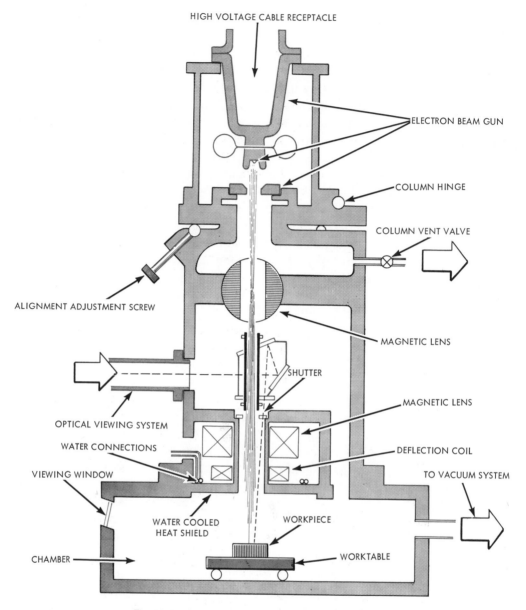

HIGH VOLTAGE CABLE RECEPTACLE

ELECTRON BEAM GUN

COLUMN HINGE

COLUMN VENT VALVE

ALIGNMENT ADJUSTMENT SCREW

MAGNETIC LENS

SHUTTER

OPTICAL VIEWING SYSTEM

MAGNETIC LENS

WATER CONNECTIONS

DEFLECTION COIL

VIEWING WINDOW

TO VACUUM SYSTEM

WATER COOLED HEAT SHIELD

WORKPIECE

CHAMBER

WORKTABLE

Fig. 16-24. Schematic of electron beam gun column.

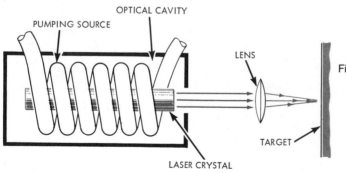

OPTICAL CAVITY

PUMPING SOURCE

LENS

Fig. 16-25. Schematic diagram of a laser welder.

TARGET

LASER CRYSTAL

When the chain reaction of collisions between atoms reaches a high enough level, a burst of red light escapes from the ruby and provides an intense heat for fusion. See Fig. 16-25.

16.4.15 Ultrasonic welding. In ultrasonic welding the joining of metal is attained by vibratory energy produced by a transducer. The vibratory energy plastically deforms the interface to produce perfectly smooth surfaces between the workpieces. This allows the atoms of one piece to readily unite with the atoms of the other piece and form a solid bond. See Fig. 16-26.

16.4.16 Plasma welding. Plasma welding utilizes a central cone of extreme temperature surrounded by a sheath of cool gas. The required heat for fusion is generated by an arc heating a gas to such a high temperature that the gas becomes ionized. As the gas is fed through the arc, it becomes heated to what is known as the plasma temperature range. The plasma jet is forced

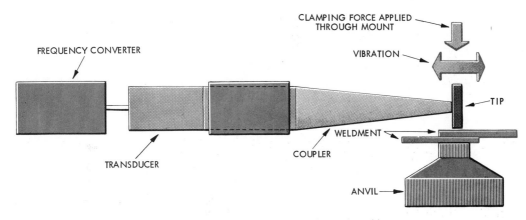

Fig. 16-26. Schematic diagram of an ultrasonic welder.

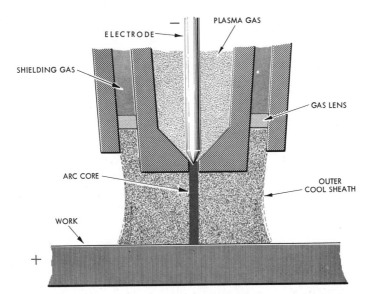

Fig. 16-27. Plasma welding uses a central core of extreme temperature surrounded by a sheath of cool gas.

through a constricting orifice of a torch at a sonic velocity (4,000 ft/sec) and produces an intense heat of approximately 30,000°F. See Fig. 16-27.

16.5 Heat Treatment[4]

Heat treatment is a heating and cooling process used to effect certain changes in the properties of steel. Heat treatment controls principally the grain structure, which in turn changes the physical properties of the steel. Changes and refinements of the grain structure take place when the steel is heated to a temperature above its critical temperature range and then cooled.

16.5.1 Heat-treating ferrous metals. The basic heat-treating operations are as follows:

Annealing is the process of heating steel for a specified period, either above or below its critical range, and then allowing it to cool slowly. Annealing is used to soften metal, remove stresses, alter such characteristics as ductility, toughness, or magnetic properties, or produce a definite grain structure.

Normalizing is practically the same as annealing, except that cooling is done in still air instead of in the furnace. Normalizing usually produces slightly higher tensile strength and hardness than does annealing.

Hardening is heating a steel above its critical temperature for a given period of time and then quenching it in liquids or gases. This action produces a metal that is extremely hard and brittle. Steels with a carbon content of less than 0.20 percent do not respond to hardening.

Tempering is a process of reheating hardened steel to a temperature below the critical range and then allowing it to cool. This treatment reduces the hardness, relieves stresses, decreases brittleness, and at the same time increases the toughness and ductility of hardened steel.

Case hardening is a process of hardening only the outer surfaces of low-carbon steels and leaving the interior soft. The result is a metal having a hard case to resist wear and abrasion and a tough core to withstand shock stresses. The common case-hardening processes are carburizing, cyaniding, nitriding, induction hardening, and flame hardening.

Carburizing is a process whereby a low-carbon steel is heated above its critical range while in contact with a carbonaceous material. The carbon—in the form of a liquid, gas, or solid—diffuses into the steel to a depth governed by the nature of carburizing material, the temperature, and the heating period.

Cyaniding is an operation in which the low-carbon steel is heated while in contact with a cyanide salt and then quenched. This process is used when only a very thin hard case is required.

Nitriding consists of heating steel in an atmosphere of ammonia vapor or placing it in contact with a nitrogenous material. No quenching follows the heating. Nitriding produces a harder case than carburizing with less distortion and cracking.

Induction hardening is the method used for case-hardening steel having not less than 0.35 percent carbon by placing it in a coil through which a high-frequency current is passed. The high-frequency current is induced in the surface of the steel, which produces an almost instantaneous heating of the case to a depth of 0.010-0.100 inch. The metal is then quenched in a water spray or oil bath. Induction hardening is used when certain hardened areas must be localized.

Flame hardening is a process of heating a metal by means of one or a series of gas-burning torches to a specified temperature and then quenching. This process is used to harden local areas of such parts as gear teeth, shafts, and rocker arms.

16.5.2 Heat-treating non-ferrous alloys. Nonferrous alloys do not respond to the same type of heat treatment as do ferrous alloys. Some materials may not be hardened by any means except cold-rolling or cold-working. Other alloys respond to precipitation hardening, often called aging. This is a process wherein one or more constituents are precipitated from a solid solution, the size and distribution of the precipitate being such that a substantial increase in strength is obtained.

Precipitation hardening is generally accomplished by quenching from a temperature slightly below the temperature at which the material be-

---

4. *General Motors Drafting Standards* (Detroit: General Motors Corp.).

gins to melt, followed by a hardening cycle ranging from room temperature to 900°F depending upon the material being hardened. This treatment may be used to harden parts ranging from small springs to large die sections.

Aluminum. Alloys of aluminum respond to a small amount of cold-working and to heat treatment. Commercially pure aluminum may be hardened by cold-working alone. In some cases both of these treatments are utilized to improve the physical properties of the alloys.

Heating aluminum alloys to 630-650°F will anneal or remove the effects of cold-working and most of the heat treatment in case of heat-treatable material.

The alloys that respond to solution treatment are heated to the required temperature, usually 950-1000°F, and held to obtain a homogeneous solid solution. They are then cooled rapidly to room temperature. In some alloys hardening begins immediately and is complete in a few hours at room temperature. It is necessary for other alloys to age at elevated temperatures for several hours in order to harden.

Copper alloys. In general, copper alloys may be hardened only by cold-working to increase the tensile strength, yield strength, and elastic limit, and to reduce the elongation and reduction in area. Most annealed copper alloys are quite ductile and respond readily to cold-working.

When the copper alloy is work-hardened, annealing is required to soften it. Annealing consists of heating the metal above the recrystallization temperature. The most common range of annealing temperature is 800°-1300°F.

16.5.3 Hardness testing.

The type of hardness test to be specified for any particular piece is governed by type of material, hardness range, area where the hardness test is to be made, surface condition of material to be tested, section size, and permissible indenter impression size.

The following hardness testers and scales are commonly used[5] (see Appendix for Table of Steel Hardness Number Conversions):

---

5. *Drafting Room Manual* (East Hartford, Conn.: Pratt and Whitney Aircraft, Division of United Aircraft Corp.).

Brinell. This is a heavy load tester, utilizing a 10-millimeter hardened steel-ball indenter, which may be used on large parts, bar stock, large heavy wall tubing, forgings, and castings where a large indenter impression is not objectionable. The Brinell is the best test for use on rough or non-homogeneous surfaces, because the large impression is less sensitive to minor variations or imperfections. It is not generally used for hardness testing of finished parts. The following scales are employed:

A *300 kilogram (kg) load* may be used for hardness values up to 444 Bhn (Brinell hardness number). For hardness values between 444 and 745 Bhn, the Rockwell C test is generally used; however, it is permissible to use Brinell in this hardness range, provided a tungsten-carbide-ball penetrator is used. This scale is most adaptable for steels but may be used for hard aluminum and copper alloys.

A *500 kilogram load* is occasionally used for hardness values below 150 Bhn and is restricted to non-ferrous alloys.

A 1000 kilogram load is occasionally used with a 10-millimeter steel ball for copper alloys and soft steels where larger indentation is undesirable.

Rockwell standard. This tester produces rapid and accurate results. See Fig. 16-28. By changing the scale (penetrator and/or major load; minor load is constant at 10 kilograms), metals of most hardness ranges can be tested. The size of the indenter impression is fairly small and decreases with increase in hardness of material tested and decrease in the major load of the Rockwell test. The Rockwell test is also applicable to checking finished parts. The following scales are used:

The *Rockwell C* (150 kg major load, diamond Brale penetrator) should only be used for hardness range *RC 20-65.* It is generally used for determining hardness of through-hardened steels and core hardness of case-hardened steels.

The *Rockwell A* (60 kg major load, diamond Brale penetrator) is generally used for checking case hardness of carburized steels. This test may be used for unusual requirements where a lighter load test than Rockwell C is required. Also it may be used in special cases for hard non-ferrous alloys.

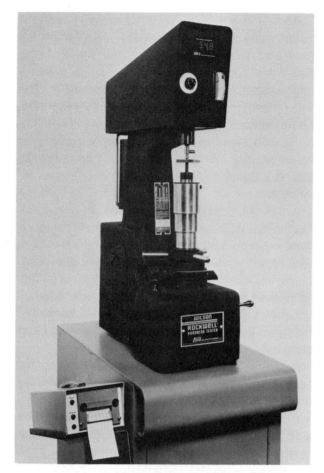

Fig. 16-28. Hardness tester.

ferrous sheet stock. Because of the shallow indentation of this test, it is often applicable to finished parts and in many cases may be used for checking the hardness of the working surfaces of parts. The following scales are employed:

The *Rockwell 30N* (30 kg major load, diamond Brale penetrator) is generally used for checking case hardness of nitrided and cyanided steels and may be used, if desired, for checking case hardness of carburized steels. It is also used for testing thin hard steels. It may be specified for finished parts when a test by a Rockwell standard scale is objectionable because of the indentation produced.

The *Rockwell 15N* (15 kg major load, diamond Brale penetrator) is similar in scope to the Rockwell 30N, but the Rockwell 30N is usually preferred. It may be desirable to stipulate Rockwell 15N for unusual requirements where a lighter load test than Rockwell 30N is required, such as for checking working surfaces of finished carburized parts.

The *Rockwell 30T* (30 kg major load, 0.06-inch steel-ball penetrator) is generally used for thin sections of non-ferrous alloys or soft steel, or on finished soft steel or non-ferrous alloy parts where small indentation is desirable.

The *Rockwell 15T* (15 kg major load, 0.06-inch steel-ball penetrator) is a lighter load test than Rockwell 30T, but similar in use.

Vickers diamond pyramid. This tester may be used for all metallic materials in any hardness range. It is not generally used as a production hardness tester but rather as a laboratory tool, and therefore

The *Rockwell B* (100 kg major load, 0.06-inch steel-ball penetrator) should not be used for hardness values greater than *RB 100*. It is generally used for soft steels and for non-ferrous alloys.

The *Rockwell F* (60 kg major load, 0.06-inch steel-ball penetrator) is generally used only on non-ferrous alloys, but for unusual requirements it may be used on soft steels where a lighter load test than Rockwell B is required.

Rockwell superficial. This tester is used where only very shallow penetration is desirable or permissible and for determining the hardness very close to the surface of the specimen. By changing the scale (penetrator and/or major load; minor load is constant at 3 kg), metals of most hardness ranges can be tested. This test may be used for checking case hardness of nitrided, cyanided, and carburized steels, and of thin ferrous and non-

NORMALIZE (OR ANNEAL) TO BRINELL ■■-■
OR (ROCKWELL C■-C■).

NORMALIZE. HARDEN AND TEMPER TO
BRINELL ■■-■■ OR (ROCKWELL C■-C■).

HARDEN AND TEMPER TO BRINELL ■■-■
OR (TEMPER TO ROCKWELL C■-C■)

CARBURIZE. .■■-.■ FINISH CASE DEPTH.
HARDEN AND TEMPER TO ROCKWELL C■■ MIN.

CYANIDE .■■ DEEP.
HARDNESS ROCKWELL SUPERFICIAL ■■.

Fig. 16-29. Types of heat treatment and hardness are shown on a drawing by means of a note.

should not be indiscriminately specified on drawings. However, for unusual requirements it may be desirable to specify the Vickers test. The loads usually used with the Vickers test are 50, 30, 10, 5, and 1 kg and the penetrator is a diamond pyramid.

16.5.4 Heat treating notes. The various heat treatment and hardness requirements are shown on a drawing in note form. See Fig. 16-29.

16.6 Surface Treatment

Most metal products, when exposed to atmospheric conditions, have a tendency to corrode unless they are treated with some protective coating. There are a number of protective measures that can be used, depending on the base material and the results desired. The following are a few of the more common methods:

Electroplating. Electroplating involves the immersion of an object in a solution containing metal or salts of metals to be deposited. As an electric current is passed through the solution, with the object to be coated serving as the negative electrode, the dissolved metal is deposited on the surface of the object. The thickness of the deposited film is governed by the amount of current, length of current flow, and shape of the object. The metals commerically used for plating are brass, cadmium, chromium, copper, gold, lead, tin, nickel, silver, and zinc.

Non-electric plating. In this process parts to be coated are left in a solution containing the coating material. The deposit is achieved through chemical electrolysis without the use of any external current. Such a process is used in tin-plating aluminum pistons and coating steel parts with copper, nickel, or cobalt from sulphates. Its advantages over electroplating include more uniform thickness of coating and better coating of internal surfaces and irregular shapes.

Hot dipping. This method consists of dipping parts in a molten solution of the plating or coating material, such as zinc, lead, or tin, and is used chiefly to coat raw sheet stock.

Sherardizing. Iron or steel is embedded in zinc powder and heated to a temperature just below the melting point of zinc. Sherardizing is used principally in coating bolts and small castings.

Anodizing. Anodizing is an electrolytic process used almost exclusively for aluminum to produce a film of oxide on the metal. The film provides a transparent, protective, anticorrosion coating that imparts a hard wear-resistant surface. The process consists of passing an electric current between the aluminum and an electrolytic bath in which the metal is immersed as an anode. The electrolytic bath may be a solution of sulfuric acid, chromic acid, boric acid, or phosphoric acid.

Parkerizing. This is a surface-plating operation in which an iron phosphate coating is applied to iron or steel parts by immersing them in a hot solution of manganese dihydrogen phosphate.

Painting. Paint is used to provide a protective finish and to serve a decorative function. Many different kinds of paints are employed. Selection of paint is based on such factors as kind and quality of finish desired, the nature and function of the fabricated product, cost, and color. Examples of paint notes used on a drawing are shown in Fig. 16-30.

whereby small pellets are blasted against a surface at high velocity. This results in a pitting action in which each shot produces a blow similar to that delivered by a small peening hammer. The effect of these shots is to release harmful stress forces which may have been generated in producing the part. In addition, shot peening also strengthens and hardens the metal surfaces. Shot peening is often used on machine parts such as camshafts, crankshafts, connecting rods, gears, and other units where it is extremely important to remove concentrated stresses.

CLEAN AND FINISH PER SPECIFICATION # ▬▬▬ TO MATCH COLOR # ▬▬▬ OR APPROVED EQUIVALENT.

ALL PAINTS TO BE SUITABLY APPLIED AND BAKED TO MEET ADHESIVE SPECIFICATIONS.

PAINT ADHESION MUST BE SATISFACTORY AFTER 24 HOURS EXPOSURE TO 100 % RELATIVE HUMIDITY AT 100° F.

Fig. 16-30. These are typical notes used on drawings to designate painting requirements.

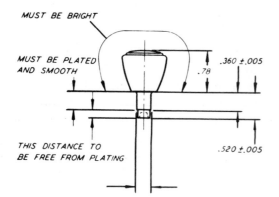

MUST BE BRIGHT

MUST BE PLATED
AND SMOOTH

.360 ±.005

.78

THIS DISTANCE TO
BE FREE FROM PLATING

.520 ±.005

.290±.001 DIA BEFORE PLATING
.292 ±.001 DIA AFTER PLATING

REMOVE ALL EXTERNAL FLASH.

CHROMIUM PLATE G.M. 4251M CODE 100.

ALL CAVITIES TO BE IDENTIFIED NUMERICALLY

MUST WITHSTAND AN OVEN TEMPERATURE OF
250° F MAX WITHOUT BLISTERING. REFER
TO G.M SPEC. 4299-P.

SURFACE AREA FOR PURPOSE OF ESTIMATING
PLATING APPROX .024 SQ FT

Fig. 16-31. This is a typical drawing of a plated part with appropriate notes.

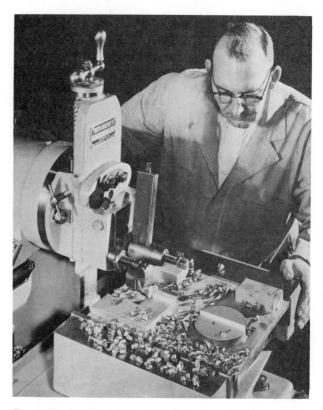

Fig. 16-32. A shaper is used for surface machining, notching, key-setting, and facing.

16.6.1 Plating specifications on a drawing. A drawing of a part that requires a metallic coating must indicate pertinent information governing the coating or plating material. The type and class of plating is generally specified by the engineer. The method of application is seldom stated.

Since plating results in increasing the thickness of the material, allowances for plating must be made on components of close fitting assemblies. A typical plating specification, including drawing and notes, is given in Fig. 16-31.

16.7 Machining[6]

Machining involves the removal of material in order to obtain the desired shape, size, and surface finish of parts being manufactured. A variety of machines and processes are used for this purpose. Here are some of the basic operations.

16.7.1 Shaping. Shaping is a surface-machining process for notching, key-seating, and facing. The cutting operation is performed by reciprocating motion of a cutting tool on a machine called a shaper. There are two basic designs of shapers: the conventional shaper and the gear shaper.

The *conventional shaper,* shown in Fig. 16-32, is extremely flexible from the point of service but requires an experienced operator. This machine is not readily adaptable to machining of production parts and is more often found in a tool room, repair shop, or job shop.

The *gear shaper* is designed primarily for use on production items. This includes the cutting or generating of gear and sprocket teeth, splines, cams, and plain or irregular outlines, both external and internal.

16.7.2 Turning. Turning is a machining process for removing material in order to produce relatively smooth and dimensionally accurate external and internal surfaces of cylindrical, conical, shouldered,

---

6. *General Motors Drafting Standards* (Detroit: General Motors Corp.).

Fig. 16-33. A lathe is used for a variety of turning operations.

16.7.3 Knurling.

Knurling is a process whereby a smooth surface or periphery of a part is machined into uniform ridges and projections. The resulting uniformly roughened surface is known as a knurl. The purpose of the knurl is to provide a suitable finger or hand grip or to restrict turning when assembled with a companion part.

The design of the knurl varies with the size and function of the part. The most common designs are the spiral and diamond-shaped knurl, which may be of varying degrees of fineness.

Another knurl design is the straight tooth type, which resembles a serration and runs parallel with the axis of the part. The dimensions and pitch of the knurl are dependent to a large extent on the size of the part and on the type of material to which the knurl is applied. Examples of diamond and straight tooth types of knurling are shown in Fig. 16-34.

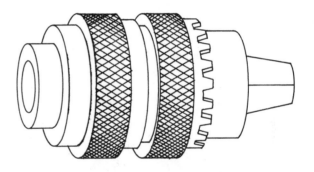

DIAMOND KNURL

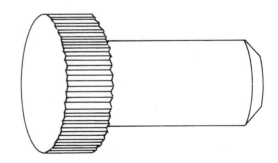

STRAIGHT TOOTH KNURL

Fig. 16-34. Knurling is a process of roughening a surface to provide a better grip or to prevent a part from turning when in contact with a companion piece.

or irregular form. Turning operations may be performed on metallic or non-metallic parts in the form of castings, forgings, moldings, bars, billets, and so on. When operations are performed internally, they are referred to as boring operations.

In performing turning operations, the work is rotated while the cutting tool is fed into or away from the work and is traversed along its axis of rotation. The work may be held in chucks or fixtures or supported on centers. The cutting tools are carried in cross slides or in a turret. Their movement may be controlled mechanically or manually.

Turning may be performed on various types of lathes. They range from the single-spindle tool room lathe, shown in Fig. 16-33, to the multiple-spindle automatic screw machine and the universal turret lathe. The single-spindle type is used for low volume or non-production work. The latter types are used for quantity production of interchangeable parts. The cutting tools vary in size, shape, and nature of cutting edges, depending upon the material being turned and the nature of the surface required.

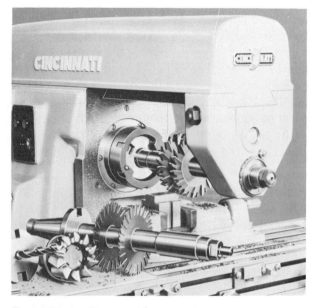

Fig. 16-35. A milling machine is used to remove material in order to produce internal or external machined surfaces.

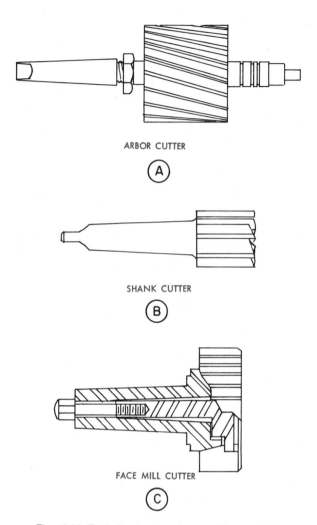

Fig. 16-36. Typical cutters used on a milling machine.

16.7.4 Milling. Milling is a machining process for removing material in order to produce internal or external machined surfaces of plain, complex, or irregular outline to close tolerances. The work is performed on a milling machine, and the cutters are circular with multiple teeth. Each tooth removes a portion of the stock as the cutter rotates about its axis and the work travels back and forth.

The milling process combines the rotation of the cutter and the feeding of the work into the path of the cutter. The cutter is supported and driven by the spindle of the machine. The work is supported on the machine table, which may be either power-driven or manually controlled. Milling machines are made with horizontal or vertical spindles and are identified as such. Fig. 16-35 shows a typical horizontal production milling machine.

Milling cutters are made in a variety of sizes and shapes. They may also be of various types, including arbor cutters, shank cutters, and face mills.

Arbor cutters are mounted on a spindle-driven arbor as shown in Fig. 16-36*A*. Shank cutters have integral shanks designed to fit directly into the machine spindles. See Fig. 16-36*B*. Face mills are designed to be attached directly to the end of the machine spindle or to a stub arbor as shown in Fig. 16-36*C*.

Milling cutters may have their cutting edges on their periphery or on the face perpendicular to their axis of rotation. The form of the milled surface is dependent on the form of the cutter.

16.7.5 Profiling. Profiling is a form of two-dimensional contour milling. The travel of the cutting tool across the work is controlled by means of a guide pin that follows the outline of a master template. See Fig. 16-37. This operation is performed on a machine known as a profiler. The machine provides power-driven spindles for the cutting tools and also blocks on which to mount the template-follower guide pins. The spindle and follower guide blocks function as a unit and are mounted on cross slides which traverse on a cross rail. The work is held in a fixture provided with a

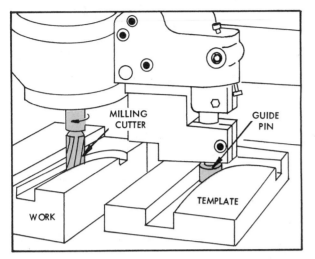

Fig. 16-37. A profiler uses a template to guide the cutter in shaping the work pieces.

Fig. 16-38. Hobbing is an automatic machining operation used in cutting gear teeth, threads, serrations, and splines.

template. The fixture is secured to the machine table, which traverses forward and backward. Both the cross slide and machine table are operated by suitable hand levers. By combining the directional travel of each, contours are machined as the guide pin, following the template, accurately guides the path of the cutter.

16.7.6 Hobbing. Hobbing is a continuous milling process. The cutter, known as a hob, and the work rotate in time relation to each other on individual spindles of a machine called a hobbing machine. In addition to the rotary motion, the hob moves across the work or across the length of the area to be hobbed.

The scope of hobbing covers a generating method of machining various types of gear teeth, threads, serrations, splines, and other special forms on external surfaces as well as threads on internal surfaces. In fact, any form or shape that is uniformly spaced on a cylindrical surface may be hobbed, providing such form or shape is of sufficient width in proportion to height in order to permit a free-rolling action of the hob.

The form of the cutting section of the hob is directly related to that of the finished gear tooth, spline, or shape to be produced. The cutting process resembles that of a worm and worm gear in mesh, with the hob representing the worm and the work representing the worm gear. Machining may be automatic, except for loading, unloading, and

starting of the machining cycle. Accordingly, hobbing is a production method of generating forms on parts of a high degree of accuracy. A typical hobbing operation is shown in Fig. 16-38.

16.7.7 Shaving. Shaving is a finishing operation which supplements general machining to obtain a higher degree of finish, improved contour, and greater accuracy of dimensions.

Shaving consists of removing a slight amount of stock and may be applied to either exterior or interior surfaces. It may be performed in several ways, depending upon the design of the part and the nature of the surfaces to be shaved.

Similar to turning, shaving is commonly performed on lathes and automatic screw machines. The stock is revolved and the shaving tool is fed into the work.

Metal stampings are very often forced through shaving dies where accuracy of dimensions, surface finish, and exacting contours are of prime importance.

Possibly the greatest application of independent shaving operations on production parts is in the final machine-finishing of spur or helical gear teeth and of splines on shafts. Basically gear and spline shaving is accomplished by means of a cutter having extremely accurate teeth conforming to the outline of the final gear tooth. Each cutter tooth is gashed or slotted at one or more points along its surface to provide multiple cutting surfaces.

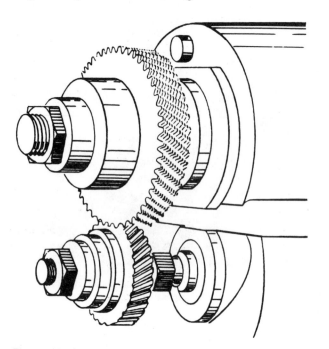

Fig. 16-39. Shaving is a finishing operation used to remove small particles of metal.

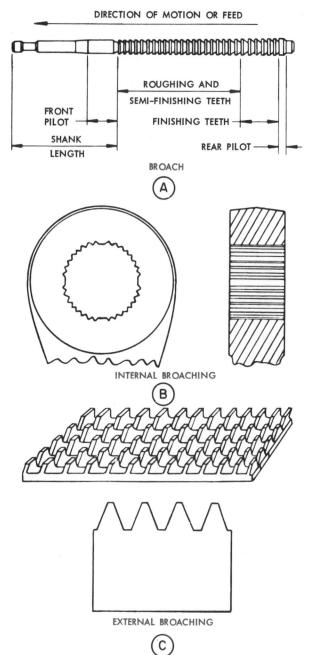

Fig. 16-40. Broaching is used in cutting holes of circular, square, or irregular outlines, keyways, internal gear teeth, splines, and flat or varied external contours.

This type of shaving may be accomplished by various methods which include (1) reciprocation of a rotary-type cutter as it engages and drives the work, (2) a rotary-type cutter in mesh with the work at crossed axes to provide an axial sliding motion, and (3) the movement of the work across a rack-type cutter which reciprocates longitudinally with the work.

Each method results in the removal of minute particles of metal to achieve a high degree of accuracy and finish. A typical example of gear shaving is shown in Fig. 16-39.

16.7.8 Broaching. Broaching is a production method of machining metal parts to a high degree of accuracy. The process employs the use of a machine-operated cutter known as a broach, which passes in a straight path through or over the stationary part to produce internal or external machined surfaces. These surfaces include holes of circular, square, or irregular outline; keyways; internal gear teeth; splines; and flat or varied external contours.

The broach, shown in Fig. 16-40*A,* is provided with several cutting teeth. These teeth are gradu-

ated in size so each tooth removes a small amount of material as the broach is passed through or over the work.

Basically there are two types of broaches: the push type and the pull type. The *push-type broach* is forced through the work, whereas the *pull type* is drawn through or over the work. Progressive broaching operations are sometimes necessary when the amount of stock to be removed exceeds the capacity of a single broach. A typical example of internal broaching is indicated in Fig. 16-40B. An example of an external broach and broached part is shown in Fig.16-40C. The conventional broaching machines are made in two principal types—horizontal and vertical. Both types have one or more rams which actuate the broaches.

16.7.9 Drilling. Drilling is a process of cutting round holes in material with a cutting tool known as a drill. The drilling operation is commonly performed on machines known as drill presses. The drills are held in a rotating spindle and fed into the work, which is supported on the machine table.

Multiple-spindle drill presses have two or more drills which are fed into the work simultaneously as shown in Fig. 16-41.

Drilling is also done on lathes, automatic screw machines, and chucking machines. In these cases the work is supported and rotated in chucking spindles, and the drill is stationary, except for feeding into the work.

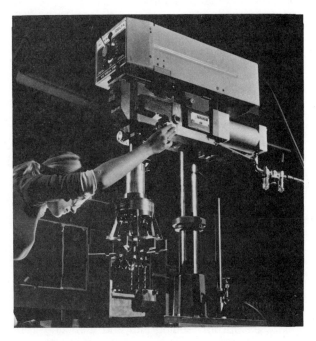

Fig. 16-41. A typical multiple spindle drill press.

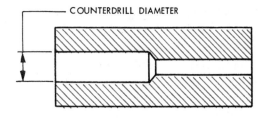

COUNTERDRILL DIAMETER

Fig. 16-42. Counterdrilling is a process of enlarging a portion of a hole.

Drilling can generally be classed as a roughing operation and is usually followed by reaming, boring, grinding, or lapping to effect a finer finish and greater accuracy of hole size. The following are typical operations related to drilling:

Counterdrilling is a drilling operation to enlarge to a given depth a portion of a hole that has previously been drilled.

In counterdrilling, the shoulder formed at the junction of the two diameters is not square with the axis of the hole but takes the conical shape of the drill point. This shoulder is usually unimportant and not intended to be a seat or bearing for another part. A typical example of counterdrilling is shown in Fig. 16-42.

Countersinking is the removal of metal around the edge of a hole with a tool having conical cutting flutes. Its purpose is (1) to provide a seat for conical screw heads and rivets, (2) to provide, for subsequent operations, a seat for supporting the work on centers of machines such as the lathe and milling machine, or (3) to remove burrs and provide a chamfered hole. A typical example of a countersinking tool and the work produced is shown in Fig. 16-43.

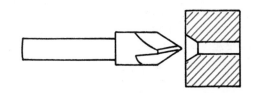

Fig. 16-43. Countersinking is a process of removing metal around the edges of a hole to form a seat for fastening devices such as screws, rivets, and bolts.

Counterboring is enlarging to a given depth a portion of a hole previously drilled or reamed, in order to accommodate a mating part having two or more diameters. The counterbore shoulder is made square with the axis of the hole to provide a seat or bearing surface for the mating part. A typical counterbore and counterboring operation are shown in Fig. 16-44. The operation is similar to drilling and is performed on the same machines.

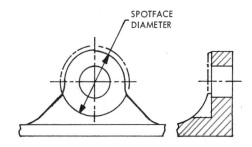

Fig. 16-46. Spotfacing is an operation that forms a seat or bearing for a bolt or nut.

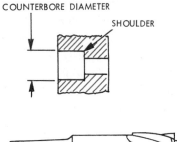

Fig. 16-44. Counterboring is a process of enlarging a portion of a hole to accommodate a mating part having two or more diameters.

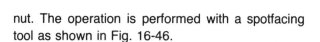

Fig. 16-47. The burnishing tool provides a high-luster finish on a surface.

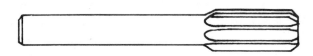

Fig. 16-45. A reamer is used to enlarge a hole for greater accuracy.

Reaming is the enlarging of a hole to obtain a higher degree of finish and accuracy of size. The process is accomplished by the use of a tool known as a reamer, which has several peripheral cutting flutes as shown in Fig. 16-45. The operation may be performed in the same manner as drilling, or it may be manually performed with the assistance of a wrench.

Spotfacing is similar to counterboring except that in spotfacing only a little metal is removed from around the top of the hole to provide a bearing surface for the head of a cap screw, bolt, or

nut. The operation is performed with a spotfacing tool as shown in Fig. 16-46.

Burnishing is a process of finish-sizing, producing an extremely smooth high-luster finish on metal surfaces previously machined. This process displaces, rather than removes, the minute surface irregularities produced with cutting tools. It may be applied to internal or external surfaces.

The burnishing tool, shown in Fig. 16-47, is provided with several annular buttons, graduated in size, so that each button displaces a small amount of material as the tool is passed through the work. External burnishing is accomplished with the use of rolls and pressure. In some instances, where the design of the part will permit, the work is forced through a burnishing die.

Precision boring, facing, and turning are strictly finishing processes whereby a small amount of stock is removed from metal parts to produce smooth, true, machined surfaces to a high degree of accuracy. These operations may be applied to internal, external, or shouldered surfaces.

The process of precision boring is accomplished with a diamond or cemented-carbide-tipped tool bit supported in a boring bar revolved at high speed by the drive spindle. The work usually remains stationary except for feeding into and retracting from the tool. Fig. 16-48 shows a typical boring operation.

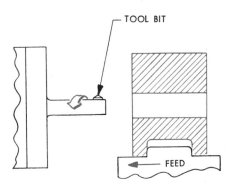

Fig. 16-48. A precision boring operation.

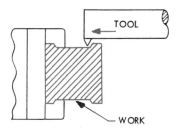

Fig. 16-49. A precision turning and facing operation.

The process of precision turning, facing, and shouldering differs slightly from precision boring. The work is held and rotated by the drive spindle at high speed, while the tool bits, supported in the machine, feed into and retract from the work. A typical turning and facing operation is shown in Fig. 16-49.

16.7.10 *Grinding.* Grinding is a process of removing material by means of a bonded-abrasive wheel mounted on a suitable machine and rotated at high speed. Each abrasive grain on the wheel can be considered a very minute sharp tool. As the wheel revolves, each grain cuts a small chip from the work, which may revolve or move transversely depending on the type of grinding operation to be performed.

Grinding wheels are classified according to abrasive material, grain size, and type of bond, the choice of which depends on the material to be worked and the surface finish required.

Grinding may be a roughing operation, as in the case of snagging; however, it is generally considered a finishing operation when applied to surfaces requiring accuracy and smooth finish. There are various classifications of grinding such as the following:

Honing is another process of removing material from surfaces by means of bonded abrasives. It is generally applied to cylindrical surfaces, although other shapes may be honed with suitable equipment and operating methods.

The tool used on cylinder bores consists of a group of equally spaced abrasive stones supported in a holder. Each contacting abrasive grain cuts a small particle from the work. The tool works with a reciprocating and rotating motion, while the work is stationary. A sectional view of a typical hone having multiple abrasive contact with the walls of a cylinder is shown in Fig. 16-50.

The tool or the work is permitted to float so that the bore and the tool may align themselves, thus maintaining the axial location of the hole. This is a contributing factor to the production of straight and round bores.

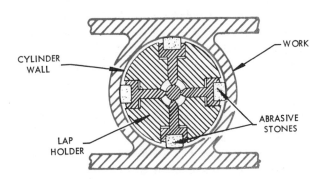

Fig. 16-50. A sectional view of a hone with multiple abrasives.

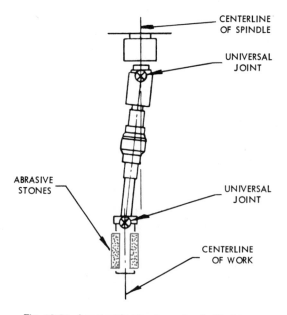

Fig. 16-51. A typical floating type of cylindrical hone.

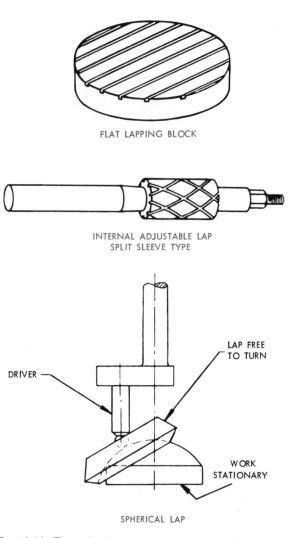

Fig. 16-52. These lapping tools are used in a final surface finishing operation.

A typical floating-type cylindrical hone is shown in Fig. 16-51.

The correction of inaccuracies in a bore sometimes necessitates removal of relatively large amounts of material. The economic range for average honing practice is 0.001 to 0.020 inch on the diameter. On parts honed for finish only, the amount of material necessary for removal of all marks left by previous operations may vary from 0.0002 to 0.001 inch on the diameter.

Honing is usually considered a precision machining operation. However, any or all of the following may be obtained: (1) rapid and economical removal of stock, (2) generation of straight round bores, and (3) any desired surface finish and dimensional accuracy.

Lapping is a process of precision finishing applied to flat, cylindrical, and spherical surfaces. It removes roughness, tool marks, and other defects left from a preceding operation.

A basic feature of lapping is the use of loose abrasive. However, modern lapping machines also use bonded-abrasive wheels and abrasive cloth or paper, all of which have been accepted commercially as lapping mediums.

The abrasive cloth or paper is attached to the lap shoes, as in the case of crankshaft journal lapping. The loose abrasive is usually mixed with a lubricant and applied between the lap and the work. The laps are the tools used for the operation and are of various designs, shapes, and sizes. Three typical designs are shown in Fig. 16-52.

Another feature of lapping is that fresh points of contact are made between the lap and the work through constantly changing relative movements. This feature is important when optical flatness or geometrical accuracy is required on the finished surface.

Lapping may be described as a final stock-removing operation producing surface quality, geometric precision, and dimensional accuracy—

all of which add to the usefulness of the part or increase its wear life.

Polishing is a process sometimes referred to as "flexible grinding." It is used to smooth a surface by the cutting action of abrasive particles bonded to the surface of resilient wheels of wool, felt, leather, canvas, or fabric. Sometimes the abrasive is bonded to belts operating over resilient wheels.

Polishing is generally a progressive operation performed with a set of wheels of different grain sizes, ranging from coarse grit to fine grit, and followed by a buffing operation. Grain progression and the number of grain sizes used depends on the work and the finish desired.

Metal polishing is usually performed for reasons of appearance rather than of accuracy. Polishing is also used on highly stressed machine parts to remove minute surface imperfections or to minimize wear on certain mating parts.

16.7.11 Related machining operations.

Other operations which are sometimes used in manufacturing parts are these:

Planing. A process similar to shaping. It is primarily intended to produce large flat machined surfaces on metals. The work is mounted on a table which reciprocates under a tool that remains stationary except for either horizontal or vertical feed.

Planing is generally identified with heavy-duty machines and is a comparatively slow operation. Therefore, planing cannot be regarded as a high production process.

Slotting. Removes material to provide a relieved longitudinal area, keyway, groove, or opening in metal parts. Slotting may be accomplished in various ways, depending on the type of slot required.

Machines used for slotting operations include shapers, key-seaters, broaching machines, spline mills, and milling machines. Hence, cutting tools for producing slots include tool bits, broaches, milling cutters, and saws.

Sandblasting. Employs the use of sharp dry sand directed against the work by air pressure. Functions of this process include removing surface oxidation; roughening metal surfaces to receive various coatings; removing hardened paint from metal surfaces; and cleaning castings, forgings, and other metal work.

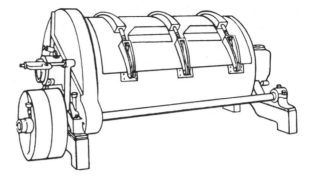

Fig. 16-53. A tumbling machine removes scale, fin projections, burrs, and tool marks from castings and other metal parts.

Grit blasting. Similar to sandblasting except that steel grit, steel shot, or a combination of both, is used instead of sand. The abrasive may be directed against the work by air pressure or by centrifugal force. The primary use of this process is in cleaning castings, forgings, and metal work.

Tumbling. For removing scale, small external fins, projections, burrs, and excessive tool marks from castings and metal parts. The work, together with the tumbling material, is placed in a closed drum, which is then rotated at slow speed. The resulting agitation cleans the work. Fig. 16-53 shows a type of tumbling machine.

Snagging. For removing sprues, gates, risers, headers, fins, and projections from castings, forgings, and raw materials when these small sections are too large for practical removal by tumbling.

Snagging is usually accomplished by rough grinding without precise limits of accuracy. The types of grinders used are dependent upon the size, shape, and weight of the work. These include swing-frame, pedestal, and portable grinders.

Sawing. For cutting, shaping, slitting, or removing material. This operation is performed with hacksaws, circular saws, or band saws. Circular sawing is considered an efficient and economical production method. Band sawing is a machine operation used for continuous rough outline cutting of irregularly shaped contours.

The type of sawing, together with the arrangement of cutting teeth in the saw, is contingent upon the character of the material that is to be processed.

Burning or torch cutting. For cutting or shaping wrought iron, rolled steel, and steel castings by heating to extremely high temperature and simultaneously oxidizing or burning away metal with a hand-operated or mechanically guided torch.

The torch is provided with two jets. One emits a flame combining a mixture of oxygen and acetylene for heating. The other introduces a large quantity of pure oxygen for oxidizing and burning away the metal to produce a rough cut.

Burning is used extensively on heavy work to produce irregular shapes, to cut extremely heavy steel plate or structural sections, and for general repair work.

16.8 Metals

Many different metals are employed for structural purposes and the manufacture of consumer products. Selection of metals is governed by many factors, such as cost, production techniques, the function of the product, the environment in which the unit is to be used, stresses imposed on the assembly, and strength requirements. Some of the more common metals are discussed here.

16.8.1 Carbon steels.
Carbon steels are those which contain iron, carbon, manganese, and silicon. These steels are classified according to their carbon content and are called low-carbon, medium-carbon, and high-carbon steels.

Plain carbon steels are made in three grades: killed, semikilled, and rimmed. A killed steel is one that is deoxidized by adding silicon or aluminum in the furnace ladle or mold to cause it to solidify quietly without evolving gases. Killed steel is homogenous, has a smooth surface, and contains no blowholes. A semikilled steel is only partially deoxidized, while a rimmed steel receives no deoxidizing treatment. Killed steels have great toughness at low temperatures. Rimmed steels are usually considered more suitable for drawing and forming.

Low-carbon steels. These have a carbon range between 0.05 and 0.30 percent. They are sometimes known as mild steels and represent the greatest bulk of steel for commercial fabrication of metal products and structures. Low-carbon steels are tough, ductile, easily machined and formed.

As a rule they do not respond to heat treatment but are readily case-hardened.

Medium-carbon steels. Those with a carbon range of 0.30 to 0.60 percent are classed as medium-carbon steels. Because of their higher carbon content, they can be heat-treated. Steels in this category are strong and hard but not nearly as ductile as the low-carbon types.

High-carbon steels. Steels with a carbon content ranging from 0.60 to 1.7 percent are classed as high-carbon. Those with a carbon range of 0.75 to 1.7 percent are often called very-high-carbon steels. Both groups respond very well to any form of heat treatment.

16.8.2 Alloy steels.
An alloy steel is one which contains one or more elements such as nickel, chromium, manganese, molybdenum, titanium, cobalt, tungsten, and vanadium. The addition of these elements gives steel greater toughness, strength, resistance to wear, and resistance to corrosion. Alloy steels are called by the predominating element that has been added.

Low-alloy steels. Low-alloy steels are those that are more ductile and easier to form than the higher carbon steels. The most common low-alloy steels are those in the 20xx series (nickel), 30xx series (nickel-chromium), and the 40xx series (molybdenum).

Nickel steels with a nickel range of 3 to 5 percent have greater elastic properties than mild steel of comparable strength. The *nickel-chromium* steels have a higher responding range to heat treatment than do straight nickel steels of the same carbon content.

Molybdenum steels are classified into three major groups: Carbon-Moly., Chrome-Moly., and Nickel-Moly. These steels are used where high strength is required at high temperatures.

High-strength low-alloy steels. High-strength low-alloy steels are classified on the basis of their mechanical properties—especially their higher yield point compared with structural carbon steels. They are intended for applications where savings in weight can be achieved because of their greater strength and durability. For example, in certain thickness ranges high-strength low-alloy steels will have a minimum yield point of 45,000 to 65,000 psi as compared with 33,000 to 36,000 psi for carbon steel. The greater mechanical properties of these steels are obtained by a combination of several elements. The blending of these elements produces a steel that is ductile and corrosion-resistant, has high strength and good forming qualities, and is readily welded. Many high-strength low-allow steels are know by such trade names as Corten, Manten, Exten, Dynalloy, and Jalten.

Stainless steels. Stainless steels are divided into general AISI groups 300 and 400. Each series includes several different kinds of steel, each of which has some special characteristic. See the Appendix.

The *300 series* is known as austenitic stainless steels, and their strength cannot be increased by heat treatment. They combine high tensile strength with exceptionally high ductility and possess the highest corrosion resistance of all stainless steels.

The *400 series* is divided into two groups known as ferritic and martensitic. Ferritic stainless steels are about 50 percent stronger than plain carbon steels and, like other austenitic steels, cannot be hardened by heat treatment. They are used considerably for decorative trim and equipment subjected to high pressures and temperatures.

The martensitic stainless steels are readily hardened by heat treatment. These steels are designed for machine parts where creep properties are very critical and high strength, corrosion resistance, and ductility are required.

16.8.3 Aluminum.
Aluminum is a very lightweight non-ferrous metal that is only about one third as heavy as steel but has an extremely high strength-to-weight ratio. It has excellent corrosion-resistance qualities, good electrical and thermal conductivity, and high reflectivity to heat and light. Aluminum has the added property of being easy to fabricate by casting, spinning, drawing, rolling, stamping, forging, machining, and extruding. Commercial aluminum is divided into three groups: pure aluminum, wrought alloys, and casting alloys.

Pure aluminum. Commercially pure aluminum has a purity of at least 99 percent. Since it lacks alloying ingredients, pure aluminum does not have a very high tensile strength. One of the chief qualities of pure aluminum is ductility, which makes the metal especially adaptable for drawing and other forming operations.

Wrought alloys. These are alloys that contain one or more elements such as copper, manganese, magnesium, silicon, chromium, zinc, and nickel. The wrought aluminums are either non-heat-treatable or heat-treatable. The non-heat-treatable are those which cannot be hardened by any form of heat treatment. Their varying degrees of hardness are controlled only by cold-working. The heat-treatable alloys are those in which hardness and strength are further improved by a heat-treating process.

Casting alloys. These alloys are used to produce aluminum castings. The molten metal is poured into sand or permanent metal molds.

Designation of aluminum. Pure aluminum and wrought aluminum alloys are designated by a four-digit index system. See Appendix. The first digit designates aluminum groups. In group I the last two digits indicate the degree of aluminum purity. The last two digits of the alloy groups have no special significance but serve only to identify the different alloys in the group. The second digit in all groups shows modifications in the production of the aluminum. A letter following the group designation indicates basic tempers. The addition of a subsequent digit refers to the specific treatment used to attain this temper condition or degree of hardness.

16.8.4. Magnesium.
Magnesium is a silvery white metal known for its extreme lightness. It weighs approximately one fourth as much as steel and two thirds as much as aluminum. Its resistance to corrosion in normal atmosphere is about equal to that of some of the aluminum alloys.

The two main magnesium groups are wrought alloys and casting alloys. Wrought alloys are in the form of sheet, plate, and extrusions.

Alloy designations for magnesium consist of one or two letters representing the alloying elements, followed by the percentages of the alloy content rounded out to whole numbers. A serial letter often is used after the coding number. Thus, AZ61A indicates 6 percent aluminum and one percent zinc. The letter representations are as follows:

A	aluminum	K	zirconium
C	copper	M	magnesium
E	rare earths	Z	zinc
H	thorium		

16.8.5 Copper. Copper is a soft, tough, and ductile metal that cannot be heat-treated but will harden when cold-worked. Commercially available coppers fall into two groups: oxygen-bearing copper and oxygen-free copper.

Oxygen-bearing copper. This is practically 99.9 percent pure and is considered to be the best conductor of heat and electricity. This copper has a small amount of oxygen in the form of copper oxide uniformly distributed throughout the metal, but this is insufficient to affect its ductility.

Oxygen-free copper. This type of copper contains a small percentage of phosphorus or some other deoxidizer, thereby leaving the metal free of oxygen. The absence of copper oxide gives the metal greater fatigue-resistance qualities and better cold-working properties than oxygen-bearing copper.

16.8.6 Copper alloys. Three copper alloys are in common use: brass, bronze, and babbitt.

Brass. Brass is an alloy of copper and zinc. By varying the amount of zinc, alloys of different characteristics can be obtained. The three most common types of brasses are known as low brasses, high brasses, and alloy brasses.

Low brasses contain 80-90 percent copper and 5-20 percent zinc. As the zinc content increases, the strength, hardness, and ductility also increase. Colors range from red to gold to green-yellows. Low brasses can be cold-worked by any commercial process such as deep drawing, rolling, spinning, and stamping.

High brasses have a copper range of 55-80 percent and a zinc content of 20-45 percent. In general, these brasses are stronger than the low brasses but not as ductile.

Alloy brasses are those which contain other elements, such as tin, manganese, aluminum, phosphorus, antimony, and iron. These additions have a marked effect on the tensile strength, ductility, hardness, and corrosion resistance. Lead is sometimes added to improve machinability.

Bronze. Bronze is an alloy of copper and tin. This metal is used for springs, hardware, bushings, bearings, tubing, gears, and many other products. The tin imparts strength, hardness, durability, and corrosion resistance to copper to a much greater extent than does zinc.

Several common bronzes are better known as commercial bronze, phosphor bronze, and manganese bronze. Aluminum bronze has aluminum instead of tin as the main alloying element, and cadmium bronze is an alloy of copper and a small amount of cadmium. Other bronzes may contain additional elements such as iron, nickel, silver, and beryllium, which produce greater strength and toughness.

Babbitt. Babbitt is an alloy of tin, copper, and antimony. The base metal is either lead or tin. The chief use of babbitt is for bearings which require especially good antifriction properties. One of its outstanding features is its low rate of shrinkage when passing from a molten to a solid state.

16.8.7 Nickel alloys. Nickel-base alloys were developed to obtain a high-strength, high-corrosion-resistant non-ferrous metal. They are often referred to by trade names such as Monel and Inconel. These metals have a high nickel content with varying percentages of copper, iron, manganese, and carbon. In general, they are as ductile as copper, brass, and aluminum and have corrosion-resistant qualities comparable to those of stainless steels.

16.8.8 Titanium. Titanium, a silver-colored metal, has become exceedingly important in the aircraft and missile industries because of its high strength-to-weight ratio. It has a melting point of approximately 3035°F with a density of about three-fifths that of carbon or stainless steels, and one and one-half that of aluminum. Another of its outstanding characteristics is its high resistance to corrosion and pitting in salt or oxidizing acid solutions. Like aluminum and magnesium, titanium depends on the formation of a passive surface film for protection from corrosion.

16.8.9 Zirconium. Zirconium has an atomic structure very much like that of titanium. Its hardness is slightly above that of aluminum alloys but below that of low-alloy steels. It is used primarily where extremely high corrosion resistance is needed. Zirconium is available in both pure and alloy forms. Pure zirconium can be hardened only by cold-working, whereas the alloys can be heat-treated and cold-worked as well.

16.8.10 Beryllium. Beryllium probably has the highest strength-to-weight ratio of any stable metal. It has a density approximating that of magnesium, a high modulus of elasticity, a melting point far higher than the melting points of aluminum and magnesium, and extreme resistance to corrosion. Another of the outstanding properties of beryllium is its ability to achieve and hold very close tolerances.

16.8.11 Cast iron. Cast iron is a term used to describe iron-base materials containing a high percentage of carbon (1.7 to 4.5 percent). There are four principal kinds of cast iron: gray, white, malleable, and nodular.

Gray cast iron. Whenever the silicon content is high and the metal is allowed to cool slowly, gray cast iron results. The carbon separates in the form of graphite. It is this separation of the carbon from the iron that makes gray cast iron brittle. Gray cast iron is used a great deal in making castings for many kinds of machine parts. It can be easily identified by its dark gray porous structure when the piece is broken. The tensile strengths of common grades of gray cast iron run from about 30,000 to 40,000 psi. Some gray cast irons are alloyed with nickel, copper, and chromium to give them higher corrosion resistance and greater strength.

White cast iron. Iron with a low silicon content and in which the carbon has united with the iron (instead of existing in a free state, as in gray cast iron) is called white cast iron. The combining of the carbon with the iron is brought about through a process of rapidly cooling the metal, leaving it very hard and brittle. In fact, it is so hard that it is exceedingly difficult to machine, and special cutting tools or grinders must be used to cut the metal. White cast iron is often used for castings having outer surfaces that must resist a great deal of wear. The structure of white cast iron will disclose a silver-white silky crystalline structure.

Malleable cast iron. This form is actually white cast iron which has been subjected to a long annealing process. The annealing treatment draws out the brittleness from the casting and leaves the metal soft but possessing considerable toughness and strength. The fracture of a piece of malleable cast iron will indicate a white rim and a dark center. Malleable irons have tensile strengths ranging from 40,000 to 100,000 psi.

Nodular iron. Nodular iron has the ductility of malleable iron, the corrosion resistance of gray iron, and a greater tensile strength than gray iron. The tensile strength will range between 60,000 and 120,000 psi, depending on the type of iron. These special qualities in nodular iron are obtained by adding a small amount of magnesium to the iron at the time of melting and by using special annealing techniques. The addition of magnesium and control of the cooling rate cause the graphite to change from a stringer structure to rounded masses in the form of spheroids or nodules. This structural change is the principal reason for the improved properties of nodular iron.

16.9 Steel Classification System

A uniform steel classification system has been adopted by the Society of Automotive Engineers (SAE) and the American Iron and Steel Institute (AISI). Indentification is based on a four- or five-digit code. The first digit indicates the type of steel; thus, 1 designates a carbon steel, 2 a nickel steel, 3 a nickel-chromium steel, and so on.

In the case of simple alloy steels, the second number of the series indicates the approximate percentage of the predominating alloying element. The last two or three digits refer to the carbon content and are expressed in hundredths of one percent. For example, a 2335 steel indicates a nickel steel of about 3 percent nickel and 0.35 percent carbon.

The following are the basic classifying numerals for various steels:

Type of Steel	Series Designation
Carbon steels	1XXX
Plain carbon	10XX
Free machining, resulfurized, rephosphorized	12XX
Free machining, resulfurized, rephosphorized	12XX
Manganese steels	13XX
High manganese carburizing steels	15XX
Nickel steels	2XXX
3.50 percent nickel	23XX
5.00 percent nickel	25XX

Nickel-chromium steels 3XXX
   1.25 percent nickel,
      0.60 percent chromium 31XX
   1.75 percent nickel,
      1.00 percent chromium 32XX
   3.50 percent nickel,
      1.50 percent chromium 33XX
Corrosion and heat resisting steels 30XXX
Molybdenum steels 4XXX
   Carbon-molybdenum 40XX
   Chromium-molybdenum 41XX
   Chromium-nickel-molybdenum 43XX
   Nickel-molybdenum ....46XX and 48XX
Chromium steels 5XXX
   Low chromium 51XX
   Medium chromium 52XXX
   Corrosion and heat resisting 51XXX

Chromium-vanadium steels 6XXX
   Chromium 1.0 percent 61XX
Nickel-chromium-molybdenum ....86XX and 87XX
Manganese-silicon 92XX
Nickel-chromium-molybdenum 93XX
Manganese-nickel-chromium-
   molybdenum 94XX
Nickel-chromium-molybdenum ....97XX and 98XX
Boron (0.0005% boron minimum) ....XXBXX

AISI also uses a prefix to indicate the steel-making process. These prefixes are as follows:

A Open-hearth alloy steel
B Acid Bessemer carbon steel
C Basic open-hearth carbon steel
D Acid open-hearth carbon steel
E Electric furnace steel of both carbon and alloy steels

Fig. 16-54. Some typical injection-molded thermoplastic items.

16.10 Plastics

There are two main families of plastics: thermosetting and thermoplastic. The *thermosetting plastics* will soften only once when exposed to heat. After they set and become hard, no subsequent heating will soften them. *Thermoplastics,* on the other hand, will soften repeatedly whenever heat is applied.

Most plastics are known by trade names or by the principal substance from which they are made. Although a great many plastics are produced, in general, they fall into the following classifications.

16.10.1 Thermoplastic family. Typical thermoplastic products are shown in Fig. 16-54. These products were formed by the injection molding process. Another group of thermoplastic products is shown in Fig. 16-55. These were made by expansion casting and are classed as structural forms.

Acrylics. These plastics are made from methyl methacrylate and are marketed under such trade names as Plexiglas, Lucite, and Lexan. They are noted for their high optical clarity and can be produced in almost any color. They are shatterproof, dimensionally stable over a wide range of temperatures, and unaffected by alkalies and weak acids. They can also withstand prolonged exposure to sunlight and weather. Acrylics are, however, attacked by oxidizing acids such as high octane gasoline, acetone, and lacquer thinner.

Polystyrenes. Polystyrenes are made principally from ethylene and are the lightest of all rigid plastics. Although they have excellent dimensional stability, they tend to be brittle and consequently

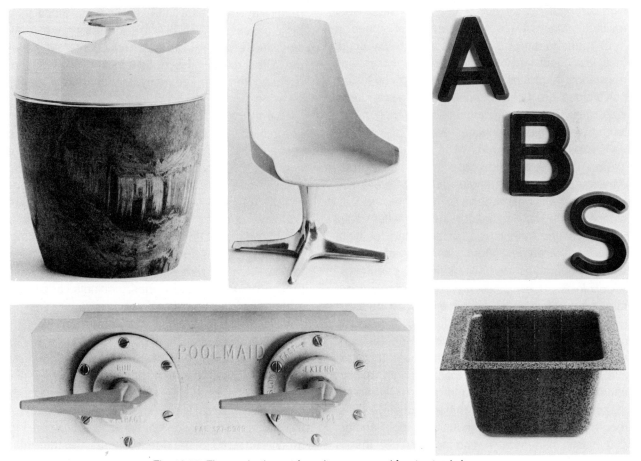

Fig. 16-55. Thermoplastic cast foam items are used for structural shapes.

are not recommended for use where severe impact or flexing is encountered. Most polystyrenes lose practically all their strength above 180°F. They make excellent electrical insulators, have low water absorption, resist acids and alkalies except oxidizing acids, and are non-toxic. As a rule polystyrenes are not suitable for outdoor use since exposure to sunlight causes them to check and to crack.

Polyethylene. This is another ethylene-base plastic which is almost transparent in film form but milky white in thicker sheets. It is quite flexible in thin sections but somewhat rigid in greater thicknesses. Due to its high folding endurance, polyethylene is used extensively for squeeze bottles. In general, its physical properties are much like those of polystyrene. A close relative of polyethylene is polypropylene, which has the same general properties except that it is more heat resistant and possesses greater strength.

Polyamide (nylon). This type has a medium-to-high strength quality but is very tough and wear resistant. Most nylons have greater resistance to heat than do other thermoplastics. Because of their low coefficient of friction, excellent corrosion resistance, high strength, and damping capacities, polyamides are used extensively for gears, bearings, dies, valves, and piping for carrying petroleum and chemical products.

Polyfluoro-hydrocarbons. This group, of which Teflon is an example, consists of fluorinated ethylene propylene materials with smooth oily surfaces that will not adhere to other sticky substances. They are noted for their excellent heat resistance and for this reason are used for high-frequency-current insulators, gaskets, packing materials, diaphragms, and self-lubricating bearings.

Cellulosics. Cellulosics are made from agricultural products such as cotton and wood pulp. Of the several types in the cellulosic group, the cellulose-nitrate type is most common for consumer products. These plastics are tough and flexible, with high tensile and impact strength, exceptional weathering qualities, and good resistance to abrasion. They are used in making adhesive films, recording tapes, electrical insulators, tool handles, and plastic hammers. A commercial fiber that is woven into cloth is known as acetate rayon.

Vinyls. These are made from ethylene gas. There are several varieties of vinyl plastics, some of which are very flexible with average compressive and tensile strength. Most vinyls have little tendency to absorb water, oil, gasoline, food, and ordinary household chemicals. Some vinyls are used for woven fabrics in upholstery and draperies. Vinyls also have wide applications as floor tile, luminous ceilings for showrooms, and as foam materials for cushions.

16.10.2 Thermosetting family. Some applications of thermosetting plastics are shown in Fig. 16-56. The widespread use of these plastics in automotive, appliance, electrical, and other industries is clearly shown.

Ureas. Ureas are translucent materials synthesized from calcium cyanamid (limestone, coke, and nitrogen) and mixed with formaldehyde. Molded urea products are not flammable but will char at about 390°F. Ureas are popular for making dishes because of their high strength. They are readily dyed to many attractive colors. Since they also possess excellent abrasion-resistance qualities, these plastics are ideal for countertop material. Urea resins have good bonding powers and, consequently, are frequently used as adhesives for plywood.

Phenolics. These plastics are produced by combining phenol and formaldehyde. They are probably the most widely used of the thermosetting plastics. Phenolics are dimensionally stable, rigid, heat- and solvent-resistant, and non-conductors of electricity. They have excellent adhesive qualities for bonding plywood and other laminated products. As foamed material, they have wide applications for making buoys and flotation tanks. Phenolics are easily molded into countless products such as telephones, cabinets, appliance handles, and electrical sockets.

Melamines. Melamines are compounds of carbon, hydrogen, and nitrogen similar to the ureas but with a larger and more complex molecular structure. They have a high heat resistance, especially when the resin is used with asbestos or glass fibers as a filler material. A great deal of melamine with cellulose fillers is used in making heavy-duty tableware for hospitals, restaurants, and hotels since these dishes are practically un-

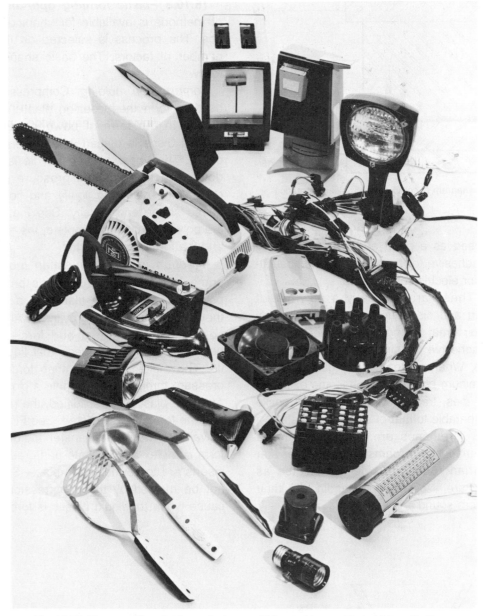

Fig. 16-56. Thermosetting plastic products are widely used in many industrial applications.

breakable and are not affected by soaps and dishwashing compounds.

Polyesters. These varieties are made from anhydride of phthalic acid and ethylene glycol. These plastics possess high-impact strength, good dielectric properties, and are dimensionally stable. They are used primarily as binders for laminates in making boat hulls, chairs, luggage, awnings, etc.

Silicone. This plastic has as its principal ingredients silicon (silica and coke) and methyl chloride. One of its outstanding qualities is the ability to withstand high heat (500°F). The plastic is used in both liquid and solid form. As a liquid, silicone is used as a water repellent on synthetic and natural fibers. It is also a good release agent for molding metal, plastic, glass, and rubber. In addition,

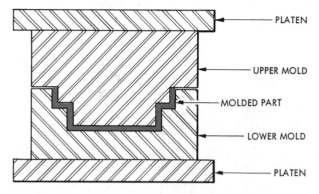

Fig. 16-57. Diagram of compression molding operation.

silicone is used as a paint additive, antifoaming compound, adhesive for tape, and an encapsulating material for electric components.

Epoxies. These are basically a polymer of epichlorhydrin and bisphenol and are particularly noted for their great strength and adhesive qualities. Many adhesive bonding materials are made from epoxies. When reinforced with glass or steel fibers or aluminum powder, they become excellent materials for dies, frequently developing a tensile strength comparable to that of structural steel.

Urethanes. Urethanes are used primarily as foaming material. These foams have a higher tear resistance than rubber but do not stretch as readily. They are especially outstanding for their ability to absorb sound and vibration.

16.10.3 Plastic forming operations.
A variety of methods is available for shaping various plastics. The process is selected on the basis of a number of factors. The basic shaping operations follow:

Compression molding. Compressing molding is used to form thermosetting plastic products. The plastic in the form of powder, pellets, or preformed discs is placed in an open heated mold cavity. The upper part of the die is then brought down and the resulting pressure plus the heat causes the plastic to liquify and flow into the required shape and density. See Fig. 16-57. When the polymerization is complete, the mold is opened and the formed part ejected.

Injection molding. Injection molding is used primarily to form thermosplastic products. Plastic powder or pellets are placed in a hopper where they drop into a heating cylinder. An injection ram compacts the material and forces it against a heated spreader or torpedo that liquifies the plastic. The melted plastic is then forced under high pressure through a nozzle into a closed cold mold. After the plastic has solidified, the mold is opened and the formed part ejected. See Fig. 16-58.

Transfer molding. Transfer molding is used to form complex and delicate parts, especially where inserts are to be molded in place. Such parts cannot be molded by the compression process because when the liquid plastic is forced around the

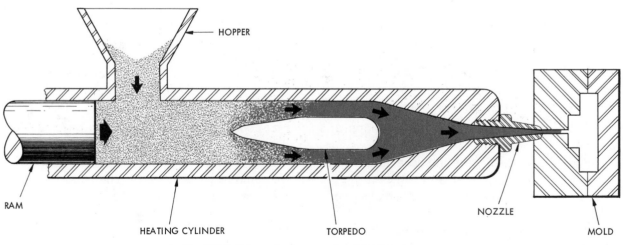

Fig. 16-58. Schematic diagram of an injection process.

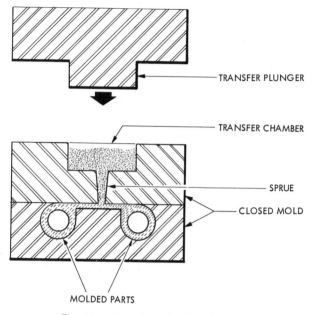

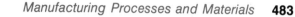

TRANSFER PLUNGER

TRANSFER CHAMBER

SPRUE

CLOSED MOLD

MOLDED PARTS

Fig. 16-59. View through a transfer mold.

mold cavity, the delicate sections may be distorted or the inserts broken. Thus, the transfer technique is designed to eliminate these shortcomings.

In this process the plastic is placed in a transfer chamber above the mold. The plastic is liquified in the chamber and forced into a heated closed mold through an opening called a sprue. From the sprue the liquified plastic flows into the mold cavity. The mold is then opened and the part ejected. See Fig. 16-59.

Extrusion molding. Extrusion molding is used to produce continuous thermoplastic lengths of tubing, rods, film, pipe, and molding. The extrusions are made by first placing plastic powder or pellets into a hopper. From the hopper the plastic drops into a heated cylinder where a rotating screw carries the liquid plastic forward and forces it through a heated die having the required cross-sectional shape. As the formed plastic leaves the die, it is cooled by water or air blasts. See Fig. 16-60.

Casting process. In the casting process liquid plastic—either thermosetting or thermoplastic—is poured without pressure into a closed mold and cured at room temperature or in a low-temperature oven.

Blow molding. This process consists of placing a softened closed-end section of thermoplastic tubing between a split mold. The mold is then closed and air pressure is forced through the open end of the plastic tubing, thereby forcing the plastic to stretch and follow the contour of the mold. See Fig. 16-61 and Fig. 16-62.

Vacuum forming. A thermoplastic sheet is clamped over a mold and the sheet heated until it becomes soft. Air is removed from the mold, creating a vacuum. Atmospheric pressure then forces the softened plastic sheet into the mold. After the plastic is cooled, the formed part is removed and trimmed. See Fig. 16-63.

Pressure forming. In pressure forming, a heated plastic sheet is formed between matched male and female dies. See Fig. 16-64.

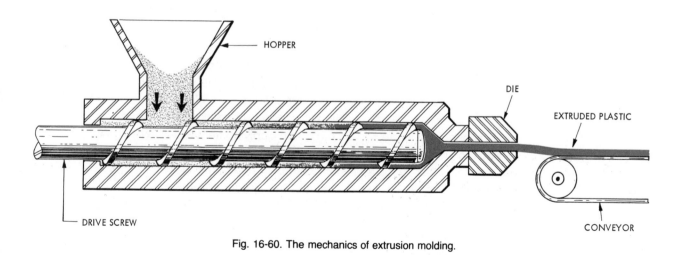

HOPPER

DIE

EXTRUDED PLASTIC

DRIVE SCREW

CONVEYOR

Fig. 16-60. The mechanics of extrusion molding.

Fig. 16-61. Some plastic products formed by blow molding.

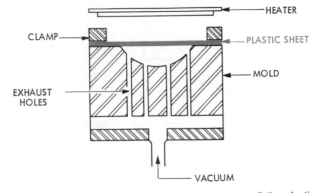

Fig. 16-63. Vacuum forming uses a vacuum to pull the plastic into the mold.

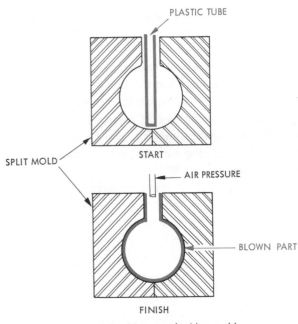

Fig. 16-62. Diagram of a blow mold.

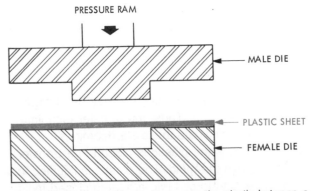

Fig. 16-64. Pressure forming squeezes the plastic between a set of dies.

Foaming. Foaming is a process of expanding particles of plastic to the required density. Plastic is foamed in two ways. In one technique a compressed gas or a volatile material is added to the plastic resin as it is made. When the plastic is heated, the gas expands each individual particle of resin into foam. With the other process the plastic resin is placed in contact with a chemical. The resulting reaction of the resin and chemical liberates a gas that changes the plastic into foam.

Forming foamed products may be done by placing the pre-expanded plastic beads in a closed mold and applying heat to further expand and fuse the beads. Another method is to place the plastic and foaming agent in a closed mold and allow the forming to take place in the mold.

UNIT 17

Product Drawings

In the fabrication of any structure or mechanism, several types of drawings are required. Some drawings provide specific information about how components of a product are to be constructed. See Fig. 17-1. Others illustrate how the parts are to be assembled. These various drawings are commonly referred to as product drawings. The types, functions, and features of product drawings are described in this unit.

Fig. 17-1. Product drawings must provide complete details for the fabrication and assembly of parts.

17.1 Types of Product Drawings

Most product drawings are based on the regular system of orthographic projection, which was described previously. The number of views shown will depend to a large extent on the complexity of the product to be manufactured. Most product drawings fall into five general classifications: design layout, detail, assembly, installation, and display. Within these groups there are special types of drawings that are associated with specific manufacturing processes such as pattern making, casting, forging, machining, stamping, and welding.

17.1.1 Design layout drawing. The development of a new tool, fixture, machine, device, appliance, or other consumer product, or the improvement of an existing structural or mechanical unit, generally starts as an idea in the mind of a designer. This idea is then placed on paper in the form of a freehand sketch. See Fig. 17-2. Once

central computer

master center transmits programs to classrooms located in Mobile Helio-craft via laser beams in computers with satellite relays

STATION 1
STATION 2
STATION 3
STATION 4
STATION 5

PROGRAMED SOFTWARE

CENTRAL COMPUTER

master tracking center is computerized

3D stimuli

laser projector

2D media television

UN 1

hologram

Helio-hover craft

Fig. 17-2. The development of a product usually starts with a freehand sketch.

the basic idea sketch is completed, other sketches may be prepared and calculations made to determine the suitability of the design.

The preliminary sketches are followed by design layout drawings, which are accurately made, often full scale, to produce the complete effects of part proportions and sizes. Some of the design details may be worked out, but no attempt is made to include the full-size description of all the component parts. Only a limited number of critical dimensions are shown. The emphasis is placed primarily on shape, how the parts go together, and how they operate. Section and auxiliary views are used if further clarity is necessary. Notes are included either on the layout itself or on separate sheets to stipulate general specifications such as materials, finishes, heat treatment, clearances, and use of standard parts.

In preparing the design layout, particular attention is given to clearances of moving parts, as-

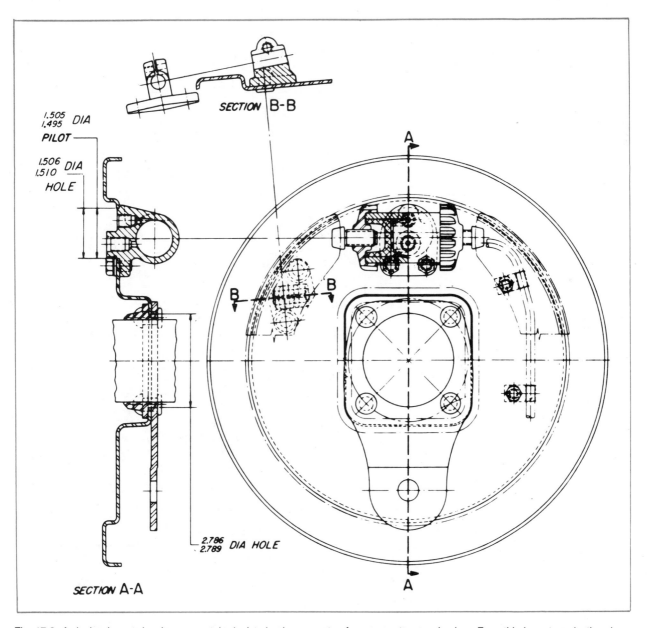

Fig. 17-3. A design layout drawing accurately depicts basic concepts of a new part or mechanism. From this layout production drawings are prepared.

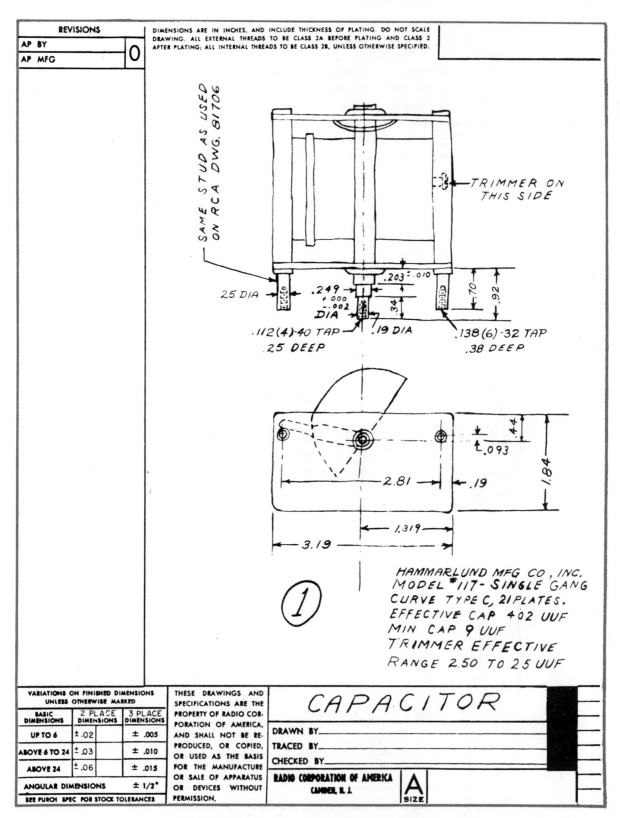

Fig. 17-4. Finished drawings are often prepared from rough freehand sketches.

sembly methods, serviceability of the unit, and functional value in terms of production costs. Fig. 17-3 illustrates a typical design layout drawing.

17.1.2 Detail drawing. After a design layout is approved, drafting personnel proceed to make the necessary production drawings. The first task involves the drawing of each individual part to be produced. These drawings are called detail drawings, and they furnish all of the essential shape and size descriptions as well as the required specifications.

Most industries use what is known as the mono-detail drawing system where a single drawing is prepared for each component part. For small quantity production where it is not feasible to have single detail drawings, the *multi-detail system* is used, which allows for more than one part to be shown on a drawing. Thus both an assembly and details of a product might be placed on one sheet.

In addition to making finished drawings of parts that are "picked-off" from new design layouts, drafting personnel are often required to prepare finished detail and assembly drawings of production components that are being modified or redesigned. Quite frequently they work directly from corrected prints or freehand sketches like the one shown in Fig. 17-4.

A detail drawing may show one or more views, depending on the shape description of the part. Section and auxiliary views are included if needed. The important consideration is to have sufficient views and enough dimensions so the shop workers will understand the shape of the piece and the nature of the work involved.

17.1.3 Pattern shop detail drawing. The pattern shop detail drawing is intended for the patternmaker and is used in making the pattern required to produce a mold for casting. A pattern

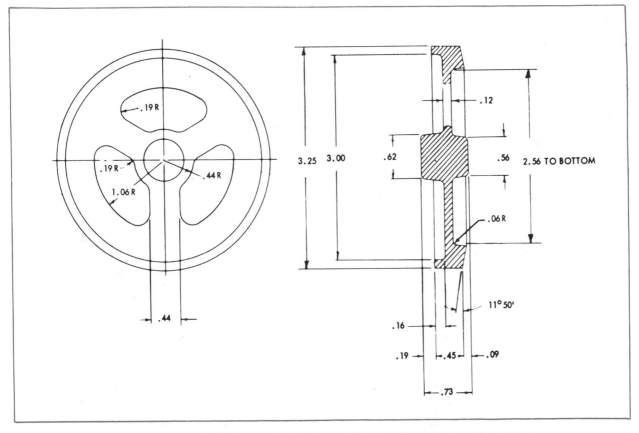

Fig. 17-5. A pattern drawing is a special drawing used by the patternmaker to make a pattern.

drawing locates the parting line of the pattern, provides for the correct amount of shrinkage, and shows how much material is to be allowed for finishing. It also stipulates the allowance for draft, and size of core prints, and the radii of fillets and rounds. See Fig. 17-5 and Unit 16.

On some pattern drawings the various allowances are omitted, in which case the patternmaker makes the necessary calculations. In other instances the engineering department makes the calculations and includes them on the drawing.

17.1.4 Casting drawing. A casting drawing gives the details of a casting. See Fig. 17-6. Two practices are often followed in the preparation of casting drawings. In one method a composite is made showing the rough and finish machined part. The other method calls for separate drawings of the rough and machined casting. Both drawings are often included on a single sheet with a vertical line between them and the rough picture located to the left of the finished part.

A casting drawing gives complete information

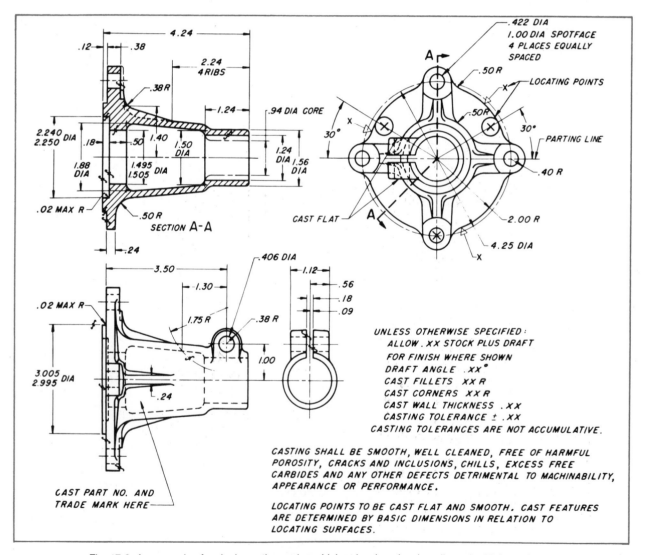

Fig. 17-6. An example of a single casting and machining drawing showing all required information.

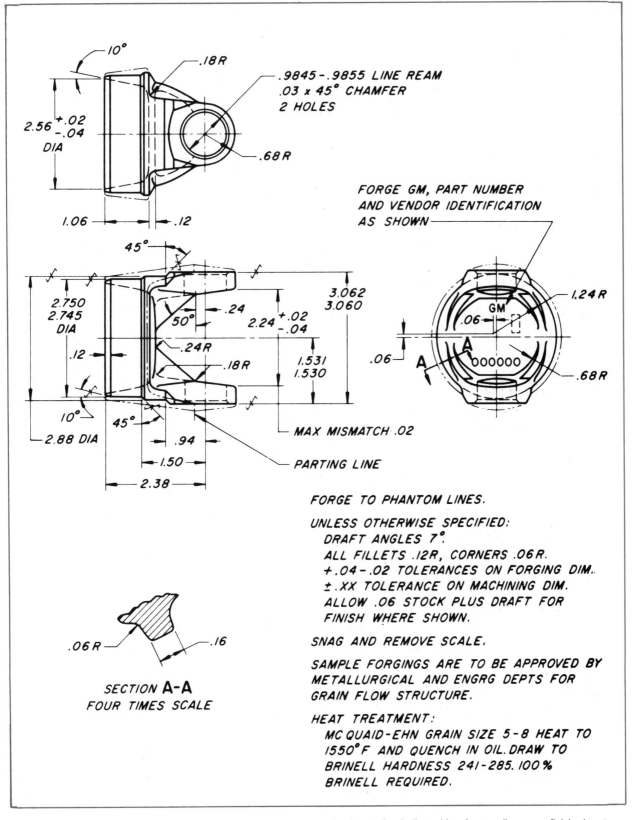

.9845-.9855 LINE REAM
.03 x 45° CHAMFER
2 HOLES

FORGE GM, PART NUMBER
AND VENDOR IDENTIFICATION
AS SHOWN

MAX MISMATCH .02

PARTING LINE

FORGE TO PHANTOM LINES.

UNLESS OTHERWISE SPECIFIED:
  DRAFT ANGLES 7°.
  ALL FILLETS .12R, CORNERS .06R.
  +.04-.02 TOLERANCES ON FORGING DIM..
  ±.XX TOLERANCE ON MACHINING DIM.
  ALLOW .06 STOCK PLUS DRAFT FOR
  FINISH WHERE SHOWN.

SNAG AND REMOVE SCALE.

SAMPLE FORGINGS ARE TO BE APPROVED BY
METALLURGICAL AND ENGRG DEPTS FOR
GRAIN FLOW STRUCTURE.

HEAT TREATMENT:
  MC QUAID-EHN GRAIN SIZE 5-8 HEAT TO
  1550°F AND QUENCH IN OIL. DRAW TO
  BRINELL HARDNESS 241-285. 100%
  BRINELL REQUIRED.

SECTION **A-A**
FOUR TIMES SCALE

Fig. 17-7. Typical single rough forging and machining drawing with forging outline indicated by phantom lines over finished part.

required to produce the casting, including sectional and auxiliary views if necessary. On rough castings dimensions will run to mold points of intersecting surfaces. In addition to regular detail information, a casting drawing will indicate allowances for machining, surface finish, draft angles, parting line, edge and corner radii, fillet radii, heat treating if required, and other essential notes.

17.1.5 Forging detail drawing. The function of a forging drawing is to show the forging operations involved in producing a rough forging. Two systems are used in preparing forging drawings. If the piece is not too complicated, the forging outline is shown in phantom lines over the finished machined part. See Fig. 17-7. When the part is so

complicated that the outline of the rough forging cannot be clearly shown if placed on the finished drawing, a separate rough forging drawing is made. It is often the practice to place both the rough forging drawing and the machining drawing on one sheet with the forging drawing to the left. See Fig. 17-8.

A rough forging drawing will always show some surface of the finished part by phantom lines with a dimension that locates it to the forged surface. If both rough forging and machining drawings are included on one sheet the notations FORGING DRAWING and MACHINING DRAWING are placed directly under their corresponding views. When separate sheets are used, the notation SEE FORGING DRAW-

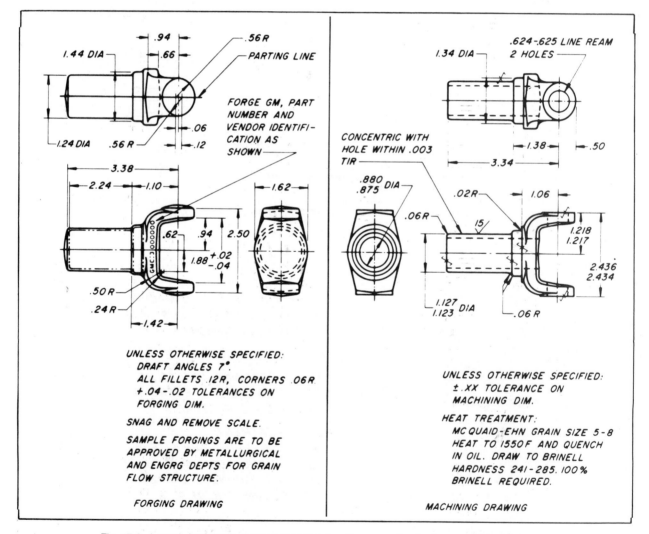

Fig. 17-8. A rough forging drawing may be placed on the same sheet with a machining drawing.

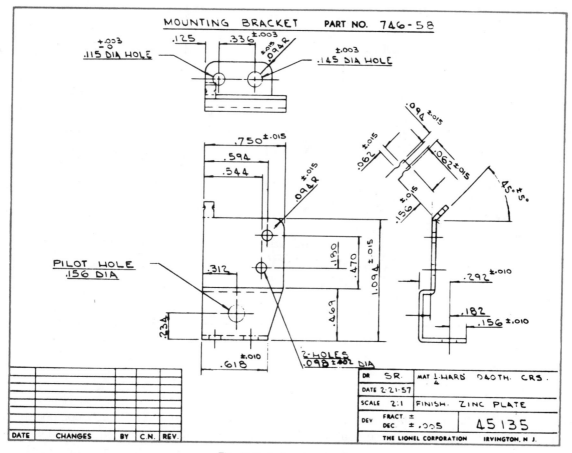

Fig. 17-9. A stamping drawing.

ING is placed on the machining drawing, and the notation SEE MACHINING DRAWING is placed on the forging drawing.

17.1.6 Machining detail drawing. The function of a machining detail drawing is to provide those dimensions, shape descriptions, and specifications involved in machining a casting or forging. The machine work may consist of drilling, reaming, broaching, or countersinking. Surfaces may have to be machined on a planer, shaper, milling machine, lathe, or grinder. Finish symbols are included to designate the kind of surface finish required. Tolerances are shown as well as the type of heat treatment necessary.

17.1.7 Stamping drawing. Stampings are products obtained by shaping sheet metal between members of a die that are under a pressure

movement. Like other types of detail drawings, a stamping drawing must accurately show how the part is to be when fabricated. Some of the basic considerations involved in stamping design and stamping drawings are discussed in Unit 15. A typical stamping drawing is illustrated in Fig. 17-9.

17.1.8 Welding drawing. A welding drawing is one that includes the essential information needed to construct a part when welding operations are involved. Welding information is designated on a drawing by symbols and notes as described in Unit 19 and is shown in Fig. 17-10.

17.1.9 Assembly drawing. An assembly drawing is a graphic presentation showing how two or more detail parts are joined to form a subassembly or complete unit. The drawing "calls out" all of the parts that are required and includes only those

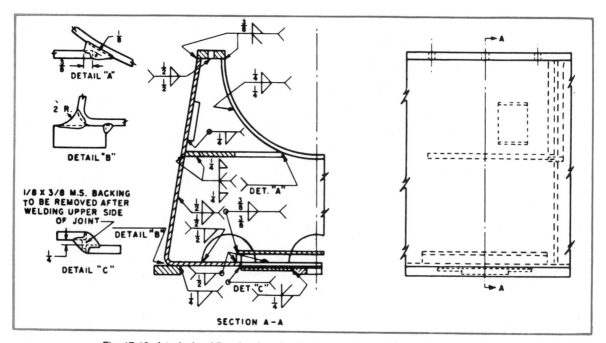

DETAIL "A"

2 R.

DETAIL "B"

1/8 X 3/8 M.S. BACKING
TO BE REMOVED AFTER
WELDING UPPER SIDE
OF JOINT

DETAIL "B"

DETAIL "C"

DET. "A"

DETAIL "B"

DET. "C"

SECTION A-A

Fig. 17-10. A typical welding drawing showing how various welding symbols are used.

dimensions needed to locate the parts with respect to each other, and overall reference sizes. See Fig. 17-11.

An assembly drawing presents only the main outline, primary movements, and relative positions of all parts of a product or product component. Here are several standard practices relating to assembly drawings:

Views. As a rule one or two views are shown—a main view and/or a view in section. Occasionally, when a mechanism is not very complicated, a single sectional view is sufficient. If the unit is symmetrical, a half section will provide the necessary appearance and the relationship of the parts. The assembly is usually completed on one sheet.

Hidden lines. In most instances hidden lines should be avoided on an assembly drawing, since they tend to interfere with the readability of the drawing. They should be included only if they help to identify or clarify unusual details.

Dimensions. Only principal reference dimensions, such as overall height, width, length, essential center distances, and working height, need to be shown. Detail dimensions of individual parts are not included.

Identification of parts. Each component of an assembly should be identified by a leader line with an arrowhead that touches a prominent part of the outline of the component and terminates on the other end with a circle. A circle 0.38 inch in diameter is generally used for this purpose. Each circle, or *balloon* as it is usually called, should contain an item number that identifies the component in the parts list or list of materials. Leaders may be drawn at any convenient angle from the identified part but must not obstruct any views or notes. Balloons should be systematically arranged either in vertical or horizontal rows and not scattered over the sheet. See Fig. 17-11.

While most industries use balloon *callouts* on assembly drawings, it is also common practice to *call out* each part name and number with just a leader, thus avoiding a listing of the parts and supplies on the face of the drawing. In such cases typewritten lists are executed and furnished with the assembly prints. See Unit 3.

To prevent leaders from becoming obscured by lines in an assembly drawing, especially on sectional views, curved leader lines are sometimes used to identify parts. These curved leaders

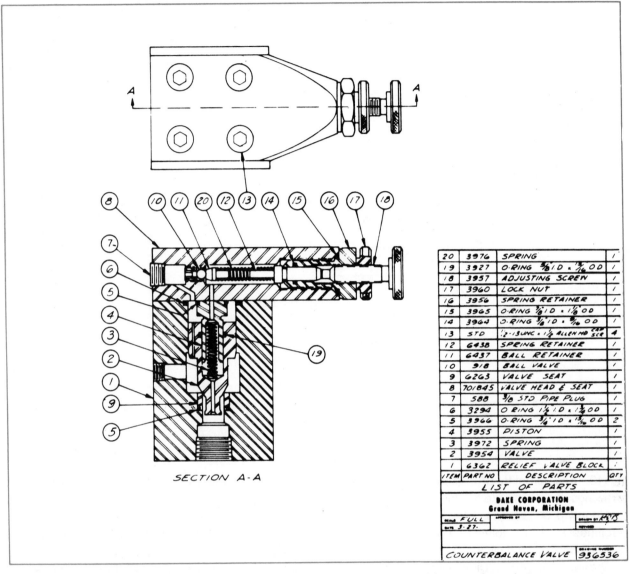

ITEM	PART NO	DESCRIPTION	QTY
20	3976	SPRING	1
19	3927	O-RING 3/8 ID x 7/16 OD	1
18	3957	ADJUSTING SCREW	1
17	3960	LOCK NUT	1
16	3956	SPRING RETAINER	1
15	3965	O-RING 7/8 ID x 1/8 OD	1
14	3964	O-RING 3/8 ID x 9/16 OD	1
13	STD	1/2-13UNC x 1 1/2 ALLEN HD CAP SCR	4
12	6438	SPRING RETAINER	1
11	6437	BALL RETAINER	1
10	918	BALL VALVE	1
9	6263	VALVE SEAT	1
8	701845	VALVE HEAD & SEAT	1
7	588	3/8 STD PIPE PLUG	1
6	3294	O-RING 1 1/2 ID x 1 1/2 OD	1
5	3966	O-RING 3/4 ID x 13/16 OD	2
4	3955	PISTON	1
3	3972	SPRING	1
2	3954	VALVE	1
1	6362	RELIEF VALVE BLOCK	1
ITEM	PART NO	DESCRIPTION	QTY

LIST OF PARTS

DAKE CORPORATION
Grand Haven, Michigan

SCALE FULL — APPROVED BY — DRAWN BY RJD
DATE 3-27- — REVISED

COUNTERBALANCE VALVE | DRAWING NUMBER 936536

SECTION A-A

Fig. 17-11. An assembly drawing shows the outline and arrangement of the parts of a product.

should not blend in with the ordinary cross section line symbols.

While there are several methods of affixing numbers to the parts, two systems are commonly used. Some industries number the parts according to the size of the pieces; that is, the largest piece is labeled 1 with higher numbers for progressively smaller parts. The other practice is to number the parts according to the sequence in which they are handled by the worker in assembling the unit. Another method is to list all purchased parts first and the items fabricated in the plant after.

17.1.10 Detail assembly drawing. Simple mechanisms are sometimes shown on one drawing that includes the assembly arrangement as well as all construction details for the various parts. All dimensions, necessary views, notes, and so on, are presented on this single drawing. See Fig. 17-12.

17.1.11 Installation drawing. An installation drawing is an outline drawing which shows general configuration and complete information required to properly install an item relative to its supporting structure or to other associated items. See Fig.

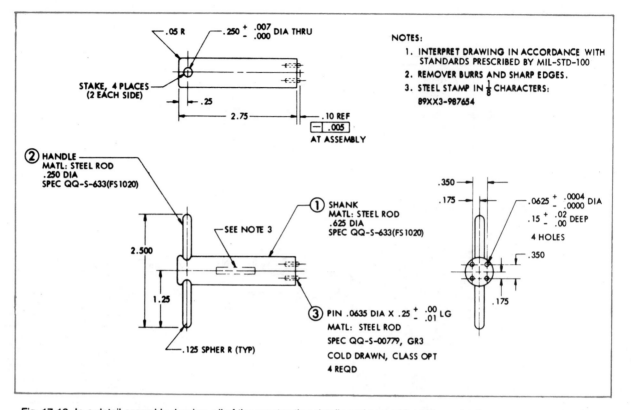

Fig. 17-12. In a detail assembly drawing, all of the construction details and assembly of the parts shown are on one drawing.

17-13. According to SAE standards, installation drawings should include the following information:[1]

Mounting dimensions. Dimensions for mounting facilities such as holes or threads for screws, studs, brackets, and clips.

Outline dimensions. Dimensions to indicate the necessary minimum space in which the product may be installed. They must show the extreme limits of operating travel and contours or surfaces from the mounting accommodation.

Servicing and functional clearance. Necessary minimum clearance past maximum outline dimensions for service removal of such items as covers, plugs, and brushes. Also extreme extended and retracted positions for angle of rotation of external mechanical actuators should be indicated.

Feature information. Any feature information pertinent to installation planning, such as the location and size of keyways and flats in respect to end stops on shafts, rotation of shafts or knobs, and the number of revolutions of the total travel for limited revolving parts. Specifications, operating and static loads, electrical ratings, lubrication, and rigging adjustments should be included where applicable on the drawing.

17.1.12 *Diagram assembly drawing (schematic).* A diagram assembly drawing is one that shows the erection or installation of equipment either in pictorial form or as a flat layout. The drawing is not made to any specific scale, and standard conventional symbols are used to represent various details. Such a drawing may illustrate, for example, the circuit of an electrical unit, a piping layout, or a fuel flow system. See Fig. 17-14. The only dimensions that are included are those that indicate distances between important points and are essential for installation. For more detailed information on this type of drawing, see Units 22, 23, and 24.

---

1. *SAE Aerospace-Automotive Drawing Standards* (New York: Society of Automotive Engineers, Inc.).

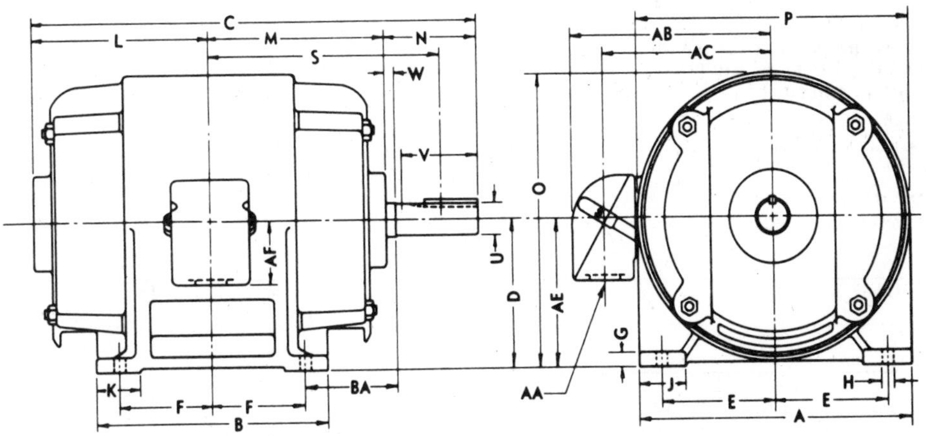

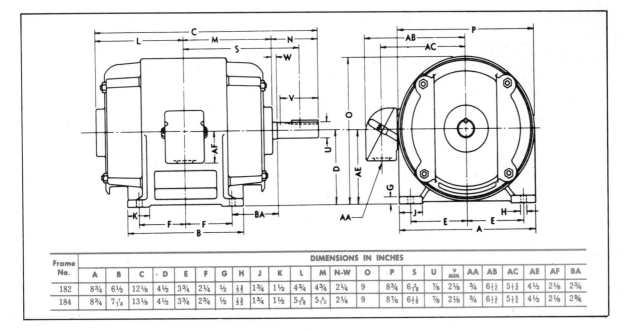

Frame No.	DIMENSIONS IN INCHES																								
	A	B	C	D	E	F	G	H	J	K	L	M	N-W	O	P	S	U	V MIN.	AA	AB	AC	AE	AF	BA
182	8¾	6½	12⅛	4½	3¾	2¼	½	13/32	1¾	1½	4¾	4¾	2¼	9	8¾	6 3/16	⅞	2⅛	¾	6 15/16	5 13/16	4½	2⅛	2¾
184	8¾	7 7/16	13⅛	4½	3¾	2¾	½	13/32	1¾	1½	5 5/16	5 5/16	2¼	9	8⅞	6 11/16	⅞	2⅛	¾	6 13/16	5 13/16	4½	2⅛	2¾

Fig. 17-13. Installation drawing.

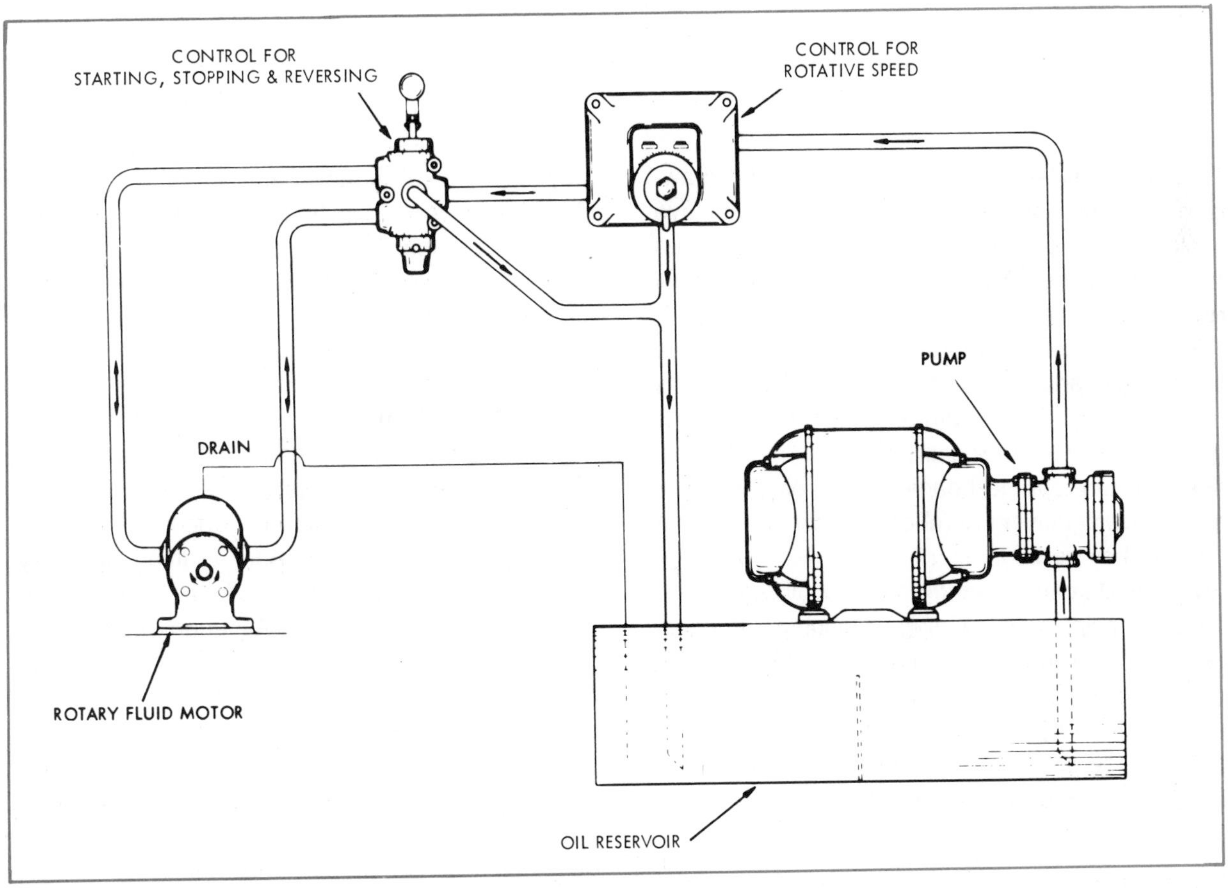

Fig. 17-14. Diagram assembly drawing showing a hydraulic circuit for a variable-horsepower type, variable-speed rotary drive.

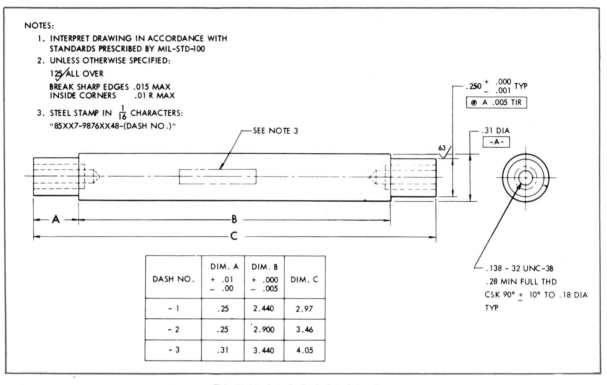

NOTES:
1. INTERPRET DRAWING IN ACCORDANCE WITH STANDARDS PRESCRIBED BY MIL-STD-100
2. UNLESS OTHERWISE SPECIFIED:
   125 ALL OVER
   BREAK SHARP EDGES .015 MAX
   INSIDE CORNERS   .01 R MAX
3. STEEL STAMP IN $\frac{1}{16}$ CHARACTERS:
   "85XX7-9876XX48-(DASH NO.)"

SEE NOTE 3

.250 $^{+\ .000}_{-\ .001}$ TYP

Ⓜ A .005 TIR

.31 DIA

-A-

.138 - 32 UNC-3B
.28 MIN FULL THD
CSK 90° $\pm$ 10° TO .18 DIA
TYP

DASH NO.	DIM. A + .01 − .00	DIM. B + .000 − .005	DIM. C
- 1	.25	2.440	2.97
- 2	.25	2.900	3.46
- 3	.31	3.440	4.05

Fig. 17-15. A typical tabulated drawing.

17.1.13 Tabulated drawing.[2] A tabulated drawing depicts similar items which as a group have constant and variable characteristics. A tabulated drawing precludes the preparation of an individual drawing for each item. See Fig. 17-15.

Normally a representation of a single item is shown with variable dimensions coded by means of letters used as headings for columns in the tabulation. The variables are entered in the table under the appropriate headings and on the same line as the identifying number or letter of the item to which they pertain.

17.1.14 Display drawing. The display drawing is used principally for catalog or display purposes. It shows the actual shape of each part of a structure, often with all of the pieces placed in their proper assembly position. See Fig. 17-16. The parts may be shaded to provide three-dimensional effects. Occasionally, different colors are em-

ployed to make certain features stand out. Each piece is properly identified, and descriptive notes are often added. Display drawings are prepared for the benefit of people who lack understanding of multiview drawings.

17.2 Drawing Controls

Before a drawing evolves into a finished print released for the manufacture of a product, it is subjected to a number of controls. These controls are designed to implement greater drafting accuracy, minimal preparation time, and more effective procedural drawing processes. Although departmental practices may vary from one industry to another, the control system will usually involve such things as progress reports, the checking of drawings, identification and classification of drawings, and procedures for revising and inactivating drawings.

17.2.1 Routing of drawings. The process of developing a new product from its design stages to final production involves many highly trained

---

2. *MIL-STD-100* (Washington: U.S. Department of Defense).

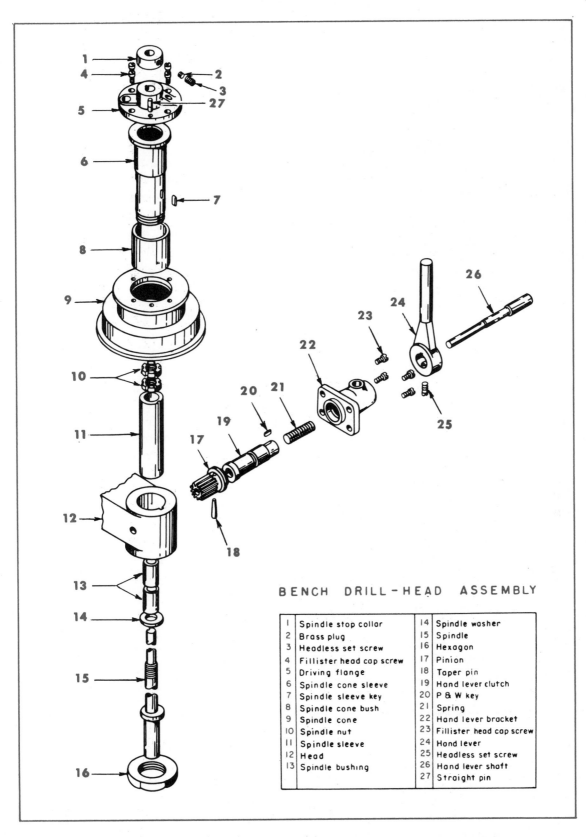

BENCH DRILL-HEAD ASSEMBLY

1	Spindle stop collar	14	Spindle washer
2	Brass plug	15	Spindle
3	Headless set screw	16	Hexagon
4	Fillister head cap screw	17	Pinion
5	Driving flange	18	Taper pin
6	Spindle cone sleeve	19	Hand lever clutch
7	Spindle sleeve key	20	P & W key
8	Spindle cone bush	21	Spring
9	Spindle cone	22	Hand lever bracket
10	Spindle nut	23	Fillister head cap screw
11	Spindle sleeve	24	Hand lever
12	Head	25	Headless set screw
13	Spindle bushing	26	Hand lever shaft
		27	Straight pin

Fig. 17-16. A display assembly drawing is used primarily for display or catalog purposes. (ANSI Y14.4)

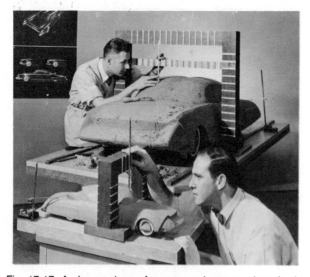

Fig. 17-17. A clay mock-up of a new car is prepared to check all the visual features of a new design.

specialists and countless hours of work. As mentioned previously, creation of a new design usually starts in the design office where experienced engineers and designers make preliminary sketches of the proposed product. The sketches are then turned over to the drafting department for preparation of preliminary layouts. From these layouts, mock-ups or a prototype of the product may be constructed. See Fig. 17-17. The prototype is then tested and necessary design changes are made.

After the changes have been incorporated and the product design has been accepted, the drafting department is called upon to prepare detailed production drawings. The manager of the drafting department, with the help of schedulers, analysts, and supervisors, breaks down the work into small blocks until each involves an individual job. These jobs are assigned to drafting personnel who work directly under a leader and prepare the required set of finished drawings.

As each drawing is completed, it is printed, and a check print is passed on to a checker for its first accuracy check. During this stage the need for certain changes may become evident—changes which are not necessarily due to inaccuracies by the person who prepared the original drawing but to design refinements in other parts or assemblies of the product. Drafting personnel take on their

greatest responsibility in bringing the drawings into final form because "checker changes" must be made on the original drawings. When these changes are completed, a second check print is sent to the checker for a final check. If the drawing is accepted by the checker, it is approved by the project engineer and is released for production.

17.2.2 Work orders, progress charts, and work logs. When a new task is undertaken, a record of its initiation and progress is maintained. A *work order* is frequently used for this purpose. The work order may contain a description of the entire job, or it may show one of several major divisions of the job. Thus, the task may be broken down into such job categories as electrical, structural, and hydraulic.

Each work order includes the name of drafting personnel assigned to the task, the estimated number of hours for completion, the date the work is scheduled to start, and the date the drawings are to be completed.

In addition to the work order, a *progress chart,* which discloses the status of the total drawing job, is often kept. This progress chart is particularly valuable since it enables the head of the drafting department to ascertain periodically whether the work is proceeding according to the established time schedule.

Some firms require drafting personnel to maintain individual work logs. These are records started at the assignment of each job and include all data concerning decisions, sketches, and calculations pertaining to it.

17.2.3 Drawing numbering systems. A great many drawings are required for the fabrication of any one product. Since most companies are involved in the manufacture of several products, a very large number of drawings will be in progress and in circulation at any given time. So that drawings do not become misplaced or lost, some system of identifying and recording them must be maintained.

Basic numbering system. No uniform system of identifying drawings is followed by all companies; each company has its own identification numbering system. Nevertheless, most numbering systems have certain principles in common. In general, the system consists of a block of numbers or a combination of numbers and letters. These may show sheet size, department or activity controlling the drawing, part number, job number, or other identifiers which a company considers pertinent in identifying a print. Thus, a drawing number may look like this:

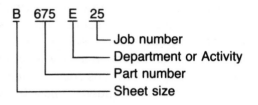

For large projects, a block of numbers such as 3,000,000 through 3,000,999 may be used. In this seven-digit system, 300 may be coded to represent machined, cast, molded, or stamped parts. In some cases the remaining numbers may reflect plating and finishing requirements, or they may indicate whether the part is a simple detail, a sub-assembly, or major assembly.

Dash numbers.[3] The dash numbering system is used to save drawing time by showing the fabrication of more than one part on the same drawing. This eliminates the necessity for separate drawings to depict individual parts, sub-assemblies, or variations on assembly and installation drawings. The name *dash number* is derived from the fact that this number is a suffix added to the basic drawing number and is always preceded by a dash whether or not the basic number is shown. Thus, all parts created in an assembly or installation drawing are assigned the number of the drawing on which they were created, with a separate dash number suffixed to the drawing number for each part, as shown here:

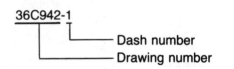

Detail dash numbers are identifying numbers used on assembly and installation drawings to identify individual parts detailed thereon, and on tabulated detail drawings to identify non-

interchangeable variations. In all cases the drawing number for a part assembly or installation is suffixed by odd dash numbers for left-hand components and even dash numbers for opposite, or right-hand, components.

Dash numbers are also used on drawings of parts identical except for one size. For example, a special screw that is manufactured in different lengths would be listed as one part number with a dash number assigned for each different length.

Dash numbers, with the exception of major assembly dash numbers, must appear on the drawing in a circle with a leader line running from the circle to the outline of the part. The title of the dash number is placed adjacent to the circle. See Fig. 17-18.

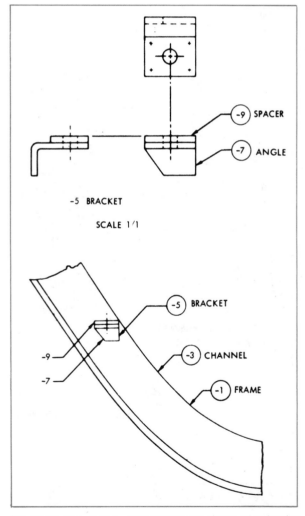

Fig. 17-18. How dash number callouts are shown on a drawing.

3. *Drafting and Design Standards* (Detroit: Chrysler Corp.).

17.2.4 Standard parts. In the design and manufacture of any product, the practice is to use as many standard parts as are available. It is not uncommon for a contracting firm to specify to a manufacturer that such parts are to be included in an assembly. The use of these parts insures a greater degree of interchangeability, faster production, more effective replacement service, and the utilization of components that meet acceptable tested specifications. It also eliminates the additional costs of preparing many drawings that are not needed.

Most standard parts have been certified by approving agencies. These parts or their controlling manufacturing specifications are identified with symbols designated by the approving agency.

17.2.5 Revision of drawings. After a drawing is released for production, drafting personnel are frequently required to make revisions to correct errors, add or delete information, clarify design, or reflect changes in manufacturing methods. Definite procedures are established by engineering drafting departments for making drawing changes. One paramount rule that is in effect everywhere is that once a drawing has been released, no changes whatsoever are to be undertaken without proper authorization and all revisions must be recorded. Authority for making changes usually is vested only in the engineering department. A drawing revision notification form called Engineering Change Order (ECO) is issued when requested changes are authorized.

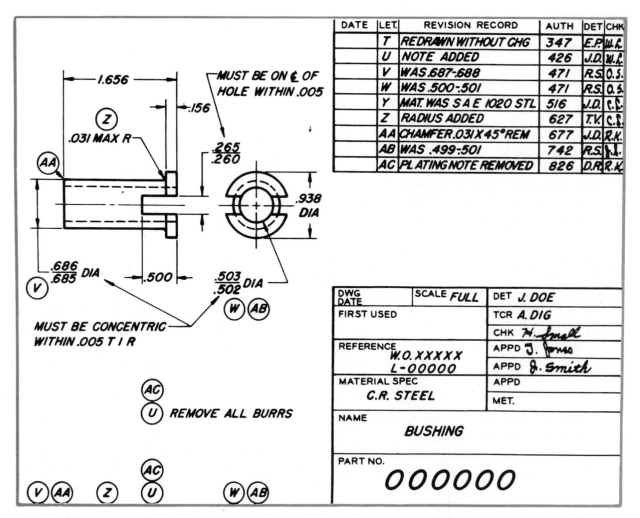

Fig. 17-19. Changes to a drawing are recorded in the revision block and noted by symbols on the drawing.

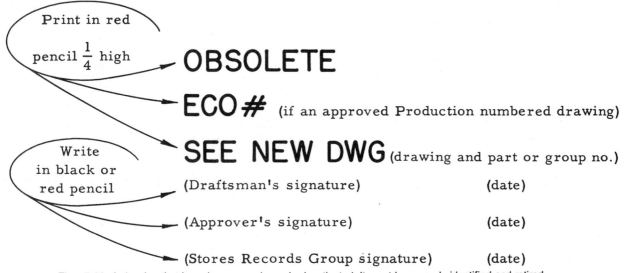

Fig. 17-20. A drawing that is no longer used may be inactivated. It must be properly identified and retired.

Information concerning the revision of a drawing is placed in the revision block. This block is usually located in the upper right-hand corner as shown in Fig. 17-19. It contains the necessary spaces to record the change symbols, dates, signatures, and descriptions of the changes. Drafting departments have their own particular revision symbols to designate drawing changes.

17.2.6 Inactivation of drawings. A drawing that is no longer used in production or service may be inactivated, or backfiled. A prescribed routine is followed to retire such a drawing. Inactivated drawings usually fall into three groups: (1) drawings that have been redrawn, with or without changes, (2) drawings that have been drawn but not released, and (3) drawings that have been replaced by other drawings.

Details of the procedure followed for inactivating a drawing will vary to some extent with different firms. Generally the first step is the issuance of an order requesting that a particular numbered drawing be inactivated. Authorization for inactivating drawings usually rests with the chief engineer.

A typical example of how an inactivated drawing is labeled is shown in Fig. 17-20. As illustrated, a note is placed on the face of the drawing above or to the left of the title block but in such a manner that it does not obstruct any views or other notes.

After the note is properly signed, the drawing is delivered to the record storage office.

17.2.7 Checking drawings. Drafting departments retain one or more experienced people known as *checkers* on their staffs. The principal job of a checker is to verify and examine all finished drawings to insure accuracy, completeness, and efficiency. Since the checker must certify the correctness of the drawings before they are released, the importance of this job is obvious.

To function effectively, a checker must be familiar with manufacturing practices and must have a thorough knowledge of drafting and drafting standards. Checkers are required not only to check those dimensions and views which are shown on a drawing but also to be constantly on the alert for omissions due to carelessness or lack of experience on the part of a detailer.

In most instances a checker makes corrections on a print of the drawing rather than on the drawing itself. Using a print avoids defacing the drawing and provides a clear record of what still needs to be done.

Checking routine. In checking a drawing, a checker uses the following questions as a guide.[4]

---

4. *Ibid.*

1. Does the general appearance of the drawing conform to drafting and design standards?

2. Is the part sufficiently strong and suitable for the function it has to perform?

3. Does the drawing represent the most economical method of manufacture?

4. Are all the necessary views and sections shown, and are they in proper relation to one another?

5. Are all necessary dimensions shown?

6. Do the dimensions agree with the layout and related parts, and are duplicate and unnecessary dimensions avoided?

7. Is the drawing to scale?

8. Is the drawing dimensioned to avoid unnecessary calculations in the shop?

9. Are stationary and operating clearances adequate?

10. Can the part or parts be assembled, disassembled, and serviced by the most economical methods?

11. Are proper limits or tolerances specified to produce the desired fits?

12. Have undesirable limit accumulations been avoided?

13. Are proper draft angles, fillets, and corner radii specified?

14. Are all necessary symbols for finishing, grinding, and so on shown?

15. Are locating points and proper finish allowances provided?

16. Are sufficient notes, including concentricity, parallelism, squareness, flatness, and so on shown?

17. Is the approximate developed length shown?

18. Is the stock size specified?

19. Are material and heat treatment specifications given?

20. Are plating and painting specifications, either for protective or decorative purposes, given?

21. Are trademark, part number, and manufacturer's identification shown according to requirements?

22. Has the title block been filled in with the correct information?

23. Has the date been entered?

24. Are primary and secondary part numbers identical?

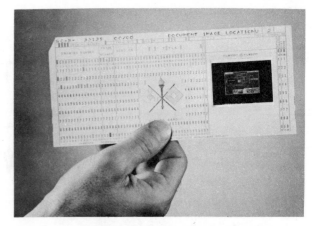

Fig. 17-21. Microfilm aperture cards eliminate hand sorting and filing of reproducibles.

25. Are necessary part numbers of detail parts and sub-assemblies shown on all assembly drawings?

26. Have original lines and drawing information damaged by erasures been properly restored?

27. Are revisions properly recorded?

28. Have all related drawings been revised in order to conform?

17.2.8 Microfilming drawings. To expedite the handling of engineering drawings and to reduce storage space, many industries are turning to microfilm. Some firms have their drawings microfilmed on frames of 35mm, 70mm, and 105mm. The 35mm frames are usually mounted on cards, while the 70mm and 105mm frames are stored in envelopes. The mounting cards are either the conventional 3 x 5 inch file type or punched coded IBM cards. See Fig. 17-21. In either case the microfilm can be used in one of two ways. The frames may be enlarged (called *blow-backs*) and used in the same way as the original tracing, or they may be placed in a table viewer. Blow-backs are made on photographic paper on enlarging machines that expose and develop them without the need for a darkroom. Table viewers or readers which produce an enlarged image on a translucent screen in ordinary room light are available. See Fig. 17-22.

Another microfilm system uses 4 x 6 inch negatives. The negatives can be enlarged to any size

Fig. 17-22. By inserting the aperture card into the viewer, a view of the microfilmed drawing is available.

up to 36 x 54 inches and working prints can be made on opaque paper or cloth by the blueprint or diazo process. The negatives can also be used to make 4 x 6 inch card prints for viewing in a table viewer, or the negatives themselves are projected onto a large screen.

17.3 Functional Drafting

Much of the emphasis in industrial drafting rooms today is on speed. Time saved in the drafting department reflects on the overall productivity of the entire organization. That is why simplification of drafting practices is encouraged by industry. Ultimately design and drafting lead to the proper fabrication and operation of those components which appear on drawings. Graphic art and art presentation are merely means to that end. To the manufacturing or engineering organization, the aesthetic aspects of drawing have little value.

As a means of achieving economies in making engineering drawings, many industries have resorted to practices known as *functional drafting*. In essence, functional drafting is any method of drawing which deviates from conventional drawing standards, yet retains all of the elements of communication necessary to convey functional information but requires a relatively shorter time to prepare.

Although there is no general uniformity among industries as to the extent these simplified practices should be utilized, there are nevertheless certain basic principles which are recognized. Briefly these are as follows:

1. Don't draw it if it can be easily and clearly described. The drawing of many simple parts such as bolts, nuts, washers, fittings, gears, tubing, gaskets, and seals can be eliminated entirely. Instead describe these units either in the list of materials or as a non-picture drawing by word description alone.

2. Eliminate unnecessary views. Time and paper are wasted frequently in drawing unnecessary views.

3. Eliminate repetitive detail. When several holes, slots, gear teeth, and so on are to be drawn in a regular pattern, include only the first two or three items and cover the remaining information with a note.

4. Use symmetry technique whenever applicable. With symmetry it is necessary to draw only one quadrant if both vertical and horizontal center lines divide symmetrical areas.

5. Avoid sectional views wherever an external view would be adequate. The time required to draw a sectional view is usually several times greater than that required for an external view showing the same internal detail with hidden lines. In cases where sectioning is necessary, the view should be broken away so that only the minimum required portion will be shown in cross section.

6. Use ordinate dimensioning when applicable. Using the ordinate system of dimensioning usually simplifies drawing. With this method, dimensions are measured from datum surfaces or center lines, thereby reducing considerably the number of conventional dimension and extension lines, which often clutter up a drawing.

7. Eliminate dashed lines wherever they are not necessary.

8. Do not show unnecessary detail on assembly drawings.

9. Make full use of standard symbols.

10. Make full use of templates.

17.4 Computer Graphics

Automated drafting is a technique of producing drawings by means of computerized drafting machines in order to achieve certain economies in the preparation of engineering documents. Documents often require a great deal of time to prepare when the work has to be done by drafting people. The potentials of computer drafting are being explored by many industries as a means of conserving human resources and reducing the time necessary to produce such documents.

Automated machines have demonstrated remarkable success in high-precision lofting work for aerospace and shipbuilding, as well as in laying out complicated circuits and templates, translating mathematical formulas and equations into drawings, converting financial and statistical data into graphic presentations, and producing perspective drawings and isometric drawings from orthographic projections.

Drawings may be made on vellum, Mylar, or steel loft plates. The Orthomat shown in Fig. 17-23 is one of several automated drafting machines. This machine has a precision table with a drawing area of from 4 by 4 feet to 6 by 20 feet, a traveling beam and carriage which moves the drawing device. Other machines use a drum plotter and servodrive motors to control the movements of the system. The drawing head holds either a single stylus or is equipped with a turret capable of holding several different styli. The stylus may be either a capillary pen, ball point pen, or a metal scriber for use on Mylar or metal. Only one stylus is in operation at any given time; selection can be programmed or made manually.

The first step in automated drawing is the visualization of the desired result. Three key factors must be understood: the specific problem, design for graphic communication, and computer programming.

The next step is to define numerically the subject to be drawn. After this information is encoded on a manuscript, it is keypunched on a tape or cards and fed into a computer. The computer assimilates the information, makes the necessary calculations, and then sends appropriate signals to the drafting machine.

17.5 Drafting for Numerical Control

Numerical control, commonly abbreviated N/C, is a method of automatically controlling the motions of machine tools in performing such operations as turning, grinding, milling, drilling, boring, and tapping, as normally used in manufacturing processes. See Fig. 17-24.

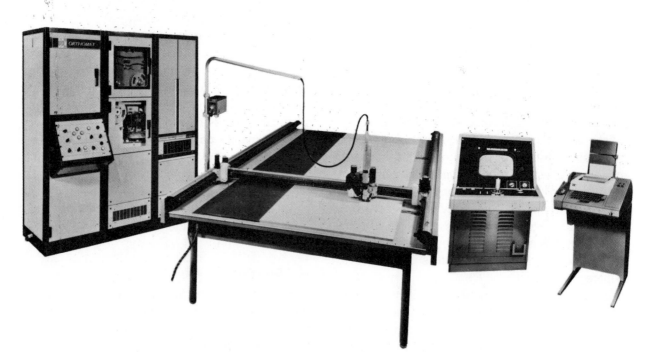

Fig. 17-23. Automated drafting machines speed up the process of preparing some types of drawings.

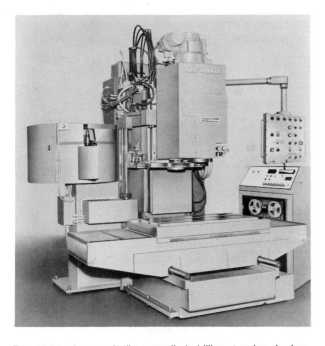

Fig. 17-24. A numerically controlled drilling, tapping, boring, and milling machine.

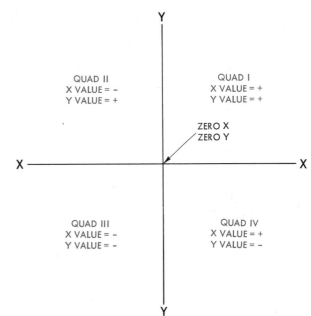

Fig. 17-25. In plane coordinates the X- and Y-axes form four quadrants in which measurements have positive or negative values as shown.

The operation of a given machine is manipulated by a control unit which commands the tool or work table to move in certain directions and through specific predetermined distances. The movements are based on input signals representing numerical values. These numerical values are indicated by linear and angular dimensions taken from a mechanical drawing or print of the part to be made.

17.5.1 Basis of numerical control. The key to numerical control is the system of rectangular, or Cartesian, coordinates. By means of these coordinates, the geometry of a part can be described by the location of points within the framework of the X-, Y-, and Z-axes. See Fig. 17-25. Thus, the true position of a feature can be expressed as a distance plotted from a single zero point along two or three mutually perpendicular axes.

Ordinarily for flat surfaces two-dimensional coordinates are used. By means of three-dimensional coordinates, it is possible to designate depth as well. For example, assume that a print calls for a hole to be drilled in a steel plate 6.250 inches to the right of the left edge, 2.750 inches in

from the front edge, and 0.50 inch deep. The x and y coordinates readily locate the position of the hole while the z coordinate indicates the depth of the hole.

Numerical control systems. The two basic numerical control systems are (1) positioning, or point-to-point, and (2) continuous path, or contouring.

The point-to-point system is the simpler since it merely involves moving the tool or work to the desired position (x and y) and, once that is reached, the vertical motion (Δz) provides the depth control.

Contouring is considered more complicated because position and motion in the X, Y, and Z dimensions must be closely synchronized in order to control the precise path of the cutting tool. The shape of the tool path may be curved or straight, in the form of a parabola, circle, circle arc, straight line, or any combination. However, since the tool itself can move only in a straight line, the straight-line distance change per pass has to be very short to produce a well blended cut that will be continuous and smooth.

Programming for numerical control. Programming for numerical control involves the preparation of a tape which describes the machining operations that are to be performed on the workpiece. The tape becomes the medium that causes the machine control gears, levers, screws, and so on to move. The person who is responsible for programming is usually referred to as the programmer

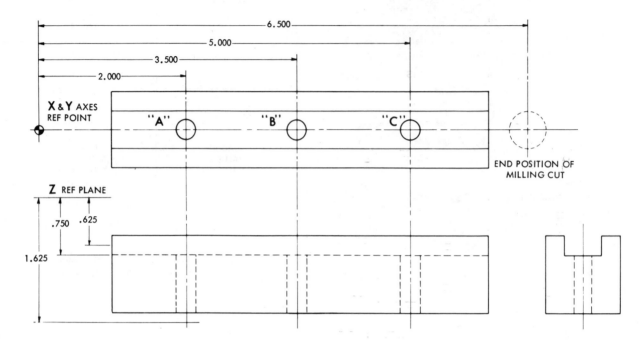

BLK	SEQ. NO. (N)	PREP. FUNC. (G)	X AXIS POSITION (X)	Y AXIS POSITION (Y)	Z AXIS FINAL DEPTH (Z)	SECONDARY FEED PT (Q)	FEED ENGAGE POINT (R)	FEED RATE (f)	SPIND SPEED (S)	TOOL (T)	MISC. FUNC. (M)	REMARKS
1	n001										SC	REWIND STOP CODE
2	n002	g89			z −750		r−625	f14	s78	t01	m03	POSITION FOR MILLING
3	n003		x+ 6500								m08	MILL SLOT
4	n004	g81	x+ 5000		z−875			f06	s38	t02		CENTER DRILL HOLE "C"
5	n005		x+ 3500									CENTER DRILL HOLE "B"
6	n006		x+ 2000									CENTER DRILL HOLE "A"
7	n007				z−1625			f10	s60	t03		DRILL HOLE "A" THRU
8	n008		x+ 3500									DRILL HOLE "B" THRU
9	n009		x+ 5000									DRILL HOLE "C" THRU
10	n010	g90			z+ 000		r+000				m09	RETURN Z TO ZERO
11	n011	g91	x+ 0000								m05	RETURN X-Y TO ZERO
12	n012										m30	REWIND TAPE

Fig. 17-26. A single sheet is often used for complete N/C planning.

or process planner. Basically, programming involves a study of the print or drawing and the conversion of the part's dimensions into coordinate dimensions, selecting appropriate cutting tools, and determining the proper sequence of machining operations. The process planner records this data on a planning sheet. Programming formats are not all alike, since no two industries produce the same design product.

A typical process-planning sheet is shown in Fig. 17-26. The programmer begins by first writing out the particular functions which the machine is to perform. After the analysis is completed, the programmer defines the operations in terms of coordinate values, which are to be used in punching the control tape. Such information is either entered on a separate planning sheet, or both the operation descriptions and their respective coding are included on a single sheet as in Fig. 17-26. Along with coordinate dimensions, auxiliary functions which are to be performed, such as turning on coolant and controlling the feed mechanism, are also recorded.

The actual punching of the control tape is accomplished with any of the commercial tape preparation devices now on the market. These devices often resemble standard electric typewriters and are operated in much the same manner. As the digits, letters, and symbols are typed on the keyboard, the machine translates these into binary coded decimal (BCD) form and punches the appropriate hole or holes in the control tape.

Although several sizes and widths of control tapes have been developed, the trend is to the use of tapes recommended by the Electronic Industries Association (EIA) and the American Standard Code for Information Interchange (ASCII). These tapes are one inch, or eight code holes, in width. Some standardization has also been achieved in defining the type of coding to be used in punching the tape, as well as in designating the symbol or function that is to be assigned to the coding combinations. See Fig. 17-27.

17.5.2 Preparation of drawings for numerical control. The primary responsibility of drafting personnel in numerical control is to prepare a drawing that will enable the process planner to correctly program the machining of a workpiece. In the preparation of such a drawing, the following practices are usually observed:[5]

1. Draw principal views as viewed by the machine cutting tools. They should contain all dimensions required to machine the features included in the view without need to refer to other views. When this is impractical, associated views should be shown as close to the principal view as possible.

2. The decimal-inch system should be used exclusively, since fractions require extensive conversions with attending errors.

3. The X-axis should be drawn parallel to the bottom edge of the drawing. It should be considered as the first and basic reference axis. The Y-axis should be drawn perpendicular to the X-axis in the plane of the drawing. The Z-axis should be drawn perpendicular to the plane of the paper. See Fig. 17-28.

4. Ordinate dimensioning should be used and all dimensions measured from two or three mutually perpendicular datum planes. These datum planes are indicated as zero coordinates, and dimensions from them are shown on extension lines

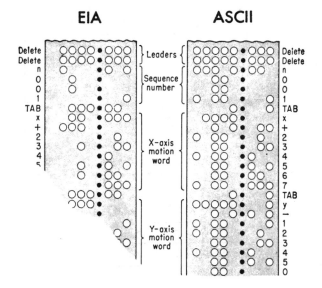

Fig. 17-27. Tapes commonly used for numerical control.

---

5. *G.E. Drafting Manual*—Sec. K4.8 (Schnectady, N.Y.: General Electric Co.).

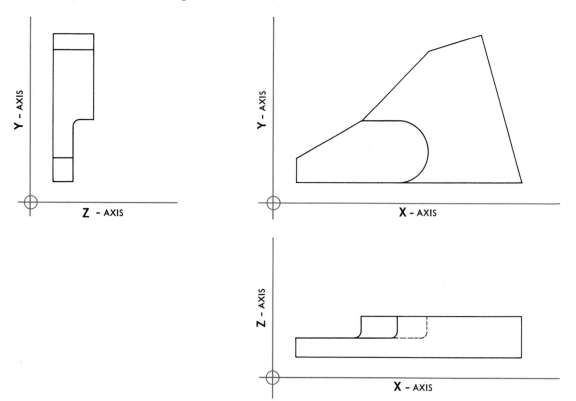

Fig. 17-28. Positions of the coordinate axes for N/C drawings.

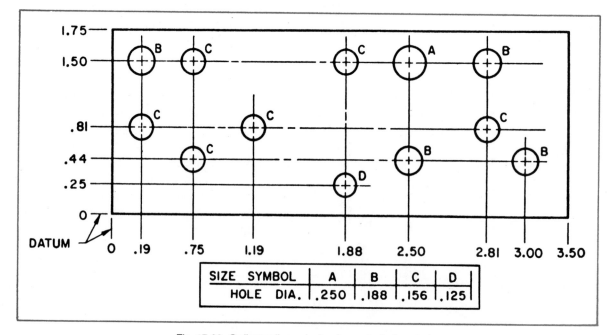

SIZE SYMBOL	A	B	C	D
HOLE DIA.	.250	.188	.156	.125

Fig. 17-29. Ordinate dimensioning for numerical control.

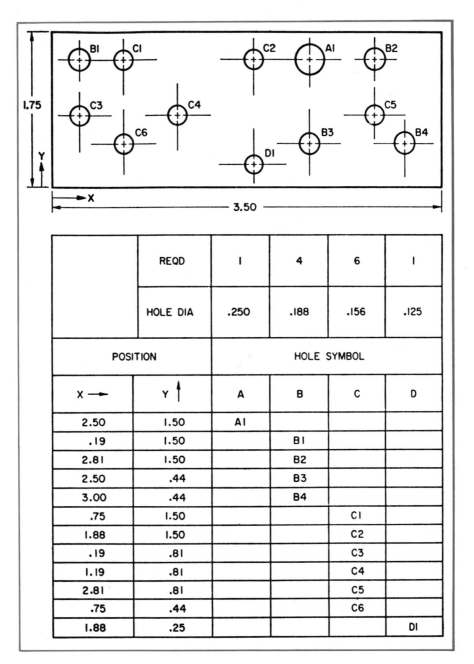

Fig. 17-30. Tabular dimensioning for numerical control.

without the use of dimension lines or arrowheads. See Fig. 17-29.

Datum dimensioning can also be in tabular form in which dimensions from mutually perpendicular datum planes are listed in a table on the drawing instead of on the view. This method may be used on drawings that require the location of a large number of similarly shaped features, as shown in Fig. 17-30.

5. Holes on a circular hole pattern should pref-

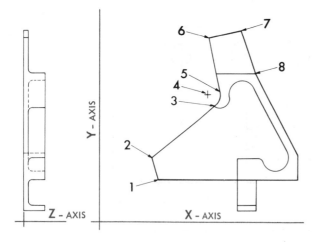

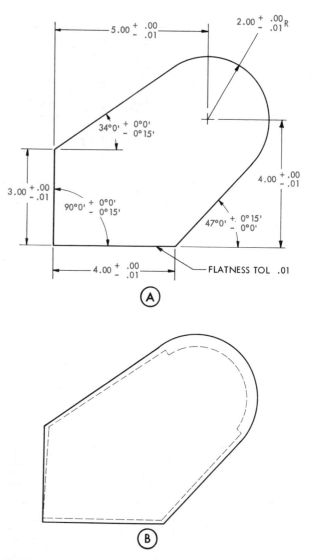

Fig. 17-32. When tolerances are included with dimensions, a programmer may misinterpret their proper function.

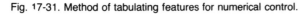

POINT NO.	X AXIS	Y AXIS	Z AXIS	DIM. INFO
1	1.401	1.000	.000	
2	1.250	1.531		
3	2.719	2.656		.500R
4	2.562	3.000		Radius Center
5	2.812	3.125		
6	2.656	4.343		
7	3.437	4.500		
8	3.781	3.500		

Fig. 17-31. Method of tabulating features for numerical control.

erably be located by polar coordinate dimensions. Linear hole patterns should be defined in the conventional way.

6. The location of tangent points and blend radii for changes in contours should be defined and given as reference information.

7. Reference planes used for establishing tabulated dimensions should be clearly marked on the delineation and in the table. The contour departures should also be identified as in Fig. 17-31.

8. For numerical control, the rule that every dimension must have a tolerance is not applicable. When the part programmer reads a drawing before preparing a process sheet, all sizes must be considered absolute. If tolerances are applied to

dimensions as shown in Fig. 17-32A, the programmer may visualize the double profile shown in Fig. 17-32B. The solid lines denote the part at maximum material condition, as reflected by the dimensions, and the dotted lines denote the part at minimum material condition, resulting from one interpretation of the tolerances. Hence, in many instances it may be impossible to determine a mean profile of a geometric shape when tolerances are associated with dimensions.

The separation of tolerances from dimensions is necessary in numerical control. This trend is evi-

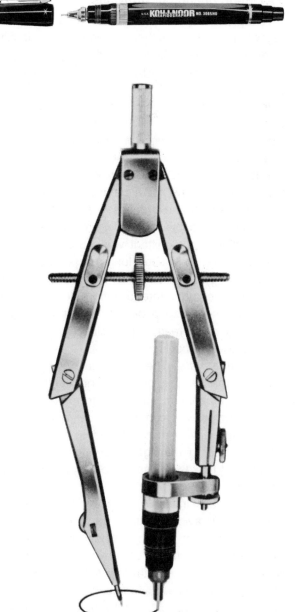

dent on all drawings which use true position dimensioning. The combination of a pictorial delineation and dimensions on the drawing is intended to define the geometry and size of a part in perfection. Dimensions provide a refined definition of specific size and location for elements of the pictorial delineation. Tolerances, or permitted deviations from perfection, apply to the physical surfaces of the part, not to the dimensions shown on the drawing.

17.6 Inking

The original working drawing is never sent to the shop. Instead, reproductions—either blueprints or whiteprints—are made and distributed to the various departments involved in the fabrication of the product. In most instances the original drawings are prepared in pencil. Rarely is ink used. The only exception is when the original must be preserved for a long time and is not subjected to frequent modifications. However, inking is often used in making drawings or sketches for illustration purposes.

Preparation of an ink tracing is accomplished by placing a tracing medium, such as vellum or drafting films, over the original penciled drawing and tracing it in ink. Sometimes the penciling is done directly on the tracing medium and then inked.

Rapidraw System. Drafting personnel now use what is known as the Rapidraw System because it eliminates all pencil work since the drawing is made directly with the technical fountain pen. The advantages of rapidraw range from the simplification of drafting techniques to savings based on perfect reproductions which eliminate the need for costly redrawings and restorations.

Rapidraw is especially suited for industrial drafting requiring precision line work for microfilming. It has wide application in product drafting, electronic schematics, and circuitry, marine design and architecture, construction detailing, cartography, and topographic profiling.

Rapidraw is accomplished with a specially designed fountain pen, shown in Fig. 17-33. This pen produces inked lines of uniform density and constant width much faster and with greater ease than does the ruling pen. The offset tip of the pen removes the danger of smearing ink lines, which is

Fig. 17-33. Special pen for rapidraw.

always a possibility with the ordinary ruling pen. Ink on film can be easily removed by means of a dampened vinyl eraser.

The pen can be used with the same ease as a pencil. Its large-capacity ink reservoir makes it possible to ink for extended periods without the time-consuming constant refilling necessary with the ruling pen.

17.7 Reproduction of Drawings

A reproduction is a copy of printed, written, or drawn material, in either a negative or positive form, to any scale—larger, smaller, or the same as the original—and done by any of several different processes. See Fig. 17-34.

In general the graphic reproduction processes used today can be divided into three major classifications: the diazo process, the silver process, and the electrostatic process.[6]

Each process has its advantages and disadvantages. Each is best for a particular application or type of work. Some processes are more permanent and less subject to soiling; others are especially suited to intermittent work. Some processes are better for continuous long-run production; still others adapt themselves to shifts in production schedules and, therefore, are more flexible.

17.7.1 The diazo process. This process produces a positive reading print directly, giving dark lines on a white background.

Dry whiteprints. The whiteprint, or dry, diazo process makes use of the principle that light-sensitive diazonium salts will form an azo dye when coupled with certain coal tar derivatives in an alkaline environment, providing the salts are not exposed to the light. Varying the azo dye component makes it possible to develop practically any color in the spectrum.

The process involves exposing the sensitized material to ultraviolet (actinic) light through a translucent material or drawing whose image prevents exposure in specified areas. See Fig. 17-35. The light decomposes the diazo in the exposed areas making it incapable of compounding into an azo dye. Alkalizing the coupler by means of an ammonia atmosphere then develops the color in the unexposed areas. This is also known as the ammonia process.

Dry sepiaprints. This process is identical to the dry whiteprint process, except that the dye color formed is a deep yellow-brown. The process is used primarily to make master copies of an original on translucent paper or other material.

Fig. 17-34. A reproduction is a copy of an original document.

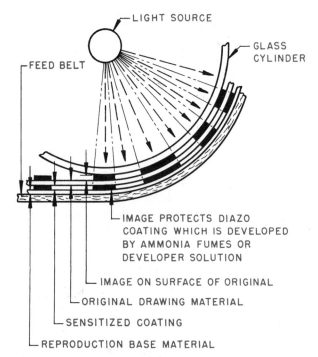

Fig. 17-35. The diazo process produces a positive print having dark lines on a white background.

Wet whiteprints. This type of diazo print is basically identical to the dry. The only difference is that the coupler is applied to the diazo as a wet solution after the diazo has been exposed to ultraviolet light. The resultant prints or reproductions are damp and have to be dried before use.

---

6. Frederick Post Co.

Wet sepia prints. These are identical to the whiteprints, except that a different coupler is used, so that the resultant positive print has deep yellow-brown lines. The base stock used in this case is usually a translucent material.

17.7.2 *The silver process.* The oldest process of reproduction is that which uses the light-sensitive properties of the silver haloids and silver nitrate compounds. This is primarily the field of photography—a process that has so expanded and become so diversified, that the modifications when used for graphic reproductions many times overshadow the basic process. Practically any base material can be coated with the sensitized solution. The process lends itself easily to reduction and enlargement of graphic illustrations and is used basically for the development of permanent reproductions, the rejuvenation of old graphic illustrations, and the development of translucent master copies from opaque originals.

Some of the sensitized coatings have to be handled under darkroom conditions, whereas others can be handled in subdued room light. Some of the sensitized materials are exposed when in direct contact with the graphic subject; others have this subject projected on them from a distance. Certain sensitized materials create a positive print from a positive original; others result in a negative print from a positive original. A number of sensitized materials are exposed by having the light pass through the original to the sensitized coating, while others are exposed by having the light pass through the sensitized coating first and bounce back from the original to the sensitized coating. Several sensitized coatings re-quire the light to pass through a yellow filter; others do not.

However, in every case the development after exposure is by standard photographic materials, water, hypo, and fixer. The reproductions must be dried before use.

The trade names under which the silver process is known are Microfilming, Autopositives, Autofax, Refax, Lithoprint, Reflex, Photocopy, Photostat, Photocontact, Rectigraph, Dupro, Cronaflex, and many others.

17.7.3 *The electrostatic process.* In this process the base material, either metal or paper, is coated with zinc oxide in an insulating resin binder. The zinc oxide is then given an electrostatic charge that is dissipated when exposed to ultraviolet light. When a graphic illustration shadows this light, either by projection or by being in direct contact with the coating, the electrostatic charge of all of the zinc oxide particles is eliminated except in the shadowed areas, and an invisible latent electrostatic image of the graphic illustration remains on the base material.

The base material is then dusted with an opposite electrostatically charged black powder that adheres to the base material only in those areas still containing charged zinc oxide, just as the north pole of a magnet adheres to the south pole of another magnet. This image is then transferred to a translucent material or offset plate and made permanent by heat fusion. If the original base material is translucent, no transfer is necessary prior to the fusion by heat.

Xerox is a trade name under which this type of process is known.

UNIT 18

Threads, Fasteners, and Springs

Structures and machine parts are held together by threaded or non-threaded fasteners or both. Their selection is governed by such factors as function of the assembly, corrosive conditions, temperature, vibration, fatigue stresses, electrical conductivity, life expectancy, and cost. Regardless of what is used, fasteners must be properly identified and specified on a drawing.

18.1 Thread Terms[1]

Threads are used for three basic purposes: fastening, adjusting, and transmitting power. Since threads perform very important functions in numerous fabricating processes, a study of the terms which follow will facilitate a better understanding of threads in general. See Fig. 18-1.

The following are terms relating to types of screw threads.

External thread. A thread on the external surface of a cylinder or cone.

Internal thread. A thread on the internal surface of a hollow cylinder or cone.

Right-hand thread. A thread which, when viewed axially, winds in a clockwise and receding direction.

Left-hand thread. A thread which, when viewed axially, winds in a counterclockwise and receding direction.

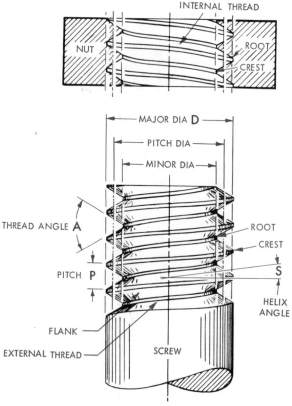

Fig. 18-1. Illustration of thread terms.

---

1. *ASME Screw Thread Manual* (New York: The American Society of Mechanical Engineers).

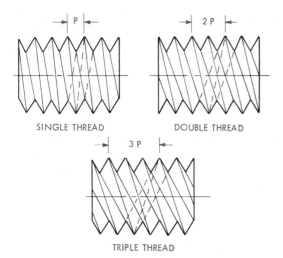

Fig. 18-2. In the single thread the lead equals the pitch; in a multiple thread the lead is greater than the pitch.

Single thread. A thread having a lead equal to the pitch; also called *single-start thread.* See also Fig. 18-2.

Multiple thread. A thread in which the lead is an integral multiple of the pitch; also called *multiple-start thread.* See also Fig. 18-2.

The following are terms relating to geometrical elements of screw threads.

Form of thread. The thread's profile in an axial plane for a length of one pitch. See also Fig. 18-3.

Flank, or side, of a thread. Either surface connecting the crest with the root, whose intersection with an axial plane is a straight line.

Crest of a thread. The surface that joins the flanks of the thread and is farthest from the cylinder or cone from which the thread projects.

Root of a thread. The surface that joins the flanks of adjacent thread forms and is identical in position with, or immediately adjacent to the cylinder or cone from which the thread projects.

Complete, or full, thread. The part of the thread having full form at both crest and root. When there is a chamfer not exceeding two pitches in length at the start of the thread, it is included within the length of complete thread.

Incomplete thread. On straight threads, the portion at the end having roots not fully formed by the lead or chamfer on threading tools; also known as the *vanish,* or *washout, thread.*

Blunt start. Designates the removal of the partial thread at the entering end of thread. This is a feature of threaded parts which are repeatedly assembled by hand, such as hose couplings and thread plug gages, to prevent cutting of hands and crossing of threads, and which was formerly known as a Higbee cut.

The following are terms relating to dimensions of screw threads.

Pitch. The distance, measured parallel to the thread's axis, between corresponding points on adjacent thread forms in the same axial plane and on the same side of the axis.

Lead. The distance a threaded part moves axially, with respect to a fixed mating part, in one complete rotation.

Threads per inch. The reciprocal of the pitch in inches.

Included angle of a thread, or angle of thread. The angle between the flanks of the thread measured in an axial plane.

Major diameter. On a straight thread, the diameter of the imaginary co-axial cylinder that bounds the crest of an external thread or the root of an internal thread. On a taper thread at a given position on the thread axis, the diameter of the major cone.

Minor diameter. On a straight thread, the diameter of the imaginary co-axial cylinder that bounds the root of an external thread or the crest of an internal thread. On a taper thread at a given position on the thread axis, the diameter of the minor cone at that position.

Pitch diameter, or simple effective diameter. On a straight thread, the diameter of the imaginary co-axial cylinder, the surface of which would pass through the thread profiles at such points as to make the width of the groove equal to one-half of the basic pitch. On a perfect thread this occurs at the point where the widths of the thread and groove are equal. On a taper thread at a given position on the thread axis, the diameter of the pitch cone at that position.

Length of thread engagement. The distance between the extreme points of contact on the pitch cylinders or cones of two mating threads measured parallel to the axis.

Crest clearance. In a thread assembly, the distance, measured perpendicular to the axis, between the crest of a thread and the root of its mating thread.

Thread class. Refers to the amount of tolerance, or of tolerance and allowance, as applied to pitch diameter. In the past, commonly called *fit of a thread,* which actually described the degree of looseness or tightness between two mating threaded parts.

18.1.1 Thread profiles. The American Standard thread shapes are identified as Unified, Knuckle, Square, Acme, and Buttress. See Fig. 18-3. The Unified thread series is the most commonly used for holding parts together. The knuckle thread is a cast or rolled thread with a rounded profile and is the one usually found on lamps and electric plugs. The square and acme threads are designed primarily for transmitting power. The buttress thread is particularly adapted for transmitting power in one direction only, such as on breech mechanisms of large guns and hubs of airplane propellers.

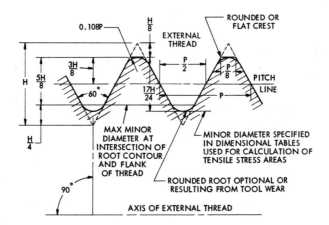

UNIFIED THREAD

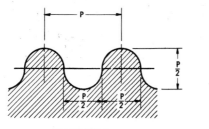

KNUCKLE

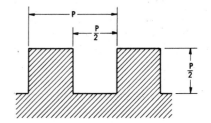

SQUARE

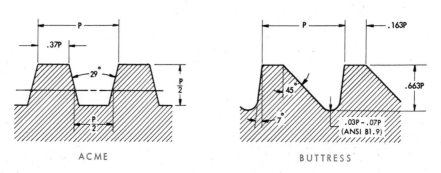

ACME

BUTTRESS

Fig. 18-3. The profile of the threads used is governed by the intended function of the thread.

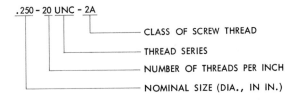

```
.250 - 20 UNC - 2A
                    CLASS OF SCREW THREAD
                    THREAD SERIES
                    NUMBER OF THREADS PER INCH
                    NOMINAL SIZE (DIA., IN IN.)
```

OR

.250 – 20 UNC – 2A
PD 0.2164 – 0.2127 (OPTIONAL)

OR

.250 – 20 UNC – 2A – LH
1.125 – 4 ACME – 2G
.625 – 20 N BUTT – 1

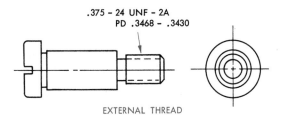

.375 – 24 UNF – 2A
PD .3468 – .3430

EXTERNAL THREAD

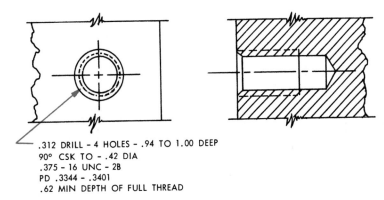

.312 DRILL – 4 HOLES – .94 TO 1.00 DEEP
90° CSK TO – .42 DIA
.375 – 16 UNC – 2B
PD .3344 – .3401
.62 MIN DEPTH OF FULL THREAD

INTERNAL THREAD

Fig. 18-4. A screw thread is designated on a drawing by a note with a leader pointing to the thread.

18.2 Thread standards[2]

Screw threads are designated on a drawing by a note with a leader and arrow pointing to the thread. See Fig. 18-4. Essential specifications include nominal size, number of threads per inch, thread type, thread class, and in some instances, the pitch diameter limits. Some examples are as follows:

---

2. *ANSI Y14.6.*

Unless otherwise specified, threads are right-hand and single-lead. Left-hand threads are designated by the letters LH following the class symbol. Double- or triple-lead threads are designated by the word DOUBLE or TRIPLE preceding the pitch diameter limits. An alternate method is to use the designation 2-START or 3-START.

For coated class *2A* external threads the pitch diameter limits should be followed by the words BEFORE COATING, and the basic pitch diameter specified as the maximum pitch diameter followed by the words AFTER COATING. Example:

.250—20 UNC—2A

PD .2164—.2127 BEFORE COATING

MAX PD .2175 AFTER COATING

18.2.1 Unified thread series.[3] The unified series is the most widely used for threaded fastening devices.

Types. The unified thread series consists of the following types: (See Appendix for additional thread sizes.)

Coarse threads, designated UNC, are used for the bulk production of screws, bolts, and nuts for general applications requiring rapid assembly or disassembly. Because of their comparatively larger thread form for a given diameter, coarse threads are less affected by corrosion, which is an important factor in equipment and structures subject to constant outdoor exposure. Coarse threads also maintain greater stripping strength in materials such as cast iron, copper alloys, aluminum, and plastics.

Fine threads, designated UNF, are used extensively on bolts and nuts employed in the automotive and aircraft industries, where great resistance to vibration is required or where great holding strength is necessary. Fine threads are also recommended where the length of engagement is short or where the wall thickness demands a fine pitch.

Extra-fine threads, designated UNEF, are used for threaded parts which require a fine adjustment,

such as bearing-retaining nuts and adjusting screws and for thin-walled tubing and thin nuts where maximum thread engagement is needed.

The *8-thread series,* designated 8 UN, (above 1 inch) is used in the utility industries for high-temperature bolting in steam-flange connections or as substitutes for coarse threads.

The *12-thread series,* designated 12 UN, is used in machine construction for thin nuts on shafts and sleeves. Twelve threads per inch is the coarsest pitch in general use that will permit a threaded collar to screw onto a threaded shoulder. The series is also used as a continuation of the fine series for diameters larger than 1.50 inches.

The *16-thread series,* designated 16 UNC, is used for adjusting collars and retaining nuts. They also serve as a continuation of the extra-fine series for diameters larger than 1.6875 inches.

The *20-, 28-, and 32-thread series* are used for adjusting nuts, screws, and collars where a fine thread is necessary.

The *4- and 6-thread series* are used primarily for heavy machine and structural applications.

Class. Threaded classes in the Unified Standard are designated by a numeral followed by the letter *A,* for external threads, or *B,* for internal threads. There are three classes of external threads: 1A, 2A, and 3A—and three classes of internal threads: 1B, 2B, and 3B.

Classes 1A and 1B have the greatest amount of allowance and are intended for ordnance use and for applications that require frequent and rapid assembly and disassembly with minimum binding even with slightly bruised or dirty threads.

Classes 2A and 2B are considered standard for general purpose threads on bolts, nuts, and screws. They provide standard allowances to insure minimum clearance between external and internal threads, which minimize galling and seizing in high-cycle wrench assembly. Because of their realistic tolerances, classes 2A and 2B are widely used for mass production purposes.

Classes 3A and 3B are suitable for applications requiring closer tolerances than those provided by classes 2A and 2B. They are designated for set screws, socket-head cap screws, and aircraft bolts or for higher strength materials where it is necessary to limit the variations of the thread elements.

---

3. *ANSI B1.1.*

The requirements for screw-thread fits depend on their end use. Combination of thread classes for components is possible. For example, a class 2A external thread may be used with a class 1B, 2B, or 3B internal thread. When selecting a class of thread, the designer must keep in mind that cost generally increases proportionately to the accuracy required. Hence, no closer thread fit should be used than is needed for the proper functioning of the components.

18.2.2 Acme threads. The American National Standards Institute lists two types of acme threads, the General Purpose and the Centralizing. The *general purpose threads* have three classes: 2G, 3G, and 4G. The classes designated as *G* (general purpose) provide ample fits for free movement of threaded parts. *Centralizing threads* have five classes: 2C, 3C, 4C, 5C, and 6C. These threads have limited clearance on major diameters in order to maintain proper alignment of the thread axes.

18.2.3 Buttress threads. Three classes have been standardized for Buttress threads: class 1 (free), class 2 (medium), and class 3 (close).

18.2.4 Pipe threads.[4] There are two standard forms of pipe threads: Regular and Dryseal. *Regular pipe thread* is the standard for the plumbing trade. *Dryseal pipe thread* is the standard for automotive, refrigeration, and hydraulic tube and pipe fittings; lubrication fittings; and drain cocks.

Regular pipe-thread forms allow crest and root clearance when the flanks contact, and unless this clearance is filled with lute or sealer, leakage occurs. With dryseal pipe thread forms there is no crest and root clearance so that a leakproof assembly is provided without the need for a sealer.

Both regular and dryseal threads come in two forms: straight and tapered. The tapered thread insures a tighter joint. Regular threads are designated as NPS (straight) and NPT (tapered). Dryseal threads are identified as NPSF (straight) and NPTF (tapered).

The designation of pipe threads on a drawing should include the nominal size, the number of threads per inch, the thread form and thread series symbols:

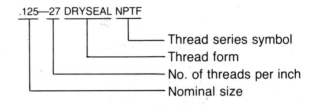

(See Appendix for data on various pipe thread sizes.)

18.2.5 Metric threads. The ISO metric thread has a profile similar to the Unified inch thread except that nominal diameters are in round or fractional millimeters. As a result, millimeter and inch threads are not interchangeable. Fig. 18-5.

There are two general series of metric threads—coarse for general purpose work and fine for precision applications. The *coarse threads* are designated by the letter *M* followed by their nominal size (basic major diameter). For example, a

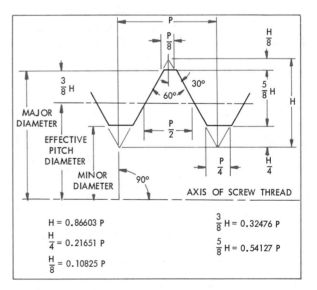

$$H = 0.86603\ P$$

$$\frac{H}{4} = 0.21651\ P$$

$$\frac{H}{8} = 0.10825\ P$$

$$\frac{3}{8}H = 0.32476\ P$$

$$\frac{5}{8}H = 0.54127\ P$$

Fig. 18-5. Metric thread profile.

---

4. *ANSI B2.1* and *B2.2*.

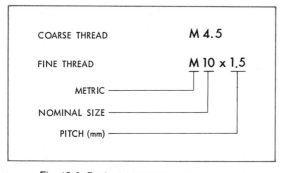

Fig. 18-6. Basic designation of metric threads.

grade 6 are classified as fine grades and are recommended for fine-quality threads or short lengths of engagement. Tolerances above grade 6 are considered coarse and are used for coarse-quality threads or long lengths of engagement.

Tolerance position. This simply identifies the thread as being internal or external. The lowercase letters *e*, *g*, and *h* are used for external threads and capital letters *H* and *G* for internal threads. These letters are designed to convey the following information: (See Fig. 18-7.)

thread specified as M4.5 means a coarse thread with a nominal diameter of 4.5 millimeters. *Fine threads* are similarly designated, but in addition, the pitch in millimeters is included. The pitch is separated from the nominal size by the sign *x*. See Fig. 18-6.

Following the basic designation, other symbols are used to indicate the tolerance class. Tolerance class refers to grade and position of tolerances.

Tolerance grade. Tolerance grade specifies the coarseness of fits for either internal or external threads. Grades are designated by numbers 3, 4, 5, 6, 7, 8, and 9. The smaller the number, the smaller the tolerance. Actually three tolerance grades are recommended—fine, medium, and coarse. Grade 6 is considered a medium class and is the most frequently used for general purpose work. This grade is comparable to the Unified thread classes 2A and 2B. Tolerances below

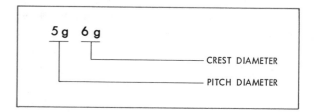

Fig. 18-8. By combining grade and position symbols, maximum and minimum tolerances can be established for a thread.

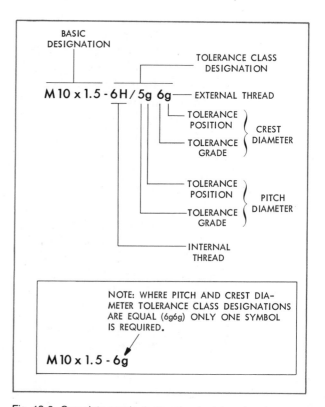

Fig. 18-9. Complete metric designation for internal and external threads.

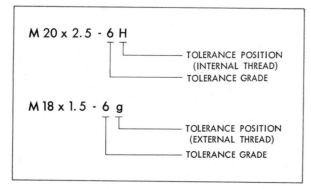

Fig. 18-7. Method of showing position and grade tolerances.

EXTERNAL THREADS (BOLTS)
    e Large allowance
    g Small allowance
    h No allowance
INTERNAL THREADS (NUTS)
    G Small allowance
    H No allowance

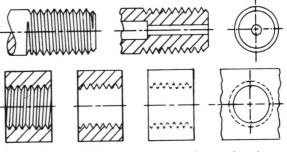

Fig. 18-11. Detailed representation of screw threads.

When extreme precision control of product tolerance limits is necessary, the class designation symbols are specified for both the pitch diameter and crest diameter. See Figs. 18-8 and 18-9. The first group represents the pitch diameter and the second group, the crest diameter. When the pitch and crest diameters are the same, for example 6g6g, only one group is given.

To show a desired fit between mating threads, the internal thread tolerance is given followed by the external class designation, the two separated by a slash. See Fig. 18-10.

When the length of a threaded fastener is to be specified, this length is separated from the rest of the designation by an additional x.

If the inch thread is to be retained even though the drawing is completely in metric, all designations should be expressed as they normally are in the inch system.

resentations. End purpose and use of drawings, general quality, drafting time, and so on influence the selection and the use of the conventions.

The *detailed representation* is a close approximation of the actual appearance of screw threads. See Fig. 18-11. The form of the thread is simplified by showing the normal helices as straight slanting lines and the truncated crests and roots as sharp V's. While the detailed rendering is comparatively difficult and time-consuming, its use is sometimes justified by such considerations as permanency of drawing, general quality of the drafting project, and the necessity for avoiding any confusion which might result from a less realistic thread representation.

The *schematic representation* is nearly as effective as the pictorial and is much easier to draw. See Fig. 18-12. The staggered lines, symbolic of the thread crests and roots, may be perpendicular to the axis of the thread or slanted to the approximate angle of the thread helix. This construction

---

M8 - 6H /6g

M10 x 1.5 6H /5g 6g

Fig. 18-10. How to show desired fit between mating threads.

18.3 Thread Representations[5]

There are three conventions in general use for depicting all forms of screw threads on drawings: the detailed, the schematic, and the simplified rep-

5. *ANSI Y14.6.*

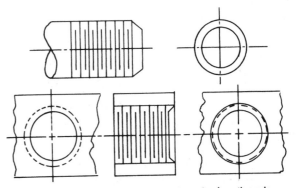

Fig. 18-12. Schematic representation of screw threads.

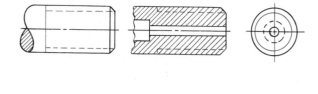

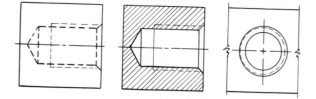

Fig. 18-13. The simplified representation of screw threads is most commonly used.

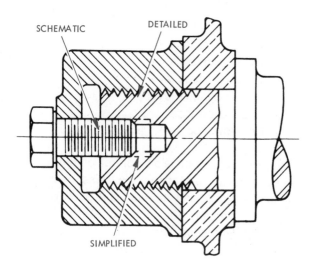

Fig. 18-15. All three thread conventions may be used on a single drawing.

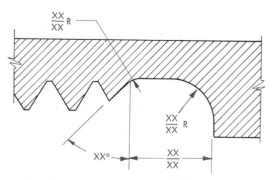

Fig. 18-14. In rare instances a representation of exact thread geometry may be used.

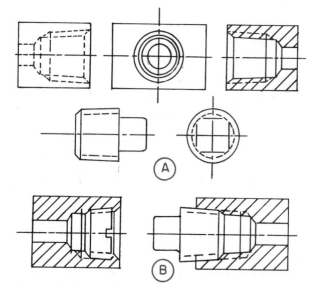

Fig. 18-16. Pipe thread is shown on a drawing by the conventional simplified representation.

should not be used for hidden internal threads or sections of external threads.

The *simplified representation,* on the grounds of economy and ease of rendering, is the most commonly used method for showing screw threads on drawings. See Fig. 18-13. It is particularly useful for indicating hidden internal threads. The simplified representation should be avoided where there is any possibility of confusion with other drawing details.

In rare instances where threads are shown as a portion of a greatly enlarged detail, the representation of exact thread geometry might be justified. See Fig. 18-14. For clarity of meaning, where good judgment dictates, all three conventions may be used on a single drawing. See Fig. 18-15.

Pipe threads should be shown on a drawing by means of the simplified method. The taper threads are drawn the same as the straight, except that the thread lines should form an angle of approximately 3° with the axis. See Fig. 18-16A. Assembled pipe-thread components should be shown as in Fig. 18-16B.

18.4 Dimensioning the Engagement Length of Threads[6]

The length of engagement required by the threaded components should be the first consideration in determination and specification of thread lengths. See Fig. 18-17. The engagement length *X* when both components are steel is equal to the nominal diameter *D* of the thread. For steel external threads assembling into cast iron, brass, or bronze, *X* is equal to 1.5*D*; and when assembled into aluminum, zinc, or plastic, *X* is equal to 2*D*.

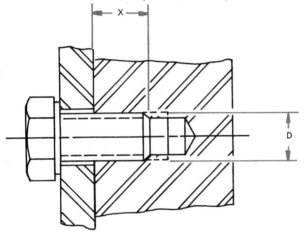

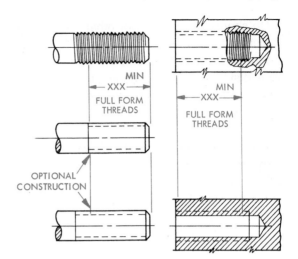

Fig. 18-18. Thread length engagement where a chamfer is involved.

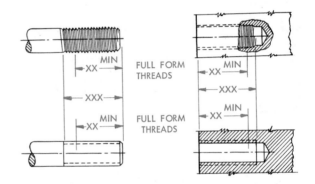

Fig. 18-17. The length of engagement depends on the function and material of the threaded component.

Fig. 18-19. When the number of partial threads is to be limited, the threaded section is dimensioned in this way.

The thread length on the drawing should be the gaging length, or the length of threads having full form, that is, the partial threads should be outside or beyond the length specified. Where there is a chamfer not exceeding two pitches in length at the start of the thread, it is included within the length of full-form thread. See Fig. 18-18.

Should there be reason to control or limit the number of partial threads, the overall thread length, including the partial threads, should be dimensioned on the drawings in addition to full-form threads. See Fig. 18-19.

6. *Ibid.*

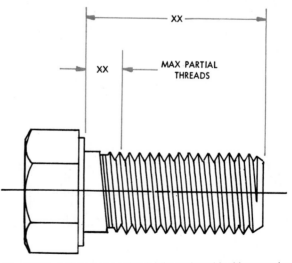

Fig. 18-20. A threaded section is dimensioned in this way when the thread must run close to the head.

On short bolts, screws, or other parts where the objective is to run the thread as close as practicable to the head or shoulder, the maximum permissible distance from the head or shoulder to the nearest thread of full form may be dimensioned on the drawing instead of the thread length. Refer to Fig. 18-20.

If a definite length of unthreaded and unscored body or shank of a threaded part is a critical functional requirement, it should be dimensioned on the drawing. See Fig. 18-21.

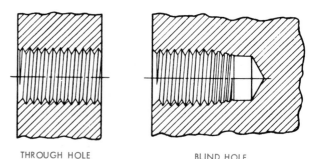

THROUGH HOLE   BLIND HOLE

Fig. 18-22. Threaded holes are either through or blind.

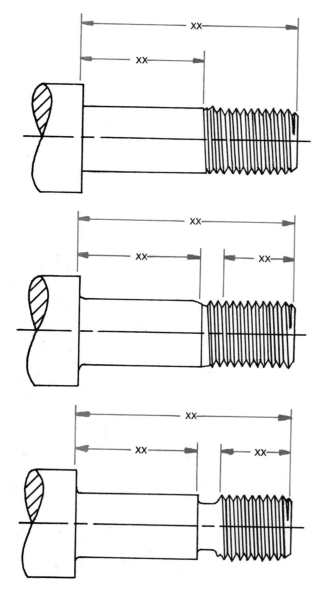

Fig. 18-21. If an unthreaded length is critical, it must be dimensioned.

The number of partial threads (called *runout*) should not be restricted unnecessarily, as it has considerable bearing on production efficiency and the life of threading tools. In general practice, allowance should be made for at least two or three partial threads on externally threaded parts.

Threaded holes are either *through* or *blind* as shown in Fig. 18-22. Through holes are preferable from a manufacturing standpoint, as they eliminate consideration of partial threads, facilitate chip disposal, and permit use of the most efficient taps.

Design permitting, through holes should be tapped their full length, except where holes are considerably deeper than the thread length required. They should break out squarely on a surface normal to the axis of the drill and not into a fillet or side wall. They should not be used in castings where removal of objectionable chips would create a problem.

The depths of drill holes for blind tapped holes should not be unnecessarily restricted. The holes should be considerably deeper than the required length of thread to afford chip clearance, speed up the tapping operation, minimize the possibility of tap breakage, and permit the use of the same taps and tapping technique as for through holes. Where hole depth is restricted by design, it should be remembered that in order to produce internal threads of full form to within one thread of the bottom of a hole, the use of a bottoming tap is necessary, often involving a second operation.

It is usual to countersink holes before threading, although this may be unnecessary if the holes are tapped with a taper tap. The included angle of countersinks should be 90°, and the diameter

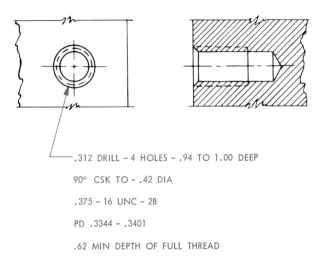

.312 DRILL – 4 HOLES – .94 TO 1.00 DEEP

90° CSK TO – .42 DIA

.375 – 16 UNC – 2B

PD .3344 – .3401

.62 MIN DEPTH OF FULL THREAD

Fig. 18-23. Typical drawing designation of an internal thread.

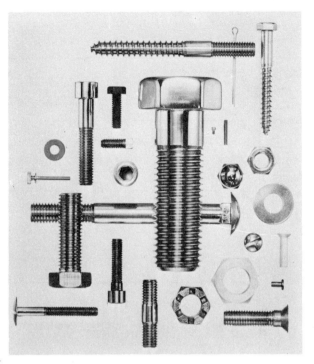

Fig. 18-24. Fasteners play important roles in the fabrication of industrial products.

should be 0.02 inch larger than the nominal diameter of the thread, with a plus tolerance of 0.02 inch. See Fig. 18-23 for a typical drawing designation of an internal thread.

18.5 Mechanical Fasteners

Countless types, sizes, and shapes of mechanical fasteners are available. See Fig. 18-24. They are made of such materials as steel, aluminum, brass, copper, nickel, stainless steel, titanium, beryllium, and plastic. Many fasteners have a special coating of zinc, cadmium, tin, nickel, or chromium. The coating is intended to improve appearance, increase corrosion resistance, and provide lubricity.

18.5.1 Metric fasteners. The fastener industry and ANSI are currently studying standards for manufacturing metric fasteners. Many fasteners already are available in millimeter sizes. Optimum conversion becomes somewhat complicated because of the many sizes of fasteners. The consensus is that when final standards have evolved, fasteners will be made in a one-type pitch series and in a limited number of sizes.

Current metric fasteners have the regular metric thread profile, which is identical to the familiar unified inch thread, but the two systems, the ISO metric and the existing inch system, are not interchangeable. Accordingly, in specifying a metric threaded fastener, it must be remembered that in the inch system pitch is identified as a number of threads per inch, while in the metric system specification is in terms of millimeter spacing of threads. See 18.2.5.

Metric fasteners are available in three fit classes: fine, medium, and coarse. Most commercial-grade fasteners for common applications are made in the medium fit class. The fine class is for fasteners to be used in precision work. The coarse class is for threads that will be surface-coated later and are intended for dirty application environments. Specifically, metric coarse threads are advantageous because they are stronger, provide more room for plating, are less affected by corrosion and high temperature, make assemblies easier and quicker, and eliminate cross threading.

18.5.2 Threaded fasteners.[7] Among threaded fasteners available are tapping screws, set

---

7. *ANSI B18.6.2* and *B13.3.*

Table 1—Tapping Screw Selection Guide

USA Type and Thread Form	Description and Recommendations
Thread-Forming Screws **AB**	Spaced thread with same pitches as Type B and with gimlet point. For sheet metal, resin-impregnated plywood, wood, and asbestos compositions. Used in pierced or punched holes where a sharp point for starting is needed. No. 6 screw for thin sheets up to and including 20 gage; larger screw sizes up to 18 gage. Joint strength of easily deformed materials can be increased with pilot holes less than root diameter of screw. Recommended hole sizes same as for Type B. Fast driving.
B	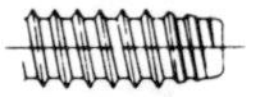 Blunt-point, spaced-thread. Can be used in heavy-gage sheetmetal and nonferrous castings. Used in assembling easily deformed materials where pilot hole is larger than root diameter of screw. Fast driving. In many applications, Type AB might be better.
BP	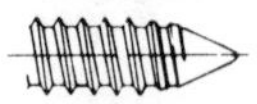 Same as Type B, but has a 45-deg included-angle, unthreaded cone point. Used for locating and aligning holes, or piercing soft materials. Industrial Fasteners Institute recommends Type AB be used instead of BP.
C	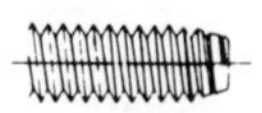 Blunt point with threads approximating machine screw threads. For applications where a machine screw thread is preferable to the spaced thread form. Unlike thread-cutting screws, Type C makes a chip-free assembly. In specific applications may require extremely high driving torques due to long thread engagement or its use in hard materials. Resists loosening by vibration since greater engaged thread surface increases frictional resistance to backing out. Smaller helix angle of Type C provides tighter clamping action than that of Type B for equivalent driving torques.
U	 Multiple threaded drive screw with steel helix angle and a blunt, unthreaded, starting pilot. Intended for permanent fastenings in metals and plastics. Are hammered or mechanically forced into the work. Should not be used in materials less than one screw diameter thick.
Thread-Cutting Screws **D**	Blunt point with single narrow flute and threads approximating machine screw threads. Flute is designed to produce a cutting edge radial to screw center. Requires less driving torque than Type C and has longer length of thread engagement. Good for low-strength metals and plastics, high-strength brittle metals, and rethreading clogged pretapped holes. Easy starting. Gives highest clamping force for a given torque of any tapping screw.
F	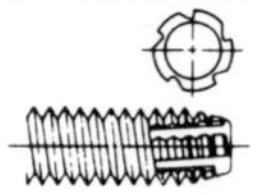 Approximate machine screw thread and blunt point. Tapered thread may be complete or incomplete. Has five evenly spaced cutting grooves and large chip cavities. Used in a wide range of materials. Fast driving. Resists vibration.
G	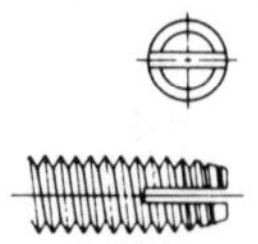 Approximate machine screw thread with single through slot that forms two cutting edges. Blunt point has incomplete tapered threads. Recommended for same general use as Type C, but requires less driving torque. Has higher percentage of thread and longer thread engagement than Type C screw. Good for low-strength metals.
T	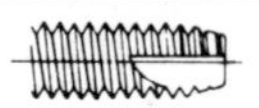 Same as Type D with single wide flute that provides more chip clearance. Cutting edge is right of the vertical center line of screw and provides an acute cutting edge that cuts easier than Type D.
BF	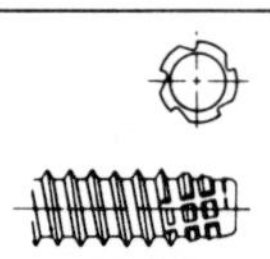 Spaced thread like Type B with blunt point and five evenly spaced cutting grooves and chip cavities. Cutting grooves remove only a small part of material, thus maintaining maximum shear strength in threaded hole wall. Wall thickness should be 1½ times major diameter of screw. Reduces stripping in brittle plastics and die castings. Good for long thread engagement, especially in blind holes. Faster driving than fine thread types.
BT	 Same as Type BF except for single wide flute, which provides room for twisted, curly chips so that binding or reaming of hole is avoided.
High-Performance Thread-Rolling Screws **SF**	Fine or coarse-thread screw with square point that forms a thread with the four rounded corners. Four-point contact provides straighter driving and permits low driving torque. Recommended for thin and heavy-gage materials.
SW	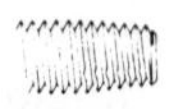 Fine or coarse-thread screw with lobes or projections spaced approximately 120 deg apart on the crest and flanks of the thread at the tapered starting end. Mating thread is formed by succession of pressures as the lobes exert three-dimensional swaging action on the crest, forward flank, and trailing-edge flank of threads. Recommended for sheetmetal, structural steel, zinc and aluminum castings, and steel, brass, and bronze forgings.
TT	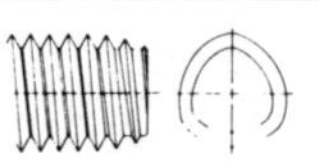 Fine or coarse-thread screw, with trilobular cross section, providing slight radial relief for its full length. Normally made with a machine screw thread and blunt point. Recommended for heavy materials and structural applications to give deep thread engagement with low driving torque. Hex-washer head style is standard.

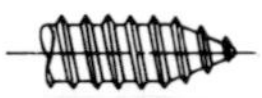

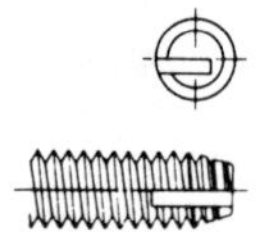

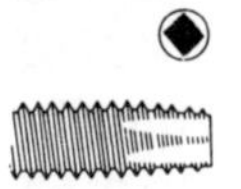

Fig. 18-25. Basic types of tapping screws.

screws, cap screws, machine screws, bolts, studs, nuts, wood screws, and drive screws.

Tapping screws. Tapping screws form or cut their own mating thread as they are driven into the material. These fasteners permit rapid installation since no nuts are required and access is necessary from only one side of the joint. They provide a close fit that keeps the screw tight even under vibrating conditions.

The three main groups of tapping screws are thread-forming, thread-cutting, and high-performance thread-rolling. See Fig. 18-25. *Thread-forming* tapping screws produce a joint by forming the material around the pilot hole so it flows around the screw thread. The various types are designated as AB, B, BP, C, and U. *Thread-cutting* tapping screws have cutting edges and chip cavities that produce a mating thread by removing material from around the edges of the pilot hole. These screws are available in both coarse- and fine-thread series and are classified as types D, F, G, T, BF, and BT. *Thread-rolling* tapping screws are used in thick-gage materials, castings, and forgings where excessive driving torques are encountered. These types are identified as SF, SW, and TT.

The sizes of tapping screws are designated by the length and wire gage number. On a drawing, they are indicated by a note:

.375—NO. 4 TAPPING SCREW

TYPE AB—NICKEL FINISH

Set screws. The function of a set screw is to prevent rotary motion between two parts, such as the hub of a pulley and shaft. It is also used to make slight adjustments between mating parts. In all cases the set screw is driven into one piece so that its point bears firmly against the other part. As a rule a set screw is rarely used where the fixed parts are subjected to heavy stresses.

Set screws are available in the following types: headless slotted, hexagon socket, fluted socket, and square head. Each of these can be obtained with these points: cup point, flat point, oval point, cone point, full dog point, and half dog point. See Fig. 18-26. Because the projecting heads often create a potentially dangerous situation, set screws with heads are being used less frequently;

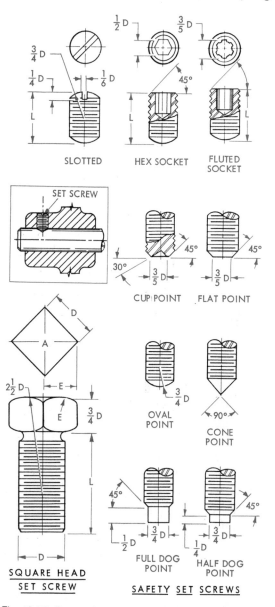

Fig. 18-26. Types of set screws and how they are used.

instead the socket, or safety headless, type is more universally recommended. The safety type has either a slotted end or an end with a fluted or hexagonal hole to receive a tightening wrench.

Specifications for set screws should include diameter, number of threads per inch, series, class of fit, type of head, type of point, and length:

.250—20 UNC—2A X .50

SLOTTED CONE PT SET SCR

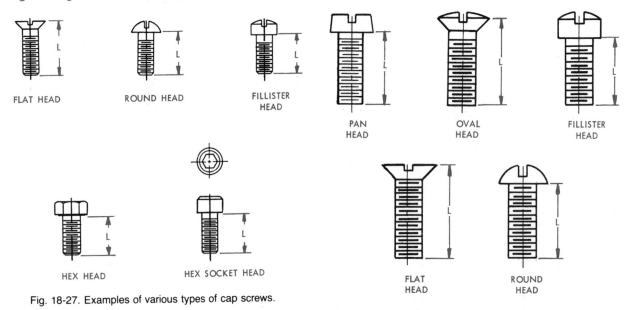

Fig. 18-27. Examples of various types of cap screws.

Fig. 18-28. Examples of different types of machine screws.

Cap screws. A cap screw passes through a clearance hole in one member of the structure and screws into a threaded or tapped hole in the other. They range in diameter from 0.25 to 1.25 inches and are made with five types of heads: hexagon, round, flat, fillister, and socket. See Fig. 18-27. The length of cap screws is not standardized; the length is measured from the largest diameter of the bearing surface of the head to the extreme point of the screw. The point of a cap screw is chamfered 45° to the flat surface.

The specifications of a cap screw should be given as:

.50—13 UNC—2A X 1.00
ROUND HD CAP SCR

Machine screws. Machine screws are similar to cap screws except that they are smaller and are used chiefly on small work having thin sections. See Fig. 18-28. Below 0.250-inch size machine screws are specified by numbers from 2 to 12. Above 0.250 inch the size is indicated by diameter. The threads run the entire length of the stem of screws 2 inches and under in length. On a drawing, machine screws are shown as follows:

NO. 10—24 UNC—2A X .38
RD HD MACH SCR

Bolts. The American National Standards Institute lists two series of square- and hexagon-head bolts known as *regular* and *heavy.* The regular bolts are recommended for general-purpose work and the heavy for use where greater bearing surface is necessary. These bolts are also classified as *finished, semifinished,* and *unfinished.* The term *finished* refers to the quality of manufacture and the closeness of tolerance. Both the semifinished and finished bolts may be obtained with a washer face. The washer face is approximately 0.02 inch thick and serves as a bearing in place of a regular washer.

The type and size of bolts to be used is determined by the engineer or designer and is largely governed by the strength requirement of the assembled unit. See Fig. 18-29. Hexagon-head bolts are the most common since they often require less head clearance. Hexagon bolts are available with either plain or slotted heads.

Bolts come in a variety of sizes. Their length is measured from the bearing surface of the head to the extreme point. In selecting thread pitches for various bolt diameters, it should be remembered that the fine-thread bolt is usually much stronger than the coarse-thread bolt and has an advantage

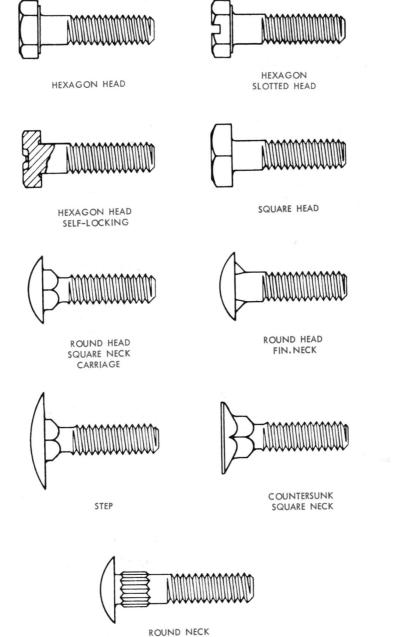

HEXAGON HEAD

HEXAGON
SLOTTED HEAD

HEXAGON HEAD
SELF-LOCKING

SQUARE HEAD

ROUND HEAD
SQUARE NECK
CARRIAGE

ROUND HEAD
FIN. NECK

STEP

COUNTERSUNK
SQUARE NECK

ROUND NECK
RIBBED NECK

Fig. 18-29. Common types of bolts.

where length of engagement is limited, where vibration may be excessive, or where thin walls may be encountered. The class of fit is also important. For general application class 2A threads are used. With automotive assemblies, such as connecting rods and main bearings, class 3A threads are generally recommended.

On a drawing the specifications of a bolt should include diameter, number of threads per inch, series, class, type of finish, type of head, name and length, as follows:

.375—16 UNC—2A X 2.50
SEMI—FIN HEX HD BOLT

Studs. A stud or stud bolt is a rod threaded on both ends. It is used when regular bolts are not suitable, especially on parts that must be removed frequently, such as cylinder heads. One end of the stud screws into a threaded or tapped hole, and the other end fits into the removable piece of the structure. A nut is used on the projecting end to hold the parts together. See Fig. 18-30. Ordinarily a stud is made with coarse threads on the stud end and fine threads on the nut end.

On a detail drawing a stud is dimensioned to show the length of thread for both ends along with an overall length. The specification is given by a note. On an assembly drawing the specifications

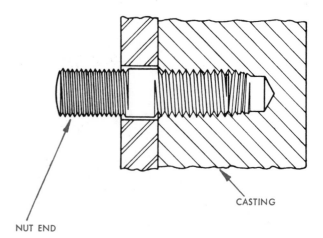

NUT END

CASTING

Fig. 18-30. This is how a stud is used.

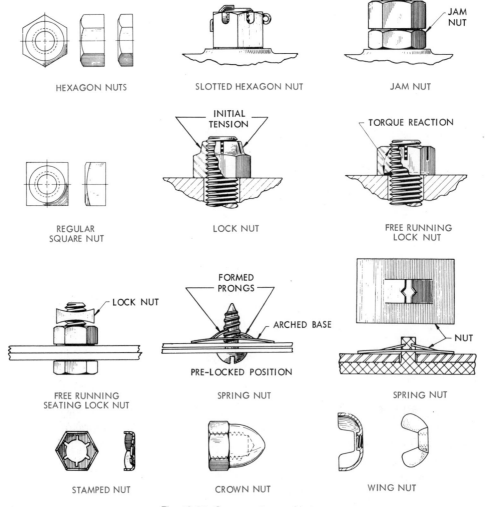

HEXAGON NUTS

SLOTTED HEXAGON NUT

JAM NUT

JAM NUT

REGULAR SQUARE NUT

INITIAL TENSION

LOCK NUT

TORQUE REACTION

FREE RUNNING LOCK NUT

LOCK NUT

FREE RUNNING SEATING LOCK NUT

FORMED PRONGS

ARCHED BASE

PRE-LOCKED POSITION

SPRING NUT

NUT

SPRING NUT

STAMPED NUT

CROWN NUT

WING NUT

Fig. 18-31. Common types of nuts.

of the stud are included in the parts list or bill of material. The specifications of a stud should be given as:

.375 X 2.50 STUD

Nuts.[8] As shown in Fig. 18-31, there are many types of nuts available to satisfy requirements. The most common of these are the following:

Hexagon nuts. Made to fit the various types of hexagon bolts. They are classified into three principal groups: regular, heavy, and light. Like hexagon bolts, they are available in the finished, semifinished, and regular finish quality.

Slotted hexagon nuts (with cotter pins or wire). Used where there is danger of the nuts coming off due to vibration or other causes.

Jam hexagon nuts. Used where height is restricted or as a means of locking the working nut.

Square nuts. Rough unfinished nuts most frequently used in conjunction with square-neck bolts and square-head bolts. Square nuts may be used with machine screws if the screws are to be driven through the head; otherwise, it is advisable to use hexagon machine screw nuts.

Lock nuts. Nuts which have a special means for gripping a threaded member or bearing surface so that relative back-off rotation between the nut and the threaded companion member is impeded. Prevailing torque-type lock nuts employ a self-contained locking feature such as deformed or undersize threads, variable lead angle, plastic or fiber washers, or plug inserts. This type of nut resists screwing on as well as unscrewing and does not depend on bolt load for locking.

Free-running lock nuts. Develop their locking action after the nut has been seated by reactive spring force against the threads or by friction against the bearing surface.

Free-running seating lock nuts. Applied over hexagon nuts. The concave surface of this type of nut, in contacting the top of the hexagon nut, tends to flatten, thereby deflecting the nut threads from their true helix and causing them to bind on the bolt or screw thread.

Spring nuts. Made of thin spring metal; have arched prongs or formed embossments to fit a single lead of a screw thread. Used extensively for sheet-metal construction where high torque is not required. A type of spring nut is available which can be used with rivets, tubing, nails, and other unthreaded parts. It is pushed on and provides a positive bite that grips securely even on very smooth surfaces.

Stamped nuts. Usually made of thin spring steel; have arched prongs formed to fit a single lead of a thread. They have the same function as spring nuts with the additional advantage of having tightening provisions by means of hexagon flanges.

Crown nuts. Generally used where the end of the external threaded part should be hidden. The crown on top of the nut provides a suitable surface to be finished for appearance.

Wing nuts. As the name implies, are provided with two wings to facilitate hand tightening and loosening. Used where high torque is not required and where the nuts are to be tightened and loosened frequently. Available in stamped, cast, or forged forms in either brass or steel.

Wood screws. Wood screws are made of steel, brass, bronze, and aluminum alloys. They are available with three types of heads: flat, oval, and round with slotted or cross recess driver provisions. See Fig. 18-32. The steel flat-head and oval-head screws are usually sold in a bright finish and the round head in a blue finish, or they may be plated. Screws are sold with either slotted or Phillips recessed-type heads.

The sizes of wood screws are designated by the length and wire gage number. On a drawing screws should be indicated as follows:

.750—NO. 6FH WOOD SCREW—STEEL

Drive screws.[9] Hardened metallic drive screws provide a permanent fastening for heavy sheet metal, castings, plastics, and so on and may be used in place of tapping screws or machine screws.

8. *General Motors Drafting Standards Manual* (Detroit: General Motors Corp.).

9. *Ibid.*

FLAT HEAD   ROUND HEAD   OVAL HEAD   WOOD DRIVE SCREW

SLOTTED TYPE HEAD

BUTTON HEAD   HIGH BUTTON HEAD (ACORN)   CONE HEAD

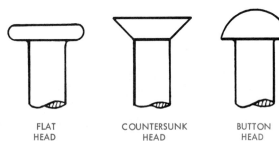

PAN HEAD   FLAT TOP COUNTERSUNK HD   ROUND TOP COUNTERSUNK HD

LARGE

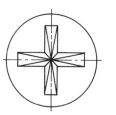
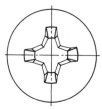

CROSS RECESSED TYPE HEAD

Fig. 18-32. Common types of wood screws.

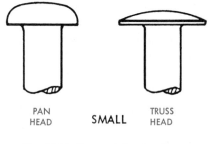

FLAT HEAD   COUNTERSUNK HEAD   BUTTON HEAD

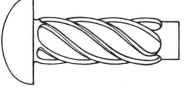

Fig. 18-33. A drive screw for metal.

PAN HEAD   **SMALL**   TRUSS HEAD

Fig. 18-34. Types of rivet heads.

Drive screws are hammered or otherwise forced into holes of suitable size. The unthreaded pilot guides the drive screw in straight, and the hardened spiral thread, which extends to the head, forms the mating thread in the hole.

Where drive screws are used in joining plastics, a clearance hole should be provided in the top piece, and if the material is brittle or friable, a chamfer should be provided to minimize spalling.

Metal drive screws are available in standard sizes, lengths, and finishes with round heads only.

The thickness of metal into which the screws are to be driven must be at least the same as the outside diameter of the drive screw to insure adequate thread engagement. An advantage of

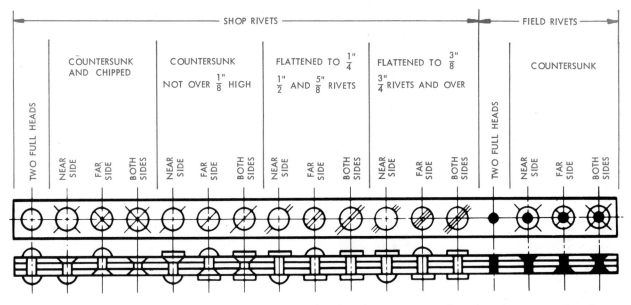

Fig. 18-35. Rivets may be indicated on a drawing by symbols that show the process as well as where the process takes place.

using these screws in place of machine screws is the elimination of tapped holes, although a pilot hole is necessary. See Fig. 18-33.

On a drawing a drive screw is identified by a note as follows:

.025 X .50 DRIVE SCREW

18.5.3 Non-threaded fasteners. Examples of non-threaded fasteners are rivets, pins, keys, nails, and washers.

Rivets. Rivets are considered permanent fastening devices and are used in joining parts constructed of sheet metal or steel plate. They are made of many different kinds of metal. The most common are wrought iron, steel, copper, brass, and aluminum.

Rivets are available in various head shapes such as cone, button, truss, countersunk, pan, and flat. See Fig. 18-34. The size of a rivet is usually indicated by the diameter and length of the stem.

On a drawing rivets are represented by circle symbols, as shown in Fig. 18-35, or simply by center lines. Rivet symbols, when used, specify the riveting procedure, location, and whether the riveting is to be done in the shop or in the field. For fabrication in the shop, the symbols are left open, whereas for field riveting the symbols are made

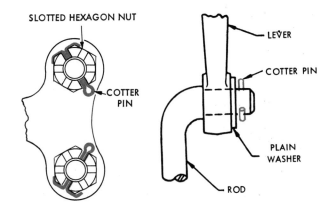

Fig. 18-36. Cotter pins are used to lock slotted nuts, rods, or movable links.

solid. When it is necessary to designate a rivet on a drawing, a note stating the size, type of head, and the length of the rivet is usually given.

Pins.[10] Pins are available in various shapes to serve specific functions. Among the most common types are these:

Cotter pins. Used for retaining or locking such devices as slotted nuts, ball sockets, movable links, and rods. See Fig. 18-36.

---

10. *Ibid.*

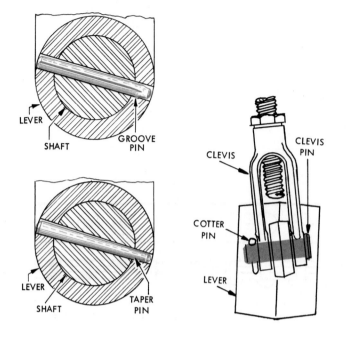

Fig. 18-37. Pins of various sorts are used to lock parts of an assembly in position.

Groove pins. Straight pins made of cold-drawn steel. They have longitudinal grooves rolled or pressed into the body to provide a locking effect when the pin is driven into a drilled hole. This type of pin makes reaming or peening unnecessary and can be disassembled a number of times without serious loss of holding power. Groove pins are used for semi-permanent fastening of levers, collars, gears, cams, and so on, to shafts. Groove pins may also be used as guides or locating pins. See Fig. 18-37.

Taper pins. Serve the same functional purpose as groove pins. However, they require taper reamed holes at assembly and depend only on taper locking, which can totally disengage when minor displacement occurs. See Fig. 18-37.

Clevis pins. Used to attach clevises to rod ends and levers and to serve as bearings. They are held in place by cotter pins. See Fig. 18-37.

Keys. Keys are used to prevent parts attached to shafts, wheels, cranks, and so on, from rotating.

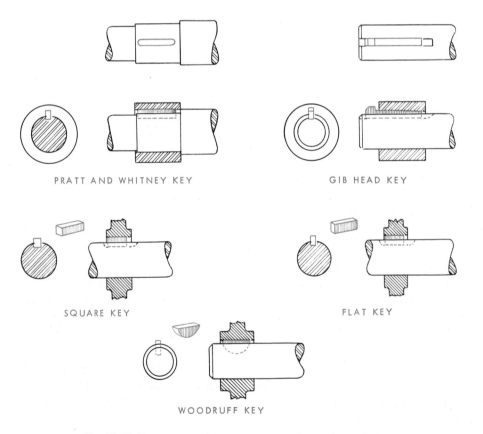

Fig. 18-38. Keys are used to secure parts against rotary motion.

Fig. 18-38 illustrates some of the more common types of keys. When force is not too severe, either a flat key or round key is used. On heavier work the rectangular key is more suitable. The Pratt and Whitney and the square keys are probably more popular for most machine designs. The gib-head key is designed so the head protrudes far enough from the hub to allow the insertion of a drift pin to remove the key. Another frequently used key is known as the Woodruff key. This key consists of a flat segmental disc with a flat or round bottom.

On drawings, flat, taper, square, and gib-head keys are specified by a note giving the width, height, and length. Pratt and Whitney keys are shown by a number. Woodruff keys are also specified by a number with the last two digits representing the nominal diameter in eighths of an inch, and the preceding digits indicating the width in thirty-seconds of an inch. (See Appendix for sizes of different keys.) The information for keys should be listed as follows:

.188 X 1.25 SQUARE KEY

.250 X .188 X 1.25 FLAT KEY

NO. 12 PRATT & WHITNEY KEY

NO. 304 WOODRUFF KEY

Keys are usually not drawn except when some special key having other than standard limits must be shown. The practice is often to dimension keyways on shafts or internal members.

Nails. Nails come in a variety of basic types and sizes. The following are the most common:

Common nails. Have larger diameters and wider heads than other types and are used mostly in rough carpentry.

Box nails. Have wide heads but are not as large in diameter as common nails. They are used extensively in box construction and in carpentry where common nails would be unsuitable.

Casing nails. Smaller in diameter and head size than box nails and are especially designed for blind nailing of flooring, ceilings, and cabinet work where large heads are undesirable.

Finishing nails. Have the smallest diameters and the smallest heads. Their chief use is in cabinet work and furniture construction where it is often necessary to sink the heads below the surface of the wood.

Sizes of nails are designated by the term *penny* (symbol *d*) with a number as a prefix, such as 4*d*, 10*d*. The term *penny* refers to the weight of the nails per thousand in quantity. Thus, a 6*d* nail means that the nails weigh six pounds per thousand. The weight has a direct relationship to the size.

Brads. The smallest type of finishing nails. The sizes of brads are indicated by their length in inches and the diameters by the gage number of the wire. The higher the gage number, the smaller the diameter.

Tacks. Used mostly for fastening material to some wooden surface. Sizes of tacks are indicated by a gage number, which in turn governs their length.

Screw nails. Used for joining sheet metal to wood. The spiral threads cut into the burr formed in the sheet metal by the pilot and form their way into the wood to provide the fastening. They are driven in a manner similar to ordinary nails. A pierced hole should be provided in heavy sheet metal to facilitate assembly. Screw nails are available with flat, countersunk, oval, and round heads in a range of sizes and lengths. See Fig. 18-39.

Helix nails. Used for the same purpose as screw nails. They are made from square stock and twisted to shape. See Fig. 18-40. Helix nails are available in flat, countersunk, and round heads and may be provided with diamond, needle, or chisel points.

Fig. 18-39. A screw nail.

Fig. 18-40. A flat-head helix nail.

On a drawing nails are shown by a note as:

8D FINISHING NAIL

1"—NO. 20 BRAD

NO. 4 CARPET TACK

Washers.[11] The three basic types of washers are plain washers, spring lock washers, and tooth lock washers. All three types are available in standard sizes to suit standard bolts and screws.

Plain washers. Annular-shaped parts, usually flat. They are used for two principal purposes. They may be used under the head of a screw or bolt or under a nut to spread a load over a greater area or they may be used to prevent the marring of the parts as a result of the turning of the screw, bolt, or nut during assembly.

Spring lock washers. Made of steel that is capable of being hardened or of bronze or aluminum alloys. They are split on one side and are helical in shape, as shown in Fig. 18-41. They have the dual function of (1) springing take-up devices to compensate for developed looseness and the loss of tension between component parts of an assembly, and (2) acting as hardened thrust bearings to facilitate assembly and disassembly of bolted fastenings by decreasing the frictional resistance between the bolted surface and the bearing face of the bolt head or nut.

Tooth lock washers. Made in three types as shown in Fig. 18-42: the external, the internal, and the internal-external. The hardened teeth of these washers are twisted offset to bite or grip both the bolt head or the nut and the respective work surface to help prevent the loosening of the assembly due to vibration. They also make good electrical contacts. Unlike the spring lock washers, they do not provide spring action to counteract wear or stretch in the parts of an assembly.

The external-tooth lock washer is the most commonly used of the tooth-type lock washers, but the internal-tooth lock washer is generally used where it is necessary to consider appearance and to insure engagement of teeth with the bearing surface of the fastener.

---

11. *Ibid.*

PLAIN   SPRING LOCK

Fig. 18-41. Spring lock washers are split and helical in shape.

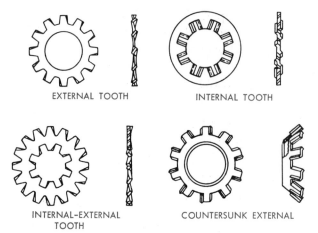

EXTERNAL TOOTH   INTERNAL TOOTH

INTERNAL-EXTERNAL TOOTH   COUNTERSUNK EXTERNAL

Fig. 18-42. Various types of tooth lock washers have teeth that are offset to bite or grip the bolt or nut and the work surface.

Where additional locking ability is required or where there is need for a large bearing surface, such as over a clearance hole, the internal-external-tooth lock washer may be used. Countersunk external-tooth lock washers are used with flat-head and oval-head machine screws.

There are many variations of special washers for specific applications. Most common among these special washers are the finish washer and the grip washer. See Fig. 18-43. *Finish washers* are used under the heads of countersunk and oval-head screws to provide proper seating of the screw heads and at the same time to eliminate countersinking of the work face. The finish washers, as the name implies, enhance the appearance of the product. *Grip washers,* both square and round, are used under bolt heads against wood. They provide a solid seat for the bolt head, and they prevent depressing or scoring of the wood.

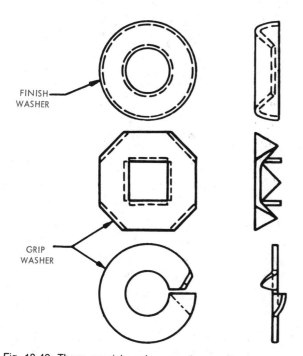

FINISH WASHER

GRIP WASHER

Fig. 18-43. These special washers are for specific applications such as under bolt heads against wood.

COMPRESSION COIL SPRING

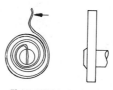

EXTENSION COIL SPRING

TORSION COIL SPRING

FLAT SPRING

VOLUTE SPRING

FLAT SPIRAL SPRING

BELLEVILLE SPRING

LEAF SPRING

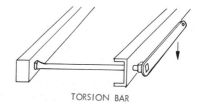

TORSION BAR

Fig. 18-44. Springs are elastic bodies designed to perform functions involving pressure and movement.

18.6 Springs

Springs are devices that store energy when distorted and then return an equivalent amount of energy upon their release. They are very important units in the operation of numerous mechanical and electrical components.

Springs are classified into three main groups: controlled action, variable action, and static. *Controlled action springs* have a regulated range of action, such as in valve or switch springs. *Variable action springs* are those which vary in their range of movement, such as those used on clutches or brakes. *Static springs* exert a constant pressure or tension between several parts.

18.6.1 Types of springs.[12] The following are the most common types of springs:

A *compression spring* is an open-coil helical spring that offers resistance to a compressive force applied axially. Compression springs are generally made cylindrical in form, although other

forms are used, such as conical, tapered, concave, or convex. See Fig. 18-44. Wire for these springs may be round, square, or of rectangular cross section.

An *extension coil spring* is a close-wound or an open-wound spring that offers resistance to an axial force tending to extend its length. When ex-

---

12. *Ibid.*

tended, it tends to draw two objects together. See Fig. 18-44. Extension springs are formed or fitted with ends which are used for attaching the spring to the assembly. They are made of wire of circular, square, or rectangular cross-section.

Torsion coil springs are springs that offer resistance to or exert a turning force in a plane at right angles to the axis of the coil. Torsion springs are generally made of circular cross-section in a variety of forms. See Fig. 18-44.

Flat springs are made of flat material formed in such a manner as to apply force in the desired direction when deflected in the opposite direction. The flat spring is often designed as an individual member of the mechanism and may perform other functions besides applying force. See Fig. 18-44.

Belleville springs are washer-shaped and made in the form of a short truncated cone. This type of spring has the ability to store a large amount of energy in a small space. It is limited to very small deflections and operates with a variable rate. See Fig. 18-44.

A *leaf spring* is composed of a series of flat leaves nested together and arranged so as to provide approximately uniform distribution of stress throughout its length. Fig. 18-44.

A *volute spring* is a conical compression-type spring produced by winding a flat material upon itself, with the wide dimension parallel to the axis of the helix, in such a manner that the inner coils telescope within the outer coils when compressed. See Fig. 18-44.

A *torsion-bar spring* is a relatively straight bar anchored at one end, on which a torque may be exerted at the other end, thus tending to twist it about its axis. See Fig. 18-44.

18.6.2 Spring glossary.[13] The following terms will help drafting personnel better understand the process of depicting springs on a drawing:

Active coils (N). Total number of coils less those rendered inactive by the nature of design or application of the spring. See Fig. 18-45.

Coil. One complete convolution, or turn, of the wire about the axis of the spring. See Fig. 18-45.

Direction of helix. A right-hand spring-coil helix runs in the same direction as the helix of a right-

hand thread of a screw. A left-hand spring-coil helix is in the opposite direction to that of a right-hand helix. See Fig. 18-45.

Free length. Length of a coil spring measured parallel to its axis in the free condition or without force applied. The measurement is taken overall on a compression spring but is usually taken inside the hooks on an extension spring.

Inside diameter (ID). The diameter of a coil spring, measured on the inside of the wire, perpendicular to the spring axis. See Fig. 18-45.

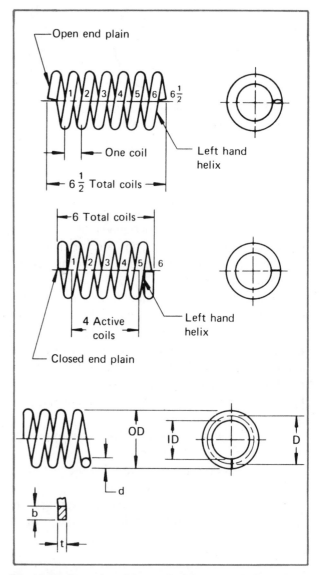

Fig. 18-45. Examples of how coil springs are designated on drawings.

13. *Ibid.*

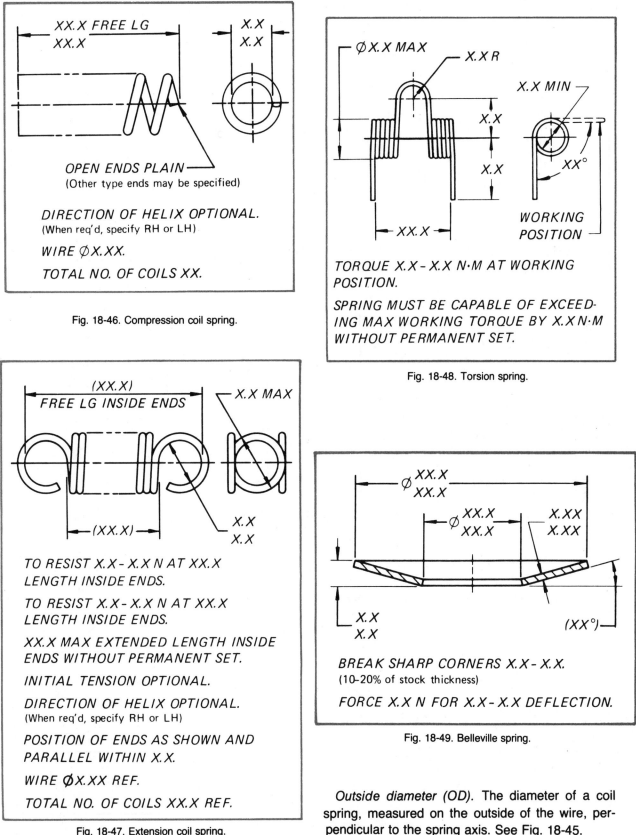

Fig. 18-46. Compression coil spring.

Fig. 18-47. Extension coil spring.

Fig. 18-48. Torsion spring.

Fig. 18-49. Belleville spring.

Outside diameter (OD). The diameter of a coil spring, measured on the outside of the wire, perpendicular to the spring axis. See Fig. 18-45.

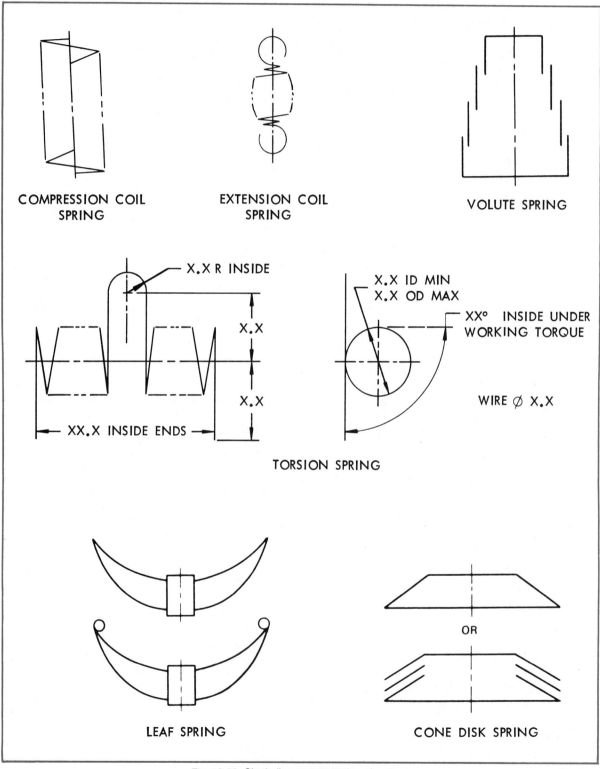

COMPRESSION COIL
SPRING

EXTENSION COIL
SPRING

VOLUTE SPRING

X.X R INSIDE

X.X

X.X

XX.X INSIDE ENDS

X.X ID MIN
X.X OD MAX

XX° INSIDE UNDER
WORKING TORQUE

WIRE ∅ X.X

TORSION SPRING

LEAF SPRING

OR

CONE DISK SPRING

Fig. 18-50. Single-line representation of springs.

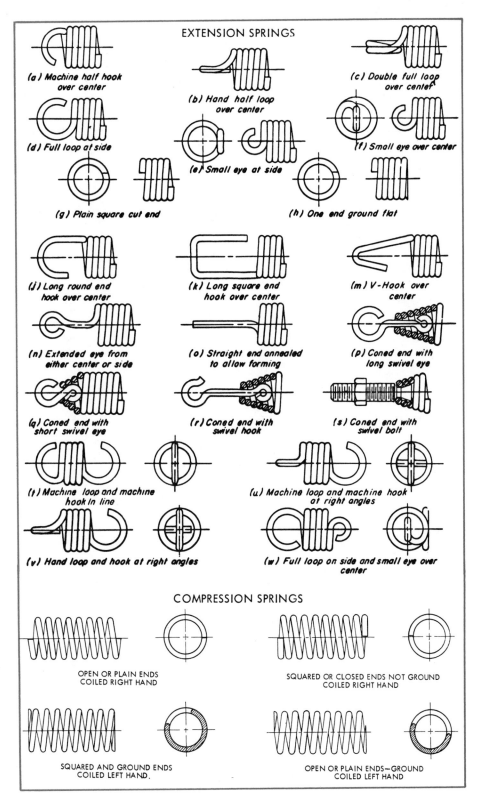

Fig. 18-51. Coil spring end types.

Rate. The ratio of change in force to the corresponding change in deflection.

Solid length. The length of a compression spring with all the coils completely compressed.

Total coils. The number of complete coils plus portions of coils. See Fig. 18-45.

Wire size (d). The cross-sectional diameter of round wire used in forming the spring. For rectangular wire it is the width and thickness (b and t, respectively). See Fig. 18-45.

18.6.3 Spring drawing practices. Typical spring drawings are shown in Figs. 18-46, 18-47, 18-48, and 18-49. An alternate method of depicting springs on detail or assembly drawings is by single-line representations as shown in Fig. 18-50. Fig. 18-51 shows various spring end types.

SECTION V
UNIT 19

Representation of Weldments

Welding requirements are indicated on a drawing by a series of symbols developed by the American Welding Society and universally adopted by industry. These symbols convey such information as what type of weld is required, where the weld is to be located, the size of the weld, and supplementary data necessary for the welding operator.

19.1 Weld Symbols

The main feature that identifies a weld is a reference line with an arrow at one end. The other data specifying various characteristics of the weld is shown by abbreviations, figures, and symbols placed around this reference line. See Fig. 19-1.

The reference line is applied to five basic joints that are used in welding: butt, corner, T, lap, and edge. See Fig. 19-2. Welds themselves are classified as fillet, plug or slot, spot, groove, flange, and seam. Each type of weld has its own specific symbol. See Fig. 19-1.

Selection of joints for various purposes is governed by four factors: (1) magnitude of the load, (2) characteristic of the load—that is, whether the load is in compression or tension, (3) application of the load—that is, whether it is steady, variable, or sudden, and (4) the cost of preparing and welding the joint.

19.2 Location of Weld Symbols

A weld is said to be either on the arrow side or other side of a joint. The arrow side is the surface in the direct line of vision, while the other side is the opposite surface of the joint. See Fig. 19-3.

Weld location is designated by running the arrowhead of the reference line to the joint. The direction of the arrow is not important; that is, it can run on either side of a joint and extend upward or downward. See Fig. 19-4. If the weld is to be made on the arrow side, the appropriate weld symbol is placed below the reference line. If the weld is to be located on the other side of the joint, the weld symbol is placed above the reference line. When both sides of the joint are to be welded, the same weld symbol appears above and below the reference line.

The only exception to this practice of indicating weld location is in seam and spot welding. With

545

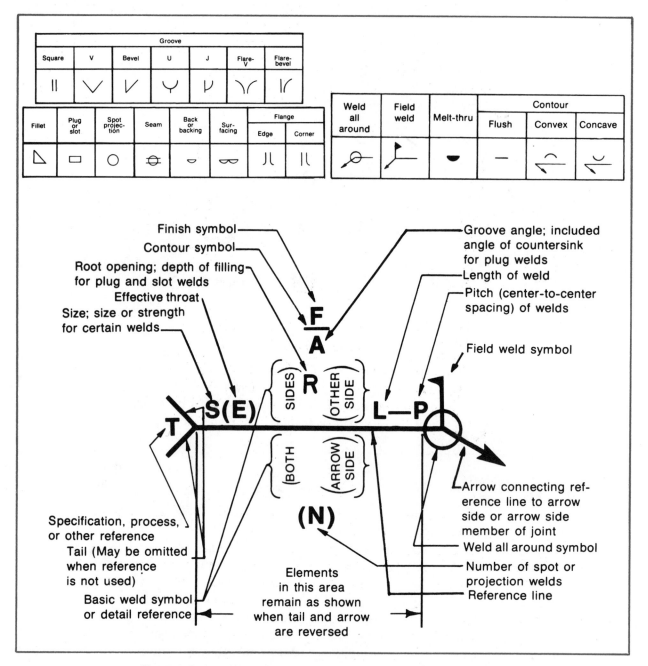

Fig. 19-1. Basic welding symbols adopted by the American Welding Society.

seam or spot welds the arrowhead is simply run to the center line of the weld seam, and the appropriate weld symbol is centered below or above the reference line. If the side is not important, the symbol is placed astride the reference line. See Fig. 19-3.

Information on weld symbols is placed to read from left to right along the reference line in accordance with the usual conventions of drafting.

Fillet, bevel-groove, J-groove, flare-bevel-groove, and corner-flange weld symbols are shown with the *perpendicular leg to the left.*

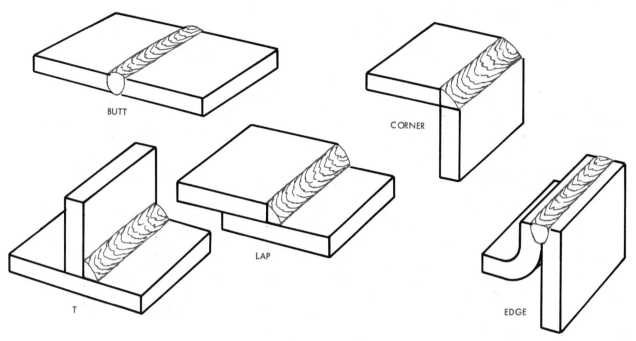

Fig. 19-2. These are the basic joints used in welding.

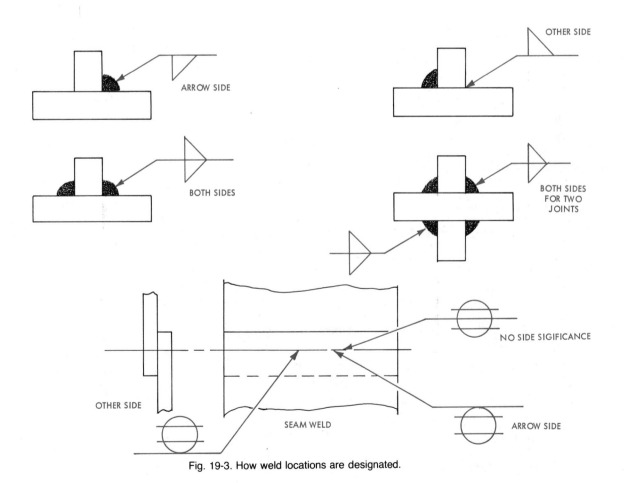

Fig. 19-3. How weld locations are designated.

LOCATION SIGNIFICANCE	FILLET	PLUG OR SLOT	SPOT OR PROJECTION	SEAM	GROOVE		
					SQUARE	V	BEVEL
ARROW SIDE							
OTHER SIDE							
BOTH SIDES		NOT USED	NOT USED	NOT USED			

LOCATION SIGNIFICANCE	GROOVE				FLANGE	
	U	J	FLARE-V	FLARE-BEVEL	EDGE	CORNER
ARROW SIDE						
OTHER SIDE						
BOTH SIDES					NOT USED	NOT USED

Fig. 19-4. Arrows may extend either upward or downward. (AWS)

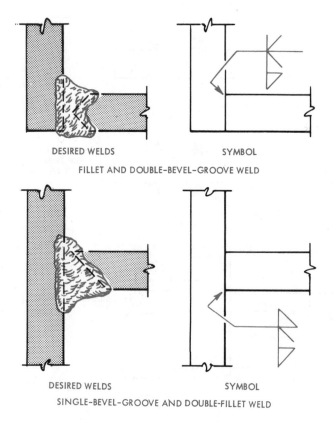

DESIRED WELDS · SYMBOL

FILLET AND DOUBLE-BEVEL-GROOVE WELD

DESIRED WELDS · SYMBOL

SINGLE-BEVEL-GROOVE AND DOUBLE-FILLET WELD

Fig. 19-5. Combined weld symbols.

19.2.1 Combined weld symbols. In the fabrication of a product, there are occasions when more than one type of weld is to be made on a joint. Thus, a joint may require both a fillet and double-bevel-groove weld. When this happens, a symbol is shown for each weld. See Fig. 19-5.

19.3 Specifying Weld Sizes

To achieve a sound welded structure, a weld must not only have the proper penetration but it also must be correct in size. To make sure that the welding operator complies with the size requirement, each weld size is specifically indicated on the weld print. The method of designating weld size is governed by the type of weld to be made.

19.3.1 Fillet welds. The width of a fillet weld is shown to the left of the weld symbol and is expressed in fractions*, decimals, or millimeters. See Fig. 19-6. When both sides of a fillet are to be welded and both welds are to have the same di-

---

*Although AWS still includes fractions for weld sizes, most industries specify sizes in decimal-inch or millimeters on drawings.

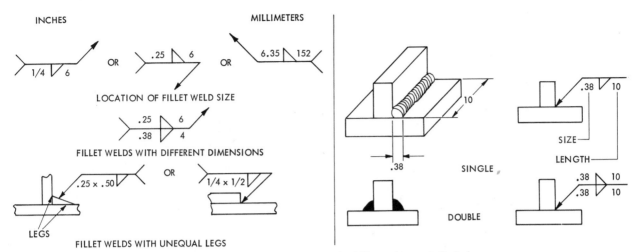

Fig. 19-6. How size and length of fillet welds are indicated.

mensions, both should be dimensioned. If the welds differ in dimensions, both must be dimensioned. Where a note that governs the size of a fillet weld appears on a drawing, no dimensions are usually shown on the symbol. In the case of a fillet weld with unequal legs, the sizes of the legs are placed in parentheses to the left of the weld symbol. See Fig. 19-6.

The length of the weld is shown to the right of the weld symbol by numerical values representing the actual required length. See Fig. 19-6.

19.3.2 Intermittent fillet welds. The length and pitch increments of intermittent welds are shown to the right of the weld symbol. The first figure represents the length of the weld section and the second figure the distance between centers of increments. See Fig. 19-7.

19.3.3 Plug welds. The size of a plug weld is shown to the left of the weld symbol, the depth when less than full on the inside of the weld symbol, the center-to-center spacing (pitch) to the right of the symbol, and the included angle of countersink below the symbol. See Fig. 19-8.

19.3.4 Slot welds. Length, width, spacing, angle of countersink, and location of slot welds are not shown on the weld symbol because it would be too cumbersome. This data is included by showing a special detail on the print. If slots are only partially filled, then the depth of filling is shown inside the weld symbol. See Fig. 19-9.

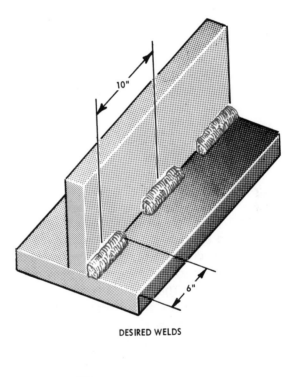

DESIRED WELDS

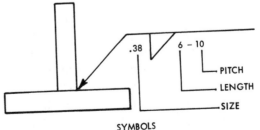

SYMBOLS

Fig. 19-7. How length and pitch of intermittent fillet welds are indicated.

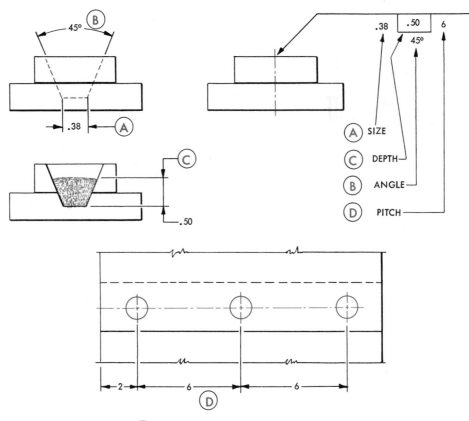

Fig. 19-8. How plug welds are indicated.

19.3.5 Spot welds. Spot, or projection, welds are dimensioned either by size or strength. Size is designated as the diameter of the weld expressed in fractions, decimals, or millimeters, and placed to the left of the symbol. The strength is also placed to the left of the symbol and expresses the required minimum shear strength in pounds per spot. The spacing of spot welds is shown to the right of the symbol. When a definite number of spot welds are needed in a joint, this number is indicated in parentheses either above or below the weld symbol. See Fig. 19-10.

19.3.6 Seam welds. Seam welds are dimensioned either by size or strength. Size is designated as the width of the weld in fractions, decimals, or millimeters, and shown to the left of the weld symbol. The length of the weld seam is placed to the right of the weld symbol. The pitch of intermittent seam welds is shown to the right of

the length dimension. See Fig. 19-11. The strength of the weld, when used, is located to the left of the symbol and is expressed as the minimum acceptable shear strength in pounds per linear inch.

19.3.7 Groove welds. There are several types of groove welds. Their sizes in fractions, decimals, or metric units are shown as follows:

1. For single-groove and symmetrical double-groove welds which extend completely through the members being joined, no size is included on the weld symbol. See Fig. 19-12.

2. For groove welds which extend only partly through the members being joined, the effective throat size is shown in parentheses on the left of the weld symbol. See Fig. 19-13.

3. A dimension not in parentheses when placed to the left of the weld symbol indicates the size of the bevel only. See Fig. 19-13. When both the ef-

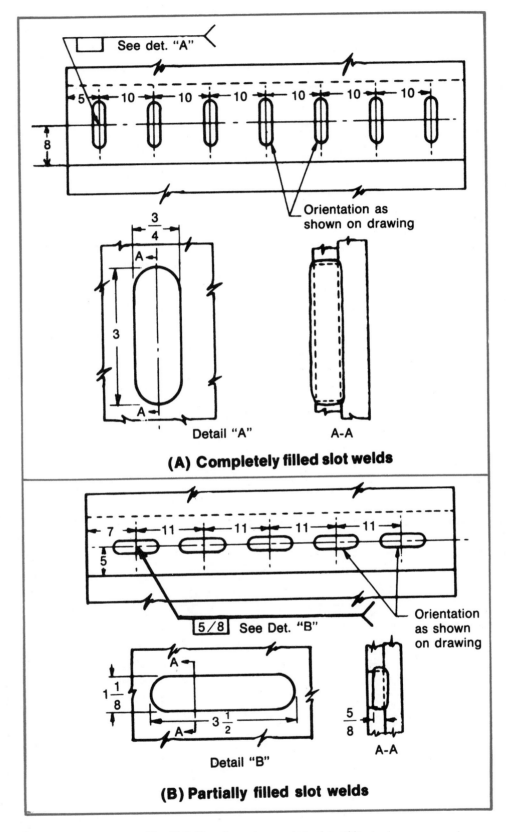

Fig. 19-9. How dimensions apply to slot welds.

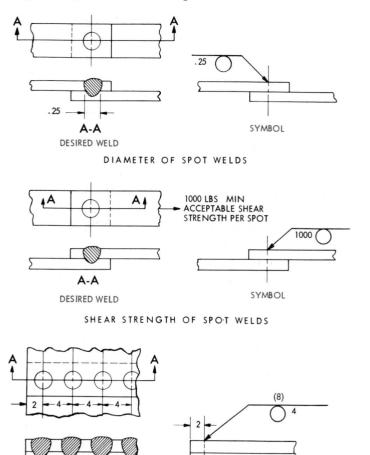

DIAMETER OF SPOT WELDS

SHEAR STRENGTH OF SPOT WELDS

Fig. 19-10. How spot welds are shown.

PITCH AND NUMBER OF SPOT WELDS SHOWN ON SYMBOL

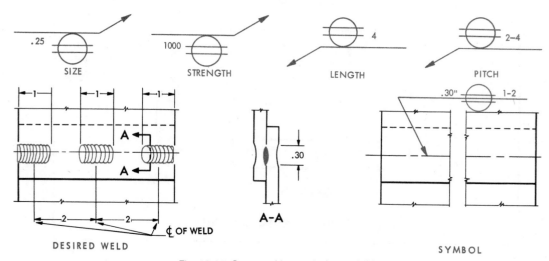

Fig. 19-11. Seam welds are designated this way.

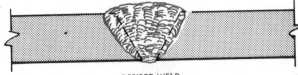

DESIRED WELD

Fig. 19-12. Size is not shown for single-groove and symmetrical double-groove welds with complete penetration.

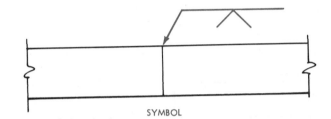

SYMBOL

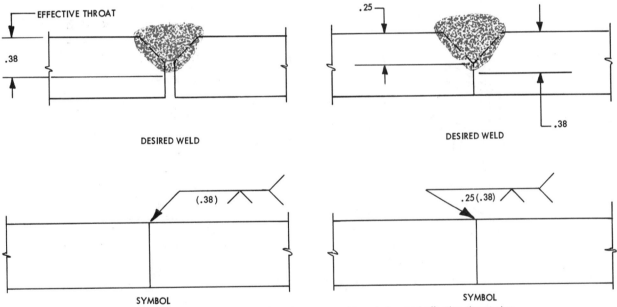

DESIRED WELD

DESIRED WELD

SYMBOL

SYMBOL

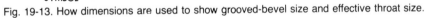

Fig. 19-13. How dimensions are used to show grooved-bevel size and effective throat size.

fective throat and bevel sizes are to be used, the groove bevel size is located to the left of the effective throat size as shown in Fig. 19-13.

4. The root opening of grooved joints is shown inside the weld symbol. See Fig. 19-14. The included angle of the bevel is placed below or above the weld symbol. See Fig. 19-14.

5. The size of flare-groove welds (V and bevel) is considered as extending only to the tangent

points as indicated by dimensional lines. Refer to Fig. 19-15.

On a bevel or J-groove joint it is often necessary to indicate which member is to be bevelled. In such cases the arrow should point with a definite break toward the member to be bevelled. See Fig. 19-16.

19.3.8 Flange welds. The radius and height of the flange are separated by a plus mark and

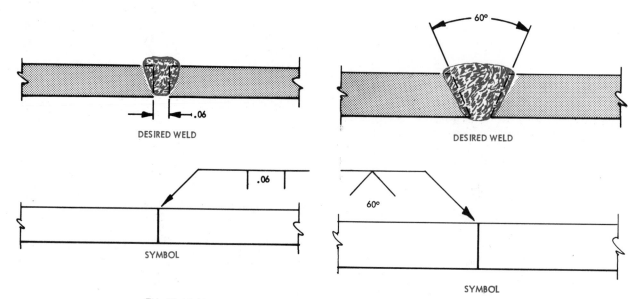

Fig. 19-14. How root opening and included angle are shown for groove welds.

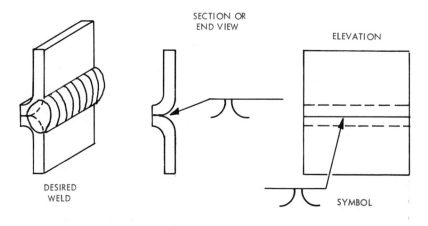

ARROW-SIDE FLARE-V-GROOVE WELDING SYMBOL

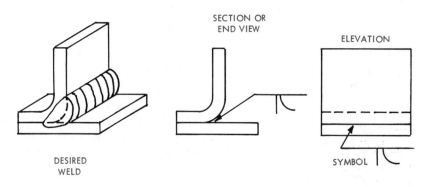

ARROW-SIDE FLARE-BEVEL-GROOVE WELDING SYMBOL

Fig. 19-15. Application of flare-groove welding symbol.

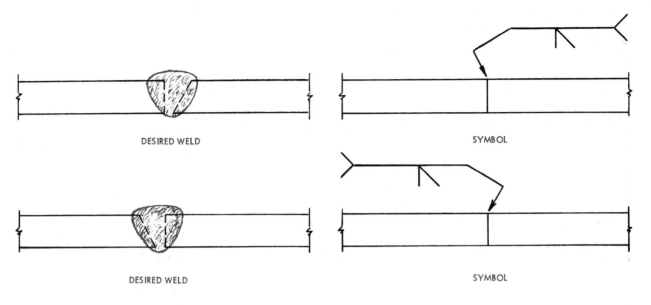

Fig. 19-16. Application of a break in arrow to show member to be bevelled.

placed to the left of the weld symbol. The size of the weld is shown by a dimension located outward of the flange dimensions. See Fig. 19-17.

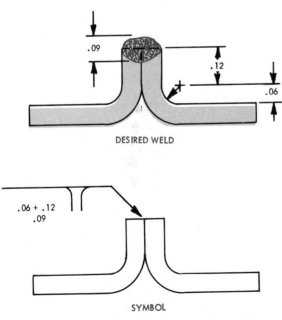

Fig. 19-17. Application of flange weld symbol.

19.4 Supplementary Welding Data

In addition to types of joints and weld sizes, other data is often included on welding prints to designate certain required weld characteristics more specifically. Supplementary information is presented by means of symbols or, in some cases, by special references or notes.

19.4.1 Reference tail. The tail is included only when some definite welding specification, procedure, reference, or welding or cutting process needs to be called out; otherwise it is omitted. This data is often in the form of symbols and is inserted in the tail, Figs. 19-18 and 19-19. Abbreviations in the tail may also call out some welding specifications which are included in more precise details on some other part of the print.

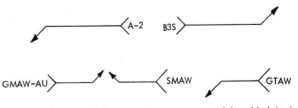

Fig. 19-18. The tail is used when some special weld data is needed.

Letter designation	Welding and allied processes	Letter designation	Welding and allied processes
AAC	air carbon arc cutting	GTAW	gas tungsten arc welding
AAW	air acetylene welding	GTAW-P	gas tungsten arc welding — pulsed arc
ABD	adhesive bonding	HFRW	high frequency resistance welding
AB	arc brazing	HPW	hot pressure welding
AC	arc cutting	IB	induction brazing
AHW	atomic hydrogen welding	INS	iron soldering
AOC	oxygen arc cutting	IRB	infrared brazing
AW	arc welding	IRS	infrared soldering
B	brazing	IS	induction soldering
BB	block brazing	IW	induction welding
BMAW	bare metal arc welding	LBC	laser beam cutting
CAC	carbon arc cutting	LBW	laser beam welding
CAW	carbon arc welding	LOC	oxygen lance cutting
CAW-G	gas carbon arc welding	MAC	metal arc cutting
CAW-S	shielded carbon arc welding	OAW	oxyacetylene welding
CAW-T	twin carbon arc welding	OC	oxygen cutting
CW	cold welding	OFC	oxyfuel gas cutting
DB	dip brazing	OFC-A	oxyacetylene cutting
DFB	diffusion brazing	OFC-H	oxyhydrogen cutting
DFW	diffusion welding	OFC-N	oxynatural gas cutting
DS	dip soldering	OFC-P	oxypropane cutting
EASP	electric arc spraying	OFW	oxyfuel gas welding
EBC	electron beam cutting	OHW	oxyhydrogen welding
EBW	electron beam welding	PAC	plasma arc cutting
ESW	electroslag welding	PAW	plasma arc welding
EXW	explosion welding	PEW	percussion welding
FB	furnace brazing	PGW	pressure gas welding
FCAW	flux cored arc welding	POC	metal powder cutting
FCAW-EG	flux cored arc welding — electrogas	PSP	plasma spraying
FLB	flow brazing	RB	resistance brazing
FLOW	flow welding	RPW	projection welding
FLSP	flame spraying	RS	resistance soldering
FOC	chemical flux cutting	RSEW	resistance seam welding
FOW	forge welding	RSW	resistance spot welding
FRW	friction welding	ROW	roll welding
FS	furnace soldering	RW	resistance welding
FW	flash welding	S	soldering
GMAC	gas metal arc cutting	SAW	submerged arc welding
GMAW	gas metal arc welding	SAW-S	series submerged arc welding
GMAW-EG	gas metal arc welding — electrogas	SMAC	shielded metal arc cutting
GMAW-P	gas metal arc welding — pulsed arc	SMAW	shielded metal arc welding
GMAW-S	gas metal arc welding — short circuiting arc	SSW	solid state welding
		SW	stud arc welding
GTAC	gas tungsten arc cutting	TB	torch brazing

Automatic	AU	Manual	MA
Machine	ME	Semiautomatic	SA

Fig. 19-19. Designation of welding and cutting processes. (AWS)

19.4.2 Surface contour of welds. When bead contour is important, a special flush, concave, or convex contour symbol is added to the weld symbol. Welds that are to be mechanically finished also carry a finish symbol along with the contour symbols. See Fig. 19-20.

19.4.3 Back welds. Back, or backing, welds refer to welds made on the opposite side of regular welds. Back welds are occasionally specified to insure adequate penetration and provide additional strength to a joint. This particular requirement is included opposite the weld symbol. No dimensions of back welds except height of reinforcement is shown on the weld symbol. See Fig. 19-21.

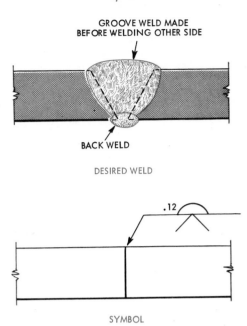

Fig. 19-21. Use of back weld symbol to indicate back weld.

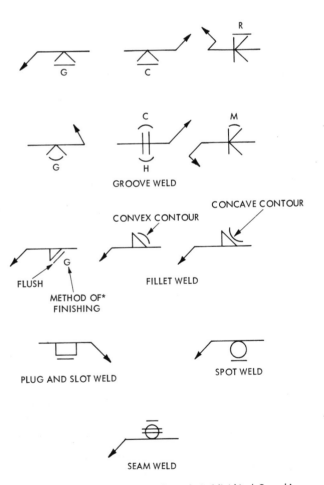

* Finish symbols used herein indicate the method of finishing(C = chipping; G = grinding; M = machining; R = rolling; H = hammering;) and not the degree of finish

Fig. 19-20. How bead contour finish is indicated.

19.4.4 Melt-thru welds. When complete joint penetration of the weld through the material is required in welds made from one side only, a special melt-thru weld symbol is placed opposite the regular weld symbol. No dimension of melt-thru, except height of reinforcement, if significant, is shown on the weld symbol. See Fig. 19-22.

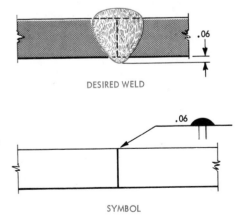

Fig. 19-22. Application of melt-thru symbol.

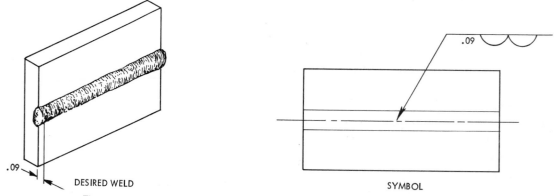

Fig. 19-23. Application of surfacing symbol to indicate surfaces built up by welding.

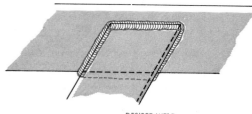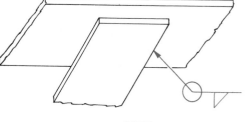

DESIRED WELD

SYMBOL

Fig. 19-24. Weld-all-around symbol.

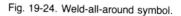

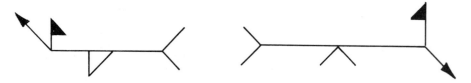

Fig. 19-25. Field weld symbol.

19.4.5 Surfacing welds. Welds whose surfaces must be built up by single- or multiple-pass welding are provided with a surfacing weld symbol. The height of the built-up surface is indicated by a dimension placed to the left of the surfacing symbol. See Fig. 19-23.

19.4.6 All-around welds. When a weld is to extend completely around a joint, a small circle is placed where the arrow connects the reference line. See Fig. 19-24.

19.4.7 Field welds. Welds that are to be made in the field (welds not made in a shop or at the place of initial construction), are indicated by a darkened triangular flag located at the junction of the reference line and arrow. The flag always points toward the tail of the arrow. See Fig. 19-25.

UNIT 20

Gears, Chain and Belt Drives

Gear, chain, and belt drives are used to transmit power and motion from one shaft to another. They also provide a means of regulating speed or altering the direction of motion. This unit presents basic information on gears, chains, and belts so that drafting personnel can design and prepare engineering drawings for power mechanical transmission systems.

20.1 Types of Gears

The following are general descriptions of the common types of gears:

Spur gears are cylindrical in shape with teeth that are formed straight across the face of the gear and parallel to the shaft axis or bore. See Fig. 20-1. They are used on shafts that are parallel to each other and transmit power or motion at the same or different speeds between shafts.

A *rack* is a type of spur gear used to transfer circular motion into straight-line motion. A rack has teeth on the surface of a straight bar instead of a cylindrical gear blank. See Fig. 20-2.

Internal gears are gears with teeth parallel to the shaft but on the inside of cylindrical forms. See

Fig. 20-1. Spur gears are the simplest of all gears.

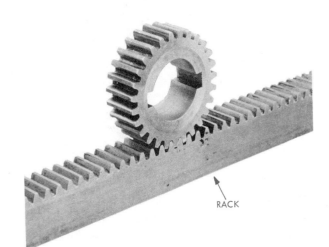

Fig. 20-2. A rack is used to transfer circular motion into straight line motion.

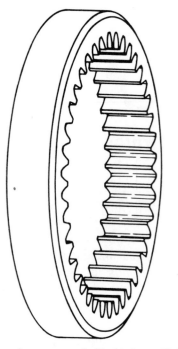

Fig. 20-3. Internal gears are cylindrical in form with teeth on the inner surface of the cylinder or ring.

Fig. 20-3. Internal gears are often called *ring gears.*

Bevel gears are shaped like cone sections and transmit power and motion between intersecting shafts. See Fig. 20-4. The angle between shafts is understood to be 90° unless otherwise specified.

Bevel gears having shafts that intersect at other than a right angle are called *angle gears.* Bevel gears in which both the driver and driven gear have the same number of teeth and with shafts at 90° are referred to as *miter gears.*

Regular bevel gears have straight teeth. Spiral bevel gears have either curved or oblique teeth. See Fig. 20-5.

Helical gears are cylindrical in form with teeth cut at an angle to the shaft axis or bore. See Fig. 20-6. Helical gears are designed to connect non-intersecting shafts, which may be at any angle with each other. Standard helical gears of 45° angle are used to connect parallel shafts or shafts at right angles to each other. These gears usually run more smoothly and quietly than spur gears.

STRAIGHT
TOOTH

Fig. 20-4. Bevel gears on shafts at right angles produce a change in speed.

SPIRAL
TOOTH

Fig. 20-5. Miter gears transmit power and motion between intersecting shafts at right angles with no change in speed.

Fig. 20-6. Helical gears with teeth at 45° are used to connect parallel shafts or shafts at 90° to each other.

WORM

Fig. 20-7. Worms and wormgears are intended for right angle shafts or non-intersecting shafts where high-ratio speed reduction is required.

They are often called *spiral gears* or *crossed axis helicals.*

Worms and wormgears are used to transmit motion or power between right-angle shafts when a high-ratio speed reduction is necessary. The worm is the small gear which drives the larger wormgear. Worm threads resemble screw threads and are made with single, double, triple, or quadruple leads. See Fig. 20-7.

A *pinion* is a small gear which meshes with the main gear and provides the essential power to run the gear. A pinion is generally referred to as the driver. See Fig. 20-1.

20.2 Spur Gears

Spur gears are the most common type of gears and, as previously described, have teeth that are straight and parallel to the shaft axis. These are simple gears, low in cost and easy to maintain. They do, however, have a lower capacity than other types, and they are sometimes noisier.

20.2.1 Spur gear terms.[1] The terms defined in

---

1. *Boston Gear Works Manual* (Quincey, Mass.: Boston Gear Works).

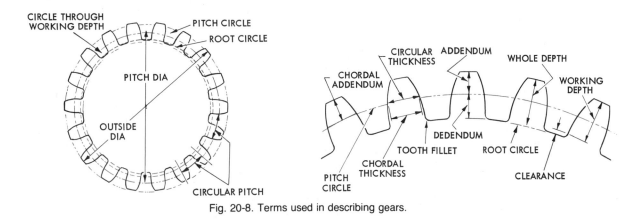

Fig. 20-8. Terms used in describing gears.

the following list are common to most spur gears. See Fig. 20-8.

Outside diameter. Diameter of the circle around the extreme outer edges of the teeth.

Pitch circle. Theoretical circle on which the teeth of the mating gears mesh.

Pitch diameter. Diameter of the pitch circle.

Root diameter. Diameter of the root circle.

Diametral pitch (DP). Refers to tooth size. It is a ratio of the number of teeth in the gear to each inch of its pitch diameter; that is, a 4-pitch gear has four teeth for each inch of pitch diameter.

Circular pitch. Distance on the circumference of the pitch circle between corresponding points of adjacent teeth. Circular pitch is measured from the center of one tooth to the center of the next tooth on the pitch circle or from the side of one tooth to the corresponding side of the next tooth on the pitch circle.

Whole depth. Distance from the top of the tooth to the root circle.

Addendum. Distance from the pitch circle to the top of the tooth.

Dedendum. Distance from the pitch circle to the root circle equals addendum plus clearance.

Working depth. Depth of engagement of two gears or the sum of their addendums.

Clearance. The amount by which the dedendum in a given gear exceeds the addendum of its mating gear. It is the space between the top of one tooth and the bottom of the mating tooth when gears are in mesh.

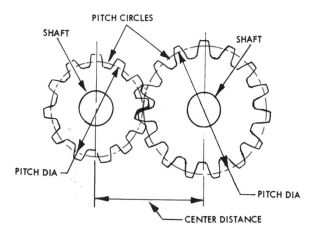

Fig. 20-9. Additional terms used with gears.

Circular thickness. Length of arc between the two sides of a gear tooth on the pitch circle.

Chordal thickness. Thickness of the gear tooth measured along a chord at the pitch circle.

Chordal addendum. Radial distance from a line representing the chordal thickness at the pitch circle to the top of the tooth.

Center distance. Distance from the center of one shaft to the center of the other. See Fig. 20-9.

Backlash. Shortest distance or play between mating teeth measured between the non-driving surfaces of adjacent teeth. See Fig. 20-10.

Pressure angle. Angle that determines the tooth shape, or tooth form, as well as the base circle. It is the angle at which pressure from the tooth of

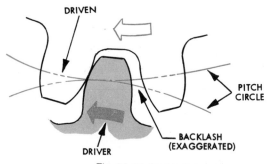

Fig. 20-10. Backlash in gears.

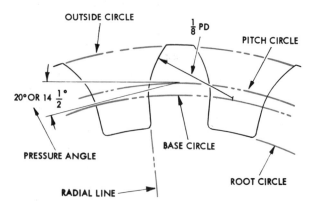

Fig. 20-12. The approximate representation of spur gear tooth profile can be produced rapidly and easily.

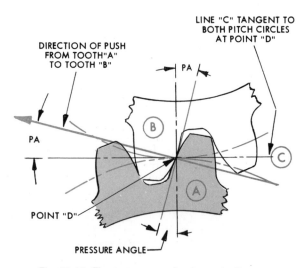

Fig. 20-11. The pressure angle of gear teeth.

one gear is passed on to the tooth of another gear. See Fig. 20-11.

Velocity. Distance that any point on the pitch circle will travel in a given period of time. It is usually stated in feet per minute (fpm).

Base circle. Circle from which the involute tooth profile is formed. See Fig. 20-12.

Gear ratio. Number of teeth on the gear divided by the number of teeth on the pinion. It is also defined as the relationship of the pitch diameter of the gear to the pitch diameter of the pinion. Ratio is further defined as the revolutions per minute (RPM) of the pinion divided by the revolutions per minute of the gear.

Face of tooth. Surface of the tooth profile that is between the pitch circle and the top of the tooth.

Flank of tooth. Surface of the tooth profile that is between the pitch circle and the bottom land including the fillet.

20.2.2 Standard spur-gear formulas. The proportion and shape of spur gear teeth are based on standardized formulas. Some of the essential spur gear formulas are included in the chart shown in Fig. 20-13.

20.2.3 Spur-gear tooth profile. A gear tooth profile must be designed so that the gears will transmit power at a constant velocity and with a minimum of vibration and noise. Various curves have been utilized to produce the type of gear-tooth profile having the correct geometric form to meet these requirements. From this group the involute curve has evolved as the most common gear-tooth profile in use today. In addition to satisfying the above requirements, involute gears can be easily manufactured by a variety of techniques. Hobbing, shaping, and shaving are the three common methods of manufacturing spur gears. Other methods used to produce spur gears are die casting for light-duty gears and stamping for thin-section gears or molding for plastic gears.

The shape of the tooth resulting from the involute system is based principally upon the pressure angle and is generated from the base circle. Therefore, the size of the base circle is governed by the magnitude of the pressure angle.

Spur gears are made with either a 14½°, 20°, or 25° pressure angle. Gears having the latter angles are gradually replacing the older 14½° angle. The

	TO FIND	HAVING	RULE	FORMULA
1	Diametral Pitch	Circular Pitch	Divide 3.1416 by Circular Pitch	$DP = \dfrac{3.1416}{CP}$
2	Diametral Pitch	Pitch Dia. and Number of Teeth	Divide the Number of Teeth by Pitch Diameter	$DP = \dfrac{N}{PD}$
3	Diametral Pitch	Outside Dia. and Number of Teeth	Divide Number of Teeth plus 2 by Outside Diameter	$DP = \dfrac{N + 2}{OD}$
4	Pitch Diameter	Number of Teeth and Diam. Pitch	Divide Number of Teeth by Diametral Pitch	$PD = \dfrac{N}{DP}$
5	Pitch Diameter	Number of Teeth and Outside Dia.	Divide the Product of Number of Teeth and Outside Diameter by Number of Teeth plus 2	$PD = \dfrac{N \times OD}{N + 2}$
6	Pitch Diameter	Outside Diameter and Diam. Pitch	Subtract from the Outside Dia. the Quotient of 2 divided by the Diametral Pitch	$PD = OD - \dfrac{2}{DP}$
7	Outside Diameter	Number of Teeth and Diam. Pitch	Divide Number of Teeth plus 2 by the Diametral Pitch	$OD = \dfrac{N + 2}{DP}$
8	Outside Diameter	Pitch Diameter and Diam. Pitch	Add to the Pitch Diameter the Quotient of 2 divided by the Diametral Pitch	$OD = PD + \dfrac{2}{DP}$
9	Outside Diameter	Pitch Diameter and No. of Teeth	Divide Product of Number of Teeth plus 2 and Pitch Diameter by Number of Teeth	$OD = \dfrac{(N+ 2) \times PD}{N}$
10	Number of Teeth	Pitch Diameter and Diam. Pitch	Multiply Pitch Diameter by Diametral Pitch	$N = PD \times DP$
11	Number of Teeth	Outside Diameter and Diam. Pitch	Multiply Outside Diameter by Diametral Pitch and Subtract 2	$N = (OD \times DP) - 2$
12	Addendum	Diametral Pitch	Divide 1 by Diametral Pitch	$A = \dfrac{1}{DP}$
13	Clearance	Diametral Pitch	Divide 0.157 by the Diametral Pitch	$K = \dfrac{0.157}{DP}$
14	Chordal Addendum	Number of Teeth, Pitch Diameter, and Addendum	Subtract the cosine of the angle determined by dividing 90° by the number of Teeth from 1; multiply this by 1/2 the Pitch Diameter and add the Addendum	$CA = A + \dfrac{PD}{2}\left(1 - \cos\dfrac{90°}{N}\right)$
15	Chordal Thickness	Number of Teeth and Pitch Diameter	Multiply the Pitch Diameter by the sine of the angle determined by dividing 90° by the number of teeth	$Th = PD \sin\dfrac{90°}{N}$

Fig. 20-13. These formulas apply to all spur gears.

20°-pressure-angle tooth shape has become the standard for new gearing because of smoother and quieter running characteristics, greater load-carrying ability, and fewer number of teeth affected by undercutting.

Two methods are used to illustrate the shape of a spur-gear tooth: approximate representation and true representation.

Approximate representation is intended for display purposes. It is also used when it is necessary to illustrate the relationship between the teeth and some special feature, such as a keyway or spline.

The construction can be executed rapidly and simply as follows: (See Fig. 20-12.)

1. Draw the required pitch circle, outside circle, and root circle, determined either by formulas or from gearing tables.

2. Draw radial lines spacing off the teeth on the pitch circle.

3. Draw a line tangent to the pitch circle.

4. Draw a second line at an angle of 14½° or 20° with the tangent. (An approximation of 15° may be used instead of 14½° to facilitate the construction.)

5. Draw the base circle tangent to the 14½° or 20° pressure-angle line.

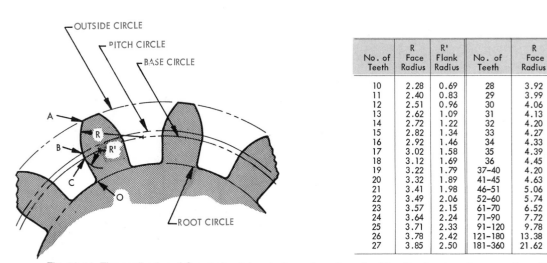

No. of Teeth	R Face Radius	R' Flank Radius	No. of Teeth	R Face Radius	R' Flank Radius
10	2.28	0.69	28	3.92	2.59
11	2.40	0.83	29	3.99	2.67
12	2.51	0.96	30	4.06	2.76
13	2.62	1.09	31	4.13	2.85
14	2.72	1.22	32	4.20	2.93
15	2.82	1.34	33	4.27	3.01
16	2.92	1.46	34	4.33	3.09
17	3.02	1.58	35	4.39	3.16
18	3.12	1.69	36	4.45	3.23
19	3.22	1.79	37–40	4.20	4.20
20	3.32	1.89	41–45	4.63	4.63
21	3.41	1.98	46–51	5.06	5.06
22	3.49	2.06	52–60	5.74	5.74
23	3.57	2.15	61–70	6.52	6.52
24	3.64	2.24	71–90	7.72	7.72
25	3.71	2.33	91–120	9.78	9.78
26	3.78	2.42	121–180	13.38	13.38
27	3.85	2.50	181–360	21.62	21.62

Fig. 20-14. The application of Grant's involute odontograph system simplifies the drawing of spur tooth profile.

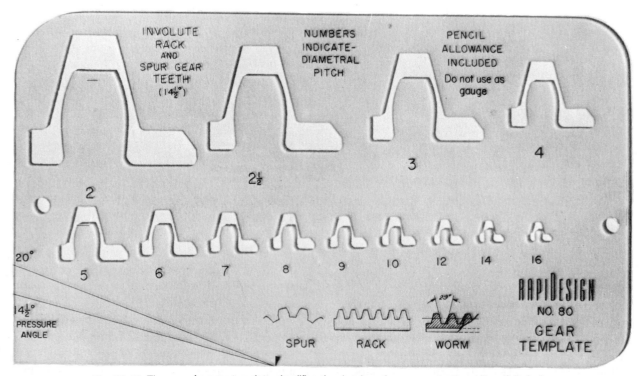

Fig. 20-15. The use of a gear template simplifies the drawing of spur gear tooth profiles (full size).

6. With centers on the base circle and radii equal to one-eighth the pitch diameter, draw circular arcs through spaced tooth points on the pitch circle. Extend the arcs from the outside circle to slightly below the base circle.

7. Draw radial lines from the base circle to the root circle and terminate with fillets to complete the tooth profile.

Another rapid approximation method of drawing tooth profiles is known as Grant's Odontograph system. Refer to Fig. 20-14. The pitch circle, addendum, root circle, and clearance circle are drawn and the teeth spaced off on the pitch circle. The face of the tooth from *A* to *B* and the flank portion *B* to *C* are drawn with two circular arcs from centers on the base circle according to the face and flank radii established in the table shown in Fig. 20-14. The tooth is then completed with a radial line *CO* and fillet. The table includes the necessary face and flank radii for gears of one diametral pitch. Other size gears can be drawn by dividing the figures in the table by the required diametral pitch.

Drawing an approximate spur-gear tooth profile is further simplified with the use of the gear template shown in Fig. 20-15. This template provides profiles of involute rack and spur gear teeth of all common pitches. All that is necessary to use the template is to draw the outer and root circles, space off the required teeth, and position the template to provide the required tooth outline.

True representation is employed only when it is necessary to lay out the true tooth profile in order to check the backlash or clearance of mating gears. The true representation of a gear tooth profile is drawn from the base circle to the outside circle as an involute of the base circle. See Fig. 20-16. To construct such a curve, proceed in the following manner:

1. Draw the base circle, outside circle, pitch circle, and root circle.

2. Divide the base circle into any convenient number of equal parts starting with point *0* and numbering them 1, 2, 3, 4, and so forth.

3. Draw a tangent *AB* to the base circle at point *0*. Divide this line into equal parts having the same lengths as the arcs on the base circle. Number them 1', 2', 3', 4', and so on.

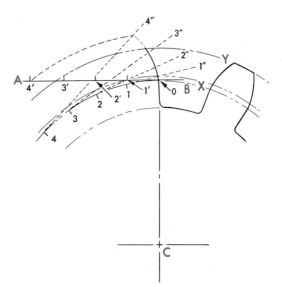

Fig. 20-16. Method of laying out a true involute tooth profile.

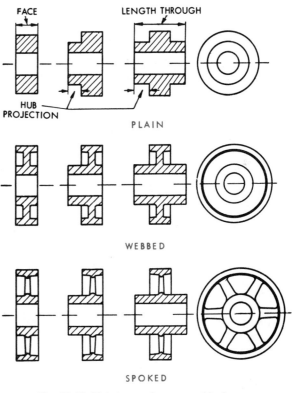

Fig. 20-17. Main types of spur gear blanks.

4. Draw tangents to points 1, 2, 3, 4, and so on, on the base circle.

5. Using the center of the base circle *C* as the pivot, draw concentric arcs from points 1', 2', 3',

4', and so on, on the base circle tangent *AB*, to intersect tangents 1", 2", 3", 4", and so on from the division points on the base circle.

6. The intersections of the arcs with the tangents represent the required points of the involute curve. These points will produce the profile of the tooth extending from *X* to *Y*. The lower portion of the tooth is completed by drawing a radial line and terminating it with the required fillet at the root circle.

20.2.4 Detail drawing of spur gears. A detail drawing of a spur gear is relatively simple. Once the style of the gear blank is selected (See Fig. 20-17.), either a single- or two-view drawing is made. See Figs. 20-18 and 20-19. In either case a cross section is used to show the web portion of the gear blank.

Notice in Fig. 20-19 that the outside circle and root circle are drawn as phantom lines and the pitch circle as a center-line circle. As a rule no teeth are included on the drawing of the blank, but one or more teeth may be shown conventionally when it is necessary to illustrate some relationship to a specific feature such as a keyway, locating-pin hole, or spline.

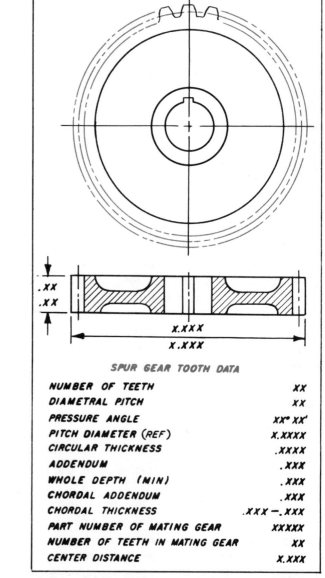

SPUR GEAR TOOTH DATA

NUMBER OF TEETH	XX
DIAMETRAL PITCH	XX
PRESSURE ANGLE	XX° XX'
PITCH DIAMETER (REF)	X.XXXX
CIRCULAR THICKNESS	.XXXX
ADDENDUM	.XXX
WHOLE DEPTH (MIN)	.XXX
CHORDAL ADDENDUM	.XXX
CHORDAL THICKNESS	.XXX — .XXX
PART NUMBER OF MATING GEAR	XXXXX
NUMBER OF TEETH IN MATING GEAR	XX
CENTER DISTANCE	X.XXX

Fig. 20-19. Teeth on a gear blank are included when a relationship must be established with a special feature of the gear. (ANSI Y14.7)

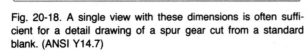

SPUR GEAR TOOTH DATA

NUMBER OF TEETH	XX
DIAMETRAL PITCH	XX
PRESSURE ANGLE	XX° XX'
PITCH DIAMETER (REF)	X.XXXX
CIRCULAR THICKNESS	.XXXX
WHOLE DEPTH (MIN)	.XXX
WORKING DEPTH	.XXX
CHORDAL ADDENDUM (MAX OD)	.XXX
CHORDAL THICKNESS	.XXX — .XXX

Fig. 20-18. A single view with these dimensions is often sufficient for a detail drawing of a spur gear cut from a standard blank. (ANSI Y14.7)

BGW Spur Service Class	Operating Conditions	Service Factor
Class I	Continuous 8 to 10 hrs., per day duty, with Smooth Load (No Shock).	1.0
Class II	Continuous 24 hr. duty, with Smooth Load, or 8 to 10 hrs. per day, with Moderate Shock.	1.2
Class III	Continuous 24 hr. duty, with Moderate Shock Load.	1.3
Class IV	Intermittent duty, not over 30 min. per hr., with Smooth Load (No Shock).	.7
Class V	Hand operation, Limited duty, with Smooth Load (No Shock).	.5
Heavy Shock loads and/or severe wear conditions require the use of higher Service Factors. Such conditions may require Factors of 1.5 to 2.0 or greater than required for Class I service		

Fig. 20-20. Types of operating service classes.

The gear-blank dimensions are shown on the given views and are generally expressed in decimal sizes in thousandths and ten-thousandths of an inch. The essential tooth data is included in chart form.

20.2.5 *Selecting a spur-gear drive.* This is the procedure for selecting a spur-gear drive:

1. Ascertain the class of service for the gear drive from the chart in Fig. 20-20.

2. Multiply the horsepower by the service factor to find the design horsepower (actual power required to drive the machine).

3. Select a spur-gear pinion equal to or greater than that required by the design horsepower from the chart in Fig. 20-21.

4. Determine the size of the driven gear.

Example. A 5-horsepower, 1200 RPM motor is to drive a machine to run 24 hours a day under a smooth load at 300 RPM. What size 20°-pressure-angle spur gears are required?

Solution. The solution is found in this way:
1. Service factor = 1.2.
2. Design horsepower = 5 x 1.2 = 6.
3. Pinion gear

   (a) From chart in Fig. 20-21, read vertically on 1200 RPM line and horizontally to horsepower

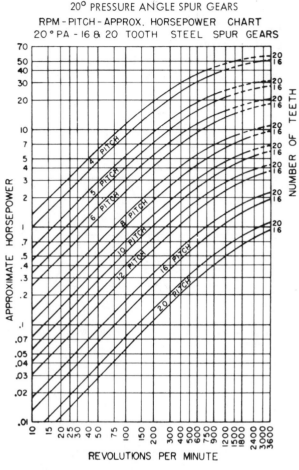

Fig. 20-21. Pitch selection chart for 20° pressure angle spur gear.

rating 6. Where these lines intersect in the pitch line area, note the recommended pitch, which is 8.

   (b) Pinion = $\frac{16}{8}$ = 2.00″

4. Driven gear

   Ratio = $\frac{1200}{300}$ = 4:1 = 16 x 4 = 64

   D = $\frac{64}{8}$ = 8.00″

20.3 Rack and Pinion[2]

As mentioned previously, a rack is essentially a spur gear with teeth spaced along a straight line and is designed for straight-line motion. See Fig.

---

2. *ANSI B6.1.*

20-22. The linear pitch of the rack must be equal to the circular pitch of the mating gear (pinion) if they are to mesh.

To obtain full involute action when the pinion is in contact with a 20° basic rack, the outside diameter of the pinion must be increased. To maintain standard center distance where a gear is substituted for the rack, the outside diameter of the mating gear must be decreased the same amount.

Tabulations of the amounts of increase and decrease in diameter and the corresponding circular tooth thickness on the pitch line of both gear and pinion are included in Table I.

Long addendum pinions in mesh with standard addendum gears will run with full involute action, but it is evident that in such cases the center distance must be increased to suit the standard size gear.

If the sum of the number of teeth in the pinion and the mating gear is less than 34, undercutting will occur in the teeth of the mating gear.

When the height of the tooth is reduced and the pressure angle increased to 20°, the teeth are known as stub teeth. Teeth of this design are

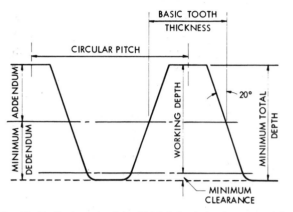

Fig. 20-22. Basic rack of the 20° full depth involute system for spur gearing.

TABLE I DATA FOR FULL 20° INVOLUTE TOOTH ACTION ON PINIONS

Number of Teeth in Pinion	Diameter Increment	Pinion Circular Tooth Thickness	Mating Gear Circular Tooth Thickness	Min Number of Teeth in Mating Gear Avoiding Undercut	Min Number of Teeth in Mating Gear for Full Involute Action
10	1.3731	1.9259	1.2157	54	27
11	1.3104	1.9097	1.2319	53	27
12	1.2477	1.8935	1.2481	52	28
13	1.1850	1.8773	1.2643	51	28
14	1.1223	1.8611	1.2805	50	28
15	1.0597	1.8449	1.2967	49	28
16	0.9970	1.8286	1.3130	48	28
17	0.9343	1.8124	1.3292	47	28
18	0.8716	1.7962	1.3454	46	28
19	0.8089	1.7800	1.3616	45	28
20	0.7462	1.7638	1.3778	44	28
21	0.6835	1.7476	1.3940	43	28
22	0.6208	1.7314	1.4102	42	27
23	0.5581	1.7151	1.4265	41	27
24	0.4954	1.6989	1.4427	40	27
25	0.4328	1.6827	1.4589	39	26
26	0.3701	1.6665	1.4751	38	26
27	0.3074	1.6503	1.4913	37	26
28	0.2447	1.6341	1.5075	36	25
29	0.1820	1.6179	1.5237	35	25
30	0.1193	1.6017	1.5399	34	24
31	0.0566	1.5854	1.5562	33	24

All dimensions in inches.

	IN TERMS OF DIAMETRAL PITCH[1] (Inches)	IN TERMS OF CIRCULAR PITCH[1] (Inches)
1. Addendum	$= \dfrac{0.8}{DP}$	$0.2546 \times CP$
2. Minimum Dedendum	$= \dfrac{1}{DP}$	$0.3183 \times CP$
3. Working Depth	$= \dfrac{1.6}{DP}$	$0.5092 \times CP$
4. Minimum Total Depth	$= \dfrac{1.8}{DP}$	$0.5729 \times CP$
5. Pitch Diameter	$= \dfrac{N}{DP}$	$0.3182 \times N \times CP$
6. Outside Diameter	$= \dfrac{N + 1.6}{DP}$	$PD + (2\ Addendums)$
7. Basic Tooth Thickness on Pitch Line	$= \dfrac{1.5708}{DP}$	$0.5 \times CP$
8. Minimum Clearance	$= \dfrac{0.2}{DP}$	$0.0637 \times CP$

N = Number of Teeth   DP = Diametral Pitch   PD = Pitch Diameter
CP = Circular Pitch

[1]Note: The term Diametral Pitch is used up to 1 DP inclusive and the term Circular Pitch is used for 3 inches CP and over.

Fig. 20-23. Stub tooth proportions for spur gears.

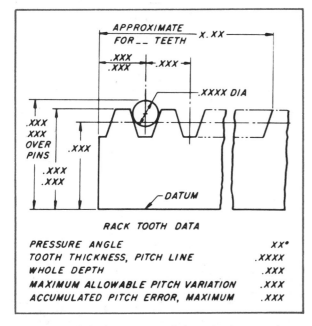

Fig. 20-24. This is the recommended way to show a rack on a drawing. (ANSI Y14.7)

stronger than the standard 14½° involute teeth. The chart in Fig. 20-23 includes the basic proportions of stub teeth for spur gears.

20.3.1 Representing racks. The prescribed method of representing a rack and of dimensioning rack teeth is shown in Fig. 20-24. Note that the teeth are dimensioned from the end of the rack blank and from some datum line. The datum line may be the bottom of the blank or any other line which provides accurate features.

20.4 Bevel and Miter Gears

Bevel and miter gears use the same involute tooth form as spur gears except that the teeth are tapered toward the apex of the cone. Regardless of the similarity in tooth form, they are not interchangeable with spur gears. Bevel and miter gears are always designed in pairs.

20.4.1 Bevel and miter gear terms. Most of the gear terms previously discussed for spur gears apply to bevel and miter gears. However, some of the definitions must be modified so they are more applicable to bevel and miter gears. The following are particularly pertinent: (See Fig. 20-25.)

Pitch diameter. Diameter of the pitch circle measured at the base of the pitch cone.

Pitch angle. Angle between an element of a pitch cone and its axis.

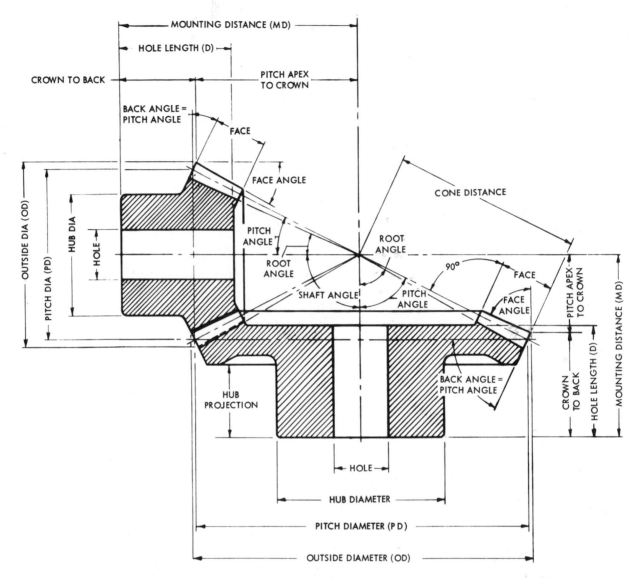

Fig. 20-25. This drawing illustrates the working relationship of two bevel gears in mesh and shows bevel gear terminology.

Cone distance. Slant length of the pitch cone.

Face. Length of the tooth.

Face angle. Angle between an element of the face cone and its axis.

Mounting distance. Distance from the pitch apex to a surface of the gear used for locating in assembly.

Root angle. Angle between an element of the root cone and its axis.

Crown backing. Distance between the rear of the hub and the outer tip of the gear tooth, measured parallel with the axis of the gear.

Crown height. Distance between the cone apex and the outer tip of the gear tooth, measured parallel with the axis of the gear.

The basic formulas used in computing bevel-gear dimensions are included in the chart shown in Fig. 20-26.

20.4.2 Detail drawing of bevel gears. A detail drawing of a bevel gear, like that of a spur gear, should contain the essential specifications for its manufacture. Two representative detail drawings are illustrated in Fig. 20-27. The gear blank dimensions are shown in a single sectional view, and

	TO FIND	RULE		TO FIND	RULE
1	Ratio	Divide the Number of Teeth in the Gear by the Number of Teeth in the Pinion	12	Dedendum Angle of <u>Pinion</u>	Divide the Dedendum of Pinion by Cone Distance. Quotient is the Tangent of the Dedendum Angle
2	Diametral Pitch (DP)	Divide 3.1416 by the Circular Pitch		Dedendum Angle of <u>Gear</u>	Divide the Dedendum of Gear by Cone Distance. Quotient is the tangent of the Dedendum Angle
3	Pitch Diameter of <u>Pinion</u>	Divide Number of Teeth in the Pinion by the D.P.	13	Root Angle of <u>Pinion</u>	Subtract the Dedendum angle of pinion from Pitch Angle of the Pinion
	Pitch Diameter of <u>Gear</u>	Divide Number of Teeth in the Gear by the D.P.		Root Angle of <u>Gear</u>	Subtract the Dedendum Angle of the Gear from Pitch Angle of the Gear
4	Whole Depth (Of Tooth)	Divide 2.188 by the Diametral Pitch and add .002	14	Face Angle of <u>Pinion</u>	Add the Addendum Angle of the Pinion to the Pitch Angle of the Pinion
5	Addendum	Divide 1 by the Diametral Pitch		Face Angle of <u>Gear</u>	Add the Addendum Angle of the <u>Gear</u> to the Pitch Angle of the <u>Gear</u>
6	Dedendum of Pinion or Gear	Divide 2.188 by the D.P. and subtract the Addendum	15	Outside Diameter of <u>Pinion</u>	Add twice the Pinion Addendum times cosine of Pinion Pitch Angle to the Pinion P.D.
7	Clearance	Divide .188 by the Diametral Pitch and add .002		Outside Diameter of <u>Gear</u>	Add twice the Gear Addendum times cosine of Gear Pitch Angle to the Gear P.D.
8	Circular Thickness Of Pinion or Gear	Divide 1.5708 by the Diametral Pitch	16	Pitch Apex to Crown of <u>Pinion</u>	Subtract the Pinion Addendum times the sine of Pinion Pitch Angle from half the Gear P.D.
9	Pitch Angle of <u>Pinion</u>	Divide No. of Teeth in Pinion by No. of Teeth in Gear. Quotient is the tangent of the Pitch Angle		Pitch Apex to Crown of <u>Gear</u>	Subtract the Gear Addendum times the sine of the Gear Pitch Angle from half the Pinion P.D.
	Pitch Angle of <u>Gear</u>	Subtract the Pitch Angle of Pinion from 90°	17	Chordal Thickness at large end of tooth in Gear or Pinion	Multiply the pitch diameter by the sine of the angle found by dividing 90° by the number of teeth
10	Cone Distance	Divide one half the P.D. by the sine of the Pitch Angle	18	Chordal Addendum at large end of tooth in Gear or Pinion	Multiply the square of the circular thickness by the cosine of the pitch angle; divide product by 4 times the pitch diameter and add the quotient to addendum
11	Addendum Angle of <u>Pinion</u>	Divide the addendum of the Pinion by the Cone Distance. Quotient is the tangent of the addendum angle			
	Addendum Angle of <u>Gear</u>	Divide the addendum of the Gear by the Cone Distance. Quotient is the tangent of the addendum angle			

Fig. 20-26. Formulas to obtain straight bevel and miter gear dimensions for 90° shaft angle.

tooth-cutting data is presented in tabular form. Note that the gear tooth profile is not generally shown. Essential mounting dimensions (bore size, keyway, or spline data) must also be included.

20.4.3 Approximate method of drawing a pair of involute bevel gears. When it is necessary to draw a pair of bevel gears in mesh, as illustrated previously in Fig. 20-25, the following procedure is suggested.

Draw the axes of the two bevel gears perpendicular (the shaft angle) to each other. Lay out in sequence the pitch angle, cone distances, pitch diameter, addendum, and dedendum. Note that both the addendum and dedendum are measured at the large end of the tooth and are perpendicular to the pitch-angle line. Using construction lines, draw lines from the pitch point to the addendum and dedendum, thus completing the root and face angles. The face of the tooth is measured along the cone distance. The lines representing the face

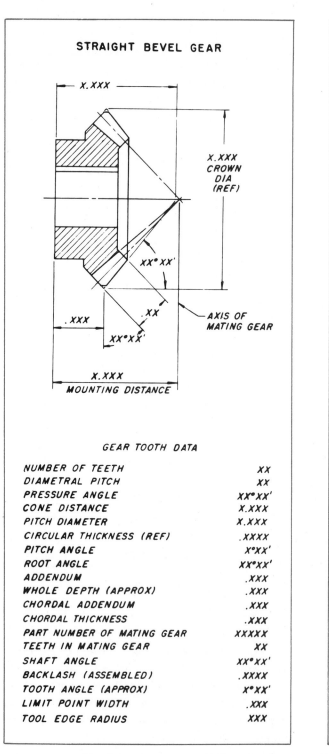

STRAIGHT BEVEL GEAR

GEAR TOOTH DATA

NUMBER OF TEETH	XX
DIAMETRAL PITCH	XX
PRESSURE ANGLE	XX°XX'
CONE DISTANCE	X.XXX
PITCH DIAMETER	X.XXX
CIRCULAR THICKNESS (REF)	.XXXX
PITCH ANGLE	X°XX'
ROOT ANGLE	XX°XX'
ADDENDUM	.XXX
WHOLE DEPTH (APPROX)	.XXX
CHORDAL ADDENDUM	.XXX
CHORDAL THICKNESS	.XXX
PART NUMBER OF MATING GEAR	XXXXX
TEETH IN MATING GEAR	XX
SHAFT ANGLE	XX°XX'
BACKLASH (ASSEMBLED)	.XXXX
TOOTH ANGLE (APPROX)	X°XX'
LIMIT POINT WIDTH	.XXX
TOOL EDGE RADIUS	XXX

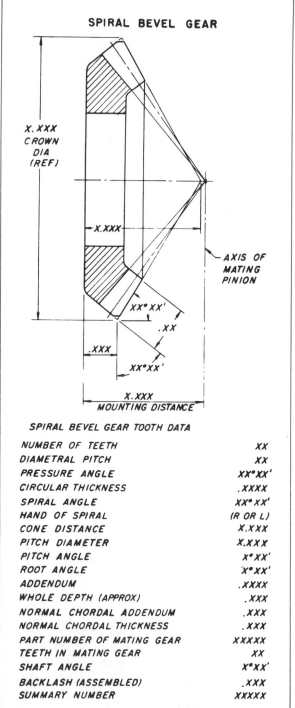

SPIRAL BEVEL GEAR

SPIRAL BEVEL GEAR TOOTH DATA

NUMBER OF TEETH	XX
DIAMETRAL PITCH	XX
PRESSURE ANGLE	XX°XX'
CIRCULAR THICKNESS	.XXXX
SPIRAL ANGLE	XX°XX'
HAND OF SPIRAL	(R OR L)
CONE DISTANCE	X.XXX
PITCH DIAMETER	X.XXX
PITCH ANGLE	X°XX'
ROOT ANGLE	X°XX'
ADDENDUM	.XXXX
WHOLE DEPTH (APPROX)	.XXX
NORMAL CHORDAL ADDENDUM	.XXX
NORMAL CHORDAL THICKNESS	.XXX
PART NUMBER OF MATING GEAR	XXXXX
TEETH IN MATING GEAR	XX
SHAFT ANGLE	X°XX'
BACKLASH (ASSEMBLED)	.XXX
SUMMARY NUMBER	XXXXX

Fig. 20-27. Typical detail drawings of bevel gears with required tooth cutting data. (ANSI Y14.7)

of one gear and a root-cone element of its mating gear are often drawn parallel to each other with the computed clearance illustrated. Information concerning the bore, hub diameter, mounting distances, and other casting information should be obtained from the gear designer.

20.5 Wormgears

Wormgears, like helical gears, are used to transmit power between non-intersecting shafts. These shafts are nearly always at 90° to each other with the worm generally acting as the driver.

A wormgear is meshed with a unit called a worm. The worm can be described as a cylinder having each tooth wrapped around the cylinder in a helical manner. The resulting configuration resembles a threaded bolt as in Fig. 20-28. Frequently the teeth of the worm are referred to as *threads* and, as such, the terms *pitch* and *lead* apply as they do with regular screws. Thus, pitch is the distance between corresponding points on

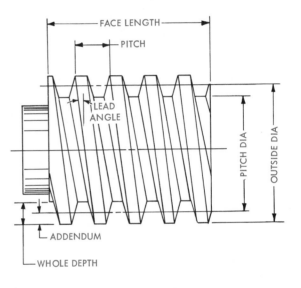

Fig. 20-28. A worm has threads which resemble those of a screw thread.

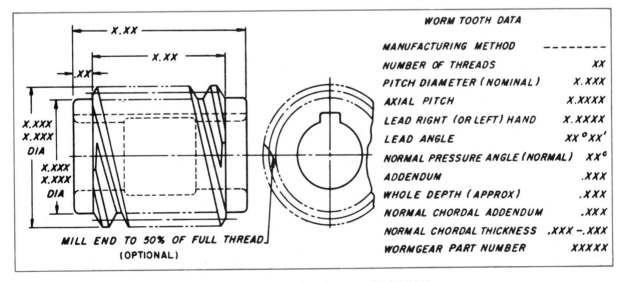

Fig. 20-29. Detail drawing of a worm. (ANSI Y14.7)

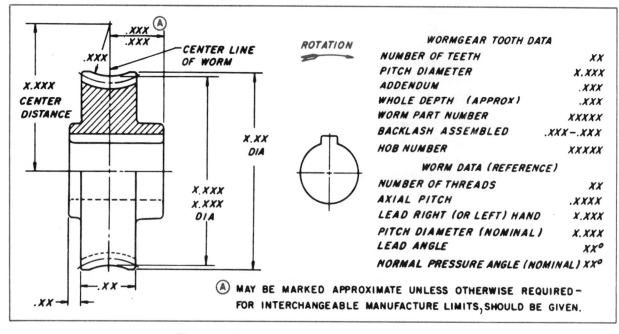

Fig. 20-30. Detail drawing of a wormgear. (ANSI Y14.7)

adjacent teeth as measured parallel to the axis of the worm. Lead is the axial distance that a thread advances in one turn of the worm. As in threads, the lead is a multiple of the pitch. In a single-thread worm, the lead equals the pitch; the lead is twice the pitch for a double-thread worm and triple the pitch for a triple-thread worm.

20.5.1 Detail drawings of wormgears. Detail drawings of the worm and wormgear are generally shown separately, as illustrated in Figs. 20-29 and 20-30. These sample drawings contain the minimum information needed by the gear manufacturer. Formulas for securing wormgear data are included in the chart in Fig. 20-31. Three types of data should be included on the detail drawing: gear blank information, tooth-cutting data, and reference data pertinent to the mating part. This last item is especially important when the mating parts are being made by different manufacturers.

A similarity exists between the drafting construction of the worm-wormgear combination and a rack and spur gear. Fig. 20-32 shows a worm and wormgear with their teeth engaged. Note that the worm, shown in section, is identical with a rack

and that the wormgear, also shown in section, is identical with a spur gear. The difference in drafting construction is illustrated in the right side view where the face radius and rim radius are constructed. These latter items alter the front view accordingly. Consequently, the construction is initiated by first laying out the center distance. The pitch diameters of the worm and wormgear, throat diameter, face radius, outside diameter, and rim radius are drawn in sequence. An approximate representation of involute spur gear teeth is generally used for the wormgear teeth. The teeth of the worm are drawn as teeth on the involute rack and are considered as threads.

20.6 Gear Trains

The *value* of any train of gears is defined as the ratio of the RPM of the first gear to the RPM of the last gear. Since most gear trains are designed to reduce speed, the value of a gear train is often known as the *speed-reduction factor*.

In the gear train in Fig. 20-33, shafts *1* and *2* are connected by gears *A* and *B*, and shafts *2* and

TO FIND	HAVING	RULE
CIRCULAR PITCH	DIAMETRAL PITCH	DIVIDE 3.1416 BY THE DIAMETRAL PITCH.
DIAMETRAL PITCH	CIRCULAR PITCH	DIVIDE 3.1416 BY THE CIRCULAR PITCH.
LEAD (OF WORM)	NUMBER OF THREADS IN WORM & CIRCULAR PITCH	MULTIPLY THE CIRCULAR PITCH BY THE NUMBER OF THREADS.
CIRCULAR PITCH OR LINEAR PITCH	LEAD AND NUMBER OF THREADS IN WORM	DIVIDE THE LEAD BY THE NUMBER OF THREADS.
ADDENDUM	CIRCULAR PITCH	MULTIPLY THE CIRCULAR PITCH BY .3183.
ADDENDUM	DIAMETRAL PITCH	DIVIDE 1 BY THE DIAMETRAL PITCH.
PITCH DIAMETER OF WORM	OUTSIDE DIAMETER AND ADDENDUM	SUBTRACT TWICE THE ADDENDUM FROM THE OUTSIDE DIAMETER.
PITCH DIAMETER OF WORM	SELECT STANDARD PITCH DIAMETER WHEN DESIGNING	WORM GEARS ARE MADE TO SUIT THE MATING WORM.
PITCH DIAMETER OF WORM GEAR	CIRCULAR PITCH AND NUMBER OF TEETH	MULTIPLY THE NUMBER OF TEETH IN THE GEAR BY THE CIRCULAR PITCH AND DIVIDE THE PRODUCT BY 3.1416.
PITCH DIAMETER OF WORM GEAR	DIAMETRAL PITCH AND NO. OF TEETH	DIVIDE THE NUMBER OF TEETH IN GEAR BY THE DIAMETRAL PITCH.
CENTER DISTANCE BETWEEN WORM AND WORM GEAR	PITCH DIAMETER OF WORM AND WORM GEAR	ADD THE PITCH DIAMETERS OF THE WORM AND WORM GEAR THEN DIVIDE THE SUM BY 2.
WHOLE DEPTH OF TEETH	CIRCULAR PITCH	MULTIPLY THE CIRCULAR PITCH BY .6866.
WHOLE DEPTH OF TEETH	DIAMETRAL PITCH	DIVIDE 2.157 BY THE DIAMETRAL PITCH.
BOTTOM DIAMETER OF WORM	WHOLE DEPTH AND OUTSIDE DIAMETER	SUBTRACT TWICE THE WHOLE DEPTH FROM THE OUTSIDE DIAMETER
THROAT DIAMETER OF WORM GEAR	PITCH DIAMETER OF WORM GEAR AND ADDENDUM.	ADD TWICE THE ADDENDUM TO THE PITCH DIAMETER OF THE WORM GEAR.
HELIX ANGLE OF WORM	PITCH DIAMETER OF THE WORM AND THE LEAD	MULTIPLY THE PITCH DIAMETER OF THE WORM BY 3.1416 AND DIVIDE THE PRODUCT BY THE LEAD, THE QUOTIENT IS THE CO-TANGENT OF THE HELIX ANGLE OF THE WORM.
RATIO	NUMBER OF STARTS (OR THREADS) IN THE WORM AND THE NUMBER OF TEETH IN THE WORM GEAR	DIVIDE THE NUMBER OF TEETH IN WORM GEAR BY NUMBER OF STARTS (OR THREADS) IN WORM.

Fig. 20-31. Worm gear formulas.

3 are connected by gears *C* and *D*. Since both gears *B* and *C* are keyed to shaft *2,* they have the same RPM. Accordingly, the ratio of gears *A* and *B* is:

$$\frac{\text{RPM of shaft 1}}{\text{RPM of shaft 2}} = \frac{N_B}{N_A}$$

where

N_A = number of teeth in gear *A*
N_B = number of teeth in gear *B*

Likewise, the ratio of gears *C* and *D* is

$$\frac{\text{RPM of shaft 2}}{\text{RPM of shaft 3}} = \frac{N_D}{N_C}$$

where

N_C = number of teeth in gear *C*
N_D = number of teeth in gear *D*

Correspondingly, the ratio of the gear train is

$$\frac{\text{RPM of shaft 1}}{\text{RPM of shaft 3}} = \frac{N_B}{N_A} \times \frac{N_D}{N_C}$$

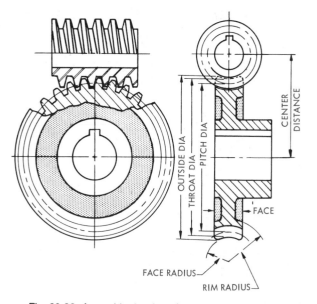

Fig. 20-32. Assembly drawing of a worm and wormgear.

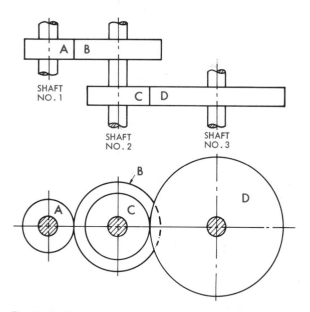

Fig. 20-33. The value of a gear train is the ratio established by the RPM of the first gear to the RPM of the last gear.

If the train contains more than two pairs of gears, the above method can be repeated. It should be noted that the number of teeth of the driven gear is always in the numerator while that of the driving gear is in the denominator.

20.7 Splines[3]

Splines are multiple keys that provide a positive connection between shafts and related members to prevent relative rotation. Spline teeth are either straight or helical in direction, straight or curved on their working profile, and straight or tapered along their length.

Since the most common splines are those having straight teeth along their length, the following discussion will be confined to this type. The involute spline has a depth equal to one-half the depth of a standard gear tooth. Standardized diametral pitches of involute splines are as follows:

1/2	4/8	8/16	16/32	32/64
2.5/5	5/10	10/20	20/40	40/80
3/6	6/12	12/24	24/48	48/96

The numerator of these pitches controls the pitch diameter and tooth thickness, and the denominator controls the addendum and dedendum.

The two basic types of splines are the flat root and fillet root. The *flat root spline* is designed for moderate stresses and thin walls, and the *fillet root spline* for highly stressed parts. A fillet-root external spline may be used with a flat-root internal spline. A flat-root external spline may be used with a fillet-root internal spline.

Splines are made with three types of fits— *sliding, close,* and *press*—which may be applied on the major diameter, minor diameter, or sides of the teeth. The fitting of parts is controlled by making the internal spline constant and the external spline varied to secure the required fit. The major diameter is most practical to maintain accuracy.

Fig. 20-34 shows composite sections of splines as recommended for various designs. Specimen drawings for internal and external drawings are illustrated in Fig. 20-35.

20.8 Serrations

Serrations, like splines, are considered as multiple keys to prevent rotary motion. However, serrations are primarily intended for parts which must

---

3. *ANSI Y14.7.*

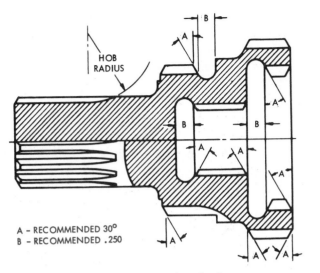

A – RECOMMENDED 30°
B – RECOMMENDED .250

Fig. 20-34. Representation of splines.

be permanently fitted together. They are made for use with uniform or tapered diameters. The standard diametral pitches of serrations are included in Table II. A typical serration drawing is shown in Fig. 20-36.

20.9 Metric Gears

Many elements of metric gears retain the same basic values as those in the decimal-inch system and, therefore, require only simple numerical conversion to change gear design data to metric. The greatest difference between the two systems is the use of the term *module* in place of diametral pitch (DP). In the inch system diametral pitch (DP) is defined as :

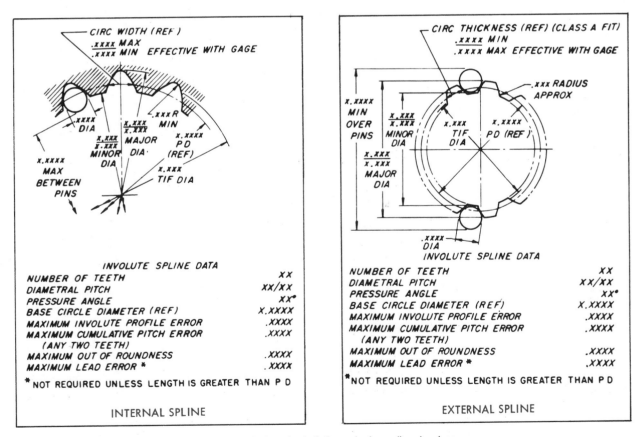

Fig. 20-35. Typical method of dimensioning spline drawings.

TABLE II DIAMETRAL PITCHES FOR STANDARD SERRATIONS

DIAMETRAL PITCH	MAJOR DIAMETER RANGE	DIAMETRAL PITCH	MAJOR DIAMETER RANGE
10/20	0.7000-10.1000	48/96	0.1459-1.2709
16/32	0.4375-6.3125	64/128	0.1875-0.5000
24/48	0.2917-4.2083	80/160	0.1500-0.3500
32/64	0.2188-1.9063	128/256	0.0937-0.1953
40/80	0.1750-1.5250		

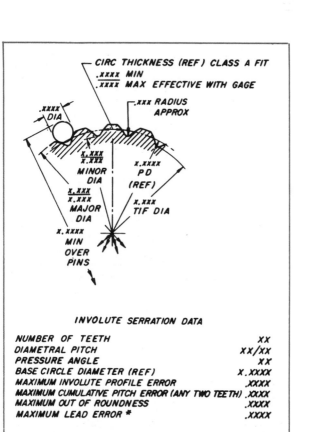

Fig. 20-36. Specimen drawing of an internal serration.

$$DP = \frac{N}{D}$$

where

DP = diametral pitch

N = number of teeth

D = pitch diameter

In the metric system, module (m) is equal to the reciprocal of diametral pitch. The module is a dimension (length of pitch diameter per tooth), whereas the diametral pitch is a ratio of the number of teeth to a unit length of pitch diameter. Thus, pitch in metric gearing relates to tooth-spacing along the pitch circle, but in the inch system this measure is known as circular pitch. Hence,

$$\text{Module (m)} = \frac{D \text{ (mm)}}{N}$$

If the diametral pitch is known, the module can be obtained by the formula

$$m = \frac{25.4}{P}$$

Most of the gear design equations are suitable for use with metric gear dimensions, providing that the module is substituted for pitch. Equations involving diametral pitch replace P with the formula:

$$P = \frac{25.4}{m}$$

Also, since circular pitch $p_c = \frac{\pi}{P}$, p_c is replaced by the formula:

$$p_c = \frac{\pi m}{25.4}$$

To Obtain	From Known	Use This Formula*
Pitch diameter, D	Module	$D = mN$
Circular pitch, p_c	Module	$p_c = m\pi = \dfrac{D}{N}\pi$
Module, m	Diametral pitch	$m = \dfrac{25.4}{P}$
No. of teeth, N	Module and pitch diameter	$N = \dfrac{D}{m}$
Addendum, a	Module	$a = m$
Dedendum, b	Module	$b = 1.25m$
Outside diameter, D_o	Module and pitch diameter or number of teeth	$D_o = D + 2m = m(N + 2)$
Root diameter, D_r	Pitch diameter and module	$D_r = D - 2.5m$
Base circle diameter, D_b	Pitch diameter and pressure angle, ϕ	$D_b = D \cos \phi$
Base pitch, p_b	Module and pressure angle	$p_b = m\pi \cos \phi$
Tooth thickness at standard pitch diameter, T_{std}	Module	$T_{std} = \dfrac{\pi}{2} m$
Center distance, C	Module and number of teeth	$C = \dfrac{m(N_1 + N_2)}{2}$
Contact ratio, m_p	Outside radii, base circle radii, center distance, pressure angle	$m_p = \dfrac{\sqrt{1R_o - 1R_b} + \sqrt{2R_o - 2R_b} - C \sin \phi}{m\pi \cos \phi}$
Backlash (linear), B	Change in center distance	$B = 2(\Delta C)\tan \phi$
Backlash (linear), B	Change in tooth thickness, T	$B = \Delta T$
Backlash (linear) along line-of-action, B_{LA}	Linear backlash above pitch circle	$B_{LA} = B \cos \phi$
Backlash (angular), B_a	Linear backlash	$B_a = 6{,}880\,\dfrac{B}{D}$ (arc minutes)
Min. number teeth for no undercutting, N_c	Pressure angle	$N_c = \dfrac{2}{\sin^2 \phi}$

*All linear dimensions in millimeters.

Fig. 20-37. Spur gear metric design formulas.

The chart in Fig. 20-37 shows conventional gear formulas that convert inch values to metric dimensions. For example, suppose a pair of 24- and 48-tooth gears have a module of 1.25. What is the center distance? Using the converted formula:

$$C = \frac{m(N_1 + N_2)}{2}$$

where

C = center distance
m = module
N_1 = number of teeth of small gear
N_2 = number of teeth of large gear

Solution.

$$C = \frac{1.25(24 + 48)}{2} = 45 \text{ mm}$$

20.10 Bearings

A bearing is a device used to reduce friction between moving elements of a mechanism.

20.10.1 Types of bearings. The following are some of the basic types of bearings:

Sleeve bearings. Sleeve bearings are hollow cylinders that support a rotating or oscillating shaft. A clearance between the bearing and shaft permits a lubricant to coat the metal surfaces and

Fig. 20-38. Sleeve bearings are used to support rotating or oscillating shafts.

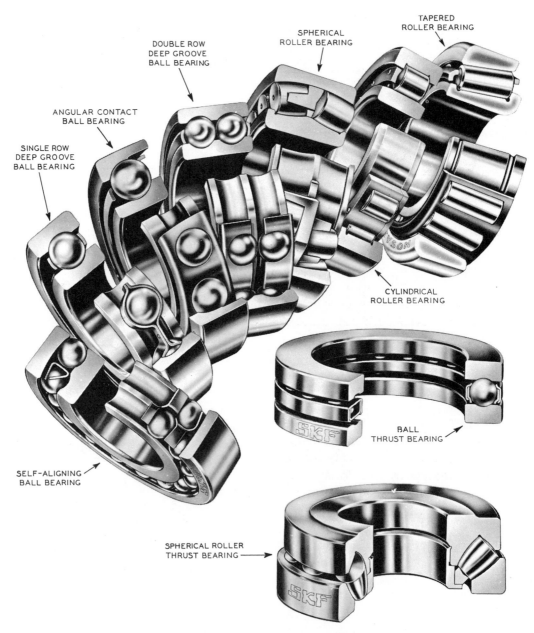

Fig. 20-39. Examples of ball and roller bearings.

·prevent scoring and seizure. The two most common types of sleeve bearings are the plain and flanged. See Fig. 20-38. Plain sleeve bearings are either full or split. Flanged bearings have one or both ends flanged and are designed for applications where both axial thrust and radial loads are encountered.

Ball bearings. Ball bearings are probably the most common antifriction bearings in use. See Fig. 20-39. These bearings consist of two grooved race rings, a set of balls, and a separator. Ball bearings are made in several types such as single-row radial, angular contact, thrust, and self-aligning. The *single-row radial bearings* are designed primarily for radial loads, although they can carry a considerable amount of thrust load as well. *Angular contact bearings* are either single- or double-row and are made specifically to support combined loads where the thrust component may be large and the axial deflection must be confined to close limits. Double-row bearings are generally preferred where maximum resistance to misalignment is necessary. *Thrust bearings* are made to support pure thrust loads only. The balls are held by two washer-type rings, one of which is usually stationary and supported by the housing, while the other revolves with the shaft on which it is mounted. *Self-aligning radial ball bearings* are used where the bearing seats cannot be maintained in line.

Ball bearings are also available in the shielded and sealed types. The *shielded bearing* has a metal shield, or plate, on one or both sides of the bearing. The chief functions of the shields are to retain the lubricant and to keep dirt and chips from getting into the bearing. *Sealed bearings* have a felt, rubber, or plastic seal mounted on the outer ring of the bearing and seals on the outside diameter of the inner ring. These bearings are filled with a special lubricant by the bearing manufacturer and require no further attention for long periods of operation.

Roller bearings. Roller bearings are particularly effective for handling heavy loads in relatively small places. The principal types are cylindrical, spherical, tapered, and thrust. See Fig. 20-39.

Pillow blocks. Pillow blocks, as shown in Fig. 20-40, consist of a housing fitted with either a ball or roller bearing. They are used on shafting or as a part of a machine.

Fig. 20-40. Pillow blocks are designed to carry light and heavy loads.

20.10.2 Selection of bearings.
Selection of bearings for any mechanism is governed by such factors as shaft size, housing design, speed, and load.

Shaft size. Shaft size is an important consideration since the bearing must be mounted firmly in place and held securely against the shaft shoulder. A loosely fitted inner ring will creep on the shaft, leading to wear. On the other hand, the shaft fit should not result in any undue tightness of the bearing. Any excessive tension in the ring would cause expansion which would disturb the internal fit of the bearing and lead to heating and increase power consumption. In general, shaft size and tolerance for seating precision bearings should be the same as the bearing bore:

Bore Size (inches)	Shaft DIA (inches)	Average Fit
Max 2.1654	Min 2.1652	Line-to-line
Min 2.1652	Max 2.1654	

Housing fits.[4] Under normal conditions of shaft rotation, the outer ring of the bearing is stationary and should be mounted with a hand push to a light tapping fit. Should the housing be the rotating member, the same fundamental considerations

---

4. *The Fafnir Bearing Company Manual* (New Britain, Conn.: The Fafnir Bearing Co.).

TABLE III BEARINGS SPECIFICATIONS (MEDIUM M300K SERIES)

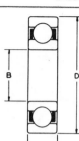

DIMENSION - TOLERANCES

bearing number	bore, B			inner ring eccen. inches	outside diameter, D			outer ring eccen. inches	width, W +.000, −.005		fillet radius ∎ inches	balls	
	mm	inches	tolerance +.0000 to minus, in.		mm	inches	tolerance +.0000 to minus, in.		mm	inches		no.	size, in.
M305K-CR	25	.9843	.0002	.0003	62	2.4409	.0004	.0005	17	.6693	.040	7	15/32
M306K-CR	30	1.1811	.0002	.0003	72	2.8346	.0004	.0005	19	.7480	.040	7	17/32
M307K-CR	35	1.3780	.0003	.0004	80	3.1496	.0004	.0005	21	.8268	.060	7	9/16
M308K-MBR♦	40	1.5748	.0003	.0004	90	3.5433	.0004	.0007	23	.9055	.060	8	5/8
M309K-MBR♦	45	1.7717	.0003	.0004	100	3.9370	.0004	.0007	25	.9843	.060	8	11/16
M310K-MBR♦	50	1.9685	.0003	.0004	110	4.3307	.0004	.0007	27	1.0630	.080	8	3/4
M311K-MBR♦	55	2.1654	.0004	.0004	120	4.7244	.0004	.0007	29	1.1417	.080	8	13/16
M312K-MBR♦	60	2.3622	.0004	.0004	130	5.1181	.0005	.0008	31	1.2205	.080	8	7/8
M313K-MBR♦	65	2.5591	.0004	.0004	140	5.5118	.0005	.0008	33	1.2992	.080	8	15/16
M314K-MBR♦	70	2.7559	.0004	.0004	150	5.9055	.0005	.0008	35	1.3780	.080	8	1

∎ Maximum shaft or housing fillet radius which bearing corners will clear.

♦ Also available as CR for high speed machines, such as routers, shapers and moulders.

LOAD RATING

Rated Radial Load Capacity in Pounds at Various rpm
Based on 500 hours minimum life—2500 hours average life

bearing number	Nd²	specific dynamic radial capacity (a 33.3 rpm	revolutions per minute											
			50	100	200	300	500	900	1200	1500	1800	2400	3600	5000
M305K-CR	1.54	4080	3560	2830	2240	1960	1650	1360	1240	1150	1080	980	855	770
M306K-CR	1.98	5160	4500	3580	2840	2480	2090	1720	1560	1450	1360	1240	1080	970
M307K-CR	2.21	5780	5050	4010	3180	2780	2340	1930	1750	1620	1530	1390	1210	1090
M308K-MBR	3.12	7660	6700	5310	4220	3680	3110	2560	2320	2160	2030	1840	1610	1440
M309K-MBR	3.78	9130	7970	6330	5020	4390	3700	3040	2760	2570	2420	2190	1920	1720
M310K-MBR	4.50	10700	9340	7410	5880	5140	4330	3560	3240	3000	2830	2570	2240	2010
M311K-MBR	5.28	12400	10800	8570	6800	5940	5010	4120	3740	3480	3270	2970	2600	...
M312K-MBR	6.12	14200	12400	9810	7790	6800	5740	4720	4290	3980	3740	3400	2970	...
M313K-MBR	7.03	16000	14000	11100	8820	7700	6500	5340	4850	4500	4240	3850	3360	...
M314K-MBR	8.00	18000	15700	12500	9900	8650	7300	6000	5450	5060	4760	4330	...	...

apply in mounting the outer race as in the case of an inner ring mounted on a rotating shaft. As a rule the minimum housing bore dimension should be 0.0001 inch less than the maximum bearing outside diameter (OD). If the bearing OD tolerance is 0.0003 inch, the maximum housing bore should be established as 0.0003 inch larger than the minimum housing bore dimension:

Bearing OD (inches)	Housing Bore (inches)	Average Fit (inches)
Max 3.5433	Min 3.5432	.0002 loose
Min 3.5430	Max 3.5435	

On exceedingly high-speed and other applications where an adjacent heat input is prevalent along the shaft, it is extremely important that the floating bearing be able to move longitudinally to compensate for thermal changes. It cannot float laterally if restricted by a tight housing bore or by the radial expansion of the bearing itself due to thermal changes. It is equally important that all shaft and housing shoulders be absolutely square and that the faces of spacers be square and parallel within 0.0001 inch.

General housing design. Housings are usually made of cast iron or steel and are generally heat-treated to lessen possible distortion. For smaller high-speed shaft applications steel housings are preferable.

In many cases of housing design, it is advantageous to employ a sub-housing or a steel sleeve between the outer ring of the bearing and the machine frame, thus allowing assembly of the bearings on the shaft and insertion of the entire unit into the machine frame. This method also provides a surface of proper hardness when machine frames are made of a material that has a low Brinell value, such as aluminum and other soft metals.

Shaft shoulders and housing shoulders should be square and true. The choice between fillets and undercut relief depends upon the individual shaft design and conditions that are surrounding its normal use.

Where screws are used to fasten parts into the main housing, adequate section should be left under the screw hole to prevent distortion of the housing bore when the screws are pulled up and the covers or other parts are pulled tightly into place.

Loads. Bearings are designed to meet practically any specified loads. Once the design load of a mechanism is established, all that needs to be done is to consult a bearing manufacturer's catalog and select from the given tables the correct type of bearing. See Table III. These tables specify the type of bearing that is required in terms of shaft size and load in pounds at various RPM levels.

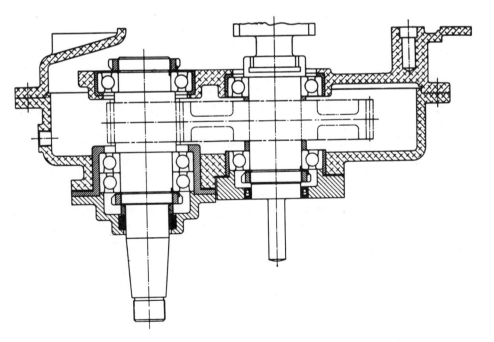

Fig. 20-41. Bearings on a component are shown in outline form.

20.10.3 Representing bearings on a drawing.
Bearings are included on the assembly drawing of a component. The bearings are shown in outline form as illustrated in Fig. 20-41. Specifications of the bearing are given in the parts list or bill of material. The bearing is often identified by the symbols designated by the bearing manufacturer. As yet, bearing symbols of different manufacturers have not been standardized.

20.11 Chain Drives

A chain drive consists of an endless chain that meshes with toothed wheels called sprockets. See Fig. 20-42. With chain drives, shaft center distances are relatively unrestricted and fixed speed ratios between two or more rotating shafts are maintained.

20.11.1 Types of chains.
A number of types, such as roller, detachable, pintle, silent, engineer-

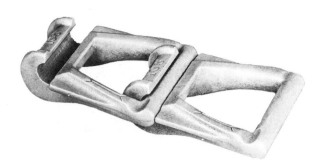

Fig. 20-44. Typical detachable chain.

The advantage of roller chains is the ability of the rollers to rotate when contacting the teeth of the sprockets. The pins that hold the link assembly together are either riveted or secured with snap rings. See Fig. 20-43.

Detachable chains. These chains have links which are easily disassembled. The links of some types have ends which hook on the bars of adjacent links. Other types have side bars and rollers which are held together by pins. Detachable chains are normally used for slower speeds and moderate loads or where chains of greater precision are not required. See Fig. 20-44.

Pintle chains. Pintle chains consist of hollow-core cylinders with offset side bars. The links are joined by pins which are inserted in the holes of the side bars and through the cored holes in the links. Lugs on the side bar prevent the pins from turning. These chains are designed for more rigorous service than the detachable chains. Refer to Fig. 20-45.

ing steel class, multiple-strand, offset, double pitch, and bead chains, are available.

Roller chains. Roller chains are probably the most widely used in industry. They consist of rollers that are evenly spaced throughout the chain.

Silent chains. Silent, or inverted-tooth, chains consist of a series of tooth link plates that operate smoothly and quietly. These chains are particularly adaptable for heavy equipment and are often used as a power take-off from the prime mover. See Fig. 20-46.

Engineering-class steel chains. These are made entirely of machined finished parts. The side bars

Fig. 20-42. A roller chain drive consists of two or more sprockets connected with one or more endless chains.

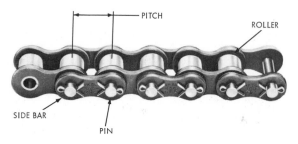

Fig. 20-43. A typical roller chain.

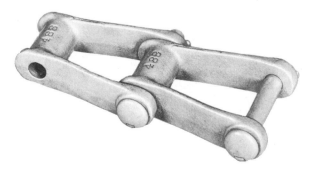

Fig. 20-45. Typical pintle chain.

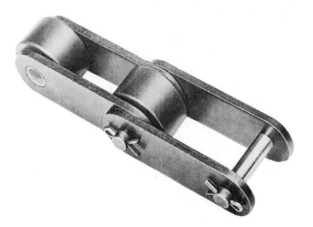

Fig. 20-47. Engineering class steel chain.

Fig. 20-46. A typical silent chain.

are attached to rollers by pins. These chains are often specified where more precision-type service is required. See Fig. 20-47.

Multiple-strand chains. Multiple-strand chains are essentially an assembly of two or more single-strand chains which are placed side by side. The pins extend through the entire width of the chains and thus maintain uniform alignment of the strands. Multiple-strand chains are used to provide greater power capacity without increasing the chain pitch or linear speed. See Fig. 20-48.

Offset chains. These have an assembly of standard roller links and offset links which are connected with riveted pins. See Fig. 20-49. These chains are able to carry heavier loads and are especially adaptable for construction machinery. Because of the offset bars, this type of chain is less affected by dirt and misalignment, which often cause binding in roller chains.

Double-pitch chains. Double-pitch chains are similar to the standard roller chains except they have a pitch that is twice as long. See Fig. 20-50. They are designed for less rigorous service than the standard chains and function well on drives with long center distances.

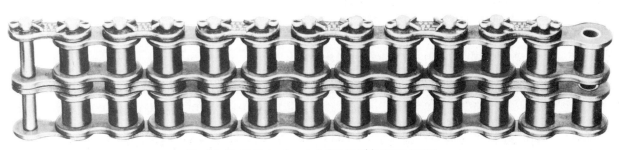

Fig. 20-48. Multiple-strand chains provide more power.

Fig. 20-49. Typical offset chain.

Fig. 20-50. Double-pitch chain.

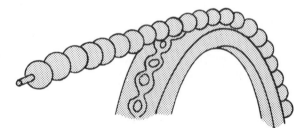

Fig. 20-51. Bead chains are used on small equipment.

Bead chains. These are light-duty chains used principally on slow drives such as those on radio and television tuners, air-conditioner controls, business machines, laboratory controls, and toys. See Fig. 20-51. These chains are made with beads of various diameters. Some types have plastic beads molded onto a flexible cord. Plastic bead chains do not carry as much load as all-metal chains, but they can run faster.

20.11.2 Sprockets. A sprocket is a toothed wheel with teeth shaped to mesh with the chain. The basic types of sprockets are plain plate with-

out hub, plate with hub on one side, plate with hub on both sides, and plate with detachable hub. See Fig. 20-52. The sprocket plate may be solid or webbed.

20.11.3 Chain-drive design factors. The principal factors involved in the design of a chain drive are the following:

1. Design horsepower
2. Chain selection
3. Sprocket sizes
4. Shaft center distance
5. Chain length

As a rule the horsepower, motor shaft speed, and shaft center distance are known. The designer must then select the chain and sprocket sizes to satisfy the necessary requirements.

The use of a specific chain-drive design problem will best illustrate the procedure involved. Let us assume that a particular machine must be fitted with a roller chain drive. The specifications established for this equipment are the following:

Problem.

1. 10-horsepower electric motor, 1200 RPM
2. Countershaft 1.9375 inches in diameter
3. Driven sprocket to produce 380 RPM
4. Shaft center distance 22 inches
5. Drive subject to heavy shock

Solution.

Step 1. Determine design horsepower. Design horsepower is the actual power required to operate the equipment efficiently. Horsepower is affected by the type of service load imposed on the chain drive. Service load must therefore be considered in order to arrive at a true horsepower

Fig. 20-52. Types of sprockets.

TABLE IV SERVICE FACTORS FOR CHAIN DRIVES

Type of Driven Load	TYPE OF INPUT POWER		
	Internal Combustion Engine with Hydraulic Drive	Electric Motor or Turbine	Internal Combustion Engine with Mechanical Drive
Smooth	1.0	1.0	1.2
Moderate Shock	1.2	1.3	1.4
Heavy Shock	1.4	1.5	1.7

rating. Since the specifications in the given problem stipulate a 10-horsepower electric motor and a drive that is subject to heavy shock loads, we must check Table IV for the service load factor. This service load factor is then multiplied by the given motor horsepower to find the actual horsepower required:

TABLE V ROLLER CHAIN SELECTION

RPM of Smaller Sprocket	DESIGN HORSEPOWER												
	½	1	1½	2	3	4	5	7½	10	15	20	25	30
	CHAIN NUMBER												
1700–2000	41	41	35	35	35	35	35	40	40	40	40–2	40–2	40–3
1400–1699	41	41	41	35	35	35	40	40	40	50	50	50–2	50–2
1150–1399	41	41	41	41	35	35	40	40	50	50	60	60	80
950–1149	41	41	41	41	35	40	40	40	50	50	60	80	80
800–949	41	41	41	41	40	40	40	40	50	50	60	80	80
650–799	41	41	41	41	40	40	40	50	50	50	60	80	80
525–649	41	41	41	40	40	40	40	50	50	60	80	80	80
425–524	41	41	40	40	40	40	50	50	60	60	80	80	80
375–424	41	41	40	40	40	40	50	50	60	60	80	80	100
325–374	41	41	40	40	40	50	50	50	60	60	80	80	100
275–324	41	40	40	40	40	50	50	60	60	80	80	100	100
225–274	41	40	40	40	40	50	60	60	60	80	80	100	100
185–224	41	40	40	50	50	50	60	60	80	80	100	100	120
160–184	41	40	40	50	50	60	60	60	80	80	100	100	120
140–159	41	40	50	50	50	60	60	80	80	80	100	120	120
120–139	40	40	50	50	60	60	80	80	80	100	120	120	140
90–119	40	50	50	60	80	80	80	80	100	100	120	120	140
75–89	40	50	60	60	80	80	80	80	100	100	120	140	160
65–74	40	50	60	60	80	80	80	100	100	120	140	140	160
55–64	40	50	60	80	80	80	100	100	120	120	140	160	. . .
45–54	50	60	60	80	80	100	100	120	120	140	160	160	. . .
35–44	50	60	80	80	100	100	100	120	120	140	160	160	. . .
31–34	50	60	80	80	100	100	120	120	140	160	160	. . .	. . .
26–30	50	80	80	80	100	100	120	140	140	160	. . .	. . .	. . .
21–25	60	80	80	100	100	120	120	140	160	160	. . .	. . .	. . .
16–20	60	80	100	100	120	120	140	160	160	. . .	. . .	. . .	. . .
11–15	80	80	100	120	140	140	140	160	. . .	. . .	. . .	. . .	. . .
5–10	80	100	120	140	160	160	. . .	. . .	. . .	. . .	. . .	. . .	. . .

TABLE VI MULTIPLE STRAND FACTORS

NUMBER OF STRANDS	MULTIPLE STRAND FACTOR
2	1.7
3	2.5
4	3.3

TABLE VII

HORSEPOWER RATINGS STANDARD SINGLE STRAND ROLLER CHAIN — NO. 50 — ⅝″ PITCH

No. of Teeth Small Spkt.	REVOLUTIONS PER MINUTE—SMALL SPROCKET																							
	10	25	50	100	200	300	400	500	700	900	1000	1200	1400	1600	1800	2100	2400	2700	3000	3500	4000	4500	5000	5500	6000
9	0.09	0.19	0.36	0.67	1.26	1.81	2.35	2.87	3.89	4.88	5.36	6.32	6.02	4.92	4.13	3.27	2.68	2.25	1.92	1.52	1.25	1.04	0.89	0.77	0.58
10	0.10	0.22	0.41	0.76	1.41	2.03	2.63	3.22	4.36	5.46	6.01	7.08	7.05	5.77	4.83	3.84	3.14	2.63	2.25	1.78	1.46	1.22	1.04	0.90	0.79
11	0.11	0.24	0.45	0.84	1.56	2.25	2.92	3.57	4.83	6.06	6.66	7.85	8.13	6.65	5.58	4.42	3.62	3.04	2.59	2.06	1.68	1.41	1.20	1.04	0.92
12	0.12	0.26	0.49	0.92	1.72	2.47	3.21	3.92	5.31	6.65	7.31	8.62	9.26	7.58	6.35	5.04	4.13	3.46	2.95	2.34	1.92	1.61	1.37	1.19	1.04
13	0.13	0.29	0.54	1.00	1.87	2.70	3.50	4.27	5.78	7.25	7.97	9.40	10.4	8.55	7.16	5.69	4.65	3.90	3.33	2.64	2.16	1.81	1.55	1.34	0
14	0.14	0.31	0.58	1.09	2.03	2.92	3.79	4.63	6.27	7.86	8.64	10.2	11.7	9.55	8.01	6.35	5.20	4.36	3.72	2.95	2.42	2.03	1.73	1.50	0
15	0.15	0.34	0.63	1.17	2.19	3.15	4.08	4.99	6.75	8.47	9.31	11.0	12.6	10.6	8.88	7.05	5.77	4.83	4.13	3.27	2.68	2.25	1.92	1.66	0
16	0.16	0.36	0.67	1.26	2.34	3.38	4.37	5.35	7.24	9.08	9.98	11.8	13.5	11.7	9.78	7.76	6.35	5.32	4.55	3.61	2.95	2.47	2.11	1.83	0
17	0.17	0.39	0.72	1.34	2.50	3.61	4.67	5.71	7.73	9.69	10.7	12.6	14.4	12.8	10.7	8.50	6.96	5.83	4.98	3.95	3.23	2.71	2.31	2.01	0
18	0.18	0.41	0.76	1.43	2.66	3.83	4.97	6.07	8.22	10.3	11.3	13.4	15.3	13.9	11.7	9.26	7.58	6.35	5.42	4.30	3.52	2.95	2.52	0	
19	0.19	0.43	0.81	1.51	2.82	4.07	5.27	6.44	8.72	10.9	12.0	14.2	16.3	15.1	12.7	10.0	8.22	6.89	5.88	4.67	3.82	3.20	2.73	0	
20	0.20	0.46	0.86	1.60	2.98	4.30	5.57	6.80	9.21	11.5	12.7	15.0	17.2	16.3	13.7	10.8	8.88	7.44	6.35	5.04	4.13	3.46	2.95	0	
21	0.21	0.48	0.90	1.69	3.14	4.53	5.87	7.17	9.71	12.2	13.4	15.8	18.1	17.6	14.7	11.7	9.55	8.01	6.84	5.42	4.44	3.72	3.18	0	
22	0.22	0.51	0.95	1.77	3.31	4.76	6.17	7.54	10.2	12.8	14.1	16.6	19.1	18.8	15.8	12.5	10.2	8.59	7.33	5.82	4.76	3.99	3.41	0	
23	0.23	0.53	1.00	1.86	3.47	5.00	6.47	7.91	10.7	13.4	14.8	17.4	20.0	20.1	16.9	13.4	11.0	9.18	7.84	6.22	5.09	4.27	0		
24	0.25	0.56	1.04	1.95	3.63	5.23	6.78	8.29	11.2	14.1	15.5	18.2	20.9	21.4	18.0	14.3	11.7	9.78	8.35	6.63	5.42	4.55	0		
25	0.26	0.58	1.09	2.03	3.80	5.47	7.08	8.66	11.7	14.7	16.2	19.0	21.9	22.8	19.1	15.2	12.4	10.4	8.88	7.05	5.77	4.83	0		
26	0.27	0.61	1.14	2.12	3.96	5.70	7.39	9.03	12.2	15.3	16.9	19.9	22.8	24.2	20.3	16.1	13.2	11.0	9.42	7.47	6.12	5.13	0		
28	0.29	0.66	1.23	2.30	4.29	6.18	8.01	9.79	13.2	16.6	18.3	21.5	24.7	27.0	22.6	18.0	14.7	12.3	10.5	8.35	6.84	5.73	0		
30	0.31	0.71	1.33	2.49	4.62	6.66	8.63	10.5	14.3	17.9	19.7	23.2	26.6	30.0	25.1	19.9	16.3	13.7	11.7	9.26	7.58	0			
32	0.33	0.76	1.42	2.66	4.96	7.14	9.25	11.3	15.3	19.2	21.1	24.9	28.6	32.2	27.7	22.0	18.0	15.1	12.9	10.2	8.35	0			
35	0.37	0.84	1.57	2.93	5.46	7.86	10.2	12.5	16.9	21.1	23.2	27.4	31.5	35.5	31.6	25.1	20.6	17.2	14.7	11.7	9.55	0			
40	0.43	0.97	1.81	3.38	6.31	9.08	11.8	14.4	19.5	24.4	26.8	31.6	36.3	41.0	38.7	30.7	25.1	21.0	18.0	14.3	0				
45	0.48	1.10	2.06	3.84	7.16	10.3	13.4	16.3	22.1	27.7	30.5	35.9	41.3	46.5	46.1	36.6	30.0	25.1	21.4	0					
	* Type A					Type B												Type C						

* Lubrication systems for chain types see 20.11.2

Motor = 10 Hp (1200 RPM)
Service factor = 1.5
Design horsepower = 15

Step 2. Chain selection. Chains are made with standard size pitches such as ¼, ⅜, ½, ⅝, ¾, and 1 inch, and up to 3 inches. These chains are identified by numbers. (Chain pitch refers to the distance between pins. See Fig. 20-43.) For example:

CHAIN NO.	CHAIN PITCH (in inches)
25	¼–.250
35	⅜–.375
40	½–.500
50	⅝–.625
60	¾–.750
80	1–1.000

To find the recommended chain number for a particular chain drive, refer to Table V. Locate the RPM of the small sprocket or drive in the left column and read horizontally to the listed design horsepower in the top column. Where the two columns intersect, the recommended chain number is given:

Given motor horsepower = 10
Design horsepower = 15
Given RPM of small sprocket = 1200
Recommended chain number from Table V = 50 (.625 inch pitch)

Table V is designed for single-strand chain drives. If multiple-chain drives are to be used, a correction in the design horsepower rating must be made. This is done by multiplying the single-strand rating by the proper factor specified for multiple-strand ratings shown in Table VI.

Step 3. Selecting the drive sprocket. The small drive sprocket must be large enough to accommodate the motor shaft. The general practice is to use a sprocket with a minimum of 17 teeth to obtain smooth operation at high speeds. For slower speeds sprockets with fewer than 17 teeth are suitable.

A No. 50 chain was selected to meet the specification established for the given problem,

and Table VII shows that for this chain at 1200 RPM and 15 Hp computed design rating, a 20 Type B tooth sprocket is necessary:

Chain required = No. 50
Design horsepower = 15
Given RPM of small sprocket = 1200
Small sprocket (from Table VII) = 20 teeth, Type B

Step 4. Selecting the driven sprocket. The number of teeth in the large sprocket is determined by the ratio of the driven and drive sprockets. This ratio for a single drive should not exceed 10:1. If a greater ratio is necessary, it is better to use two drive shafts in series—a drive or input to a countershaft and another drive from the countershaft to the driven or output shaft. Usually large sprockets should have not more than 120 teeth, although it is possible to have wheels with 150 teeth or more. The derived ratio is then multiplied by the number of teeth in the small sprocket.

Driver = 1200 RPM
Driven = 380 RPM
Small sprocket = 20 teeth
Ratio $= \dfrac{1200}{380} \approx 3{:}1$
Number of teeth = 20 x 3 = 60

Step 5. Checking center distance. In general, the shaft center distance should be within 30 to 50 chain pitches. See Fig. 20-53. Minimum center distance must be sufficient to provide clearance between the teeth of the two sprockets and also to insure a chain wrap of 120° around the small

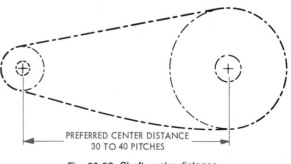

PREFERRED CENTER DISTANCE
30 TO 40 PITCHES

Fig. 20-53. Shaft center distance.

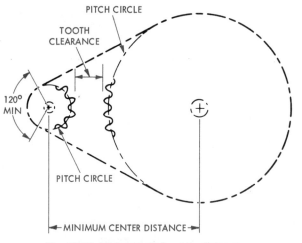

Fig. 20-54. Minimum shaft center distance.

sprocket. See Fig. 20-54. To have an effective 120° wrap, sprockets should have not less than a 3:1 ratio. For ratios greater than 3:1, the center distance should be no less than the difference in sprocket diameters. Maximum spacing should be about 80 pitches. Longer distances will often result in excessive chain tension unless special support guides or rollers are provided. Effective wraps for greater distances should be no less than the difference between sprocket diameters.

The given problem has specified a 22-inch center distance, and an analysis of the factors governing center distances indicates that the center distance of 22 inches meets these conditions.

Step 6. Finding chain length. Chain length is computed in terms of chain pitches. The result is then multiplied by the selected chain pitch to obtain the length in inches.

Actual chain length usually can be found by referring to a manufacturer's catalog. Without a manufacturer's chain length table, chain length can be calculated by the formula

$$CL = \left(\frac{2 \times CD}{CP}\right) + \left(\frac{N + n}{2}\right)$$

where
CL = chain length in pitches
CD = center distance in inches
CP = chain pitch

N = number of teeth in large sprocket
n = number of teeth in small sprocket

In our case the chain length can be found using these computations:

Center distance = 22″   N = 60
Chain pitch = .625   n = 20

$$CL = \frac{2 \times 22}{.625} + \frac{60 + 20}{2}$$
$$= 70.4 + 40$$
$$= 110.4 \text{ or } 111 \text{ pitches (111 pitches increased to even number = 112 pitches)}$$
$$= 112 \times .625 = 70″$$

Since the chain cannot have a fractional part of a pitch, it is necessary to increase or decrease the pitch to the next whole number (preferably the next even number) and then correct for distance.

Step 7. Correcting center distance. After the chain length is found, a correction is necessary to establish the exact center distance of the shafts. This is done by using the formula

$$CD_c = \frac{CL - \frac{N + n}{2} + \sqrt{\left(CL - \frac{N + n}{2}\right)^2 - 8\frac{(N - n)^2}{4\pi^2}}}{4}$$

where
CD_c = corrected center distance
CL = chain length in pitches
N = number of teeth in large sprocket
n = number of teeth in small sprocket.

The solution is found through these computations:

$$CD_c = \frac{112 - \frac{60 + 20}{2} + \sqrt{\left(112 - \frac{60 + 20}{2}\right)^2 - 8\frac{(60 - 20)^2}{39.48}}}{4}$$

$$= 35.42 \text{ pitches}$$
$$= 35.42 \times .625 = 22.14″$$

20.11.4 Additional design factors. Additional factors which should be considered in selecting an appropriate chain drive are chain lubricating methods and shaft positions.

Lubricating methods. Three types of lubricating systems are in general use. These are

Type A   Manual and drip lubrication
Type B   Oil bath and disc lubrication, and
Type C   Oil stream or pressure spray lubrication.

The horsepower rating tables will usually indicate the desired lubricating system to be used, which is influenced by chain speed and amount of power transmitted. See Table VII.

The *manual or drip lubricating system* is generally recommended for low- and medium-speed drives. With the manual method oil is periodically applied with a brush or spout can. Drip lubrication involves the use of a feed pipe from which oil drops on the chain while the drive is in motion. See Fig. 20-55.

The *oil bath system* has a casing where the lower strands of the chain run through a sump of

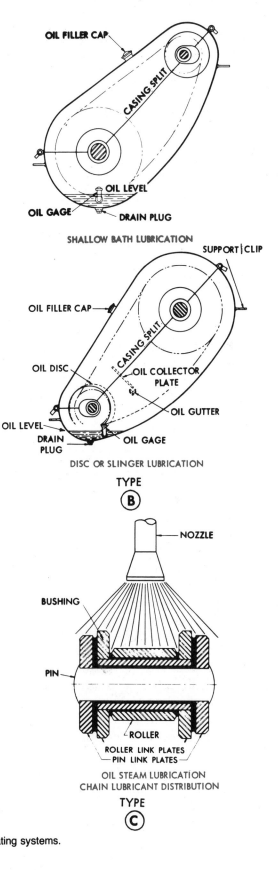

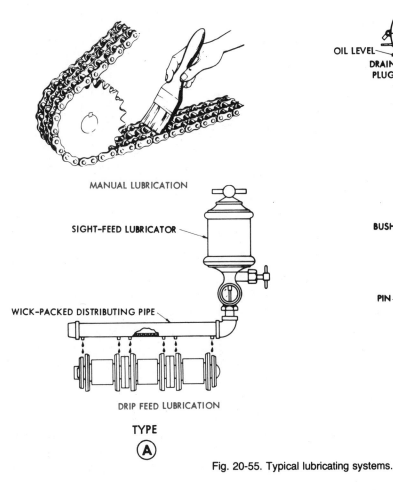

Fig. 20-55. Typical lubricating systems.

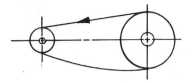

Satisfactory arrangement for
drives with short centers

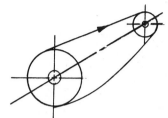

Satisfactory arrangement for
drives with short centers

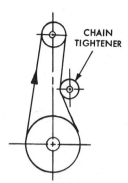

CHAIN
TIGHTENER

Best arrangement for vertical drives where
means for adjusting slack is possible

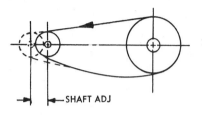

SHAFT ADJ

It is best to have one shaft adjustable
as shown directly above, or use chain
tightener as shown in lower arrangement

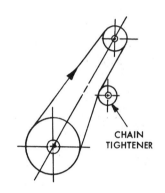

CHAIN
TIGHTENER

For drives on steep inclines some means must
be provided to adjust slack side tension

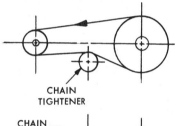

CHAIN
TIGHTENER

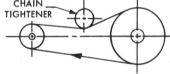

CHAIN
TIGHTENER

When slack side is on top some means must
be provided to adjust slack side tension

Unsatisfactory arrangement
(no adjustment is provided)

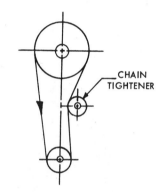

CHAIN
TIGHTENER

This arrangement, while sometimes uses,
is not satisfactory as that shown above

Fig. 20-56. Drive chain arrangements.

oil. *Disc lubrication* is a slight variation of the oil bath. In this system the chain operates above the oil level and discs pick up oil from the sump and deposit it on the chain. The use of casings also protects the chain from dusty atmospheres, high ambient temperatures, and corrosive fumes—all of which tend to shorten the life of chain drives. See Fig. 20-55.

Oil stream or pressure spray lubrication is the most satisfactory system for chain drives operating at relatively high speeds and loads. In this system a pump delivers an oil spray under pressure onto the chain. See Fig. 20-55.

Shaft positions. Location of shaft position is a necessary consideration to obtain maximum operating efficiency. The most favorable operating condition is to have the drives in an approximately horizontal position. If a vertical or steeply inclined drive is necessary, there should be less chain slack in order to keep the chain in proper alignment with the lower sprocket. See Fig. 20-56.

20.12 Belt Drives

Belt drives consist of endless flexible strands made of leather, rubberized cord, or fabric, or of other reinforced materials such as rayon, nylon,

Fig. 20-57. Typical belt drive.

Belt sizes have been standardized and are specified by a code which indicates cross-sectional shape and length. Cross-sectional shapes are classified as Conventional *(A, B, C, D, E)*, Narrow *(V)*, and Light Duty *(L)*. Each type has a special width and thickness. For example, a 90B V-belt is classified as a heavy-duty type belt having a width of 0.656 inch, a thickness of 0.406 inch, and a length of 90 inches. See Fig. 20-58. Other class designations are similarly coded except that for narrow belts, the length represents tenths of an inch. Thus, 5V1400 is a narrow belt with a 5V-type shape and 140 inches in length.

steel, or glass fiber. See Fig. 20-57. Belt drives transmit power at the lowest cost; they are smooth-running and shock-absorbing but are not as strong or as durable as gear or chain drives. Since belts are subject to creep, there must be some provision to adjust the equipment so belts can be tightened.

20.12.1 Types of belts. The following are the common types of belts used for power transmission:

V-Belts. V-belts have the widest application because they provide the best traction force, operating speed, and service life. V-belts function most effectively at speeds between 1500 and 6500 fpm. Most V-belts are designed for normal ambient temperatures. Any temperatures above 185°F or below −30°F will shorten belt life.

Fig. 20-59. Typical positive-drive belt.

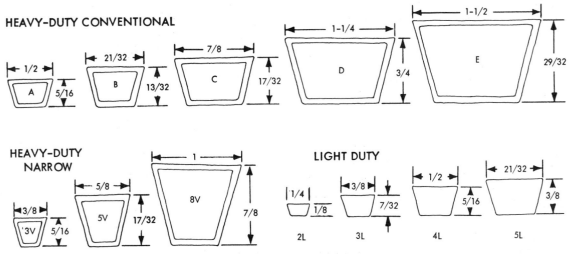

Fig. 20-58. Standard V-belt sections.

Agricultural belts have the same cross sections as conventional belts but are identified by the letter *H* before the section designation. Automotive belts are available in six SAE sections and are specified by nominal top widths of 0.380, 0.500, $^{11}/_{16}$, ¾, ⅞, and 1 inch.

Flat belts. These belts are used where high speed rather than power is the primary requisite. These belts do not provide the grip of V-belts and therefore tend to slip more. As a rule flat belts

have comparatively low efficiency at moderate speeds and are much noisier than other belt drives.

Positive-drive belts. These belts combine the advantages of the chain and gear drives and are very versatile on drives up to 600 horsepower and speeds up to and over 10,000 fpm. The belts have a flat outer surface and a ribbed underside. See Fig. 20-59. With positive-drive belts there is no slippage or speed variation. They are generally

	DRIVER					
DRIVEN MACHINE ■	AC Normal Torque **Electric Motor** □ (NEMA Design A-B)			AC High Torque Electric Motor ▲ (NEMA Design C-D)		
	Intermittent Service ①	Normal Service ②	Continuous Service ③	Intermittent Service ①	Normal Service ②	Continuous Service ③
Agitators for Liquids............ Blowers and Exhausters........... Centrifugal Pumps and Compressors........... Conveyors (Light Duty)............ Fans (up to 10 H.P.)................	1.0	1.1	1.2	1.1	1.2	1.3
Belt Conveyors for Sand, Grain, etc............ Fans (over 10 H.P.)................ Generators................ Laundry Machinery............ Line Shafts................ Machine Tools................ Mixers (Dough)................ Positive Displacement Rotary Pumps............ Printing Machinery................ Punches-Presses-Shears................ Revolving and Vibrating Screens..............	1.1	1.2	1.3	1.2	1.3	1.4
Blowers (Positive Displacement)............ Brick Machinery................ Compressors (Piston)............ Conveyors (Drag-Pan-Screw)............ Elevators (Bucket)................ Exciters................ Hammer Mills................ Paper Mill Beaters................ Pulverizers................ Pumps (Piston)................ Saw Mill and Woodworking Machinery........ Textile Machinery................	1.2	1.3	1.4	1.4	1.5	1.6
Crushers (Giratory-Jaw-Roll)............ Mills (Ball-Rod-Tube)............ Hoists................ Rubber Calenders-Extruders-Mills..............	1.3	1.4	1.5	1.5	1.6	1.8

① Intermittent Service refers to 3–5 hours of daily or seasonal operation.
② Normal Service indicates 8–10 hours of daily operation.
③ Continuous Service refers to 16–24 hours of daily operation.

Fig. 20-60. Service factors for belt drives.

recommended for drives requiring synchronization, high mechanical efficiency, and constant angular velocity.

20.12.2 Belt-drive design factors. Selection of a proper belt drive involves the following factors:

1. Horsepower and speed of driving shaft
2. Type of duty service
3. Speed of driven unit
4. True center distance between shafts
5. Belt length
6. Number of belts required

Although these selection factors normally apply to many belt types, further discussion of belt-drive design will be confined to V-belts, since they are most commonly used.

To carry out a typical belt-selection procedure, let us turn to a problem:

Problem. A 2-horsepower 1750 RPM electric motor is to operate a blower at approximately 660 RPM under a normal duty service load. The center distance between motor and blower shaft is to be

about 24 inches. Both motor and blower shafts are 1.375 inches in diameter with standard keysets.

Solution.

Step 1. Find design horsepower. Determine the proper service factor from the chart in Fig. 20-60. Notice that three kinds of services are listed for different driven machines: intermittent, normal, and continuous. From this table we find that for the blower, a service factor of 1.1 is specified. Multiply the given rated horsepower by this service factor:

Motor = 2 HP
Service factor = 1.1
Design horsepower = 2.2

Step 2. Select belt cross section. Use the chart in Fig. 20-61 to find the appropriate belt cross section. On the bottom line of this chart, locate the design horsepower. Then read up to the RPM of the faster shaft. The area opposite the RPM column will indicate the desired belt cross section:

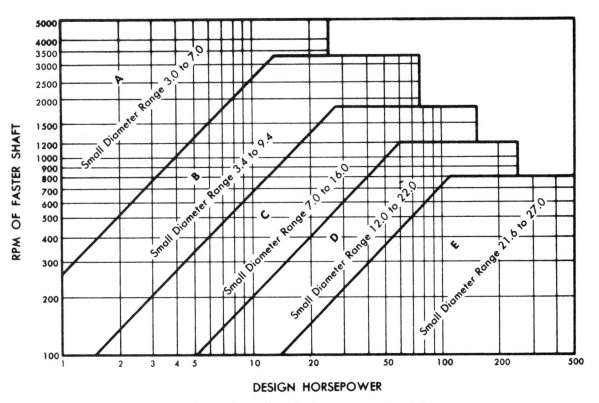

Fig. 20-61. Belt selection and outside diameter range of small sheave.

SHEAVE PITCH DIAMETER		APPROXIMATE DRIVEN SPEEDS AND HORSEPOWER PER BELT												NOMINAL CENTER DISTANCES—USING SURE-GRIP "A" BELTS (MULTIPLY H.P. FROM TABLE BY NEAREST COMBINED ARC & LENGTH CORRECTION FACTOR SHOWN IN BOLD FIGURES)																		
		1160 RPM DRIVER				1750 RPM DRIVER				3500 RPM DRIVER				V-BELT NUMBER																		
▲ Driver	▲ Driven	Driven Speed	H.P. Standard belt	H.P. Super belt	H.P. Steel belt	Driven Speed	H.P. Standard belt	H.P. Super belt	H.P. Steel belt	Driven Speed	H.P. Standard belt	H.P. Super belt	H.P. Steel belt	A26	A31	A35	A38	A42	A46	A51	A55	A60	A68	A75	A80	A85	A90	A96	A105	A112	A120	A128
																.82	**.83**	**.84**	**.85**	**.88**	**.91**	**.92**	**.96**	**.98**	**1.01**	**1.02**	**1.03**	**1.05**	**1.08**	**1.09**	**1.11**	**1.12** ■
6.4	15.6	475	2.92	3.33	3.47	716	3.95	4.54	4.73	1432	4.97*	6.09*	6.37*	...	...	...	...	...	...	...	...	...	16.6	20.2	22.8	25.3	27.9	30.9	35.4	39.0	43.0	47.0
3.6	9.0	464	1.46	1.71	1.79	700	2.00	2.37	2.50	1400	3.12	3.83	4.05	...	...	...	9.4	11.4	13.5	16.0	18.1	20.6	24.6	28.1	30.6	33.2	35.7	38.7	43.2	46.7	50.7	54.7
4.8	12.0	464	2.30	2.61	2.72	700	3.18	3.64	3.79	1400	4.83	5.70	5.96	...	...	...	...	...	...	12.4	14.5	17.1	21.2	24.7	27.2	29.7	32.3	35.3	39.8	43.3	47.3	51.3
6.0	15.0	464	2.79	3.17	3.30	700	3.80	4.35	4.54	1400	5.08*	6.13*	6.41*	...	...	...	...	...	...	...	...	13.4	17.6	21.2	23.7	26.3	28.8	31.8	36.4	39.9	43.9	47.9
6.2	15.6	462	2.86	3.25	3.39	698	3.87	4.44	4.64	1392	5.04*	6.12*	6.40*	...	...	...	...	...	...	...	...	...	16.8	20.4	23.0	25.5	28.0	31.1	35.6	39.1	43.1	47.2
																.82	**.83**	**.84**	**.85**	**.88**	**.91**	**.92**	**.96**	**.98**	**1.01**	**1.02**	**1.03**	**1.05**	**1.08**	**1.09**	**1.11**	**1.12** ■
4.2	10.6	460	1.88	2.16	2.26	694	2.60	3.01	3.16	1389	4.03	4.82	5.06	...	...	...	...	...	11.6	14.2	16.2	18.8	22.8	26.3	28.9	31.4	33.9	36.9	41.4	44.9	48.9	52.9
3.0	7.6	458	1.02	1.24	1.32	692	1.38	1.71	1.81	1382	2.11	2.74	2.93	...	...	9.6	11.2	13.2	15.1	17.3	19.4	21.9	26.0	29.6	32.1	34.6	37.1	40.1	44.6	48.2	52.2	56.2
5.2	13.2	456	2.51	2.84	2.96	689	3.46	3.95	4.11	1376	5.11	6.03	6.29	...	...	...	...	...	...	...	12.9	16.0	19.6	23.2	25.7	28.3	30.8	34.1	38.4	41.9	45.9	49.9
3.2	8.2	453	1.17	1.40	1.48	684	1.59	1.93	2.04	1367	2.46	3.11	3.32	...	...	8.8	10.4	12.4	14.5	17.0	19.0	21.6	25.6	29.1	31.6	34.1	36.6	39.6	44.1	47.6	51.6	55.6
7.0	18.0	451	3.12	3.55	3.71	681	4.16	4.79	5.01	1362	...	...	...	...	...	...	...	...	...	...	...	...	...	17.7	20.3	22.9	25.4	28.5	33.1	36.6	40.6	44.7
																.79	**.83**	**.84**	**.87**	**.88**	**.91**	**.92**	**.96**	**.98**	**1.01**	**1.02**	**1.03**	**1.05**	**1.08**	**1.09**	**1.11**	**1.12** ■
5.8	15.0	448	2.72	3.09	3.22	676	3.72	4.25	4.44	1351	5.11*	6.13*	6.41*	...	...	...	...	...	...	...	...	13.5	17.7	21.3	23.9	26.4	29.0	32.0	36.5	40.1	44.1	48.1
6.0	15.6	446	2.79	3.17	3.30	672	3.80	4.35	4.54	1345	5.08*	6.13*	6.41*	...	...	...	...	...	...	...	...	12.3	16.5	20.2	22.9	25.4	27.9	31.0	35.6	39.1	43.1	47.2
4.6	12.0	444	2.16	2.46	2.57	670	2.99	3.43	3.58	1341	4.58	5.42	5.67	...	...	...	...	...	...	12.6	14.7	17.2	21.3	24.8	27.4	29.9	32.4	35.4	39.9	43.5	47.5	51.5
5.0	13.2	439	2.44	2.76	2.87	662	3.37	3.84	4.00	1325	5.07	5.96	6.23	...	...	...	...	...	...	...	13.4	16.1	20.0	23.6	26.1	28.5	31.0	34.0	38.8	42.0	46.0	50.0
3.4	9.0	438	1.31	1.55	1.64	660	1.80	2.15	2.27	1321	2.80	3.48	3.69	...	...	...	9.5	11.6	13.6	16.2	18.2	20.7	24.8	28.3	30.8	33.3	35.8	38.8	43.3	46.8	50.8	54.8
																.79	**.82**	**.84**	**.87**	**.88**	**.91**	**.92**	**.96**	**.98**	**1.01**	**1.02**	**1.03**	**1.05**	**1.08**	**1.09**	**1.11**	**1.12** ■
4.0	10.6	438	1.74	2.01	2.11	660	2.40	2.80	2.94	1321	3.74	4.50	4.74	...	...	...	...	...	11.7	14.3	16.3	18.9	22.9	26.5	29.0	31.5	34.0	37.0	41.5	45.1	49.1	53.1
5.6	15.0	433	2.65	3.01	3.13	653	3.63	4.15	4.33	1306	5.12*	6.11*	6.39*	...	...	...	...	...	...	...	...	13.7	17.9	21.5	24.0	26.6	29.1	32.1	36.7	40.2	44.2	48.3
5.8	15.6	431	2.72	3.09	3.22	650	3.72	4.25	4.44	1300	5.11*	6.13*	6.41*	...	...	...	...	...	...	...	...	...	16.9	20.5	23.0	25.6	28.6	31.5	36.2	39.8	43.7	47.3
3.0	8.2	425	1.02	1.24	1.32	641	1.38	1.71	1.81	1282	2.11	2.74	2.93	...	...	9.0	10.5	12.6	14.6	17.2	19.2	21.7	25.7	29.2	31.8	34.3	36.8	39.8	44.3	47.8	51.8	55.8
4.4	12.0	425	2.02	2.32	2.42	641	2.80	3.22	3.37	1282	4.31	5.12	5.37	...	...	...	...	...	...	12.7	14.8	17.4	21.4	25.0	27.5	30.0	32.6	35.6	40.1	43.6	47.6	51.6
																.79	**.81**	**.84**	**.86**	**.87**	**.88**	**.90**	**.94**	**.96**	**.99**	**1.00**	**1.02**	**1.04**	**1.06**	**1.08**	**1.10**	**1.12** ■
6.6	18.0	425	2.99	3.40	3.55	641	4.02	4.62	4.83	...	...	...	...	...	...	...	...	...	...	...	...	...	14.7	18.2	20.7	23.3	25.8	28.8	33.3	36.8	40.8	44.8
4.8	13.2	422	2.30	2.61	2.72	636	3.18	3.64	3.79	1272	4.83	5.70	5.96	...	...	...	...	...	...	11.1	13.2	16.2	19.9	23.5	26.0	28.6	31.1	34.1	38.7	42.2	46.2	50.3
5.4	15.0	417	2.58	2.93	3.05	629	3.55	4.05	4.22	1259	5.12	6.08	6.35	...	...	...	...	...	...	...	...	13.8	18.0	21.6	24.2	26.7	29.2	32.3	36.8	40.4	44.4	48.4
3.8	10.6	416	1.60	1.86	1.95	627	2.20	2.59	2.72	1254	3.44	4.17	4.40	...	...	...	...	...	11.9	14.4	16.5	19.0	23.1	26.6	29.1	31.7	34.2	37.2	41.7	45.2	49.2	53.2
5.6	15.6	416	2.65	3.01	3.13	627	3.63	4.15	4.33	1254	5.12*	6.11*	6.39*	...	...	...	...	...	...	...	...	13.2	17.3	21.0	23.6	26.1	28.8	31.4	36.0	39.5	43.5	47.6
																	.80	**.84**	**.86**	**.87**	**.88**	**.90**	**.94**	**.96**	**.99**	**1.00**	**1.02**	**1.04**	**1.06**	**1.08**	**1.10**	**1.12** ■
7.0	19.6	414	3.12	3.55	3.71	625	4.16	4.79	5.01	...	...	...	...	...	...	...	...	...	...	...	...	...	...	...	18.7	21.3	23.9	27.0	31.6	35.2	39.3	43.3
3.2	9.0	413	1.17	1.40	1.48	623	1.59	1.93	2.04	1246	2.46	3.11	3.32	...	...	...	9.6	11.7	13.8	16.3	18.4	20.9	24 9	28.4	30.9	33.5	36.0	39.0	43.5	47.0	51.0	55.0
6.4	18.0	413	2.92	3.33	3.47	623	3.95	4.54	4.73	1246	4.97*	6.09*	6.37*	...	...	...	...	...	...	...	...	...	...	18.1	20.7	23.3	25.8	28.9	33.5	37.0	41.1	45.1
4.2	12.0	406	1.88	2.16	2.26	612	2.60	3.01	3.16	1224	4.03	4.82	5.06	...	...	...	...	...	...	12.8	14.9	17.5	21.6	25.1	27.7	30.2	32.7	35.7	40.3	43.8	47.8	51.8
4.6	13.2	404	2.16	2.46	2.57	610	2.99	3.43	3.58	1219	4.58	5.42	5.67	...	...	...	...	...	...	...	13.6	16.3	20.3	23.9	26.2	28.5	31.3	34.3	39.0	42.3	46.3	50.3
																	.80	**.82**	**.84**	**.86**	**.88**	**.90**	**.94**	**.96**	**.99**	**1.00**	**1.02**	**1.04**	**1.06**	**1.08**	**1.10**	**1.12** ■
5.2	15.0	403	2.51	2.84	2.96	608	3.46	3.95	4.11	1215	5.11	6.03	6.29	...	...	...	...	...	...	...	...	13.9	18.1	21.7	24.3	26.9	29.4	32.4	37.0	40.5	44.5	48.6
5.4	15.6	401	2.58	2.93	3.05	605	3.55	4.05	4.22	1210	5.12	6.08	6.35	...	...	...	...	...	...	...	...	13.4	17.4	21.1	23.7	26.2	28.9	31.5	36.1	39.6	43.6	47.7
6.2	18.0	400	2.86	3.25	3.39	603	3.87	4.44	4.64	1207	5.04*	6.12*	6.40*	...	...	...	...	...	...	...	...	...	...	18.2	20.8	23.4	26.0	29.1	33.6	37.2	41.2	45.3
3.6	10.6	395	1.46	1.71	1.79	595	2.00	2.37	2.50	1190	3.12	3.83	4.05	...	...	...	...	9.9	12.0	14.6	16.6	19.2	23.2	26.8	29.3	31.8	34.3	37.3	41.9	45.4	49.4	53.4
6.6	19.6	391	2.99	3.40	3.55	588	4.02	4.62	4.83	...	...	...	...	...	...	...	...	...	...	...	...	...	...	16.2	18.8	21.4	24.0	27.1	31.6	35.2	39.2	43.3
																	.78	**.80**	**.84**	**.86**	**.88**	**.90**	**.94**	**.96**	**.99**	**1.00**	**1.01**	**1.04**	**1.06**	**1.08**	**1.10**	**1.12** ■
3.0	9.0	387	1.02	1.24	1.32	583	1.38	1.71	1.81	1167	2.11	2.74	2.93	...	...	...	9.8	11.9	13.9	16.5	18.5	21.0	25.1	28.6	31.1	33.6	36.1	39.1	43.6	47.1	51.2	55.2
4.0	12.0	387	1.74	2.01	2.11	583	2.40	2.80	2.94	1167	3.74	4.50	4.74	...	...	...	...	...	...	13.0	15.1	17.6	21.7	25.3	27.8	30.3	32.9	35.9	40.4	43.9	47.9	51.9
4.4	13.2	387	2.02	2.32	2.42	583	2.80	3.22	3.37	1167	4.31	5.12	5.37	...	...	...	...	...	...	...	13.7	16.4	20.5	24.0	26.4	29.0	31.5	34.5	39.2	42.5	46.5	50.5
5.0	15.0	387	2.44	2.76	2.87	583	3.37	3.84	4.00	1167	5.07	5.96	6.23	...	...	...	...	...	...	...	...	14.1	18.3	21.9	24.4	27.0	29.5	32.6	37.1	40.7	44.7	48.7
5.2	15.6	387	2.51	2.84	2.96	583	3.46	3.95	4.11	1167	5.11	6.03	6.29	...	...	...	...	...	...	...	...	...	17.5	20.9	23.5	26.1	28.7	31.8	36.3	40.1	44.0	48.0
																	.78	**.78**	**.84**	**.86**	**.88**	**.90**	**.94**	**.96**	**.99**	**1.00**	**1.01**	**1.04**	**1.06**	**1.08**	**1.10**	**1.12** ■
6.0	18.0	387	2.79	3.17	3.30	583	3.80	4.35	4.54	1167	5.08*	6.13*	6.41*	...	...	...	...	...	...	...	...	...	...	18.3	21.0	23.6	26.1	29.2	33.8	37.3	41.4	45.4
6.4	19.6	379	2.92	3.33	3.47	572	3.95	4.54	4.73	1144	4.97*	6.09*	6.37*	...	...	...	...	...	...	...	...	...	...	16.4	19.1	21.7	24.3	27.4	32.1	35.6	39.7	43.7
5.8	18.0	374	2.72	3.09	3.22	565	3.72	4.25	4.44	1129	5.11*	6.13*	6.41*	...	...	...	...	...	...	...	...	...	14.7	18.5	21.1	23.7	26.3	29.3	33.9	37.5	41.5	45.6
3.4	10.6	372	1.31	1.55	1.64	561	1.80	2.15	2.27	1122	2.80	3.48	3.69	...	...	...	...	10.0	12.1	14.7	16.8	19.3	23.4	26.9	29.4	32.0	34.5	37.5	42.0	45.5	49.5	53.5
5.0	15.6	372	2.44	2.76	2.87	561	3.37	3.84	4.00	1120	5.07	5.96	6.23	...	...	...	...	...	...	...	...	13.5	17.7	21.3	23.9	26.5	29.1	32.1	36.2	39.8	43.8	47.8
																		.78	**.84**	**.85**	**.88**	**.90**	**.94**	**.96**	**.98**	**.99**	**1.00**	**1.03**	**1.05**	**1.07**	**1.10**	**1.12** ■
4.8	15.0	371	2.30	2.61	2.72	559	3.18	3.64	3.79	1118	4.83	5.70	5.96	...	...	...	...	...	...	...	...	14.2	18.4	22.0	24.6	27.1	29.7	32.7	37.3	40.8	44.8	48.8
4.2	13.2	369	1.88	2.16	2.26	557	2.60	3.01	3.16	1115	4.03	4.82	5.06	...	...	...	...	...	...	...	13.7	16.2	20.4	24.1	26.6	29.0	31.5	34.5	39.1	42.7	46.7	50.7
3.8	12.0	367	1.60	1.86	1.95	554	2.20	2.59	2.72	1108	3.44	4.17	4.40	...	...	...	...	...	...	13.1	15.2	17.8	21.9	25.4	27.9	30.5	33.0	36.0	40.5	44.0	48.1	52.1
6.2	19.6	367	2.86	3.25	3.39	554	3.87	4.44	4.64	1108	5.04*	6.12*	6.40*	...	...	...	...	...	...	...	...	...	...	16.5	19.2	21.9	24.5	27.6	32.2	35.8	39.8	43.9
5.6	18.0	361	2.65	3.01	3.13	545	3.63	4.15	4.33	1090	5.12*	6.11*	6.39*	...	...	...	...	...	...	...	...	...	14.8	18.6	21.2	23.8	26.4	29.5	34.1	37.6	41.7	45.7
																				.84	**.88**	**.90**	**.94**	**.96**	**.98**	**.99**	**1.00**	**1.03**	**1.05**	**1.07**	**1.10**	**1.12** ■

■ This line of figures shows the combined Arc & Length Correction Factors.

Fig. 20-62. Stock drive selection table for conventional A-type V-belt drives.

Given RPM of small sprocket = 1750
Design HP = 2.2
Belt cross section = Type A

Step 3. Select pulley diameters. Turn to the drive selection chart, Fig. 20-62, for the A-type belt and read down the column headed Approximate Driven Speed to the nearest desired driven speed. On this line move across to the column headed Sheave (pulley) Pitch Diameter and find the diameters of the driver and driven pulleys:

Motor = 1750 RPM
Driven speed = 660 RPM
Driven pulley = 10.6″ dia.
Driver pulley = 4.0″ dia.

Step 4. Select belt length. From the location of the required pulley diameters in the drive selection

*For design problems requiring other V-belt sizes and types, consult manufacturer's (T. B. Wood's Sons Co.) catalog.

chart, Fig. 20-62, read across and find the nearest specified center distance. Then move upward to the top column which gives the belt length:

Nearest center distance = 22.9″
V-belt (No. A68) length = 68″
Adjusted center distance = 22.9″

Step 5. Determine number of belts required. From the drive selection chart in Fig. 20-62, find the rated horsepower per belt listed under the column headed 1750 RPM Driver and corresponding to the required sheave pitch diameters. Multiply the rated horsepower per belt by the combined Arc & Length Correction Factor which appears in bold type and is further identified by a black square at the extreme right of the table. Then divide the corrected horsepower per belt into the design horsepower found in step 1. If the answer contains a fraction, use the next larger number of belts:

Horsepower per belt = 2.40
Arc & Length correction factor = .96
Corrected horsepower = .96 × 2.40 = 2.30

$$\text{Belts required} = \frac{\text{Design HP}}{\text{Corrected HP per belt}} = \frac{2.2}{2.3} = .96$$

Belts needed = 1

20.12.3 Design of light-duty belt drives. For light-duty belt drives in the L-type classification (see Fig. 20-58), the following is all that is required in order to design the required drive:

1. Speed (RPM) and horsepower of motor or drive unit
2. Speed (RPM) of driven shaft
3. Space available for the belt drive

Problem. Design the proper belt drive to run a ventilating fan intermittently at a speed of 550 RPM by an electric motor rated ½ horsepower and 1750 RPM. The center distance of shafts is to be approximately 10 inches.

TABLE VIII HORSEPOWER RATING

RPM of small pulley	outside diameter of small v-pulley—inches														
	1.50	1.75	2.00	2.25	2.50	2.75	3.00	3.25	3.50	3.75	4.00	4.25	4.50	4.75	5.00
200	...	...	...	...	...	...	...	...	...	...	0.18	0.22	0.24	0.28	0.29
400	...	...	0.06	0.08	0.12	0.15	0.18	0.22	0.25	0.31	0.35	0.42	0.46	0.52	0.56
600	0.04	0.07	0.08	0.12	0.18	0.22	0.27	0.32	0.36	0.44	0.51	0.58	0.66	0.73	0.81
800	0.05	0.08	0.11	0.15	0.22	0.28	0.34	0.41	0.45	0.55	0.64	0.74	0.81	0.93	1.00
1000	0.06	0.10	0.12	0.18	0.26	0.33	0.42	0.48	0.55	0.64	0.75	0.86	0.99	1.10	1.21
1160	0.07	0.11	0.15	0.21	0.29	0.38	0.46	0.54	0.62	0.69	0.84	0.98	1.07	1.23	1.35
1400	0.08	0.12	0.17	0.23	0.33	0.43	0.53	0.64	0.74	0.84	0.96	1.10	1.25	1.42	1.55
1600	0.08	0.14	0.19	0.25	0.36	0.48	0.58	0.69	0.80	0.90	1.02	1.20	1.36	1.53	1.68
1750	0.08	0.15	0.20	0.25	0.38	0.51	0.63	0.74	0.85	0.96	1.08	1.25	1.43	1.61	1.78
2000	0.09	0.16	0.22	0.28	0.41	0.55	0.68	0.81	0.92	1.05	1.17	1.35	1.54	1.73	1.90
2200	0.09	0.17	0.24	0.31	0.44	0.58	0.72	0.86	0.99	1.12	1.25	1.41	1.61	1.80	1.99
2400	0.10	0.18	0.25	0.32	0.45	0.61	0.76	0.91	1.05	1.19	1.32	1.45	1.65	1.86	2.02
2600	0.10	0.19	0.26	0.35	0.47	0.64	0.79	0.96	1.09	1.24	1.38	1.48	1.69	1.89	2.09
2800	0.11	0.19	0.28	0.36	0.48	0.66	0.83	0.99	1.14	1.28	1.42	1.48	1.71	1.91	2.11
3000	0.11	0.21	0.29	0.39	0.49	0.68	0.85	1.02	1.18	1.32	1.46	1.48	1.69	1.89	2.08
3200	0.11	0.21	0.30	0.39	0.51	0.70	0.88	1.05	1.20	1.36	1.50	1.50	1.67	1.86	2.03
3450	0.12	0.22	0.32	0.41	0.51	0.71	0.90	1.07	1.23	1.38	1.52	1.52	1.61	1.78	1.94
3600	0.12	0.22	0.33	0.42	0.52	0.72	0.91	1.09	1.25	1.40	1.54	1.54	1.54	1.71	1.85
3800	0.12	0.22	0.33	0.42	0.52	0.72	0.92	1.09	1.25	1.41	1.54	1.54	1.54	1.59	1.72
4000	0.12	0.22	0.34	0.44	0.53	0.72	0.92	1.10	1.26	1.40	1.52	1.52	1.52	1.52	1.55

For [ ] Background use a 3L Section   For [ ] Background use a 4L Section   For [ ] Background use a 5L Section

NOTE: This table incorporates a service factor of 1.3. For heavy duty, multiply normal duty horsepower rating by .85. For light duty, multiply normal duty horsepower rating by 1.20.

Solution.

1. Read down left column of Table VIII to the RPM figure nearest to that specified for the motor.

2. Read across this line to the figure closest to the design horsepower of the drive.

3. See bottom of Table VIII for the recommended type belt section.

4. From Table VIII read up from the design horsepower rating to find the motor pulley diameter in the top horizontal column.

5. Refer to Table IX and read across the top to the figure nearest the small pulley size. Read down this column to the figure nearest the desired speed of driven shaft and follow this line to the extreme left column. The figure shown is the diameter of the driven or large pulley.

6. Add diameters of the two pulleys and select the number in the top row of Table X that is nearest to this sum.

7. Read down this column below the shaded area. The given figure indicates the required center distance in inches.

8. Using the figure given for the corrected center distance, follow the line to the left to the column marked Belt Length. The given figure represents the required belt length.

Motor = 1750 RPM

Design HP = HP × service factor
= .50 × 1.0 = .50
(See Fig. 20-60.)

Belt = 4L

Motor pulley = 2.75″ dia.

Desired speed of driven
pulley = 550 RPM

Closest given speed = 564 RPM

Driven pulley size = 8.0″

DIA of small pulley = 2.75″

DIA of large pulley = 8.00″

Total = 10.75″

Closest table figure = 11

Specified center distance = 10″

Corrected center distance = 11.1″

Belt length = 40″

Required belt type = 4L400

TABLE IX DRIVEN SPEEDS FOR 1750 RPM MOTORS

DriveN V-Pulley O.D. inches	DriveR V-Pulley O.D.—inches													
	1.50	1.75	2.00	2.25	2.50	2.75	3.00	3.25	3.50	3.75	4.00	4.25	4.50	4.75	5.00
1.5	1750	2100	2450	2800	3150	3500	3850	...	...	...	...	...	...	...
2.0	1250	1500	1750	2000	2250	2500	2750	3000	3250	3500	3750	4000	...	...	...
2.5	974	1167	1360	1555	1750	1945	2140	2330	2530	2725	2915	3110	3305	3500	3690
3.0	797	955	1113	1272	1431	1590	1750	1910	2070	2225	2385	2545	2700	2865	3022
3.5	674	808	942	1077	1210	1346	1480	1615	1750	1885	2020	2155	2290	2420	2555
4.0	584	700	817	935	1050	1168	1283	1400	1518	1634	1750	1865	1985	2100	2215
4.5	516	618	720	824	926	1030	1131	1235	1339	1440	1543	1650	1750	1850	1955
5.0	462	554	646	737	830	922	1013	1105	1198	1290	1382	1473	1568	1660	1750
5.5	417	500	584	667	750	834	917	1000	1082	1167	1250	1333	1417	1500	1581
6.0	381	456	533	610	685	760	837	913	990	1065	1140	1217	1290	1370	1445
6.5	350	420	490	560	630	700	771	840	910	980	1050	1120	1190	1260	1330
7.0	324	389	454	518	584	648	713	778	843	907	973	1039	1102	1168	1231
8.0	282	339	394	451	507	564	620	676	734	789	845	902	959	1016	1072
9.0	250	300	350	400	450	500	550	600	650	700	750	800	850	900	950
10.0	224	270	315	360	405	450	495	540	585	630	675	720	765	810	855
11.0	203	244	285	326	366	407	448	488	530	570	610	652	692	733	774
12.0	186	224	261	298	336	373	410	446	485	522	560	596	634	671	708
13.0	172	206	240	275	309	343	378	412	447	480	515	549	584	618	652
14.0	159	191	223	255	286	318	350	382	414	445	477	509	540	573	605
15.0	148	178	208	247	267	297	326	358	386	415	445	475	505	534	564

TABLE X APPROXIMATE CENTER DISTANCES (IN INCHES)

Installation allowance	take-up	belt length	sum of both v-belt pulley diameters																																		
			4	4½	5	5½	6	6½	7	7½	8	8½	9	9½	10	10½	11	11½	12	12½	13	13½	14	14½	15	15½	16	16½	17	17½	18	18½	19	19½	20	20½	21
½	½	16	4.9	4.5	4.1																																
⅝	½	18	5.9	5.5	5.1	4.6																															
⅝	½	20	6.9	6.5	6.1	5.6	5.2																														
⅝	½	22	7.9	7.5	7.1	6.6	6.2	5.8																													
⅝	½	24	8.9	8.5	8.1	7.6	7.2	6.8	6.3	5.8																											
⅝	½	26	9.9	9.5	9.1	8.6	8.2	7.8	7.3	6.9	6.5																										
⅝	½	28	10.9	10.5	10.1	9.6	9.2	8.8	8.4	7.9	7.6	7.1	6.6																								
⅝	½	30	11.9	11.5	11.1	10.6	10.2	9.8	9.4	8.9	8.6	8.1	7.7	7.3																							
⅝	½	32	12.9	12.5	12.1	11.6	11.2	10.8	10.4	10.0	9.6	9.1	8.7	8.4	8.0																						
⅝	½	34	13.9	13.5	13.1	12.7	12.2	11.8	11.4	11.0	10.6	10.2	9.7	9.4	9.0	8.6																					
⅝	½	36	14.9	14.5	14.1	13.7	13.2	12.8	12.4	12.0	11.6	11.2	10.7	10.4	10.0	9.6	9.0																				
⅝	½	38	15.9	15.5	15.1	14.7	14.2	13.8	13.4	13.0	12.6	12.2	11.8	11.4	11.0	10.6	10.0	9.7	9.1																		
¾	½	40	16.9	16.5	16.1	15.7	15.3	14.8	14.4	14.0	13.6	13.2	12.8	12.4	12.0	11.6	11.1	10.7	10.1	9.8																	
¾	½	42	17.9	17.5	17.1	16.7	16.3	15.8	15.4	15.0	14.6	14.2	13.8	13.4	13.1	12.6	12.1	11.7	11.2	10.8	10.2																
¾	½	44	18.9	18.5	18.1	17.7	17.3	16.8	16.4	16.0	15.6	15.2	14.8	14.4	14.1	13.6	13.1	12.8	12.2	11.9	11.2	10.9															
¾	½	46	19.9	19.5	19.1	18.7	18.3	17.9	17.4	17.0	16.6	16.2	15.8	15.4	15.1	14.6	14.1	13.8	13.2	12.9	12.3	12.0	10.9	10.5													
¾	½	48	20.9	20.5	20.1	19.7	19.3	18.9	18.4	18.0	17.7	17.2	16.8	16.4	16.1	15.6	15.1	14.8	14.3	13.9	13.3	13.0	12.0	11.6	11.3												
¾	½	50	21.9	21.5	21.1	20.7	20.3	19.9	19.4	19.0	18.7	18.2	17.8	17.4	17.1	16.7	16.2	15.8	15.3	14.9	14.4	14.0	13.1	12.7	12.4	12.1	11.7										
¾	½	52	22.9	22.5	22.1	21.7	21.3	20.9	20.4	20.0	19.7	19.2	18.8	18.4	18.1	17.7	17.2	16.8	16.3	15.9	15.4	15.0	14.1	13.8	13.5	13.1	12.8	12.5									
¾	½	54	23.9	23.5	23.1	22.7	22.3	21.9	21.4	21.0	20.7	20.2	19.8	19.4	19.1	18.7	18.2	17.8	17.3	17.0	16.4	16.1	15.2	14.8	14.5	14.2	13.8	13.5	13.2	12.9							
¾	½	56	24.9	24.5	24.1	23.7	23.3	22.9	22.4	22.0	21.7	21.2	20.8	20.4	20.1	19.7	19.2	18.8	18.3	18.0	17.4	17.1	16.2	15.9	15.6	15.2	14.9	14.6	14.2	13.9							
¾	½	58	25.9	25.5	25.1	24.7	24.3	23.9	23.4	23.0	22.7	22.2	21.8	21.4	21.1	20.7	20.2	19.8	19.3	19.0	18.5	18.1	17.3	16.9	16.6	16.3	15.9	15.6	15.2	14.9	13.5	13.2					
⅞	¾	60	26.9	26.5	26.1	25.7	25.3	24.9	24.5	24.0	23.7	23.2	22.8	22.4	22.1	21.7	21.2	20.8	20.4	20.0	19.5	19.1	18.3	18.0	17.6	17.3	17.0	16.6	16.3	15.9	14.6	14.3	14.0				
⅞	¾	62	27.9	27.5	27.1	26.7	26.3	25.9	25.5	25.0	24.7	24.3	23.8	23.4	23.1	22.7	22.2	21.8	21.4	21.0	20.5	20.1	19.4	19.0	18.7	18.3	18.0	17.6	17.3	16.9	15.7	15.4	15.1	14.8			
⅞	¾	64	28.9	28.5	28.1	27.7	27.3	26.9	26.5	26.0	25.7	25.3	24.8	24.4	24.1	23.7	23.2	22.9	22.4	22.0	21.5	21.1	20.4	20.0	19.7	19.4	19.0	18.7	18.3	18.0	16.8	16.5	16.1	15.8	15.5		
⅞	¾	66	29.9	29.5	29.1	28.7	28.3	27.9	27.5	27.0	26.7	26.3	25.9	25.4	25.1	24.7	24.2	23.9	23.4	23.0	22.5	22.2	21.4	21.1	20.7	20.4	20.0	19.7	19.3	19.0	17.8	17.5	17.2	16.9	16.5	16.2	14.9
⅞	¾	68	30.9	30.5	30.1	29.7	29.3	28.9	28.5	28.1	27.7	27.3	26.9	26.4	26.1	25.7	25.2	24.9	24.4	24.0	23.5	23.2	22.4	22.1	21.7	21.4	21.0	20.7	20.3	20.0	18.9	18.6	18.2	17.9	17.6	17.2	16.0
⅞	¾	70	31.9	31.5	31.1	30.7	30.3	29.9	29.5	29.1	28.7	28.3	27.9	27.4	27.1	26.7	26.2	25.9	25.4	25.0	24.5	24.2	23.5	23.1	22.8	22.4	22.1	21.7	21.4	21.0	20.0	19.6	19.3	19.0	18.6	18.3	17.1
⅞	¾	72	32.9	32.5	32.1	31.7	31.3	30.9	30.5	30.1	29.7	29.3	28.9	28.4	28.1	27.7	27.2	26.9	26.4	26.0	25.5	25.2	24.5	24.1	23.8	23.4	23.1	22.7	22.4	22.0	21.0	20.7	20.3	20.0	19.6	19.3	18.2
⅞	¾	74	33.9	33.5	33.1	32.7	32.3	31.9	31.5	31.1	30.7	30.3	29.9	29.4	29.1	28.7	28.2	27.9	27.4	27.0	26.5	26.2	25.5	25.1	24.8	24.4	24.1	23.7	23.4	23.0	22.0	21.7	21.3	21.0	20.6	20.3	19.2
⅞	¾	76	34.9	34.5	34.1	33.7	33.3	32.9	32.5	32.1	31.7	31.3	30.9	30.4	30.1	29.7	29.2	28.9	28.4	28.0	27.6	27.2	26.5	26.2	25.8	25.5	25.1	24.7	24.4	24.0	23.1	22.7	22.4	22.1	21.7	21.4	20.3
⅞	¾	78	35.9	35.5	35.1	34.7	34.2	33.9	33.5	33.1	32.7	32.3	31.9	31.4	31.1	30.7	30.2	29.9	29.4	29.0	28.6	28.2	27.5	27.2	26.8	26.5	26.1	25.7	25.4	25.0	24.1	23.7	23.4	23.1	22.7	22.4	21.3
⅞	¾	80	36.9	36.5	36.1	35.7	35.3	34.9	34.5	34.1	33.7	33.3	32.9	32.4	32.1	31.7	31.3	30.9	30.4	30.0	29.6	29.2	28.6	28.2	27.9	27.5	27.1	26.8	26.4	26.1	25.1	24.8	24.5	24.1	23.8	23.4	22.4
⅞	¾	82	37.6	37.1	36.7	36.3	35.9	35.5	35.1	34.7	34.3	33.9	33.5	33.0	32.7	32.3	31.9	31.5	31.0	30.7	30.2	29.8	29.2	28.8	28.5	28.1	27.8	27.4	27.0	26.7	25.8	25.5	25.1	24.8	24.4	24.1	23.1
⅞	¾	84	38.9	38.5	38.1	37.7	37.3	36.9	36.5	36.1	35.7	35.3	34.9	34.4	34.1	33.7	33.3	32.9	32.4	32.1	31.6	31.2	30.6	30.2	29.9	29.5	29.2	28.8	28.4	28.1	27.2	26.9	26.5	26.2	25.8	25.5	24.5
⅞	¾	86	39.6	39.1	38.7	38.3	37.9	37.5	37.1	36.7	36.3	35.9	35.5	35.0	34.7	34.3	33.9	33.5	33.0	32.7	32.2	31.8	31.2	30.9	30.5	30.1	29.8	29.4	29.0	28.7	27.8	27.5	27.1	26.8	26.4	26.1	25.2
⅞	¾	88	40.9	40.5	40.1	39.7	39.3	38.9	38.5	38.1	37.7	37.3	36.9	36.4	36.1	35.7	35.3	34.9	34.4	34.1	33.6	33.2	32.6	32.3	31.9	31.5	31.2	30.8	30.4	30.1	29.2	28.9	28.5	28.2	27.8	27.5	26.6
⅞	¾	90	41.9	41.5	41.1	40.7	40.3	39.9	39.5	39.1	38.7	38.3	37.9	37.4	37.1	36.7	36.3	35.9	35.5	35.1	34.6	34.2	33.6	33.3	32.9	32.5	32.2	31.8	31.4	31.1	30.2	29.9	29.5	29.2	28.8	28.5	27.6
⅞	¾	92	42.9	42.5	42.1	41.7	41.3	40.9	40.5	40.1	39.7	39.3	38.9	38.4	38.1	37.7	37.3	36.9	36.5	36.1	35.6	35.2	34.6	34.3	33.9	33.6	33.2	32.8	32.5	32.1	31.3	30.9	30.6	30.2	29.9	29.5	28.7
⅞	¾	94	43.9	43.5	43.1	42.7	42.3	41.9	41.5	41.1	40.7	40.3	39.9	39.4	39.1	38.7	38.3	37.9	37.5	37.1	36.6	36.2	35.6	35.3	34.9	34.6	33.2	33.8	33.5	33.1	32.3	31.9	31.6	31.2	30.9	30.5	29.7
⅞	¾	96	44.9	44.5	44.1	43.7	43.3	42.9	42.5	42.1	41.7	41.3	40.9	40.4	40.1	39.7	39.3	38.9	38.5	38.1	37.6	37.3	36.7	36.3	35.9	35.6	35.2	34.8	34.5	34.1	33.3	33.0	32.6	32.3	31.9	31.5	30.7
⅞	¾	98	45.9	45.5	45.1	44.7	44.3	43.9	43.5	43.1	42.7	42.3	41.9	41.4	41.1	40.7	40.3	39.9	39.5	39.1	38.6	38.3	37.7	37.3	36.9	36.6	35.2	35.8	35.5	35.1	34.3	34.0	33.6	33.3	32.9	32.5	31.7
⅞	¾	100	46.9	46.5	46.1	45.7	45.3	44.9	44.5	44.1	43.7	43.3	42.9	42.4	42.1	41.7	41.3	40.9	40.5	40.1	39.6	39.3	38.7	38.3	37.9	37.6	37.2	36.8	36.5	36.1	35.3	35.0	34.6	34.3	33.9	33.6	32.8

The centers shown in this shaded area are below the recommended minimum.

This table provides approximate center distances for the great majority of V-Belt Drives. It is calculated for 1 to 3 pulley diameters, but gives useful approximations for all other ratios.

UNIT 21

Cam Development

Cams are machine elements in the shape of plates or cylinders, which impart predetermined motion to other machine members. By means of cams, rotary motion can be changed to reciprocating motion. Cams in the shape of plates or cylinders impart motion to other machine members.

Today cam design is often accomplished with the aid of a computer. By feeding data into the computer, design solutions are readily ascertained without resorting to time-consuming manual calculations. The material in this unit is intended primarily to give drafting personnel a better understanding of design information when preparing cam drawings.

21.1 Function of Cams

A cam is designed primarily to convert constant rotary motion into timed irregular motion. This basic action is illustrated in Fig. 21-1, where an irregularly shaped disk revolves on a constant-speed shaft and imparts irregular motion to a follower. The follower is a plunger that is held in contact with the profile of the cam by means of gravity or a spring. The special movements made possible by the use of cams generally are not easily obtained by any other mechanical means. Evidence of this is the universal use of cams in automatic screw machines, sewing machines, textile machinery, and engines.

21.1.1 Cams. The cam mechanism consists of a cam and a follower. Each of these components has many possible design variations. However, the two main types of cams are *plate*, or *radial*, and *cylindrical*. See Fig. 21-2. Both types rotate with their respective drive shafts but differ in the action of their followers. The follower of the radial cam operates in a plane perpendicular to the axis of the camshaft, whereas the follower of the cylindrical cam oscillates in a plane parallel with the axis of the cam-shaft.

21.1.2 Cam followers. Cam followers may have a variety of physical forms. Four common cam fol-

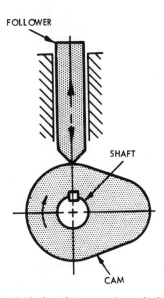

Fig. 21-1. A cam is designed to convert constant rotary motion into timed irregular motion.

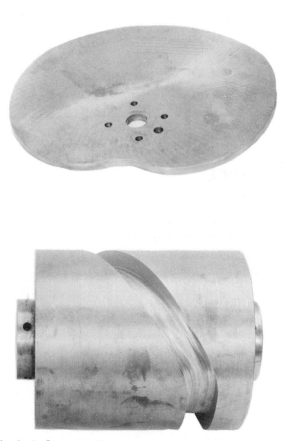

Fig. 21-2. Cams as plates or cylinders impart motion to other machine elements.

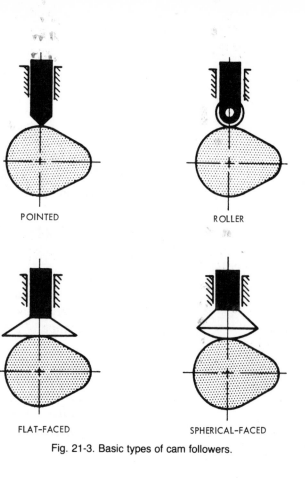

POINTED

ROLLER

FLAT-FACED

SPHERICAL-FACED

Fig. 21-3. Basic types of cam followers.

Fig. 21-4. The position of cam follower may have to be modified to meet special requirements.

lower types are *knife-edge,* or *pointed, roller, flat-faced,* and *spherical-faced.* See Fig. 21-3. The type used depends upon such factors as mechanism speed, cam profile, possible misalignment, and stresses involved. Space requirements may also influence the designer to modify one of the basic followers. A pivoted, or swinging, follower is one example of such a modification. See Fig. 21-4. The design requirements often necessitate the follower's being offset from the center line common with the cam.

21.2 Motion Diagrams

A cam may be used to produce nearly any motion even if the resulting motion defies definition. However, three definable motions are frequently used: *uniform,* or *constant velocity, har-* *monic,* and *uniform accelerated and retarded.* These motions may be used separately or in any combination. Normally the motion of the follower is specified by the designer. Drafting personnel must

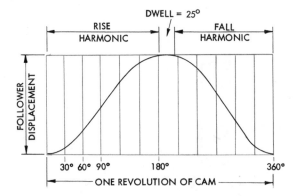

Fig. 21-5. An example of a motion diagram showing cam rise and fall.

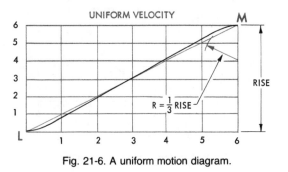

Fig. 21-6. A uniform motion diagram.

then develop the type of cam profile curve that will result in the desired follower motion.

Motion diagrams, often termed displacement diagrams, are used to graphically analyze cam curvature and layout of the cam disk. The diagram is generally drawn prior to laying out the cam curve. This permits a modification, if necessary, of the theoretical curve to insure a smooth follower action. Later, points may be transmitted from the diagram to a drawing of the disk cam.

As shown in Fig. 21-5, the displacement diagram is constructed on the abscissa (horizontal base line), which represents one revolution of the cam. The abscissa may be any length. However, this length is sometimes drawn equal to the circumference of a circle having a radius equal to the distance from the center of the camshaft to the highest point in the cam rise. The ordinate (vertical scale) is drawn to scale equal to the maximum rise of the cam follower. The abscissa, or horizontal base line, can be divided to represent time intervals or angle of cam rotation. The motion diagram illustrated in Fig. 21-5 is for a follower which rises with harmonic motion in a 180° revolution, dwells for 25°, and returns to its starting point by harmonic motion. Single-motion diagrams illustrating uniform motion, harmonic motion, and constant acceleration are drawn with the abscissa at any convenient length, and the ordinate should be equal to the rise.

21.2.1 Uniform motion.
This is constant velocity of the follower, or straight-line motion. The motion is such that the follower movement is directly

proportional to the time of movement. To construct a uniform motion curve as shown in Fig. 21-6, divide the ordinate and the abscissa into any number of equal parts. Connect points *L* and *M* with a straight line. Note that line *LM* is modified by an arc at each end. This is done to eliminate abrupt starting at the beginning and ending of the motion interval. Any convenient curve could be used which would be tangent to the rise line, but a radius of one-third the rise is recommended. This curve permits the follower to rise equal distances in each equal time interval.

21.2.2 Harmonic motion.
With this type of motion, the velocity of the follower is increased from zero at the initial position of the follower to a maximum at the mid-position and is then decreased to zero at the final position. See Fig. 21-7. The abscissa and ordinate are drawn as for the uniform-motion diagram. Then construct a semicircle having a radius equal to one-half the rise to join the abscissa and ordinate. Next divide the semicircle, as illustrated, into the same number of equal parts as the abscissa. Draw horizontal lines from the points in the semicircle to intersect the vertical lines drawn from the division points on the

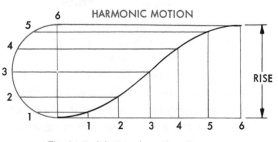

Fig. 21-7. A harmonic motion diagram.

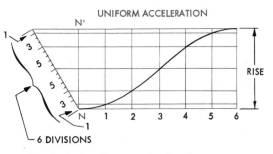

Fig. 21-8. A uniform acceleration diagram.

abscissa. Draw a curved line through the intersections to complete the construction of the displacement curve.

21.2.3 Uniform accelerated and retarded motion. This is constant acceleration (change of velocity) of the follower, or parabolic motion. The motion is characterized by the follower's having a constant acceleration for the first half of the rise and then a constant retardation (deceleration) for the remaining half of the rise. See Fig. 21-8. Again the abscissa and the ordinate are drawn to represent the position in degrees and the rise, respectively. The problem is to divide lines *NN'* and *N6* into the same number of parts. However, each unit of line *NN'* must be proportional to the square of the time interval. Thus, line *NN'* will be divided into increments of 1, 3, 5, and so on. Horizontal lines are drawn through these divisions and intersect the vertical lines drawn from the time intervals on the base line. A line is drawn through the intersecting points, completing the uniform accelerated phase. A change in velocity is now indicated by uniform retarded motion constructed by reversing the increments, as 5, 3, 1, from this point to the end of the cycle.

21.3 Constructing Plate-cam Profiles

The general procedures for constructing profiles of plate cams for use with the most common types of followers are as follows:

21.3.1 Roller follower. To construct a cam profile with a roller-type follower, the cam must be considered as being stationary and the follower as moving around the cam in a direction opposite to the true cam rotation. This method is illustrated in

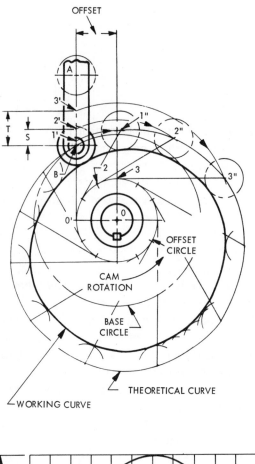

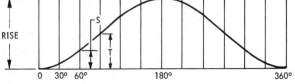

Fig. 21-9. Construction of a plate-cam profile having an offset roller-type follower.

Fig. 21-9, where a profile is developed for a plate cam having an offset roller follower. Here is the basic procedure:

The displacement diagram is constructed as a preliminary step to the actual layout of the cam profile, whereas the construction of the cam can be considered to initiate with the drawing of the base circle.

The base circle is constructed with the follower shown in its lowest position. Next, using the center of the cam *O* as the center, draw the offset circle with a radius equal to the offset distance *OO'*. The offset circle is then divided into the same number of divisions as were used in the displacement diagram. In the example shown, twelve divisions are used. Lines are drawn from each of these divisions tangent to the offset circle. Note that these divisions are numbered in a direction opposite to the cam rotation. The numbering of these points aids in the construction only and will not appear on the finished drawing.

The ordinate, or vertical, distances *S* and *T* are transferred from the displacement diagram to the center line of the follower. These distances *A1'*, *A2'*, *A3'*, and so on represent successive positions of the follower. Consequently, in the example shown, the follower will have risen *T* distance from its initial position when the cam has rotated 90°.

Points 1', 2', 3', and so on are next successively revolved in a direction opposite to cam rotation. This is accomplished by using the cam center *O* and a radius equal to *O1'*, *O2'*, and so on. These arcs intersect their corresponding tangent lines (tangents to the offset circle), resulting in intersecting points which are the locations of the center of the follower at these various positions.

Using a radius equal to the radius of the follower and the centers just found, draw arcs as indicated. The line drawn tangent to these arcs will be the true cam profile.

A smoother cam profile will result if a theoretical, or pitch, curve is utilized as a center line for closely spaced roller circles as illustrated. The pitch curve is the smooth curve drawn through points 1", 2", 3", and so on. Generally neither the pitch curve nor the alternate positions of the cam follower are shown on a detail drawing. When they are shown, however, they are represented by phantom outlines.

21.3.2 Pivoted follower. The profile of a plate cam having a pivoted follower is constructed similarly to that of a plate cam with a roller-type follower. The exact procedure for drawing the profile is illustrated in Fig. 21-10.

Construction of the displacement diagram precedes the cam layout. Note that the true path of the follower is an arc. Thus, the height of the ordinate distance in the displacement diagram will equal the rectified length of arc *AB*.

First draw the base circle with the roller follower

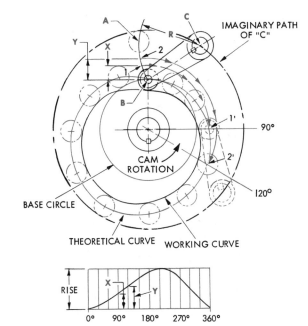

Fig. 21-10. Construction of a profile of a plate-cam having a pivoted follower.

in its lowest position. Next locate *C*, the center of the follower pivot. The base circle is then divided into the same number of divisions as were utilized in the displacement diagram. Radius *R* is drawn using *C* as a center.

Now transfer distances *X*, *Y*, and so on from the displacement diagram to the cam drawing, as illustrated. These distances are then rotated into position; that is, radius *O1* is rotated to the 90° radial position, radius *O2* is rotated to the 120° position, and so forth. Small arcs representing the roller follower are drawn from points 1', 2', and so on. The cam profile is then drawn tangent to these small arcs. As was previously suggested, a smoother and more accurate curve will result if the pitch curve is drawn first, permitting the centers for the roller-arcs to be located.

21.3.3 Flat-face follower. The procedure for constructing the profile of a plate cam having a flat-face follower employs methods previously discussed. The construction of the displacement diagram, base circle, offset circle, and location of points is identical to the procedure described in subsection 21.3.2.

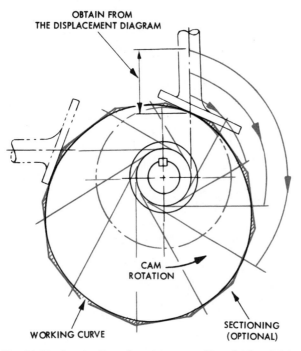

OBTAIN FROM
THE DISPLACEMENT DIAGRAM

CAM
ROTATION

WORKING CURVE

SECTIONING
(OPTIONAL)

Fig. 21-11. Construction of a plate-cam with a flat-faced follower.

Note in Fig. 21-11 that the initial contact point between the cam profile and the follower changes as the follower rises. The cam profile is drawn tangent to the face of the follower in the positions previously established.

21.4 Cylindrical Cam Development

The procedure for drawing a cylindrical cam follows the general method of plate-cam layout, much as if the follower were moving in a direction opposite to the cam's rotation with the cam stationary. See Fig. 21-12. The actual construction starts with the displacement diagram, which is laid out as a plane development of the cylindrical surface. That is, the length of the diagram representing 360° is accurately laid out equal to πD, and the ordinate distance is drawn equal to the height of the cylinder. Thus, the resulting curves will be an accurate development of the outer surface of the cam cylinder.

The displacement diagram is drawn adjacent to the space reserved for the front view. This location

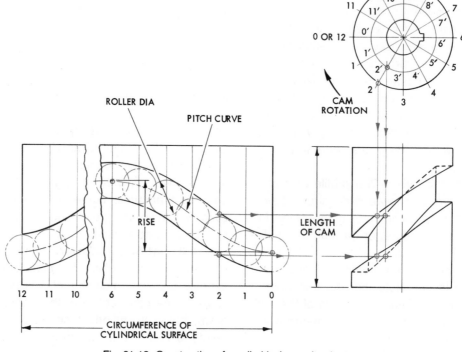

ROLLER DIA

PITCH CURVE

RISE

CAM
ROTATION

LENGTH
OF CAM

CIRCUMFERENCE OF
CYLINDRICAL SURFACE

Fig. 21-12. Construction of a cylindrical cam development.

permits the direct projection of points from the displacement diagram to the front view.

The theoretical, or pitch, curve is constructed for the desired motion. Circles equal to the roller diameter are then drawn using the pitch curve as a location for the roller centers. Curves are drawn tangent to these roller circles with an irregular curve.

The top view of the cam is drawn next. This permits the projection of points from both the displacement diagram and the top view. For example, in Fig. 21-12, points 2 and 2' are easily located in the front view by projection. Note that a separate inner cylindrical layout would be needed if a true picture of the inner curve were desired. Since this information is seldom required and the development of the outer cylinder provides the necessary information for the cam manufacturer, such a curve is usually omitted in the front view.

21.4.1 Circular arc cams.

Cams composed of tangent arcs of circles have the advantage of being comparatively easy to manufacture. To lay out such a cam, assume, for example, that the cam in Fig. 21-13 rotates about point O. Arc A_1A_2 has a radius R_1, arc A_1B_1 has a radius R_3, arc B_1B_2 a

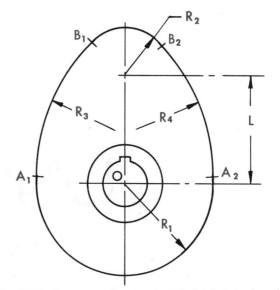

Fig. 21-13. Cams are often more easily fabricated when the design incorporates tangent circular arcs.

radius R_2, and arc B_2A_2 a radius R_4. The lift of the cam is equal to $L + R_2 - R_1$. Usually R_3 is made equal to R_4.

A cam similar to the one in Fig. 21-13 but constructed using tangent straight lines instead of the arcs formed by R_3 and R_4 is a tangential cam.

SECTION V
UNIT 22

Piping Drawings

Special drawings are prepared to depict piping systems. As a rule engineering personnel will design a piping system and then turn over design sketches with specifications to the drafting department for preparing finished piping drawings. This unit contains basic information and introduces practices that will be found useful in preparing piping drawings.

22.1 Kinds of Pipe

Pipe is made of various materials, the most common being cast iron, steel, wrought iron, copper, brass, lead, and plastic. The type of pipe used depends on its intended function.

Cast-iron pipe is particularly suitable for underground installations to convey gas, water, or steam. It has high corrosion-resistance qualities and can withstand considerable stresses of expansion, contraction, and vibration.

Steel pipe is utilized for a great many purposes, not only in carrying fluids of different kinds but in structural work as well. Very often pipe is substituted for steel structural shapes to function as columns or in constructing rail guards, scaffolds, frames, and so forth. Steel pipe is resistant to high temperatures and pressures and has an outstanding advantage in that connections can be made easily by welding. When steel pipe is dipped in molten zinc to prevent rust, it is called galvanized pipe. Galvanized pipe is intended primarily for lines carrying drinking water.

Wrought-iron pipe originally was considered best in situations where high corrosive conditions pre-vailed. However, installation records and tests show that steel pipe is equal to wrought-iron pipe for general-purpose work. Consequently, steel pipe is used more extensively today, especially since it is less expensive. Where extreme corrosive conditions exist, wrought iron, stainless steel, or other non-corrosive pipes are necessary.

Brass and copper pipe are utilized for general plumbing and heating purposes, gas, steam, and oil lines. Brass pipe is extremely resistant to corrosion except under severe acid conditions. Copper pipe is not suitable where continuous high temperatures or repeated severe stresses occur, because under such conditions it deteriorates rapidly. Copper pipe also has wide applications for radiant heating and air conditioning installations. The use of copper or brass pipe in these applications minimizes installation time since they can be shaped readily without the need for special fittings; connections are easily and quickly made by soldering operations.

Lead pipe and lead-lined pipe are widely used for conveying chemicals or in piping systems subjected to acid conditions. Lead-lined steel pipe is

610

designed for situations where severe corrosion conditions and higher pressures are encountered.

Many state plumbing codes now permit the use of plastic pipe in some areas of new plumbing installation. The advantage of plastic pipe is ease of cutting, joining, and bending.

22.1.1 Sizes of pipe. There are three standard weights of steel pipe and wrought-iron pipe: *standard, extra-strong,* and *double extra-strong.* For any given diameter, each of the three types has a different wall thickness, but for any given nominal size, all have the same outside diameter. When reference is made to a specified inside diameter, it does not imply that this size is actually as stated. Thus, the inside diameter of a 1-inch nominal standard pipe is 1.049 inches, of an extra-strong is 0.957 inch, and of a double extra-strong is 0.599 inch.

Commercial sizes of wrought-iron and steel pipe ranging from 0.125 to 12 inches are designated by their nominal inside diameters (ID). Pipe of 14 inches diameter and larger is indicated by its outside diameter (OD). The general practice in specifying the size of steel and wrought-iron pipe for 12-inch and smaller is to indicate the desired weight by the terms standard, extra-strong, and double extra-strong. For pipe 14-inch or larger the desired wall thickness is given instead of the weight terminology.

The required volume of flow usually governs the inside diameter of pipe to be used. Wall thickness depends on such factors as internal pressure, external pressure, and the amount of expansion stress encountered.

The size of cast-iron pipe is designated by the nominal inside diameter, wall thickness, and quite often, the strength or working pressure class. (See Appendix.) The wall thickness of the pipe varies according to its strength and diameter.

Cast-iron pipe is available either with flanged ends or with bell-and-spigot ends. The greatest use of the bell-and-spigot is for underground waterlines. This joint when properly calked and leaded makes a very tight joint. See Fig. 22-1A.

Flanged pipe has higher strength values and is used to convey fluids under greater pressure. See Fig. 22-1B.

Another type of cast-iron pipe, known as Universal pipe, is available for gas and water lines of all pressures. The pipe is made with a hub-and-spigot end, and the joint is held together by two or more bolts. The pipe ends are tapered, providing a tight iron-to-iron contact joint. See Fig. 22-1C.

Brass and copper pipe is manufactured in two standard weights—regular and extra-strong—and in sizes ranging from 0.125 to 12 inches in diameter. Size is specified by both inside and outside diameters. Outside diameters are always the same as corresponding nominal sizes of steel pipe.

Lead pipe is supplied in straight lengths, reels, or coils. The size of both lead and lead-lined pipe is indicated by inside and outside diameters.

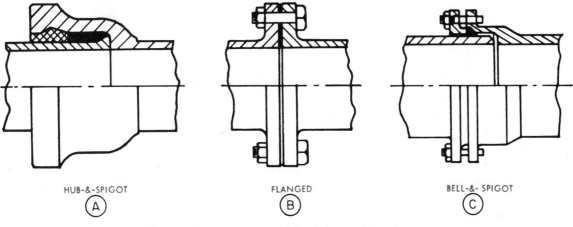

HUB-&-SPIGOT   Ⓐ   FLANGED   Ⓑ   BELL-&-SPIGOT   Ⓒ

Fig. 22-1. Common types of joints used on cast iron pipes.

22.1.2 Tubing. Generally speaking, when piping does not correspond to standard steel-pipe wall thicknesses and diameters, it is referred to as tubing. Tubing is manufactured in round, square, rectangular, hex, and octagonal shapes. Steel, copper, brass, aluminum, and stainless steel are the common materials used for tubing. Copper and brass tubing are especially adaptable for gas, air, or fluid lines in heating, refrigeration, air conditioning, and plumbing installations. Stainless steel and aluminum tubing have wide applications in high-pressure hydraulic lines.

Copper and brass tubing are furnished in three basic weights designated as types *K* and *L*, in hard and soft tempers, and type *M* in hard temper. Aluminum tubing is specified according to the standard aluminum classification symbols, which designate the type of alloy and tempers, such as 3003-*O* and 6061-*F*. Stainless steel tubing is available in a variety of types in the 300 or 400 classification series.

In all cases the size of round tubing is indicated by the outside diameter and wall thickness. For square, hex, and octagonal tubing the distance across flats and the wall thickness are given.

22.1.3 Pipe and tube fittings. Fittings are designed to make connections in piping systems and to change the direction of flow. They are made of cast iron, malleable iron, steel, brass, copper, aluminum, and other special alloys. Pipe fittings fall into four general categories: screwed, flanged, welded, and soldered.

Screwed fittings are used primarily in small piping systems such as house plumbing, oil, and hydraulic lines. *Soldered fittings* are found where connections must be permanently and tightly sealed, especially in refrigeration units, radiant heating systems, and other small low-pressure fluid lines. *Flanged* and *welded fittings* are employed in large piping systems where connections must be strong enough to carry the weight of pipes and withstand high pressures.

Tube fittings are either of the *soldered, welded, flared,* or *threaded-sleeve* type. Flared tube fittings are widely used for connecting lines which carry liquids or gases at relatively high pressures, such as in hydraulic and air braking systems. See Fig. 22-2. Soldered, welded, and threaded-sleeve tube

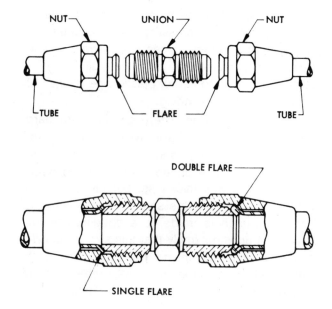

FLARED TYPE CONNECTION

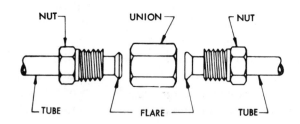

INVERTED FLARED TYPE CONNECTION

Fig. 22-2. Flared tube fittings are used on high-pressure lines.

fittings are intended mostly for connecting lines that will carry either gases or liquids at medium pressures.

In most instances fittings should have the same qualities as those of the pipe or tubing they join.

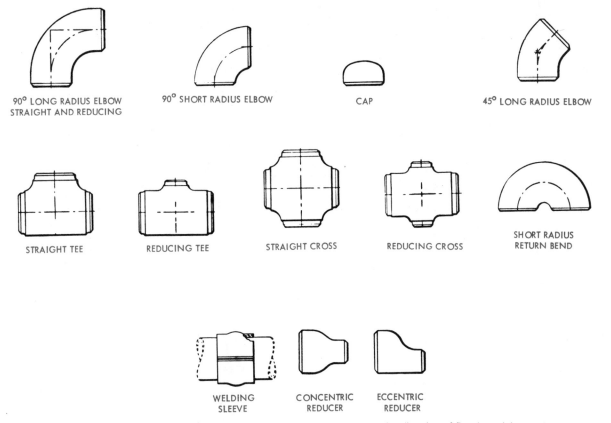

Fig. 22-3. These fittings are used to connect various units and to change the direction of flow in a piping system.

Hence, fittings are usually specified according to the nominal piping size, material, and strength factor or pressure rating. Standard pressure ratings for cast-iron fittings for low- and medium-pressure lines are 25, 125, and 250 pounds. Steel fittings with high pressure ratings 150 to 2,500 pounds are available for high-pressure lines. Thus, a fitting may be specified as

2"—125 LB 90° ELBOW, CAST IRON

Some of the more common fittings are illustrated in Fig. 22-3. Included are the following:

Coupling. For connecting straight sections.

Cap. Fits on the end of pipe to close it.

Plug. Used to close an opening in a fitting.

Nipple. Short piece of pipe for making connection to fittings.

Bushing. Reduces the size of an opening in a fitting.

Union. Used to close a piping system and to connect pipes that must be disconnected occasionally for repairs.

Tees, crosses, and laterals. Form the connections for lines branching off the piping system.

Ells. Used to change the direction of pipe lines.

Reducer. Permits the use of different pipe sizes in a piping system.

22.1.4 Pipe bends. Pipe bends are formed lengths of pipe obtainable in practically any size and shape to meet existing situations arising in a piping system. Specifically they are used to (1) eliminate the need for a number of joints in a pipe line, (2) compensate for expansion and contraction, (3) permit installation of a pipe line around obstructions, and (4) reduce friction in the piping. Some of the more common pipe bends are shown in Fig. 22-4, as well as the method of dimensioning them. Table I contains the recommended radii and tangents of pipe bends for pipe of various sizes.

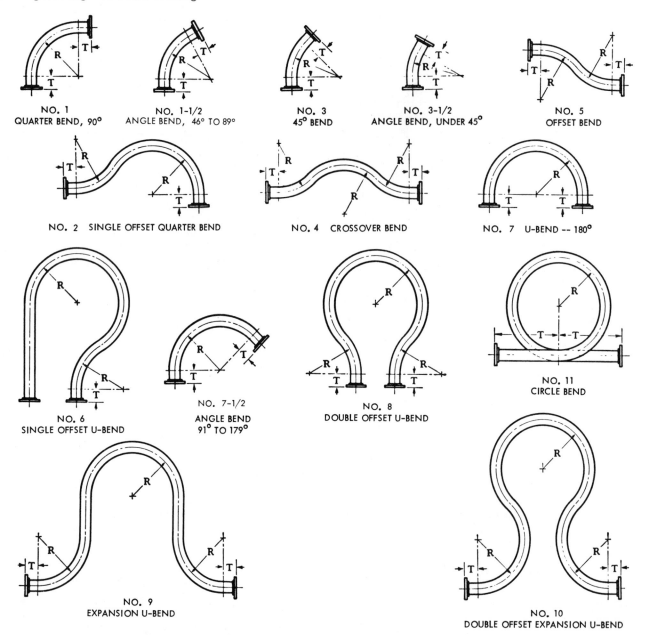

NO. 1
QUARTER BEND, 90°

NO. 1-1/2
ANGLE BEND, 46° TO 89°

NO. 3
45° BEND

NO. 3-1/2
ANGLE BEND, UNDER 45°

NO. 5
OFFSET BEND

NO. 2   SINGLE OFFSET QUARTER BEND

NO. 4   CROSSOVER BEND

NO. 7   U-BEND -- 180°

NO. 6
SINGLE OFFSET U-BEND

NO. 7-1/2
ANGLE BEND
91° TO 179°

NO. 8
DOUBLE OFFSET U-BEND

NO. 11
CIRCLE BEND

NO. 9
EXPANSION U-BEND

NO. 10
DOUBLE OFFSET EXPANSION U-BEND

Fig. 22-4. Specially formed pipe bends can be obtained to meet any piping requirements.

22.2 Valves

The function of a valve is to control the quantity or the direction of flow in a piping system. There are many different types of valves, and they are made of several kinds of metal. Usually valves on small-size pipe lines are of brass or bronze. Large piping systems with low and intermediate pres-sures are equipped with cast-iron valves. Pipe lines with high-pressure flows have either cast-steel or cast-alloy valves.

The *gate valve* is probably the most frequently used valve for water and oil lines. This valve is operated by turning a handwheel which raises or lowers a disk or wedge into the flow stream. See Fig. 22-5A. The quick-opening gate valve is de-

TABLE I RADII AND TANGENTS OF PIPE BENDS, IN INCHES

Size of Pipe	Column "A" R Minimum Recommended Radius Steel or Wrought Iron Pipe — Standard or Extra Strong Weight	Column "B" R Shortest Radii To Which Pipe Can Be Bent — Steel Pipe Only — Threaded Ends, Screwed, or Welded Flanges — Standard Weight Pipe	Extra Strong Weight Pipe	Wrought Iron Pipe — Threaded Ends, Screwed, or Welded Flanges — Standard Weight Pipe	Extra Strong Weight Pipe	Column "C" T Minimum Length of Tangent Steel or Wrought Iron Pipe Standard or Extra Strong — Threaded Ends or Screwed Flanges
1/4	1 1/4	1	5/8	1 1/4	1	1
3/8	1 7/8	1 1/4	3/4	1 7/8	1 1/2	1 1/4
1/2	2 1/2	1 1/2	1	2 1/2	2	1 1/2
3/4	3 3/4	1 3/4	1 1/4	3	2 1/2	1 3/4
1	5	2	1 1/2	4	3	2
1 1/4	6 1/4	2 1/4	1 3/4	5	4	2
1 1/2	7 1/2	2 1/2	2	6	5	2 1/2
2	10	3	2 1/2	8	5	3
2 1/2	12 1/2	5	4 1/4	10	8	4
3	15	8	6	12	10	4
3 1/2	17 1/2	10	8	14	12	5
4	20	12	10	16	12	5
5	25	18	14	20	15	6
6	30	22	15	26	18	7
8	40	30	23	30	28	9
10	50	36	30	36	32	12
12	60	46	36	46	42	14
*14 OD	70	60	48	60	54	16
*16 OD	96	80	60	80	70	18
*18 OD	108	90	66	90	80	18
*20 OD	120	100	72	100	90	18
*24 OD	144	144	108	144	122	18

*For sizes 14-inch O.D. and larger, the radii shown are based upon pipe with a wall thickness of 7/16-inch or lighter under the "Standard Weight Pipe" column, and a wall thickness of 1/2-inch or heavier under the "Extra Strong Weight Pipe" column.

signed so it can be used on lines which must be opened or closed instantly.

The *globe valve* is especially adaptable for throttling steam in low- and high-pressure lines. Control is effected by raising or lowering a plug or disk into a seat. See Fig. 22-5*B*.

The *needle valve* is found in small piping systems where close control is necessary.

The *check valve* is constructed to permit flow of a fluid in one direction only. This valve closes automatically if the flow should reverse its direction. See Fig. 22-5*C*.

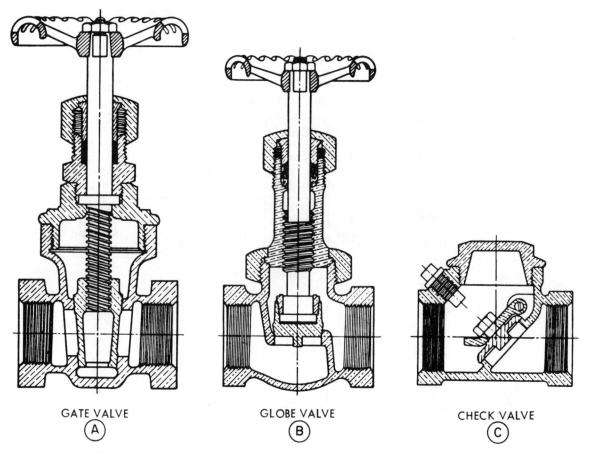

GATE VALVE
Ⓐ

GLOBE VALVE
Ⓑ

CHECK VALVE
Ⓒ

Fig. 22-5. Valves are used in a piping system to control the quantity or direction of flow.

The *diaphragm valve* is designed for piping systems conveying acids, alkalies, and volatile substances. The valve has no metal-to-metal seat; closure is actuated by a rubber diaphragm.

22.3 Piping Identification

A scheme of identifying a piping system for industrial plants has been adopted by the American National Standards Institute. This scheme consists of a series of colors which classifies the piping system according to the nature of materials carried. The approved colors are painted over the entire lengths of pipe, or several bands are painted at various intervals throughout the piping system. In some instances the contents of the pipe are actually stenciled in abbreviated form over the color bands. The color scheme is shown in Table II.

TABLE II COLOR SCHEME FOR A PIPING SYSTEM

Class	Color
Fire Protection Equipment	Red
Dangerous Materials	Yellow or Orange
Safe Materials	Green, White, Black Gray, or Aluminum
Protective Materials	Bright Blue
Extra Valuable Materials	Deep Purple

In preparing a piping drawing or sketch, all fittings, fixtures, valves, and other units are shown by means of graphic symbols. The basic symbols that have been approved by ANSI are included in the Appendix. The use of these symbols simplifies considerably the preparation of piping drawings and conserves a great deal of time and effort.

22.4 Piping Drawings

The function of a piping drawing is to show the location, the type, and the position of various units in a pipe line and to specify the sizes and descriptions of all parts used in the piping system. The drawing may be a freehand sketch or a finished mechanical drawing.

The actual views of a piping system may consist of a single orthographic view or an isometric or oblique pictorial representation. Occasionally a two- or three-view orthographic projection is prepared especially for systems which are basic in design and are installed frequently, or when special valve designs or other units in the piping system must be clearly shown. See Fig. 22-6.

The pictorial drawing of a piping system has a very decided advantage over orthographic presentations in that a pictorial can reveal changes in direction and differences in installation levels. See Fig. 22-7.

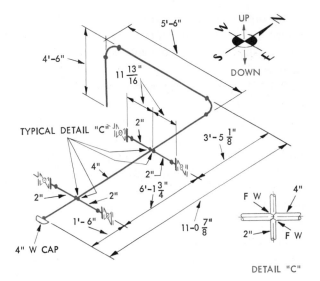

Fig. 22-7. A pictorial drawing clearly shows the arrangement of components in a piping installation.

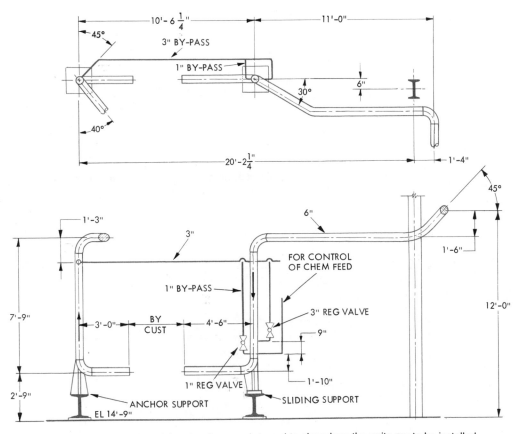

Fig. 22-6. A multiview piping drawing may be used to show how the units are to be installed.

The presentation of a piping layout in either orthographic or pictorial form is accomplished by what is known as a single-line or double-line representation.

22.4.1 Single-line representation. The single-line drawing is commonly used when installation of small-size pipe is involved as well as for laying out small piping systems and making preliminary layouts and calculations. In a single-line drawing single lines are employed to designate pipe regardless of the pipe size. Conventional symbols are included for all fittings, valves, and fixtures. See Fig. 22-8.

The center line of the pipe is drawn as a heavy line and the fitting symbols are attached to it.

Sizes of the symbols are variable, depending on the scale of the layout. A piping template, as shown in Fig. 22-9, is often used to draw symbols.

22.4.2 Double-line representation. The double-line drawing is more often prepared by manufacturers of piping equipment when a drawing is to be used repeatedly for similar installations on numerous projects. A double-line drawing is also quite common for large piping systems. See Fig. 22-10. It is also used when lengths of pipe or distances are critical or when pipe is pre-cut and shipped before assembly.

Pipe and fittings are usually drawn to scale. The drawing is generally an orthographic projection unless it depicts a piping system for a catalog or

Fig. 22-8. A conventional single-line piping drawing.

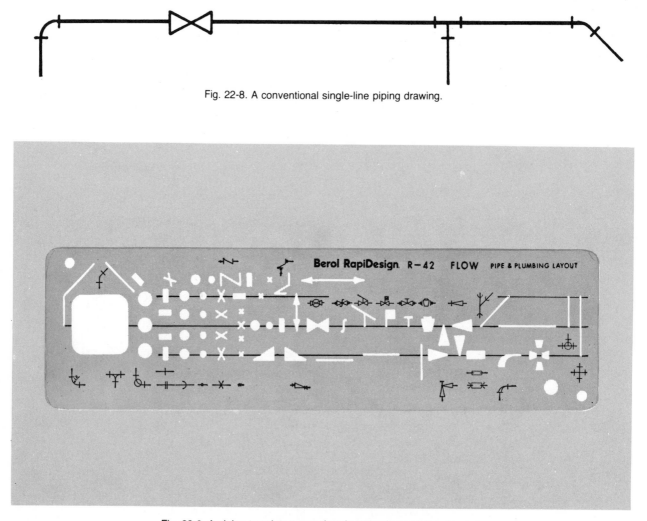

Fig. 22-9. A piping template saves time in preparing a piping drawing.

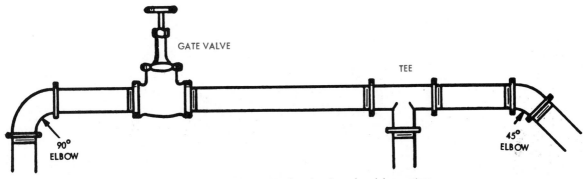

Fig. 22-10. A typical double-line drawing of a piping system.

service manual, in which case a pictorial is often preferred.

22.4.3 Dimensioning a piping drawing.
Generally speaking, the same rules of dimensioning that govern orthographic and pictorial drawings also apply to pipe drawings. In every instance it is important to supply sufficient information for the pipe fitter to proceed with the installation without any guesswork.

The following rules are particularly significant in dimensioning a piping drawing:

1. All straight lengths of pipe should be dimensioned.

2. Location dimensions must be clearly indicated.

3. Pipe fittings, valves, and other units are located by center-to-center distances.

4. Wherever possible, dimensions should be placed on a solid dimension line instead of in between breaks of lines.

5. Size and kind of pipe fittings and valves are identified at their location position or in a bill of material.

6. For some drawings it will be necessary to supply a bill of material specifying quantity and kinds of material needed for the piping system.

SECTION V
UNIT 23

Fluid Power Diagrams

Fluid power, often referred to as fluidics, is a form of power that utilizes liquids to transfer energy from one location to another. Fluid-power systems are used to perform essential tasks in many important work areas. Thus, we see the use of fluid power in hydraulic presses for drawing or compressing metals, in hoists and jacks for lifting heavy objects, in brake systems, and in operating a great variety of mechanisms.

Drawings of one kind or another are used in the design, operation, and maintenance of hydraulic systems. The purpose of this unit is to provide some basic information about fluid-power units, the standard symbols used in fluid-power diagrams, and the various types of drawings that are characteristic of fluid-power systems.

23.1 Basic Fluid-Power Units

In the preparation of fluid-power diagrams, the various fluid-power components are shown by graphical symbols, pictorial symbols, cutaway symbols, or a combination of the three types.

Graphical symbols consist of simple geometric figures that identify the function and method of operation of each component.[1] *Pictorial symbols* are miniature drawings that show the physical outline of hydraulic units. *Cutaway symbols* are miniature section drawings that present internal parts of hydraulic components.

The following are the basic components with their representative symbols used in fluid-power diagrams:

Pumps. Pumps are used to develop pressure in the hydraulic fluid. They create a partial vacuum at their inlet, thereby permitting the atmospheric pressure above the liquid in the reservoir to force liquid into the pump and deliver it into the hydraulic system.

Pumps are either of the rotary or reciprocating type, although in some systems centrifugal pumps are also used. A *rotary pump* uses a rotary motion to convey the fluid from the pump inlet to the pump outlet. The elements that propel or transmit the liquid may be gears, lobes, vanes, or pistons enclosed in a casing. A *reciprocating pump* uses a reciprocating motion to transmit liquid. These pumps may have single or multiple cylinders arranged either vertically or horizontally. Fig. 23-1 illustrates the symbols that are used to depict hydraulic pumps.

Hydraulic cylinders. Hydraulic cylinders, sometimes called linear or reciprocating motors, are used to transform the flow of pressurized fluid in a system into an applied work force through the push or pull action of pistons.

The three common types of hydraulic cylinders

---

1. See *ANSI Y32.10.*

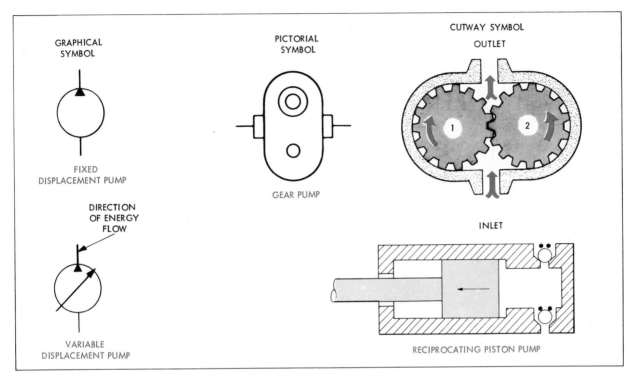

Fig. 23-1. Types of symbols used to show pumps in a fluid-power drawing.

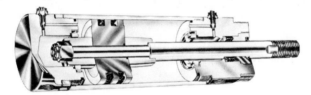

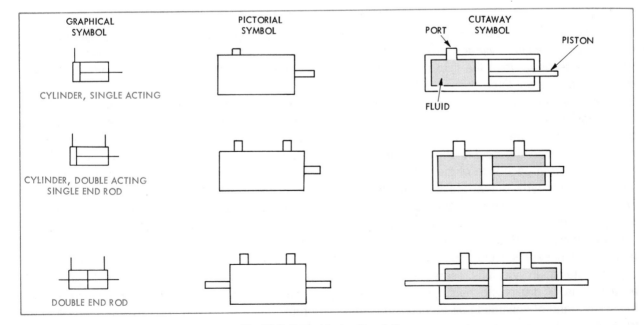

Fig. 23-2. Typical hydraulic cylinders.

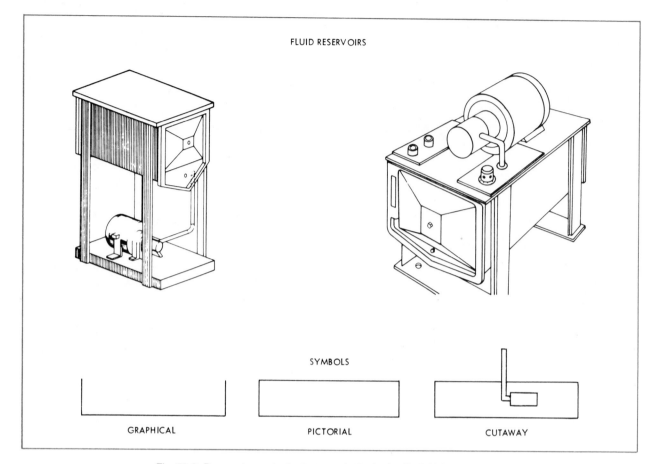

Fig. 23-3. Reservoirs are tanks that supply the hydraulic fluid in a system.

are single-acting, double-acting, and double-end rod. The *single-acting cylinder* powers a stroke in only one direction. When the fluid drains from the cylinder, an external force pushes the piston back to its starting position. The *double-acting cylinder* is the most common and contains two fluid chambers so that pressure can be used to extend and retract the rod. In the *double-end rod cylinder* a rod extends from each side of the piston. The advantage of this type is that the working areas of both sides of the cylinder are equal. See Fig. 23-2 for typical symbols.

Fluid reservoirs. Reservoirs are tanks or containers which supply the hydraulic fluid in fluid-power systems. In addition, reservoirs also serve as a means of dissipating heat generated within the system and of separating air and contaminants from the fluid. Tank sizes and shapes are gov-

erned by the type and size of the hydraulic system. A square or rectangular tank with the pump located on top or bottom is a common arrangement. See Fig. 23-3.

Filters. To insure long life and trouble-free performance of hydraulic components, the system is often equipped with some filtering device. Filters usually consist of 60-mesh or 100-mesh screens, which may be located in the reservoir or between the reservoir and the pump. See Fig. 23-4. Notice in Fig. 23-4 the standard symbol for filtering devices and how this symbol is used in a fluid-power diagram.

Fluid-temperature controls. If a fluid-power system is to operate efficiently, the fluid temperature must be kept within prescribed limits. Each hydraulic fluid—oil, water, or synthetic—has a specified operating temperature. When fluid tempera-

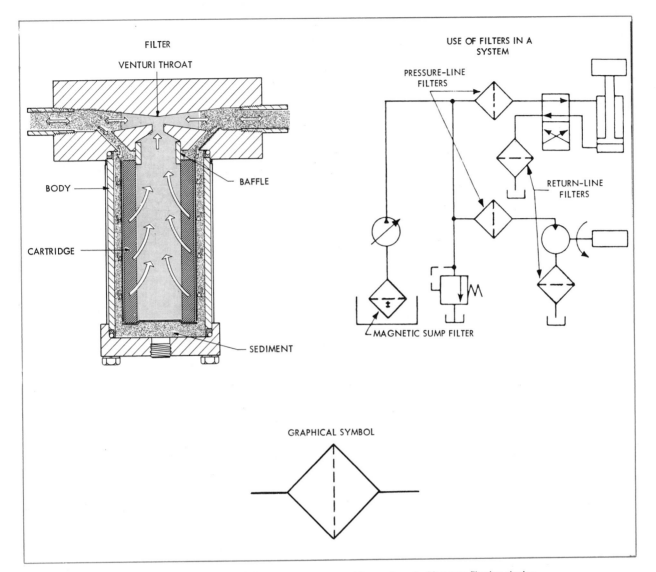

Fig. 23-4. For long life a fluid-power system must be equipped with some filtering device.

tures are below the required range, there is a slower movement of the fluid through the orifices and piping system, thus resulting in incorrect time responses of the equipment. If the temperatures are too high, the lubricating characteristics of the fluids are reduced. In some instances the fluids may actually break down, forming sludge or other contaminants which may lead to plugged orifices, pipes, or control valves. In some fluid-power systems temperature control is achieved by the use of a special heat exchanger that will either cool or heat the fluid. See Fig. 23-5.

Accumulators. Accumulators are chambers where fluid can be stored under pressure and from which fluid may be withdrawn when needed. They are used to supply peak demands in a system having an intermittent-duty cycle where power output requirements are both low and high. See Fig. 23-6.

Valves. Valves are devices in a circuit that control pressure or flow rate or that direct fluid flow. The three basic valve types are known as pressure-control, flow-control, and directional-control valves.

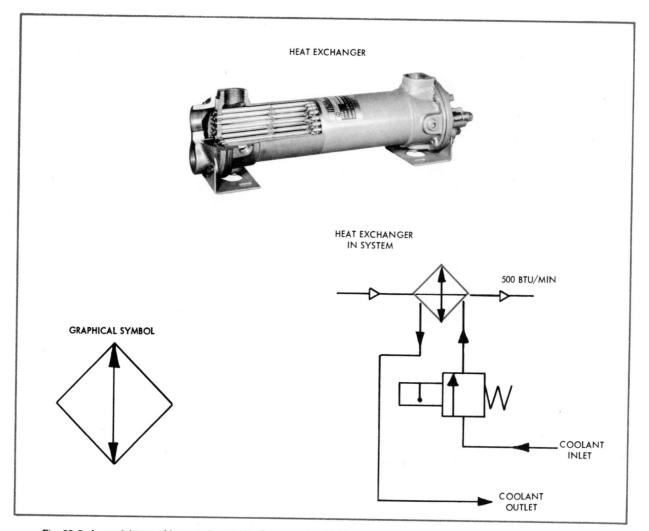

HEAT EXCHANGER

HEAT EXCHANGER
IN SYSTEM

500 BTU/MIN

GRAPHICAL SYMBOL

COOLANT
INLET

COOLANT
OUTLET

Fig. 23-5. A special type of heat exchanger is often used in a fluid-power system for exacting control of fluid temperature.

Pressure-control valves function to maintain fluid pressure at the required levels in various parts of the system. Such valves are particularly important in a system where precise control of pressure is mandatory in order to perform types of work that demand the application of constant forces. See Fig. 23-7.

Flow-control valves are used to regulate the rate of fluid flow in a system. See Fig. 23-8. The correct flow rate is essential since the effective movement of machine elements is governed by the flow of the pressurized hydraulic fluid. Some

hydraulic systems have compensated types of control valves. Ordinary simple types of valves do not compensate for variation of fluid flow due to changes in fluid temperature or pressure. Compensating flow-control valves will automatically change valve adjustments to insure a constant flow at an established setting.

Directional-control valves are designed to regulate the direction of fluid flow in a system so that the pressurized fluid will cause the desired machine movements. These valves are also used to accelerate, decelerate, stop, and control the di-

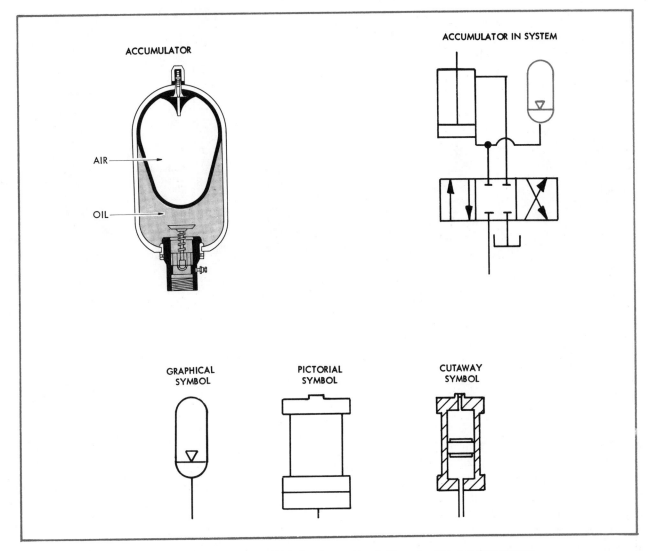

Fig. 23-6. Accumulators are special chambers where fluid can be stored under pressure.

rection of motion in hydraulic cylinders. Directional-control valves are classified according to the number of flow paths within the valve; that is, two, three, or four. See Fig. 23-9.

Electrohydraulic servovalves provide automatic control of machine functions by means of an electrical signal to a hydraulic output. The signal may activate a control to start a sequence of events, establish a pressure level or a stop position, or move a piston or cylinder.

The various symbols used to designate valves in a fluid-power system are shown in Fig. 23-10.

23.2 Hydraulic Plumbing

In any hydraulic system every effort is made to keep lines short. This is done by incorporating bends wherever it is feasible. Bends make it possible to eliminate angular fittings, absorb strain caused by vibrations, and compensate for expansion and contraction.

Individual tubing sections are kept symmetrical so they can be assembled from either end. Since sharp bends effect pressure loss in a line, bend radii are made so they are not less than 3 to 3.5 times the diameter of the line. The lines are in-

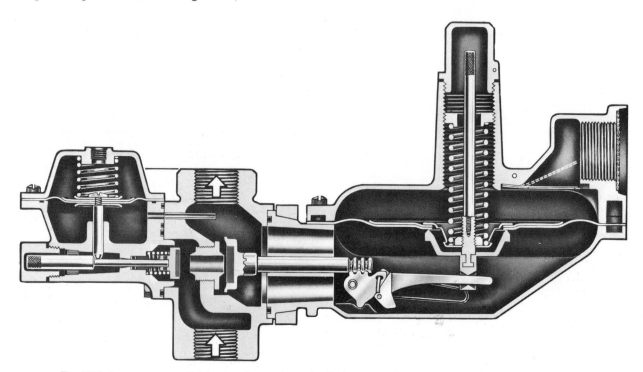

Fig. 23-7. A pressure control valve may be used to maintain the correct fluid pressure in all parts of a system.

stalled so they can be readily removed without dismantling any of the circuit components.

23.2.1 Pipe and tubing. Hydraulic plumbing may be either of pipe or tubing with appropriate fittings. The primary considerations in a plumbing system are material, diameter, and wall thickness. Although pipe is used in some hydraulic installations, tubing is more common because it can be bent easily, thereby reducing the number of fittings. Steel pipe is used mainly where disassembly is infrequent, where large volumes of fluid are moved, or where the line is long and straight. Recommended sizes for hydraulic pipe are included in the Appendix.

Tubing for hydraulic systems may be of copper, steel, aluminum, or plastic. The Joint Industrial Council (JIC), an industrial association which recommends industrial standards to promote safety, long life, and maintenance ease of tools and equipment, restricts the use of copper to stationary systems with low pressures. Copper has poor resistance qualities to vibrations. Seamless, cold-drawn, dead soft SAE 1010 steel tubing is the only

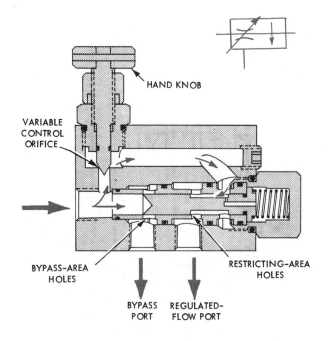

Fig. 23-8. Flow-control valves are usually a combination of needle valve and check valve, which provides free flow in one direction and restricted flow in the other.

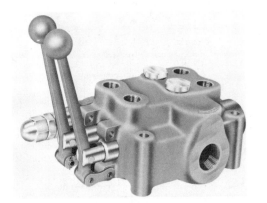

MANUALLY OPERATED

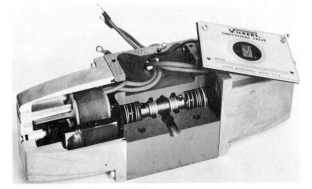

SOLENOID OPERATED

Fig. 23-9. Directional-control valves.

VALVES IN SYSTEM

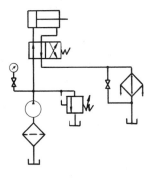

PICTORIAL VALVE SYMBOLS

CUTAWAY VALVE SYMBOLS

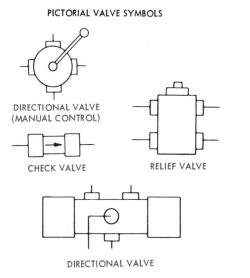

DIRECTIONAL VALVE
(MANUAL CONTROL)

CHECK VALVE

RELIEF VALVE

DIRECTIONAL VALVE

DIRECTIONAL VALVE

CHECK VALVE

RELIEF VALVE

DIRECTIONAL VALVE

Fig. 23-10. Basic valve symbols. (Continued on next page.)

GRAPHIC SYMBOLS	
VALVE, CHECK	
VALVE, MANUAL SHUTOFF	
VALVE, MAXIMUM PRESSURE (RELIEF)	
VALVE, BASIC SYMBOL SINGLE FLOW PATH IS MODIFIED.	
VALVE, BASIC SYMBOL MULTIPLE FLOW PATHS ARE CHANGED	
VALVE, SINGLE FLOW PATH, NORMALLY CLOSED	
VALVE, SINGLE FLOW PATH, NORMALLY OPEN	
VALVE, MULTIPLE FLOW PATHS, BLOCKED	
VALVE, MULTIPLE FLOW PATHS, OPEN (ARROWS DENOTE DIRECTION OF FLOW)	

GRAPHIC SYMBOLS	
VALVE, RELIEF REMOTELY OPERATED (UNLOADING VALVE)	
VALVE, SEQUENCE, DIRECTLY OPERATED	
VALVE, PRESSURE REDUCING	
VALVE, FLOW-RATE CONTROL, VARIABLE	
VALVE, DIRECTIONAL, 2 POSITION 3 CONNECTION	
VALVE, DIRECTIONAL, 3 POSITION 4 CONNECTION OPEN CENTER	
VALVE, DIRECTIONAL, 3 POSITION 4 CONNECTION CLOSED CENTER	

Fig. 23-10 (Continued). Basic valve symbols.

material approved by JIC standards without restrictions. Aluminum seamless tubing is approved for low-pressure application. Greater use is now being made of plastic tubing for many fluid-power systems. Types of plastic tubing that are currently used are made of polyvinyl chloride, polyethylene, nylon, and tetra-fluoroethylene.

The inside diameter of a fluid line is important because if the diameter is too small, excessive turbulence and friction heat are created. The result is considerable power loss. (See Appendix for hydraulic tubing sizes.)

Wall thickness governs the bursting pressure of a line. As a rule a wall thickness is selected that provides a safety factor of eight. This means that the rated bursting pressure is at least eight times greater that the maximum working pressure in the system.

23.2.2 Fittings. Hydraulic lines are connected with various types of fittings. Pipe fittings are either threaded, brazed, or welded. Several typical hydraulic pipe fittings are shown in Fig. 23-11.

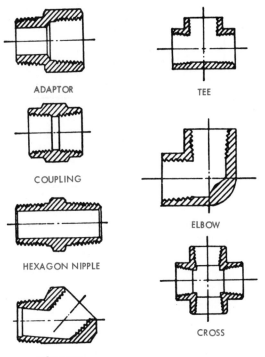

ADAPTOR

TEE

COUPLING

HEXAGON NIPPLE

ELBOW

CROSS

45° STREET

Fig. 23-11. Typical hydraulic pipe fittings. All threads are tapered, preferably NPTF.

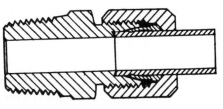

FLARE TYPE FITTING

FLARELESS TYPE FITTING

Fig. 23-12. Typical fittings used for connecting hydraulic tubing.

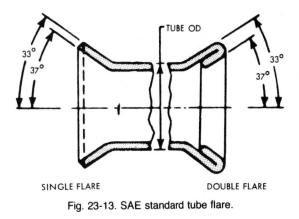

TUBE OD

33°
37°

33°
37°

SINGLE FLARE

DOUBLE FLARE

Fig. 23-13. SAE standard tube flare.

These fittings are made of brass for low- and medium-pressure application, cast iron for large sizes, and steel for high-pressure systems.

Tubing is connected by flaring or brazing. In flaring, the flared end of the tubing presses against the mating surface of the fitting. Pressure against the flare is provided by a nut. See Fig. 23-12. The SAE standard for tubing flares is shown in Fig. 23-13. Flareless-type fittings are also used on tubing. Here a ferrule, or sleeve, is forced into the tubing and forms a seal. A nut applies the necessary pressure to the sleeve. See Fig. 23-12.

23.3 Types of Fluid-power Diagrams[2]

Four types of diagrams are used to show fluid-power layouts: pictorial, cutaway, graphical, and combination.

23.3.1 Pictorial diagrams. Diagrams of this type provide a simplified method of showing piping between components. See Fig. 23-14. General piping arrangement is emphasized. Circuit analysis is difficult since only external features of components are shown.

Extensive use is made of pictorial symbols with interconnecting lines in the preparation of pictorial diagrams. Pictorial symbols depict configurations of various components in miniature. Sufficient detail is included so that the unit is readily identifiable. Each symbol usually is drawn with thick line width and made to indicate relative size.

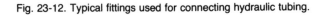

2. *ANSI Y14.17.*

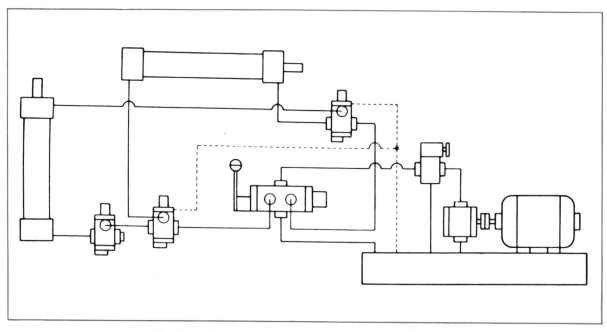

Fig. 23-14. Pictorial type of fluid-power diagram. (ANSI Y14.17)

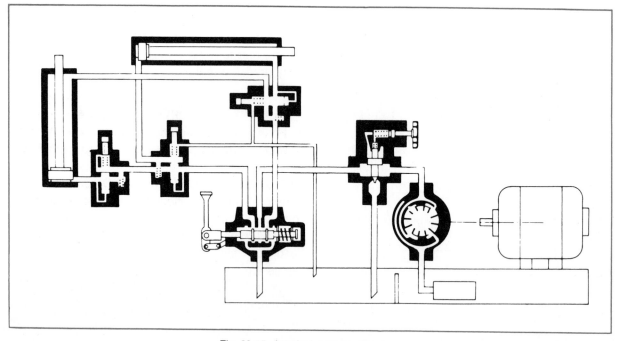

Fig. 23-15. A typical cutaway diagram.

23.3.2 Cutaway diagrams. Diagrams of this type provide a means of showing principal internal working parts. See Fig. 23-15. Function of individual components is emphasized. Sometimes multiple cutaway drawings are used to show positions of movable parts and resulting flow paths during each phase of system operation.

Cutaway symbols with interconnecting lines are

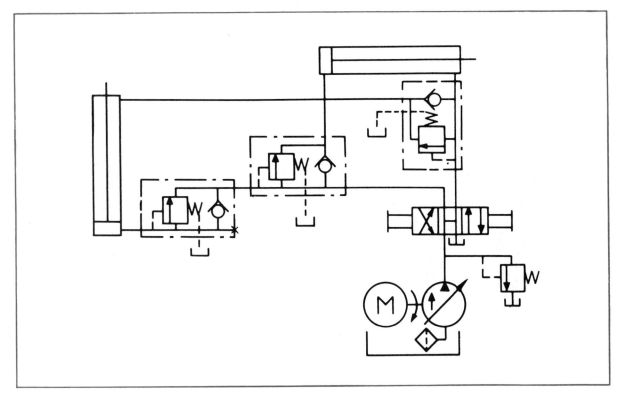

Fig. 23-16. A typical graphical fluid-power diagram.

used to make cutaway diagrams. These symbols show miniature cross-sectional shapes of hydraulic units, which clearly illustrate internal and external parts, indicate flow paths, show control elements such as levers and springs, and include physical characteristics to permit easy identification. Drawing cutaway symbols requires some artistic ability to make them look realistic. Line widths are varied for emphasis and clarity. Components are usually drawn to indicate relative size.

23.3.3 Graphical diagrams. Graphical diagrams are most ideally suited to the drafting function. Diagrams of this type provide a simple method for emphasizing functions of the circuit and function and method of operation of each component. Graphical diagrams and symbols can be standardized in many countries and can therefore promote the universal understanding of fluid-power systems. Fig. 23-16 shows how graphical symbols of components are drawn with interconnecting lines to depict a complete fluid-power circuit.

23.3.4 Combination diagrams. A combination of pictorial, cutaway, and graphical symbols is drawn with interconnecting lines as in Fig. 23-17. Diagrams of this type provide a method of showing piping, function, or flow paths for each component as best suits the purpose of the diagram. The symbols are used to emphasize a portion or feature of an assembly.

23.4 Preparing Fluid-Power Diagrams[3]

To be effective, any fluid-power diagram must clearly fulfill its basic function. Whether the diagram is of the pictorial, cutaway, graphical, or combination type, the symbols representative of the various units must be clearly outlined and correctly located in the system. Sufficient descriptive data and specifications must be included to properly identify all components.

---

3. *Ibid.*

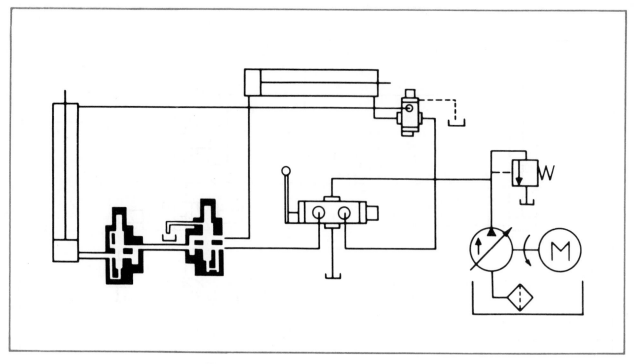

Fig. 23-17. A combination fluid-power diagram.

Some of the more significant rules governing the preparation of fluid-power diagrams as recommended by ANSI are included in the paragraphs to follow.

23.4.1 Arrangement of symbols. Symbols should be arranged in a diagram to facilitate use of direct and straight interconnecting lines. Where components have definite mechanical, functional, or otherwise important relationships to one another, their symbols should be so placed in the diagram. Where a component requires a specific mounting position, its symbol should be so drawn and noted. Spacing should provide room for adjacent data without crowding.

23.4.2 Uses of lines. As with other types of drawings, certain line conventions are employed in fluid-power diagrams.

Interconnecting lines. Interconnecting lines between various components in a diagram represent means of conducting fluid or transmitting signals. The lines constitute a very important portion of the system and should be drawn carefully. For clarity, lines should be direct and straight. They do not in-

dicate actual installation of piping and fittings. Horizontal and vertical lines are preferred.

Types of lines. Interconnecting lines may be either single or double, depending upon the type of diagram. Single lines are used in graphical diagrams. Double lines are used in cutaway diagrams. Pictorial and combination diagrams may use single or double lines or both. See Fig. 23-18.

Single lines. Single lines conveying power-actuating fluid, either pressure or return, are called working lines and should be drawn as single unbroken lines. Lines that convey fluid used to control components are called pilot lines and should be drawn as a series of long dashes. Sensing lines, such as gage lines, should be drawn the same as the line or lines to which they connect. Lines which indicate internal seepage of components or exhaust pilot fluid being returned to the tank by a drain should be drawn as a series of short dashes. Working, pilot, and drain lines should be drawn thick line-width. Sharp angles, preferably of 90°, should be used when lines drawn between symbols change direction. Dashes should

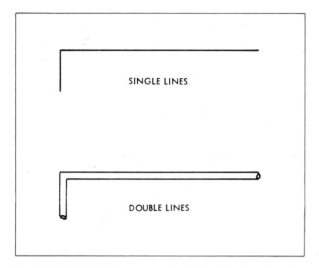

Fig. 23-18. Conductors are represented by single and double lines.

join at corners. Where more than one fluid must be shown on a single diagram, lines should be coded for differentiation.

Double lines. Double lines should be used to depict the interconnecting lines in cutaway diagrams. Double lines are sometimes used in pictorial diagrams to show the piping arrangements more clearly. Included space between double lines may indicate relative pipe size and may also be used to indicate functions such as working, pilot, drain, bleed, and fluids.

Double lines should be drawn thick line-width. Inside radius of double-line bends should be drawn approximately equal to space between lines.

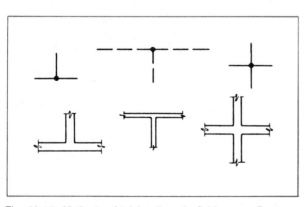

Fig. 23-19. Methods of joining lines in fluid-power diagrams. (ANSI Y14.17)

Joining lines. Joining lines should be drawn as indicated in Fig. 23-19. Single lines joining symbols should terminate without space or dot. A dot should be used at the joint of two or more single lines and the diameter should be approximately five line-widths. Dash lines should begin and end with a dash of full length at junction with other lines. Double lines forming a junction should end at square corners.

Crossing lines. Crossing lines should be drawn as indicated in Fig. 23-20. A loop should be used in one of two single lines which cross but do not join. The loop should be unbroken and should join its line with sharp corners. The radius of loops should be uniform and of such size as to be clearly recognized. Arcs should preferably loop upward from the horizontal. Where one line crosses several closely spaced parallel lines, a single loop (often flat-topped) may be used.

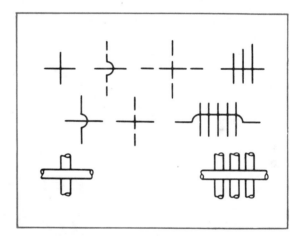

Fig. 23-20. Method of crossing lines in fluid-power diagrams.

Additional line functions. Fig. 23-21 illustrates several other functions of lines as used in fluid-power diagrams.

23.4.3 Color or pattern code. Interconnecting lines are sometimes colored or patterned to show pressure, flow, special functions, or different fluids, during a phase or each phase of operation. See Fig. 23-22. A note, preferably located in the lower left corner of the sheet, serves to identify and illustrate the code for each condition used.

LINE, WORKING	—
LINE, PILOT (L>20W)	– – –
LINE, DRAIN (L<5W)	- - - -
CONNECTOR (DOT TO BE 5X WIDTH OF LINES)	●
LINE, FLEXIBLE	⌣
DIRECTON OF FLOW	→
LINE TO RESERVOIR ABOVE FLUID LEVEL BELOW FLUID LEVEL	
LINE TO VENTED MANIFOLD	
PLUG OR PLUGGED CONNECTION	✕
RESTRICTION, FIXED	
RESTRICTION, VARIABLE	

Fig. 23-21. These lines are used to indicate certain functions in fluid-power diagrams.

FUNCTION	COLOR	PATTERN
INTENSIFIED PRESSURE	VIOLET	
SUPPLY PRESSURE	RED	
CHARGING PRESSURE REDUCED PRESSURE PILOT PRESSURE	ORANGE	
METERED OR BLOCKED FLOW	YELLOW	
EXHAUST	BLUE	
INTAKE DRAIN	GREEN	
INACTIVE	BLANK	

Fig. 23-22. Color and pattern coding used in fluid-power diagrams. (ANSI Y14.17)

When applicable, the color and pattern code illustrated should be used. Only those lines showing active functions for the phase shown should be coded. When more than one function is appropriate, the code which gives the greatest emphasis and clarity should be selected.

23.5 Design Specifications of Hydraulic Units[4]

In the design of a fluid-power system, not only must the various hydraulic components be clearly identified but other relevant specifications such as size, capacity, rate flow, and function must be provided. Examples of essential design specifications for fluid-power units are included in the paragraphs to follow.

Power supply. Horsepower, speed, and direction of rotation should be shown for each pump or compressor drive. Direction of rotation should be indicated by an arrow or an appropriate note. If an

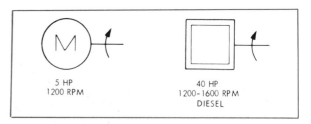

5 HP
1200 RPM

40 HP
1200-1600 RPM
DIESEL

Fig. 23-23. Method of designating horsepower, speed, and rotation in a power supply unit. (ANSI Y14.17)

arrow is used on the symbol of the pump shaft, it is understood that the arrow is on the near side of the shaft to denote direction of rotation. Where speed of drive is relatively constant, as with an AC motor, the nominal speed should be given. Where speed varies considerably, as with an engine, the speed range should be given. Type of drive should also be indicated, such as electric motor, gas engine, machine power-take-off, or mill shafting. Horsepower, speed, and rotation should be placed adjacent to the symbol in the diagram or appear in the component list if the drive component is listed. See Fig. 23-23.

---

4. *Ibid.*

Pump or compressor rating. In a hydraulic circuit the nominal delivery rate of each constant displacement pump should be given and, where applicable, range of delivery used should be stated. Individual delivery rates should be given for pumps with more than one outlet. Each delivery rate should be given for pumps with more than one nominal delivery from a single outlet.

Information should appear adjacent to the symbol in the diagram or in the component list with manufacturer's model number of the component. See Fig. 23-24.

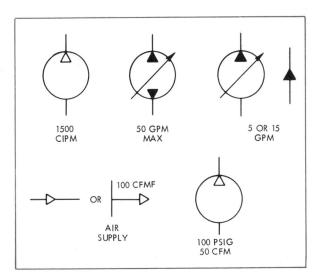

Fig. 23-24. Methods of indicating pump or compressor rating.

Pressure controls. The operating pressure of controls for proper circuit function should appear as close as practicable to the appropriate symbol.

SET PSI indicates a factory-set pressure of non-adjustable controls and a field-set pressure of adjustable controls. The range of adjustable controls may appear next to the symbol or in the component list. The words SET and RANGE and standard abbreviations for units of measurement should be used with numerical values, as shown in Fig. 23-25.

Directional controls. Solenoids or solenoid-operated or controlled directional valves should be identified. Solenoid identification on the diagram should be by consecutive letters, omitting *I, O,* and

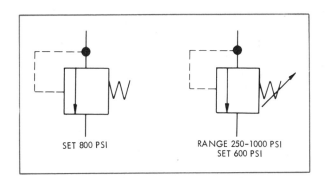

Fig. 23-25. Method of showing operating pressure of control valves.

Reservoir or receiver capacity. The nominal liquid capacity of each hydraulic reservoir should be given. Capacity of each air receiver should be stated. Information should be placed within or adjacent to the symbol in the diagram or in the component list.

Heat transfer devices. The rating of each heat transfer device, such as a cooler or heater, should be indicated either by its model number in the component list or by an appropriate note adjacent to its symbol in the diagram. See Fig. 23-5.

Filters. Rated flow capacity of all filters should be given. Information should be placed adjacent to the symbol of the component in the diagram or in the component list.

Q. The letter *A* should be used as a prefix for the second progression through the alphabet, that is, *AA, AB, AC,* and so on. The letter *B* should be used as a prefix for the third progression.

Control positions of manually- and mechanically-operated directional valves should be identified. Identification should agree with markings on the component or with the manufacturer's installation drawing of the component and should be as close as practicable to the relevant symbol. For examples of solenoid identification and control positions, see Fig. 23-26.

Flow controls. Appropriate data required for circuit analysis should be given at symbols of flow controls. Examples of this data are shown in Fig.

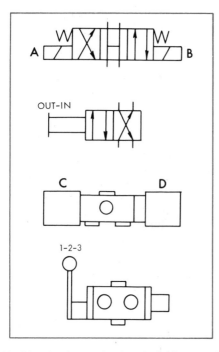

Fig. 23-26. Directional-control valves in fluid-power diagrams. (ANSI Y14.17)

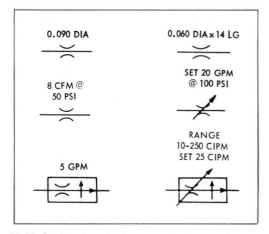

Fig. 23-27. Designations for flow-control valves. (ANSI Y14.17)

23-27 and may include the fixed flow rate of pressure-compensated restrictions; set flow rate and range of adjustment, where pertinent, of adjustable pressure-compensated controls; and flow rate and pressure drop through non-compensated restrictions, either adjustable or fixed. Data may

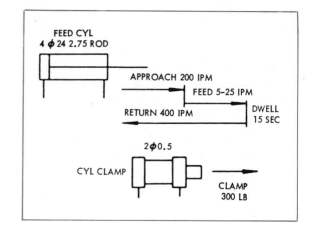

Fig. 23-28. Cylinder component data in fluid-power diagrams. (ANSI Y14.17)

be dimensional, as the area of an orifice or the diameter and length of a choke. Setting, range, rate, dimensions, and units of measurement should be identified by word or abbreviation. Information should appear as close as practicable to the symbol.

Linear motors (cylinders). Linear motors are devices which produce straight-line motion from fluid flow and force from fluid pressure. Dimensions, bore, stroke, and, where applicable, rod or ram diameters should be given for each cylinder. The model number of each cylinder listed in the component list may convey this information. Otherwise the information should appear as close as practicable to the symbol. See Fig. 23-28.

Function should be given where cylinders or groups of cylinders serve various purposes. CLAMP CYLINDER, FEED CYLINDER, or other appropriate name should appear as close as practicable to the symbol.

Information such as speed and force resulting from cylinder operation should be noted so that circuit analysis can be made without calculating these values from flow rate, pressure, and area. Speed, range of speed, cycle time, and units of measurement should be identified by word or abbreviation. Direction of function or operation should be indicated by arrows and named. Information should appear as close as practicable to the relevant symbol.

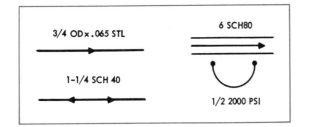

Fig. 23-29. Tube, pipe, and hose information is shown in these ways. (ANSI Y14.17)

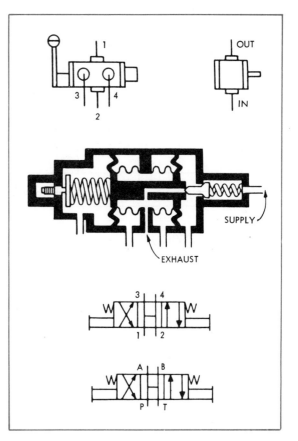

Fig. 23-30. Port identification of components in a fluid-power system. (ANSI Y14.17)

Tube, pipe, and hose. Appropriate symbols, notes, or coding should be used to indicate physical characteristics, direction of flow, and identification of all interconnecting lines. Outside diameter and wall thickness should be given for all types of tubing used. Tubing material, such as steel, copper alloy, or plastic, should also be identified. Where iron or steel pipe is used, the nominal pipe size and schedule number should be given. Inside diameter and working pressure rating of hose should be designated. The length of lines should be given when the information is important to circuit analysis. Information should appear adjacent to each interconnecting line as shown in Fig. 23-29, or it may be adequately covered by note or code. If lines are coded, a code note should be shown elsewhere on the drawing.

Direction of flow in interconnecting lines may be shown by arrowheads on single lines or by arrows within double lines.

Port identification. Functional ports of components should be clearly identified in the diagram. Identification should agree with port designation on the component and its installation drawing. The identification should appear adjacent to or within the symbol and should be as close as practicable to the port being identified. See Fig. 23-30.

Operation notes. Explanatory notes or charts should be used as required to explain circuit operation. Where various operations of circuit follow a prescribed pattern, each action should be explained in order of occurrence. Each phase of operation should be numbered or lettered and followed by a brief description of initiating and resulting action. The notation should be titled SEQUENCE

OF OPERATIONS and located in the upper left corner of the sheet. See Fig. 23-31.

Functional description. A description should be provided to briefly explain the function and purpose of each individual component if this is required for circuit analysis. Key numbers of components may be used in this description for clarity and brevity. The note should be titled FUNCTIONAL DESCRIPTION and located directly below sequence of operation, allowing reasonable space for additions or changes. See Fig. 23-31.

Component list. Name, model number, and supplier, if given adjacent to the symbol for each component, would clutter up the diagram and make it difficult to follow. This information should be given in a column headed COMPONENT LIST in the upper right corner of the drawing. Listing components in this manner has other advantages,

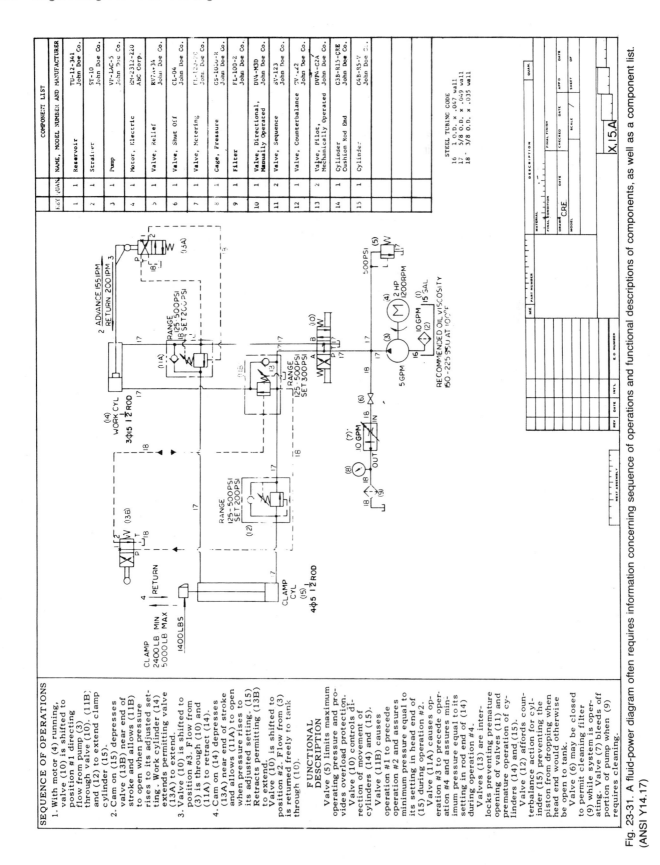

Fig. 23-31. A fluid-power diagram often requires information concerning sequence of operations and functional descriptions of components, as well as a component list. (ANSI Y14.17)

such as providing readily available information for quotations or orders. See Fig. 23-31.

Each component in the diagram should be assigned a key arabic number for reference purposes, such as in the component list, notes on the drawing, and functional description. Each key number should be placed in parentheses adjacent to component symbol and in a prominent location so it may be easily found. Key numbers should be assigned in sequence, beginning with the reservoir or compressor and progressing to the output motor in numerical order. Key numbers should determine the order of items in the component list.

Where two or more identical components are used, they should be assigned the same key number. Letters should be used as a suffix to each key number to identify each identical component, for example (1A), (1B), and (1C). The arabic number without a suffix letter should appear in the component list. Key numbers with suffix letters placed adjacent to symbols will also be used in the functional description and sequence of operations to identify individual components.

A column should be provided in the component list to show the required quantity of each component listed.

The name of component, its model number, and manufacturer's name should be given in the component list. Other necessary procurement information should also be given.

The component list may include pressure or flow ranges, motor displacement, or information previously specified in the component data.

UNIT 24

Electrical and Electronics Diagrams

Electrical and electronic circuits are generally designed by electrical engineers. As a rule these designs are in the form of freehand sketches which show the essential components and circuit path by means of standard symbols. Sketches are then reproduced as finished diagrams by drafting personnel.

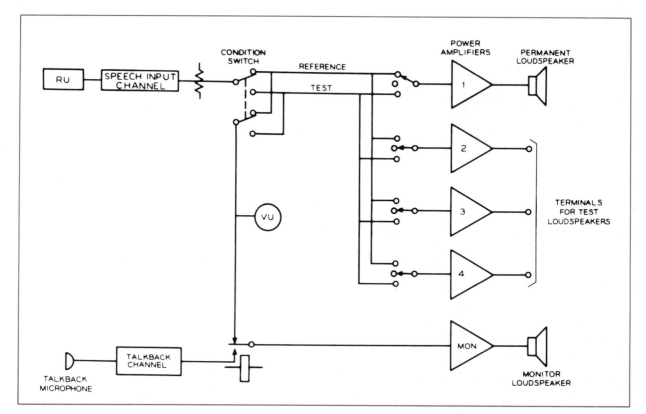

Fig. 24-1. Typical single-line electronics diagram. (ANSI Y14.15)

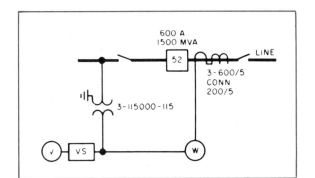

Fig. 24-2. Portion of a typical single-line diagram using symbols. (ANSI Y14.15)

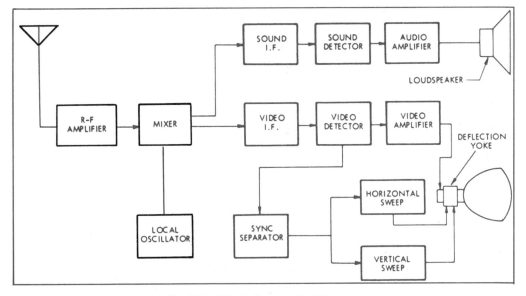

Fig. 24-3. A block diagram of a TV receiver.

Basic electrical and electronics diagrams are classified into five main groups: single-line, schematic, connection, interconnection, and terminal.[1]

24.1.1 Single-line diagrams. The single-line, or one-line, diagram shows by means of single lines and graphical symbols the path of an electrical circuit and the component devices or parts used in the system. See Fig. 24-1. Primary circuits are indicated by thick connecting lines, and connections to the circuit and potential sources are shown by

medium lines. Relays, meters, and instruments are shown by circles with appropriate abbreviations. Rectangles with abbreviations are used to depict resistors and switches. Power circuit devices are identified by graphical symbols. See Fig. 24-2.

A single-line diagram often appears as a block diagram. Each unit is laid out beginning from a point where the signal is introduced, progressing through the various stages, and finally terminating at the output. See Fig. 24-3.

Whenever possible, the circuit path of a block diagram is drawn to run from left to right or from the upper left corner to the lower right corner. The blocks representing the units or stages are shown as rectangles, squares, or triangles with one line

1. *ANSI Y14.15.*

connecting them. Arrows are used on the connecting line to designate the direction of flow. Identification of each unit is included in the respective block.

24.1.2 Schematic diagrams. A schematic diagram shows by means of graphical symbols the electrical connections and functions of a specific circuit arrangement. The diagram is prepared so

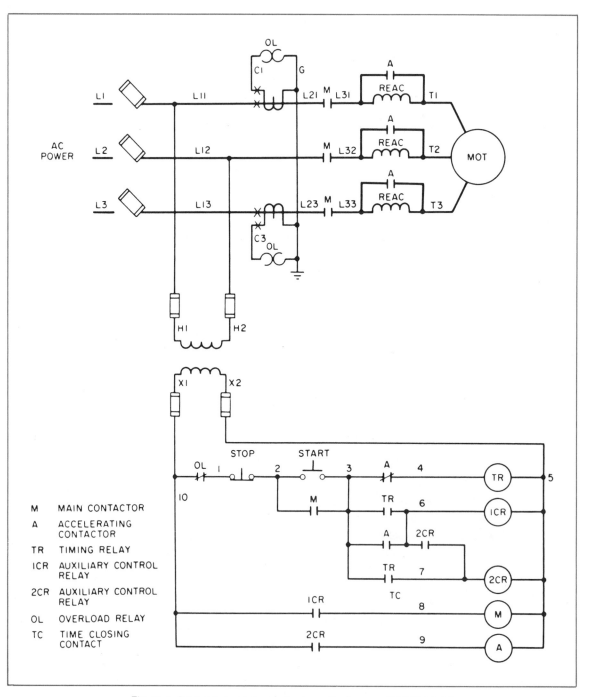

Fig. 24-4. Typical industrial control schematic diagram. (ANSI Y14.15)

that functions can be followed from left to right or from upper left to lower right. The overall circuit layout traces the signal or transmission path from input to output or in the order of functional sequence. See Fig. 24-4.

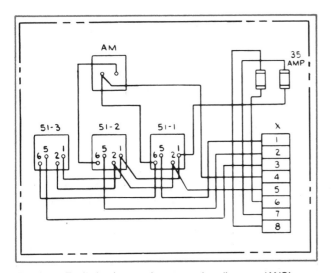

Fig. 24-5. Typical point-to-point connection diagram. (ANSI Y14.15)

24.1.3 Connection diagrams. A connection, or wiring, diagram includes all of the devices in an electrical system, arranged to show their physical relations to each other. Thus, poles, terminals, coils, and other units are drawn in their correct places on each device. Diagrams of this type give the necessary information for actually wiring a group of control devices and are particularly useful for tracing wires when troubleshooting.

Connection diagrams are classified as: continuous-line, interrupted-line, and tabular.

Continuous-line diagrams are either of the point-to-point type or the highway type. The point-to-point is used primarily for simple equipment requiring only a few items. See Fig. 24-5. The highway, or cable connection, diagram is similar to the point-to-point except that groups of connecting lines are merged into lines called highways rather than shown as separate lines from terminal to terminal. See Fig. 24-6. A highway diagram is particularly useful in illustrating the wiring of switchboards and panels. One or more highways may be shown, depending on the wire routing necessary

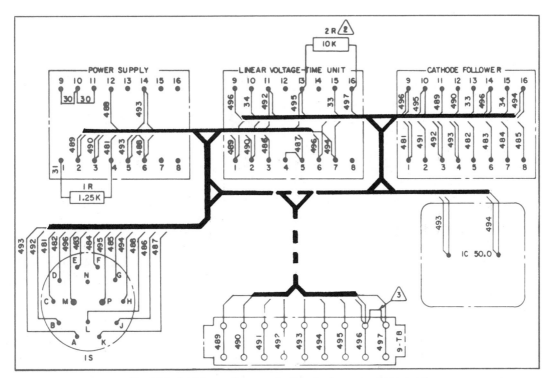

Fig. 24-6. A highway diagram of a timed reference voltage supply.

for the components, with short feed lines running to terminals. The direction a feed line takes when it joins the highway is indicated by a small arc or slanted line.

In highway diagrams each wire is coded to identify component, terminal destination, type and size of wire, and wire color. Thus, the label E12-3-B6 indicates the following:

E12-3-B6
— color
— type and size of wire
— terminal
— component

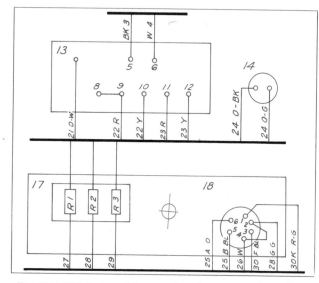

Fig. 24-7. This is a partial connection diagram using base lines.

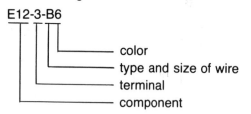

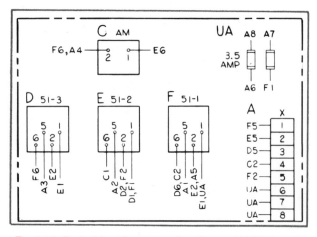

Fig. 24-8. Typical feed-line connection diagram (ANSI Y14.15)

An *interrupted-line diagram* is one where connecting lines start at a symbol and are interrupted a short distance away. The two somewhat similar methods used for this type of diagram are referred to as base-line and feed-line. In the base-line method, lines are terminated in a base line for convenience of alignment, while in the feed-line method no base line is used. See Figs. 24-7 and 24-8. Each component is properly identified and

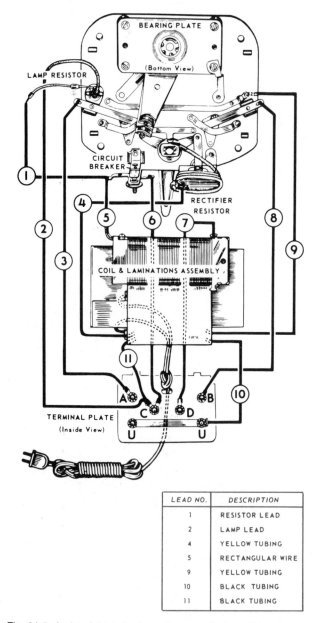

LEAD NO.	DESCRIPTION
1	RESISTOR LEAD
2	LAMP LEAD
4	YELLOW TUBING
5	RECTANGULAR WIRE
9	YELLOW TUBING
10	BLACK TUBING
11	BLACK TUBING

Fig. 24-9. A pictorial tabular-type diagram of a transformer assembly.

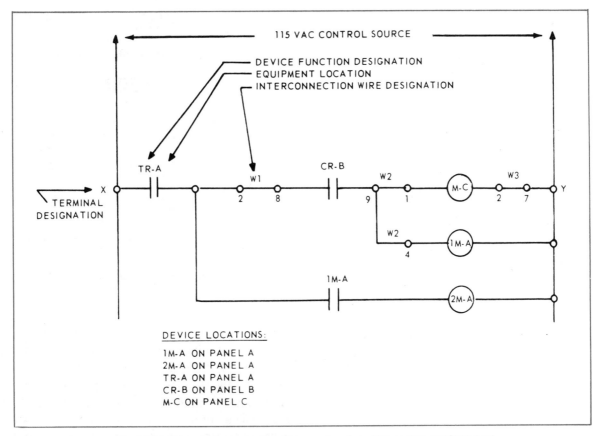

Fig. 24-10. Typical interconnection diagram of an industrial control. (ANSI Y14.15)

each wire correctly coded to simplify the task of tracing circuit paths.

In the *tabular-type diagram,* wiring information is arranged in tabular form rather than shown on feed lines. This tabular information is usually accompanied by a pictorial drawing. See Fig. 24-9.

24.1.4 Interconnection diagrams. Interconnection diagrams are those which show only external connections between unit assemblies or equipment. Internal connections of the unit assemblies are usually omitted. Interconnection diagrams may be prepared in continuous-line, point-to-point, highway, interrupted-line, or tabular form. See Fig. 24-10.

24.1.5 Terminal diagrams. A terminal diagram depicts the internal circuits of an item or device to its terminal physical configuration. Its function is to supply information specifically applicable to enclosed or sealed devices, such as electron tubes, semiconductor devices, and packaged circuits.

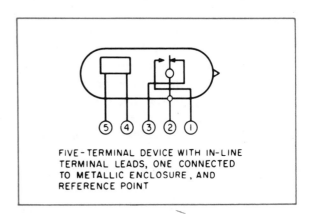

Fig. 24-11. Typical relay terminal diagram. (ANSI Y14.15)

The usual graphical symbols are employed to show the units in a circuit. Refer to Fig. 24-11 and Fig. 24-12.

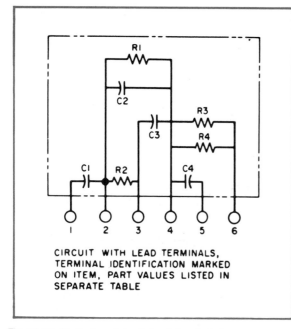

Fig. 24-12. Typical packaged circuit terminal diagram. (ANSI Y14.15)

Fig. 24-13. Line conventions for diagrams.

24.2 Preparing Electrical and Electronics Diagrams[2]

The following are some of the recommended practices for preparing electrical and electronics diagrams:

1. Use medium-weight lines for wire, leads, and symbols. A thin line may be used for brackets, leader lines, and so forth. When emphasis on special features such as main or transmission paths is necessary, heavy lines are advisable to provide the desired contrast. See Fig. 24-13.

2. Arrange the layout so that the main features are prominently shown. The parts of the diagram should be spaced to provide an even balance between blank spaces and lines. Sufficient blank area should be provided around symbols to avoid crowding of notes or reference information.

3. When a circuit contains parts which need to be shown grouped, the grouping may be indicated

by means of a boundary (phantom) line enclosure. See Fig. 24-12. The phantom line enclosure may be omitted if sufficient space is provided between parts. Typical groupings are unit assemblies, subassemblies, printed circuits, hermetically sealed units, contactor parts, and relays. The dash line used to indicate shielding also implies that the parts enclosed by the dash line are grouped.

4. In a block type of drawing, the blocks should be drawn with heavy lines. Connecting lines should be light. The exception is when the path of signal travel is to be emphasized, in which case the connecting lines are drawn heavier than the blocks. Once the size of the block is established, it should be used throughout the drawing. No attempt should be made to indicate the relative importance of the components by varying the block size. The description or identity of the components should be placed in the blocks. The direction of signal flow should be designated by arrows. The signal path should travel from left to right. Whenever possible, the input should be located at the left and the output at the upper right corner.

5. There are no specific requirements governing the size of the symbols to be used in a drawing. They may be drawn vertically or horizontally and to any proportion; however, they should produce a balanced appearance. Care should be taken to avoid congested areas and numerous large white

---

2. *American Standard Drafting Manual,* Electrical Diagrams—*ANSI Y14.15* (New York: The American Society of Mechanical Engineers).

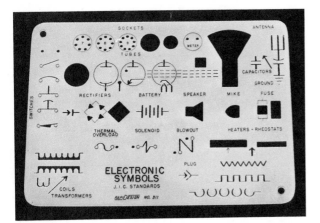

Fig. 24-14. Templates simplify the task of drawing electronics symbols.

areas. ANSI recommends that in most diagrams intended for manufacturing purposes or for ultimate use in a reduced form the symbols be drawn approximately 1.5 times the size of those shown in ANSI Standard Y32.2.

Construction of symbols is simplified by the use of templates. See Fig. 24-14.

6. The layout of a schematic diagram should produce a circuit, signal, or transmission path which follows from input to output, source to load, or in the order of functional sequence. Every effort should be made to avoid long interconnecting lines between parts of the circuit. The diagram should be arranged so it will read functionally from left to right; that is, the source or input is at the left and the output at the right. For some industrial types of electronic circuits, it is often the practice to place the power input at the top and the signal at any convenient location on the drawing. Complex drawings may be laid out in two or more layers with each layer reading from left to right.

7. Connecting lines should preferably be drawn horizontally or vertically and with as few bends and crossovers as possible. Connection of four or more lines at one point should be avoided when it is equally convenient to use an alternate arrangement. Parallel lines should be arranged in groups, preferably three, with approximately double spacing between groups of lines. Spacing between parallel lines should be a minimum of 0.06 inch.

8. Connecting lines, whether single or in groups, may be interrupted at convenient points and identified, with the destination indicated. Letters, numbers, abbreviations, or other identifiers for interrupted lines must be located as close as possible to the point of interruption.

For single interrupted lines the line identification may also serve to indicate destination as shown in Fig. 24-15. In general, identification practice for single interrupted lines should be the same as for grouped and bracketed lines described in the next paragraph.

When interrupted lines are grouped and bracketed, depending on whether the lines are horizon-

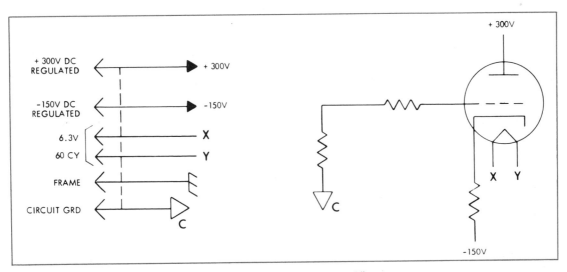

Fig. 24-15. Identification of interrupted lines.

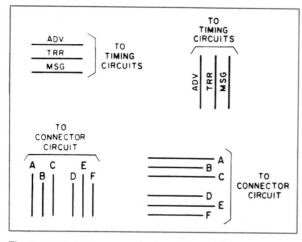

Fig. 24-16. Typical arrangements of line identification and destinations. (ANSI Y14.15)

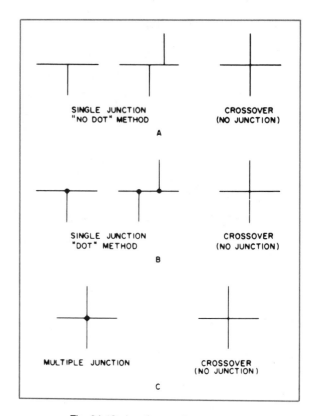

Fig. 24-18. Junctions and crossovers.

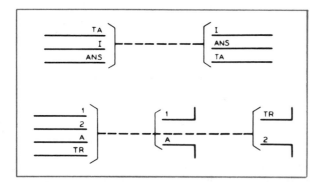

Fig. 24-17. Typical interrupted lines interconnected by dash lines. (ANSI Y14.15)

tal or vertical, line identification should be indicated as shown in Fig. 24-16. Bracket designations or connections may be indicated either by means of notations outside the brackets, as shown in Fig. 24-16, or as shown in Fig. 24-17, by means of a dash line. When the dash line is used to connect brackets, it must be drawn so that it will not be mistaken for a continuation of one of the bracketed lines. The dash line originates in one bracket and terminates in no more than two brackets.

9. All junctions of connecting lines should be shown as single junctions, as in Fig. 24-18A, the preferred method, or as in Fig. 24-18B. When lay-

out considerations prevent the exclusive use of the single junction methods, multiple junctions may be shown as in Fig. 24-18C.

10. Terminal identifications may be added to graphical symbols to indicate actual physical markings which appear on or near part terminations. When the terminals of parts such as relay windings, switches, and transformers are not shown or marked on the part, number or letter identifications may be arbitrarily assigned. When terminal identifications are arbitrarily assigned, the diagram should include an explanatory note and a simplified terminal orientation diagram that relates assigned symbol terminal nomenclature to relative physical locations on the part. For examples of terminal numbering and simplified terminal orientation diagrams, see Fig. 24-19.

When terminals or leads of multilead parts are identified on the part by a wire color code, a letter, number, or geometric symbol, this identification

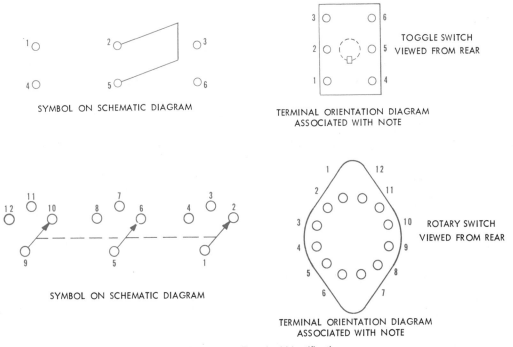

SYMBOL ON SCHEMATIC DIAGRAM

TOGGLE SWITCH
VIEWED FROM REAR

TERMINAL ORIENTATION DIAGRAM
ASSOCIATED WITH NOTE

SYMBOL ON SCHEMATIC DIAGRAM

ROTARY SWITCH
VIEWED FROM REAR

TERMINAL ORIENTATION DIAGRAM
ASSOCIATED WITH NOTE

Fig. 24-19. Terminal identification.

should be shown on or near the connecting line adjacent to the symbol.

11. Wire colors may be indicated by giving either color designations or numerical color codes. Indication of color designations is preferable when many colors and color combinations such as BK-W are to be shown. When numerical color codes are used, care should be taken to avoid confusion with other numerical references. Recommended single and two-letter color designations for use specifically on diagrams and corresponding color identifications by numerical code are:

Wire Color	Designation (ANSI Z32.13)	Numerical Code (ANSI C83.1)
Black	BK	0
Brown	BR	1
Red	R	2
Orange	O	3
Yellow	Y	4
Green	G	5

Wire Color	Designation (ANSI Z32.13)	Numerical Code (ANSI C83.1)
Blue	BL	6
Violet (Purple)	V (PR)	7
Gray (Slate)	GY (S)	8
White	W	9

12. In the specification of resistance, capacitance, and inductance values, the overall objective should be to use forms of expression which will require the identification of the fewest ciphers.

13. When an explanatory note is needed to supplement a given symbol, the note is placed at some convenient distance from the symbol. An arrow is then drawn from the note to the component. The arrowhead should either touch or point to the symbol.

To avoid repeating abbreviations of units of measurement which are generally applicable throughout the diagram, a general drawing note may be used and only the numerical value need be specified on the diagram. A recommended form of the note is

UNLESS OTHERWISE SPECIFIED:
RESISTANCE VALUES ARE IN OHMS.
CAPACITANCE VALUES ARE IN MICROFARADS
    (or PICAFARADS)

24.3 Electrical and Electronic Components and Symbols[3]

The majority of electrical and electronics diagrams incorporate standardized graphical symbols representing a variety of components. The following are the ones most frequently used:

Antenna. The antenna is an arrangement of conductors for the reception or transmission of electromagnetic waves. Symbols for three basic types are shown in Fig. 24-20.

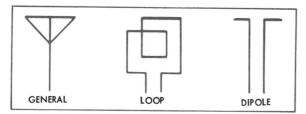

Fig. 24-20. Symbols for antennas.

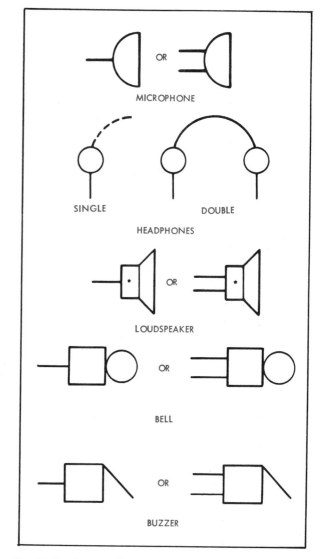

Fig. 24-21. Symbols for devices which convert electrical impulses into sound or sound into impulses.

Audio transducers. These are devices such as microphones, loudspeakers, headphones, bells, and buzzers. Their primary function is to convert electrical impulses into sound or sound into electrical signals. Symbols for these devices are shown in Fig. 24-21.

Battery. A cell is a single unit consisting of dissimilar electrodes and the intervening materials, which converts chemical energy into electrical energy. When cells are combined, they are known as a battery. The symbols for cells and batteries are illustrated in Fig. 24-22.

Capacitor. A capacitor, or condenser, is a set of electrodes, usually in the form of plates, for storing electricity. It is used in a circuit to pass rapidly varying currents or to stabilize the fluctuation of voltage which may be high at one moment and low the next. A capacitor is either of the fixed or variable type, depending on whether one of the sets of plates may be moved in relation to another. The symbols for fixed and variable capacitors are illustrated in Fig. 24-23.

Coil. A coil, referred to as an inductance or inductor when connected to a circuit to increase op-

---

3. *ANSI Y32.2.*

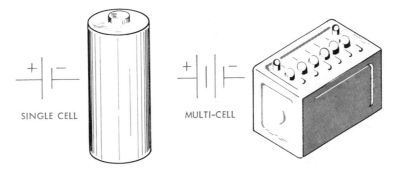

Fig. 24-22. Symbols for cells and batteries.

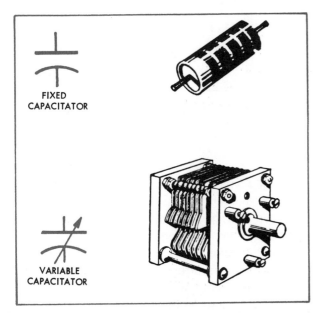

Fig. 24-23. Symbols for fixed and variable capacitors.

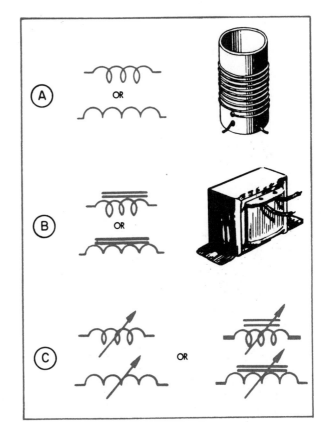

Fig. 24-24. Inductance coil symbols.

position to the flow of alternating current in that circuit, is a spiral loop of conducting material designated to concentrate the magnetic field established by the flow of current through the conductor. It tends to prevent the applied current flowing through a circuit from undergoing variations. Fig. 24-24*A* shows the basic symbol for a coil. Iron, which carries magnetism better than air, is used in some inductors to increase their inductance. Fig. 24-24*B* shows the symbol for an iron-core inductance. Occasionally, instead of a fixed inductance, a variable inductance may be specified. Fig. 24-24*C* shows the symbols for variable inductances.

Conductors. A conductor is a material that allows free passage of electricity. Included in this category are solid and stranded wire, shielded wire, coaxial cable, and wave guides. Symbols for conductors are shown in Fig. 24-25.

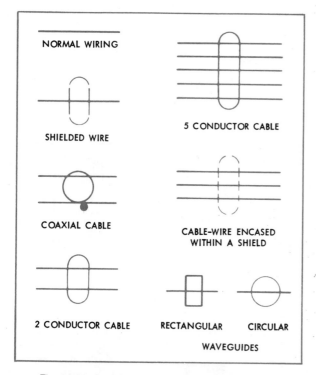

Fig. 24-25. Symbols for some types of conductors.

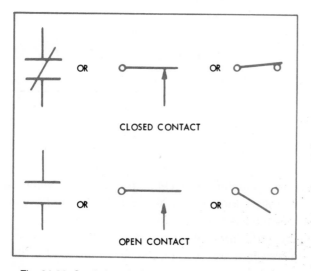
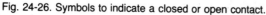

Fig. 24-26. Symbols to indicate a closed or open contact.

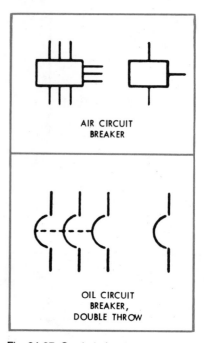

Fig. 24-27. Symbols for circuit breakers.

Circuit contact. When it is necessary to show whether a circuit is in an operative or a non-operative condition, the symbols in Fig. 24-26 are incorporated into the diagram.

Circuit breaker. A circuit breaker is a device used to open a current-carrying circuit automatically under abnormal conditions. Circuit breaker symbols are shown in Fig. 24-27.

Connector. A connector is any device that joins one circuit to another. These devices may be in the form of separate connectors, terminal boards, jacks, plugs, or receptacles. See Fig. 24-28.

Direction of current flow. The direction of current flow or signal is designated by an arrow as in Fig. 24-29.

Feeder Circuit. A feeder circuit connects generating stations or substations, lines from a main board to a distribution panel, or an antenna to a radio transmitter. Rating and type of load is included for each feeder circuit as shown in Fig. 24-30.

Fuse. A fuse is a safety device to prevent overloading a circuit. See Fig. 24-31 for fuse symbols.

Generator. A generator is a rotary machine for converting mechanical energy into electrical power

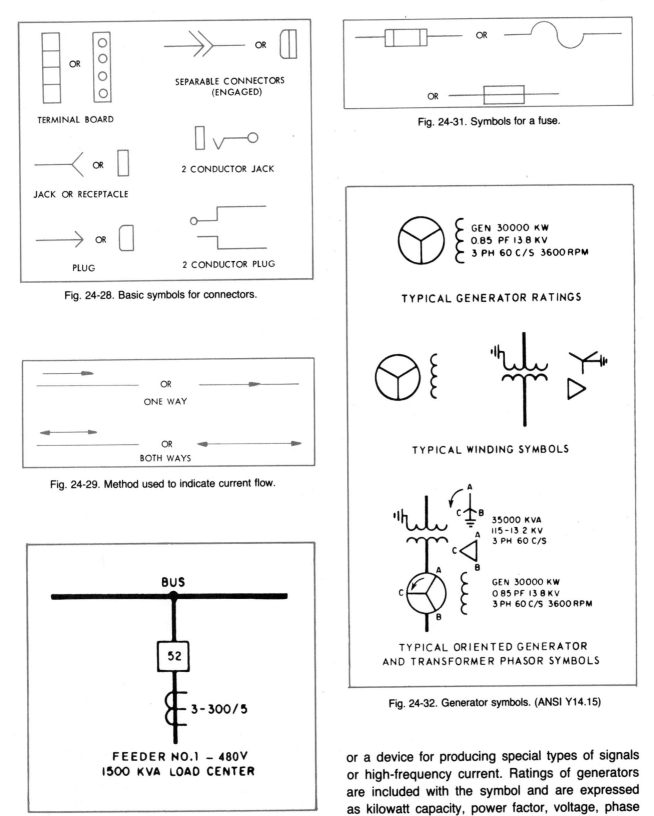

Fig. 24-28. Basic symbols for connectors.

Fig. 24-29. Method used to indicate current flow.

Fig. 24-30. Typical marking of feeder-circuit load. (ANSI Y14.15)

Fig. 24-31. Symbols for a fuse.

TYPICAL GENERATOR RATINGS

GEN 30000 KW
0.85 PF 13.8 KV
3 PH 60 C/S 3600 RPM

TYPICAL WINDING SYMBOLS

TYPICAL ORIENTED GENERATOR
AND TRANSFORMER PHASOR SYMBOLS

35000 KVA
115-13 2 KV
3 PH 60 C/S

GEN 30000 KW
0 85 PF 13 8 KV
3 PH 60 C/S 3600 RPM

Fig. 24-32. Generator symbols. (ANSI Y14.15)

or a device for producing special types of signals or high-frequency current. Ratings of generators are included with the symbol and are expressed as kilowatt capacity, power factor, voltage, phase windings, cycles, and revolutions per minute. See Fig. 24-32.

Ground connections. The two types of ground connections are the signal ground and the chassis. The signal ground indicates an earth, or signal-frequency, ground. The chassis ground refers to the metal structure on which the different components are mounted. See Fig. 24-33 for appropriate symbols.

Grouped leads. When several leads must be joined into a cable, the practice is to show them as in Fig. 24-34.

Instrument, meter, and associated switches. These show the operating source, current or potential, or both, by connecting lines for each device. See Fig. 24-35.

Lamp. A lamp is either of the light-producing type such as the incandescent and fluorescent or of the signaling type. Symbols for various lamps are illustrated in Fig. 24-36.

Meter. A meter is an electrical measuring instrument used to determine electrical quantities such as voltage, current, and power. The basic

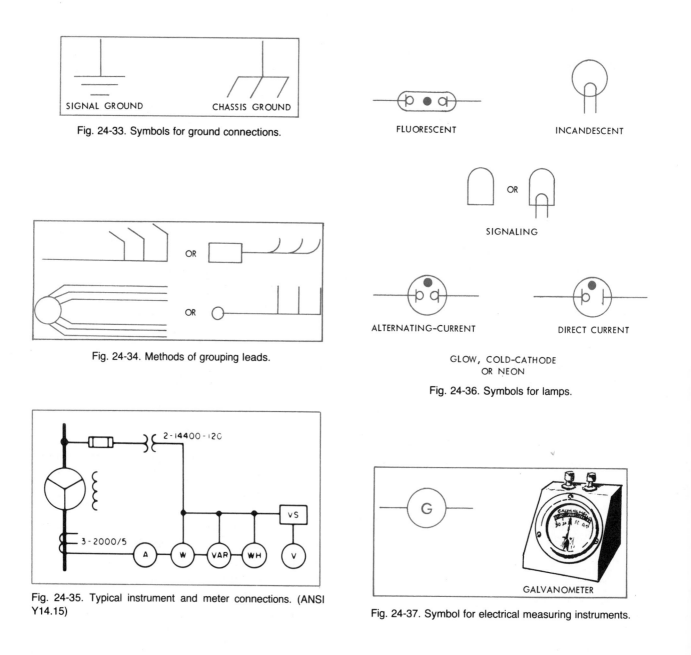

Fig. 24-33. Symbols for ground connections.

Fig. 24-34. Methods of grouping leads.

Fig. 24-35. Typical instrument and meter connections. (ANSI Y14.15)

Fig. 24-36. Symbols for lamps.

Fig. 24-37. Symbol for electrical measuring instruments.

symbol is a circle, which is the same for all meters. The appropriate letter designating the instrument is placed in the circle. See Fig. 24-37.

Polarity. Inclusion of some electronic components on a drawing requires that the circuit polarity at the point of inclusion be known. Current flow through some components is in one direction only, and in certain cases care must be taken to draw the symbol of a component in such a way that the component can function properly if installed according to the drawing. A small plus sign (+) or minus sign (−) near one terminal of a component shows that the polarity there must be as indicated. Typical examples are shown in Fig. 24-38.

Rectifier. A rectifier is used to convert alternating current into direct current. This is accomplished by various devices, including vacuum tubes (diode), mercury-vapor tubes (diode), or solid state units of selenium, copper oxide, germanium, and silicon. The basic symbols for rectifiers are shown in Fig. 24-39.

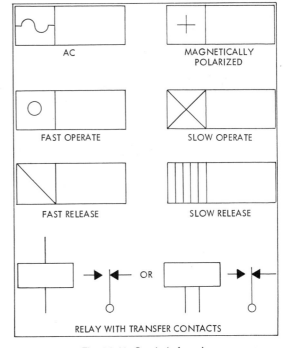

Fig. 24-40. Symbols for relays.

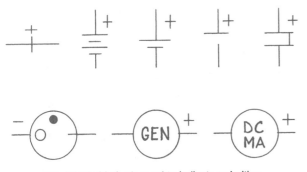

Fig. 24-38. Methods used to indicate polarities.

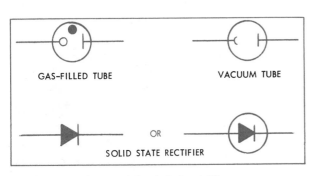

Fig. 24-39. Symbols for rectifiers.

Relay. A relay is an electromagnetic device which, when energized by an electrical current, will force a pivoted contact-bearing arm to make or break a connection in a circuit by an electrical pulse. See Fig. 24-40 for the basic symbols.

Resistors. A resistor is a device that resists, or retards, the flow of current in an electronic circuit. It is used for control, operation, and protection of electrical and electronic circuits. It is either of the fixed or variable type. A fixed resistor provides a fixed value of resistance which cannot be changed. With a variable resistor it is possible to vary the flow of current in a circuit from zero to full value.

If a fixed resistor has one or more additional connections, it is called a tapped resistor. A variable resistor with taps is usually referred to as a rheostat and is found in audio circuits. Symbols for resistors are found in Fig. 24-41. Symbols for two other types of resistance units are shown in Fig. 24-41 also. These are meter shunt and heater elements.

When rotary-type adjustable resistors are used on schematic diagrams, the direction of rotation (clockwise or counterclockwise) is frequently indi-

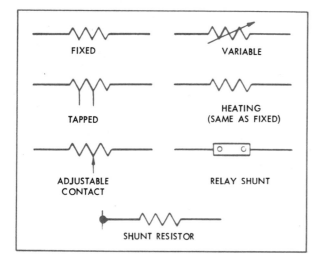

Fig. 24-41. Symbols for resistors.

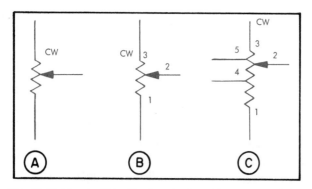

Fig. 24-42. Terminal identification of adjustable resistors. (ANSI Y14.15)

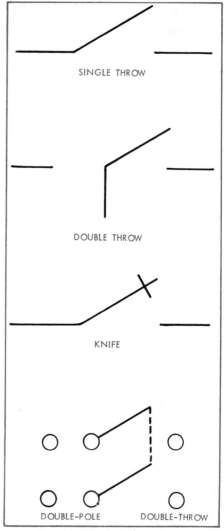

Fig. 24-43. Symbols for switches.

cated. Thus, for a clockwise position the designation CW is placed near the terminal as shown in Fig. 24-42. Numbers may be used with the resistor symbols to show the number of taps. See Figs. 24-42*B* and 24-42*C*.

Switch. The functions of a switch are to open, close, or change connections of a circuit, and to alter the voltage or signals in a circuit. A variety of switches are made: the single-pole single-throw, single-pole double-throw, and others. See Fig. 24-43 for switch symbols.

On schematic diagrams it is necessary to show the relation of switch position to circuit function. For simple toggle switches this may be done with a notation such as ON-OFF. When complex switches are used, position-function relation is shown near the switch symbol or in tabular form at some other convenient location on the drawing. See Fig. 24-44.

Transformer. A transformer consists of two closely coupled coils, primary and secondary, which increase or decrease the voltage in a circuit.

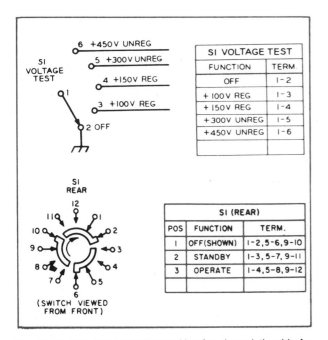

Fig. 24-44. Method of showing position-function relationship for rotary switches. (ANSI Y14.15)

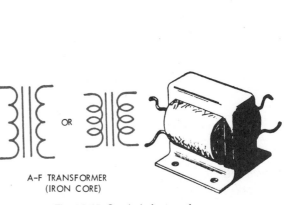

Fig. 24-45. Symbols for transformers.

In electronics the air-core type is used where high-frequency currents are flowing and the iron-core type is used in low-frequency circuits. In order to keep losses at a minimum in intermediate-frequency circuits, transformers with powdered iron cores are often used. Such cores are made by molding processes similar to those employed in powder metallurgy. Symbols for transformers are shown in Fig. 24-45.

Transistor. A transistor functions in much the same way as does a vacuum tube. Due to its smaller size, greater resistance to shock and vibration, and longer life expectancy, the transistor provides a great improvement in reliability over the vacuum tube. Some of the basic symbols currently used for transistors are illustrated in Fig. 24-46.

Vacuum tube. Vacuum tubes serve as signal amplifiers for transmitting and receiving and as rectifiers. There are many varieties of tubes to perform various functions in a circuit. Tubes are often described according to the number of elements they contain. Thus, a diode has two elements, a

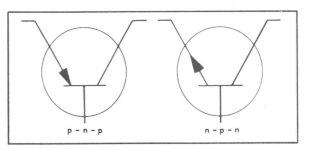

Fig. 24-46. Symbols for transistors.

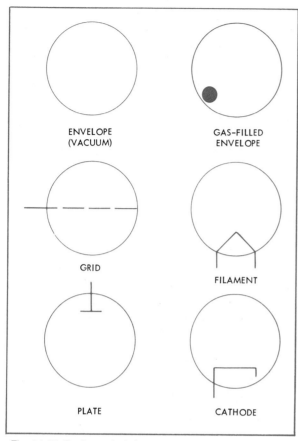

Fig. 24-47. Basic symbols for several vacuum tube elements.

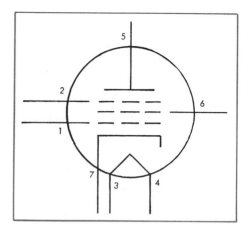

Fig. 24-48. Electronic tube pin identification. (ANSI Y14.15)

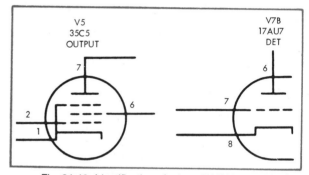

Fig. 24-49. Identification of tubes. (ANSI Y14.15)

triode three elements, a tetrode four elements, and a pentode five elements. Symbols for several tube elements are illustrated in Fig. 24-47.

Electronic tube pin numbers are often shown outside the tube envelope and immediately above or to one side of the connecting line. See Fig. 24-48.

Tubes are also identified by means of letters and numbers. In addition, the circuit function of the tube is sometimes included. This information is placed adjacent to the tube symbol, preferably above it. See Fig. 24-49.

Problems for Section V
Industrial Production Drafting

The problems in this section deal with drafting principles and practices involving

Unit 15. *Product Design and Development*
Unit 16. *Manufacturing Processes and Materials*
Unit 17. *Product Drawing*
Unit 18. *Threads, Fasteners and Springs*
Unit 19. *Representation of Weldments*
Unit 20. *Gears, Chain and Belt Drives*
Unit 21. *Cam Development*
Unit 22. *Piping Drawings*
Unit 23. *Fluid Power Diagrams*
Unit 24. *Electrical and Electronic Diagrams*

Problems for Section V include

Problems 1-24 Product Drawings
Problems 25-39 Threads, Fasteners and Spring Drawings
Problems 40-48 Welding Drawings
Problems 49-64 Gear, Chain and Belt Drive Drawings
Problems 65-73 Cam Drawings
Problems 74-78 Piping Drawings
Problems 79-85 Fluid Power Diagrams
Problems 86-90 Electrical and Electronics Diagrams
Problems 91-108 Product Design Problems

Problems having decimal-inch values may be converted to equivalent metric sizes by using the millimeter conversion chart or the problems may be redesigned by assigning new millimeter dimensional values more compatible with metric production.

Problems 1-24 Product Drawings

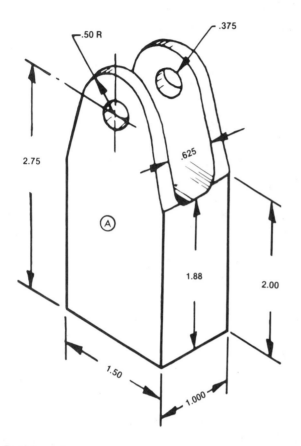

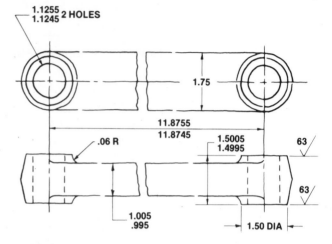

Problem 2
Make a combination forging and machining drawing of the lever arm shown. Select a suitable steel. Make holes parallel to within .002″ and perpendicular to a flat machined surface to within .001″.

Problem 1
Make a detail drawing of the pivot mount to be fabricated from 1.00″ x 1.50″ commercial steel bar, SAE 1020. Analyze the project to determine the machining processes involved. Make the two holes in line and perpendicular to surface *A* to .002″ TIR and bottom surface flat to within .003″.

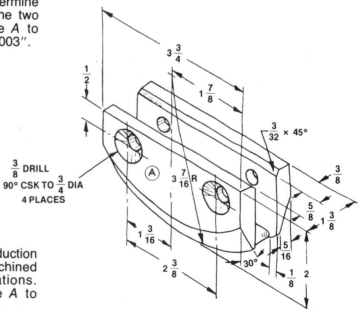

Problem 3
Redesign the counterweight for metric production and prepare a detail drawing. Identify machined surfaces with non-control finish designations. Make holes perpendicular to flat surface *A* to within .001″.

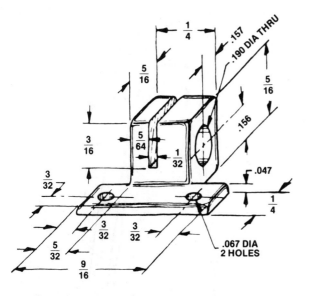

Problem 4

Make a detail drawing of the plastic die-cast gyro bracket shown. The 1/32″ slot is to be machined in the casting. Convert all fractional sizes to decimal-inch values.

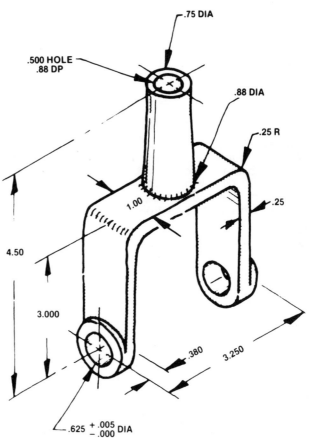

Problem 6

Make a complete detail drawing of the die-cast aluminum rod yoke. Choose suitable fillets and rounds. Supply any necessary sizes. Add geometric tolerancing as deemed applicable.

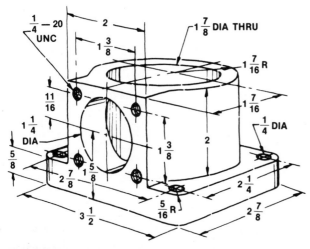

Problem 5

Make both a casting and a machining drawing of the gyro mounting. Material: aluminum. Supply any needed size information. Convert all dimensions to decimal-inch values of two or three places. Make the base lower surface flat to within .002″ and the front flat to within .001″. Show the threaded holes to be at true position within .005″ DIA.

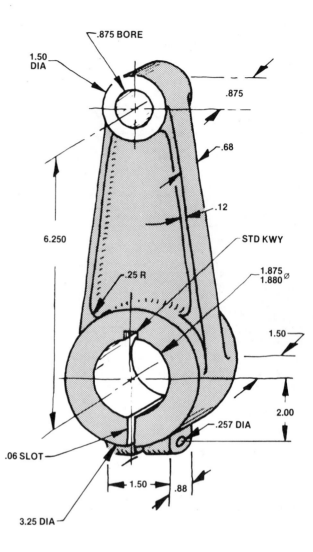

.875 BORE

1.50 DIA

.875

.68

.12

STD KWY

6.250

.25 R

1.875 1.880 ⌀

1.50

2.00

.257 DIA

.06 SLOT

1.50

.88

3.25 DIA

Problem 7
Make a rough casting drawing and a finished machining drawing of the arbor support. Material: steel, SAE 030. Web thickness: 0.25″. Supply any necessary size and surface finish information and apply desired feature controls.

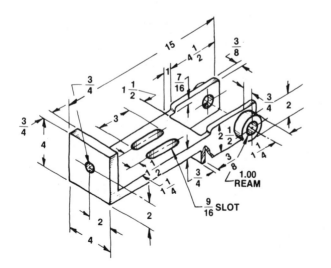

15

4 1/2

3/8

3/4

1 1/2

1

7/16

3/4

2

3

3

8

1.00 REAM

3/4

2

1 1/2

1 1/4

4

3/4

2

4

2

9/16 SLOT

Problem 8
Redesign the pulley bracket for metric production. Dimension completely in mm. Assign surface finishes and any desired feature controls.

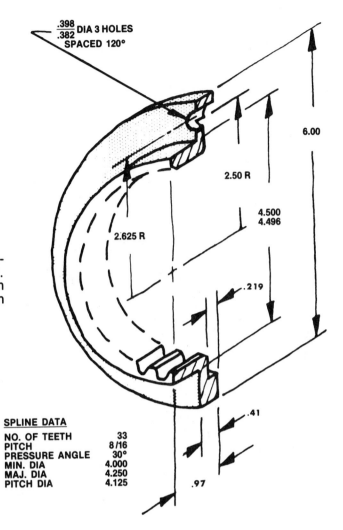

Problem 9

Prepare a forging drawing and a machining drawing for the steel armature mounting hub shown. The spline is a standard SAE involute type. Finish to be 125/ on machined surfaces except 63/ on splines. ∨

SPLINE DATA

NO. OF TEETH	33
PITCH	8/16
PRESSURE ANGLE	30°
MIN. DIA	4.000
MAJ. DIA	4.250
PITCH DIA	4.125

Problem 10

Make stamping drawings of the guide bracket in both flat blank and finish formed conditions. Punch holes while piece is in the flat. Material: ½ hard cold-rolled steel, AISI C1020. Revise all dimensional data for metric production.

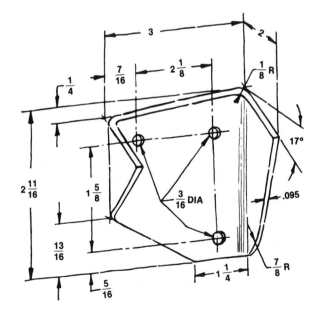

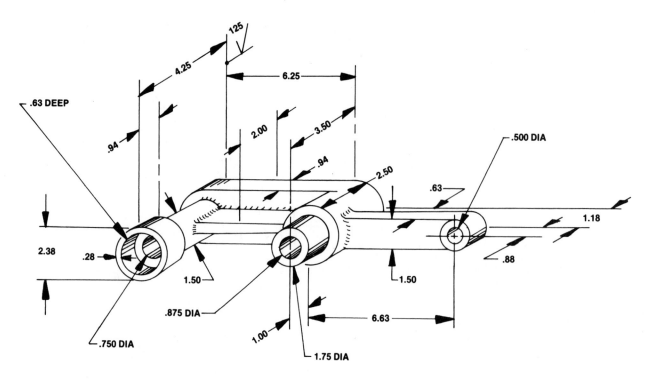

Problem 11
Make a rough casting drawing and a finish machining drawing of the wheel arm. Material: brass, SAE 40.

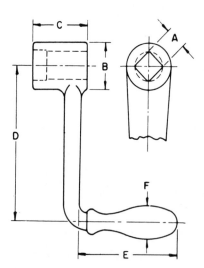

PART No. FINISHED	A	B	C	D	E	F
HC-501	½	1	1¼	1¾	2⅜	¾
HC-502	½	1	1¼	2¼	2½	¾
HC-503	½	1¹⁄₁₆	1⁵⁄₁₆	3½	2¾	⅞
HC-564	⁹⁄₁₆	1¼	1½	3	2⅞	⅞
HC-565	⁹⁄₁₆	1¼	1⅜	4	3⅛	1
HC-626	⅝	1¼	1¹³⁄₁₆	5	3⅛	1
HC-687	¹¹⁄₁₆	1¼	1¹⁵⁄₁₆	6	3⅛	1
HC-758	¾	1⅜	2¹⁄₁₆	7	3⅜	1⅛
HC-879	⅞	1⁹⁄₁₆	2⁷⁄₁₆	8	3⅝	1⅛
HC-880	⅞	1½	2½	9⅛	3¾	1⅛
HC-1011	1	1¾	3	10	4	1¼

Problem 12
Using the tabular data as assigned by instructor, make detail forging and machining drawings of a crank handle. Assume any sizes not indicated. Finish non-machined surfaces with a gray enamel finish.

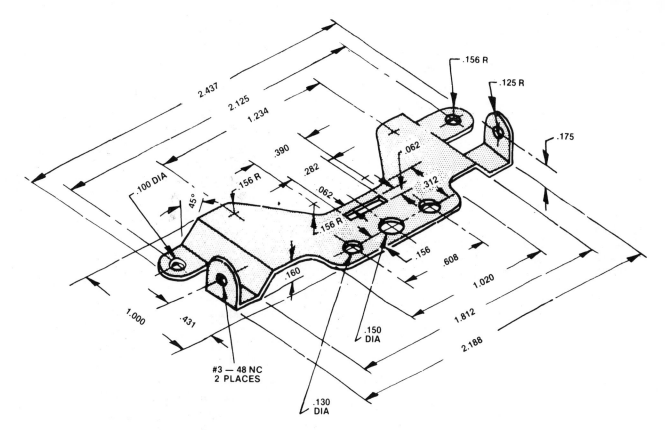

Problem 13
Make a complete detail drawing of the valve
hanger shown. Determine and specify the flat
blank size before forming. Material: .040″ ¼ hard
cold-rolled steel. Finish: black oxide.

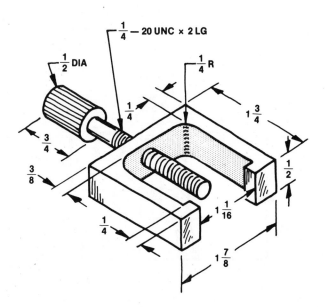

Problem 14
Make a complete set of detail and assembly drawings of the gear puller. Use a straight knurl,
medium pitch, on the screw knob. Choose appropriate materials. Redesign all parts for metric production. Use metric values nearest the fractional
sizes shown. Change unified thread to an equivalent metric thread.

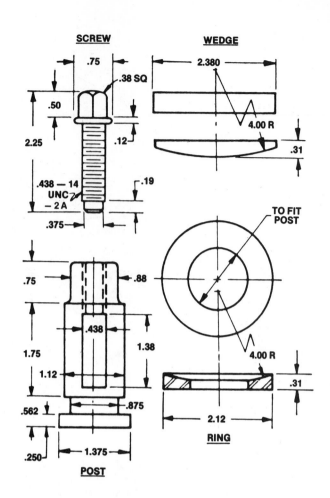

SCREW

.75

.38 SQ

.50

.12

2.25

.438 — 14
UNC
— 2 A

.19

.375

WEDGE

2.380

4.00 R

.31

TO FIT
POST

.75

.88

.438

1.38

1.75

1.12

.562

.875

4.00 R

.31

2.12

RING

.250

1.375

POST

Problem 15
Make an assembly drawing of the tool post with a standard tool bit holder in proper position shown in phantom.

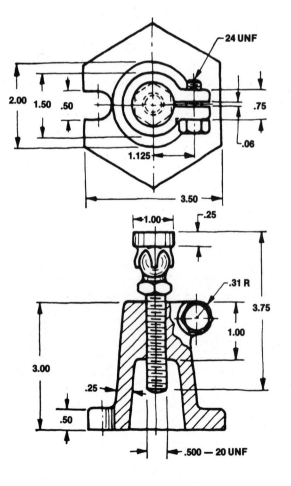

24 UNF

2.00 1.50 .50

.75

.06

1.125

3.50

1.00

.25

.31 R

3.75

1.00

3.00

.25

.50

.500 — 20 UNF

Problem 16
Make detail and assembly drawings of the leveling jack shown. Provide all information needed.

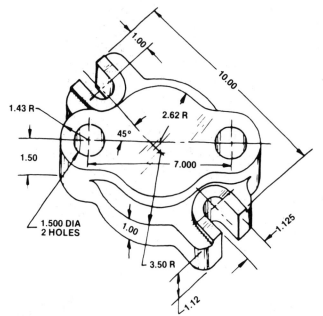

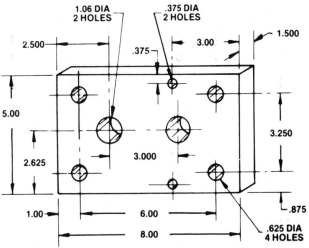

Problem 18

Make a drawing of the plate. Dimension the piece according to the system shown in Fig. 17-29 or 17-30 as assigned by your instructor.

Problem 17

Make both casting and machining drawings, as assigned, of the die shoe. Assign surface finishes and tolerances to machining drawing and design the two holes for accepting force fit. Dowels: each 1.500″ φ. Material: semi-steel casting.

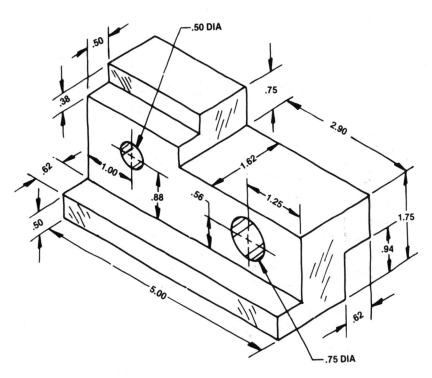

Problem 19

Make a drawing for producing the index block by N/C machining.

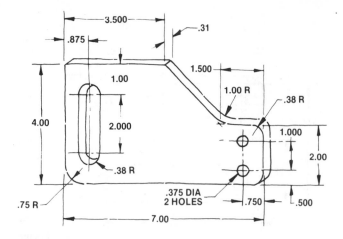

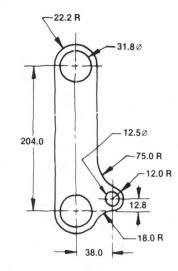

Problem 20
Make a drawing of the gage plate showing dimensions to locate all features for N/C machining.

Problem 22
Make a drawing of the level link for N/C manufacture. Use center line of 12.5 φ hole as O point. All dimensions in mm.

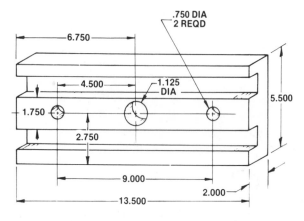

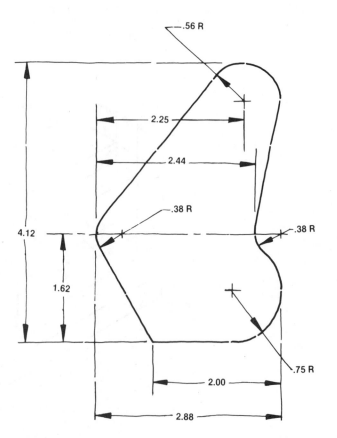

Problem 21
Make a detailed drawing of the slide guide to be produced on an N/C machine tool. All surfaces machined. Datum point 0 established 1.500″ in front and 2.125″ to the left of the workpiece. The two slots parallel to the part's center line, 1.125″ wide and .94″ deep. Three holes .88″ deep.

Problem 23
Provide N/C drawing and program plan for machining contour of blank using .375″ DIA cutter. Choose zero point to suit.

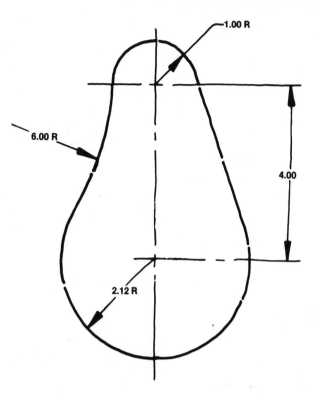

Problem 24

Make a drawing of the cam blank for N/C machining. Prepare a program planning sheet for the center line plan of a .50″ DIA cutting tool. Choose zero point to suit or as designated by instructor.

Problems 25-39 Threads, Fasteners and Spring Drawings

Problem 25

Complete a two-view drawing showing holes for fastenings according to the types and representations indicated below. Determine tap drill sizes, numbers of threads for the series listed, and other machining data as required. Use Appendix tables. Letter proper callouts for each hole.

A—.500-UNC Thru—schematic representation

B—.250-UNF 1″ deep—schematic representation

C—.750-UNC Thru—simplified representation

D—.312-UNF 0.75″ deep—simplified representation

E—Clearance hole for a No. 10-24 flat head machine screw

F—Clearance hole for a flush mounted .375-UNC hex socket head cap screw.

Problem 26

Make two-view drawings of the following standard fasteners, showing thread representations as assigned by your instructor. Include correct nomenclature.

A—.750-10UNC-2A semi-finished hex head bolt 4″ long with assembled spring lock washer and hex nut

B—1.000-8UNC square head bolt 4″ long with plain washer and square nut

C—.375-24UNF-3A hex socket head cap screw 2″ long

D—.250-20UNC-2A round head machine screw, 1.50″ long

E—.500-20UNF-2A slotted set screw, 1.50″ long, cone point

F—.750-10UNC-2A flat head cap screw, 2.00″ long.

Problem 27

Make a two-view dimensioned drawing of a steel tapping pad, ½″ x 2½″ x 6″ showing two holes, threaded .625-18UNF-2B on center line spaces 1.50″ from end and 3.000″ between centers. Dimension completely including tap drill size.

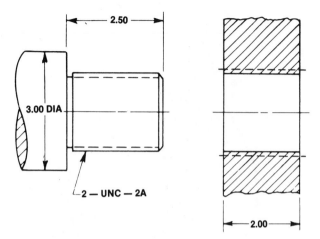

2.50

3.00 DIA

2 — UNC — 2A

2.00

Problem 28

Redesign the plug and threaded hole shown for the nearest size standard metric thread. Show threads schematically. Include all dimensions on the drawing.

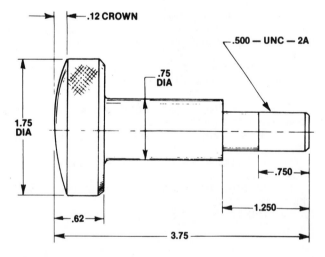

.12 CROWN

.500 — UNC — 2A

.75 DIA

1.75 DIA

.750

1.250

.62

3.75

Problem 29

Draw the necessary views of the hand knob, giving complete dimensional data. Provide a gripping surface on the knob by providing a 21 P diamond knurl as shown. Material: steel, SAE 1115. Show thread simplified.

Problem 30

Make a two-view drawing of a steel shaft 1.50″ DIA, 18″ long having .125″ x 45° chamfered ends. Show a keyway for a #608 woodruff key located 3.500″ from one end. Include a hole for a #6 taper pin to be inserted through a 2.50″ O.D. collar positioned 3.50″ from the opposite end and in line with the keyway.

Problem 31

Make two views of a cast steel collar 4″ O.D. x 2″ long having a 2.502″/2.498″ bore. Show a keyway for a standard square key for this size bore. Show a .625-18UNF-2A square head set screw centered on collar opposite the keyway. Rounds .25″ R. Include all dimensions.

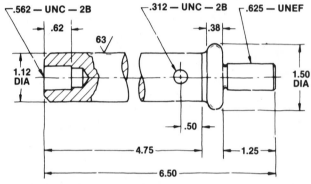

.562 — UNC — 2B

.312 — UNC — 2B

.625 — UNEF

.62

63

1.12 DIA

1.50 DIA

.50

4.75

1.25

6.50

Problem 32

Prepare a dimensioned drawing of the shaft showing all threads by simplified representation. Supply any dimensions and information needed for production. Material: steel, SAE 1115.

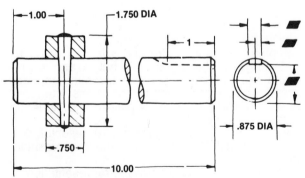

1.00

1.750 DIA

1

.750

.875 DIA

10.00

Problem 33

Make detail drawings of the shaft and collar. Supply data for a keyseat for a standard square key. Determine the proper size taper pin and dimension the hole to suit. Material: SAE 1040 steel.

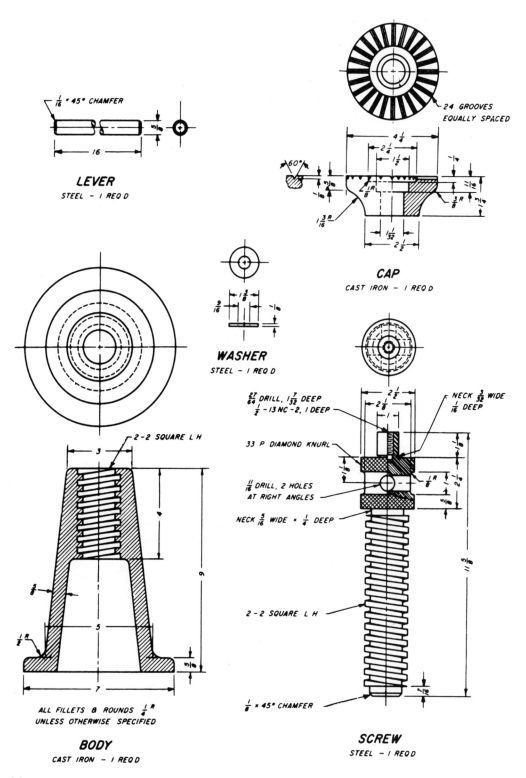

$\frac{1}{16}$ × 45° CHAMFER

16

LEVER
STEEL – 1 REQ D

24 GROOVES
EQUALLY SPACED

60°

$4\frac{1}{4}$

$2\frac{1}{4}$

$1\frac{1}{2}$

$\frac{1}{4}$

$\frac{3}{8}$R

$1\frac{3}{16}$R

$1\frac{1}{32}$

$2\frac{1}{2}$

CAP
CAST IRON – 1 REQ D

$\frac{9}{16}$ $1\frac{3}{8}$ $\frac{1}{8}$

WASHER
STEEL – 1 REQ D

$2\frac{1}{2}$

$2\frac{1}{8}$

1

NECK $\frac{3}{32}$ WIDE
$\frac{1}{16}$ DEEP

$\frac{27}{64}$ DRILL, $1\frac{7}{32}$ DEEP
$\frac{1}{2}$ – 13 NC – 2, 1 DEEP

33 P DIAMOND KNURL

$\frac{11}{16}$ DRILL, 2 HOLES
AT RIGHT ANGLES

$\frac{3}{8}$R

$2\frac{1}{4}$

NECK $\frac{5}{16}$ WIDE × $\frac{1}{4}$ DEEP

2 – 2 SQUARE L H

$\frac{1}{8}$ × 45° CHAMFER

$\frac{7}{16}$

$11\frac{5}{8}$

SCREW
STEEL – 1 REQ D

2 – 2 SQUARE L H

3

4

9

5

7

$\frac{1}{2}$R

ALL FILLETS 8 ROUNDS $\frac{1}{4}$R
UNLESS OTHERWISE SPECIFIED

BODY
CAST IRON – 1 REQ D

Problem 34
Make detail drawings of each of the screw jack components. Indicate finished surfaces with appropriate symbols. In addition make a half-section assembly drawing of the completed project with the screw approximately 1 inch above its lowest position. Provide a standard hex head cap screw to fasten the cap to the screw. Change all dimensions to decimal values.

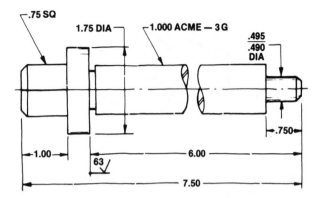

Problem 35

Redesign the lead screw for metric production. Maintain the standard Acme thread which should be shown by detailed representation. Dimension completely.

Problem 36

Make a two-view drawing of a coil compression spring to these specifications:
O.D.: 1.500″
Free length: 3.50″
No of coils: 8
Type of ends: plain ground
Finish: plain
Material: No. 10 gage (.135) hard drawn steel spring wire.

Problem 37

Make a two-view drawing of a coil compression spring having 10 active coils 2.750″ free length, closed and ground ends, 1.125 O.D. of .105″ hard drawn steel spring wire.

Problem 38

Make a two-view drawing of a coil extension spring having the following specifications:
O.D.: .934″
24 coils
Ends: single loop over center
Material: #18 W & M Ga. (.048) oil-tempered steel spring wire
Finish: cadmium plate .0003″ thick. Provide complete dimensions and notes.

Problem 39

Make a two-view drawing of a coil extension spring having the following specifications:
O.D.: .750″
14.5 coils, wound LH
Ends: single loop over center
Material: .080 oil-tempered steel spring wire
Finish: black Japan
Provide complete dimensions and notes.

Problems 40–48 Welding Drawings

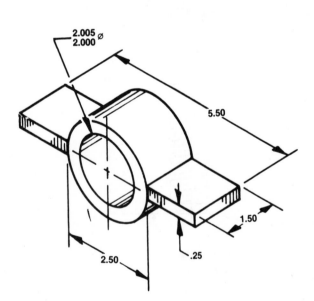

Problem 40

Make the necessary views of the anchor collar. Weld all joints using fillet welds. Material: SAE 1020 steel. Show all dimensions.

Problem 41

Make a two-view drawing of the angle bracket. Material: steel plate. Choose appropriate welded joints. Dimension completely.

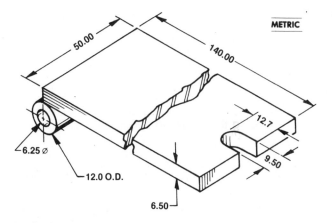

METRIC

Problem 42
Make a dimensioned drawing of the jig lid. Use 12 mm DIA cold-rolled steel shafting for hinge and 6.5 mm mild steel for lid. Choose proper welded joints.

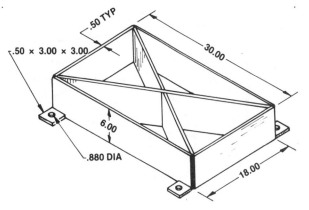

Problem 44
Make a dimensioned drawing of the welded machine base. Provide tabs for anchoring base with .75 DIA bolts. Show symbols for all welds.

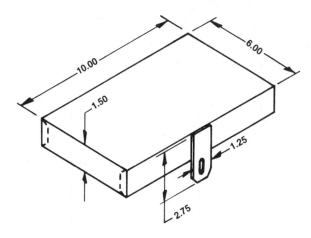

Problem 43
Make a two-view drawing of the cover and hasp assembly. Spot weld as needed. Show all weld symbols. Assume all sizes not given.

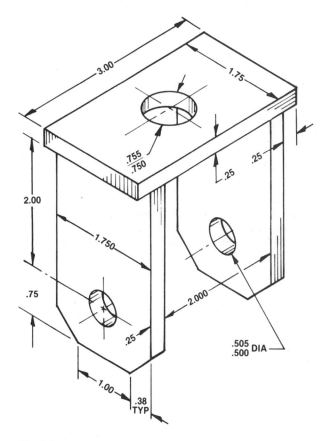

Problem 45
Design a caster wheel bracket as shown. Show all dimensions and weld symbols. Choose appropriate weld joints. Material: SAE 1020 steel.

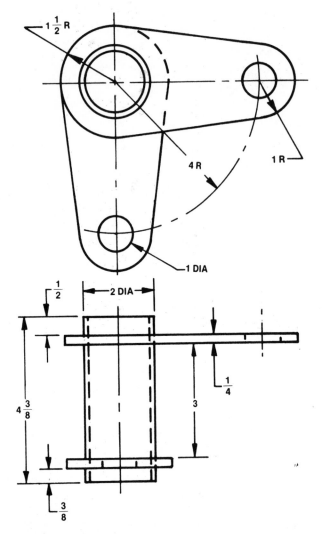

Problem 46
Make a welded assembly drawing of the lever arm shown. Use steel tubing having a 0.125″ wall and mild steel plate. Choose welds. Change all sizes to mm.

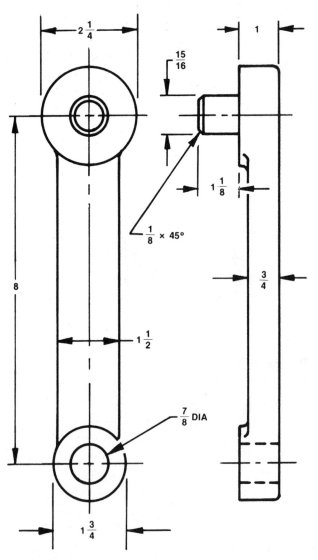

Problem 47
Redesign the forged steel link shown for welded fabrication. Use HR steel bar and steel shafting. Change all sizes to mm.

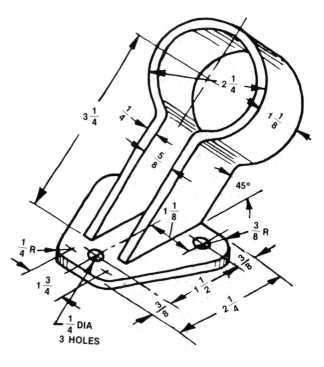

Problem 48
Make a drawing showing both an assembly of the welded bracket shown as well as individual detail drawings of the two parts. Choose suitable weld joints and provide any fabrication dimensions not shown. Material: steel, SAE 1010-1020. Convert all dimensions to two-place decimals.

Problems 49-64 Gear, Chain and Belt Drive Drawings

Problem 49
Make a detail drawing of a 4-pitch, 14½° involute full depth tooth spur gear having the following specifications: 12 teeth, 1.50″ face, 2.00″ diameter hub, and 1.188″ diameter bore projecting .50″ on one side. Material: cast steel SAE 3120.

Problem 50
Make complete detail drawings of the problems given in Tables I and II as assigned. Each gear is an involute full depth tooth spur gear with a pressure angle of either 14½° or 20°. Supply any design data not specified.

TABLE 1
Spur Gears (inch)

Problem	Number of Teeth	Diametral Pitch	Face	Style	Bore	HUB O.D.	Proj.
A	16	5	2.50	Plain	1.062	2.38	.50
B	24	3	3.00	Webbed	1.437	3.00	.75
C	35	5	2.50	Webbed	1.250	2.75	0
D	20	8	1.25	Plain	.875	1.50	.12
E	32	16	.31	Spoked	.312	.75	.12
F	18	8	1.50	Plain	.937	1.50	0
G	96	10	1.12	Spoked	1.125	2.25	.38

TABLE II
Spur Gears (metric)*

Problem	Number	Module	Face	Style	Bore	HUB	
						O.D.	Proj.
A	24	1.5	12	Plain	14	–	–
B	20	1.5	18	Plain	14	25	4.4
C	48	2.0	20	Webbed	28	50	8
D	25	2.5	28	Plain	20	–	–
E	16	2.5	28	Plain	12	30	20
F	40	3.0	30	Webbed	35	60	15

* dimensions in mms

Problem 51

Make a single-view layout drawing of an involute spur gear rack and pinion in mesh. Show a minimum of 6 teeth on each part. Use the approximate method of constructing the involute gear tooth. Given: 20° pressure angle, 5-pitch, 32 gear teeth. Calculate the addendum, whole depth dimensions, in addition to the linear pitch of the rack. Choose suitable bore and hub sizes with standard keyway for pinion.

Problem 52

Make a single-view drawing of a 3-pitch spur gear, pinion and rack in mesh. Place the large gear between the pinion and the rack. Develop the involute tooth profile accurately and show all teeth. A template may be used to transfer tooth outlines. The gear and pinion have 24 and 15 full depth teeth respectively and 20° involute form. Use plain style gears with 1.438″ diameter bores and standard keyways.

Problem 53

Make complete detail drawings of the problems in Table III as assigned by your instructor. Supply all needed data and specifications not given.

Problem 54

Make a single-view drawing of a pair of miter gears in mesh. The gears are of steel and have the following specifications: 6 diametral pitch, 20° pressure angle, 1.188″ bore, 42 teeth, 3.75″ mounting distance, 3.50″ diameter hub projecting approximately 1.50″. Show all dimensions.

Problem 55

Construct a detail drawing of a right-hand, 12-pitch, 14½° bronze wormgear having 80 teeth. The face is 1.50″ wide and the bore 1.188″. A 2.00″ diameter hub projects .38″.

Problem 56

Make a detail drawing of a worm for the wormgear in Problem 55. The worm has a 1.75″ pitch diameter and a .500″ bore.

TABLE III
Bevel Gears

Problem	Pitch	P.A.	Number of Teeth	Bore	Mounting Distance	HUB	
						Ø	Proj.
A	4	20°	42	1.125	4.00	3.75	1.50
B	4	20	14	1.125	7.25	3.25	1.94
C	6	20°	48	1.125	3.44	3.25	1.00
D	10	20°	30	.750	2.25	2.50	1.00
E	12	20°	18	.50	1.875	1.25	.656
F	8	20°	72	1.125	3.25	3.00	1.69

Problem 57

Make a two-view assembly drawing of a worm and wormgear with the following specifications: 32-pitch, double thread, .218″ face, 100 teeth on the wormgear. The wormgear is spoked having a 1.06″ diameter hub that projects .31″ on each side. The gear is bored to .312″ diameter. The worm has .438″ pitch diameter and is bored to .218″ diameter.

Problem 58

Using Fig. 20-33 as a guide, find the RPM of shaft 3 if gear *A* has 28 teeth and rotates at 2400 RPM, gear *B* has 96 teeth, gear *C* has 32 teeth and gear *D* has 112 teeth. What is the value of the gear train? Is the direction of shaft 3 opposite or the same as that of shaft 1?

Problem 59

Using Table III in Unit 20, determine the shaft outside diameter and housing inside diameter with tolerances for the following bearings: (1) Bearing No. M307K-CR and (2) Bearing No. M312K-MBR.

Problem 60

A bearing is required to have a radial load capacity of 2030 pounds at 1800 RPM. Using Table III in Unit 20, select a bearing to meet this condition. Design a partial end section of a shaft for mounting this bearing. Show complete dimensions for the diameter on which the bearing is to be mounted, shoulder radius, and shoulder diameter.

Problem 61

Determine the RPM of gear *D* in Fig. 20-33. Gear *A*, 24 teeth, turns at 20 rpm; gear *B* has 30 teeth; gear *C*, 18 teeth; gear *D*, 60 teeth.

Problem 62

Determine the length in pitches of a .75″ pitch roller chain in a drive having a center distance of 36″ and sprockets with 24 and 36 teeth.

Problem 63

A 1750 RPM electric motor powers a V-belt drive having an approximate center distance of 22″. The speed of the driven sheave is to be 650 RPM. Determine the pitch diameters for both sheaves, the proper type V-belt, the exact center distance and the ratio of the drive.

Problem 64

A speed reducer having an output of 55 RPM and driven by an electric motor is to be mounted on a metal plate so the reducer is stationary and the motor can be adjusted for slack take-up in the V-belt. Specifications for a commercially produced speed reducer, motor, V-belt with sheaves, and a roller chain sprocket which will be placed on the reducer output shaft should be obtained from a vender's current catalog.

Required: 1. Make a dimensioned drawing of the drive plate showing reducer, motor, etc., in phantom. 2. Design a belt guard to be mounted on the drive plate as a safety device.

Given Specifications:

   Drive plate: .50″ thick steel, length and width to be determined

   Speed reducer: helical gear, single reduction, 23.95:1 ratio, horizontal parallel shafts

   Motor: ¾ HP, 1725 RPM, single-phase, 60 cycles, 115/230 V, totally enclosed, fan cooled; frame size to be determined

   V-Belt (B section): length and center distance to be determined

   Sheaves: driver—3″ pitch diameter, type B, single-groove; driven—to be determined

   Sprocket (reducer output): 5.101 PD, type B single strand for No. 40, 0.50″ pitch American Standard roller chain.

Problems 65-73 Cam Drawings

Problem 65

For each cam assigned in the problems shown in Table IV on the next page, make a displacement diagram and a complete detail drawing. In each case provide a standard keyway and lightening holes as needed for balancing. Style of cam may be cast or welded construction depending upon size.

Problem 66

Construct the pitch curve and working curve for a plate cam having a flat-face follower; the face is perpendicular to the center line of the follower. The base circle diameter is 3.500″. The follower is offset .50″ to the left of the center of the camshaft. Select the diameter of the hub, camshaft and size of the keyway. Follower movement outward is 1.250″ with uniform accelerated and retarded motion 120° of cam rotation. Return in 150° with harmonic motion and dwell for 90°.

TABLE IV

Prob	Base Cir Dia	Hub O.D.	Bore Ø	Hub Project.	Cam Rot.	FOLLOWER MOTION			Follow
						Rise	Dwell	Fall	
A	3.25	1.75	1.00	.50 B.S.		1.125 0° – 120° Parabolic	120°–165°	120° .25 str. Drop 165° – 360° Harmonic	Pointed
B	3.50	–	1.25	–		1.50 0° – 180° Modified Uniform	None	180° – 360° Harmonic	.75 O
C	3.50	1.88	1.125	.75 B.S.		1.50 0° – 90° Modified Uniform 1.00 180° –240° Harmonic	90°–180°	240°–360° Parabolic	1.00 O
D	3.75	2.00	1.25	.75 B.S.		.875 30° – 120° Harmonic .750 165°–225° Parabolic	0° – 30° 120°–165°	225°–360° Harmonic	.875 O
E	3.00	–	1.00	–		1.75 0° – 120° Modified Uniform	120°–180°	180°–360° Harmonic	.75 O
F	3.25	1.88	1.00	.50 B.S.		1.50 0° – 180° Modified Uniform	180°–200°	180° .50 str. Drop 200 – 360 Modified Uniform	Pointed

Problem 67
Construct the front and top views of a cylindrical cam with a 3.500″ diameter cylinder, 2.75″ high, roller .500″ diameter, roller depth .375″, camshaft .750″ diameter, follower rod .62″ wide and .31″ thick. Motion: up 1.500″ in 180° with harmonic motion, return to initial position in 180° using the same motion. The cam rotates clockwise.

Problem 68
Same as Problem 67 except up motion to be 1.750″ using parabolic motion in 120°, dwell 60°, down with harmonic motion in 180°.

Problem 69
Design a disk edge cam that will move a roller follower 1.375″ by uniform motion and return in one revolution. Diameter of shaft 1.250″, diameter of hub 2.50″, roller follower 1.250″ diameter, hub length 1.62″, cam thickness .75″. Use a standard flat key. See figure.

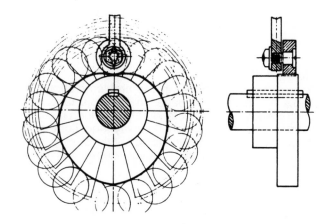

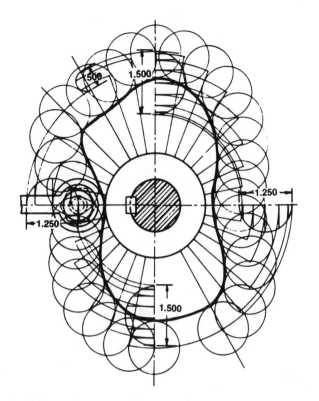

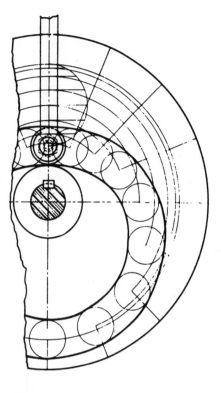

Problem 70

Design a disk edge cam that will move a recip- rocating roller follower by harmonic motion through a 1.250″ rise in 45°, dwell 15°, .500″ rise in 30°, 1.500″ drop in 90°, 1.250″ rise in 60°, dwell 30°, and return to origin through the remainder of the revolution. Shaft diameter 1.250″, hub diameter 2.500″, hub length 1.75″, cam thickness .750″. Use a standard flat key. See figure.

Problem 72

Design a disk path cam that will move a roller fol- lower through 2.250″ in 180° by harmonic motion and return to origin. Shaft diameter 1.00″, hub diameter 2.000″, roller diameter 1.00″, plate diameter 9.12″. Draw plan view only. See figure.

Problem 71

Design a uniform motion plate cam to be welded to a 2.50″ diameter x 1.125″ long hub that will ac- tuate a 1.00″ diameter x .375″ thick roller follower to rise .750″ in 60°, dwell 25°, rise 1.000″ in 75°, dwell 20°, fall 1.250″ in 45°, dwell 15°, rise 1.00″ in 30°, dwell 20°, drop through the remaining angle to the starting point. Cam thickness .38″, shaft diameter 1.250″. See figure.

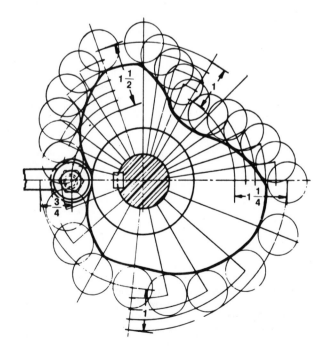

Problem 73

Design a cam to move a 1.250″ diameter roller follower through 1.500″ uniform rise in 45°, 1.000″ harmonic rise through 30°, dwell through 15°, .750″ uniform rise in 45°, dwell 30°, harmonic drop through the remaining angle to origin. Draw plan view only. Shaft diameter 1.250″ and hub diameter 2.500″.

Problems 74-78 Piping Drawings

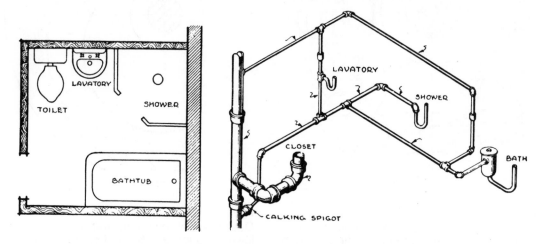

Problem 74

Prepare a single-line plumbing layout in isometric for the bathroom shown. Use standard symbols.

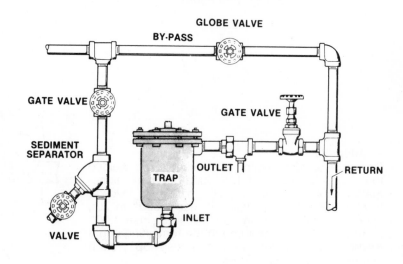

Problem 75

Make a single-line diagram of the piping arrangement for an inverted open float steam trap as shown. Details with no standard symbols should be made in outline and noted.

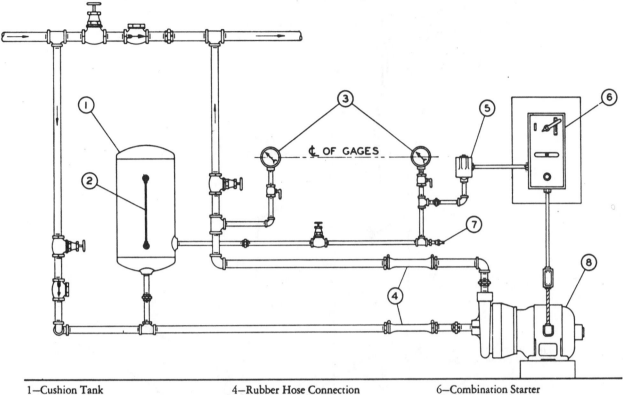

1—Cushion Tank	4—Rubber Hose Connection	6—Combination Starter
2—Water Gauge	5—Pressure Switch	7—Tank Valve
3—Pressure Gauge		8—Centrifugal Pump

Problem 76

Make a single-line schematic drawing of the assembly shown. Use standard symbols where applicable.

Problem 77

From the engineer's proposal sketch shown, prepare an assembly drawing of the piping for the tubular conveyor system. Omit receiving and discharge units. Make a detail drawing of the pipe bends, which have a typical radius of 24 inches with 6 inch tangent lengths. Determine pipe lengths and prepare a list of pipe sections, bends, flanges, bolts and nuts. All pipe to be 4″ std, and all flanges 4″ std, 125 lbs, slip-on type.

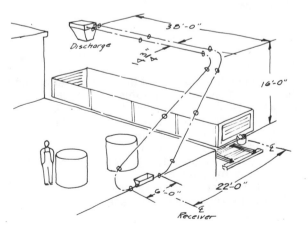

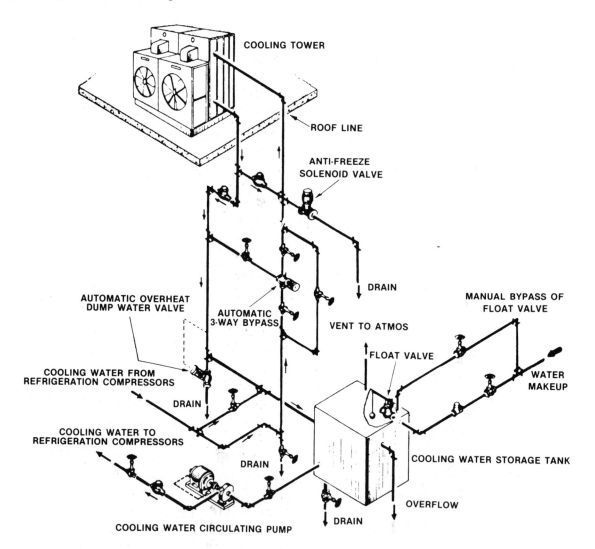

COOLING TOWER

ROOF LINE

ANTI-FREEZE
SOLENOID VALVE

DRAIN

AUTOMATIC OVERHEAT
DUMP WATER VALVE

AUTOMATIC
3-WAY BYPASS

VENT TO ATMOS

MANUAL BYPASS OF
FLOAT VALVE

FLOAT VALVE

COOLING WATER FROM
REFRIGERATION COMPRESSORS

WATER
MAKEUP

DRAIN

COOLING WATER TO
REFRIGERATION COMPRESSORS

DRAIN

COOLING WATER STORAGE TANK

COOLING WATER CIRCULATING PUMP

OVERFLOW

DRAIN

Problem 78

Prepare a single-line isometric diagram of the piping connection for a non-freeze cooling tower shown. Use standard symbols and choose appropriate distances to clearly illustrate the system's arrangement.

Problems 79-85 Fluid Power Diagrams

Problems 79-82

Make complete diagrams of the fluid power circuits shown. Use standard graphical symbols for all components, lines, etc. according to the following item identification code. Select suitable scale.

Item	Component
A	Cylinder, single acting
B	Cylinder, double acting, single end rod
C	Valve, manual shutoff
D	Motor, electric, fixed displacement
E	Pump, single, fixed displacement
F	Filter

G	Reservoir
H	Valve, check
J	Valve, maximum pressure (relief)
K	Valve, directional, 2-position, 4 connection, manual
L	Valve, directional, 3-position, 4-connection, closed center, manual
M	Valve, flow-rate control, variable pressure compensated
N	Accumulator
P	Pressure Gage

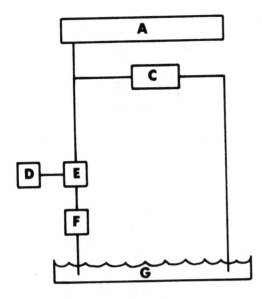

Problem 79.

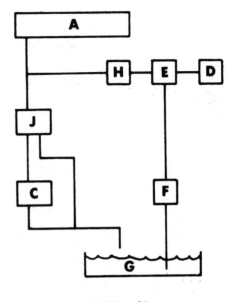

Problem 81.

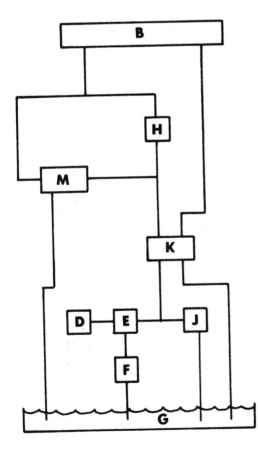

Problem 80.

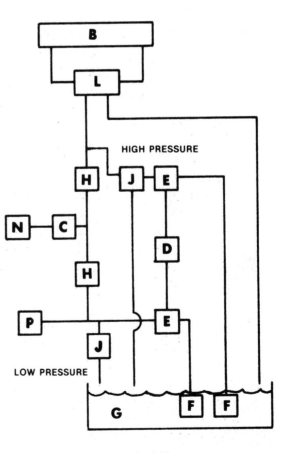

Problem 82.

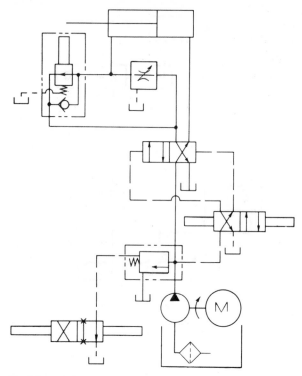

Problem 83
Identify the units used in the circuit shown. Determine the shapes of these units from catalogs of fluid-power components and make a pictorial diagram of the circuit.

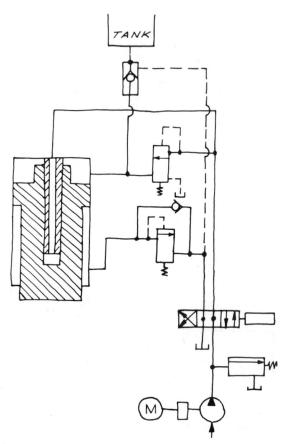

Problem 85
Make a diagram of the hydraulic circuit for the press shown.

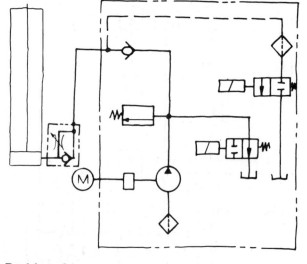

Problem 84
Reproduce the sketch of the hydraulic circuit for the industrial lift truck shown.

Problems 86-90 Electrical and Electronics Diagrams

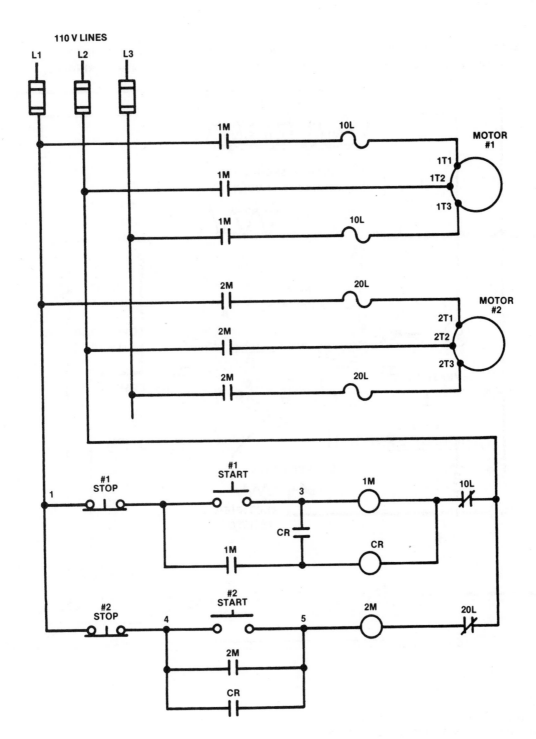

Problem 86
Make a diagram of the motor control unit shown.

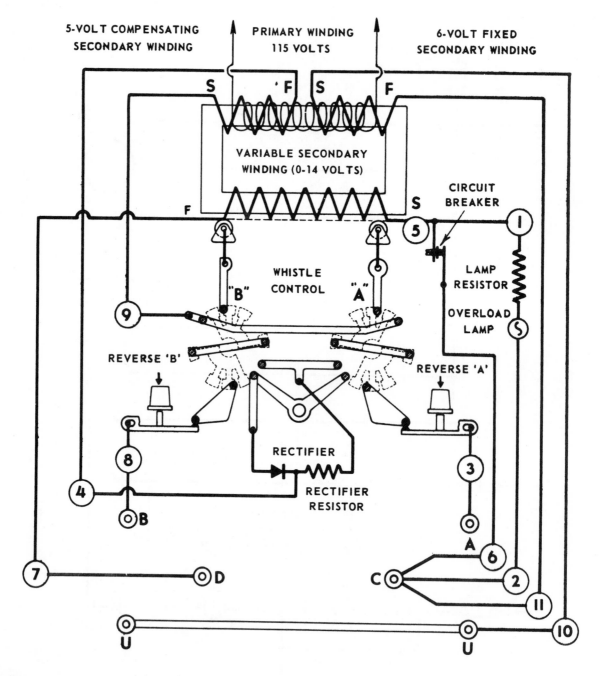

5-VOLT COMPENSATING SECONDARY WINDING

PRIMARY WINDING 115 VOLTS

6-VOLT FIXED SECONDARY WINDING

VARIABLE SECONDARY WINDING (0-14 VOLTS)

CIRCUIT BREAKER

WHISTLE CONTROL

"B"

"A"

LAMP RESISTOR

OVERLOAD LAMP

REVERSE 'B'

REVERSE 'A'

RECTIFIER

RECTIFIER RESISTOR

Problem 87
Prepare a drawing suitable for publication in an instruction manual of the wiring for a *KW* transformer unit. Finished size of the illustration is to be 4.50″ x 4.75″.

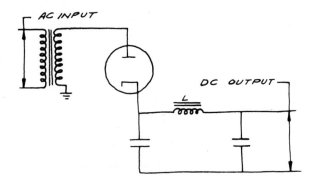

Problem 88
Make an instrument schematic drawing of the DC power supply circuit shown.

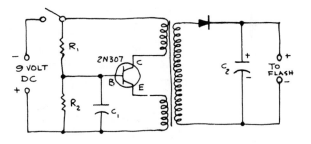

Problem 89
Make an instrument drawing of the photo-flash power supply unit shown.

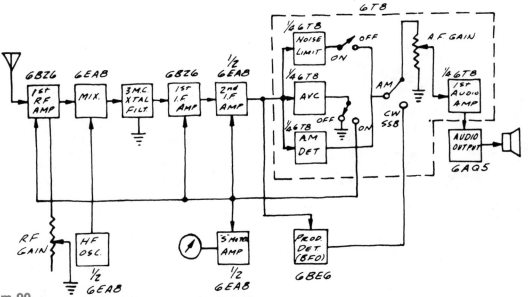

Problem 90
Draw an instrument block diagram of the 8-tube mobile receiver shown.

Problems 91-108 Product Design Problems

In working out each of the listed problems, students are expected to show some degree of resourcefulness in applying the basic design principles described in this chapter. A necessary requirement is the preparation of essential design sketches and/or complete detail and assembly drawings as specified by the instructor.

For many of the problems, considerable research will be required to arrive at a successful solution. Special consideration should be given to the most economical means of manufacturing the design product.

Students undertaking a design project should attempt to channel their efforts through four stages, namely:

1. Identification of the problem—clearly define the objectives, that is, identify the needs and function of the product.

2. Research—collection of ideas.

3. Critical analysis—review and evaluation of the design in terms of market appeal, function, strength, economy of production, and so on.

4. Preparation of final drawings and specifications for manufacturing purposes.

Problem 91

Design a child's toy similar to the cycle shown for a 1½- to 4-year-old. The seat is 8″ from the floor. Overall sizes not to exceed 18″ x 18″ x 15¼″ for packaging purposes. Use commercially produced wheels, handle grips, and fastenings. Make detail and assembly drawings and parts list.

Problem 92

Design a drill press vise similar to that shown in the sketch and prepare complete detail and assembly drawings. Choose all design sizes not indicated. Provide hardened steel jaw plates as extra equipment. Specify materials.

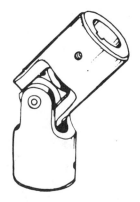

Problem 93

Design and prepare a set of drawings for a universal joint similar to the one shown in the sketch capable of accommodating .812″ DIA shafts.

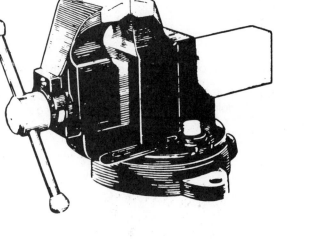

Problem 94

Design and prepare complete working drawings including assemblies of a machinist's vise similar to the one shown in the sketch. Provide a jaw width of 3.50″ with a maximum opening of 5.62″. Use an Acme thread on the lead screw.

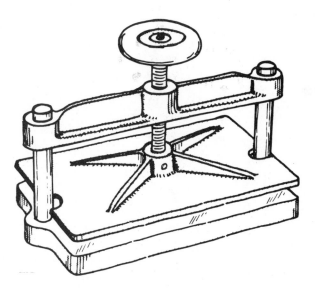

Problem 95

Design the components for a modelmaker's press, as shown in the sketch having a platen 6.38″ x 10.50″ and a lead screw 0.625″ DIA. All castings gray iron. Prepare a complete set of detail and assembly drawings.

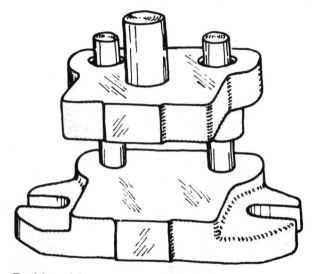

Problem 96
Design a production-type die set, such as the one shown in the sketch having a die space left to right of 4.50″ and front to back of 3.25″; the punch holder 1″. Make hardened steel guide pins 1″ DIA x 5.50″ long. Prepare detail and assembly drawings.

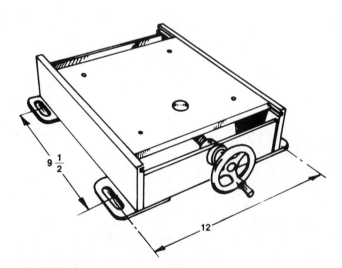

Problem 97
Design and prepare complete working drawings and assemblies for an adjustable motor base similar to that shown. The motor base is to accommodate an electric motor having NEMA frame #184. Make the welded base of steel plate 4.25″ high, 14.00″ long and 10.50″ wide and a 6″ adjustment for the slide plate.

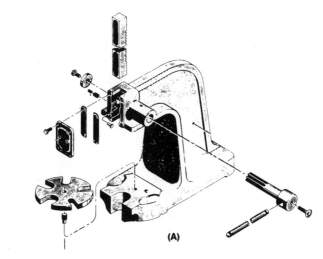

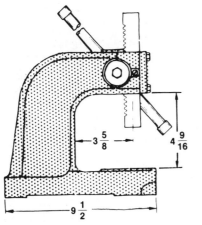

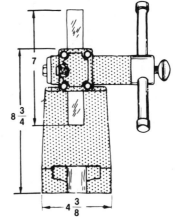

(A)

(B)

Problem 98

Design and prepare complete detail and assembly drawings of an arbor press using the basic dimensions as indicated in the sketch. (Note to Instructor: this project may be assigned to a design group of three or four students under a group leader or undertaken on an individual basis.) Convert all dimensions to mm.

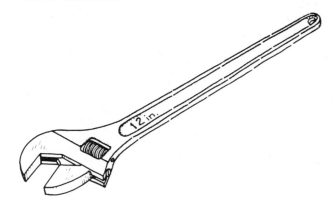

Problem 99

Design a 304mm forged steel wrench similar to that shown in the sketch having a capacity of 34.0mm. Make complete detail and assembly drawings. Add to the product line by designing two other wrenches of identical design but different sizes to make a set of three wrenches. Design a portable metal carrying case for the three wrenches.

Problem 100
Design a barbell set having 12 weights (4 each, of 10, 5, and 2.50 lbs) on a 5½′ bar, 1″ in diameter. Provide a revolving chrome sleeve separator 32″ long.

Problem 101
Under the direction of your instructor, set up a design group to prepare design layouts for a racing kart similar to that shown. The racing kart is to have a wheelbase of 48 inches with a tread width of 29 inches. Use commercially produced components whenever possible. Select a suitable drive arrangement and braking mechanism. Prepare detail, sub-assembly and final assembly drawings together with parts list for the complete vehicle. Make an inked assembly drawing in pictorial, exploded or orthographic for catalog or advertising purposes.

Problem 102
Design and prepare suitable detail and assembly drawings of a rack to hold 20 liter-size wine bottles in an inclined position so their corks are kept moist.

Problem 103
Design a rack to hold four water skis which can be installed in a variety of outboard runabouts and removed for storage during the off season.

Problem 104
Canvass the market and prepare design drawings of an outdoor hammock stand. Determine material costs.

Problem 105
Design a carrier for transporting one 16 HP and two 40 HP outboard motors. Provide arrangements for carrying a standard 6 gallon gasoline can.

Problem 106
Design a fiberglass covering for a standard 8′ fleetside box on a ½-ton pickup truck.

Problem 107
Design a trailer for transporting two snowmobiles.

Problem 108
Design a launching ramp for beaching outboard runabouts of up to 16′ lengths.

APPENDIX

Abbreviations Used on Drawings

A

Absolute	ABS
Accelerate	ACCEL
Accessory	ACCESS.
Accumulator	ACC
Actual	ACT.
Adapter	ADPT
Addendum	ADD.
Adjust	ADJ
Advance	ADV
Aeronautic	AERO
After	AFT.
Aggregate	AGGR
Air Condition	AIR COND
Aircraft	ACFT
Air Force—Navy	AN
Allowance	ALLOW.
Alloy	ALY
Alternate	ALT
Alternating Current	AC
Altitude	ALT
Aluminum	AL
American Standard	AMER STD
American Wire Gage	AWG
Ammeter	AM.
Amount	AMT
Ampere	AMP
Amplifier	AMPL
Amplitude Modulation	AM
Anneal	ANL
Antenna	ANT.
Anti-Aircraft	AA
Anti-Friction Bearing	AFB
Apparatus	APP, APPAR
Approved	APPD
Approximate	APPROX
Arc Weld	ARC/W
Area	A
Armature	ARM.
Armour Plate	ARM PL
Arrangement	ARR
Asbestos	ASB
Asphalt	ASPH
Assemble	ASSEM
Assembly	ASSY
Assistant	ASST
Associate	ASSOC
Association	ASSN
Atomic	AT
Atomic Hydrogen Weld	AT/W
Audible	AUD
Audio Frequency	AF
Authorized	AUTH
Automatic	AUTO.
Automatic Frequency Control	AFC
Automatic Voltage Control	AVC
Auxiliary	AUX
Average	AVG
Aviation	AVN

B

Babbit	BAB
Back Feed	BF
Back Pressure	BP
Back to Back	B to B
Backgear	BG
Balance	BAL
Ball Bearing	BB
Barrel	BBL
Base Line	BL
Base Plate	BP
Battery	BAT. or BT
Bearing	BRG
Bell and Flange	B&F
Bell and Spigot	B&S
Bench Mark	BM
Bending Moment	M
Between	BET.
Between Centers	BC
Bevel	BEV
Bill of Material	B/M
Birmingham Wire Gage	BWG
Blank	BLK
Board	BD
Bolt Circle	BC
Both Faces	BF
Both Sides	BS
Both Ways	BW
Bottom	BOT
Bottom Chord	BC

(B continued)

Bottom Face	BF
Bracket	BRKT
Brake	BK
Brake Horsepower	BHP
Brass	BRS
Brazing	BRZG
Break	BRK
Brinell Hardness	BH
British Standard	BR STD
British Thermal Units	BTU
Broach	BRO
Bronze	BRZ
Brown & Sharpe	B&S
Building	BLDG
Bulkhead	BHD
Bureau of Standards	BU STD
Bushing	BUSH.
Button	BUT.
By-Pass	BYP

C

Cabinet	CAB.
Cadmium Plate	CD PL
Calculate	CALC
Caliber	CAL
Candlepower	CP
Capacitance	C
Capacity	CAP.
Cap Screw	CAP SCR
Carbon	C
Carburetor, Carburize	CARB
Carriage	CRG
Carton	CTN
Case Harden	CH
Cast Iron	CI
Cast Iron Pipe	CIP
Cast Steel	CS
Casting	CSTG
Castle Nut	CAS NUT
Catalogue	CAT.
Cathode Ray Tube	CRT
Celsius	C
Cement	CEM
Center	CTR
Center Line	CL
Center of Gravity	CG
Center Punch	CP
Center to Center	C to C
Centigram	cg
Centiliter	cL
Centimeter	cm
Centrifugal	CENT.
Ceramic	CER
Chamfer	CHAM
Change	CHG
Change Notice	CN
Channel	CHAN
Check	CHK
Chemical	CHEM
Chord	CHD
Chrome Molybdenum	CR MOLY
Chromium Plate	Cr PL
Chrome Vanadium	CR VAN.
Circle	CIR
Circular Pitch	CP
Circumference	CIRC
Cleanout	CO
Clearance	CL
Clockwise	CW
Coated	CTD
Coaxial	COAX
Cold Drawn	CD
Cold Drawn Steel	CDS
Cold Finish	CF
Cold Punched	CP
Cold Rolled Steel	CRS
Combination	COMB.
Commercial	COML
Commutator	COMM
Companion	COMP
Compensator	COMP
Complete	COMPL
Compressor	COMPR
Concentric	CONC
Concrete	CONC
Condition	COND
Constant	CONST
Construction	CONSTR
Container	CNTR
Continue	CONT
Contract, Contractor	CONTR
Contractor Furnished Equipment	CFE

(C continued)

Control	CONT
Conveyor	CNVR
Copper Plate	COP PL
Correct	CORR
Corrosion Resistant	CRE
Corrosion Resistant Steel	CRES
Corrugate	CORR
Cotter	COT.
Cotton Webbing	COT. WEB.
Counter Clockwise	CCW
Counterbore	CBORE
Counterdrill	CDRILL
Counterpunch	CPUNCH
Countersink	CSK
Coupling	CPLG
Cowling	COWL.
Cross Section	XSECT
Cubic	CU
Cubic Foot	CU FT
Cubic Inch	CU IN.
Cubic Inches per Minute	CIPM
Cubic Yard	CU YD
Current	CUR.
Cyanide	CYN
Cycle	CY
Cylinder	CYL

D

Decalcomania	DECAL
Decigram	dg
Deciliter	dL
Decimal	DEC
Decimeter	dm
Dedendum	DED
Deep Drawn	DD
Degree	(°)DEG
Dekagram	dag
Dekaliter	daL
Dekameter	dam
Density	D
Department	DEPT
Design	DSGN
Detail	DET
Develop	DEV
Diagonal, Diagram	DIAG
Diameter	DIA
Diametral Pitch	DP
Diaphragm	DIAPH
Direct Current	DC
Discharge	DISCH
Ditto	DO.
Division	DIV
Dovetail	DVTL
Dowel	DWL
Drafting	DFTG
Draftsman	DFTSMN
Drawing	DWG
Drill, Drill Rod	DR
Drive	DR
Drive Fit	DF
Drop Forge	DF
Duplicate	DUP

E

Each	EA
Eccentric	ECC
Effective	EFF
Elastic Limit	EL
Elbow	ELL.
Electric	ELEC
Electromotive Force	EMF
Elementary	ELEM
Elevate	ELEV
Elevation	EL
Elongation	ELONG
Enclose	ENCL
Engine	ENG
Engineer	ENGR
Engineering	ENGRG
Engineering Change Order	ECO
Entrance	ENT
Envelope	ENV
Equal	EQ
Equipment	EQUIP
Equivalent	EQUIV
Estimate	EST
Exchange	EXCH
Executive	EXEC
Exhaust	EXH
Existing	EXIST.
Explosion Proof	EP
Extension, Exterior	EXT

(E continued)

Extra Heavy	X HVY
Extra Strong	X STR
Extrude	EXTR

F

Fabricate	FAB
Face to Face	F to F
Fahrenheit	F
Far Side	FS
Federal	FED.
Federal Specification	FS
Federal Stock Number	FSN
Feeder	FDR
Feet	(') FT
Feet Board Measure	FBM
Fiber	FBR
Figure	FIG.
Fillet, Fillister	FIL
Filter	FLT
Finish	FIN.
Finish All Over	FAO
Fireproof	FPRF
Fitting	FTG
Fixture	FIX.
Flange	FLG
Flat	F
Flat Head	FH
Flexible	FLEX.
Fluid	FL
Fluorescent	FLUOR
Foot	(') FT
Forged Steel	FST
Forging	FORG
Forward	FWD
Foundation	FDN
Foundry	FDRY
Fractional	FRAC
Fractional Horsepower	FHP
Frame	FR
Framework	FRWK
Frequency	FREQ
Frequency Modulation	FM
Front	FR
Furnish	FURN

G

Gage	GA
Gallon	GAL
Galvanize	GALV
Galvanized Iron	GI
Galvanized Steel	GS
Gasket	GSKT
General, Generator	GEN
Glass	GL
Government	GOVT
Grade	GR
Gram	g
Graphic	GRAPH.
Graphite	GPH
Grind	GRD
Grommet	GROM
Groove	GRV
Ground	GND or GRD

H

Half-Hard	½H
Half-Round	½RD
Handle	HDL
Hanger	HGR
Hard	H
Hard-Drawn	HD
Harden	HDN
Hardware	HDW
Head	HD
Headless	HDLS
Heat	HT
Heat Exchanger	HE
Heat Treat	HT TR
Heavy	HVY
Height	HGT
Hexagon	HEX, HEX.
High Frequency	HF
High-Pressure	HP
High-Speed	HS
High-Speed Steel	HSS
High-Tensile Cast Iron	HTCI
High-Tensile Steel	HTS
Horizontal	HORIZ
Horsepower	HP
Hot Rolled	HR
Hot Rolled Steel	HRS

Hour ... HR
Hundredweight ... CWT
Hydraulic ... HYD
Hydrostatic ... HYDRO

I

Ignition ... IGN
Impregnate ... IMPG
Inboard ... INBD
Inch ... (") IN.
Indicate, Industrial ... IND
Inside Diameter ... ID
Inspect ... INSP
Install ... INSTL
Insulate ... INS
Interior, Internal, Intersect ... INT
Intermediate Frequency ... IF
International Pipe Standard ... IPS
Iron ... I
Irregular ... IRREG

J

Job Order ... JO
Joint ... JT
Journal ... JNL
Junction ... JCT

K

Kelvin ... K
Key ... K
Keyseat ... KST
Keyway ... KWY
Kiln-Dried ... KD
Kilocycle ... KC
Kilogram ... kg
Kiloliter ... kL
Kilometer ... km
Kilowatt Hour ... KWH
Kilovolt ... KV
Kip (1000 lb) ... K

L

Laboratory ... LAB
Lacquer ... LAQ
Laminate ... LAM
Lateral ... LAT.
Left ... L
Left Hand ... LH
Length ... LG
Length Over All ... LOA
Letter ... LTR
Light ... LT
Limit ... LIM
Line ... L
Linear ... LIN
Liquid ... LIQ
Locate ... LOC
Long ... LG
Low Pressure ... LP
Lubricate, Lubricator ... LUB
Lumber ... LBR

M

Machine ... MACH.
Machine Steel ... MS
Maintenance ... MAINT
Malleable ... MALL.
Malleable Iron ... MI
Manual ... MAN.
Manufacture ... MFR
Manufactured ... MFD
Manufacturing ... MFG
Material ... MATL
Maximum ... MAX
Mechanical ... MECH
Median ... MED
Memorandum ... MEMO
Metal ... MET.
Meter (Instrument or Measure of Length) ... M
Micro (10^{-6}) ... μ or U
Micrometer ... MIC
Mile ... MI
Miles per Gallon ... MPG
Miles per Hour ... MPH
Military ... MIL
Milligram ... mg
Milliliter ... mL
Millimeter ... mm
Minimum ... MIN
Minute ... (') MIN
Miscellaneous ... MISC
Mixture ... MIX.
Model ... MOD
Molded ... MLD
Motor ... MOT
Mounted ... MTD
Mounting ... MTG
Multiple ... MULT
Music Wire Gage ... MWG

N

National ... NATL
National Electrical Code ... NEC
Natural ... NAT
Near Face ... NF
Near Side ... NS
Negative ... NEG
Nipple ... NIP.
Nominal ... NOM
Normal ... NORM.
Not to Scale ... NTS
Number ... NO

O

Obsolete ... OBS
Octagon ... OCT
Office ... OFF.
Ohms ... Ω
On Center ... OC
One Pole ... 1 P
Opening ... OPNG
Opposite ... OPP
Ordnance ... ORD
Original ... ORIG
Oscillator ... OSC
Outboard ... OUTBD
Outlet ... OUT.
Outside Diameter ... OD
Overall ... OA
Overflow ... OVFL
Overhead ... OVHD
Oxidized ... OXD

P

Pack ... PK
Packing ... PKG
Page ... P
Pair ... PR
Panel ... PNL
Paragraph ... PARA
Parallel ... PAR.
Part ... PT
Passage ... PASS.
Patent ... PAT.
Pattern ... PATT
Peck ... PK
Penny (Nails, etc) ... d
Permanent ... PERM
Perpendicular ... PERP
Phenolic ... PHEN
Phosphor Bronze ... PH BRZ
Piece ... PC
Pipe Tap ... PT
Pitch ... P
Pitch Circle ... PC
Pitch Diameter ... PD
Plastic ... PLSTC
Plate ... PL
Plumbing ... PLMB
Pneumatic ... PNEU
Point ... PT
Pole ... P
Polish ... POL
Porcelain ... PORC
Position ... POS
Potential ... POT.
Potentiometer ... POT
Pound ... LB
Pounds per Square Inch ... PSI
Power ... PWR
Power Factor ... PF
Precast ... PRCST
Prefabricated ... PREFAB
Preferred ... PFD
Prepare ... PREP
Pressure ... PRESS.
Pressure Angle ... PA.
Pressure Controlled ... PC
Pressure Switch ... PS
Primary ... PRIM.
Process ... PROC
Production ... PROD.
Production Order ... PO
Profile ... PF
Project ... PROJ
Proposed ... PROP.
Publication ... PUB.
Pull Switch ... PS
Pump, Fixed Displacement ... PF
Pump, Variable Displacement ... PV
Punch ... PCH
Purchase ... PUR
Push Button ... PB

Q

Quadrant ... QUAD.
Quality ... QUAL
Quantity ... QTY
Quarter-Hard ... ¼ H
Quarter-Round ... ¼ RD

R

Radial ... RAD
Radio Frequency ... RF
Radius ... R
Railroad ... RR
Ream ... RM
Received ... RECD
Receptacle ... RECP
Record ... REC
Rectangle, Rectifier ... RECT
Reduce, Reducer ... RED.
Reference ... REF
Regulator ... REG
Reinforce ... REINF
Relative Humidity ... RH
Relay, Relief ... REL
Remove ... REM
Reproduce ... REPRO
Require ... REQ
Required ... REQD
Requisition ... REQ
Resistance ... R
Retard, Return ... RET
Reverse, Revise, Revolution ... REV
Revolutions per Minute ... RPM
Right ... R
Right Hand ... RH
Rockwell Hardness ... RH
Roller Bearing ... RB
Root Diameter ... RD
Root Mean Square ... RMS
Round ... RD

S

Schedule ... SCH
Schematic ... SCHEM
Screw ... SCR
Seamless ... SMLS
Section ... SECT.
Semi-Finished ... SF
Separator ... SEP
Serial ... SER
Servo ... SERV
Set Screw ... SS
Shaft ... SFT
Sheet ... SH
Short Wave ... SW
Shoulder ... SHLD
Shut Off Valve ... SOV
Side ... S
Similar ... SIM
Sketch ... SK
Sleeve ... SLV
Sleeve Bearing ... SB
Slotted ... SLOT.
Small ... SM
Socket ... SOC
Soft ... S
Solenoid ... SOL
Space, Spare ... SP
Special ... SPL
Special Treatment Steel ... STS
Specific Gravity ... SP GR
Specification ... SPEC
Speed ... SP
Spherical ... SPHER
Spot Faced ... SF
Spring ... SPG
Square ... SQ
Squirrel Cage ... SQ CG
Stainless ... STN
Stainless Steel ... SST
Standard ... STD
Station ... STA
Steel ... STL
Stock ... STK
Storage ... STOR
Straight ... STR
Street ... ST
Stress Anneal ... SA
Strip, Structural, Strainer ... STR
Substitute ... SUB.
Summary ... SUM.
Superintendent ... SUPT
Supersede ... SUPSD
Superstructure ... SUPRSTR
Supervise ... SUPV
Supplement ... SUPP
Supply ... SUP.
Surface ... SURF.
Switch ... SW
Switchgear ... SWGR
Symbol, Symmetrical ... SYM
Synchronous, Synthetic ... SYN
System ... SYS

T

Tabulate ... TAB
Tangent ... TAN.
Taper ... TPR
Technical ... TECH
Tee, Teeth ... T
Teeth per Inch ... TPI
Temperature, Template ... TEMP
Tensile Strength ... TS
Tension ... TENS.
Terminal ... TERM.
Thick ... THK
Thousand ... M
Thousand Foot Pounds ... KIP-FT
Thousand Pound ... KIP
Thread ... THD
Threads per Inch ... TPI
Through ... THRU
Time Delay ... TD
Tobin-Bronze ... TOB BRZ
Toggle ... TGL
Tolerance ... TOL
Tool Steel ... TS
Tooth ... T
Total ... TOT.
Total Indicator Reading ... TIR
Transfer, Transformer ... TRANS
Transmission ... XMSN
Transportation ... TRANS
Tubing ... TUB.
Typical ... TYP

U

Ultimate ... ULT
Ultra-High Frequency ... UHF
Universal ... UNIV

V

Vacuum ... VAC
Vacuum Tube ... VT
Valve ... V
Vapor Proof ... VAP PRF
Varnish ... VARN
Velocity ... V
Vent Pipe ... VP
Ventilate ... VENT.
Vertical ... VERT
Very-High Frequency ... VHF
Vibrate ... VIB
Volt ... V
Volume ... VOL

W

Wall ... W
Washer ... WASH.
Watt ... W
Week ... WK
Weight ... WT
Wheel Base ... WB
Width ... W
Wire ... W
Without ... W/O
Woodruff ... WDF
Wrought ... WRT
Wrought Iron ... WI

X Y Z

Yard ... YD
Year ... YR

Note: For additional standard abbreviations, see MIL-STD-12 and ANSI Y1.1-1972.

DUAL DIMENSIONING

CONVERSION TABLE INCH/MM

Group 1

Drill No. or Letter	Inch	mm
	.001	0.0254
	.002	0.0508
	.003	0.0762
	.004	0.1016
	.005	0.1270
	.006	0.1524
	.007	0.1778
	.008	0.2032
	.009	0.2286
	.010	0.2540
	.011	0.2794
	.012	0.3048
	.013	0.3302
80 .0135	.014	0.3556
79 .0145	.015	0.3810
1/64 .0156	.016	0.4064
78	.017	0.4318
	.018	0.4572
77	.019	0.4826
	.020	0.5080
76	.021	0.5334
75	.022	0.5588
74 .0225	.023	0.5842
73	.024	0.6096
72	.025	0.6350
71	.026	0.6604
	.027	0.6858
70	.028	0.7112
69 .0292	.029	0.7366
68	.030	0.7620
	.031	0.7874
1/32 .0312	.032	0.8128
67	.033	0.8382
66	.034	0.8636
65	.035	0.8890
64	.036	0.9144
63	.037	0.9398
62	.038	0.9652
61	.039	0.9906
60	.040	1.0160
59	.041	1.0414
58	.042	1.0668
57	.043	1.0922
	.044	1.1176
	.045	1.1430
56 .0465	.046	1.1684
3/64 .0469	.047	1.1938
	.048	1.2192
	.049	1.2446
	.050	1.2700
	.051	1.2954
55	.052	1.3208
	.053	1.3462
	.054	1.3716
54	.055	1.3970
	.056	1.4224
	.057	1.4478
	.058	1.4732
53 .0595	.059	1.4986
	.060	1.5240
	.061	1.5494
1/16 .0625	.062	1.5748
52 .0635	.063	1.6002
	.064	1.6256
	.065	1.6510
	.066	1.6764
51	.067	1.7018
	.068	1.7272
	.069	1.7526
50	.070	1.7780
	.071	1.8034
	.072	1.8288
49	.073	1.8542
	.074	1.8796
	.075	1.9050
48	.076	1.9304
	.077	1.9558
47 .0785 / 5/64 .0781	.078	1.9812
	.079	2.0066
	.080	2.0320
46	.081	2.0574
45	.082	2.0828
	.083	2.1082
	.084	2.1336
	.085	2.1590
44	.086	2.1844
	.087	2.2098
	.088	2.2352
43	.089	2.2606
	.090	2.2860
	.091	2.3114
	.092	2.3368
42 .0935 / 3/32 .0937	.093	2.3622
	.094	2.3876
	.095	2.4130
41	.096	2.4384
	.097	2.4638
40	.098	2.4892
	.099	2.5146
39 .0995	.100	2.5400

Group 2

Drill No. or Letter	Inch	mm
38 .1015	.101	2.5654
	.102	2.5908
37	.103	2.6162
	.104	2.6416
	.105	2.6670
36 .1065	.106	2.6924
	.107	2.7178
	.108	2.7432
7/64 .1094	.109	2.7686
35	.110	2.7940
34	.111	2.8194
33	.112	2.8448
	.113	2.8702
	.114	2.8956
32	.115	2.9210
	.116	2.9464
	.117	2.9718
	.118	2.9972
	.1181	3.0000
31	.119	3.0226
	.120	3.0480
	.121	3.0734
	.122	3.0988
	.123	3.1242
	.124	3.1496
1/8 .125	.125	3.1750
	.126	3.2004
	.127	3.2258
30 .1285	.128	3.2512
	.129	3.2766
	.130	3.3020
	.131	3.3274
	.132	3.3528
	.133	3.3782
	.134	3.4036
29	.135	3.4290
	.136	3.4544
	.137	3.4798
	.138	3.5052
	.139	3.5306
28 .1405 / 9/64 .1406	.140	3.5560
	.141	3.5814
	.142	3.6068
27	.143	3.6322
	.144	3.6576
	.145	3.6830
26	.146	3.7084
	.147	3.7338
	.148	3.7592
25 .1495	.149	3.7846
	.150	3.8100
	.151	3.8354
24	.152	3.8608
	.153	3.8862
23	.154	3.9116
	.155	3.9370
	.156	3.9624
5/32 .1562	.157	3.9878
22	.1575	4.0000
	.158	4.0132
21	.159	4.0386
	.160	4.0640
20	.161	4.0894
	.162	4.1148
	.163	4.1402
	.164	4.1656
	.165	4.1910
19	.166	4.2164
	.167	4.2418
	.168	4.2672
18 .1695	.169	4.2926
	.170	4.3180
	.171	4.3434
11/64 .1719	.172	4.3688
17	.173	4.3942
	.174	4.4196
	.175	4.4450
16	.176	4.4704
	.177	4.4958
	.178	4.5212
	.179	4.5466
15	.180	4.5720
	.181	4.5974
14	.182	4.6228
	.183	4.6482
	.184	4.6736
13	.185	4.6990
	.186	4.7244
	.187	4.7498
3/16 .1875		4.7625
	.188	4.7752
12	.189	4.8006
	.190	4.8260
11	.191	4.8514
	.192	4.8768
	.193	4.9022
10 .1935	.194	4.9276
	.195	4.9530
9	.196	4.9784
	.1969	5.0000
	.197	5.0038
	.198	5.0292
	.199	5.0546
8	.200	5.0800

Group 3

Drill No. or Letter	Inch	mm
7	.201	5.1054
	.202	5.1308
	.203	5.1562
13/64 .2031	.204	5.1816
6	.205	5.2070
5 .2055	.206	5.2324
	.207	5.2578
	.208	5.2832
4	.209	5.3086
	.210	5.3340
	.211	5.3594
3	.212	5.3848
	.213	5.4102
	.214	5.4356
	.215	5.4610
	.216	5.4864
7/32 .2187	.217	5.5118
	.218	5.5372
	.219	5.5626
	.220	5.5880
2	.221	5.6134
	.222	5.6388
	.223	5.6642
	.224	5.6896
	.225	5.7150
	.226	5.7404
	.227	5.7658
1	.228	5.7912
	.229	5.8166
	.230	5.8420
	.231	5.8674
	.232	5.8928
	.233	5.9182
A	.234	5.9436
15/64 .2344	.235	5.9690
	.236	5.9944
	.2362	6.0000
	.237	6.0198
B	.238	6.0452
	.239	6.0706
	.240	6.0960
	.241	6.1214
C	.242	6.1468
	.243	6.1722
	.244	6.1976
	.245	6.2230
D	.246	6.2484
	.247	6.2738
	.248	6.2992
	.249	6.3246
E 1/4 .250	.250	6.3500
	.251	6.3754
	.252	6.4008
	.253	6.4262
	.254	6.4516
	.255	6.4770
F	.256	6.5024
	.257	6.5278
	.258	6.5532
	.259	6.5786
	.260	6.6040
G	.261	6.6294
	.262	6.6548
	.263	6.6802
	.264	6.7056
	.265	6.7310
17/64 .2656	.266	6.7564
H	.267	6.7818
	.268	6.8072
	.269	6.8326
	.270	6.8580
	.271	6.8834
I	.272	6.9088
	.273	6.9342
	.274	6.9596
	.275	6.9850
	.2756	7.0000
	.276	7.0104
J	.277	7.0358
	.278	7.0612
	.279	7.0866
	.280	7.1120
K 9/32 .2812	.281	7.1374
	.282	7.1628
	.283	7.1882
	.284	7.2136
	.285	7.2390
	.286	7.2644
	.287	7.2898
	.288	7.3152
	.289	7.3406
L	.290	7.3660
	.291	7.3914
	.292	7.4168
	.293	7.4422
	.294	7.4676
	.295	7.4930
M	.296	7.5184
19/64 .2969	.297	7.5438
	.298	7.5692
	.299	7.5946
	.300	7.6200

Group 4

Drill No. or Letter	Inch	mm
N	.301	7.6454
	.302	7.6708
	.303	7.6962
	.304	7.7216
	.305	7.7470
	.306	7.7724
	.307	7.7978
	.308	7.8232
	.309	7.8486
	.310	7.8740
	.311	7.8994
	.312	7.9248
5/16 .3125		7.9375
	.313	7.9502
	.314	7.9756
	.3150	8.0000
O	.315	8.0010
	.316	8.0264
	.317	8.0518
	.318	8.0772
	.319	8.1026
	.320	8.1280
P	.321	8.1534
	.322	8.1788
	.323	8.2042
	.324	8.2296
	.325	8.2550
	.326	8.2804
	.327	8.3058
	.328	8.3312
21/64 .3281		8.3344
	.329	8.3566
	.330	8.3820
	.331	8.4074
	.332	8.4328
Q	.333	8.4582
	.334	8.4836
	.335	8.5090
	.336	8.5344
	.337	8.5598
	.338	8.5852
R	.339	8.6106
	.340	8.6360
	.341	8.6614
	.342	8.6868
11/32 .3437		8.7312
	.343	8.7122
	.344	8.7376
	.345	8.7630
	.346	8.7884
	.347	8.8138
S	.348	8.8392
	.349	8.8646
	.350	8.8900
	.351	8.9154
	.352	8.9408
	.353	8.9662
	.354	8.9916
	.3543	9.0000
	.355	9.0170
	.356	9.0424
	.357	9.0678
T	.358	9.0932
	.359	9.1186
23/64 .3594		9.1281
	.360	9.1440
	.361	9.1694
	.362	9.1948
	.363	9.2202
	.364	9.2456
	.365	9.2710
	.366	9.2964
	.367	9.3218
U	.368	9.3472
	.369	9.3726
	.370	9.3980
	.371	9.4234
	.372	9.4488
	.373	9.4742
	.374	9.4996
3/8 .375	.375	9.5250
V	.376	9.5504
	.377	9.5758
	.378	9.6012
	.379	9.6266
	.380	9.6520
	.381	9.6774
	.382	9.7028
	.383	9.7282
	.384	9.7536
	.385	9.7790
W	.386	9.8044
	.387	9.8298
	.388	9.8552
	.389	9.8806
	.390	9.9060
25/64 .3906		9.9219
	.391	9.9314
	.392	9.9568
	.393	9.9822
	.3937	10.0000
	.394	10.0076
	.395	10.0330
	.396	10.0584
X	.397	10.0838
	.398	10.1092
	.399	10.1346
	.400	10.1600

Group 5

Drill No. or Letter	Inch	mm
	.401	10.1854
	.402	10.2108
	.403	10.2362
Y	.404	10.2616
	.405	10.2870
	.406	10.3124
13/32 .4062		10.3187
	.407	10.3378
	.408	10.3632
	.409	10.3886
	.410	10.4140
	.411	10.4394
	.412	10.4648
Z	.413	10.4902
	.414	10.5156
	.415	10.5410
	.416	10.5664
	.417	10.5918
	.418	10.6172
	.419	10.6426
	.420	10.6680
	.421	10.6934
27/64 .4219		10.7156
	.422	10.7188
	.423	10.7442
	.424	10.7696
	.425	10.7950
	.426	10.8204
	.427	10.8458
	.428	10.8712
	.429	10.8966
	.430	10.9220
	.431	10.9474
	.432	10.9728
	.433	10.9982
	.4331	11.0000
	.434	11.0236
	.435	11.0490
	.436	11.0744
	.437	11.0998
7/16 .4375		11.1125
	.438	11.1252
	.439	11.1506
	.440	11.1760
	.441	11.2014
	.442	11.2268
	.443	11.2522
	.444	11.2776
	.445	11.3030
	.446	11.3284
	.447	11.3538
	.448	11.3792
	.449	11.4046
	.450	11.4300
	.451	11.4554
	.452	11.4808
	.453	11.5062
29/64 .4531		11.5094
	.454	11.5316
	.455	11.5570
	.456	11.5824
	.457	11.6078
	.458	11.6332
	.459	11.6586
	.460	11.6840
	.461	11.7094
	.462	11.7348
	.463	11.7602
	.464	11.7856
	.465	11.8110
	.466	11.8364
	.467	11.8618
	.468	11.8872
15/32 .4687		11.9062
	.469	11.9126
	.470	11.9380
	.471	11.9634
	.472	11.9888
	.4724	12.0000
	.473	12.0142
	.474	12.0396
	.475	12.0650
	.476	12.0904
	.477	12.1158
	.478	12.1412
	.479	12.1666
	.480	12.1920
	.481	12.2174
	.482	12.2428
	.483	12.2682
	.484	12.2936
31/64 .4844		12.3031
	.485	12.3190
	.486	12.3444
	.487	12.3698
	.488	12.3952
	.489	12.4206
	.490	12.4460
	.491	12.4714
	.492	12.4968
	.493	12.5222
	.494	12.5476
	.495	12.5730
	.496	12.5984
	.497	12.6238
	.498	12.6492
	.499	12.6746
1/2 .500	.500	12.7000

CONVERSION TABLE INCH/MM (Continued)

Inch	mm
.501	12.7254
.502	12.7508
.503	12.7762
.504	12.8016
.505	12.8270
.506	12.8524
.507	12.8778
.508	12.9032
.509	12.9286
.510	12.9540
.511	12.9794
.5118	13.0000
.512	13.0048
.513	13.0302
.514	13.0556
.515	13.0810
33/64 .5156	13.0968
.516	13.1064
.517	13.1318
.518	13.1572
.519	13.1826
.520	13.2080
.521	13.2334
.522	13.2588
.523	13.2842
.524	13.3096
.525	13.3350
.526	13.3604
.527	13.3858
.528	13.4112
.529	13.4366
.530	13.4620
.531	13.4874
17/32 .5312	13.4937
.532	13.5128
.533	13.5382
.534	13.5636
.535	13.5890
.536	13.6144
.537	13.6398
.538	13.6652
.539	13.6906
.540	13.7160
.541	13.7414
.542	13.7668
.543	13.7922
.544	13.8176
.545	13.8430
.546	13.8684
35/64 .5469	13.8906
.547	13.8938
.548	13.9192
.549	13.9446
.550	13.9700
.551	13.9954
.5512	14.0000
.552	14.0208
.553	14.0462
.554	14.0716
.555	14.0970
.556	14.1224
.557	14.1478
.558	14.1732
.559	14.1986
.560	14.2240
.561	14.2494
.562	14.2748
9/16 .5625	14.2875
.563	14.3002
.564	14.3256
.565	14.3510
.566	14.3764
.567	14.4018
.568	14.4272
.569	14.4526
.570	14.4780
.571	14.5034
.572	14.5288
.573	14.5542
.574	14.5796
.575	14.6050
.576	14.6304
.577	14.6558
.578	14.6812
37/64 .5781	14.6844
.579	14.7066
.580	14.7320
.581	14.7574
.582	14.7828
.583	14.8082
.584	14.8336
.585	14.8590
.586	14.8844
.587	14.9098
.588	14.9352
.589	14.9606
.590	14.9860
.5906	15.0000
.591	15.0114
.592	15.0368
.593	15.0622
19/32 .5937	15.0812
.594	15.0876
.595	15.1130
.596	15.1384
.597	15.1638
.598	15.1892
.599	15.2146

Inch	mm
.600	15.2400
.601	15.2654
.602	15.2908
.603	15.3162
.604	15.3416
.605	15.3670
.606	15.3924
.607	15.4178
.608	15.4432
.609	15.4686
39/64 .6094	15.4781
.610	15.4940
.611	15.5194
.612	15.5448
.613	15.5702
.614	15.5956
.615	15.6210
.616	15.6464
.617	15.6718
.618	15.6972
.619	15.7226
.620	15.7480
.621	15.7734
.622	15.7988
.623	15.8242
.624	15.8496
5/8 .625	15.8750
.626	15.9004
.627	15.9258
.628	15.9512
.629	15.9766
.6299	16.0000
.630	16.0020
.631	16.0274
.632	16.0528
.633	16.0782
.634	16.1036
.635	16.1290
.636	16.1544
.637	16.1798
.638	16.2052
.639	16.2306
.640	16.2560
41/64 .6406	16.2719
.641	16.2814
.642	16.3068
.643	16.3322
.644	16.3576
.645	16.3830
.646	16.4084
.647	16.4338
.648	16.4592
.649	16.4846
.650	16.5100
.651	16.5354
.652	16.5608
.653	16.5862
.654	16.6116
.655	16.6370
.656	16.6624
21/32 .6562	16.6687
.657	16.6878
.658	16.7132
.659	16.7386
.660	16.7640
.661	16.7894
.662	16.8148
.663	16.8402
.664	16.8656
.665	16.8910
.666	16.9164
.667	16.9418
.668	16.9672
.669	16.9926
.6693	17.0000
.670	17.0180
.671	17.0434
43/64 .6719	17.0656
.672	17.0688
.673	17.0942
.674	17.1196
.675	17.1450
.676	17.1704
.677	17.1958
.678	17.2212
.679	17.2466
.680	17.2720
.681	17.2974
.682	17.3228
.683	17.3482
.684	17.3736
.685	17.3990
.686	17.4244
.687	17.4498
11/16 .6875	17.4625
.688	17.4752
.689	17.5006
.690	17.5260
.691	17.5514
.692	17.5768
.693	17.6022
.694	17.6276
.695	17.6530
.696	17.6784
.697	17.7038
.698	17.7292
.699	17.7546
.700	17.7800

Inch	mm
.701	17.8054
.702	17.8308
.703	17.8562
45/64 .7031	17.8594
.704	17.8816
.705	17.9070
.706	17.9324
.707	17.9578
.708	17.9832
.7087	18.0000
.709	18.0086
.710	18.0340
.711	18.0594
.712	18.0848
.713	18.1102
.714	18.1356
.715	18.1610
.716	18.1864
.717	18.2118
.718	18.2372
23/32 .7187	18.2562
.719	18.2626
.720	18.2880
.721	18.3134
.722	18.3388
.723	18.3642
.724	18.3896
.725	18.4150
.726	18.4404
.727	18.4658
.728	18.4912
.729	18.5166
.730	18.5420
.731	18.5674
.732	18.5928
.733	18.6182
.734	18.6436
47/64 .7344	18.6532
.735	18.6690
.736	18.6944
.737	18.7198
.738	18.7452
.739	18.7706
.740	18.7960
.741	18.8214
.742	18.8468
.743	18.8722
.744	18.8976
.745	18.9230
.746	18.9484
.747	18.9738
.748	18.9992
.7480	19.0000
.749	19.0246
3/4 .750	19.0500
.751	19.0754
.752	19.1008
.753	19.1262
.754	19.1516
.755	19.1770
.756	19.2024
.757	19.2278
.758	19.2532
.759	19.2786
.760	19.3040
.761	19.3294
.762	19.3548
.763	19.3802
.764	19.4056
.765	19.4310
49/64 .7656	19.4469
.766	19.4564
.767	19.4818
.768	19.5072
.769	19.5326
.770	19.5580
.771	19.5834
.772	19.6088
.773	19.6342
.774	19.6596
.775	19.6850
.776	19.7104
.777	19.7358
.778	19.7612
.779	19.7866
.780	19.8120
.781	19.8374
25/32 .7812	19.8433
.782	19.8628
.783	19.8882
.784	19.9136
.785	19.9390
.786	19.9644
.787	19.9898
.7874	20.0000
.788	20.0152
.789	20.0406
.790	20.0660
.791	20.0914
.792	20.1168
.793	20.1422
.794	20.1676
.795	20.1930
.796	20.2184
51/64 .7969	20.2402
.797	20.2438
.798	20.2692
.799	20.2946

Inch	mm
.800	20.3200
.801	20.3454
.802	20.3708
.803	20.3962
.804	20.4216
.805	20.4470
.806	20.4724
.807	20.4978
.808	20.5232
.809	20.5486
.810	20.5740
.811	20.5994
.812	20.6248
13/16 .8125	20.6375
.813	20.6502
.814	20.6756
.815	20.7010
.816	20.7264
.817	20.7518
.818	20.7772
.819	20.8026
.820	20.8280
.821	20.8534
.822	20.8788
.823	20.9042
.824	20.9296
.825	20.9550
.826	20.9804
.8268	21.0000
.827	21.0058
.828	21.0312
53/64 .8281	21.0344
.829	21.0566
.830	21.0820
.831	21.1074
.832	21.1328
.833	21.1582
.834	21.1836
.835	21.2090
.836	21.2344
.837	21.2598
.838	21.2852
.839	21.3106
.840	21.3360
.841	21.3614
.842	21.3868
.843	21.4122
27/32 .8437	21.4312
.844	21.4376
.845	21.4630
.846	21.4884
.847	21.5138
.848	21.5392
.849	21.5646
.850	21.5900
.851	21.6154
.852	21.6408
.853	21.6662
.854	21.6916
.855	21.7170
.856	21.7424
.857	21.7678
.858	21.7932
.859	21.8186
55/64 .8594	21.8281
.860	21.8440
.861	21.8694
.862	21.8948
.863	21.9202
.864	21.9456
.865	21.9710
.866	21.9964
.8661	22.0000
.867	22.0218
.868	22.0472
.869	22.0726
.870	22.0980
.871	22.1234
.872	22.1488
.873	22.1742
.874	22.1996
7/8 .875	22.2250
.876	22.2504
.877	22.2758
.878	22.3012
.879	22.3266
.880	22.3520
.881	22.3774
.882	22.4028
.883	22.4282
.884	22.4536
.885	22.4790
.886	22.5044
.887	22.5298
.888	22.5552
.889	22.5806
.890	22.6060
57/64 .8906	22.6219
.891	22.6314
.892	22.6568
.893	22.6822
.894	22.7076
.895	22.7330
.896	22.7584
.897	22.7838
.898	22.8092
.899	22.8346
.900	22.8600

Inch	mm
.901	22.8854
.902	22.9108
.903	22.9362
.904	22.9616
.905	22.9870
.9055	23.0000
.906	23.0124
29/32 .9062	23.0187
.907	23.0378
.908	23.0632
.909	23.0886
.910	23.1140
.911	23.1394
.912	23.1648
.913	23.1902
.914	23.2156
.915	23.2410
.916	23.2664
.917	23.2918
.918	23.3172
.919	23.3426
.920	23.3680
.921	23.3934
59/64 .9219	23.4156
.922	23.4188
.923	23.4442
.924	23.4696
.925	23.4950
.926	23.5204
.927	23.5458
.928	23.5712
.929	23.5966
.930	23.6220
.931	23.6474
.932	23.6728
.933	23.6982
.934	23.7236
.935	23.7490
.936	23.7744
.937	23.7998
15/16 .9375	23.8125
.938	23.8252
.939	23.8506
.940	23.8760
.941	23.9014
.942	23.9268
.943	23.9522
.944	23.9776
.9449	24.0000
.945	24.0030
.946	24.0284
.947	24.0538
.948	24.0792
.949	24.1046
.950	24.1300
.951	24.1554
.952	24.1808
.953	24.2062
61/64 .9531	24.2094
.954	24.2316
.955	24.2570
.956	24.2824
.957	24.3078
.958	24.3332
.959	24.3586
.960	24.3840
.961	24.4094
.962	24.4348
.963	24.4602
.964	24.4856
.965	24.5110
.966	24.5364
.967	24.5618
.968	24.5872
31/32 .9687	24.6062
.969	24.6126
.970	24.6380
.971	24.6634
.972	24.6888
.973	24.7142
.974	24.7396
.975	24.7650
.976	24.7904
.977	24.8158
.978	24.8412
.979	24.8666
.980	24.8920
.981	24.9174
.982	24.9428
.983	24.9682
.984	24.9936
.9843	25.0000
63/64 .9844	25.0031
.985	25.0190
.986	25.0444
.987	25.0698
.988	25.0952
.989	25.1206
.990	25.1460
.991	25.1714
.992	25.1968
.993	25.2222
.994	25.2476
.995	25.2730
.996	25.2984
.997	25.3238
.998	25.3492
.999	25.3746
1.000	25.4000

Distance Across Corners of Hexagons and Squares

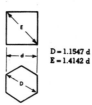

DISTANCE ACROSS CORNERS OF HEXAGON; AND SQUARES

D = 1.1547 d
E = 1.4142 d

d	D	E	d	D	E	d	D	E
¼	0.2886	0.3535	1¼	1.4434	1.7677	2⁵/₁₆	2.6702	3.2703
⁹/₃₂	0.3247	0.3977	1⁹/₃₂	1.4794	1.8119	2⅜	2.7424	3.3587
⁵/₁₆	0.3608	0.4419	1⁵/₁₆	1.5155	1.8561	2⁷/₁₆	2.8145	3.4471
¹¹/₃₂	0.3968	0.4861	1¹¹/₃₂	1.5516	1.9003	2½	2.8867	3.5355
⅜	0.4329	0.5303	1⅜	1.5877	1.9445	2⁹/₁₆	2.9583	3 6239
¹³/₃₂	0.4690	0.5745	1¹³/₃₂	1.6238	1.9887	2⅝	3.0311	3.7123
⁷/₁₆	0.5051	0.6187	1⁷/₁₆	1.6598	2.0329	2¹¹/₁₆	3.1032	3.8007
¹⁵/₃₂	0.5412	0.6629	1¹⁵/₃₂	1.6959	2.0771	2¾	3.1754	3.8891
½	0.5773	0.7071	1½	1.7320	2.1213	2¹³/₁₆	3.2476	3.9794
¹⁷/₃₂	0.6133	0.7513	1¹⁷/₃₂	1.7681	2.1655	2⅞	3.3197	4.0658
⁹/₁₆	0.6494	0.7955	1⁹/₁₆	1.8042	2.2097	2¹⁵/₁₆	3.3919	4.1542
¹⁹/₃₂	0.6855	0.8397	1¹⁹/₃₂	1.8403	2.2539	3	3.4641	4.2426
⅝	0.7216	0.8839	1⅝	1.8764	2.2981	3¹/₁₆	3.5362	4.3310
²¹/₃₂	0.7576	0.9281	1²¹/₃₂	1.9124	2.3423	3⅛	3.6084	4.4194
¹¹/₁₆	0.7937	0.9723	1¹¹/₁₆	1.9485	2.3865	3³/₁₆	3.6806	4.5078
²³/₃₂	0.8298	1.0164	1²³/₃₂	1.9846	2.4306	3¼	3.7527	4.5962
¾	0.8659	1.0606	1¾	2.0207	2.4708	3⁵/₁₆	3.8249	4.6846
²⁵/₃₂	0.9020	1.1048	1²⁵/₃₂	2.0568	2.5190	3⅜	3.8971	4.7729
¹³/₁₆	0.9380	1.1490	1¹³/₁₆	2.0929	2.5632	3⁷/₁₆	3.9692	4.8613
²⁷/₃₂	0.9741	1.1932	1²⁷/₃₂	2.1289	2.6074	3½	4.0414	4.9497
⅞	1.0102	1.2374	1⅞	2.1650	2.6516	3⁹/₁₆	4.1136	5.0381
²⁹/₃₂	1.0463	1.2816	1²⁹/₃₂	2.2011	2.6958	3⅝	4.1857	5.1265
¹⁵/₁₆	1.0824	1.3258	1¹⁵/₁₆	2.2372	2.7400	3¹¹/₁₆	4.2579	5.2149
³¹/₃₂	1.1184	1.3700	1³¹/₃₂	2.2733	2.7842	3¾	4.3301	5.3033
1	1.1547	1.4142	2	2.3094	2.8284	3¹³/₁₆	4.4023	5.3917
1¹/₃₂	1.1907	1.4584	2¹/₃₂	2.3453	2.8726	3⅞	4.4744	5.4801
1¹/₁₆	1.2268	1.5026	2¹/₁₆	2.3815	2.9168	3¹⁵/₁₆	4.5466	5.5684
1³/₃₂	1.2629	1.5468	2³/₃₂	2.4176	2.9610	4	4.6188	5.6568
1⅛	1.2990	1.5910	2⅛	2.4537	3.0052	4⅛	4.7631	5.8336
1⁵/₃₂	1.3351	1.6352	2⁵/₃₂	2.4898	3.0494	4¼	4.9074	6.0104
1³/₁₆	1.3712	1.6793	2³/₁₆	2.5259	3.0936	4⅜	5.0518	6.1872
1⁷/₃₂	1.4073	1.7235	2¼	2.5981	3.1820	4½	5.1961	6.3639

Standard Sheet Gauges and Weights

Ga. No.	Carbon Sheets (To USS or Mfrs. ga.)		Galv. Sheets (To galv. sht. ga.)	Stainless Sheets (To Stainless Sheet Gauge)		
	Thickness in Inches	Weight per sq. Ft. in Lbs.	Weight per Sq. Ft. in Lbs.	Thickness in Inches	Wt. per Sq. Ft. in Lbs.	
					Straight Chrome	Chrome Nickel
	¼", & over are plates					
7	.1793	7.500	....	....	....	....
8	.1644	6.875	....	....	....	....
9	.1494	6.250	....	....	....	....
10	.1345	5.625	5.7812	.140625	5.7937	5.9062
11	.1196	5.000	5.1562	.125000	5.1500	5.2500
12	.1046	4.375	4.5312	.109375	4.5063	4.5937
13	.0897	3.750	3.9062	.093750	3.8625	3.9375
14	.0747	3.125	3.2812	.078125	3.2187	3.2812
15	.0673	2.812	2.9687	.070312	2.8968	2.9531
16	.0598	2.500	2.6562	.062500	2.5750	2.6250
17	.0538	2.250	2.4062	.056250	2.3175	2.3625
18	.0478	2.000	2.1562	.050000	2.0600	2.1000
19	.0418	1.750	1.9062	.043750	1.8025	1.8375
20	.0359	1.500	1.6562	.037500	1.5450	1.5750
21	.0329	1.375	1.5312	.034375	1.4160	1.4437
22	.0299	1.250	1.4062	.031250	1.2875	1.3125
23	.0269	1.125	1.2812	.028125	1.1587	1.1813
24	.0239	1.000	1.1562	.025000	1.0300	1.0500
25	.0209	.875	1.0312	.021875	.9013	.9187
26	.0179	.750	.9062	.018750	.7725	.7875
27	.0164	.6875	.8437	.017187	.7081	.7218
28	.0149	.625	.7812	.015625	.6438	.6562
29	.0135	.5625	.7187	.014062	.5794	.5906
30	.0120	.500	.6562	.012500	.5150	.5250

Weights of Materials

Material	Avg. Wt. per Cu. Ft.
Aluminum	167.1
Brass, cast	519
Brass, rolled	527
Brick, common and hard	125
Bronze, copper 8, tin 1	546
Cement, Portland, 376 lbs. net per bbl	110–115
Concrete, conglomerate, with Portland cement	150
Copper, cast	542
Copper, rolled	555
Fibre, hard	87
Fir, Douglas	31
Glass, window or plate	162
Gravel, round	100–125
Iron, cast	450
Iron, wrought	480
Lead, commercial	710
Mahogany, Honduras, dry	35
Manganese	465
Masonry, granite or limestone	165
Nickel, rolled	541
Oak, live, perfectly dry, .88 to 1.02	59.3
Pine, white, perfectly dry	25
Pine, yellow, southern dry	45
Plastics, molded	74–137
Rubber, manufactured	95
Slate, granulated	95
Snow, freshly fallen	5–15
Spruce, dry	29
Steel	489.6
Tin, cast	459
Walnut, black, perfectly dry	38
Water, distilled or pure rain	62.4
Zinc or spelter, cast	443

Steel Wire Gauges and Decimal Equivalents

Ga. No.	Birmingham Wire Gauge or Stubs Ga. Thickness In.	Wt. per Sq. Ft.	Brown & Sharpe Gauge or American Wire	Steel Wire Gauge (Washburn & Moen)	Ga. No.	Birmingham Wire Gauge or Stubs Ga. Thickness In.	Wt. per Sq. Ft.	Brown & Sharpe Gauge or American Wire	Steel Wire Gauge (Washburn & Moen)
3	.259	10.567	.2294	.2437	17	.058	2.366	.0453	.0540
4	.238	9.710	.2043	.2253	18	.049	1.999	.0403	.0475
5	.220	8.976	.1819	.2070	19	.042	1.714	.0359	.0410
6	.203	8.282	.1620	.1920	20	.035	1.428	.0320	.0348
7	.180	7.344	.1443	.1770	21	.032	1.306	.0285	.0317
8	.165	6.732	.1285	.1620	22	.028	1.142	.0253	.0286
9	.148	6.038	.1144	.1483	23	.025	1.020	.0226	.0258
10	.134	5.467	.1019	.1350	24	.022	.898	.0201	.0230
11	.120	4.896	.0907	.1205	25	.020	.816	.0179	.0204
12	.109	4.447	.0808	.1055	26	.018	.734	.0159	.0181
13	.095	3.876	.0720	.0915	27	.016	.653	.0142	.0173
14	.083	3.386	.0641	.0800	28	.014	.571	.0126	.0162
15	.072	2.938	.0571	.0720	29	.013	.530	.0113	.0150
16	.065	2.652	.0508	.0625	30	.012	.490	.0100	.0140

Hardness Conversion Numbers for Steel
Based on Rockwell Hardness

C Scale 150 Kg., 120° Cone	B Scale 100 Kg., 1/16" Ball	15-N Scale 15 Kg., Superficial Brale Penetrator	30-N Scale 30 Kg., Superficial Brale Penetrator	Hardness No.	Diam. 3000 Kg., 10 mm. Ball	Vickers	Shore Scleroscope	Tensile Strength 1000 Lb./Sq. Inch	C Scale 150 Kg., 120° Cone	B Scale 100 Kg., 1/16" Ball	15-N Scale 15 Kg., Superficial Brale Penetrator	30-N Scale 30 Kg., Superficial Brale Penetrator	Hardness No.	Diam. 3000 Kg., 10 mm. Ball	Vickers	Shore Scleroscope	Tensile Strength 1000 Lb./Sq. Inch
68	...	93.2	84.4	...	...	940	97	...	44	...	82.5	63.1	409	3.02	434	58	206
67	...	92.9	83.6	...	...	900	95	...	43	...	82.0	62.2	400	3.05	423	57	201
66	...	92.5	82.8	...	...	865	92	...	42	...	81.5	61.3	390	3.09	412	56	196
65	...	92.2	81.9	739	...	832	91	...	41	...	80.9	60.4	381	3.12	402	55	191
64	...	91.8	81.1	722	2.28	800	88	...	40	...	80.4	59.5	371	3.16	392	54	186
63	...	91.4	80.1	705	2.31	772	87	...	39	...	79.9	58.6	362	3.19	382	52	181
62	...	91.1	79.3	688	2.33	746	85	...	38	...	79.4	57.7	353	3.24	372	51	176
61	...	90.7	78.4	670	2.36	720	83	...	37	...	78.8	56.8	344	3.28	363	50	172
60	...	90.2	77.5	654	2.40	697	81	...	36	(109)	78.3	55.9	336	3.32	354	49	168
59	...	89.8	76.6	634	2.43	674	80	326	35	...	77.7	55.0	327	3.37	345	48	163
58	...	89.3	75.7	615	2.47	653	78	315	34	(108)	77.2	54.2	319	3.41	336	47	159
57	...	88.9	74.8	595	2.51	633	76	305	33	...	76.6	53.3	311	3.45	327	46	154
56	...	88.3	73.9	577	2.55	613	75	295	32	(107)	76.1	52.1	301	3.51	318	44	150
55	...	87.9	73.0	560	2.58	595	74	287	31	(106)	75.6	51.3	294	3.54	310	43	146
54	...	87.4	72.0	543	2.63	577	72	278	30	...	75.0	50.4	286	3.59	302	42	142
53	...	86.9	71.2	525	2.67	560	71	269	29	...	74.5	49.5	279	3.63	294	41	138
52	...	86.4	70.2	512	2.71	544	69	262	28	(104)	73.9	48.6	271	3.69	286	41	134
51	...	85.9	69.4	496	2.75	528	68	253	27	(103)	73.3	47.7	264	3.74	279	40	131
50	...	85.5	68.5	481	2.79	513	67	245	26	...	72.8	46.8	258	3.78	272	39	127
49	...	85.0	67.6	469	2.83	498	66	239	25	...	72.2	45.9	253	3.81	266	38	124
48	...	84.5	66.7	455	2.87	484	64	232	24	(101)	71.6	45.0	247	3.84	260	37	121
47	...	83.9	65.8	443	2.91	471	63	225	23	100	71.0	44.0	243	3.88	254	36	118
46	...	83.5	64.8	432	2.94	458	62	219	22	99	70.5	43.2	237	3.93	248	35	115
45	...	83.0	64.0	421	2.98	446	60	212	21	...	69.9	42.3	231	3.98	243	35	113
44	...	82.5	63.1	409	3.02	434	58	206	20	98	69.4	41.5	226	4.02	238	34	110

Types of Stainless Steels

CHROMIUM-NICKEL TYPES
Austenitic Non-Magnetic Non-Hardenable by Heat Treatment

AISI Type Number	Carbon %	Manganese Max.%	Phosphorus Max.%	Sulfur Max.%	Silicon Max.%	Chromium %	Nickel %	Other Elements %
301	Over 0.08 to 0.15	2.00	0.04	0.03	1.00	16.0-18.0	6.0-8.0	
302	Over 0.08 to 0.15	2.00	0.04	0.03	1.00	17.0-19.0	8.0-10.0	
302B	Over 0.08 to 0.15	2.00	0.04	0.03	2.0-3.0	17.0-19.0	8.0-10.0	
303	0.15 max.	2.00	X	X	1.00	17.0-19.0	8.0-10.0	
304	0.08 max.	2.00	0.04	0.03	1.00	18.0-20.0	8.0-11.0	
304L	0.03 max.	2.00	0.04	0.03	1.00	18.0-20.0	8.0-11.0	
305	0.12 max.	2.00	0.04	0.03	1.00	17.0-19.0	10.0-13.0	
308	0.08 max.	2.00	0.04	0.03	1.00	19.0-21.0	10.0-12.0	
309	0.20 max.	2.00	0.04	0.03	1.00	22.0-24.0	12.0-15.0	
309S	0.08 max.	2.00	0.04	0.03	1.00	22.0-24.0	12.0-15.0	
310	0.25 max.	2.00	0.04	0.03	1.50	24.0-26.0	19.0-22.0	
310S	0.08 max.	2.00	0.04	0.03	1.50	24.0-26.0	19.0-22.0	
314	0.25 max.	2.00	0.04	0.03	1.5-3.0	23.0-26.0	19.0-22.0	
316	0.10 max.	2.00	0.04	0.03	1.00	16.0-18.0	10.0-14.0	Mo-1.75-2.50
316L	0.03 max.	2.00	0.04	0.03	1.00	16.0-18.0	10.0-14.0	Mo-1.75-2.50
317	0.10 max.	2.00	0.04	0.03	1.00	18.0-20.0	11.0-14.0	Mo-3.00-4.00
317L	0.03 max.	2.00	0.04	0.03	1.00	18.0-20.0	11.0-14.0	Mo-3.00-4.00
330	0.25 max.	2.00	0.04	0.04	1.00	14.0-16.0	33.0-36.0	
321	0.08 max.	2.00	0.04	0.03	1.00	17.0-19.0	8.0-11.0	Ti-5 X C min.
347	0.08 max.	2.00	0.04	0.03	1.00	17.0-19.0	9.0-12.0	Cb-Ta-10XC min.
347F	0.08 max.	2.00	X	X	1.00	17.0-19.0	9.0-12.0	Cb-Ta-10XC min.

CHROMIUM TYPES
Martensitic Magnetic Hardenable by Heat Treatment

	Carbon %	Manganese Max.%	Phosphorus Max.%	Sulfur Max.%	Silicon Max.%	Chromium %	Nickel %	Other Elements %
403	0.15 max.	1.00	0.04	0.03	0.50	11.5-13.0		
410	0.15 max.	1.00	0.04	0.03	1.00	11.5-13.5		
414	0.15 max.	1.00	0.04	0.03	1.00	11.5-13.5	1.25-2.50	
416	0.15 max.	1.25	X	X	1.00	12.0-14.0		X
420	Over 0.15	1.00	0.04	0.03	1.00	12.0-14.0		
420F	Over 0.15	1.00	X	X	1.00	12.0-14.0		X
431	0.20 max.	1.00	0.04	0.03	1.00	15.0-17.0	1.25-2.50	
440A	0.60-0.75	1.00	0.04	0.03	1.00	16.0-18.0		Mo-0.75 max.
440B	0.75-0.95	1.00	0.04	0.03	1.00	16.0-18.0		Mo-0.75 max.
440C	0.95-1.20	1.00	0.04	0.03	1.00	16.0-18.0		Mo-0.75 max.
440F	0.95-1.20	1.00	X	X	1.00	16.0-18.0		X

CHROMIUM TYPES
Ferritic Magnetic Non-Hardenable by Heat Treatment

	Carbon %	Manganese Max.%	Phosphorus Max.%	Sulfur Max.%	Silicon Max.%	Chromium %	Nickel %	Other Elements %
405	0.08 max.	1.00	0.04	0.03	1.00	11.5-13.5		Al-0.10-0.30
430	0.12 max.	1.00	0.04	0.03	1.00	14.0-18.0		
430 Ti	0.12 max.	1.00	0.04	0.03	1.00	14.0-18.0		Ti-6 X C min.
430F	0.12 max.	1.25	X	X	1.00	14.0-18.0		X
442	0.20 max.	1.00	0.04	0.03	1.00	18.0-23.0		
446	0.35 max.	1.50	0.04	0.03	1.00	23.0-27.0		

X Free-Machining Stainless Steels—Phosphorus, Sulfur or Selenium—0.07% min., Zr or Mo—0.60% max.

Designations for Aluminum

		AA Number
Aluminum — 99.00% minimum and greater		1xxx

	Major Alloying Element	
Aluminum Alloys grouped by major Alloying Elements	Copper	2xxx
	Manganese	3xxx
	Silicon	4xxx
	Magnesium	5xxx
	Magnesium and Silicon	6xxx
	Zinc	7xxx
	Other Element	8xxx

	2017	5056
1160	2117	X5356
1175	2018	5357
1187	2218	6061
EC	2618	6062
1095	2024	6063
1099	2025	6066
1197	2225	7070
1085	4032	7072
1187	4043	7075
1090	4343	7277
1070	X4543	X7178
1080	4343	X8280
1180	4045	8112
1075	5050	4343
1050	5005	4043
1060	X5405	5005
1160	6151	5357
1150	X6251	6951
1030	6951	6003
1145	5052	5083
1100	5652	5086
3003	6053	2014
3004	6253	5005
X3005	X6453	6003
2011	6553	1130
2014	5154	8099
X2214	5254	1235
X2316	X5055	1180

Locational Clearance Fits
Limits are in thousandths of an inch.

Nominal Size Range Inches Over	To	Class LC 1 Limits of Clearance	Hole H6	Shaft h5	Class LC 2 Limits of Clearance	Hole H7	Shaft h6	Class LC 3 Limits of Clearance	Hole H8	Shaft h7	Class LC 4 Limits of Clearance	Hole H10	Shaft h9	Class LC 5 Limits of Clearance	Hole H7	Shaft g6
0 —	0.12	0 / 0.45	+0.25 / -0	+0 / -0.2	0 / 0.65	+0.4 / -0	+0 / -0.25	0 / 1	+0.6 / -0	+0 / -0.4	0 / 2.6	+1.6 / -0	+0 / -1.0	0.1 / 0.75	+0.4 / -0	-0.1 / -0.35
0.12—	0.24	0 / 0.5	+0.3 / -0	+0 / -0.2	0 / 0.8	+0.5 / -0	+0 / -0.3	0 / 1.2	+0.7 / -0	+0 / -0.5	0 / 3.0	+1.8 / -0	+0 / -1.2	0.15 / 0.95	+0.5 / -0	-0.15 / -0.45
0.24—	0.40	0 / 0.65	+0.4 / -0	+0 / -0.25	0 / 1.0	+0.6 / -0	+0 / -0.4	0 / 1.5	+0.9 / -0	+0 / -0.6	0 / 3.6	+2.2 / -0	+0 / -1.4	0.2 / 1.2	+0.6 / -0	-0.2 / -0.6
0.40—	0.71	0 / 0.7	+0.4 / -0	+0 / -0.3	0 / 1.1	+0.7 / -0	+0 / -0.4	0 / 1.7	+1.0 / -0	+0 / -0.7	0 / 4.4	+2.8 / -0	+0 / -1.6	0.25 / 1.35	+0.7 / -0	-0.25 / -0.65
0.71—	1.19	0 / 0.9	+0.5 / -0	+0 / -0.4	0 / 1.3	+0.8 / -0	+0 / -0.5	0 / 2	+1.2 / -0	+0 / -0.8	0 / 5.5	+3.5 / -0	+0 / -2.0	0.3 / 1.6	+0.8 / -0	-0.3 / -0.8
1.19—	1.97	0 / 1.0	+0.6 / -0	+0 / -0.4	0 / 1.6	+1.0 / -0	+0 / -0.6	0 / 2.6	+1.6 / -0	+0 / -1	0 / 6.5	+4.0 / -0	+0 / -2.5	0.4 / 2.0	+1.0 / -0	-0.4 / -1.0
1.97—	3.15	0 / 1.2	+0.7 / -0	+0 / -0.5	0 / 1.9	+1.2 / -0	+0 / -0.7	0 / 3	+1.8 / -0	+0 / -1.2	0 / 7.5	+4.5 / -0	+0 / -3	0.4 / 2.3	+1.2 / -0	-0.4 / -1.1
3.15—	4.73	0 / 1.5	+0.9 / -0	+0 / -0.6	0 / 2.3	+1.4 / -0	+0 / -0.9	0 / 3.6	+2.2 / -0	+0 / -1.4	0 / 8.5	+5.0 / -0	+0 / -3.5	0.5 / 2.8	+1.4 / -0	-0.5 / -1.4

Nominal Size Range Inches Over	To	Class LC 6 Limits of Clearance	Hole H9	Shaft f8	Class LC 7 Limits of Clearance	Hole H10	Shaft e9	Class LC 8 Limits of Clearance	Hole H10	Shaft d9	Class LC 9 Limits of Clearance	Hole H11	Shaft c10	Class LC 10 Limits of Clearance	Hole H12	Shaft	Class LC 11 Limits of Clearance	Hole H13	Shaft
0 —	0.12	0.3 / 1.9	+1.0 / 0	-0.3 / -0.9	0.6 / 3.2	+1.6 / 0	-0.6 / -1.6	1.0 / 3.6	+1.6 / -0	-1.0 / -2.0	2.5 / 6.6	+2.5 / -0	-2.5 / -4.1	4 / 12	+4 / -0	-4 / -8	5 / 17	+6 / -0	-5 / -11
0.12—	0.24	0.4 / 2.3	+1.2 / 0	-0.4 / -1.1	0.8 / 3.8	+1.8 / 0	-0.8 / -2.0	1.2 / 4.2	+1.8 / -0	-1.2 / -2.4	2.8 / 7.6	+3.0 / -0	-2.8 / -4.6	4.5 / 14.5	+5 / -0	-4.5 / -9.5	6 / 20	+7 / -0	-6 / -13
0.24—	0.40	0.5 / 2.8	+1.4 / 0	-0.5 / -1.4	1.0 / 4.6	+2.2 / 0	-1.0 / -2.4	1.6 / 5.2	+2.2 / -0	-1.6 / -3.0	3.0 / 8.7	+3.5 / -0	-3.0 / -5.2	5 / 17	+6 / -0	-5 / -11	7 / 25	+9 / -0	-7 / -16
0.40—	0.71	0.6 / 3.2	+1.6 / 0	-0.6 / -1.6	1.2 / 5.6	+2.8 / 0	-1.2 / -2.8	2.0 / 6.4	+2.8 / -0	-2.0 / -3.6	3.5 / 10.3	+4.0 / -0	-3.5 / -6.3	6 / 20	+7 / -0	-6 / -13	8 / 28	+10 / -0	-8 / -18
0.71—	1.19	0.8 / 4.0	+2.0 / 0	-0.8 / -2.0	1.6 / 7.1	+3.5 / 0	-1.6 / -3.6	2.5 / 8.0	+3.5 / -0	-2.5 / -4.5	4.5 / 13.0	+5.0 / -0	-4.5 / -8.0	7 / 23	+8 / -0	-7 / -15	10 / 34	+12 / -0	-10 / -22
1.19—	1.97	1.0 / 5.1	+2.5 / 0	-1.0 / -2.6	2.0 / 8.5	+4.0 / 0	-2.0 / -4.5	3.0 / 9.5	+4.0 / -0	-3.0 / -5.5	5 / 15	+6 / -0	-5 / -9	8 / 28	+10 / -0	-8 / -18	12 / 44	+16 / -0	-12 / -28
1.97—	3.15	1.2 / 6.0	+3.0 / 0	-1.2 / -3.0	2.5 / 10.0	+4.5 / 0	-2.5 / -5.5	4.0 / 11.5	+4.5 / -0	-4.0 / -7.0	6 / 17.5	+7 / -0	-6 / -10.5	10 / 34	+12 / -0	-10 / -22	14 / 50	+18 / -0	-14 / -32
3.15—	4.73	1.4 / 7.1	+3.5 / 0	-1.4 / -3.6	3.0 / 11.5	+5.0 / 0	-3.0 / -6.5	5.0 / 13.5	+5.0 / -0	-5.0 / -8.5	7 / 21	+9 / -0	-7 / -12	11 / 39	+14 / -0	-11 / -25	16 / 60	+22 / -0	-16 / -38

Locational Transition Fits
Limits are in thousandths of an inch.
"Fit" represents the masimum interference (minus values) and the maximum clearance (plus values).

Nominal Size Range Inches Over	To	Class LT 1 Fit	Hole H7	Shaft js6	Class LT 2 Fit	Hole H8	Shaft js7	Class LT 3 Fit	Hole H7	Shaft k6	Class LT 4 Fit	Hole H8	Shaft k7	Class LT 5 Fit	Hole H7	Shaft n6	Class LT 6 Fit	Hole H7	Shaft n7
0 —	0.12	-0.10 / +0.50	+0.4 / -0	+0.10 / -0.10	-0.2 / +0.8	+0.6 / -0	+0.2 / -0.2							-0.5 / +0.15	+0.4 / -0	+0.5 / +0.25	-0.65 / +0.15	+0.4 / -0	+0.65 / +0.25
0.12 —	0.24	-0.15 / +0.65	+0.5 / -0	+0.15 / -0.15	-0.25 / +0.95	+0.7 / -0	+0.25 / -0.25							-0.6 / +0.2	+0.5 / -0	+0.6 / +0.3	-0.8 / +0.2	+0.5 / -0	+0.8 / +0.3
0.24 —	0.40	-0.2 / +0.8	+0.6 / -0	+0.2 / -0.2	-0.3 / +1.2	+0.9 / -0	+0.3 / -0.3	-0.5 / +0.5	+0.6 / -0	+0.5 / +0.1	-0.7 / +0.8	+0.9 / -0	+0.7 / +0.1	-0.8 / +0.2	+0.6 / -0	+0.8 / +0.4	-1.0 / +0.2	+0.6 / -0	+1.0 / +0.4
0.40 —	0.71	-0.2 / +0.9	+0.7 / -0	+0.2 / -0.2	-0.35 / +1.35	+1.0 / -0	+0.35 / -0.35	-0.5 / +0.6	+0.7 / -0	+0.5 / +0.1	-0.8 / +0.9	+1.0 / -0	+0.8 / +0.1	-0.9 / +0.2	+0.7 / -0	+0.9 / +0.5	-1.2 / +0.2	+0.7 / -0	+1.2 / +0.5
0.71 —	1.19	-0.25 / +1.05	+0.8 / -0	+0.25 / -0.25	-0.4 / +1.6	+1.2 / -0	+0.4 / -0.4	-0.6 / +0.7	+0.8 / -0	+0.6 / +0.1	-0.9 / +1.1	+1.2 / -0	+0.9 / +0.1	-1.1 / +0.2	+0.8 / -0	+1.1 / +0.6	-1.4 / +0.2	+0.8 / -0	+1.4 / +0.6
1.19 —	1.97	-0.3 / +1.3	+1.0 / -0	+0.3 / -0.3	-0.5 / +2.1	+1.6 / -0	+0.5 / -0.5	-0.7 / +0.9	+1.0 / -0	+0.7 / +0.1	-1.1 / +1.5	+1.6 / -0	+1.1 / +0.1	-1.3 / +0.3	+1.0 / -0	+1.3 / +0.7	-1.7 / +0.3	+1.0 / -0	+1.7 / +0.7
1.97 —	3.15	-0.3 / +1.5	+1.2 / -0	+0.3 / -0.3	-0.6 / +2.4	+1.8 / -0	+0.6 / -0.6	-0.8 / +1.1	+1.2 / -0	+0.8 / +0.1	-1.3 / +1.7	+1.8 / -0	+1.3 / +0.1	-1.5 / +0.4	+1.2 / -0	+1.5 / +0.8	-2.0 / +0.4	+1.2 / -0	+2.0 / +0.8
3.15 —	4.73	-0.4 / +1.8	+1.4 / -0	+0.4 / -0.4	-0.7 / +2.9	+2.2 / -0	+0.7 / -0.7	-1.0 / +1.3	+1.4 / -0	+1.0 / +0.1	-1.5 / +2.1	+2.2 / -0	+1.5 / +0.1	-1.9 / +0.4	+1.4 / -0	+1.9 / +1.0	-2.4 / +0.4	+1.4 / -0	+2.4 / +1.0

Locational Interference Fits
Limits are in thousandths of an inch.

Nominal Size Range Inches		Class LN 1			Class LN 2			Class LN 3		
		Limits of Interference	Standard Limits		Limits of Interference	Standard Limits		Limits of Interference	Standard Limits	
Over	To		Hole H6	Shaft n5		Hole H7	Shaft p6		Hole H7	Shaft r6
0	0.12	0 / 0.45	+0.25 / −0	+0.45 / +0.25	0 / 0.65	+0.4 / −0	+0.65 / +0.4	0.1 / 0.75	+0.4 / −0	+0.75 / +0.5
0.12	0.24	0 / 0.5	+0.3 / −0	+0.5 / +0.3	0 / 0.8	+0.5 / −0	+0.8 / +0.5	0.1 / 0.9	+0.5 / −0	+0.9 / +0.6
0.24	0.40	0 / 0.65	+0.4 / −0	+0.65 / +0.4	0 / 1.0	+0.6 / −0	+1.0 / +0.6	0.2 / 1.2	+0.6 / −0	+1.2 / +0.8
0.40	0.71	0 / 0.8	+0.4 / −0	+0.8 / +0.5	0 / 1.1	+0.7 / −0	+1.1 / +0.7	0.3 / 1.4	+0.7 / −0	+1.4 / +1.0
0.71	1.19	0 / 1.0	+0.5 / −0	+1.0 / +0.5	0 / 1.3	+0.8 / −0	+1.3 / +0.8	0.4 / 1.7	+0.8 / −0	+1.7 / +1.2
1.19	1.97	0 / 1.1	+0.6 / −0	+1.1 / +0.6	0 / 1.6	+1.0 / −0	+1.6 / +1.0	0.4 / 2.0	+1.0 / −0	+2.0 / +1.4
1.97	3.15	0.1 / 1.3	+0.7 / −0	+1.3 / +0.7	0.2 / 2.1	+1.2 / −0	+2.1 / +1.4	0.4 / 2.3	+1.2 / −0	+2.3 / +1.6
3.15	4.73	0.1 / 1.6	+0.9 / −0	+1.6 / +1.0	0.2 / 2.5	+1.4 / −0	+2.5 / +1.6	0.6 / 2.9	+1.4 / −0	+2.9 / +2.0

Solution of Right-angled Traingles

As shown in the illustration, the sides of the right-angled triangle are designated a and b and the hypotenuse, c. The angles opposite each of these sides are designated A and B respectively.

Angle C, opposite the hypotenuse c is the right angle, and is therefore always one of the known quantities.

Sides and Angles Known	Formulas for Sides and Angles to be Found		
Side a; side b	$c = \sqrt{a^2 + b^2}$	$\tan A = \dfrac{a}{b}$	$B = 90° - A$
Side a; hypotenuse c	$b = \sqrt{c^2 - a^2}$	$\sin A = \dfrac{a}{c}$	$B = 90° - A$
Side b; hypotenuse c	$a = \sqrt{c^2 - b^2}$	$\sin B = \dfrac{b}{c}$	$A = 90° - B$
Hypotenuse c; angle B	$b = c \times \sin B$	$a = c \times \cos B$	$A = 90° - B$
Hypotenuse c; angle A	$b = c \times \cos A$	$a = c \times \sin A$	$B = 90° - A$
Side b; angle B	$c = \dfrac{b}{\sin B}$	$a = b \times \cot B$	$A = 90° - B$
Side b; angle A	$c = \dfrac{b}{\cos A}$	$a = b \times \tan A$	$B = 90° - A$
Side a; angle B	$c = \dfrac{a}{\cos B}$	$b = a \times \tan B$	$A = 90° - B$
Side a; angle A	$c = \dfrac{a}{\sin A}$	$b = a \times \cot A$	$B = 90° - A$

Force and Shrink Fits
Limits are in thousandths of an inch.

Nominal Size Range Inches		Class FN 1			Class FN 2			Class FN 3			Class FN 4			Class FN 5		
		Limits of Interference	Standard Limits		Limits of Interference	Standard Limits		Limits of Interference	Standard Limits		Limits of Interference	Standard Limits		Limits of Interference	Standard Limits	
Over	To		Hole H6	Shaft		Hole H7	Shaft s6		Hole H7	Shaft t6		Hole H7	Shaft u6		Hole H8	Shaft x7
0	0.12	0.05 / 0.5	+0.25 / −0	+0.5 / +0.3	0.2 / 0.85	+0.4 / −0	+0.85 / +0.6				0.3 / 0.95	+0.4 / −0	+0.95 / +0.7	0.3 / 1.3	+0.6 / −0	+1.3 / +0.9
0.12	0.24	0.1 / 0.6	+0.3 / −0	+0.6 / +0.4	0.2 / 1.0	+0.5 / −0	+1.0 / +0.7				0.4 / 1.2	+0.5 / −0	+1.2 / +0.9	0.5 / 1.7	+0.7 / −0	+1.7 / +1.2
0.24	0.40	0.1 / 0.75	+0.4 / −0	+0.75 / +0.5	0.4 / 1.4	+0.6 / −0	+1.4 / +1.0				0.6 / 1.6	+0.6 / −0	+1.6 / +1.2	0.5 / 2.0	+0.9 / −0	+2.0 / +1.4
0.40	0.56	0.1 / 0.8	+0.4 / −0	+0.8 / +0.5	0.5 / 1.6	+0.7 / −0	+1.6 / +1.2				0.7 / 1.8	+0.7 / −0	+1.8 / +1.4	0.6 / 2.3	+1.0 / −0	+2.3 / +1.6
0.56	0.71	0.2 / 0.9	+0.4 / −0	+0.9 / +0.6	0.5 / 1.6	+0.7 / −0	+1.6 / +1.2				0.7 / 1.8	+0.7 / −0	+1.8 / +1.4	0.8 / 2.5	+1.0 / −0	+2.5 / +1.8
0.71	0.95	0.2 / 1.1	+0.5 / −0	+1.1 / +0.7	0.6 / 1.9	+0.8 / −0	+1.9 / +1.4				0.8 / 2.1	+0.8 / −0	+2.1 / +1.6	1.0 / 3.0	+1.2 / −0	+3.0 / +2.2
0.95	1.19	0.3 / 1.2	+0.5 / −0	+1.2 / +0.8	0.6 / 1.9	+0.8 / −0	+1.9 / +1.4	0.8 / 2.1	+0.8 / −0	+2.1 / +1.6	1.0 / 2.3	+0.8 / −0	+2.3 / +1.8	1.3 / 3.3	+1.2 / −0	+3.3 / +2.5
1.19	1.58	0.3 / 1.3	+0.6 / −0	+1.3 / +0.9	0.8 / 2.4	+1.0 / −0	+2.4 / +1.8	1.0 / 2.6	+1.0 / −0	+2.6 / +2.0	1.5 / 3.1	+1.0 / −0	+3.1 / +2.5	1.4 / 4.0	+1.6 / −0	+4.0 / +3.0
1.58	1.97	0.4 / 1.4	+0.6 / −0	+1.4 / +1.0	0.8 / 2.4	+1.0 / −0	+2.4 / +1.8	1.2 / 2.8	+1.0 / −0	+2.8 / +2.2	1.8 / 3.4	+1.0 / −0	+3.4 / +2.8	2.4 / 5.0	+1.6 / −0	+5.0 / +4.0
1.97	2.56	0.6 / 1.8	+0.7 / −0	+1.8 / +1.3	0.8 / 2.7	+1.2 / −0	+2.7 / +2.0	1.3 / 3.2	+1.2 / −0	+3.2 / +2.5	2.3 / 4.2	+1.2 / −0	+4.2 / +3.5	3.2 / 6.2	+1.8 / −0	+6.2 / +5.0
2.56	3.15	0.7 / 1.9	+0.7 / −0	+1.9 / +1.4	1.0 / 2.9	+1.2 / −0	+2.9 / +2.2	1.8 / 3.7	+1.2 / −0	+3.7 / +3.0	2.8 / 4.7	+1.2 / −0	+4.7 / +4.0	4.2 / 7.2	+1.8 / −0	+7.2 / +6.0
3.15	3.94	0.9 / 2.4	+0.9 / −0	+2.4 / +1.8	1.4 / 3.7	+1.4 / −0	+3.7 / +2.8	2.1 / 4.4	+1.4 / −0	+4.4 / +3.5	3.6 / 5.9	+1.4 / −0	+5.9 / +5.0	4.8 / 8.4	+2.2 / −0	+8.4 / +7.0
3.94	4.73	1.1 / 2.6	+0.9 / −0	+2.6 / +2.0	1.6 / 3.9	+1.4 / −0	+3.9 / +3.0	2.6 / 4.9	+1.4 / −0	+4.9 / +4.0	4.6 / 6.9	+1.4 / −0	+6.9 / +6.0	5.8 / 9.4	+2.2 / −0	+9.4 / +8.0

Unified Screw Thread Standard Series

Nominal Size — Preferred	Nominal Size — Secondary	Basic Major (Nominal) Diameter	Coarse UNC	Fine UNF	Extra-Fine UNEF	4 UN	6 UN	8 UN	12 UN	16 UN	20 UN	28 UN	32 UN	Nominal Size
0		0.0600	—	80	—	—	—	—	—	—	—	—	—	0
	1	0.0730	64	72	—	—	—	—	—	—	—	—	—	1
2		0.0860	56	64	—	—	—	—	—	—	—	—	—	2
	3	0.0990	48	56	—	—	—	—	—	—	—	—	—	3
4		0.1120	40	48	—	—	—	—	—	—	—	—	—	4
5		0.1250	40	44	—	—	—	—	—	—	—	—	—	5
6		0.1380	32	40	—	—	—	—	—	—	—	—	UNC	6
8		0.1640	32	36	—	—	—	—	—	—	—	—	UNC	8
10		0.1900	24	32	—	—	—	—	—	—	—	UNF	UNF	10
	12	0.2160	24	28	32	—	—	—	—	—	—	UNF	UNEF	12
1/4		0.2500	20	28	32	—	—	—	—	—	UNC	UNF	UNEF	1/4
5/16		0.3125	18	24	32	—	—	—	—	—	20	28	UNEF	5/16
3/8		0.3750	16	24	32	—	—	—	—	UNC	20	28	UNEF	3/8
7/16		0.4375	14	20	28	—	—	—	—	16	UNF	UNEF	32	7/16
1/2		0.5000	13	20	28	—	—	—	—	16	UNF	UNEF	32	1/2
9/16		0.5625	12	18	24	—	—	—	UNC	16	20	28	32	9/16
5/8		0.6250	11	18	24	—	—	—	12	16	20	28	32	5/8
	11/16	0.6875	—	—	24	—	—	—	12	16	20	28	32	11/16
3/4		0.7500	10	16	20	—	—	—	12	UNF	UNEF	28	32	3/4
	13/16	0.8125	—	—	20	—	—	—	12	16	UNEF	28	32	13/16
7/8		0.8750	9	14	20	—	—	—	12	16	UNEF	28	32	7/8
	15/16	0.9375	—	—	20	—	—	—	12	16	UNEF	28	32	15/16
1		1.0000	8	12	20	—	—	UNC	UNF	16	UNEF	28	32	1
	1-1/16	1.0625	—	—	18	—	—	8	12	16	20	28	—	1-1/16
1-1/8		1.1250	7	12	18	—	—	8	UNF	16	20	28	—	1-1/8
	1-3/16	1.1875	—	—	18	—	—	8	12	16	20	28	—	1-3/16
1-1/4		1.2500	7	12	18	—	—	8	UNF	16	20	28	—	1-1/4
	1-5/16	1.3125	—	12	18	—	—	8	12	16	20	28	—	1-5/16
1-3/8		1.3750	6	12	18	—	UNC	8	UNF	16	20	28	—	1-3/8
	1-7/16	1.4375	—	—	18	—	6	8	12	16	20	28	—	1-7/16
1-1/2		1.5000	6	12	18	—	UNC	8	UNF	16	20	28	—	1-1/2
	1-9/16	1.5625	—	—	18	—	6	8	12	16	20	—	—	1-9/16
1-5/8		1.6250	—	—	18	—	6	8	12	16	20	—	—	1-5/8
	1-11/16	1.6875	—	—	18	—	6	8	12	16	20	—	—	1-11/16
1-3/4		1.7500	5	—	—	—	6	8	12	16	20	—	—	1-3/4
	1-13/16	1.8125	—	—	—	—	6	8	12	16	20	—	—	1-13/16
1-7/8		1.8750	—	—	—	—	6	8	12	16	20	—	—	1-7/8
	1-15/16	1.9375	—	—	—	—	6	8	12	16	20	—	—	1-15/16
2		2.0000	4½	—	—	—	6	8	12	16	20	—	—	2
	2-1/8	2.1250	—	—	—	—	6	8	12	16	20	—	—	2-1/8
2-1/4		2.2500	4½	—	—	—	6	8	12	16	20	—	—	2-1/4
	2-3/8	2.3750	—	—	—	—	6	8	12	16	20	—	—	2-3/8
2-1/2		2.5000	4	—	—	UNC	6	8	12	16	20	—	—	2-1/2
	2-5/8	2.6250	—	—	—	4	6	8	12	16	20	—	—	2-5/8
2-3/4		2.7500	4	—	—	UNC	6	8	12	16	20	—	—	2-3/4
	2-7/8	2.8750	—	—	—	4	6	8	12	16	20	—	—	2-7/8
3		3.0000	4	—	—	UNC	6	8	12	16	20	—	—	3
	3-1/8	3.1250	—	—	—	4	6	8	12	16	—	—	—	3-1/8
3-1/4		3.2500	4	—	—	UNC	6	8	12	16	—	—	—	3-1/4
	3-3/8	3.3750	—	—	—	4	6	8	12	16	—	—	—	3-3/8
3-1/2		3.5000	4	—	—	UNC	6	8	12	16	—	—	—	3-1/2
	3-5/8	3.6250	—	—	—	4	6	8	12	16	—	—	—	3-5/8
3-3/4		3.7500	4	—	—	UNC	6	8	12	16	—	—	—	3-3/4
	3-7/8	3.8750	—	—	—	4	6	8	12	16	—	—	—	3-7/8
4		4.0000	4	—	—	UNC	6	8	12	16	—	—	—	4
	4-1/8	4.1250	—	—	—	4	6	8	12	16	—	—	—	4-1/8
4-1/4		4.2500	—	—	—	4	6	8	12	16	—	—	—	4-1/4
	4-3/8	4.3750	—	—	—	4	6	8	12	16	—	—	—	4-3/8
4-1/2		4.5000	—	—	—	4	6	8	12	16	—	—	—	4-1/2
	4-5/8	4.6250	—	—	—	4	6	8	12	16	—	—	—	4-5/8
4-3/4		4.7500	—	—	—	4	6	8	12	16	—	—	—	4-3/4
	4-7/8	4.8750	—	—	—	4	6	8	12	16	—	—	—	4-7/8
5		5.0000	—	—	—	4	6	8	12	16	—	—	—	5
	5-1/8	5.1250	—	—	—	4	6	8	12	16	—	—	—	5-1/8
5-1/4		5.2500	—	—	—	4	6	8	12	16	—	—	—	5-1/4
	5-3/8	5.3750	—	—	—	4	6	8	12	16	—	—	—	5-3/8
5-1/2		5.5000	—	—	—	4	6	8	12	16	—	—	—	5-1/2
	5-5/8	5.6250	—	—	—	4	6	8	12	16	—	—	—	5-5/8
5-3/4		5.7500	—	—	—	4	6	8	12	16	—	—	—	5-3/4
	5-7/8	5.8750	—	—	—	4	6	8	12	16	—	—	—	5-7/8
6		6.0000	—	—	—	4	6	8	12	16	—	—	—	6

(Graded Pitch Series: Coarse UNC, Fine UNF, Extra-Fine UNEF; Constant Pitch Series: 4 UN through 32 UN. All values are Threads Per Inch.)

Taper Pipe Threads
(All dimensions in inches.)

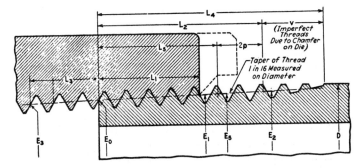

Basic Dimensions, American Standard Taper Pipe Thread

Nominal Pipe Size	Outside Diameter of Pipe D	Threads per Inch n	Pitch of Thread p	Pitch Diameter at Beginning of External Thread E_0	Hand-Tight Engagement			Effective Thread, External		
					Length L_1 In.	Thds	Diam E_1 In.	Length L_4 In.	Thds	Diam E_2 In.
1/16	0.3125	27	0.03704	0.27118	0.160	4.32	0.28118	0.2611	7.05	0.28750
1/8	0.405	27	0.03704	0.36351	0.180	4.86	0.37476	0.2639	7.12	0.38000
1/4	0.540	18	0.05556	0.47739	0.200	3.60	0.48989	0.4018	7.23	0.50250
3/8	0.675	18	0.05556	0.61201	0.240	4.32	0.62701	0.4078	7.34	0.63750
1/2	0.840	14	0.07143	0.75843	0.320	4.48	0.77843	0.5337	7.47	0.79179
3/4	1.050	14	0.07143	0.96768	0.339	4.75	0.98887	0.5457	7.64	1.00179
1	1.315	11 1/2	0.08696	1.21363	0.400	4.60	1.23863	0.6828	7.85	1.25630
1 1/4	1.660	11 1/2	0.08696	1.55713	0.420	4.83	1.58338	0.7068	8.13	1.60130
1 1/2	1.900	11 1/2	0.08696	1.79609	0.420	4.83	1.82234	0.7235	8.32	1.84130
2	2.375	11 1/2	0.08696	2.26902	0.436	5.01	2.29627	0.7565	8.70	2.31630
2 1/2	2.875	8	0.12500	2.71953	0.682	5.46	2.76216	1.1375	9.10	2.79062
3	3.500	8	0.12500	3.34062	0.766	6.13	3.38850	1.2000	9.60	3.41562
3 1/2	4.000	8	0.12500	3.83750	0.821	6.57	3.88881	1.2500	10.00	3.91562
4	4.500	8	0.12500	4.33438	0.844	6.75	4.38712	1.3000	10.40	4.41562
5	5.563	8	0.12500	5.39073	0.937	7.50	5.44929	1.4063	11.25	5.47862
6	6.625	8	0.12500	6.44609	0.958	7.66	6.50597	1.5125	12.10	6.54062
8	8.625	8	0.12500	8.43359	1.063	8.50	8.50003	1.7125	13.70	8.54062
10	10.750	8	0.12500	10.54531	1.210	9.68	10.62094	1.9250	15.40	10.66562
12	12.750	8	0.12500	12.53281	1.360	10.88	12.61781	2.1250	17.00	12.66562
14 OD	14.000	8	0.12500	13.77500	1.562	12.50	13.87262	2.2500	18.90	13.91562
16 OD	16.000	8	0.12500	15.76250	1.812	14.50	15.87575	2.4500	19.60	15.91562
18 OD	18.000	8	0.12500	17.75000	2.000	16.00	17.87500	2.6500	21.20	17.91562
20 OD	20.000	8	0.12500	19.73750	2.125	17.00	19.87031	2.8500	22.80	19.91562
24 OD	24.000	8	0.12500	23.71250	2.375	19.00	23.86094	3.2500	26.00	23.91562

Tap Drill Sizes for Unified and American Threads

SCREW THREAD		TAP DRILL
No. or Diameter	Threads per Inch	Size or Number
0	80	3/64
1	64 72	53
2	56 64	50
3	48 56	47 45
4	40 48	43 42
5	40 44	38 37
6	32 40	36 33
8	32 36	29
10	24 32	25 21
12	24 28	16 14
1/4	20 28	7 3
5/16	18 24	F I
3/8	16 24	5/16 Q
7/16	14 20	U 25/64
1/2	13 20	27/64 29/64
9/16	12 18	31/64 33/64
5/8	11 18	17/32 37/64
3/4	10 16	21/32 11/16
7/8	9 14	49/64 13/16
1	8 12	7/8 59/64
1 1/8	7 12	63/64 1 3/64
1 1/4	7 12	1 7/64 1 11/64

Fractional, Number, and Letter Sizes for Twist Drills

DRILL NO.	FRAC.	DECI.	DRILL NO.	FRAC.	DECI.	DRILL NO.	FRAC.	DECI.	DRILL NO.	FRAC.	DECI.	DRILL NO.	FRAC.	DECI.	DRILL NO.	FRAC.	DECI.	DRILL NO.	FRAC.	DECI.	DRILL NO.	FRAC.	DECI.
80	--	.0135	60	--	.0400	--	3/32	.0938	24	--	.152	6	--	.204	L	--	.290	--	13/32	.406	--	47/64	.734
79	--	.0145	59	--	.0410	41	--	.0960	23	--	.154	5	--	.206	M	--	.295	Z	--	.413	--	3/4	.750
--	1/64	.0156	58	--	.0420	40	--	.0980	--	5/32	.156	4	--	.209	--	19/64	.297	--	27/64	.422	--	49/64	.766
78	--	.0160	57	--	.0430	39	--	.0995	22	--	.157	3	--	.213	N	--	.302	--	7/16	.438	--	25/32	.781
77	--	.0180	56	--	.0465	38	--	.1015	21	--	.159	--	7/32	.219	--	5/16	.313	--	29/64	.453	--	51/64	.797
76	--	.0200	--	3/64	.0469	37	--	.1040	20	--	.161	2	--	.221	O	--	.316	--	15/32	.469	--	13/16	.813
75	--	.0210	55	--	.0520	36	--	.1065	19	--	.166	1	--	.228	P	--	.323	--	31/64	.484	--	53/64	.828
74	--	.0225	54	--	.0550	--	7/64	.1094	18	--	.170	A	--	.234	--	21/64	.328	--	1/2	.500	--	27/32	.844
73	--	.0240	53	--	.0595	35	--	.1100	--	11/64	.172	--	15/64	.234	Q	--	.332	--	33/64	.516	--	55/64	.859
72	--	.0250	--	1/16	.0625	34	--	.1110	17	--	.173	B	--	.238	R	--	.339	--	17/32	.531	--	7/8	.875
71	--	.0260	52	--	.0635	33	--	.1130	16	--	.177	C	--	.242	--	11/32	.344	--	35/64	.547	--	57/64	.891
70	--	.0280	51	--	.0670	32	--	.116	15	--	.180	D	--	.246	S	--	.348	--	9/16	.562	--	29/32	.906
69	--	.0292	50	--	.0700	31	--	.120	14	--	.182	--	1/4	.250	T	--	.358	--	37/64	.578	--	59/64	.922
68	--	.0310	49	--	.0730	--	1/8	.125	13	--	.185	E	--	.250	--	23/64	.359	--	19/32	.594	--	15/16	.938
--	1/32	.0313	48	--	.0760	30	--	.129	--	3/16	.188	F	--	.257	U	--	.368	--	39/64	.609	--	61/64	.953
67	--	.0320	--	5/64	.0781	29	--	.136	12	--	.189	G	--	.261	--	3/8	.375	--	5/8	.625	--	31/32	.969
66	--	.0330	47	--	.0785	--	9/64	.140	11	--	.191	--	17/64	.266	V	--	.377	--	41/64	.641	--	63/64	.984
65	--	.0350	46	--	.0810	28	--	.141	10	--	.194	H	--	.266	W	--	.386	--	21/32	.656	--	1	1.000
64	--	.0360	45	--	.0820	27	--	.144	9	--	.196	I	--	.272	--	25/64	.391	--	43/64	.672			
63	--	.0370	44	--	.0860	26	--	.147	8	--	.199	J	--	.277	X	--	.397	--	11/16	.688			
62	--	.0380	43	--	.0890	25	--	.150	7	--	.201	--	9/32	.281	Y	--	.404	--	45/64	.703			
61	--	.0390	42	--	.0935				--	13/64	.203	K	--	.281				--	23/32	.719			

Dimensions of Machine Screws
(All dimensions in inches.)

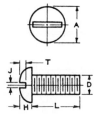

Round Head

Nominal Size	D — Max Diameter of Screw	A — Head Diameter		H — Height of Head		J — Width of Slot		T — Depth of Slot	
		Max	Min	Max	Min	Max	Min	Max	Min
0	0.060	0.113	0.099	0.053	0.043	0.023	0.016	0.039	0.029
1	0.073	0.138	0.122	0.061	0.051	0.026	0.019	0.044	0.033
2	0.086	0.162	0.146	0.069	0.059	0.031	0.023	0.048	0.037
3	0.099	0.187	0.169	0.078	0.067	0.035	0.027	0.053	0.040
4	0.112	0.211	0.193	0.086	0.075	0.039	0.031	0.058	0.044
5	0.125	0.236	0.217	0.095	0.083	0.043	0.035	0.063	0.047
6	0.138	0.260	0.240	0.103	0.091	0.048	0.039	0.068	0.051
8	0.164	0.309	0.287	0.120	0.107	0.054	0.045	0.077	0.058
10	0.190	0.359	0.334	0.137	0.123	0.060	0.050	0.087	0.065
12	0.216	0.408	0.382	0.153	0.139	0.067	0.056	0.096	0.072
¼	0.250	0.472	0.443	0.175	0.160	0.075	0.064	0.109	0.082
⁵⁄₁₆	0.3125	0.590	0.557	0.216	0.198	0.084	0.072	0.132	0.099
³⁄₈	0.375	0.708	0.670	0.256	0.237	0.094	0.081	0.155	0.117
⁷⁄₁₆	0.4375	0.750	0.707	0.328	0.307	0.094	0.081	0.196	0.148
½	0.500	0.813	0.766	0.355	0.332	0.106	0.091	0.211	0.159
⁹⁄₁₆	0.5625	0.938	0.887	0.410	0.385	0.118	0.102	0.242	0.183
⅝	0.625	1.000	0.944	0.438	0.411	0.133	0.116	0.258	0.195
¾	0.750	1.250	1.185	0.547	0.516	0.149	0.131	0.320	0.242

Flat Head

Nominal Size	D — Max Diameter of Screw	A — Head Diameter			H — Height of Head		J — Width of Slot		T — Depth of Slot	
		Max Sharp	Min Sharp	Absolute Min with Max S	Max	Min	Max	Min	Max	Min
0	0.060	0.119	0.105	0.101	0.035	0.026	0.023	0.016	0.015	0.010
1	0.073	0.146	0.130	0.126	0.043	0.033	0.026	0.019	0.019	0.012
2	0.086	0.172	0.156	0.150	0.051	0.040	0.031	0.023	0.023	0.015
3	0.099	0.199	0.181	0.175	0.059	0.048	0.035	0.027	0.027	0.017
4	0.112	0.225	0.207	0.200	0.067	0.055	0.039	0.031	0.030	0.020
5	0.125	0.252	0.232	0.225	0.075	0.062	0.043	0.035	0.034	0.022
6	0.138	0.279	0.257	0.249	0.083	0.069	0.048	0.039	0.033	0.024
8	0.164	0.332	0.308	0.300	0.100	0.084	0.054	0.045	0.045	0.029
10	0.190	0.385	0.359	0.348	0.116	0.098	0.060	0.050	0.053	0.034
12	0.216	0.438	0.410	0.397	0.132	0.112	0.067	0.056	0.060	0.039
¼	0.250	0.507	0.477	0.462	0.153	0.131	0.075	0.064	0.070	0.046
⁵⁄₁₆	0.3125	0.635	0.600	0.581	0.191	0.165	0.084	0.072	0.088	0.058
³⁄₈	0.375	0.762	0.722	0.700	0.230	0.200	0.094	0.081	0.106	0.070
⁷⁄₁₆	0.4375	0.812	0.771	0.743	0.223	0.190	0.094	0.081	0.103	0.066
½	0.500	0.875	0.831	0.802	0.223	0.186	0.106	0.091	0.103	0.065
⁹⁄₁₆	0.5625	1.000	0.950	0.919	0.260	0.220	0.118	0.102	0.120	0.077
⅝	0.625	1.125	1.069	1.035	0.298	0.253	0.133	0.116	0.137	0.088
¾	0.750	1.375	1.306	1.267	0.372	0.319	0.149	0.131	0.171	0.111

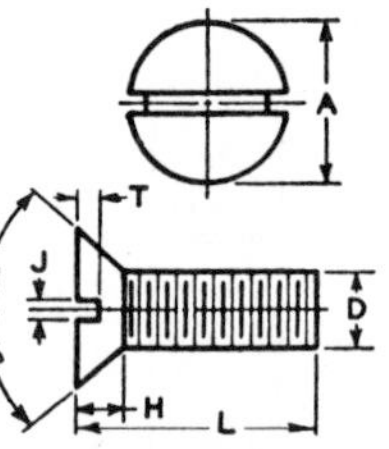

Fillister Head

Nominal Size	D — Max Diameter of Screw	A — Head Diameter		H — Height of Head		O — Total Height of Head		J — Width of Slot		T — Depth of Slot	
		Max	Min	Max	Min	Max	Min	Max	Min	Max	Min
0	0.060	0.096	0.083	0.045	0.037	0.059	0.043	0.023	0.016	0.025	0.015
1	0.073	0.118	0.104	0.053	0.045	0.071	0.055	0.026	0.019	0.031	0.020
2	0.086	0.140	0.124	0.062	0.053	0.083	0.066	0.031	0.023	0.037	0.025
3	0.099	0.161	0.145	0.070	0.061	0.095	0.077	0.035	0.027	0.043	0.030
4	0.112	0.183	0.166	0.079	0.069	0.107	0.088	0.039	0.031	0.048	0.035
5	0.125	0.205	0.187	0.088	0.078	0.120	0.100	0.043	0.035	0.054	0.040
6	0.138	0.226	0.208	0.096	0.086	0.132	0.111	0.048	0.039	0.060	0.045
8	0.164	0.270	0.250	0.113	0.102	0.156	0.133	0.054	0.045	0.071	0.054
10	0.190	0.313	0.292	0.130	0.118	0.180	0.156	0.060	0.050	0.083	0.064
12	0.216	0.357	0.334	0.148	0.134	0.205	0.178	0.067	0.056	0.094	0.074
¼	0.250	0.414	0.389	0.170	0.155	0.237	0.207	0.075	0.064	0.109	0.087
⁵⁄₁₆	0.3125	0.518	0.490	0.211	0.194	0.295	0.262	0.084	0.072	0.137	0.110
³⁄₈	0.375	0.622	0.590	0.253	0.233	0.355	0.315	0.094	0.081	0.164	0.133
⁷⁄₁₆	0.4375	0.625	0.589	0.265	0.242	0.368	0.321	0.094	0.081	0.170	0.135
½	0.500	0.750	0.710	0.297	0.273	0.412	0.362	0.106	0.091	0.190	0.151
⁹⁄₁₆	0.5625	0.812	0.768	0.336	0.308	0.466	0.410	0.118	0.102	0.214	0.172
⅝	0.625	0.875	0.827	0.375	0.345	0.521	0.461	0.133	0.116	0.240	0.193
¾	0.750	1.000	0.945	0.441	0.406	0.612	0.542	0.149	0.131	0.281	0.226

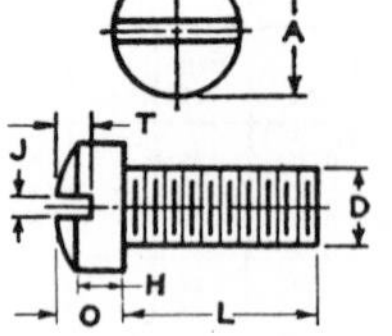

Dimensions of Machine Screws
(All dimensions in inches.)

Oval Head

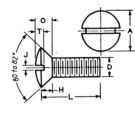

Nominal Size	D Max Diameter of Screw	A Head Diameter			H Height of Head		O Total Height of Head		J Width of Slot		T Depth of Slot	
		Max Sharp	Min Sharp	Absolute Min with Max S	Max	Min	Max	Min	Max	Min	Max	Min
0	0.060	0.119	0.105	0.101	0.035	0.026	0.056	0.041	0.023	0.016	0.030	0.025
1	0.073	0.146	0.130	0.126	0.043	0.033	0.068	0.052	0.026	0.019	0.038	0.031
2	0.086	0.172	0.156	0.150	0.051	0.040	0.080	0.063	0.031	0.023	0.045	0.037
3	0.099	0.199	0.181	0.175	0.059	0.048	0.092	0.073	0.035	0.027	0.052	0.043
4	0.112	0.225	0.207	0.200	0.067	0.055	0.104	0.084	0.039	0.031	0.059	0.049
5	0.125	0.252	0.232	0.225	0.075	0.062	0.116	0.095	0.043	0.035	0.067	0.055
6	0.138	0.279	0.257	0.249	0.083	0.069	0.128	0.105	0.048	0.039	0.074	0.060
8	0.164	0.332	0.308	0.300	0.100	0.084	0.152	0.126	0.054	0.045	0.088	0.072
10	0.190	0.385	0.359	0.348	0.116	0.098	0.176	0.148	0.060	0.050	0.103	0.084
12	0.216	0.438	0.410	0.397	0.132	0.112	0.200	0.169	0.067	0.056	0.117	0.096
1/4	0.250	0.507	0.477	0.462	0.153	0.131	0.232	0.197	0.075	0.064	0.136	0.112
5/16	0.3125	0.635	0.600	0.581	0.191	0.165	0.290	0.249	0.084	0.072	0.171	0.141
3/8	0.375	0.762	0.722	0.700	0.230	0.200	0.347	0.300	0.094	0.081	0.206	0.170
7/16	0.4375	0.812	0.771	0.743	0.223	0.190	0.345	0.295	0.094	0.081	0.210	0.174
1/2	0.500	0.875	0.831	0.802	0.223	0.186	0.354	0.299	0.106	0.091	0.216	0.176
9/16	0.5625	1.000	0.950	0.919	0.260	0.220	0.410	0.350	0.118	0.102	0.250	0.207
5/8	0.625	1.125	1.069	1.035	0.298	0.253	0.467	0.399	0.133	0.116	0.285	0.235
3/4	0.750	1.375	1.306	1.267	0.372	0.319	0.578	0.497	0.149	0.131	0.353	0.293

Truss Head

Nominal Size	D Max Diameter of Screw	A Head Diameter		H Height of Head		J Width of Slot		T Depth of Slot		R Radius
		Max	Min	Max	Min	Max	Min	Max	Min	Max
2	0.086	0.194	0.180	0.053	0.044	0.031	0.023	0.031	0.022	0.129
3	0.099	0.226	0.211	0.061	0.051	0.035	0.027	0.036	0.026	0.151
4	0.112	0.257	0.241	0.069	0.059	0.039	0.031	0.040	0.030	0.169
5	0.125	0.289	0.272	0.078	0.066	0.043	0.035	0.045	0.034	0.191
6	0.138	0.321	0.303	0.086	0.074	0.048	0.039	0.050	0.037	0.211
7	0.151	0.352	0.333	0.094	0.081	0.048	0.039	0.054	0.041	0.231
8	0.164	0.384	0.364	0.102	0.088	0.054	0.045	0.058	0.045	0.254
10	0.190	0.448	0.425	0.118	0.103	0.060	0.050	0.068	0.053	0.283
12	0.216	0.511	0.487	0.134	0.118	0.067	0.056	0.077	0.061	0.336
1/4	0.250	0.573	0.546	0.150	0.133	0.075	0.064	0.087	0.070	0.375
5/16	0.3125	0.698	0.666	0.183	0.162	0.084	0.072	0.106	0.085	0.457
3/8	0.375	0.823	0.787	0.215	0.191	0.094	0.081	0.124	0.100	0.538
7/16	0.4375	0.948	0.907	0.248	0.221	0.094	0.081	0.142	0.116	0.619
1/2	0.500	1.073	1.028	0.280	0.250	0.106	0.091	0.161	0.131	0.701
9/16	0.5625	1.198	1.149	0.312	0.279	0.118	0.102	0.179	0.146	0.783
5/8	0.625	1.323	1.269	0.345	0.309	0.133	0.116	0.196	0.162	0.863
3/4	0.750	1.573	1.511	0.410	0.368	0.149	0.131	0.234	0.182	1.024

Binding Head

Nominal Size	D Max Diameter of Screw	A Head Diameter		O Total Height of Head		J Width of Slot		T Depth of Slot		F Height of Oval		U Diameter of Undercut		X Depth of Undercut	
		Max	Min	Max	Min	Max	Min	Max	Min	Max	Min	Max	Min	Max	Min
2	0.086	0.181	0.171	0.046	0.041	0.031	0.023	0.030	0.024	0.018	0.013	0.141	0.124	0.010	0.005
3	0.099	0.208	0.197	0.054	0.048	0.035	0.027	0.036	0.029	0.022	0.016	0.162	0.143	0.011	0.006
4	0.112	0.235	0.223	0.063	0.056	0.039	0.031	0.042	0.034	0.025	0.018	0.184	0.161	0.012	0.007
5	0.125	0.263	0.249	0.071	0.064	0.043	0.035	0.048	0.039	0.029	0.021	0.205	0.180	0.014	0.009
6	0.138	0.290	0.275	0.080	0.071	0.048	0.039	0.053	0.044	0.032	0.024	0.226	0.199	0.015	0.010
8	0.164	0.344	0.326	0.097	0.087	0.054	0.045	0.065	0.054	0.039	0.029	0.269	0.236	0.017	0.012
10	0.190	0.399	0.378	0.114	0.102	0.060	0.050	0.077	0.064	0.045	0.034	0.312	0.274	0.020	0.015
12	0.216	0.454	0.430	0.130	0.117	0.067	0.056	0.089	0.074	0.052	0.039	0.354	0.311	0.023	0.018
1/4	0.250	0.513	0.488	0.153	0.138	0.075	0.064	0.105	0.088	0.061	0.046	0.410	0.360	0.026	0.021
5/16	0.3125	0.641	0.609	0.193	0.174	0.084	0.072	0.134	0.112	0.077	0.059	0.513	0.450	0.032	0.027
3/8	0.375	0.769	0.731	0.234	0.211	0.094	0.081	0.163	0.136	0.094	0.071	0.615	0.540	0.039	0.034

Dimensions of Machine Screws
(All dimensions in inches.)

Pan Head

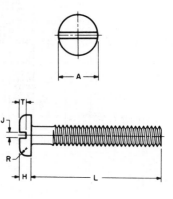

Nominal Size or Basic Screw Diameter		A Head Diameter		H Head Height		R Head Radius	J Slot Width		T Slot Depth	
		Max	Min	Max	Min	Max	Max	Min	Max	Min
0000	0.0210	0.042	0.036	0.016	0.010	0.007	0.008	0.004	0.008	0.004
000	0.0340	0.066	0.060	0.023	0.017	0.010	0.012	0.008	0.012	0.008
00	0.0470	0.090	0.082	0.032	0.025	0.015	0.017	0.010	0.016	0.010
0	0.0600	0.116	0.104	0.039	0.031	0.020	0.023	0.016	0.022	0.014
1	0.0730	0.142	0.130	0.046	0.038	0.025	0.026	0.019	0.027	0.018
2	0.0860	0.167	0.155	0.053	0.045	0.035	0.031	0.023	0.031	0.022
3	0.0990	0.193	0.180	0.060	0.051	0.037	0.035	0.027	0.036	0.026
4	0.1120	0.219	0.205	0.068	0.058	0.042	0.039	0.031	0.040	0.030
5	0.1250	0.245	0.231	0.075	0.065	0.044	0.043	0.035	0.045	0.034
6	0.1380	0.270	0.256	0.082	0.072	0.046	0.048	0.039	0.050	0.037
8	0.1640	0.322	0.306	0.096	0.085	0.052	0.054	0.045	0.058	0.045
10	0.1900	0.373	0.357	0.110	0.099	0.061	0.060	0.050	0.068	0.053
12	0.2160	0.425	0.407	0.125	0.112	0.078	0.067	0.056	0.077	0.061
1/4	0.2500	0.492	0.473	0.144	0.130	0.087	0.075	0.064	0.087	0.070
5/16	0.3125	0.615	0.594	0.178	0.162	0.099	0.084	0.072	0.106	0.085
3/8	0.3750	0.740	0.716	0.212	0.195	0.143	0.094	0.081	0.124	0.100
7/16	0.4375	0.863	0.837	0.247	0.228	0.153	0.094	0.081	0.142	0.116
1/2	0.5000	0.987	0.958	0.281	0.260	0.175	0.106	0.091	0.161	0.131
9/16	0.5625	1.041	1.000	0.315	0.293	0.197	0.118	0.102	0.179	0.146
5/8	0.6250	1.172	1.125	0.350	0.325	0.219	0.133	0.116	0.197	0.162
3/4	0.7500	1.435	1.375	0.419	0.390	0.263	0.149	0.131	0.234	0.192

Plain and Slotted Regular and Large Hex Head

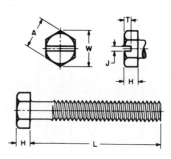

Nominal Size or Basic Screw Diameter		A^3	$W^{3,4}$	A^3	$W^{3,4}$		H		J^5		$T^{5,6}$	
		Regular Head[7]		Large Head[7,8]			Head Height		Slot Width		Slot Depth	
		Width Across Flats	Across Corners	Width Across Flats		Across Corners						
		Max	Min	Min	Max	Min	Min	Max	Min	Max	Min	Max	Min
1	0.0730	0.125	0.120	0.134	—	—	—	0.044	0.036	—	—	—	—
2	0.0860	0.125	0.120	0.134	—	—	—	0.050	0.040	—	—	—	—
3	0.0990	0.188	0.181	0.202	—	—	—	0.055	0.044	—	—	—	—
4	0.1120	0.188	0.181	0.202	0.219	0.213	0.238	0.060	0.049	0.039	0.031	0.036	0.025
5	0.1250	0.188	0.181	0.202	0.250	0.244	0.272	0.070	0.058	0.043	0.035	0.042	0.030
6	0.1380	0.250	0.244	0.272	—	—	—	0.093	0.080	0.048	0.039	0.046	0.033
8	0.1640	0.250	0.244	0.272	0.312	0.305	0.340	0.110	0.096	0.054	0.045	0.066	0.052
10	0.1900	0.312	0.305	0.340	—	—	—	0.120	0.105	0.060	0.050	0.072	0.057
12	0.2160	0.312	0.305	0.340	0.375	0.367	0.409	0.155	0.139	0.067	0.056	0.093	0.077
1/4	0.2500	0.375	0.367	0.409	0.438	0.428	0.477	0.190	0.172	0.075	0.064	0.101	0.083
5/16	0.3125	0.500	0.489	0.545	—	—	—	0.230	0.208	0.084	0.072	0.122	0.100
3/8	0.3750	0.562	0.551	0.614	—	—	—	0.295	0.270	0.094	0.081	0.156	0.131

Cross Recessed Pan Head

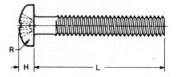

Nominal Size or Basic Screw Diameter		A Head Diameter		H Head Height		R Head Radius	M Recess Diameter		T Recess Depth		N Recess Width	Driver Size
		Max	Min	Max	Min	Min	Max	Min	Max	Min	Min	
0	0.0600	0.116	0.104	0.044	0.036	0.005	0.067	0.054	0.039	0.021	0.013	0
1	0.0730	0.142	0.130	0.053	0.044	0.005	0.074	0.061	0.045	0.025	0.014	0
2	0.0860	0.167	0.155	0.062	0.053	0.010	0.104	0.091	0.059	0.041	0.017	1
3	0.0990	0.193	0.180	0.071	0.062	0.010	0.112	0.099	0.068	0.050	0.019	1
4	0.1120	0.219	0.205	0.080	0.070	0.010	0.122	0.109	0.078	0.060	0.019	1
5	0.1250	0.245	0.231	0.089	0.079	0.015	0.158	0.145	0.083	0.057	0.028	2
6	0.1380	0.270	0.256	0.097	0.087	0.015	0.166	0.153	0.091	0.066	0.028	2
8	0.1640	0.322	0.306	0.115	0.105	0.015	0.182	0.169	0.108	0.082	0.030	2
10	0.1900	0.373	0.357	0.133	0.122	0.020	0.199	0.186	0.124	0.100	0.031	2
12	0.2160	0.425	0.407	0.141	0.139	0.025	0.259	0.246	0.141	0.115	0.034	3
1/4	0.2500	0.492	0.473	0.175	0.162	0.035	0.281	0.268	0.161	0.135	0.036	3
5/16	0.3125	0.615	0.594	0.218	0.203	0.040	0.350	0.337	0.193	0.169	0.059	4
3/8	0.3750	0.740	0.716	0.261	0.244	0.040	0.389	0.376	0.233	0.210	0.065	4
7/16	0.4375	0.863	0.837	0.305	0.284	0.050	0.413	0.400	0.259	0.234	0.068	4
1/2	0.5000	0.987	0.958	0.348	0.325	0.055	0.435	0.422	0.280	0.255	0.071	4
9/16	0.5625	1.041	1.000	0.391	0.366	0.065	0.470	0.447	0.312	0.288	0.076	4
5/8	0.6250	1.172	1.125	0.434	0.406	0.075	0.587	0.564	0.343	0.314	0.081	5
3/4	0.7500	1.435	1.375	0.521	0.488	0.085	0.633	0.610	0.382	0.355	0.086	5

Dimensions of Cap Screws
(All dimensions in inches)

Flat Countersunk Head

Nominal Size¹ or Basic Screw Diameter	E Body Diameter		A Head Diameter		H² Head Height	J Slot Width		T Slot Depth		U Fillet Radius
	Max	Min	Max, Edge Sharp	Min, Edge Rounded or Flat	Ref	Max	Min	Max	Min	Max
1/4 0.2500	0.2500	0.2450	0.500	0.452	0.140	0.075	0.064	0.068	0.045	0.100
5/16 0.3125	0.3125	0.3070	0.625	0.567	0.177	0.084	0.072	0.086	0.057	0.125
3/8 0.3750	0.3750	0.3690	0.750	0.682	0.210	0.094	0.081	0.103	0.068	0.150
7/16 0.4375	0.4375	0.4310	0.812	0.736	0.210	0.094	0.081	0.103	0.068	0.175
1/2 0.5000	0.5000	0.4930	0.875	0.791	0.210	0.106	0.091	0.103	0.068	0.200
9/16 0.5625	0.5625	0.5550	1.000	0.906	0.244	0.118	0.102	0.120	0.080	0.225
5/8 0.6250	0.6250	0.6170	1.125	1.020	0.281	0.133	0.116	0.137	0.091	0.250
3/4 0.7500	0.7500	0.7420	1.375	1.251	0.352	0.149	0.131	0.171	0.115	0.300

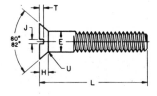

Round Head

Nominal Size¹ or Basic Screw Diameter	E Body Diameter		A Head Diameter		H Head Height		J Slot Width		T Slot Depth		U Fillet Radius	
	Max	Min	Max	Min	Max	Min	Max	Min	Max	Min	Max	Min
1/4 0.2500	0.2500	0.2450	0.437	0.418	0.191	0.175	0.075	0.064	0.117	0.097	0.031	0.016
5/16 0.3125	0.3125	0.3070	0.562	0.540	0.245	0.226	0.084	0.072	0.151	0.126	0.031	0.016
3/8 0.3750	0.3750	0.3690	0.625	0.603	0.273	0.252	0.094	0.081	0.168	0.138	0.031	0.016
7/16 0.4375	0.4375	0.4310	0.750	0.725	0.328	0.302	0.094	0.081	0.202	0.167	0.047	0.016
1/2 0.5000	0.5000	0.4930	0.812	0.786	0.354	0.327	0.106	0.091	0.218	0.178	0.047	0.016
9/16 0.5625	0.5625	0.5550	0.937	0.909	0.409	0.378	0.118	0.102	0.252	0.207	0.047	0.016
5/8 0.6250	0.6250	0.6170	1.000	0.970	0.437	0.405	0.133	0.116	0.270	0.220	0.062	0.031
3/4 0.7500	0.7500	0.7420	1.250	1.215	0.546	0.507	0.149	0.131	0.338	0.278	0.062	0.031

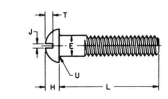

Fillister Head

Nominal Size¹ or Basic Screw Diameter	E Body Diameter		A Head Diameter		H Head Side Height		O Total Head Height		J Slot Width		T Slot Depth		U Fillet Radius	
	Max	Min	Max	Min	Max	Min	Max	Min	Max	Min	Max	Min	Max	Min
1/4 0.2500	0.2500	0.2450	0.375	0.363	0.172	0.157	0.216	0.194	0.075	0.064	0.097	0.077	0.031	0.016
5/16 0.3125	0.3125	0.3070	0.437	0.424	0.203	0.186	0.253	0.230	0.084	0.072	0.115	0.090	0.031	0.016
3/8 0.3750	0.3750	0.3690	0.562	0.547	0.250	0.229	0.314	0.284	0.094	0.081	0.142	0.112	0.031	0.016
7/16 0.4375	0.4375	0.4310	0.625	0.608	0.297	0.274	0.368	0.336	0.094	0.081	0.168	0.133	0.047	0.016
1/2 0.5000	0.5000	0.4930	0.750	0.731	0.328	0.301	0.413	0.376	0.106	0.091	0.193	0.153	0.047	0.016
9/16 0.5625	0.5625	0.5550	0.812	0.792	0.375	0.346	0.467	0.427	0.118	0.102	0.213	0.168	0.047	0.016
5/8 0.6250	0.6250	0.6170	0.875	0.853	0.422	0.391	0.521	0.478	0.133	0.116	0.239	0.189	0.062	0.031
3/4 0.7500	0.7500	0.7420	1.000	0.976	0.500	0.466	0.612	0.566	0.149	0.131	0.283	0.223	0.062	0.031

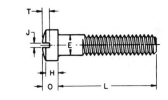

Hexagonal Socket Head

D Body Diameter			A Head Diameter		H Head Height		S Head Side-Height			J Socket Width Across Flats		T Key Engagement
Nom	Max	Min	Max	Min	Max	Min	Nom	Max	Min	Max	Min	Min
0	0.060	0.0583	0.0960	0.0926	0.0600	0.0574	0.055	0.056	0.054	0.051	0.050	0.025
1	0.0730	0.0711	0.1180	0.1142	0.0730	0.0702	0.067	0.068	0.066	0.051	0.050	0.031
2	0.0860	0.0840	0.140	0.136	0.086	0.083	0.079	0.081	0.078	0.0635	1/16	0.038
3	0.0990	0.0968	0.161	0.157	0.099	0.096	0.091	0.093	0.089	0.0791	5/64	0.044
4	0.1120	0.1096	0.183	0.178	0.112	0.109	0.103	0.105	0.101	0.0791	5/64	0.051
5	0.1250	0.1226	0.205	0.200	0.125	0.122	0.115	0.117	0.113	0.0947	3/32	0.057
6	0.1380	0.1353	0.226	0.221	0.138	0.134	0.127	0.129	0.125	0.0947	3/32	0.064
8	0.1640	0.1613	0.270	0.265	0.164	0.160	0.150	0.152	0.148	0.1270	1/8	0.077
10	0.1900	0.1867	5/16	0.306	0.190	0.185	0.174	0.176	0.172	0.1582	5/32	0.090
12	0.2160	0.2127	11/32	0.337	0.216	0.211	0.198	0.200	0.196	0.1582	5/32	0.103
1/4		0.2464	3/8	0.367	1/4	0.244	0.229	0.232	0.226	0.1895	3/16	0.120
5/16	0.3125	0.3084	7/16	0.429	5/16	0.306	0.286	0.289	0.283	0.2207	7/32	0.151
3/8	0.3750	0.3705	9/16	0.553	3/8	0.368	0.344	0.347	0.341	0.3155	5/16	0.182
7/16	0.4375	0.4326	5/8	0.615	7/16	0.430	0.401	0.405	0.397	0.3155	5/16	0.213
1/2	0.5000	0.4948	3/4	0.739	1/2	0.492	0.458	0.462	0.454	0.3780	3/8	0.245
9/16	0.5625	0.5569	13/16	0.801	9/16	0.554	0.516	0.520	0.512	0.3780	3/8	0.276
5/8	0.6250	0.6191	7/8	0.863	5/8	0.616	0.573	0.577	0.569	0.5030	1/2	0.307
3/4	0.7500	0.7436	1	0.987	3/4	0.741	0.688	0.693	0.684	0.5655	9/16	0.370
7/8	0.8750	0.8680	1 1/8	1.111	7/8	0.865	0.802	0.807	0.797	0.5655	9/16	0.432

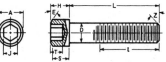

Dimensions of Set Screws
(All dimensions in inches.)

Square Head

Nominal Size[1] or Basic Screw Diameter	F Width Across Flats		G Width Across Corners		H Head Height		K Neck Relief Diameter		S Neck Relief Fillet Radius	U Neck Relief Width	V Head Radius
	Max	Min	Max	Min	Max	Min	Max	Min	Max	Max	Min
10 0.1900	0.188	0.180	0.265	0.247	0.148	0.134	0.145	0.140	0.027	0.083	0.48
1/4 0.2500	0.250	0.241	0.354	0.331	0.196	0.178	0.185	0.170	0.032	0.100	0.62
5/16 0.3125	0.312	0.302	0.442	0.415	0.245	0.224	0.240	0.225	0.036	0.111	0.78
3/8 0.3750	0.375	0.362	0.530	0.497	0.293	0.270	0.294	0.279	0.041	0.125	0.94
7/16 0.4375	0.438	0.423	0.619	0.581	0.341	0.315	0.345	0.330	0.046	0.143	1.09
1/2 0.5000	0.500	0.484	0.707	0.665	0.389	0.361	0.400	0.385	0.050	0.154	1.25
9/16 0.5625	0.562	0.545	0.795	0.748	0.437	0.407	0.454	0.439	0.054	0.167	1.41
5/8 0.6250	0.625	0.606	0.884	0.833	0.485	0.452	0.507	0.492	0.059	0.182	1.56
3/4 0.7500	0.750	0.729	1.060	1.001	0.582	0.544	0.620	0.605	0.065	0.200	1.88

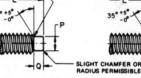

FLAT POINT

DOG POINT

HALF DOG POINT

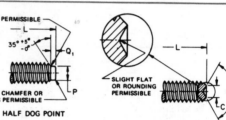

SLIGHT FILLET PERMISSIBLE

SLIGHT CHAMFER OR RADIUS PERMISSIBLE

SLIGHT FLAT OR ROUNDING PERMISSIBLE

CUP POINT

SLIGHT FLAT PERMISSIBLE

OVAL POINT

Square Head Set Screws (continued)

Nominal Size[1] or Basic Screw Diameter	C Cup and Flat Point Diameters		P Dog and Half Dog Point Diameters		Q Point Length Dog		Q₁ Point Length Half Dog		R Oval Point Radius +0.031 −0.000	Y Cone Point Angle 90°±2° For These Nominal Lengths or Longer; 118°±2° For Shorter Screws
	Max	Min	Max	Min	Max	Min	Max	Min		
10 0.1900	0.102	0.088	0.127	0.120	0.095	0.085	0.050	0.040	0.142	1/4
1/4 0.2500	0.132	0.118	0.156	0.149	0.130	0.120	0.068	0.058	0.188	5/16
5/16 0.3125	0.172	0.156	0.203	0.195	0.161	0.151	0.083	0.073	0.234	3/8
3/8 0.3750	0.212	0.194	0.250	0.241	0.193	0.183	0.099	0.089	0.281	7/16
7/16 0.4375	0.252	0.232	0.297	0.287	0.224	0.214	0.114	0.104	0.328	1/2
1/2 0.5000	0.291	0.270	0.344	0.334	0.255	0.245	0.130	0.120	0.375	9/16
9/16 0.5625	0.332	0.309	0.391	0.379	0.287	0.275	0.146	0.134	0.422	5/8
5/8 0.6250	0.371	0.347	0.469	0.456	0.321	0.305	0.164	0.148	0.469	3/4
3/4 0.7500	0.450	0.425	0.562	0.549	0.383	0.367	0.196	0.180	0.562	7/8

SLIGHT FLAT OR ROUNDING PERMISSIBLE

CONE POINT

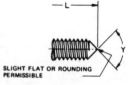

FLAT POINT

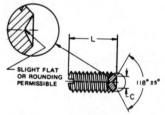

SLIGHT FLAT OR ROUNDING PERMISSIBLE

CUP POINT

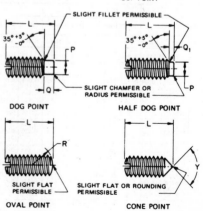

SLIGHT FILLET PERMISSIBLE

DOG POINT

HALF DOG POINT

SLIGHT CHAMFER OR RADIUS PERMISSIBLE

SLIGHT FLAT PERMISSIBLE

OVAL POINT

SLIGHT FLAT OR ROUNDING PERMISSIBLE

CONE POINT

Slotted Headless

Nominal Size[1] or Basic Screw Diameter	I[2] Crown Radius	J Slot Width		T Slot Depth		C Cup and Flat Point Diameters		P Dog Point Diameters		Q Point Length Dog		Q₁ Half Dog		R[2] Oval Point Radius	Y Cone Point Angle 90°±2° For These Nominal Lengths or Longer; 118°±2° For Shorter Screws
	Basic	Max	Min	Max	Min	Max	Min	Max	Min	Max	Min	Max	Min	Basic	
0 0.0600	0.060	0.014	0.010	0.020	0.016	0.033	0.027	0.040	0.037	0.032	0.028	0.017	0.013	0.045	5/64
1 0.0730	0.073	0.016	0.012	0.020	0.016	0.040	0.033	0.049	0.045	0.040	0.036	0.021	0.017	0.055	3/32
2 0.0860	0.086	0.018	0.014	0.025	0.019	0.047	0.039	0.057	0.053	0.046	0.042	0.024	0.020	0.064	7/64
3 0.0990	0.099	0.020	0.016	0.028	0.022	0.054	0.045	0.066	0.062	0.052	0.048	0.027	0.023	0.074	1/8
4 0.1120	0.112	0.024	0.018	0.031	0.025	0.061	0.051	0.075	0.070	0.058	0.054	0.030	0.026	0.084	5/32
5 0.1250	0.125	0.026	0.020	0.036	0.026	0.067	0.057	0.083	0.078	0.063	0.057	0.033	0.027	0.094	3/16
6 0.1380	0.138	0.028	0.022	0.040	0.030	0.074	0.064	0.092	0.087	0.073	0.067	0.038	0.032	0.104	3/16
8 0.1640	0.164	0.032	0.026	0.046	0.036	0.087	0.076	0.109	0.103	0.083	0.077	0.043	0.037	0.123	1/4
10 0.1900	0.190	0.035	0.029	0.053	0.043	0.102	0.088	0.127	0.120	0.095	0.085	0.050	0.040	0.142	1/4
12 0.2160	0.216	0.042	0.035	0.061	0.051	0.115	0.101	0.144	0.137	0.105	0.105	0.060	0.050	0.162	5/16
1/4 0.2500	0.250	0.049	0.041	0.068	0.058	0.132	0.118	0.156	0.149	0.130	0.120	0.068	0.058	0.188	5/16
5/16 0.3125	0.312	0.055	0.047	0.083	0.073	0.172	0.156	0.203	0.195	0.161	0.151	0.083	0.073	0.234	3/8
3/8 0.3750	0.375	0.068	0.060	0.099	0.089	0.212	0.194	0.250	0.241	0.193	0.183	0.099	0.089	0.281	7/16
7/16 0.4375	0.438	0.076	0.068	0.114	0.104	0.252	0.232	0.297	0.287	0.224	0.214	0.114	0.104	0.328	1/2
1/2 0.5000	0.500	0.086	0.078	0.130	0.120	0.291	0.270	0.344	0.334	0.255	0.245	0.130	0.120	0.375	9/16
9/16 0.5625	0.562	0.096	0.086	0.146	0.136	0.332	0.309	0.391	0.379	0.287	0.275	0.146	0.134	0.422	5/8
5/8 0.6250	0.625	0.107	0.097	0.161	0.151	0.371	0.347	0.469	0.456	0.321	0.305	0.164	0.148	0.469	3/4
3/4 0.7500	0.750	0.134	0.124	0.193	0.183	0.450	0.425	0.562	0.549	0.383	0.367	0.196	0.180	0.562	7/8

Dimensions of Bolts
(All dimensions in inches.)

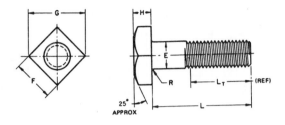

Square

Nominal Size or Basic Product Dia		E Body Dia.	F Width Across Flats			·G Width Across Corners		H Height			R Radius of Fillet		L_T Thread Length For Bolt Lengths	
													6 in. and shorter	Over 6 in.
		Max	Basic	Max	Min	Max	Min	Basic	Max	Min	Max	Min	Basic	Basic
1/4	0.2500	0.260	3/8	0.375	0.362	0.530	0.498	11/64	0.188	0.156	0.03	0.01	0.750	1.000
5/16	0.3125	0.324	1/2	0.500	0.484	0.707	0.665	13/64	0.220	0.186	0.03	0.01	0.875	1.125
3/8	0.3750	0.388	9/16	0.562	0.544	0.795	0.747	1/4	0.268	0.232	0.03	0.01	1.000	1.250
7/16	0.4375	0.452	5/8	0.625	0.603	0.884	0.828	19/64	0.316	0.278	0.03	0.01	1.125	1.375
1/2	0.5000	0.515	3/4	0.750	0.725	1.061	0.995	21/64	0.348	0.308	0.05	0.01	1.250	1.500
5/8	0.6250	0.642	15/16	0.938	0.906	1.326	1.244	27/64	0.444	0.400	0.06	0.01	1.500	1.750
3/4	0.7500	0.768	1 1/8	1.125	1.088	1.591	1.494	1/2	0.524	0.476	0.06	0.02	1.750	2.000
7/8	0.8750	0.895	1 5/16	1.312	1.269	1.856	1.742	19/32	0.620	0.568	0.06	0.02	2.000	2.250

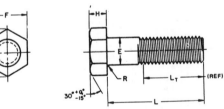

Hex

Nominal Size or Basic Product Dia		E Body Dia	F Width Across Flats			G Width Across Corners		H Height			R Radius of Fillet		L_T Thread Length For Bolt Lengths	
													6 in. and Shorter	Over 6 in.
		Max	Basic	Max	Min	Max	Min	Basic	Max	Min	Max	Min	Basic	Basic
1/4	0.2500	0.260	7/16	0.438	0.425	0.505	0.484	11/64	0.188	0.150	0.03	0.01	0.750	1.000
5/16	0.3125	0.324	1/2	0.500	0.484	0.577	0.552	7/32	0.235	0.195	0.03	0.01	0.875	1.125
3/8	0.3750	0.388	9/16	0.562	0.544	0.650	0.620	1/4	0.268	0.226	0.03	0.01	1.000	1.250
7/16	0.4375	0.452	5/8	0.625	0.603	0.722	0.687	19/64	0.316	0.272	0.03	0.01	1.125	1.375
1/2	0.5000	0.515	3/4	0.750	0.725	0.866	0.826	11/32	0.364	0.302	0.03	0.01	1.250	1.500
5/8	0.6250	0.642	15/16	0.938	0.906	1.063	1.033	27/64	0.444	0.378	0.06	0.02	1.500	1.750
3/4	0.7500	0.768	1 1/8	1.125	1.088	1.299	1.240	1/2	0.524	0.455	0.06	0.02	1.750	2.000
7/8	0.8750	0.895	1 5/16	1.312	1.269	1.516	1.447	37/64	0.604	0.531	0.06	0.02	2.000	2.250
1	1.0000	1.022	1 1/2	1.500	1.450	1.732	1.653	43/64	0.700	0.591	0.09	0.03	2.250	2.500
1 1/8	1.1250	1.149	1 11/16	1.688	1.631	1.949	1.859	3/4	0.780	0.658	0.09	0.03	2.500	2.750
1 1/4	1.2500	1.277	1 7/8	1.875	1.812	2.165	2.066	27/32	0.876	0.749	0.09	0.03	2.750	3.000
1 3/8	1.3750	1.404	2 1/16	2.062	1.994	2.382	2.273	29/32	0.940	0.810	0.09	0.03	3.000	3.250
1 1/2	1.5000	1.531	2 1/4	2.250	2.175	2.598	2.480	1	1.036	0.902	0.09	0.03	3.250	3.500
1 3/4	1.7500	1.785	2 5/8	2.625	2.538	3.031	2.893	1 5/32	1.196	1.054	0.12	0.04	3.750	4.000

Dimensions of Nuts
(All dimensions in inches.)

Square Nuts

Nominal Size or Basic Major Dia of Thread		F Width Across Flats			G Width Across Corners		H Thickness		
		Basic	Max	Min	Max	Min	Basic	Max	Min
1/4	0.2500	7/16	0.438	0.425	0.619	0.584	7/32	0.235	0.203
5/16	0.3125	9/16	0.562	0.547	0.795	0.751	17/64	0.283	0.249
3/8	0.3750	5/8	0.625	0.606	0.884	0.832	21/64	0.346	0.310
7/16	0.4375	3/4	0.750	0.728	1.061	1.000	3/8	0.394	0.356
1/2	0.5000	13/16	0.812	0.788	1.149	1.082	7/16	0.458	0.418
5/8	0.6250	1	1.000	0.969	1.414	1.330	35/64	0.569	0.525
3/4	0.7500	1 1/8	1.125	1.088	1.591	1.494	21/32	0.680	0.632
7/8	0.8750	1 5/16	1.312	1.269	1.856	1.742	49/64	0.792	0.740
1	1.0000	1 1/2	1.500	1.450	2.121	1.991	7/8	0.903	0.847
1 1/8	1.1250	1 11/16	1.688	1.631	2.386	2.239	1	1.030	0.970
1 1/4	1.2500	1 7/8	1.875	1.812	2.652	2.489	1 3/32	1.126	1.062
1 3/8	1.3750	2 1/16	2.062	1.994	2.917	2.738	1 13/64	1.237	1.169
1 1/2	1.5000	2 1/4	2.250	2.175	3.182	2.986	1 5/16	1.348	1.276

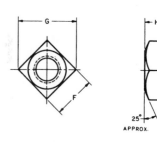

Hex and Hex Jam

Nominal Size or Basic Major Dia of Thread		F Width Across Flats			G Width Across Corners		H Thickness Hex Nuts			H₁ Thickness Hex Jam Nuts		
		Basic	Max	Min	Max	Min	Basic	Max	Min	Basic	Max	Min
1/4	0.2500	7/16	0.438	0.428	0.505	0.488	7/32	0.226	0.212	5/32	0.163	0.150
5/16	0.3125	1/2	0.500	0.489	0.577	0.557	17/64	0.273	0.258	3/16	0.195	0.180
3/8	0.3750	9/16	0.562	0.551	0.650	0.628	21/64	0.337	0.320	7/32	0.227	0.210
7/16	0.4375	11/16	0.688	0.675	0.794	0.768	3/8	0.385	0.365	1/4	0.260	0.240
1/2	0.5000	3/4	0.750	0.736	0.866	0.840	7/16	0.448	0.427	5/16	0.323	0.302
9/16	0.5625	7/8	0.875	0.861	1.010	0.982	31/64	0.496	0.473	5/16	0.324	0.301
5/8	0.6250	15/16	0.938	0.922	1.083	1.051	35/64	0.559	0.535	3/8	0.387	0.363
3/4	0.7500	1 1/8	1.125	1.088	1.299	1.240	41/64	0.665	0.617	27/64	0.446	0.398
7/8	0.8750	1 5/16	1.312	1.269	1.516	1.447	3/4	0.776	0.724	31/64	0.510	0.458

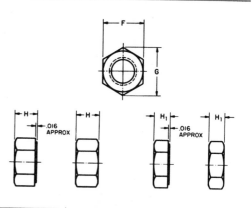

Hex Slotted Nuts

Nominal Size or Basic Major Dia of Thread		F Width Across Flats			G Width Across Corners		H Thickness			T Unslotted Thickness		S Width of Slot	
		Basic	Max	Min	Max	Min	Basic	Max	Min	Max	Min	Max	Min
1/4	0.2500	7/16	0.438	0.428	0.505	0.488	7/32	0.226	0.212	0.14	0.12	0.10	0.07
5/16	0.3125	1/2	0.500	0.489	0.577	0.557	17/64	0.273	0.258	0.18	0.16	0.12	0.09
3/8	0.3750	9/16	0.562	0.551	0.650	0.628	21/64	0.337	0.320	0.21	0.19	0.15	0.12
7/16	0.4375	11/16	0.688	0.675	0.794	0.768	3/8	0.385	0.365	0.23	0.21	0.15	0.12
1/2	0.5000	3/4	0.750	0.736	0.866	0.840	7/16	0.448	0.427	0.29	0.27	0.18	0.15
9/16	0.5625	7/8	0.875	0.861	1.010	0.982	31/64	0.496	0.473	0.31	0.29	0.18	0.15
5/8	0.6250	15/16	0.938	0.922	1.083	1.051	35/64	0.559	0.535	0.34	0.32	0.24	0.18
3/4	0.7500	1 1/8	1.125	1.088	1.299	1.240	41/64	0.665	0.617	0.40	0.38	0.24	0.18
7/8	0.8750	1 5/16	1.312	1.269	1.516	1.447	3/4	0.776	0.724	0.52	0.49	0.24	0.18

Hex Castle

Nominal Size or Basic Major Dia of Thread		F Width Across Flats			G Width Across Corners		H Thickness			T Unslotted Thickness and Height of Flats			S Width of Slot		R Radius of Fillet ±0.010	U Dia of Cylindrical Part Min
		Basic	Max	Min	Max	Min	Basic	Max	Min	Nom	Max	Min	Max	Min		
1/4	0.2500	7/16	0.438	0.428	0.505	0.488	9/32	0.288	0.274	3/16	0.20	0.18	0.10	0.07	0.094	0.371
5/16	0.3125	1/2	0.500	0.489	0.577	0.557	21/64	0.336	0.320	15/64	0.24	0.22	0.12	0.09	0.094	0.425
3/8	0.3750	9/16	0.562	0.551	0.650	0.628	13/32	0.415	0.398	9/32	0.29	0.27	0.15	0.12	0.094	0.478
7/16	0.4375	11/16	0.688	0.675	0.794	0.768	29/64	0.463	0.444	19/64	0.31	0.29	0.15	0.12	0.094	0.582
1/2	0.5000	3/4	0.750	0.736	0.866	0.840	9/16	0.573	0.552	13/32	0.42	0.40	0.18	0.15	0.125	0.637
9/16	0.5625	7/8	0.875	0.861	1.010	0.982	39/64	0.621	0.598	27/64	0.43	0.41	0.18	0.15	0.156	0.744
5/8	0.6250	15/16	0.938	0.922	1.083	1.051	23/32	0.731	0.706	1/2	0.51	0.49	0.24	0.18	0.156	0.797
3/4	0.7500	1 1/8	1.125	1.088	1.299	1.240	13/16	0.827	0.798	9/16	0.57	0.55	0.24	0.18	0.188	0.941
7/8	0.8750	1 5/16	1.312	1.269	1.516	1.447	29/32	0.922	0.890	21/32	0.67	0.64	0.24	0.18	0.188	1.097

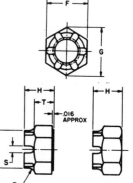

Dimensions of Washers
(All dimensions in inches.)

PLAIN WASHERS

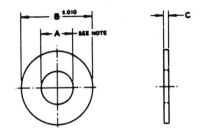

Inside Diameter A	Outside Diameter B	Thickness, C Nom	Inside Diameter A	Outside Diameter B	Thickness, C Nom
3/64	3/16	0.020	5/8	1 1/2	0.109
3/32	7/32	0.020	5/8	2 1/4	0.134
3/32	1/4	0.020	21/32	1 5/16	0.095
3/32	1/4	0.022	11/16	1 1/2	0.134
1/8	1/4	0.032	11/16	1 3/4	0.134
1/8	5/16				
5/32	5/16	0.035	11/16	2 3/8	0.165
5/32	3/8	0.049	13/16	1 1/2	0.134
11/64	11/32	0.049	13/16	1 3/4	0.148
3/16	3/8	0.049	13/16	2	0.148
3/16	7/16	0.049	13/16	2 7/8	0.165
13/64	15/32	0.049	15/16	1 3/4	0.134
7/32	7/16	0.049	15/16	2	0.165
7/32	1/2	0.049	15/16	2 1/4	0.165
15/64	17/32	0.049	15/16	3 3/8	0.180
1/4	1/2	0.049	1 1/16	2	0.134
1/4	9/16	0.049	1 1/16	2 1/4	0.165
1/4	9/16	0.065	1 1/16	2 1/2	0.165
17/64	5/8	0.049	1 1/16	3 7/8	0.238
9/32	5/8	0.065	1 3/16	2 1/2	0.165
5/16	3/4	0.065	1 1/4	2 3/4	0.165
5/16	7/8	0.065	1 5/16	2 3/4	0.165
11/32	11/16	0.065	1 3/8	3	0.165
3/8	3/4	0.065	1 7/16	3	0.180
3/8	7/8	0.083	1 1/2	3 1/4	0.180
3/8	1 1/16	0.065	1 9/16	3 1/4	0.180
13/32	13/16	0.065	1 5/8	3 1/2	0.180
7/16	7/8	0.083	1 11/16	3 1/2	0.180
7/16	1	0.083	1 3/4	3 3/4	0.180
7/16	1 1/4	0.083	1 13/16	3 3/4	0.180
15/32	1 9/64	0.065	1 7/8	4	0.180
1/2	1 1/8	0.083	1 15/16	4	0.180
1/2	1 1/4	0.083	2	4 1/4	0.180
1/2	1 5/8	0.083			
17/32	1 1/16	0.095			
9/16	1 1/4	0.109			
9/16	1 3/8	0.109			
9/16	1 7/8	0.109			
19/32	1 3/16	0.095			
5/8	1 3/8	0.109			

MEDIUM SPRING LOCK WASHERS

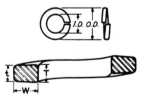

Nominal Size	Inside Diameter Min	Washer Sections (Min) Width W	Washer Sections (Min) Thickness $\frac{T+t}{2}$	Outside Diam Max*
0.086 (No. 2)	0.088	0.035	0.020	0.175
0.099 (No. 3)	0.102	0.040	0.025	0.198
0.112 (No. 4)	0.115	0.040	0.025	0.212
0.125 (No. 5)	0.128	0.047	0.031	0.239
0.138 (No. 6)	0.141	0.047	0.031	0.253
0.164 (No. 8)	0.168	0.055	0.040	0.296
0.190 (No. 10)	0.194	0.062	0.047	0.337
0.216 (No. 12)	0.221	0.070	0.056	0.380
1/4	0.255	0.109	0.062	0.493
5/16	0.319	0.125	0.078	0.591
3/8	0.382	0.141	0.094	0.688
7/16	0.446	0.156	0.109	0.784
1/2	0.509	0.171	0.125	0.879
9/16	0.573	0.188	0.141	0.979
5/8	0.636	0.203	0.156	1.086
11/16	0.700	0.219	0.172	1.184
3/4	0.763	0.234	0.188	1.279
13/16	0.827	0.250	0.203	1.377
7/8	0.890	0.266	0.219	1.474
15/16	0.954	0.281	0.234	1.570
1	1.017	0.297	0.250	1.672
1 1/16	1.081	0.312	0.266	1.768
1 1/8	1.144	0.328	0.281	1.865
1 3/16	1.208	0.344	0.297	1.963
1 1/4	1.271	0.359	0.312	2.058
1 5/16	1.335	0.375	0.328	2.156
1 3/8	1.398	0.391	0.344	2.253
1 7/16	1.462	0.406	0.359	2.349
1 1/2	1.525	0.422	0.375	2.446

Dimensions of Woodruff Keys and Keyseats
(All dimensions in inches.)

USA Standard Woodruff Keys (USAS B17.2-1967)

Key No.	Nominal Key Size W × B	Actual Length F +0.000 -0.010	Height of Key C Max.	Height of Key C Min.	Height of Key D Max.	Height of Key D Min.	Distance Below Center E
202	1/16 × 1/4	0.248	0.109	0.104	0.109	0.104	1/64
202.5	1/16 × 5/16	0.311	0.140	0.135	0.140	0.135	1/64
302.5	3/32 × 5/16	0.311	0.140	0.135	0.140	0.135	1/64
203	1/16 × 3/8	0.374	0.172	0.167	0.172	0.167	1/64
303	3/32 × 3/8	0.374	0.172	0.167	0.172	0.167	1/64
403	1/8 × 3/8	0.374	0.172	0.167	0.172	0.167	1/64
204	1/16 × 1/2	0.491	0.203	0.198	0.194	0.188	3/64
304	3/32 × 1/2	0.491	0.203	0.198	0.194	0.188	3/64
404	1/8 × 1/2	0.491	0.203	0.198	0.194	0.188	3/64
405	1/8 × 5/8	0.612	0.250	0.245	0.240	0.234	1/16
505	5/32 × 5/8	0.612	0.250	0.245	0.240	0.234	1/16
605	3/16 × 5/8	0.612	0.250	0.245	0.240	0.234	1/16
406	1/8 × 3/4	0.740	0.313	0.308	0.303	0.297	1/16
506	5/32 × 3/4	0.740	0.313	0.308	0.303	0.297	1/16
606	3/16 × 3/4	0.740	0.313	0.308	0.303	0.297	1/16
507	5/32 × 7/8	0.866	0.375	0.370	0.365	0.359	1/16
607	3/16 × 7/8	0.866	0.375	0.370	0.365	0.359	1/16
707	7/32 × 7/8	0.866	0.375	0.370	0.365	0.359	1/16
608	3/16 × 1	0.992	0.438	0.433	0.428	0.422	1/16
708	7/32 × 1	0.992	0.438	0.433	0.428	0.422	1/16
808	1/4 × 1	0.992	0.438	0.433	0.428	0.422	1/16
1008	5/16 × 1	0.992	0.438	0.433	0.428	0.422	1/16
1208	3/8 × 1	0.992	0.438	0.433	0.428	0.422	1/16
609	3/16 × 1 1/8	1.114	0.484	0.479	0.475	0.469	5/64
709	7/32 × 1 1/8	1.114	0.484	0.479	0.475	0.469	5/64
809	1/4 × 1 1/8	1.114	0.484	0.479	0.475	0.469	5/64
1009	5/16 × 1 1/8	1.114	0.484	0.479	0.475	0.469	5/64
610	3/16 × 1 1/4	1.240	0.547	0.542	0.537	0.531	5/64
710	7/32 × 1 1/4	1.240	0.547	0.542	0.537	0.531	5/64
810	1/4 × 1 1/4	1.240	0.547	0.542	0.537	0.531	5/64
1010	5/16 × 1 1/4	1.240	0.547	0.542	0.537	0.531	5/64
1210	3/8 × 1 1/4	1.240	0.547	0.542	0.537	0.531	5/64
811	1/4 × 1 3/8	1.362	0.594	0.589	0.584	0.578	3/32
1011	5/16 × 1 3/8	1.362	0.594	0.589	0.584	0.578	3/32
1211	3/8 × 1 3/8	1.362	0.594	0.589	0.584	0.578	3/32
812	1/4 × 1 1/2	1.484	0.641	0.636	0.631	0.625	7/64
1012	5/16 × 1 1/2	1.484	0.641	0.636	0.631	0.625	7/64
1212	3/8 × 1 1/2	1.484	0.641	0.636	0.631	0.625	7/64

USA Standard Woodruff Keys (USAS B17.2-1967)

Key No.	Nominal Size Key	Keyseat — Shaft Width A Min.	Width A Max.	Depth B +0.005 -0.000	Diameter F Min.	Diameter F Max.	Key Above Shaft Height C +0.005 -0.005	Keyseat — Hub Width D +0.002 -0.000	Depth E +0.005 -0.000
202	1/16 × 1/4	0.0615	0.0630	0.0728	0.250	0.268	0.0312	0.0635	0.0372
202.5	1/16 × 5/16	0.0615	0.0630	0.1038	0.312	0.330	0.0312	0.0635	0.0372
302.5	3/32 × 5/16	0.0928	0.0943	0.0882	0.312	0.330	0.0469	0.0948	0.0529
203	1/16 × 3/8	0.0615	0.0630	0.1358	0.375	0.393	0.0312	0.0635	0.0372
303	3/32 × 3/8	0.0928	0.0943	0.1202	0.375	0.393	0.0469	0.0948	0.0529
403	1/8 × 3/8	0.1240	0.1255	0.1045	0.375	0.393	0.0625	0.1260	0.0685
204	1/16 × 1/2	0.0615	0.0630	0.1668	0.500	0.518	0.0312	0.0635	0.0372
304	3/32 × 1/2	0.0928	0.0943	0.1511	0.500	0.518	0.0469	0.0948	0.0529
404	1/8 × 1/2	0.1240	0.1255	0.1355	0.500	0.518	0.0625	0.1260	0.0685
305	3/32 × 5/8	0.0928	0.0943	0.1981	0.625	0.643	0.0469	0.0948	0.0529
405	1/8 × 5/8	0.1240	0.1255	0.1825	0.625	0.643	0.0625	0.1260	0.0685
505	5/32 × 5/8	0.1553	0.1568	0.1669	0.625	0.643	0.0781	0.1573	0.0841
605	3/16 × 5/8	0.1863	0.1880	0.1513	0.625	0.643	0.0937	0.1885	0.0997
406	1/8 × 3/4	0.1240	0.1255	0.2455	0.750	0.768	0.0625	0.1260	0.0685
506	5/32 × 3/4	0.1553	0.1568	0.2299	0.750	0.768	0.0781	0.1573	0.0841
606	3/16 × 3/4	0.1863	0.1880	0.2143	0.750	0.768	0.0937	0.1885	0.0997
806	1/4 × 3/4	0.2487	0.2505	0.1830	0.750	0.768	0.1250	0.2510	0.1310
507	5/32 × 7/8	0.1553	0.1568	0.2919	0.875	0.895	0.0781	0.1573	0.0841
607	3/16 × 7/8	0.1863	0.1880	0.2763	0.875	0.895	0.0937	0.1885	0.0997
707	7/32 × 7/8	0.2175	0.2193	0.2607	0.875	0.895	0.1093	0.2198	0.1153
807	1/4 × 7/8	0.2487	0.2505	0.2450	0.875	0.895	0.1250	0.2510	0.1310
608	3/16 × 1	0.1863	0.1880	0.3393	1.000	1.020	0.0937	0.1885	0.0997
708	7/32 × 1	0.2175	0.2193	0.3237	1.000	1.020	0.1093	0.2198	0.1153
808	1/4 × 1	0.2487	0.2505	0.3080	1.000	1.020	0.1250	0.2510	0.1310
1008	5/16 × 1	0.3111	0.3130	0.2768	1.000	1.020	0.1562	0.3135	0.1622
1208	3/8 × 1	0.3735	0.3755	0.2455	1.000	1.020	0.1875	0.3760	0.1935
609	3/16 × 1 1/8	0.1863	0.1880	0.3853	1.125	1.145	0.0937	0.1885	0.0997
709	7/32 × 1 1/8	0.2175	0.2193	0.3697	1.125	1.145	0.1093	0.2198	0.1153
809	1/4 × 1 1/8	0.2487	0.2505	0.3540	1.125	1.145	0.1250	0.2510	0.1310
1009	5/16 × 1 1/8	0.3111	0.3130	0.3228	1.125	1.145	0.1562	0.3135	0.1622
610	3/16 × 1 1/4	0.1863	0.1880	0.4483	1.250	1.273	0.0937	0.1885	0.0997
710	7/32 × 1 1/4	0.2175	0.2193	0.4327	1.250	1.273	0.1093	0.2198	0.1153
810	1/4 × 1 1/4	0.2487	0.2505	0.4170	1.250	1.273	0.1250	0.2510	0.1310
1010	5/16 × 1 1/4	0.3111	0.3130	0.3858	1.250	1.273	0.1562	0.3135	0.1622
1210	3/8 × 1 1/4	0.3735	0.3755	0.3545	1.250	1.273	0.1875	0.3760	0.1935
811	1/4 × 1 3/8	0.2487	0.2505	0.4640	1.375	1.398	0.1250	0.2510	0.1310
1011	5/16 × 1 3/8	0.3111	0.3130	0.4328	1.375	1.398	0.1562	0.3135	0.1622
1211	3/8 × 1 3/8	0.3735	0.3755	0.4015	1.375	1.398	0.1875	0.3760	0.1935

All dimensions are given in inches.

The key numbers indicate nominal key dimensions. The last two digits give the nominal diameter B in eighths of an inch and the digits preceding the last two give the nominal width W in thirty-seconds of an inch.

Dimensions of Taper Pins

(All dimensions in inches.)

American Standard Taper Pins

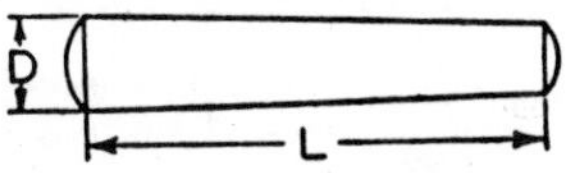

Taper 1/4 inch per foot

No. of Taper Pin	Diam. Large End D	Approx. Size D	Range of Lengths L†	No. of Taper Pin	Diam. Large End D	Approx. Size D	Range of Lengths L†
7/0	0.0625	1/16	3/8 to 5/8	3	0.219	7/32	3/4 to 1 3/4
6/0	0.078	5/64	3/8 to 3/4	4	0.250	1/4	3/4 to 2
5/0	0.094	3/32	1/2 to 1	5	0.289	19/64	1 to 2 1/4
4/0	0.109	7/64	1/2 to 1	6	0.341	11/32	1 1/4 to 3
3/0	0.125	1/8	1/2 to 1	7	0.409	13/32	2 to 3 3/4
2/0	0.141	9/64	1/2 to 1 1/4	8	0.492	1/2	2 to 4 1/2
0	0.156	5/32	1/2 to 1 1/4	9	0.591	19/32	2 3/4 to 5 1/4
1	0.172	11/64	5/8 to 1 1/4	10	0.706	45/64	3 1/2 to 6
2	0.193	3/16	3/4 to 1 1/2				

† These lengths L are suitable for use with the standard reamers listed. Longer lengths available for Nos. 1 to 9, incl.

Special sizes No. 11 (0.860), No. 12 (1.032), No. 13 (1.241), and No. 14 (1.523) are also available. Their lengths are special.

Tolerance on diameter is +0.0013 −0.0007 for all sizes.

To find diameter at small end of pin, multiply length L by 0.0208 and subtract product from large end diameter D.

Diameter at Small Ends of Standard Taper Pins

Pin Length in Inches	Pin Number and Small End Diam. for Given Length										
	0	1	2	3	4	5	6	7	8	9	10
3/4	0.140	0.156	0.177	0.203	0.235	0.273	0.325	0.393	0.476	0.575	0.690
1	0.135	0.151	0.172	0.198	0.230	0.268	0.320	0.388	0.471	0.570	0.685
1 1/4	0.130	0.146	0.167	0.192	0.224	0.263	0.315	0.382	0.466	0.565	0.680
1 1/2	0.125	0.141	0.162	0.187	0.219	0.258	0.310	0.377	0.460	0.560	0.675
1 3/4	0.120	0.136	0.157	0.182	0.214	0.252	0.305	0.372	0.455	0.554	0.669
2	0.114	0.130	0.151	0.177	0.209	0.247	0.299	0.367	0.450	0.549	0.664
2 1/4	0.109	0.125	0.146	0.172	0.204	0.242	0.294	0.362	0.445	0.544	0.659
2 1/2	0.104	0.120	0.141	0.166	0.198	0.237	0.289	0.356	0.440	0.539	0.654
2 3/4	0.099	0.115	0.136	0.161	0.193	0.232	0.284	0.351	0.434	0.534	0.649
3	0.094	0.110	0.131	0.156	0.188	0.227	0.279	0.346	0.429	0.528	0.643
3 1/4	...	...	...	0.151	0.182	0.221	0.273	0.340	0.424	0.523	0.638
3 1/2	...	...	...	0.146	0.177	0.216	0.268	0.335	0.419	0.518	0.633
3 3/4	...	...	...	0.141	0.172	0.211	0.263	0.330	0.414	0.513	0.628
4	...	...	...	0.136	0.167	0.206	0.258	0.326	0.409	0.508	0.623
4 1/4	...	...	...	0.131	0.162	0.201	0.253	0.321	0.403	0.502	0.617
4 1/2	...	...	...	0.125	0.156	0.195	0.247	0.315	0.398	0.497	0.612
5	...	...	...	...	0.146	0.185	0.237	0.305	0.389	0.487	0.602
5 1/2	...	...	...	...	...	...	...	0.294	0.377	0.476	0.591
6	...	...	...	...	...	...	...	0.284	0.367	0.466	0.581

American Standard Welded and Seamless Steel Pipe Data

Nominal Pipe Size	Outside Diameter	WALL THICKNESS			WEIGHT PER FOOT		
		Standard Weight *	Extra Strong **	Double Extra Strong	Standard Weight +	Extra Strong	Double Extra Strong
1/8	.405	.068	.095	-	.244	.314	-
1/4	.540	.088	.119	-	.424	.535	-
3/8	.675	.091	.126	-	.567	.738	-
1/2	.840	.109	.147	.294	.850	1.087	1.714
3/4	1.050	.113	.154	.308	1.130	1.473	2.440
1	1.315	.133	.179	.358	1.678	2.171	3.659
1-1/4	1.660	.140	.191	.382	2.272	2.996	5.214
1-1/2	1.900	.145	.200	.400	2.717	3.631	6.408
2	2.375	.154	.218	.436	3.652	5.022	9.029
2-1/2	2.875	.203	.276	.552	5.79	7.66	13.70
3	3.500	.216	.300	.600	7.58	10.25	18.58
3-1/2	4.000	.226	.318	-	9.11	12.51	-
4	4.500	.237	.337	.674	10.79	14.98	27.54
5	5.563	.258	.375	.750	14.62	20.78	38.55
6	6.625	.280	.432	.864	18.97	28.57	53.16
8	8.625	.322	.500	.875	28.55	43.39	72.42
10	10.750	.365	.500	-	40.48	54.74	-
12	12.750	.375	.500	-	49.56	65.42	-

Dimensions in inches, weights in pounds.

*Same as ASA B36.10 "Schedule 40" Pipe, except 12-inch diameter.

**Same as ASA B36.10 "Schedule 80" Pipe, except 10 and 12-inch diameter.

+ Plain Ends

Dimensions and Characteristics of Pipe for Hydraulic Systems

NOMINAL PIPE SIZE (IN.)	PIPE OD (IN.)	THREADS PER INCH	SCHEDULE 40 STANDARD		SCHEDULE 80 EXTRA HEAVY		SCHEDULE 160		DOUBLE EXTRA HEAVY	
			PIPE ID (IN.)	BURSTING PRESSURE (PSI)	PIPE ID (IN.)	BURSTING PRESSURE (PSI)	PIPE ID (IN.)	BURSTING PRESSURE (PSI)	PIPE ID (IN.)	BURSTING PRESSURE (PSI)
1/4	0.540	18	0.364	16,000	0.302	22,000	....	...	...	....
3/8	0.675	18	0.493	13,500	0.423	19,000	....	....	....	....
1/2	0.840	14	0.622	13,200	0.546	17,500	0.466	21,000	0.252	35,000
3/4	1.050	14	0.824	11,000	0.742	15,000	0.614	21,000	0.434	30,000
1	1.315	11 1/2	1.049	10,000	0.957	13,600	0.815	19,000	0.599	27,000
1 1/4	1.660	11 1/2	1.380	8,400	1.278	11,500	1.160	15,000	0.896	23,000
1 1/2	1.900	11 1/2	1.610	7,600	1.500	10,500	1.388	14,800	1.100	21,000
2	2.375	11 1/2	2.067	6,500	1.939	9,100	1.689	14,500	1.503	19,000
2 1/2	2.875	8	2.469	7,000	2.323	9,600	2.125	13,000	1.771	18,000
3	3.500	8	3.068	6,100	2.900	8,500	2.624	12,500	—	—

Basic Dimensions for Commercially Available Tubing

OUTSIDE DIAMETER (IN.)	WALL THICKNESS (IN.)	INSIDE DIAMETER (IN.)	OUTSIDE DIAMETER (IN.)	WALL THICKNESS (IN.)	INSIDE DIAMETER (IN.)	OUTSIDE DIAMETER (IN.)	WALL THICKNESS (IN.)	INSIDE DIAMETER (IN.)	OUTSIDE DIAMETER (IN.)	WALL THICKNESS (IN.)	INSIDE DIAMETER (IN.)
1/8	0.028	0.069	1/2	0.035	0.430	7/8	0.049	0.777	1 1/4	0.120	1.010
	.032	.061		.042	.416		.058	.759			
	.035	.055		.049	.402		.065	.745	1 1/2	0.065	1.370
				.058	.384		.072	.731		.072	1.356
3/16	0.032	0.1235		.065	.370		.083	.709		.083	1.334
	.035	.1175		.072	.358		.095	.685		.095	1.310
				.083	.334		.109	.657		.109	1.282
1/4	0.035	0.180								.120	1.260
	.042	.166	5/8	0.035	0.555	1	0.049	0.902			
	.049	.152		.042	.541		.058	.884	1 3/4	0.065	1.620
	.058	.134		.049	.527		.065	.870		.072	1.606
	.065	.120		.058	.509		.072	.856		.083	1.584
				.065	.495		.083	.834		.095	1.560
5/16	0.035	0.2425		.072	.481		.095	.810		.109	1.532
	.042	.2285		.083	.459		.109	.782		.120	1.510
	.049	.2145		.095	.435		.120	.760		.134	1.482
	.058	.1965									
	.065	.1825	3/4	0.049	0.652	1 1/4	0.049	1.152	2	0.065	1.870
				.058	.634		.058	1.134		.072	1.856
3/8	0.035	0.305		.065	.620		.065	1.120		.083	1.834
	.042	.291		.072	.606		.072	1.106		.095	1.810
	.049	.277		.083	.584		.083	1.064		.109	1.782
	.058	.259		.095	.560		.095	1.060		.120	1.760
	.065	.245		.109	.532		.109	1.032		.134	1.732

Cast-Iron Fittings
(All dimensions in inches.)

90° ELBOW   TEE   CROSS   45° ELBOW

125-Lb Screwed Fittings

Nominal Pipe Size	Center to End, Elbows, Tees, and Crosses A	Center to End, 45-Deg Elbows C	Length of Thread, Min B	Width of Band, Min E	Inside Diameter of Fitting P Max	Inside Diameter of Fitting P Min	Metal Thickness G	Outside Diameter of Band, Min H
1/4	0.81	0.73	0.32	0.38	0.584	0.540	0.110	0.93
3/8	0.95	0.80	0.36	0.44	0.719	0.675	0.120	1.12
1/2	1.12	0.88	0.43	0.50	0.897	0.840	0.130	1.34
3/4	1.31	0.98	0.50	0.56	1.107	1.050	0.155	1.63
1	1.50	1.12	0.58	0.62	1.385	1.315	0.170	1.95
1 1/4	1.75	1.29	0.67	0.69	1.730	1.660	0.185	2.39
1 1/2	1.94	1.43	0.70	0.75	1.970	1.900	0.200	2.68
2	2.25	1.68	0.75	0.84	2.445	2.375	0.220	3.28
2 1/2	2.70	1.95	0.92	0.94	2.975	2.875	0.240	3.86
3	3.08	2.17	0.98	1.00	3.600	3.500	0.260	4.62
3 1/2	3.42	2.39	1.03	1.06	4.100	4.000	0.280	5.20
4	3.79	2.61	1.08	1.12	4.600	4.500	0.310	5.79
5	4.50	3.05	1.18	1.18	5.663	5.563	0.380	7.05
6	5.13	3.46	1.28	1.28	6.725	6.625	0.430	8.28
8	6.56	4.28	1.47	1.47	8.725	8.625	0.550	10.63
10	8.08	5.16	1.68	1.68	10.850	10.750	0.690	13.12
12	9.50	5.97	1.88	1.88	12.850	12.750	0.800	15.47

250-Lb Screwed Fittings

Nominal Pipe Size	Center to End, Elbows, Tees, and Crosses A	Center to End, 45-Deg Elbows C	Length of Thread, Min B	Width of Band, Min E	Inside Diameter of Fitting P Max	Inside Diameter of Fitting P Min	Metal Thickness G	Outside Diameter of Band, Min H
1/4	0.94	0.81	0.43	0.49	0.584	0.540	0.18	1.17
3/8	1.06	0.88	0.47	0.55	0.719	0.675	0.18	1.36
1/2	1.25	1.00	0.57	0.60	0.897	0.840	0.20	1.59
3/4	1.44	1.13	0.64	0.68	1.107	1.050	0.23	1.88
1	1.63	1.31	0.75	0.76	1.385	1.315	0.28	2.24
1 1/4	1.94	1.50	0.84	0.88	1.730	1.660	0.33	2.73
1 1/2	2.13	1.69	0.87	0.97	1.970	1.900	0.35	3.07
2	2.50	2.00	1.00	1.12	2.445	2.375	0.39	3.74
2 1/2	2.94	2.25	1.17	1.30	2.975	2.875	0.43	4.60
3	3.38	2.50	1.23	1.40	3.600	3.500	0.48	5.36
3 1/2	3.75	2.63	1.28	1.49	4.100	4.000	0.52	5.98
4	4.13	2.81	1.33	1.57	4.600	4.500	0.56	6.61
5	4.88	3.19	1.43	1.74	5.663	5.563	0.66	7.92
6	5.63	3.50	1.53	1.91	6.725	6.625	0.74	9.24
8	7.00	4.31	1.72	2.24	8.725	8.625	0.90	11.73
10	8.63	5.19	1.93	2.58	10.850	10.750	1.08	14.37
12	10.00	6.00	2.13	2.91	12.850	12.750	1.24	16.84

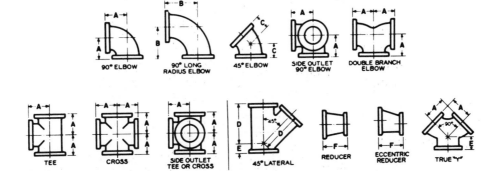

90° ELBOW   90° LONG RADIUS ELBOW   45° ELBOW   SIDE OUTLET 90° ELBOW   DOUBLE BRANCH ELBOW

TEE   CROSS   SIDE OUTLET TEE OR CROSS   45° LATERAL   REDUCER   ECCENTRIC REDUCER   TRUE "Y"

Dimensions of Elbows, Double Branch Elbows, Tees, Crosses, Laterals, True Y's (Straight Sizes), and Reducers

Nominal Pipe Size	Inside Diam of Fittings	Center to Face 90 Deg Elbow Tees, Crosses True "Y" and Double Branch Elbow A	Center to Face 90 Deg Long Radius Elbow B	Center to Face 45 Deg Elbow C	Center to Face Lateral D	Short Center to Face True "Y" and Lateral E	Face to Face Reducer F	Diam of Flange	Thickness of Flange (Min)	Wall Thickness
1	1	3 1/2	5	1 1/4	5 3/4	1 1/4	....	4 1/4	7/16	5/16
1 1/4	1 1/4	3 3/4	5 1/2	2	6 1/4	1 1/4	....	4 5/8	1/2	5/16
1 1/2	1 1/2	4	6	2 1/4	7	2	....	5	9/16	5/16
2	2	4 1/2	6 1/2	2 1/2	8	2 1/2	5	6	5/8	5/16
2 1/2	2 1/2	5	7	3	9 1/2	2 1/2	5 1/2	7	11/16	5/16
3	3	5 1/2	7 3/4	3	10	3	6	7 1/2	3/4	3/8
3 1/2	3 1/2	6	8 1/2	3 1/2	11 1/2	3	6 1/2	8 1/2	13/16	7/16
4	4	6 1/2	9	4	12	3	7	9	15/16	1/2
5	5	7 1/2	10 1/4	4 1/2	13 1/2	3 1/2	8	10	15/16	9/16
6	6	8	11 1/2	5	14 1/2	3 1/2	9	11	1	7/16

Graphical Symbols for Piping

PLUMBING

1 ACID WASTE		ACID
2 COLD WATER		
3 COMPRESSED AIR		A
4 DRINKING-WATER FLOW		

5 DRINKING-WATER RETURN		
6 FIRE LINE		F
7 GAS		G
8 HOT WATER		
9 HOT-WATER RETURN		
10 SOIL, WASTE OR LEADER (ABOVE GRADE)		

11 SOIL, WASTE OR LEADER (BELOW GRADE)		
12 VACUUM CLEANING		V
13 VENT		
PNEUMATIC TUBES 14 TUBE RUNS		
SPRINKLERS 15 BRANCH AND HEAD		
16 DRAIN		S
17 MAIN SUPPLIES		S

Basic Elements of Metrics

Metric Unit Prefixes

Value	Multiples and Submultiples	Prefixes	Symbols
1000	10^3	kilo	k
100	10^2	hecto	h
10	10^1	deka	da
0.1	10^{-1}	deci	d
0.01	10^{-2}	centi	c
0.001	10^{-3}	milli	m

Linear Measurements

One thousand meters (10^3 meters) is a kilometer - km

One hundred meters (10^2 meters) is a hectometer - hm

Ten meters (10^1 meters) is a dekameter - dam

A meter (10^0 meters) is a meter - m

One tenth of a meter (10^{-1} meters) is a decimeter - dm

One hundredth of a meter (10^{-2} meters) is a centimeter - cm

One thousandth of a meter (10^{-3} meters) is a millimeter - mm

Mass (weight) Measurements

One thousand grams (10^3 grams) is a kilogram - kg

One hundred grams (10^2 grams) is a hectogram - hg

Ten grams (10^1 grams) is a dekagram - dag

A gram (10^0 grams) is a gram - g

One tenth of a gram (10^{-1} grams) is a decigram - dg

One hundredth of a gram (10^{-2} grams) is a centrigram - cg

One thousandth of a gram (10^{-3} grams) is a milligram - mg

Volume (liquid) Measurements

One thousand liters (10^3 liters) is a kiloliter - kl

One hundred liters (10^2 liters) is a hectoliter - hl

Ten liters (10^1 liters) is a deckaliter - dal

A liter (10^0 liters) is a liter - 1

One tenth of a liter (10^{-1} liters) is a deciliter - dl

One hundredth of a liter (10^{-2} liters) is a centiliter - cl

One thousandth of a liter (10^{-3} liters) is a milliliter - ml

Meter Conversions

Metric values to English system

millimeters $\times$ 0.039 = inches
centimeters $\times$ 0.39 = inches
meters $\times$ 39.4 = inches
centimeters $\times$ 0.33 = feet
meters $\times$ 3.28 = feet
meters $\times$ 1.09 = yards
kilometers $\times$ 0.62 = miles

English system to metric values

inch $\times$ 25.4 = mm
inch $\times$ 2.5 = cm
inch $\times$ 0.025 = m
feet $\times$ 30.5 = cm
feet $\times$ 0.305 = m
yard $\times$ 0.91 = m
mile $\times$ 1.6 = km

Gram Conversions

1 lb = 0.453 kg
1 kg = 2.20 lb
1 oz = 28.34 g
1 g = 0.035 oz
1 lb = 453.59 g
1 g = 0.002 lb

Liter Conversions

1 pint = 0.47 liters
1 liter = 2.1 pints
1 quart = 0.95 liters
1 liter = 1.06 quarts
1 gallon = 3.8 liters
1 liter = 0.26 gallons

Area Conversions

in² $\times$ 6.5 = cm²
ft² $\times$ 0.09 = m²
yd² $\times$ 0.8 = m²
mi² $\times$ 2.6 = km²
cm². $\times$ 0.16 = in²
m² $\times$ 1.2 = yd²
km² $\times$ 0.4 = mi²

Volume Conversions

ft³ $\times$ 0.03 = cubic meters (m³)
yd³ $\times$ 0.76 = cubic meters (m³)
m³ $\times$ 35 = cubic feet (ft³)
m³ $\times$ 1.3 = cubic yards (yd³)

Temperature Conversions

To convert from	To	Use formula
Celsius (C)	Kelvin (K)	$K = C + 273.15$
Fahrenheit (F)	Kelvin (K)	$K = (F + 459.67) \div 1.8$
Celsius (C)	Fahrenheit (F)	$F = 1.8C + 32$
Fahrenheit (F)	Celsius (C)	$C = (F - 32) \div 1.8$
Kelvin (K)	Celsius (C)	$C = K - 273.15$
Kelvin (K)	Fahrenheit (F)	$F = 1.8K - 459.67$

Metric Drill Sizes (mm)[1]

Preferred	Decimal Equivalent in Inches (Ref)	Preferred	Decimal Equivalent in Inches (Ref.)	Preferred	Decimal Equivalent in Inches (Ref)
.50	.0197	2.20	.0866	9.00	.3543
.55	.0217	2.40	.0945	9.50	.3740
.60	.0236	2.50	.0984	10.00	.3937
.65	.0256	2.60	.1024	10.50	.4134
.70	.0276	2.80	.1102	11.00	.4331
.75	.0295	3.00	.1181		
.80	.0315	3.20	.1260	12.00	.4724
.85	.0335	3.40	.1339	12.50	.4921
.90	.0354	3.60	.1417	13.00	.5118
.95	.0374			14.00	.5512
1.00	.0394	3.80	.1496		
1.05	.0413	4.00	.1575	15.00	.5906
1.10	.0433	4.20	.1654	16.00	.6299
1.20	.0472	4.50	.1772	17.00	.6693
1.25	.0492	4.80	.1890	18.00	.7087
1.30	.0512	5.00	.1969	19.00	.7480
1.40	.0551	5.30	.2087	20.00	.7874
1.50	.0591	5.60	.2205	21.00	.8268
1.60	.0630	6.00	.2362	22.00	.8661
1.70	.0669	6.30	.2480	24.00	.9449
1.80	.0709	6.70	.2638	25.00	.9842
1.90	.0748	7.10	.2795	26.00	1.0236
2.00	.0787	7.50	.2953	28.00	1.1024
2.10	.0827	8.00	.3150	30.00	1.1811
		8.50	.3346	32.00	1.2598

[1] Metric drill sizes listed in the "Preferred" column are based on the R′40 series of preferred numbers shown in the ISO Standard R497.

Metric Tap Drill Sizes

METRIC TAP SIZE	RECOMMENDED METRIC DRILL				CLOSEST RECOMMENDED INCH DRILL			
	DRILL SIZE mm	Inch Equiv.	PROBABLE HOLE SIZE (Inches)	PROBABLE PERCENT OF THREAD	DRILL SIZE	Inch Equiv.	PROBABLE HOLE SIZE (Inches)	PROBABLE PERCENT OF THREAD
M1.6 × 0.35	1.25	0.0492	0.0507	69	—	—	—	—
M1.8 × 0.35	1.45	0.0571	0.0586	69	—	—	—	—
M2 × 0.4	1.60	0.0630	0.0647	69	#52	0.0635	0.0652	66
M2.2 × 0.45	1.75	0.0689	0.0706	70	—	—	—	—
M2.5 × 0.45	2.05	0.0807	0.0826	69	#46	0.0810	0.0829	67
M3 × 0.5	2.50	0.0984	0.1007	68	#40	0.0980	0.1003	70
M3.5 × 0.6	2.90 .	0.1142	0.1168	68	#33	0.1130	0.1156	72
M4 × 0.7	3.30	0.1299	0.1328	69	#30	0.1285	0.1314	73
M4.5 × 0.75	3.70	0.1457	0.1489	74	#26	0.1470	0.1502	70
M5 × 0.8	4.20	0.1654	0.1686	69	#19	0.1660	0.1692	68
M6 × 1	5.00	0.1968	0.2006	70	#9	0.1960	0.1998	71
M7 × 1	6.00	0.2362	0.2400	70	15/64	0.2344	0.2382	73
M8 × 1.25	6.70	0.2638	0.2679	74	17/64	0.2656	0.2697	71
M8 × 1	7.00	0.2756	0.2797	69	J	0.2770	0.2811	66
M10 × 1.5	8.50	0.3346	0.3390	71	Q	0.3320	0.3364	75
M10 × 1.25	8.70	0.3425	0.3471	73	11/32	0.3438	0.3483	71
M12 × 1.75	10.20	0.4016	0.4063	74	Y	0.4040	0.4087	71
M12 × 1.25	10.80	0.4252	0.4299	67	27/64	0.4219	0.4266	72
M14 × 2	12.00	0.4724	0.4772	72	15/32	0.4688	0.4736	76
M14 × 1.5	12.50	0.4921	0.4969	71	—	—	—	—
M16 × 2	14.00	0.5512	0.5561	72	35/64	0.5469	0.5518	76
M16 × 1.5	14.50	0.5709	0.5758	71	—	—	—	—
M18 × 2.5	15.50	0.6102	0.6152	73	39/64	0.6094	0.6144	74
M18 × 1.5	16.50	0.6496	0.6546	70	—	—	—	—
M20 × 2.5	17.50	0.6890	0.6942	73	11/16	0.6875	0.6925	74
M20 × 1.5	18.50	0.7283	0.7335	70	—	—	—	—
M22 × 2.5	19.50	0.7677	0.7729	73	49/64	0.7656	0.7708	75
M22 × 1.5	20.50	0.8071	0.8123	70	—	—	—	—
M24 × 3	21.00	0.8268	0.8327	73	53/64	0.8281	0.8340	72
M24 × 2	22.00	0.8661	0.8720	71	—	—	—	—
M27 × 3	24.00	0.9449	0.9511	73	15/16	0.9375	0.9435	78
M27 × 2	25.00	0.9843	0.9913	70	63/64	0.9844	0.9914	70
M30 × 3.5	26.50	1.0433						
M30 × 2	28.00	1.1024						
M33 × 3.5	29.50	1.1614						
M33 × 2	31.00	1.2205						
M36 × 4	32.00	1.2598						
M36 × 3	33.00	1.2992						
M39 × 4	35.00	1.3780						
M39 × 3	36.00	1.4173						

FORMULA FOR METRIC TAP DRILL SIZE

$$\text{Basic Major Dia.}_{(mm)} - \frac{\% \text{ Thread} \times \text{Pitch (mm)}}{76.980} = \text{DRILLED HOLE SIZE (mm)}$$

FORMULA FOR PERCENT OF THREAD

$$\frac{76.980}{\text{Pitch (mm)}} \times \left[\text{Basic Major Dia.}_{(mm)} - \text{Drilled Hole Size}_{(mm)} \right] = \text{Percent of Thread}$$

Dimensions of Standard Series Threads for Commercial ISO Metric Screws, Bolts and Nuts (Inches)

Nominal Size Diam. (mm)	Pitch P (mm)	Basic Thread Designation	Tol Class	Allowance	Major Diameter Max	Major Diameter Min	Pitch Diameter Max	Pitch Diameter Min	Pitch Diameter Tol	Minor Diameter Max	Minor Diameter Min b	Tol Class	Minor Diameter Min	Minor Diameter Max	Pitch Diameter Min	Pitch Diameter Max	Pitch Diameter Tol	Major Diam. Min
1.6	0.35	M1.6	6g	0.0008	0.0622	0.0589	0.0533	0.0509	0.0024	0.0453	0.0419	6H	0.0481	0.0520	0.0541	0.0574	0.0033	0.0630
1.8	0.35	M1.8	6g	0.0008	0.0701	0.0668	0.0611	0.0588	0.0023	0.0531	0.0498	6H	0.0560	0.0598	0.0620	0.0652	0.0032	0.0709
2	0.4	M2	6g	0.0009	0.0779	0.0743	0.0677	0.0652	0.0025	0.0586	0.0549	6H	0.0617	0.0661	0.0686	0.0720	0.0034	0.0788
2.2	0.45	M2.2	6g	0.0009	0.0858	0.0819	0.0743	0.0716	0.0027	0.0640	0.0601	6H	0.0675	0.0723	0.0752	0.0788	0.0033	0.0867
2.5	0.45	M2.5	6g	0.0009	0.0976	0.0938	0.0861	0.0834	0.0027	0.0759	0.0719	6H	0.0793	0.0841	0.0870	0.0906	0.0036	0.0985
3	0.5	M3	6g	0.0009	0.1173	0.1132	0.1045	0.1016	0.0029	0.0931	0.0889	6H	0.0969	0.1023	0.1054	0.1092	0.0038	0.1182
3.5	0.6	M3.5	6g	0.0009	0.1369	0.1321	0.1216	0.1183	0.0033	0.1079	0.1030	6H	0.1123	0.1185	0.1225	0.1268	0.0043	0.1378
4	0.7	M4	6g	0.0009	0.1566	0.1512	0.1387	0.1352	0.0034	0.1227	0.1173	6H	0.1277	0.1347	0.1396	0.1442	0.0046	0.1575
4.5	0.75	M4.5	6g	0.0010	0.1762	0.1708	0.1571	0.1536	0.0035	0.1400	0.1345	6H	0.1452	0.1526	0.1580	0.1626	0.0046	0.1772
5	0.8	M5	6g	0.0010	0.1959	0.1900	0.1754	0.1717	0.0037	0.1572	0.1513	6H	0.1628	0.1706	0.1764	0.1812	0.0048	0.1969
6	1	M6	6g	0.0012	0.2351	0.2282	0.2096	0.2052	0.0044	0.1868	0.1797	6H	0.1936	0.2028	0.2107	0.2165	0.0058	0.2363
7	1	M7	6g	0.0011	0.2745	0.2675	0.2489	0.2446	0.0043	0.2262	0.2191	6H	0.2330	0.2422	0.2500	0.2559	0.0059	0.2756
8	1.25	M8	6g	0.0012	0.3138	0.3056	0.2818	0.2773	0.0045	0.2535	0.2454	6H	0.2617	0.2721	0.2830	0.2892	0.0062	0.3150
	1	M8 x 1	6g	0.0011	0.3139	0.3069	0.2883	0.2840	0.0043	0.2656	0.2584	6H	0.2724	0.2816	0.2894	0.2952	0.0058	0.3150
10	1.5	M10	6g	0.0013	0.3924	0.3832	0.3540	0.3489	0.0051	0.3199	0.3102	6H	0.3298	0.3415	0.3554	0.3624	0.0070	0.3937
	1.25	M10 x 1.25	6g	0.0012	0.3925	0.3843	0.3606	0.3560	0.0046	0.3322	0.3241	6H	0.3404	0.3508	0.3618	0.3680	0.0062	0.3937
12	1.75	M12	6g	0.0014	0.4711	0.4607	0.4263	0.4205	0.0058	0.3865	0.3758	6H	0.3979	0.4110	0.4277	0.4355	0.0078	0.4725
	1.25	M12 x 1.25	6g	0.0012	0.4713	0.4630	0.4393	0.4342	0.0051	0.4109	0.4023	6H	0.4192	0.4295	0.4405	0.4475	0.0070	0.4725
14	2	M14	6g	0.0016	0.5496	0.5387	0.4985	0.4923	0.0062	0.4530	0.4412	6H	0.4660	0.4807	0.5001	0.5083	0.0082	0.5512
	1.5	M14 x 1.5	6g	0.0013	0.5499	0.5407	0.5115	0.5061	0.0054	0.4774	0.4677	6H	0.4873	0.4990	0.5129	0.5203	0.0074	0.5512
16	2	M16	6g	0.0016	0.6284	0.6175	0.5772	0.5710	0.0062	0.5318	0.5199	6H	0.5447	0.5594	0.5788	0.5871	0.0083	0.6300
	1.5	M16 x 1.5	6g	0.0014	0.6286	0.6194	0.5903	0.5894	0.0054	0.5561	0.5465	6H	0.5660	0.5777	0.5916	0.5990	0.0074	0.6300
18	2.5	M18	6g	0.0017	0.7070	0.6939	0.6430	0.6364	0.0066	0.5862	0.5725	6H	0.6022	0.6198	0.6448	0.6535	0.0087	0.7087
	1.5	M18 x 1.5	6g	0.0013	0.7074	0.6982	0.6690	0.6636	0.0054	0.6349	0.6252	6H	0.6448	0.6565	0.6704	0.6777	0.0073	0.7087
20	2.5	M20	6g	0.0018	0.7857	0.7726	0.7218	0.7152	0.0066	0.6649	0.6513	6H	0.6809	0.6985	0.7235	0.7322	0.0087	0.7875
	1.5	M20 x 1.5	6g	0.0014	0.7861	0.7769	0.7477	0.7423	0.0054	0.7136	0.7039	6H	0.7235	0.7352	0.7491	0.7565	0.0074	0.7875
22	2.5	M22	6g	0.0018	0.8644	0.8513	0.8005	0.7939	0.0066	0.7437	0.7300	6H	0.7597	0.7773	0.8023	0.8110	0.0087	0.8662
	1.5	M22 x 1.5	6g	0.0014	0.8648	0.8556	0.8265	0.8211	0.0054	0.7924	0.7827	6H	0.8023	0.8140	0.8278	0.8352	0.0074	0.8662
24	3	M24	6g	0.0020	0.9429	0.9283	0.8662	0.8584	0.0078	0.7980	0.7817	6H	0.8171	0.8366	0.8682	0.8785	0.0103	0.9449
	2	M24 x 2	6g	0.0016	0.9433	0.9324	0.8922	0.8856	0.0066	0.8467	0.8345	6H	0.8597	0.8744	0.8938	0.9025	0.0087	0.9449
27	3	M27	6g	0.0019	1.0611	1.0464	0.9843	0.9765	0.0078	0.9161	0.8999	6H	0.9352	0.9548	0.9863	0.9966	0.0103	1.0630
	2	M27 x 2	6g	0.0016	1.0614	1.0505	1.0103	1.0037	0.0066	0.9648	0.9526	6H	0.9778	0.9925	1.0119	1.0206	0.0087	1.0630
30	3.5	M30	6g	0.0022	1.1790	1.1623	1.0895	1.0812	0.0083	1.0099	0.9917	6H	1.0320	1.0539	1.0917	1.1026	0.0109	1.1812
	2	M30 x 2	6g	0.0016	1.1796	1.1686	1.1284	1.1218	0.0066	1.0829	1.0707	6H	1.0959	1.1106	1.1300	1.1387	0.0087	1.1812
33	3.5	M33	6g	0.0022	1.2971	1.2804	1.2076	1.1993	0.0083	1.1280	1.1099	6H	1.1501	1.1720	1.2098	1.2207	0.0109	1.2993
	2	M33 x 2	6g	0.0016	1.2977	1.2867	1.2465	1.2399	0.0066	1.2011	1.1888	6H	1.2140	1.2287	1.2481	1.2568	0.0087	1.2993
36	4	M36	6g	0.0025	1.4149	1.3963	1.3126	1.3039	0.0087	1.2217	1.2017	6H	1.2469	1.2704	1.3151	1.3268	0.0117	1.4174
	3	M36 x 3	6g	0.0020	1.4154	1.4007	1.3386	1.3309	0.0077	1.2705	1.2542	6H	1.2895	1.3091	1.3406	1.3510	0.0104	1.4174
39	4	M39	6g	0.0025	1.5330	1.5144	1.4307	1.4220	0.0087	1.3398	1.3198	6H	1.3650	1.3885	1.4332	1.4449	0.0117	1.5355
	3	M39 x 3	6g	0.0020	1.5335	1.5188	1.4568	1.4490	0.0078	1.3886	1.3723	6H	1.4076	1.4272	1.4587	1.4691	0.0104	1.5355

Metric Machine Screws

SLOTTED FLAT COUNTERSUNK HEAD

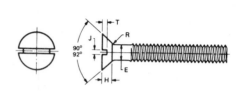

Nom Screw Size and Thread Pitch	E Body Dia		A Head Diameter			H Head Height	R Fillet Radius		J Slot Width		T Slot Depth	
			Theoretical Sharp		Actual							
	Max	Min	Max	Min	Min	Max Ref	Max	Min	Max	Min	Max	Min
M2x0.4	2.00	1.65	4.40	3.90	3.60	1.20	0.8	0.2	0.7	0.5	0.6	0.4
M2.5x0.45	2.50	2.12	5.50	4.90	4.60	1.50	1.0	0.3	0.8	0.6	0.7	0.5
M3x0.5	3.00	2.58	6.60	5.80	5.50	1.80	1.2	0.3	1.0	0.8	0.9	0.6
M3.5x0.6	3.50	3.00	7.70	6.80	6.44	2.10	1.4	0.4	1.2	1.0	1.0	0.7
M4x0.7	4.00	3.43	8.65	7.80	7.44	2.32	1.6	0.4	1.4	1.2	1.1	0.8
M5x0.8	5.00	4.36	10.70	9.80	9.44	2.85	2.0	0.5	1.5	1.2	1.4	1.0
M6.3x1	6.30	5.51	13.50	12.30	11.87	3.60	2.5	0.6	1.9	1.6	1.8	1.3
M8x1.25	8.00	7.04	16.80	15.60	15.17	4.40	3.2	0.8	2.3	2.0	2.1	1.6
M10x1.5	10.00	8.86	20.70	19.50	18.98	5.35	4.0	1.0	2.8	2.5	2.6	2.0
M12x1.75	12.00	10.68	24.70	23.50	22.88	6.35	4.8	1.2	2.8	2.5	3.1	2.5

SLOTTED OVAL COUNTERSUNK HEAD

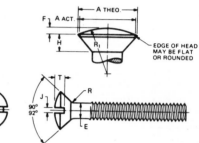

Nom Screw Size and Thread Pitch	E Body Dia		A Head Diameter			H Head Side Height	F Raised Head Height	R₁ Head Radius	R Fillet Radius		J Slot Width		T Slot Depth	
			Theoretical Sharp		Actual									
	Max	Min	Max	Min	Min	Max Ref	Max	Approx	Max	Min	Max	Min	Max	Min
M2x0.4	2.00	1.65	4.40	3.90	3.60	1.20	0.50	3.8	0.8	0.2	0.7	0.5	1.0	0.8
M2.5x0.45	2.50	2.12	5.50	4.90	4.60	1.50	0.60	5.0	1.0	0.3	0.8	0.6	1.2	1.0
M3x0.5	3.00	2.58	6.60	5.80	5.50	1.80	0.75	5.7	1.2	0.3	1.0	0.8	1.5	1.2
M3.5x0.6	3.50	3.00	7.70	6.80	6.44	2.10	0.90	6.5	1.4	0.4	1.2	1.0	1.7	1.4
M4x0.7	4.00	3.43	8.65	7.80	7.44	2.32	1.00	7.8	1.6	0.4	1.4	1.2	1.9	1.6
M5x0.8	5.00	4.36	10.70	9.80	9.44	2.85	1.25	9.9	2.0	0.5	1.5	1.2	2.3	2.0
M6.3x1	6.30	5.51	13.50	12.30	11.87	3.60	1.60	12.2	2.5	0.6	1.9	1.6	3.0	2.6
M8x1.25	8.00	7.04	16.80	15.60	15.17	4.40	2.00	15.8	3.2	0.8	2.3	2.0	3.7	3.2
M10x1.5	10.00	8.86	20.70	19.50	18.98	5.35	2.50	19.8	4.0	1.0	2.8	2.5	4.5	4.0
M12x1.75	12.00	10.68	24.70	23.50	22.88	6.35	3.00	23.8	4.8	1.2	2.8	2.5	5.3	4.8

SLOTTED AND RECESSED PAN HEAD

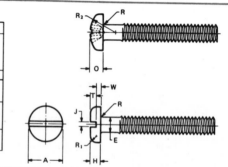

Nom Screw Size and Thread Pitch	E Body Diameter		A Head Diameter		H Head Height				R₁ Head Radius (Slttd)	R₂ Head Radius (Rcsed)	R Fillet Radius		J Slot Width		T Slot Depth	W Un-slotted Thickness
					Slotted Head		Recessed Head									
	Max	Min	Max	Min	Max	Min	Max	Min	Max	Ref	Max	Min	Max	Min	Min	Min
M2x0.4	2.00	1.65	3.90	3.60	1.35	1.15	1.60	1.40	0.8	4	0.3	0.1	0.7	0.5	0.55	0.44
M2.5x0.45	2.50	2.12	4.90	4.60	1.65	1.45	1.95	1.75	1.0	5	0.4	0.1	0.8	0.6	0.73	0.55
M3x0.5	3.00	2.58	5.80	5.50	1.90	1.65	2.30	2.05	1.2	6	0.5	0.2	1.0	0.8	0.80	0.66
M3.5x0.6	3.50	3.00	6.80	6.44	2.25	2.00	2.50	2.25	1.4	7	0.5	0.2	1.2	1.0	0.95	0.77
M4x0.7	4.00	3.43	7.80	7.44	2.55	2.30	2.80	2.55	1.6	8	0.6	0.2	1.4	1.2	1.15	0.88
M5x0.8	5.00	4.36	9.80	9.44	3.10	2.85	3.50	3.25	2.0	10	0.8	-0.3	1.5	1.2	1.35	1.10
M6.3x1	6.30	5.51	12.00	11.57	3.90	3.50	4.30	4.00	2.5	13	1.0	0.3	1.9	1.6	1.70	1.36
M8x1.25	8.00	7.04	15.60	15.17	5.00	4.60	5.60	5.20	3.2	16	1.2	0.4	2.3	2.0	2.20	1.76
M10x1.5	10.00	8.86	19.50	18.98	6.20	5.70	7.00	6.50	4.0	20	1.5	0.5	2.8	2.5	2.70	2.20
M12x1.75	12.00	10.68	23.40	22.88	7.50	6.90	8.30	7.80	4.8	24	1.8	0.6	2.8	2.5	3.20	2.70

CROSS RECESS DIMENSIONS OF PAN HEAD

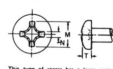

Nom Screw Size	Type 1								Type 1A							
	M Recess Dia		T Recess Depth		N Recess Width	Driver Size	Recess Penetration Gaging Depth		M Recess Dia		T Recess Depth		N Recess Width	Driver Size	Recess Penetration Gaging Depth	
	Max	Min	Max	Min	Min		Max	Min	Max	Min	Max	Min	Min		Max	Min
M 2	1.88	1.55	1.14	0.64	0.36	0	1.02	0.56	2.31	1.98	1.60	1.19	0.46	0	1.42	1.02
M 2.5	2.84	2.51	1.73	1.27	0.48	1	1.55	1.09	3.10	2.77	2.03	1.63	0.74	1	1.78	1.37
M 3	3.10	2.77	1.98	1.52	0.48	1	1.80	1.35	3.35	3.02	2.31	1.91	0.74	1	2.06	1.65
M 3.5	4.22	3.89	2.31	1.68	0.71	2	2.03	1.40	4.11	3.78	2.34	1.88	1.02	2	1.93	1.47
M 4	4.62	4.29	2.74	2.08	0.76	2	2.46	1.80	4.50	4.17	2.74	2.29	1.04	2	2.34	1.88
M 5	5.05	4.72	3.15	2.54	0.79	2	2.87	2.26	4.90	4.57	3.15	2.69	1.04	2	2.74	2.29
M 6.3	7.14	6.81	4.09	3.43	0.91	3	3.66	3.00	6.93	6.60	4.04	3.58	1.45	3	3.48	3.02
M 8	8.89	8.56	4.90	4.29	1.50	4	4.39	3.78	8.66	8.33	4.85	4.39	2.18	4	4.17	3.71
M10	10.49	10.16	6.58	5.94	1.73	4	6.07	5.44	10.13	9.80	6.38	5.92	2.18	4	5.69	5.23
M12	11.05	10.72	7.11	6.48	1.80	4	6.60	5.97	10.67	10.34	6.93	6.48	2.18	4	6.25	5.79

This type of recess has a large center opening, tapered wings, and blunt bottom, with all edges relieved or rounded.

TYPE 1

This type of recess has a large center opening, wide straight wings, and blunt bottom, with all edges relieved or rounded.

TYPE 1A

HEX HEAD

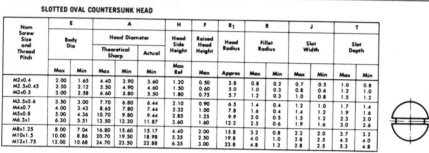

SHAPE OF INDENTATION OPTIONAL

INDENTED HEAD

Nom Screw Size and Thread Pitch	E Body Diameter		A Width Across Flats		W Width Across Corners	H Head Height		R Fillet Radius		F Protrusion Beyond Gaging Ring
	Max	Min	Max	Min	Min	Max	Min	Max	Min	Min
M2x0.4	2.00	1.65	3.20	3.02	3.36	1.27	1.02	0.3	0.1	0.61
M2.5x0.45	2.50	2.12	4.00	3.82	4.25	1.40	1.12	0.4	0.1	0.67
M3x0.5	3.00	2.58	5.00	4.82	5.36	1.52	1.24	0.5	0.2	0.74
M3.5x0.6	3.50	3.00	5.50	5.32	5.92	2.36	2.03	0.5	0.2	1.22
M4x0.7	4.00	3.43	7.00	6.78	7.55	2.79	2.44	0.6	0.2	1.46
M5x0.8	5.00	4.36	8.00	7.78	8.66	3.05	2.67	0.8	0.3	1.60
M6.3x1	6.30	5.51	10.00	9.78	10.89	4.83	4.37	1.0	0.3	2.62
M8x1.25	8.00	7.04	13.00	12.73	14.17	5.84	5.28	1.2	0.4	3.17
M10x1.5	10.00	8.86	15.00	14.73	16.41	7.49	6.86	1.5	0.5	4.12
M12x1.75	12.00	10.68	18.00	17.73	19.75	9.50	8.66	1.8	0.6	5.20

Metric Hex Bolts

Nominal Bolt Size & Thread Pitch	E Body Diameter		F Width Across Flats		G Width Across Corners		H Head Height	
	Max	Min	Max	Min	Max	Min	Max	Min
M5x0.8	5.48	4.52	8.00	7.75	9.24	8.84	3.88	3.35
M6.3x1	6.78	5.72	10.00	9.69	11.55	11.05	4.70	4.13
M8x1.25	8.58	7.42	13.00	12.60	15.01	14.36	5.73	5.10
M10x1.5	10.58	9.42	15.00	14.50	17.32	16.53	6.86	6.17
M12x1.75	12.70	11.30	18.00	17.40	20.78	19.84	7.99	7.24
M14x2	14.70	13.30	21.00	20.30	24.25	23.14	9.32	8.51
M16x2	16.70	15.30	24.00	23.20	27.71	26.45	10.56	9.68
M20x2.5	20.84	19.16	30.00	29.00	34.64	33.06	13.12	12.12
M24x3	24.84	23.16	36.00	35.00	41.57	39.67	15.68	14.56
M30x3.5	30.84	29.16	46.00	44.50	53.12	50.73	19.48	17.92
M36x4	37.00	35.00	55.00	53.20	63.51	60.65	23.38	21.72
M42x4.5	43.00	41.00	65.00	62.90	75.06	71.71	26.97	25.03
M48x5	49.00	47.00	75.00	72.60	86.60	82.76	31.07	28.93
M56x5.5	57.20	54.80	85.00	82.20	98.15	93.71	36.20	33.80
M64x6	65.52	62.80	95.00	91.80	109.70	104.65	41.32	38.68
M72x6	73.84	70.80	105.00	101.40	121.24	115.60	46.45	43.55
M80x6	82.16	78.80	115.00	111.00	132.79	126.54	51.58	48.42
M90x6	92.48	88.60	130.00	125.50	150.11	143.07	57.74	54.26
M100x6	102.80	98.60	145.00	140.00	167.43	159.60	63.90	60.10

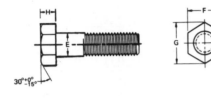

30°+0°/-15°

Metric Hex Nuts

REGULAR

Nominal Nut Size and Thread Pitch	F Width Across Flats		G Width Across Corners		O Bearing Face Dia	H Nut Thickness Style 1		H₁ Style 2	
	Max	Min	Max	Min	Min	Max	Min	Max	Min
M1.6x0.35	3.20	3.02	3.70	3.44	2.5	—	—	1.3	1.1
M2x0.4	4.00	3.82	4.62	4.35	3.1	—	—	1.6	1.3
M2.5x0.45	5.00	4.82	5.77	5.49	4.1	—	—	2.0	1.7
M3x0.5	5.50	5.32	6.35	6.06	4.6	—	—	2.4	2.1
M3.5x0.6	7.00	6.78	8.08	7.73	6.0	—	—	2.8	2.5
M4x0.7	7.00	6.78	8.08	7.73	6.0	—	—	3.2	2.9
M5x0.8	8.00	7.78	9.24	8.87	7.0	4.5	4.2	5.3	5.0
M6.3x1	10.00	9.76	11.55	11.13	8.9	5.6	5.3	6.5	6.2
M8x1.25	13.00	12.73	15.01	14.51	11.6	6.6	6.2	7.8	7.4
M10x1.5	15.00	14.70	17.32	16.76	13.6	9.0	8.5	10.7	10.2
M12x1.75	18.00	17.67	20.78	20.14	16.6	10.7	10.2	12.8	12.3
M14x2	21.00	20.64	24.25	23.53	19.4	12.5	11.9	14.9	14.3
M16x2	24.00	23.61	27.71	26.92	22.4	14.5	13.9	17.4	16.8
M20x2.5	30.00	29.00	34.64	33.06	27.6	18.4	17.4	21.2	20.2
M24x3	36.00	34.80	41.57	39.67	32.9	22.0	20.9	25.4	24.3
M30x3.5	46.00	44.50	53.12	50.73	42.5	26.7	25.4	31.0	29.7
M36x4	55.00	53.20	63.51	60.65	50.8	32.0	30.5	37.6	36.1

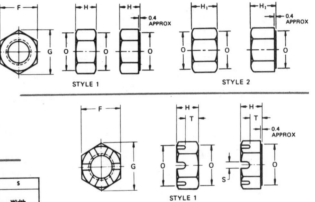

STYLE 1 STYLE 2

SLOTTED

Nominal Nut Size and Thread Pitch	F Width Across Flats		G Width Across Corners		O Bearing Face Dia	H Nut Thickness Style 1		H₁ Style 2		T Unslotted Thickness Style 1		T₁ Style 2		S Width of Slot	
	Max	Min	Max	Min	Min	Max	Min	Max	Min	Max	Min	Max	Min	Max	Min
M5x0.8	8.00	7.78	9.24	8.87	7.0	4.5	4.2	5.3	5.0	3.2	2.7	3.7	3.2	2.2	1.4
M6.3x1	10.00	9.76	11.55	11.13	8.9	5.6	5.3	6.5	6.2	3.9	3.4	4.5	4.0	2.8	2.0
M8x1.25	13.00	12.73	15.01	14.51	11.6	6.6	6.2	7.8	7.4	4.5	4.0	5.3	4.8	3.3	2.5
M10x1.5	15.00	14.70	17.32	16.76	13.6	9.0	8.5	10.7	10.2	6.0	5.5	7.1	6.6	3.6	2.8
M12x1.75	18.00	17.67	20.78	20.14	16.6	10.7	10.2	12.8	12.3	7.1	6.6	8.5	8.0	4.3	3.5
M14x2	21.00	20.64	24.25	23.53	19.4	12.5	11.9	14.9	14.3	8.2	7.7	9.8	9.3	4.3	3.5
M16x2	24.00	23.61	27.71	26.92	22.4	14.5	13.9	17.4	16.8	9.5	9.0	11.4	10.9	6.0	4.5
M20x2.5	30.00	29.00	34.64	33.06	27.6	18.4	17.4	21.2	20.2	12.1	11.3	13.9	13.1	6.0	4.5
M24x3	36.00	34.80	41.57	39.67	32.9	22.0	20.9	25.4	24.3	14.4	13.6	16.6	15.8	7.0	5.5
M30x3.5	46.00	44.50	53.12	50.73	42.5	26.7	25.4	31.0	29.7	17.3	16.5	20.1	19.3	9.3	7.0
M36x4	55.00	53.20	63.51	60.65	50.8	32.0	30.5	37.6	36.1	20.8	19.8	24.5	23.5	9.3	7.0

STYLE 1 STYLE 2

Metric Washers

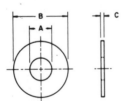

PLAIN CIRCULAR WASHERS

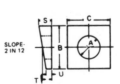

SLOPE 2 IN 12

BEVELED SQUARE WASHERS

Nom Bolt Size	B Outside Dia		A Hole Dia		T Thickness		C Width		A Hole Dia		S Thick Side ±0.5	T Mean Nom	U Thin Side ±0.5
	Max	Min	Max	Min	Max	Min	Max	Min	Max	Min			
M16x2	34.0	33.0	19.0	18.0	4.5	3.1	45.0	43.0	19.0	18.0	11.7	8	4.3
M20x2.5	41.0	40.0	23.0	22.0	4.5	3.5	45.0	43.0	23.0	22.0	11.7	8	4.3
M24x3	49.0	48.0	27.0	26.0	4.5	3.5	45.0	43.0	27.0	26.0	11.7	8	4.3
M30x3.5	60.0	59.0	34.0	33.0	4.5	3.5	59.0	57.0	34.0	33.0	12.8	8	3.2
M36x4	72.0	71.0	40.0	39.0	4.5	3.5	59.0	57.0	40.0	39.0	12.8	8	3.2

Conversion Table
Fraction – Decimal – Millimeter

4ths	8ths	16ths	32nds	64ths	to 2 places	to 3 place	mm	4ths	8ths	16ths	32nds	64ths	to 2 places	to 3 places	mm
				1/64	0.02	0.016	0.3969					33/64	0.52	0.516	13.0968
			1/32		0.03	0.031	0.7937				17/32		0.53	0.531	13.4937
				3/64	0.05	0.047	1.1906					35/64	0.55	0.547	13.8906
		1/16			0.06	0.062	1.5875			9/16			0.56	0.562	14.2875
				5/64	0.08	0.078	1.9844					37/64	0.58	0.578	14.6844
			3/32		0.09	0.094	2.3812				19/32		0.59	0.594	15.0812
				7/64	0.11	0.109	2.7781					39/64	0.61	0.609	15.4781
	1/8				0.12	0.125	3.1750		5/8				0.62	0.625	15.8750
				9/64	0.14	0.141	3.5719					41/64	0.64	0.641	16.2719
			5/32		0.16	0.156	3.9687				21/32		0.66	0.656	16.6687
				11/64	0.17	0.172	4.3656					43/64	0.67	0.672	17.0656
		3/16			0.19	0.188	4.7625			11/16			0.69	0.688	17.4625
				13/64	0.20	0.203	5.1594					45/64	0.70	0.703	17.8594
			7/32		0.22	0.219	5.5562				23/32		0.72	0.719	18.2562
				15/64	0.23	0.234	5.9531					47/64	0.73	0.734	18.6532
1/4					0.25	0.250	6.3500	3/4					0.75	0.750	19.0500
				17/64	0.27	0.266	6.7469					49/64	0.77	0.766	19.4469
			9/32		0.28	0.281	7.1437				25/32		0.78	0.781	19.8433
				19/64	0.30	0.297	7.5406					51/64	0.80	0.797	20.2402
		5/16			0.31	0.312	7.9375			13/16			0.81	0.812	20.6375
				21/64	0.33	0.328	8.3344					53/64	0.83	0.828	21.0344
			11/32		0.34	0.344	8.7312				27/32		0.84	0.844	21.4312
				23/64	0.36	0.359	9.1281					55/64	0.86	0.859	21.8281
	3/8				0.38	0.375	9.5250		7/8				0.88	0.875	22.2250
				25/64	0.39	0.391	9.9219					57/64	0.89	0.891	22.6219
			13/32		0.41	0.406	10.3187				29/32		0.91	0.906	23.0187
				27/64	0.42	0.422	10.7156					59/64	0.92	9.922	23.4156
		7/16			0.44	0.438	11.1125			15/16			0.94	0.938	23.8125
				29/64	0.45	0.453	11.5094					61/64	0.95	0.953	24.2094
			15/32		0.47	0.469	11.9062				31/32		0.94	0.969	24.6062
				31/64	0.48	0.484	12.3031					63/64	0.98	0.984	25.0031
1/2					0.50	0.500	12.7000	1					1.00	1.000	25.4000

INDEX

Numbers appearing in **bold type** refer to illustrations.